Energy Systems in Electrical Engineering

Series Editor

Muhammad H. Rashid, Florida Polytechnic University, Lakeland, USA

Energy Systems in Electrical Engineering is a unique series that aims to capture advances in electrical energy technology as well as advances electronic devices and systems used to control and capture other sources of energy. Electric power generated from alternate energy sources is getting increasing attention and supports for new initiatives and developments in order to meet the increased energy demands around the world. The availability of computer–based advanced control techniques along with the advancement in the high-power processing capabilities is opening new doors of opportunity for the development, applications and management of energy and electric power. This series aims to serve as a conduit for dissemination of knowledge based on advances in theory, techniques, and applications in electric energy systems. The Series accepts research monographs, introductory and advanced textbooks, professional books, reference works, and select conference proceedings. Areas of interest include, electrical and electronic aspects, applications, and needs of the following key areas:

- Biomass and Wastes Energy
- Carbon Management
- Costs and Marketing
- Diagnostics and Protections
- Distributed Energy Systems
- Distribution System Control and Communication
- Electric Vehicles and Tractions Applications
- Electromechanical Energy Conversion
- Energy Conversion Systems
- Energy Costs and Monitoring
- Energy Economics
- Energy Efficiency
- Energy and Environment
- Energy Management, and Monitoring
- Energy Policy
- Energy Security
- Energy Storage and Transportation
- Energy Sustainability
- Fuel Cells
- Geothermal Energy
- Hydrogen, Methanol and Ethanol Energy
- Hydropower and Technology
- Intelligent Control of Power and Energy Systems
- Nuclear Energy and Technology
- Ocean Energy
- Power and Energy Conversions and Processing
- Power Electronics and Power Systems
- Renewable Energy Technologies
- Simulation and Modeling for Energy Systems
- Superconducting for Energy Applications
- Tidal Energy
- Transport Energy

Farzin Asadi

Essential Circuit Analysis Using Proteus®

Farzin Asadi
Department of Electrical and Electronics Engineering
Maltepe University
Istanbul, Turkey

ISSN 2199-8582 ISSN 2199-8590 (electronic)
Energy Systems in Electrical Engineering
ISBN 978-981-19-4355-3 ISBN 978-981-19-4353-9 (eBook)
https://doi.org/10.1007/978-981-19-4353-9

This Springer imprint is published by the registered company Springer Nature Singapore Pte Ltd.
The registered company address is: 152 Beach Road, #21-01/04 Gateway East, Singapore 189721, Singapore

In loving memory of my father Moloud Asadi
and my mother Khorshid Tahmasebi,
always on my mind, forever in my heart.

Preface

A computer simulation is an attempt to model a real-life or hypothetical situation on a computer so that it can be studied to see how the system works. By changing variables in the simulation, predictions may be made about the behavior of the system. So, computer simulation is a tool to virtually investigate the behavior of the system under study.

Computer simulation has many applications in science, engineering, education and even in entertainment. For instance, pilots use computer simulations to practice what they learned without any danger or loss of life.

A circuit simulator is a computer program which permits us to see the circuit behavior, i.e., circuit voltages and currents, without making it. Using a circuit simulator is a cheap, efficient and safe way to study the behavior of circuits. A circuit simulator even saves your time and energy. It permits you to test your ideas before you go wasting all that time building it with a breadboard or hardware, just to find out it doesn't really work.

This book shows how a circuit can be simulated in a Proteus® Design Suite environment. Proteus Design Suite (designed by Labcenter Electronics Ltd.) is a software toolset, mainly used for creating schematics, simulating electronics and embedded circuits and designing PCB layouts.

Proteus ISIS is used by engineering students and professionals to create schematics and simulations of different electronic circuits. Proteus ARES is used for designing PCB layouts of electronic circuits. ARES is not studied in this book.

This book contains 125 sample simulations. A brief summary of book chapters is given below:

Chapter 1 introduces Proteus and shows how it can be used to analyze electric circuits. Students who take/took electric circuits I/II courses can use this chapter as a reference to learn how to solve an electric circuit problem with the aid of a computer. This chapter has 61 sample simulations.

Chapter 2 focuses on the simulation of electronic circuits (i.e., circuits which contain diodes, transistors, ICs, etc.) with Proteus. Students who take/took electronics I/II courses can use this chapter as a reference to learn how to analyze an electronic circuit with the aid of a computer. This chapter has 39 sample simulations.

Chapter 3 focuses on the simulation of digital circuits with Proteus. Students who take/took digital design course can use this chapter as a reference to learn how to simulate a digital circuit with the aid of a computer. This chapter has 12 sample simulations.

Chapter 4 focuses on the simulation of power electronics circuits with Proteus. Students who take/took power electronics/industrial electronics course can use this chapter as a reference to learn how to simulate a power electronic circuit with the aid of a computer. This chapter has 13 sample simulations.

Engineering students (for instance, electrical, biomedical, mechatronics and robotic to name a few), engineers who work in industry and anyone who want to learn the art of circuit simulation with Proteus can benefit from this book. I hope that this book will be useful to the readers, and I welcome comments on the book.

Istanbul, Turkey Farzin Asadi

Contents

1 Simulation of Electric Circuits with Proteus® 1
1.1 Introduction 1
1.2 Example 1: Simple Resistive Voltage Divider 1
1.3 Example 2: Project Documentation and Reporting 33
1.4 Example 3: Addition of Text to Schematic 35
1.5 Example 4: Defining Variables 36
1.6 Example 5: Removing the Unused Components from the Component List 41
1.7 Example 6: V_{source} Block 43
1.8 Example 7: C_{source} Block 44
1.9 Example 8: V_{sine} Block 45
1.10 Example 9: Exporting the Drawn Schematic as a Graphical File 50
1.11 Example 10: Measurement with Probes (I) 52
1.12 Example 11: Measurement with Probes (II) 56
1.13 Example 12: Measurement with Probes (III) 59
1.14 Example 13: Junction Dot Mode 62
1.15 Example 14: Excluding a Component from Simulation 65
1.16 Example 15: Potentiometer Block 68
1.17 Example 16: Measurement with AC Voltmeter/Ammeter (I) 73
1.18 Example 17: Measurement with AC Voltmeter/Ammeter (II) . . . 84
1.19 Example 18: Wattmeter Block (I) 90
1.20 Example 19: Wattmeter Block (II) 104
1.21 Example 20: Measurement of Power Factor 108
1.22 Example 21: Power Factor Correction 113
1.23 Example 22: Measurement of Phase Difference 119
1.24 Example 23: Giving a Name to Oscilloscope Blocks 127
1.25 Example 24: I_{sine} Block 130
1.26 Example 25: Grounded Current Sources 134
1.27 Example 26: Thevenin Equivalent Circuit 137

1.28 Example 27: Making Connections Without Using the Wire (I) 141
1.29 Example 28: Making Connections Without Using the Wire (II) 147
1.30 Example 29: Current Controlled Voltage Source Block 150
1.31 Example 30: Current Sensor 152
1.32 Example 31: Voltage Controlled Current Source Block 155
1.33 Example 32: Three-Phase Voltage Source Block 158
1.34 Example 33: Voltage Difference Measurement (I) 161
1.35 Example 34: Voltage Difference Measurement (II) 162
1.36 Example 35: Transient Analysis (I) 165
1.37 Example 36: Transient Analysis (II) 180
1.38 Example 37: Transient Analysis (III) 182
1.39 Example 38: Transient Analysis (IV) 186
1.40 Example 39: Transient Analysis (V) 193
1.41 Example 40: Increasing the Accuracy of Transient Analysis Graph 194
1.42 Example 41: Copying the Waveform Graph into the Clipboard Memory 196
1.43 Example 42: Exporting the Waveforms into MATLAB® 197
1.44 Example 43: Multiplier Block 200
1.45 Example 44: Gain Block 205
1.46 Example 45: Coupled Inductors (I) 206
1.47 Example 46: Coupled Inductors (II) 214
1.48 Example 47: Single-Phase Transformer 223
1.49 Example 48: Single-Phase Transformer with Two Outputs 227
1.50 Example 49: Center Tap Transformer 232
1.51 Example 50: Three-Phase Transformer 237
1.52 Example 51: Impulse Response of a RLC Circuit (I) 242
1.53 Example 52: Impulse Response of an RLC Circuit (II) 249
1.54 Example 53: Step Response of a RC Circuit 254
1.55 Example 54: Pulse Response of a RC Circuit 259
1.56 Example 55: Frequency Response of Electric Circuits (I) 261
1.57 Example 56: Frequency Response of Electric Circuits (II) 274
1.58 Example 57: Input Impedance of Electric Circuits (I) 280
1.59 Example 58: Input Impedance of Electric Circuits (II) 291
1.60 Example 59: Input Impedance of Electric Circuits (III) 294
1.61 Example 60: AC Sweep Analysis 305
1.62 Example 61: Samples Simulations 311
1.63 Exercises 314
References for Further Study 317

2 Simulation of Electronic Circuits with Proteus® 319
2.1 Introduction 319
2.2 Example 1: DC Sweep Analysis 319
2.3 Example 2: Diode IV Characteristic 330
2.4 Example 3: DC Transfer Curve Analysis 335
2.5 Example 4: Small Signal Resistance of Diode 343
2.6 Example 5: Doing the Simulation at a Specific Temperature 349
2.7 Example 6: LED and Push Button Blocks 351
2.8 Example 7: Different Kinds of Mechanical Switches 356
2.9 Example 8: Turning on and off a Lamp 357
2.10 Example 9: Turning on and off a Lamp from Two Different Points 359
2.11 Example 10: Measurement of Output Voltage Ripple for Half Wave Diode Rectifier 360
2.12 Example 11: Input Current of Half Wave Rectifier 363
2.13 Example 12: Full Wave Rectifier 365
2.14 Example 13: Measurement of Average Value of Output Voltage for Full Wave Rectifier 369
2.15 Example 14: Current Passed from Rectifier Diodes 373
2.16 Example 15: Bridge Block 378
2.17 Example 16: Fourier Analysis of Output Voltage of Full Wave Rectifier 379
2.18 Example 17: Harmonic Content of a Triangular Waveform 386
2.19 Example 18: Voltage Regulator (I) 396
2.20 Example 19: Voltage Regulator (II) 399
2.21 Example 20: Voltage Regulator (III) 401
2.22 Example 21: Common Emitter Amplifier 403
2.23 Example 22: Signal Generator Block 414
2.24 Example 23: Input Impedance of Common Emitter Amplifier . . . 417
2.25 Example 24: Frequency Response of Input Impedance of Common Emitter Amplifier 419
2.26 Example 25: Output Impedance of Common Emitter Amplifier 424
2.27 Example 26: Frequency Response of Output Impedance of Common Emitter Amplifier 426
2.28 Example 27: Frequency Response of Amplifier 430
2.29 Example 28: Modeling Custom Semiconductor Devices 438
2.30 Example 29: Bill of Material 443
2.31 Example 30: Common Mode Rejection Ratio (CMRR) of Difference Amplifier 446
2.32 Example 31: CMRR of Differential Pair 463
2.33 Example 32: Measurement of Differential Mode Input Impedance of Differential Pair 475
2.34 Example 33: Astable Oscillator with 555 481
2.35 Example 34: Colpitts Oscillator 484

2.36 Example 35: Total Harmonic Distortion (THD) of Colpitts Oscillator 487
2.37 Example 36: Wien Bridge Oscillator 489
2.38 Example 37: Optocoupler Block 493
2.39 Example 38: Relay Block 496
2.40 Example 39: Simulation of Control Systems 501
2.41 Exercises 508
References for Further Study 510

3 Simulation of Digital Circuits with Proteus® 511
3.1 Introduction 511
3.2 Example 1: Full Adder Circuit 511
3.3 Example 2: Logic Probe Block 518
3.4 Example 3: Decade Counter 520
3.5 Example 4: Dclock Block 523
3.6 Example 5: Frequency Divider Circuit 525
3.7 Example 6: Frequency Meter Block 528
3.8 Example 7: Two-Bit Binary Counter 530
3.9 Example 8: Generating of Desired Digital Pulses 532
3.10 Example 9: Digital Graph 535
3.11 Example 10: Boolean Block 543
3.12 Example 11: Bus 547
3.13 Example 12: Simulation of Circuits Contains a Microcontroller 553
3.14 Exercises 560
References for Further Study 562

4 Simulation of Power Electronics Circuits with Proteus® 563
4.1 Introduction 563
4.2 Example 1: Buck Converter Circuit 563
4.3 Example 2: Operating Mode of Converter 569
4.4 Example 3: Efficiency of the Converter 573
4.5 Example 4: Dimmer Circuit 577
4.6 Example 5: Single-Phase Half Wave Controlled Rectifier 581
4.7 Example 6: Single-Phase Full Wave Controlled Rectifier 586
4.8 Example 7: Three-Phase Controlled Rectifier (I) 590
4.9 Example 8: Three-Phase Controlled Rectifier (II) 606
4.10 Example 9: Three-Phase Controlled Rectifier (III) 612
4.11 Example 10: Three-Phase Controlled Rectifier (IV) 618
4.12 Example 11: Harmonic Content of Output Voltage of a Rectifier 621
4.13 Example 12: Single-Phase Inverter 626
4.14 Example 13: Three-Phase Inverter 642
4.15 Exercises 654
References for Further Study 659

Index 661

About the Author

Farzin Asadi received his B.Sc. in Electronics Engineering, his M.Sc. in Control Engineering, and his Ph.D. in Mechatronics Engineering. Currently, he is with the Department of Electrical and Electronics Engineering at the Maltepe University, Istanbul, Turkey. Dr. Asadi has published more than 50 international papers and 20 books. He is on the editorial board of 7 scientific journals as well. His research interests include switching converters, control theory, robust control of power electronics converters, and robotics.

Chapter 1
Simulation of Electric Circuits with Proteus®

Abstract This chapter shows how an electric circuit can be analyzed in Proteus. In this chapter you will learn how to measure voltage, current and power, how to observe a waveform withoscilloscope, how to model coupled inductors, how to obtain transient response and how to obtain impulse response, step response and frequency response of an electric circuit.

Keywords Capacitor · Inductor · Resistor · Proteus · Thevenin's theorem · Norton theorem · First order circuit · Second order circuit · Frequency response · Transient analysis · Single phase transformer · Three phase transformer · Voltage controlled voltage source · Current controlled current source · Voltage controlled current source · Current controlled voltage source · Dependent source · Independent source

1.1 Introduction

In this chapter you will learn how to analyze electric circuits in Proteus. The theory behind the studied circuits can be found in any standard circuit theory text book (Alexander and Sadiku 2016; Hayt et al. 2021; Nilsson and Riedel 2018; Thomas et al. 2020). It is a good idea to do some hand calculations for the circuits that are given and compare them with Proteus results.

1.2 Example 1: Simple Resistive Voltage Divider

In this example, we want to simulate the resistive voltage divider shown in Fig. 1.1. From basic circuit theory we know that $V_{R_1} = \frac{R_1}{R_1 + R_L} \times V_{\text{in}} = \frac{1k}{1k + 2.2k} \times 10 = 3.13\,\text{V}$ and $V_{R_L} = \frac{R_L}{R_1 + R_L} \times V_{\text{in}} = \frac{2.2k}{1k + 2.2k} \times 10 = 6.87\,\text{V}$.

F. Asadi, *Essential Circuit Analysis Using Proteus®*, Energy Systems in Electrical Engineering, https://doi.org/10.1007/978-981-19-4353-9_1

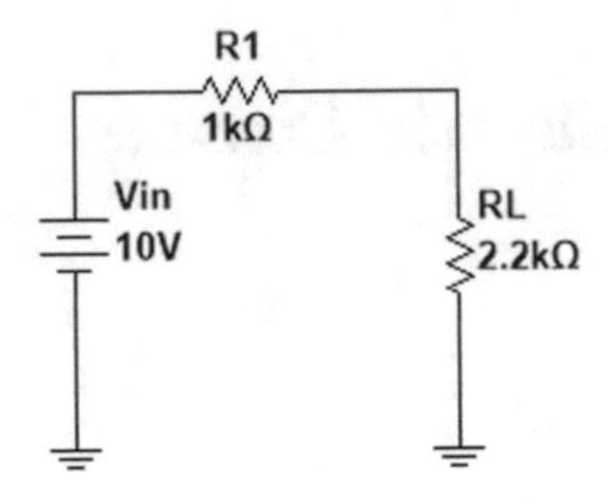

Fig. 1.1 Circuit for Example 1

Let's analyze the circuit with Proteus. Run Proteus (Fig. 1.2).

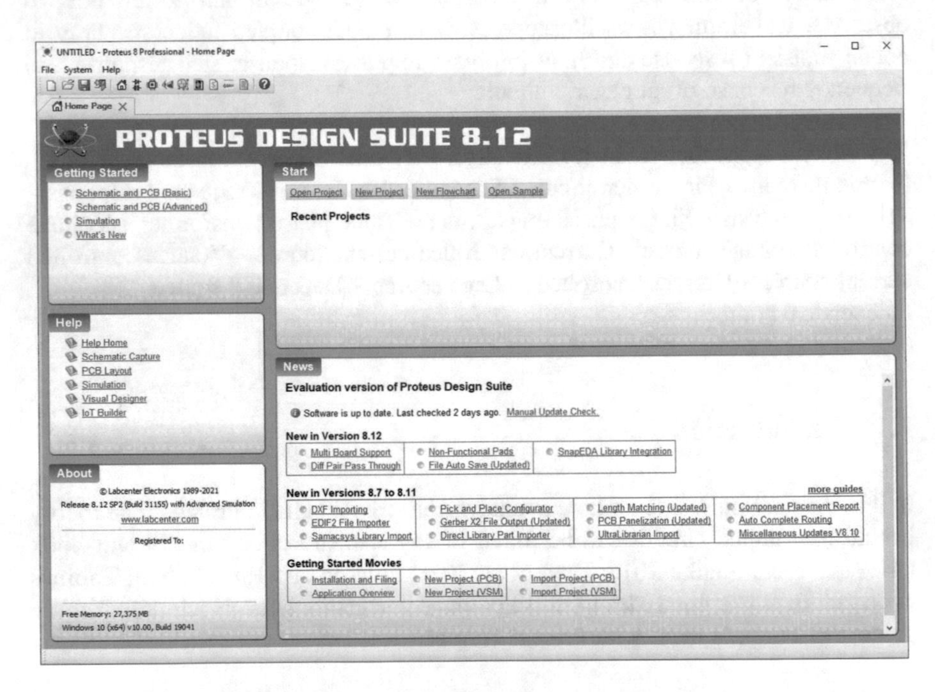

Fig. 1.2 Proteus welcome page

Click the New Project icon (Fig. 1.3) to start a new project. After clicking the New Project icon, the New Project Wizard window appears on the screen (Fig. 1.4). Use this window to set the project name and path (Fig. 1.5). After entering the project name and path, click the Next button.

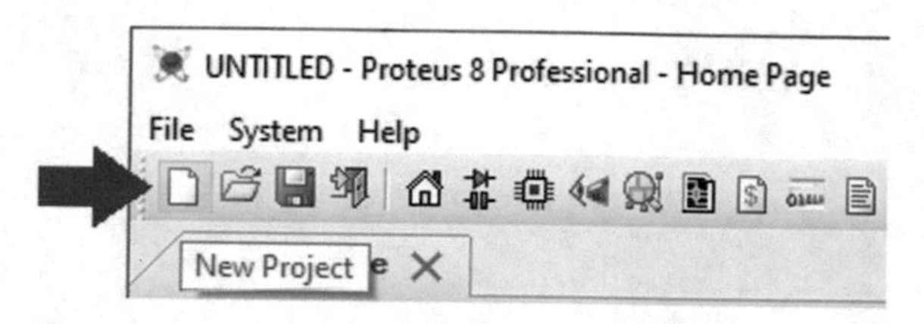

Fig. 1.3 New project icon

Fig. 1.4 New project wizard window

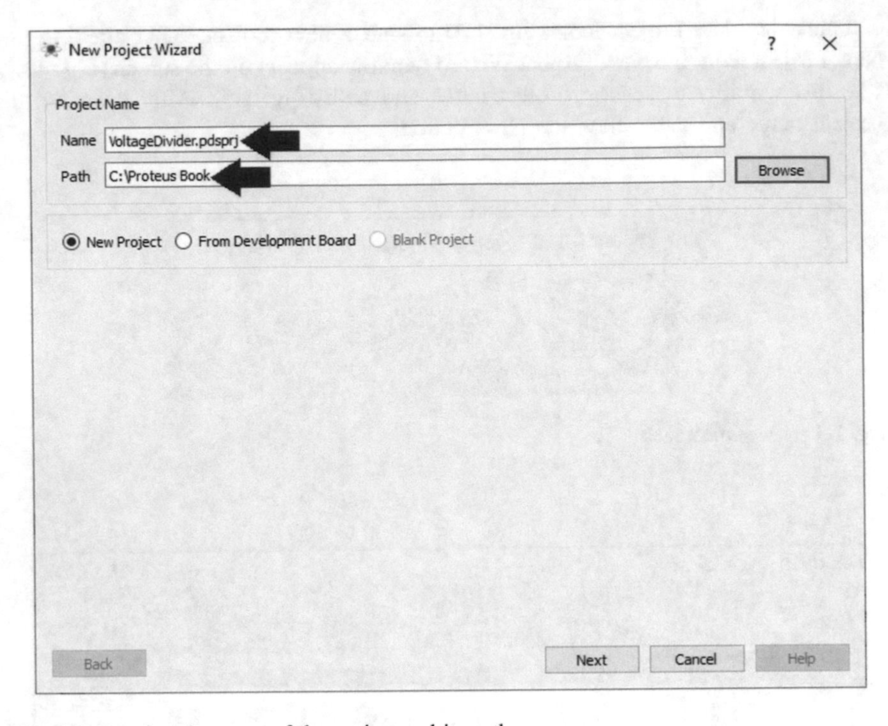

Fig. 1.5 Entering the name of the project and its path

After clicking the Next button, the window shown in Fig. 1.6 appears. You can determine the size of schematic in this window. DEFAULT is enough for schematics in this book. After selecting the desired schematic size, click the Next button.

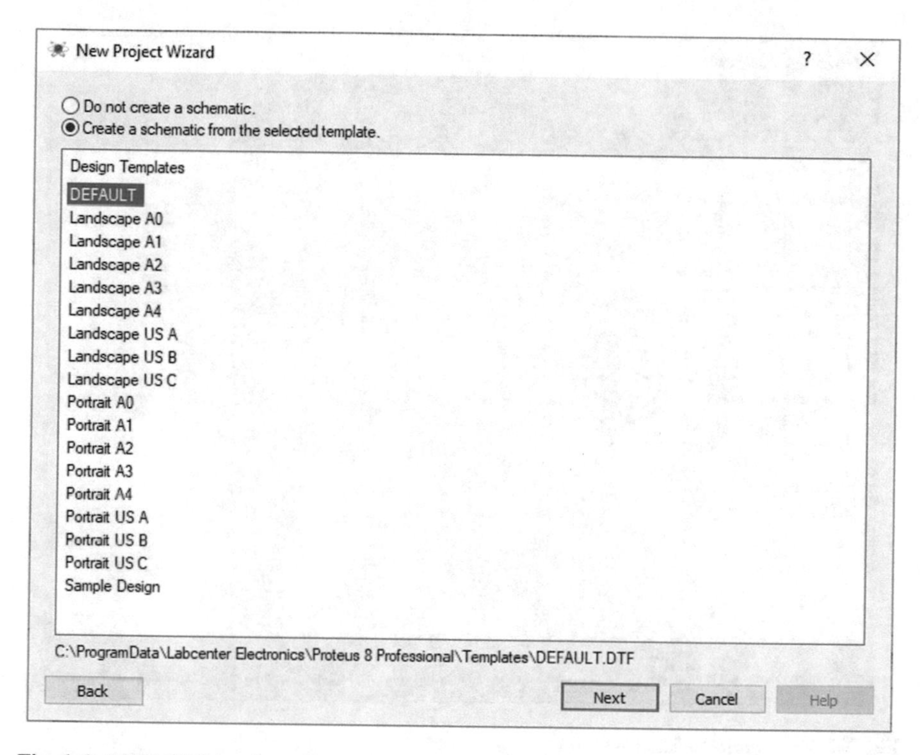

Fig. 1.6 DEFAULT is selected

After clicking the Next button, the window shown in Fig. 1.7 appears on the screen. We don't want to design a PCB for the schematic that we will draw. Therefore, select the Don't create a PCB layout (Fig. 1.7) and click the Next button.

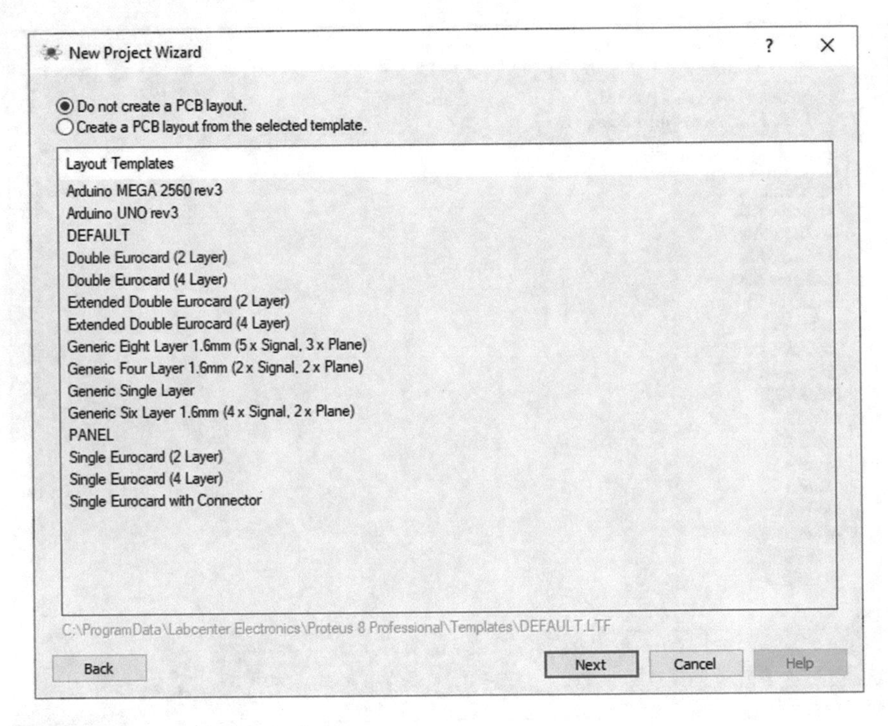

Fig. 1.7 New project wizard window

After clicking the Next button, the window shown in Fig. 1.8 appears on the screen. Select the No Firmware Project and click the Next button.

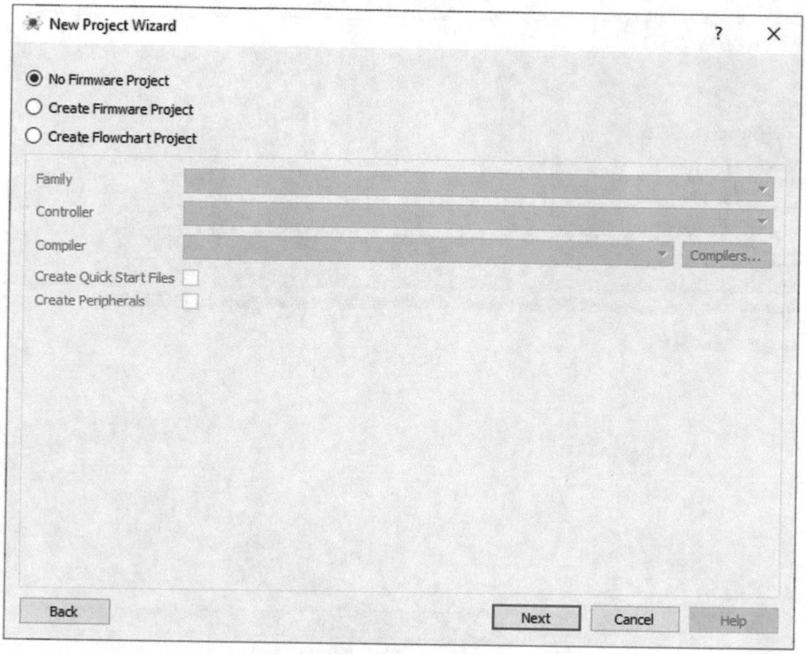

Fig. 1.8 New project wizard window

After clicking the Next button, the window shown in Fig. 1.9 appears on the screen. Click the Finish button to enter the Proteus environment.

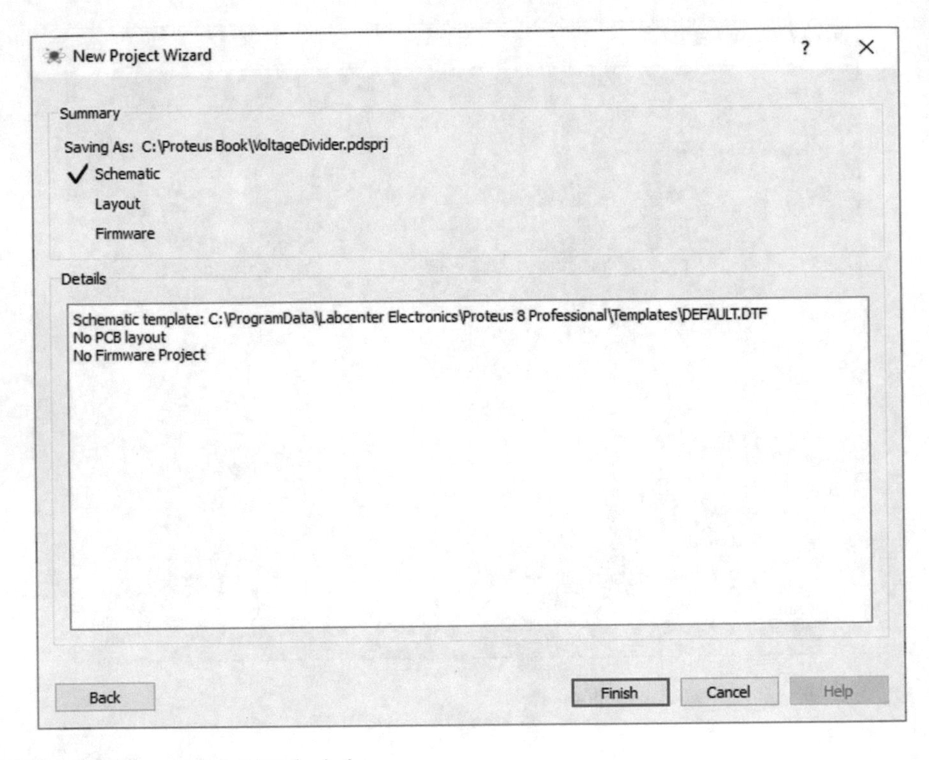

Fig. 1.9 New project wizard window

After clicking the Finish button, the Proteus environment appears on the screen (Fig. 1.10). Proteus schematic editor has some grid lines by default. You can press the keyboard *G* key to change the grid lines to grid dots (Fig. 1.11) or remove them completely (Fig. 1.12).

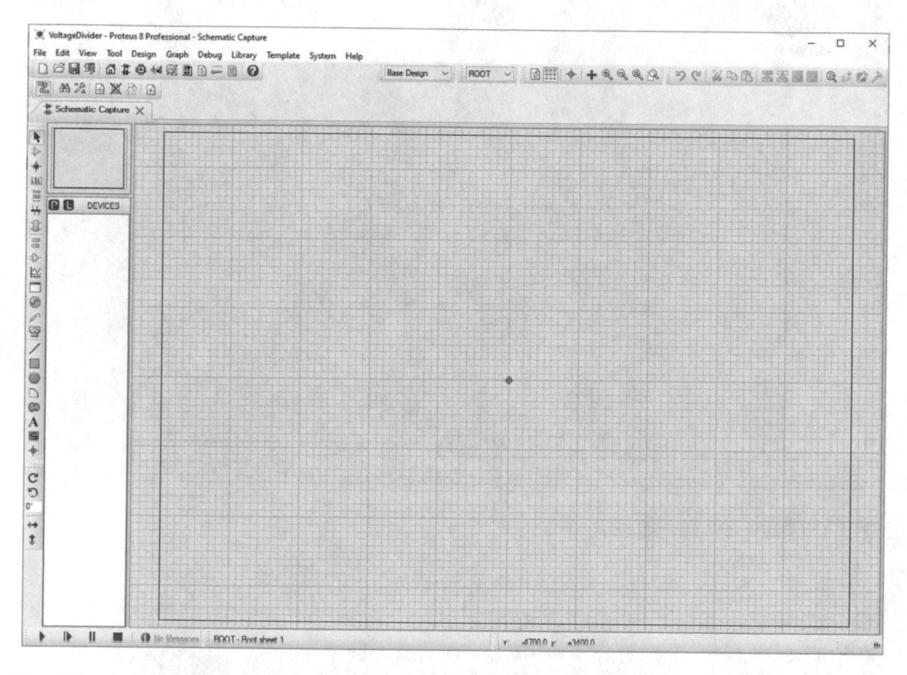

Fig. 1.10 Proteus environment

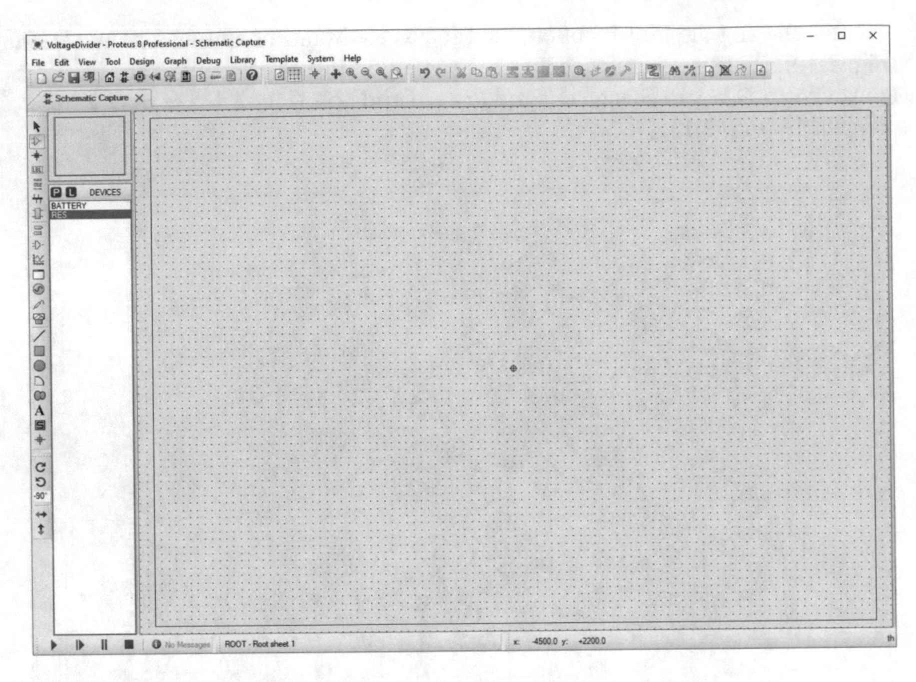

Fig. 1.11 Grids are changed into dots

Fig. 1.12 Grids are removed

After selecting the desired theme for schematic editor, click the *P* icon (Fig. 1.13) to select the required components. After clicking the *P* icon, the Pick Devices window (Fig. 1.14) appears on the screen.

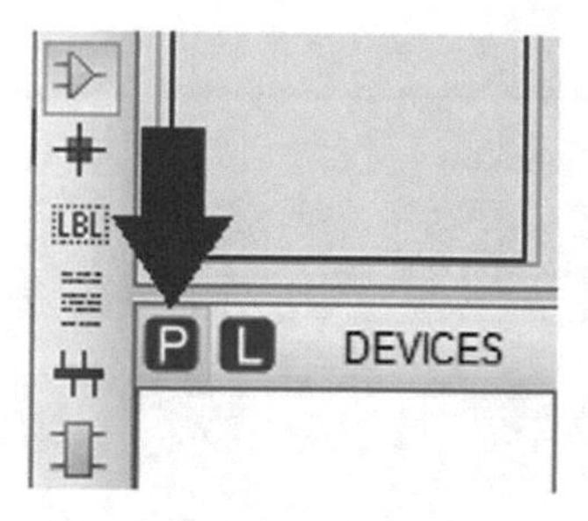

Fig. 1.13 *P* icon

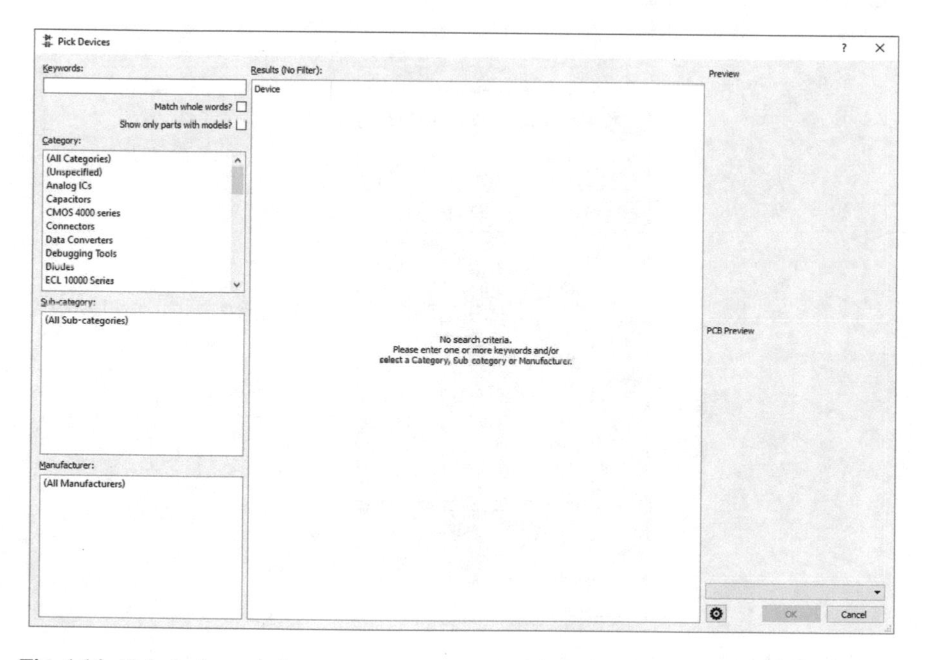

Fig. 1.14 Pick devices window

Enter "res" to Keywords box (Fig. 1.15) and press the Enter key of your keyboard. Note that Proteus is not case sensitive. So, you can enter "RES" to Keywords box and press the Enter key as well. After pressing the Enter key, a resistor is added to DEVICES section (Fig. 1.16).

Pick Devices

Keywords:

res

Match whole words?

Show only parts with models?

Fig. 1.15 Searching for resistor

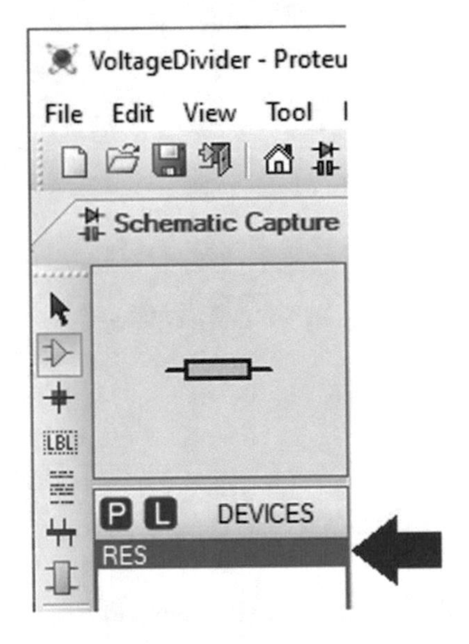

Fig. 1.16 Resistor is added to the DEVICES section

The circuit shown in Fig. 1.1 has a DC voltage source. Let's add the DC voltage source to the component list. Click the P icon again and enter "battery" to the Keywords box (Fig. 1.17). Then press the Enter key. After pressing the Enter key, the battery block is added to the component list (Fig. 1.18).

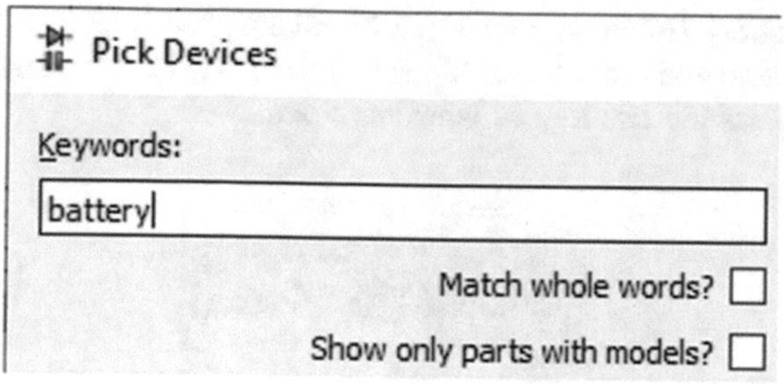

Fig. 1.17 Searching for battery

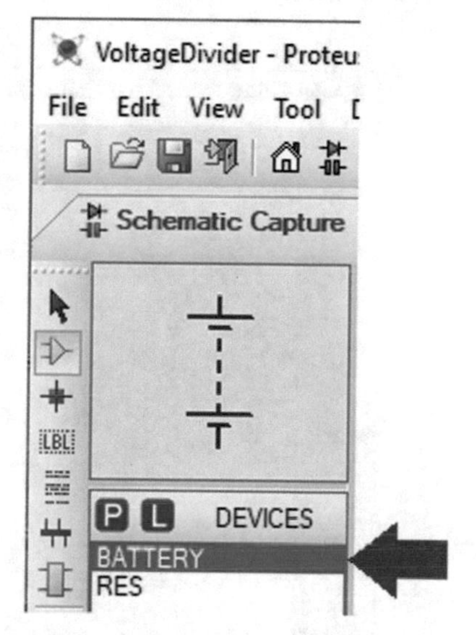

Fig. 1.18 Battery is added to the DEVICES section

You can remove a component from component list by right clicking on it and click the Delete (Fig. 1.19).

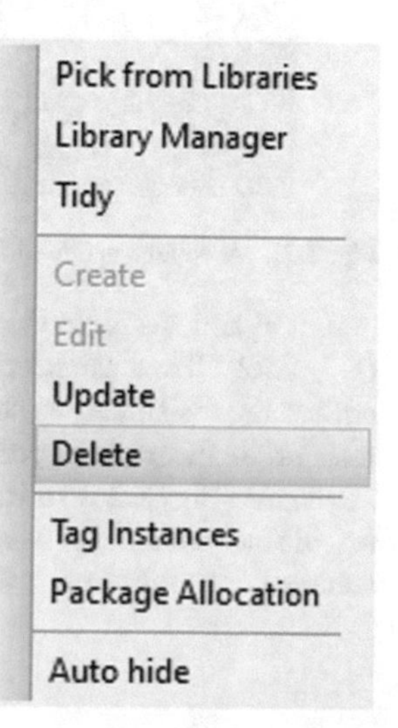

Fig. 1.19 Click delete to remove a component

Select the battery from component list by clicking on it (Fig. 1.20). Then click on the schematic to add the selected block to it (Fig. 1.21). After adding the battery to schematic, press the Esc key of your keyboard.

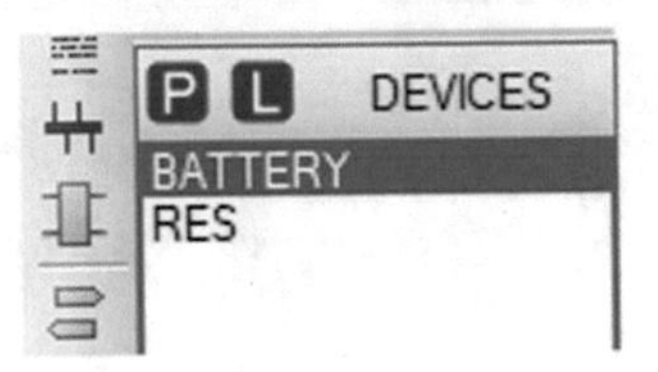

Fig. 1.20 Battery is selected

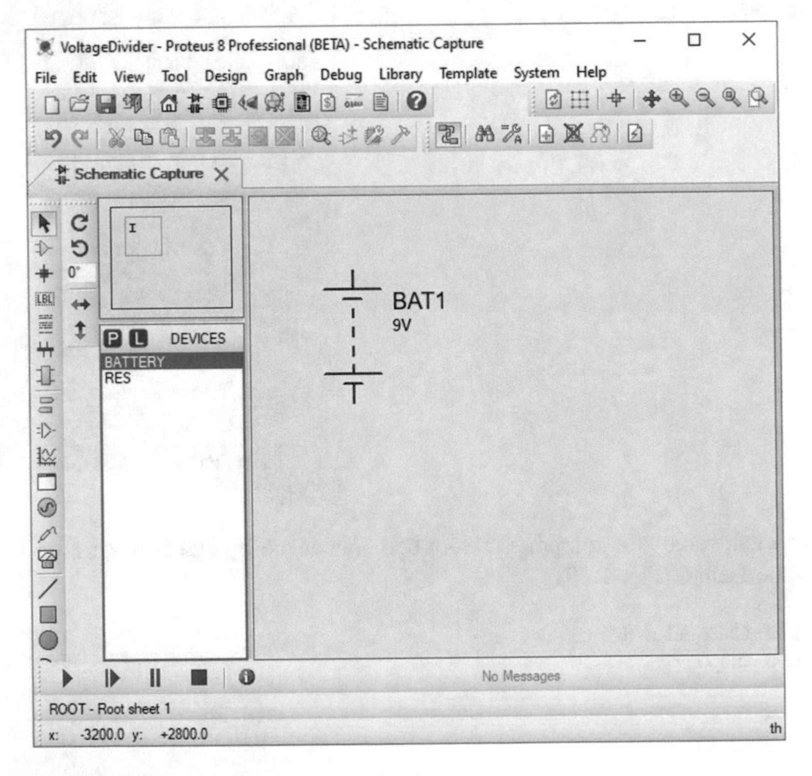

Fig. 1.21 A battery is added to the schematic

Let's add resistors to the schematic. Select the resistor by clicking on it (Fig. 1.22). Then click on the schematic to add a resistor to it (Fig. 1.23). After adding the resistor R_1 to schematic, click on the schematic again. This adds another resistor to the mouse pointer. Press the + or − key of numeric section of your keyboard (Fig. 1.24) to rotate the component. After rotating the component, click on the schematic to place it (Fig. 1.25). After placing the resistor R_2 on the schematic, press the Esc key of your keyboard.

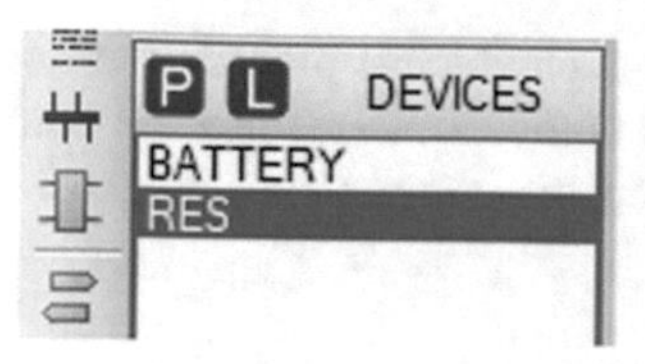

Fig. 1.22 Resistor is selected

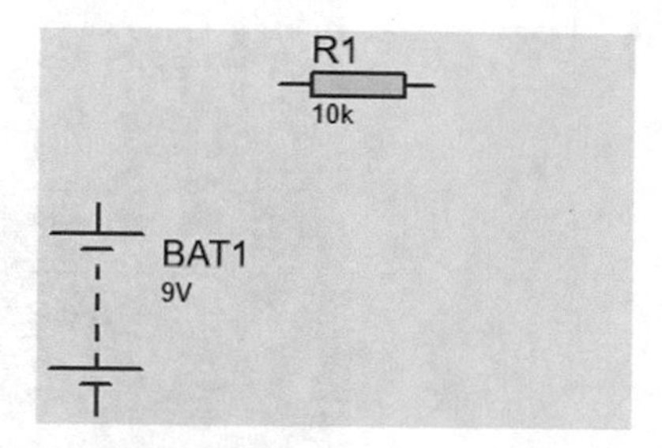

Fig. 1.23 A resistor is added to the schematic

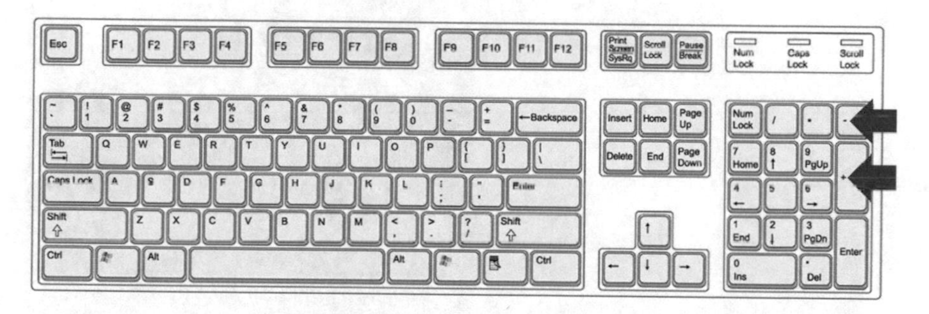

Fig. 1.24 + and − keys of the keyboard

Fig. 1.25 Second resistor is added to the schematic

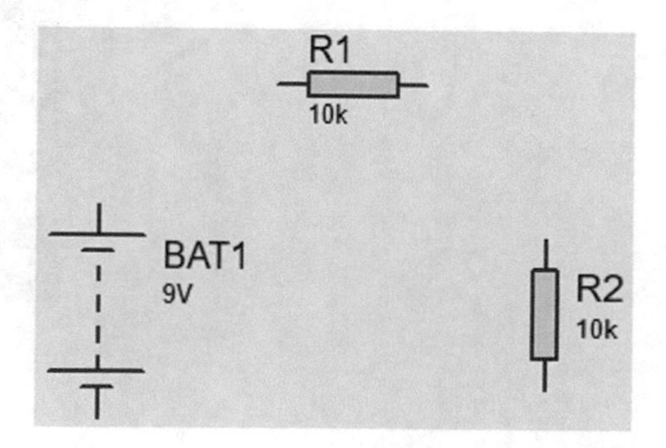

You can rotate a component which is placed on the schematic as well. In order to do this, right click on the component and use the commands shown in Fig. 1.26. Another way is to click on the component and press the + or − key of numeric section of your keyboard.

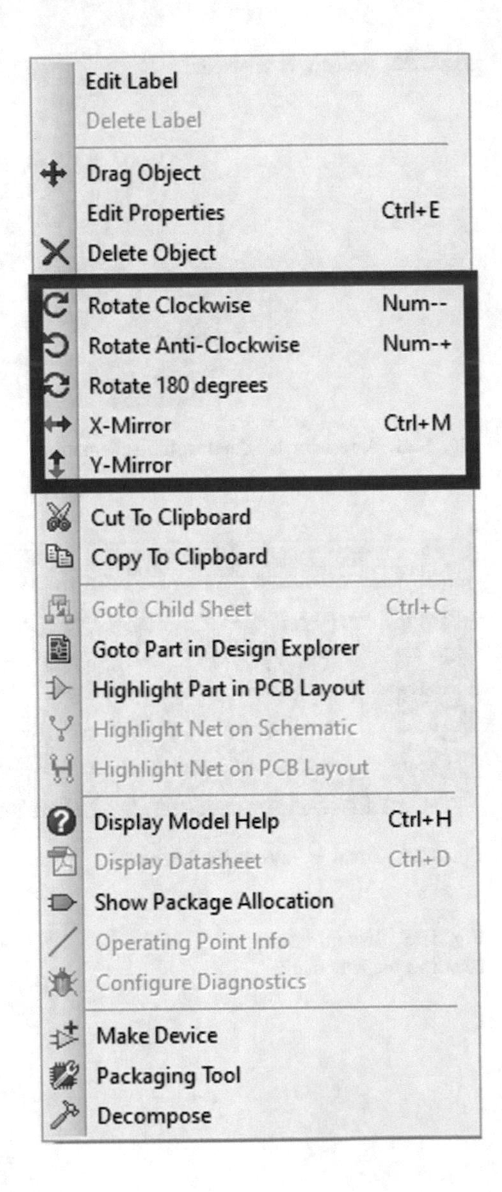

Fig. 1.26 Different rotate commands

Let's determine the value of components. Double click on the battery block and enter "V_1" and "10 V" to the Part Reference and Voltage boxes, respectively (Fig. 1.27). Then click the OK button. After clicking the OK button, the schematic changes to what is shown in Fig. 1.28.

Edit Component
Part Reference: V1 Hidden:
Voltage: 10V Hidden:
Element: New
Simulator Primitive Type: ANALOGUE Hide All
Other Properties:
Exclude from Simulation
Exclude from PCB Layout
Exclude from Current Variant
Attach hierarchy module
Hide common pins
Edit all properties as text
OK
Cancel

Fig. 1.27 Edit component window

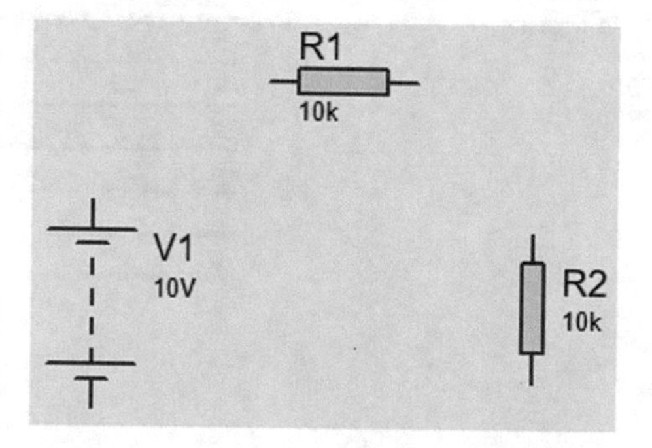

Fig. 1.28 Voltage of battery is changed to 10 V

Double click the R_1 and enter "1*k*" to the Resistance box (Fig. 1.29). Then click the OK button. "1*k*" means 1 kΩ. Table 1.1 shows the list of prefixes that you can use in Proteus.

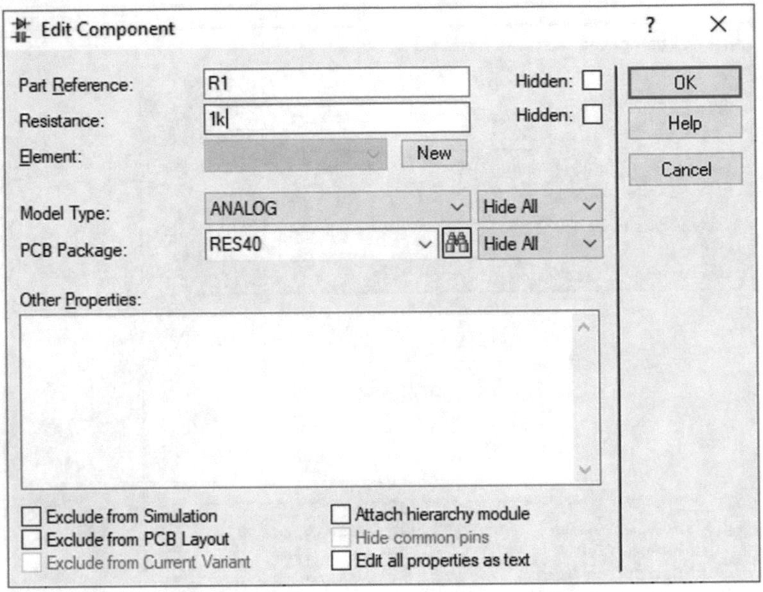

Fig. 1.29 Edit component window

Table 1.1 Available prefixes in Proteus

Prefix	Meaning
M	Mega
k	Kilo
m	Mili
μ	Micro
n	Nano
p	Pico

Double click the R_2 and enter "2.2*k*" to the Resistance box (Fig. 1.30). Then click the OK button.

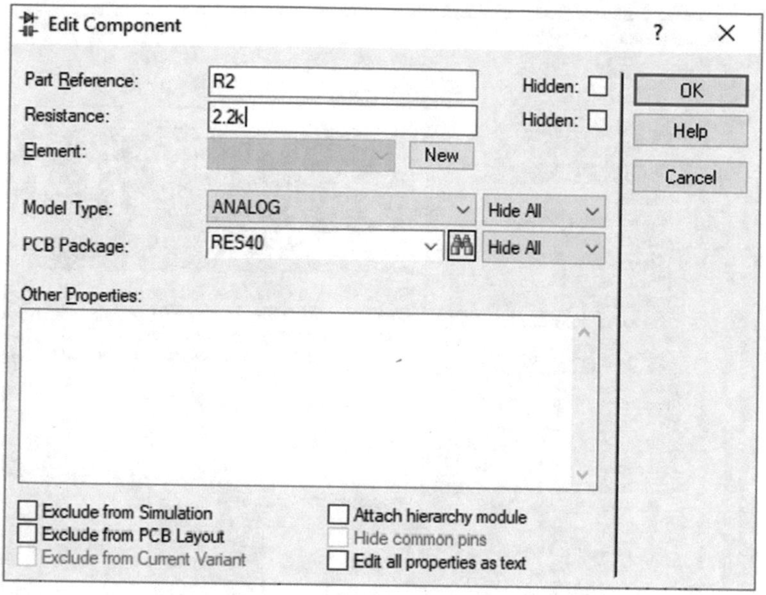

Fig. 1.30 Edit component window

If you click the Help button in Fig. 1.30, the window shown in Fig. 1.31 appears and gives more information about the resistor component.

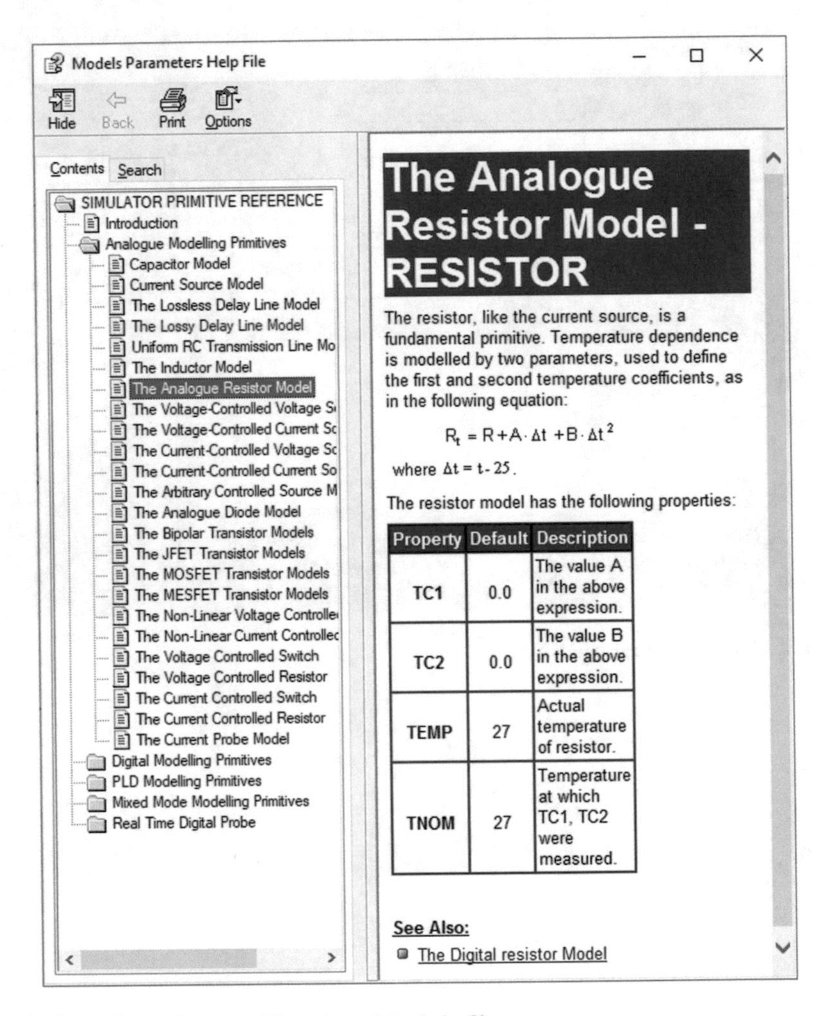

Fig. 1.31 Analog resistor model section of the help file

After entering the component values, the schematic changes to what is shown in Fig. 1.32. It's time to connect the components together. If you bring the mouse pointer close to the component terminals, it changes to a pencil. After you saw the pencil, click on the source terminal. This connects a wire to the source terminal. Then move the wire toward the destination terminal. After you clicked the destination terminal, a wire is drawn between the source and destination terminals. Use the aforementioned method to connect the component together (Fig. 1.33).

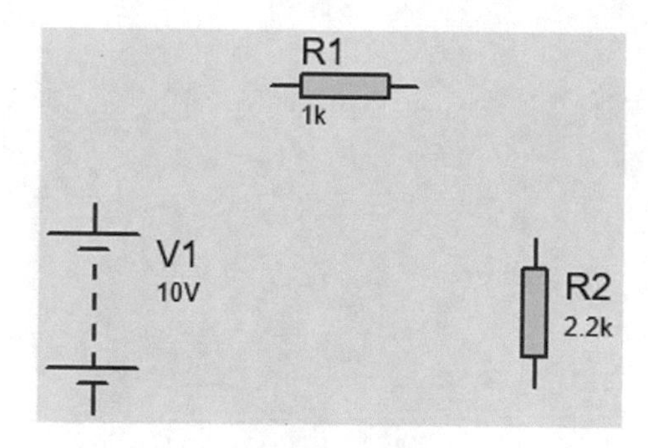

Fig. 1.32 Values of components are entered to the schematic

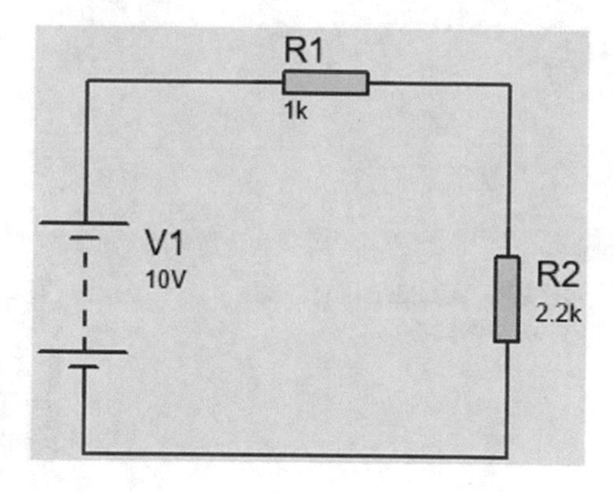

Fig. 1.33 Components are connected to each other

All the circuits must have a ground. Add a ground block (Fig. 1.34) to the schematic (Fig. 1.35).

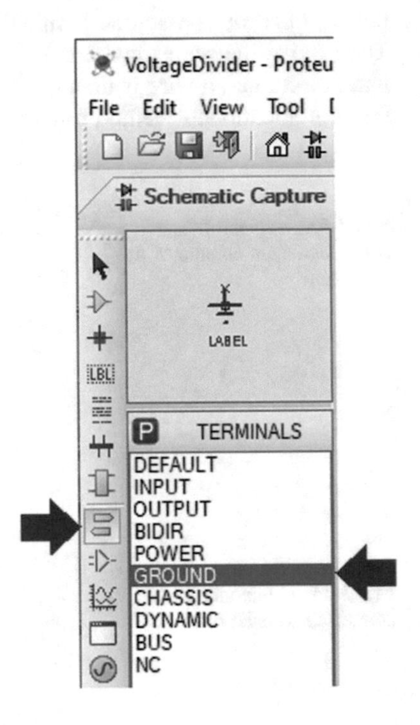

Fig. 1.34 Ground block

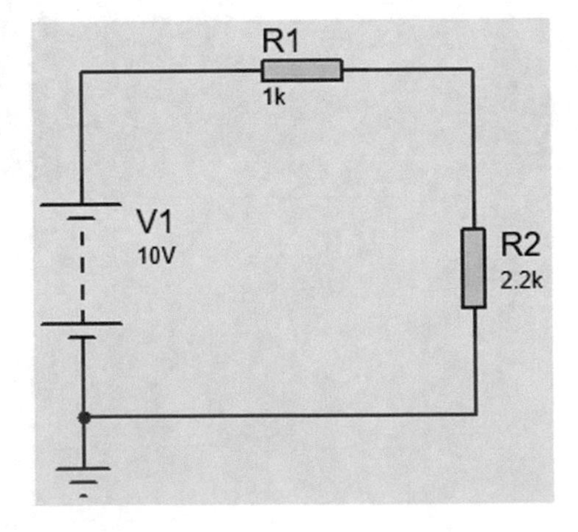

Fig. 1.35 Addition of ground to the circuit

We need a DC voltmeter (Fig. 1.36) to measure the voltage of resistor R_2 (Fig. 1.37).

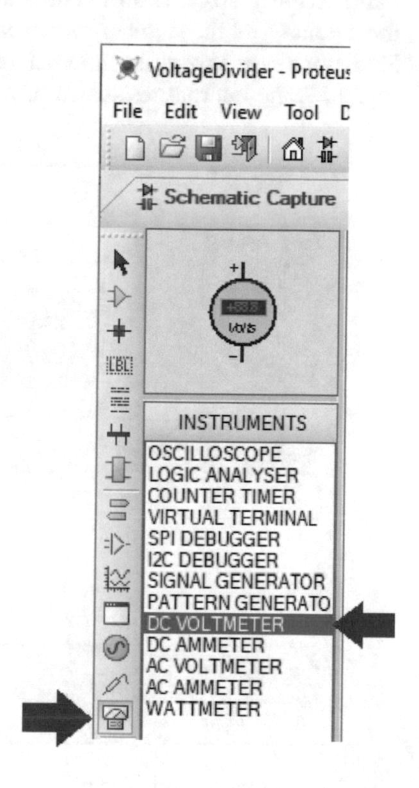

Fig. 1.36 DC voltmeter block

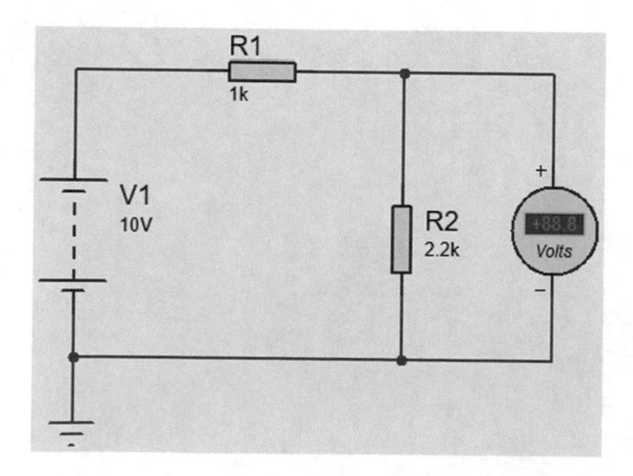

Fig. 1.37 Addition of DC voltmeter to the schematic

If you double click the DC voltmeter, the window shown in Fig. 1.38 appears. A descriptive text can be entered to the Part Reference box. The text entered to the Part Reference box is shown behind the DC voltmeter. So, it helps the user to know the meaning of the number which is displayed on the DC voltmeter. The Load Resistance box shows the internal resistance of the DC voltmeter. According to Fig. 1.38, the internal resistance of the DC voltmeter is 100 MΩ.

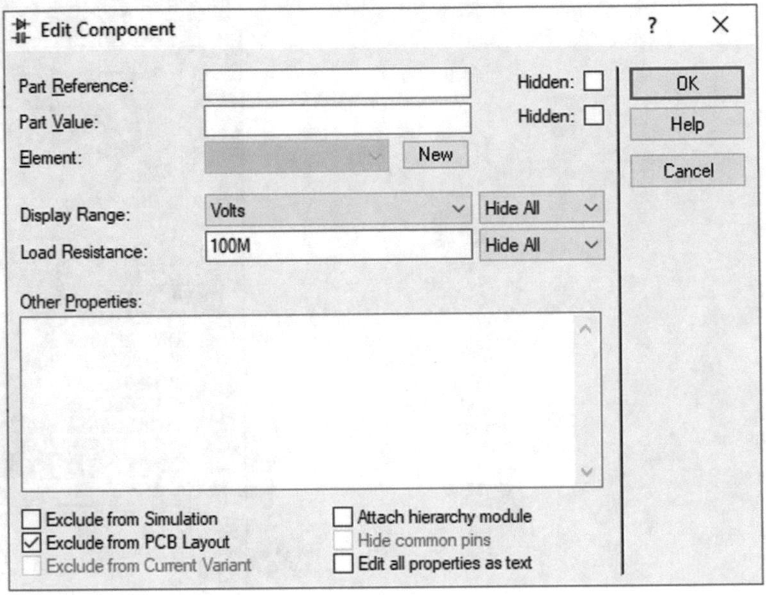

Fig. 1.38 Edit component window

The Display Range drop down list (Fig. 1.39) sets the accuracy of the voltmeter. Table 1.2 shows the possible measurement range for each of the options. DC voltmeter shows + MAX or − MAX for values outside the given ranges. For instance, you cannot measure 1.2 V with Millivolts. Select Volts for this example.

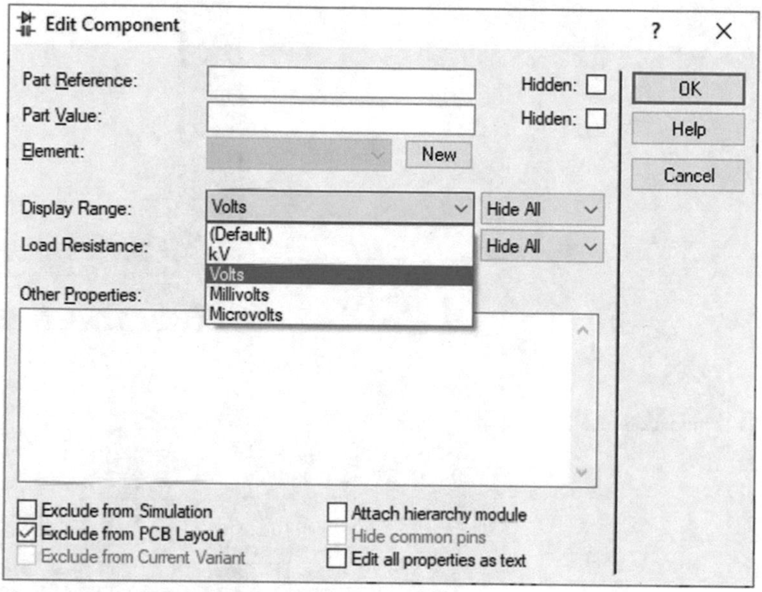

Fig. 1.39 Edit component window

Table 1.2 Possible measurement ranges for each of the display range options

Display range	Possible measurement range
kV	[− 999 kV, + 999 kV]
Volts	[− 999 V, + 999 V]
Millivolts	[− 999 mV, + 999 mV]
Microvolts	[− 999 μV, + 999 μV]

Click the icon shown in Fig. 1.40 to run the simulation. Simulation result is shown in Fig. 1.41.

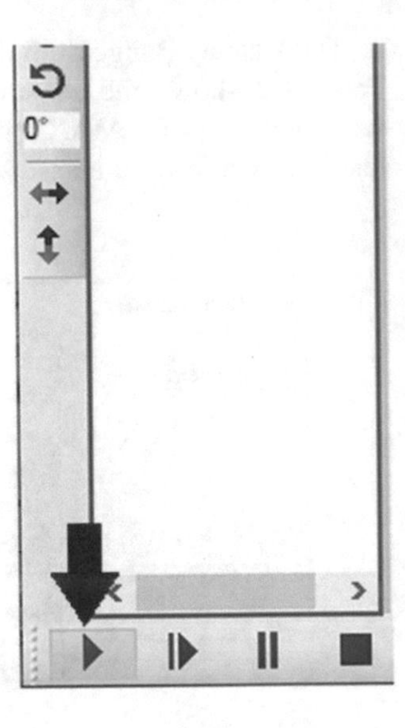

Fig. 1.40 Icon to run the simulation

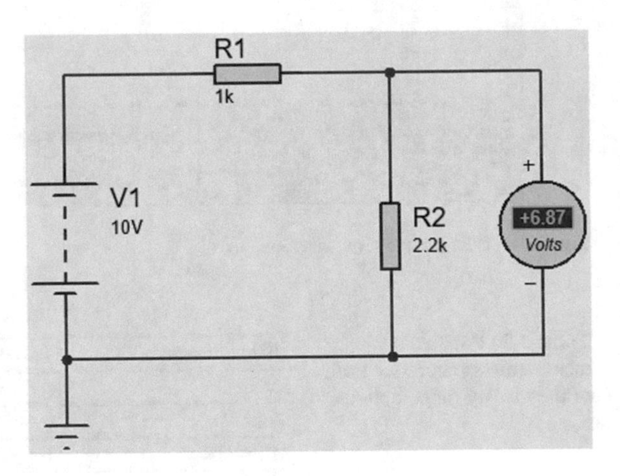

Fig. 1.41 Simulation result

Click the icon shown in Fig. 1.42 to stop the simulation.

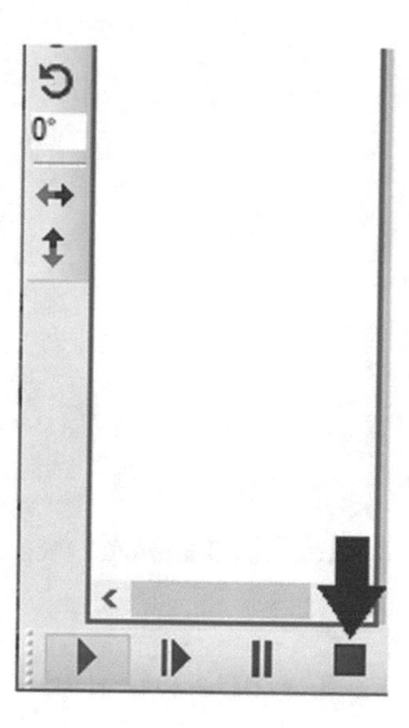

Fig. 1.42 Icon to stop the simulation

Let's measure the circuit current. A DC ammeter can be used to measure the circuit current. Right click on the wire which connects the voltage source to the resistor R_1. Then click the Delete Wire (Fig. 1.43). This removes the wire (Fig. 1.44). Another way to remove the wire between voltage source and resistor R_1 is double right click on the wire.

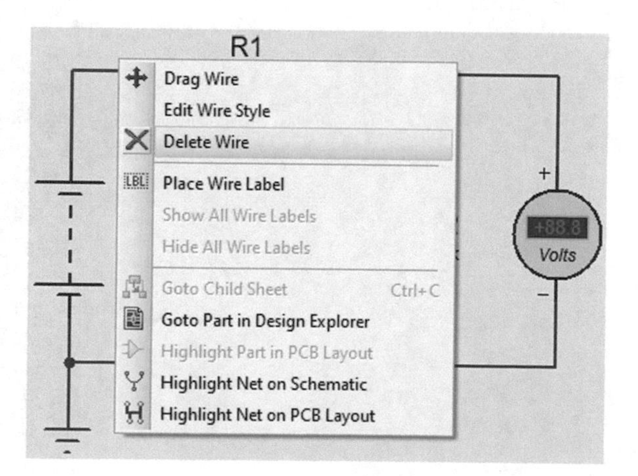

Fig. 1.43 Deleting a wire

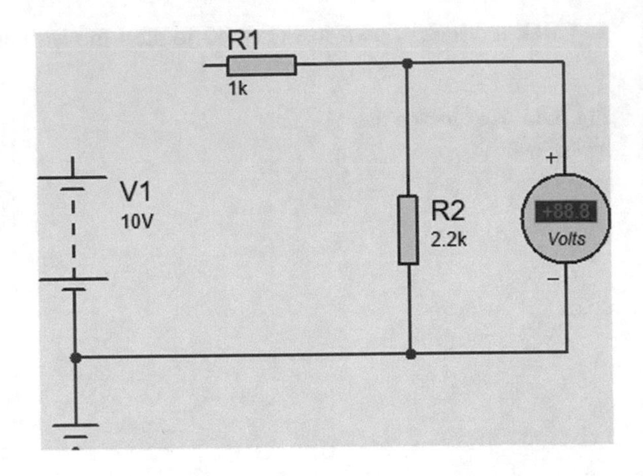

Fig. 1.44 Wire between V_1 and R_1 is deleted

Add a DC ammeter (Fig. 1.45) to the schematic (Fig. 1.46).

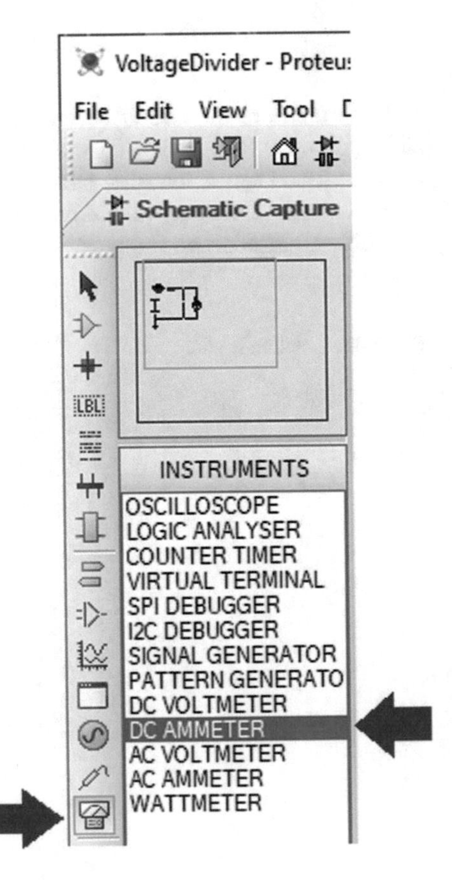

Fig. 1.45 DC ammeter block

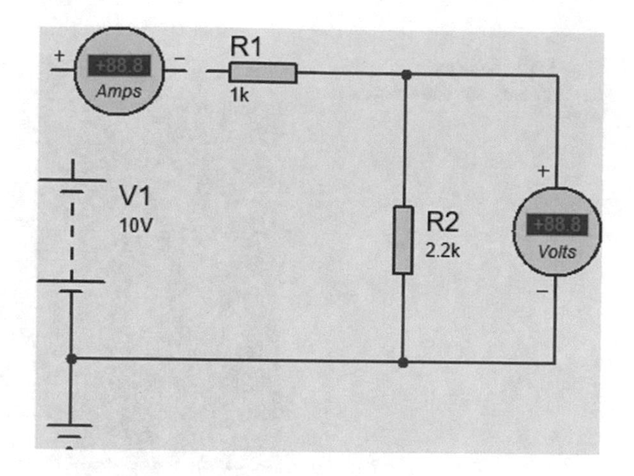

Fig. 1.46 Addition of a DC ammeter block to the schematic

Click on an empty point of the schematic and without releasing the mouse left button, draw a rectangle around the resistor R_1, the resistor R_2 and the DC voltmeter (Fig. 1.47). After drawing the rectangle around the aforementioned components, release the mouse left button. After releasing the mouse left button, the color of components inside the rectangle change to red (Fig. 1.48). Now left click on the rectangle and drag the rectangle toward right (Fig. 1.49). This makes space for the ammeter (Fig. 1.50).

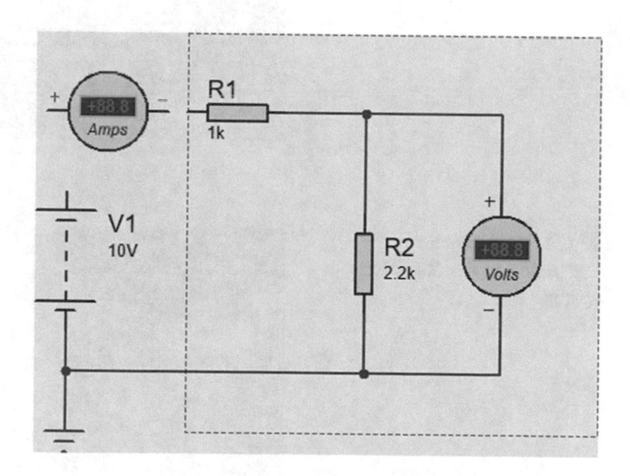

Fig. 1.47 Selecting the R_1, R_2 and DC voltmeter

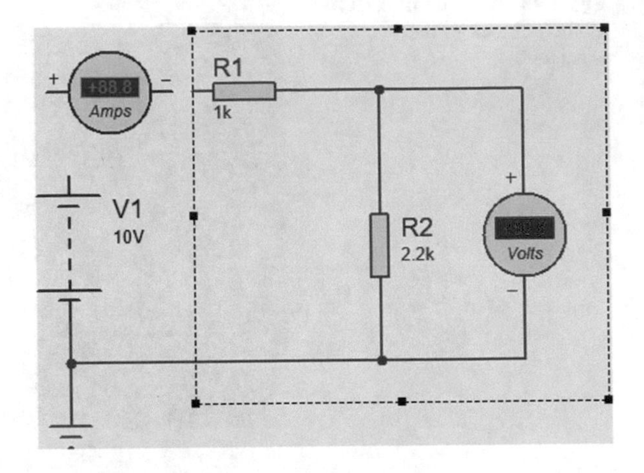

Fig. 1.48 Selected components are shown with red

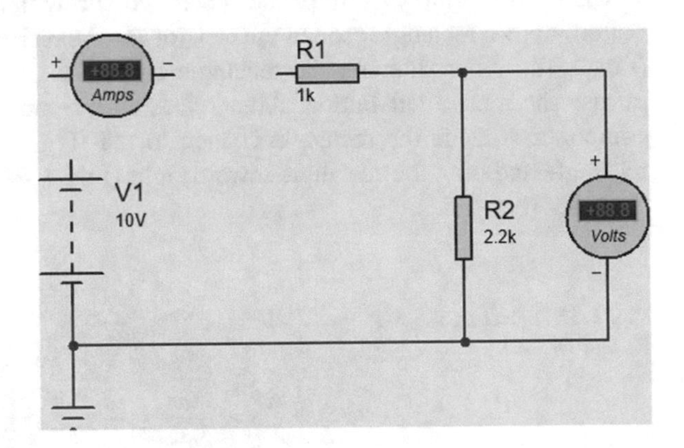

Fig. 1.49 Selected components are moved to left

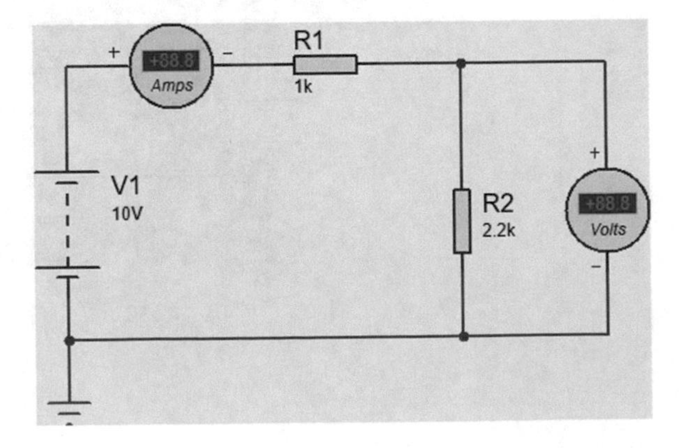

Fig. 1.50 Ammeter is connected to the rest of the system

Run the simulation. The simulation result is shown in Fig. 1.51. The ammeter shows the 0.00 A. Let's increase the accuracy of the ammeter. Stop the simulation and double click the ammeter. Then select the Milliamps for Display Range (Fig. 1.52). After selecting the Milliamps for Display Range, the schematic changes to what is shown in Fig. 1.53.

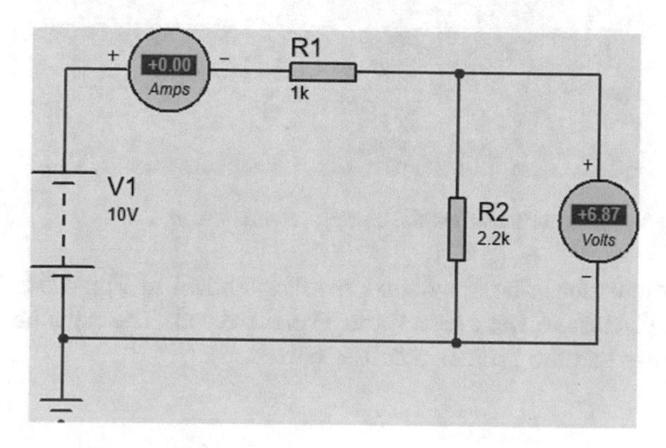

Fig. 1.51 Simulation result

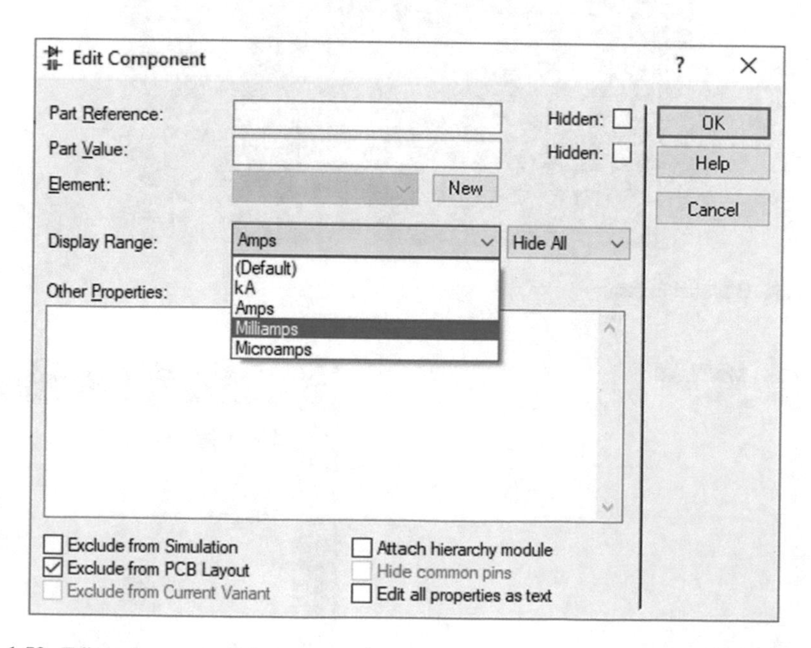

Fig. 1.52 Edit component window

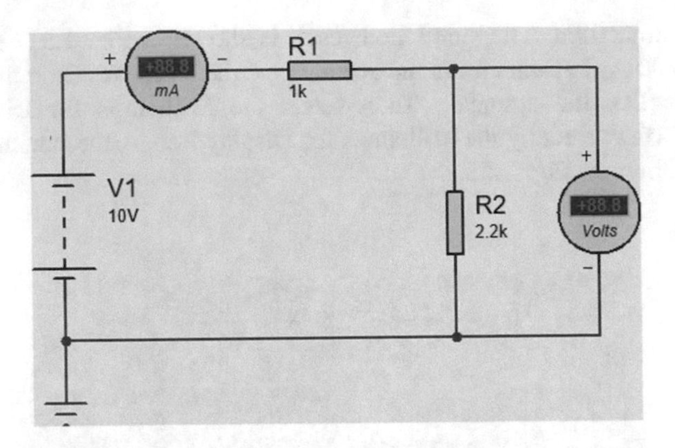

Fig. 1.53 Now the ammeter measures the current with unit of mA

Run the simulation. The simulation result is shown in Fig. 1.54. Now we can read the circuit current. Let's check the Proteus result. The calculation shown in Fig. 1.55 shows that the Proteus result is correct.

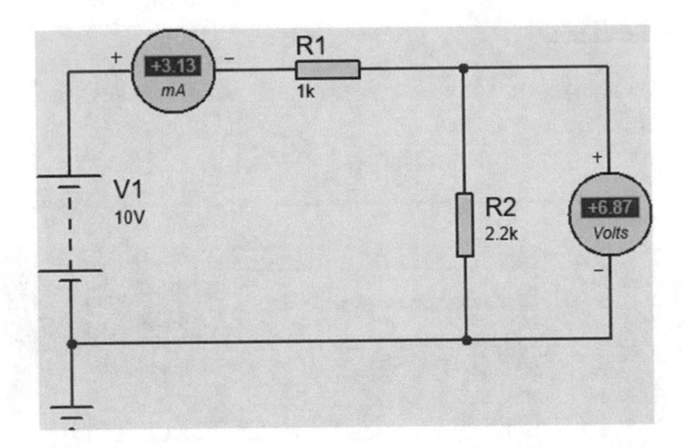

Fig. 1.54 Simulation result

Fig. 1.55 MATLAB calculations

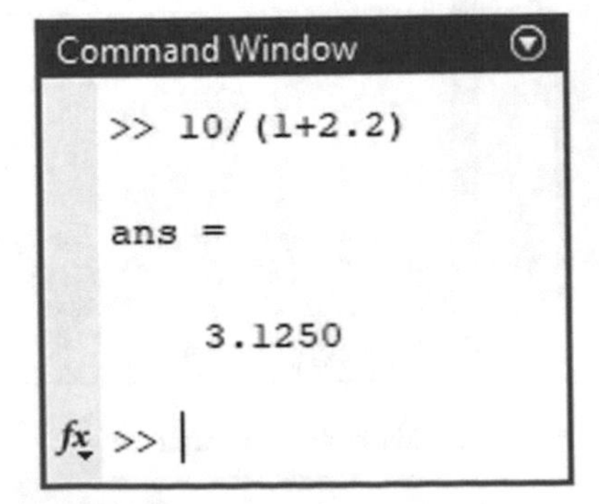

1.3 Example 2: Project Documentation and Reporting

Proteus has a text editing tool that permits you to prepare documents and reports for your design. In this example, we study this tool. Click the Project Notes icon (Fig. 1.56) to run the text editing tool.

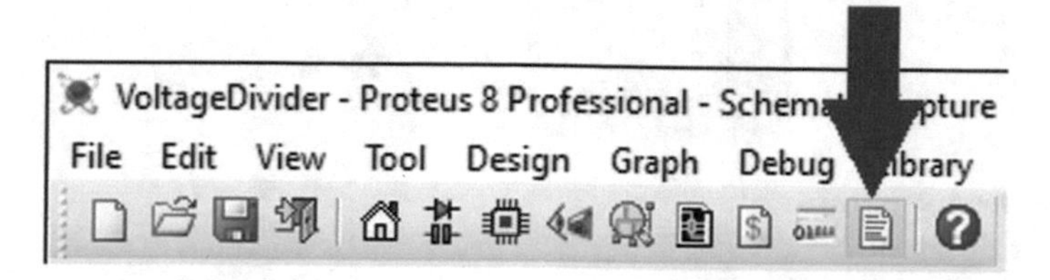

Fig. 1.56 Project notes icon

After clicking the Project Notes icon, the environment shown in Fig. 1.57 appears and you can write what you want. Use the Insert menu to add image to the document (Fig. 1.58). You can save what you wrote by clicking the Save icon (Fig. 1.59).

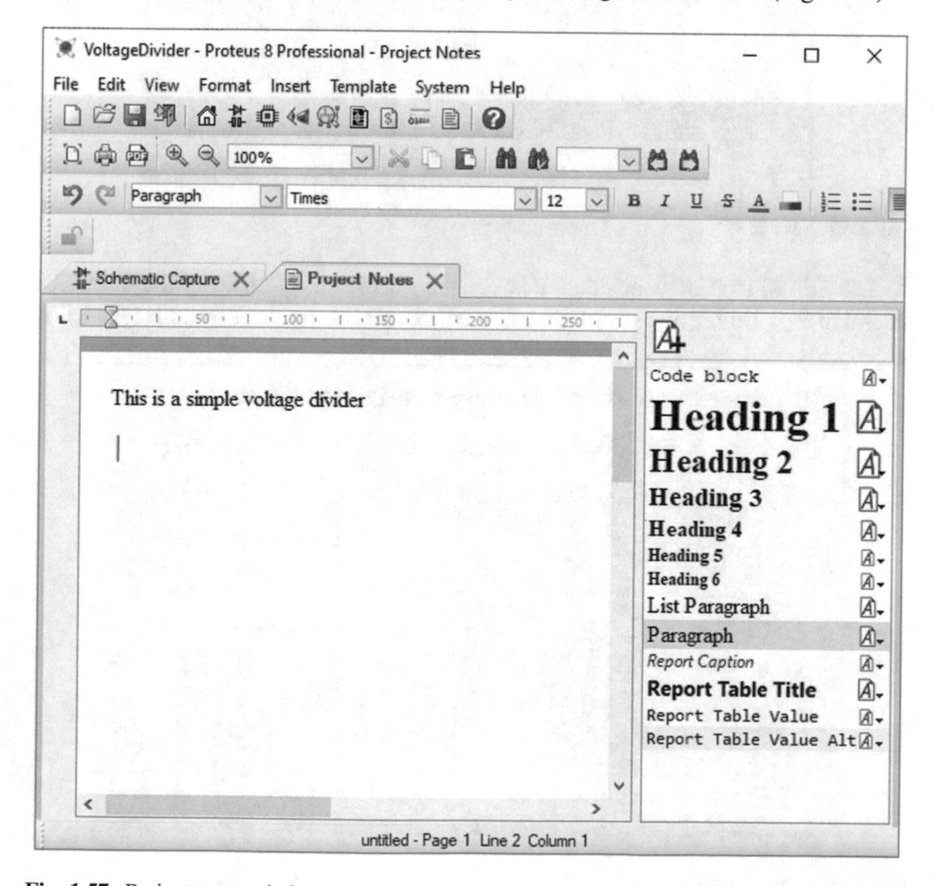

Fig. 1.57 Project notes window

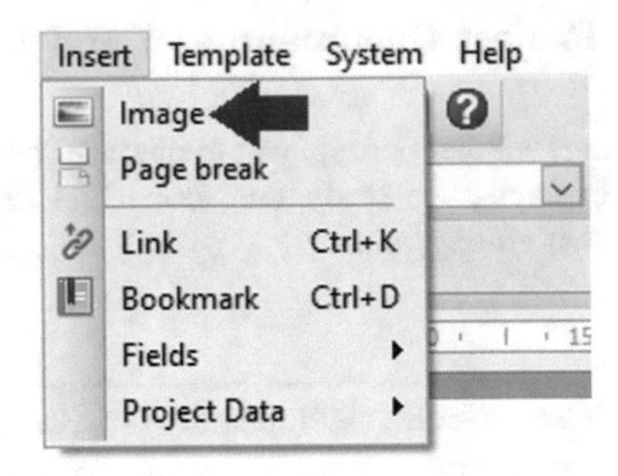

Fig. 1.58 Insert > image

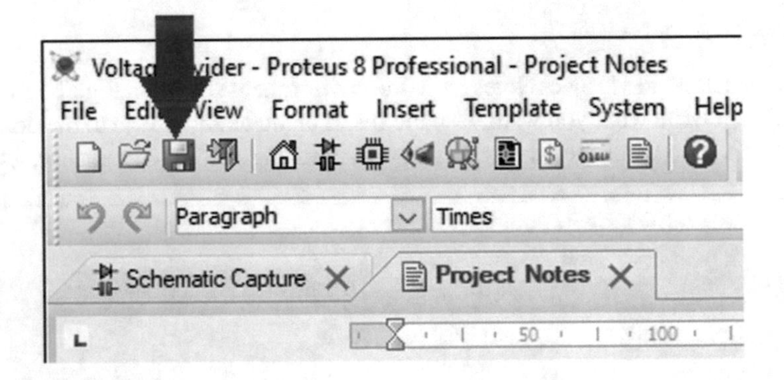

Fig. 1.59 Save icon

The Pdf version of your document can be generated by clicking the PDF icon shown in Fig. 1.60. You can close the text editor by clicking the *X* button (Fig. 1.60).

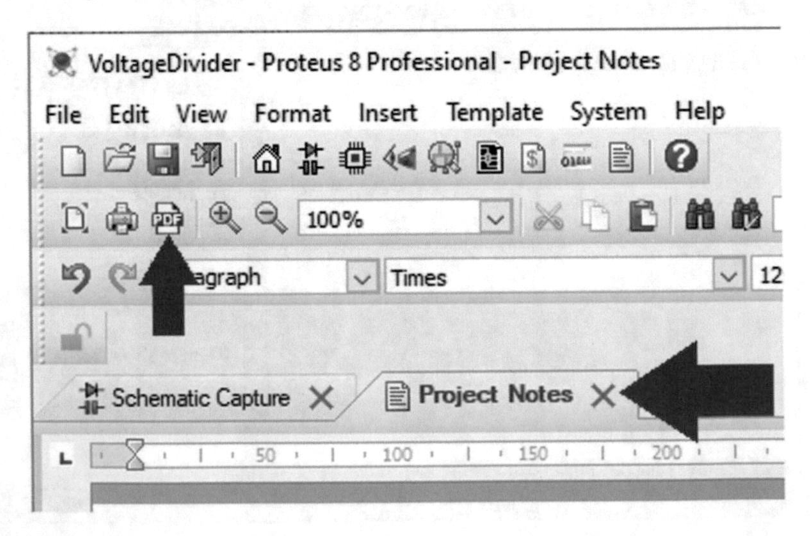

Fig. 1.60 PDF and close icons

The next time you open the text editor, you see a red padlock icon (Fig. 1.61) in the toolbar. When the padlock is locked, you cannot edit the document. You need to unlock the padlock by clicking on it (Fig. 1.62). After unlocking the padlock, you can apply changes to what you wrote before.

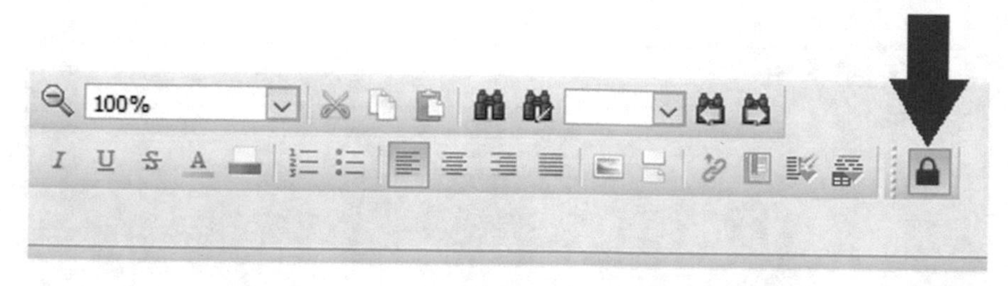

Fig. 1.61 Red (locked) padlock icon

Fig. 1.62 Green (unlocked) padlock icon

1.4 Example 3: Addition of Text to Schematic

In this example, we will see how to add text to the schematic. You can add text to the schematic with the aid of 2D Graphics Text Mode icon (Fig. 1.63).

Fig. 1.63 2D graphics text mode icon

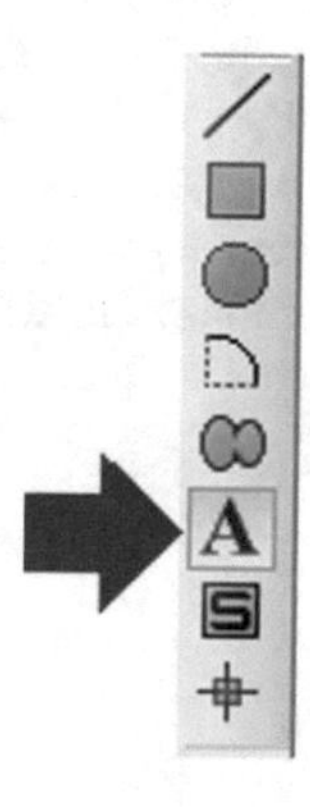

After clicking the 2D Graphics Text Mode icon, the mouse pointer changes to a pencil. Click the desired location of the schematic that you want to add the text there. After clicking, the window shown in Fig. 1.64 appears on the screen. Enter the desired text into the String box and click the OK button. After clicking the OK button, the entered text is added to the schematic.

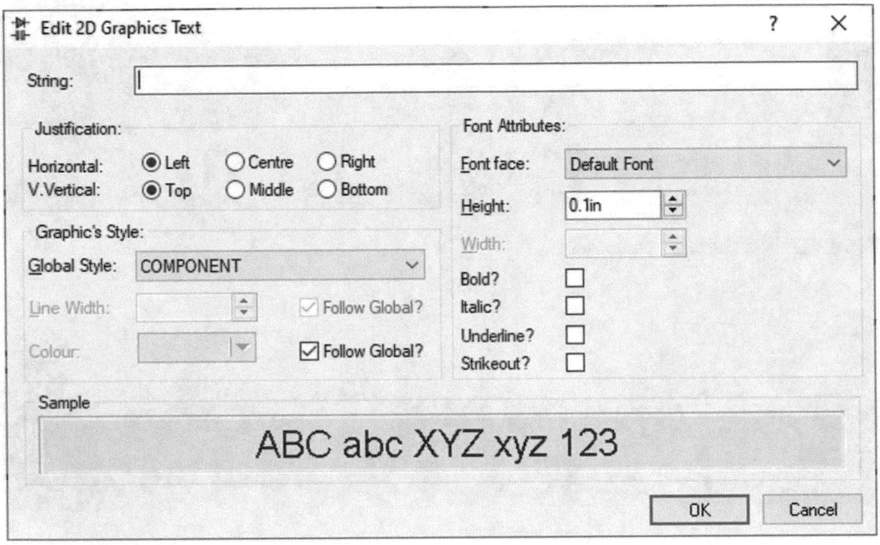

Fig. 1.64 Edit 2D graphics text window

1.5 Example 4: Defining Variables

In the Example 1, we entered the values of components directly to the components. In this example, we will see how to define variables in Proteus environment.

Open the schematic of Example 1 and click the Text Script Mode icon (Fig. 1.65). After clicking the Text Script Mode icon, the window shown in Fig. 1.66 appears on the screen. Enter the commands shown in Fig. 1.67 and click the OK button. These commands define two variables: R_1 with value of $1k$ (= 1000) and R_2 with value of $2.2k$ (= 2200). After clicking the OK button, the schematic changes to what is shown in Fig. 1.68.

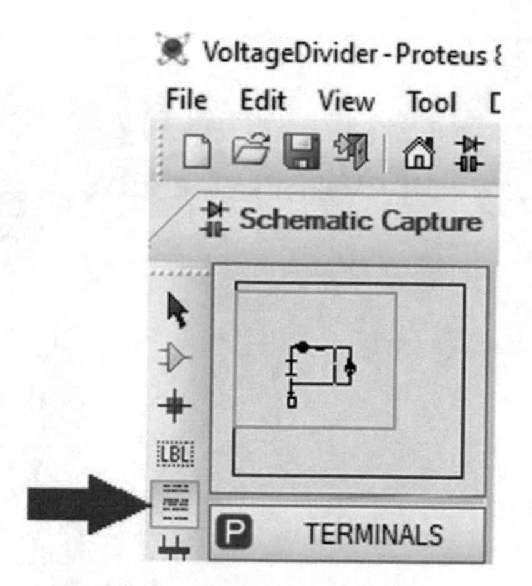

Fig. 1.65 Text script mode icon

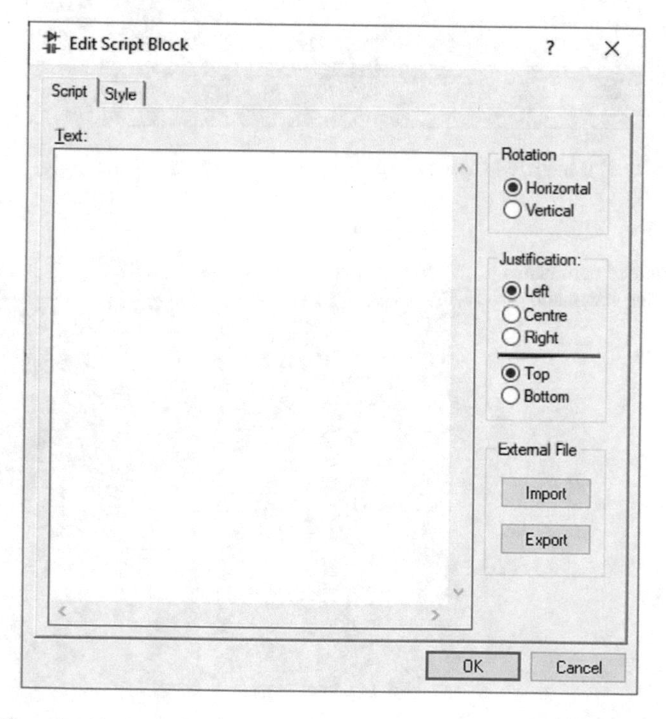

Fig. 1.66 Edit script block window

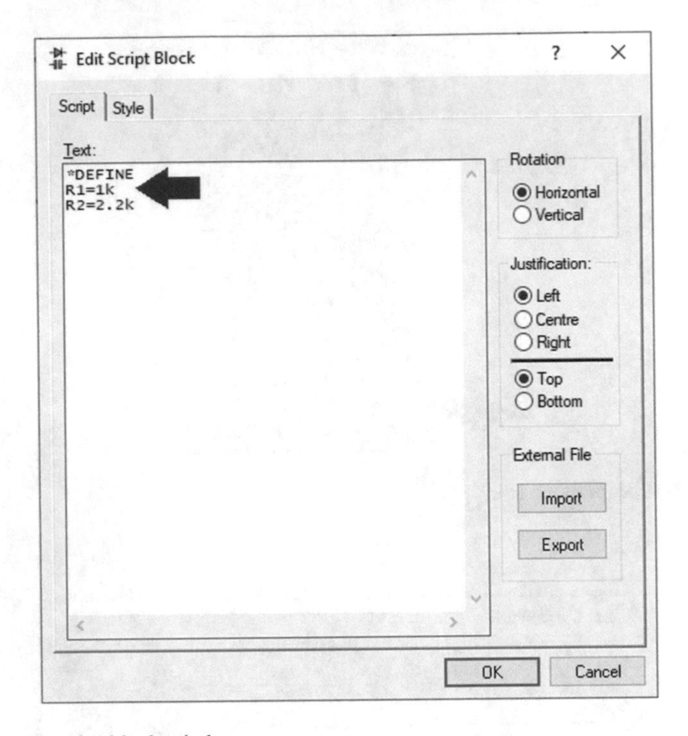

Fig. 1.67 Edit script block window

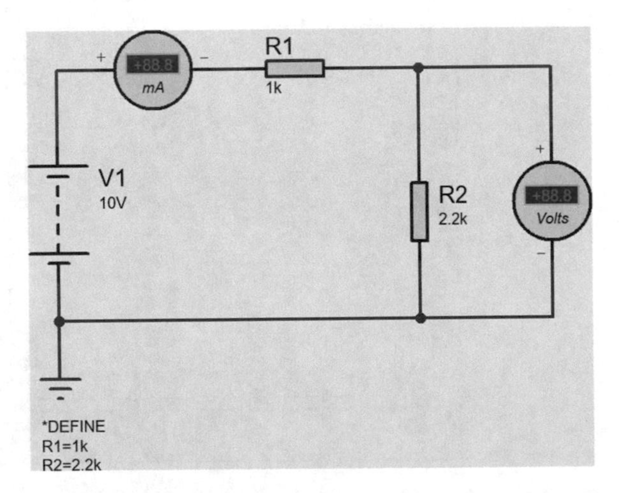

Fig. 1.68 Entered commands are added to the schematic

Double click the resistor R_1 and enter $\langle R_1 \rangle$ to the Resistance box (Fig. 1.69). After clicking the OK button, the schematic changes to what is shown in Fig. 1.70.

Double click the resistor R_2 and enter $\langle R_2 \rangle$ to the Resistance box (Fig. 1.71). After clicking the OK button, the schematic changes to what is shown in Fig. 1.72.

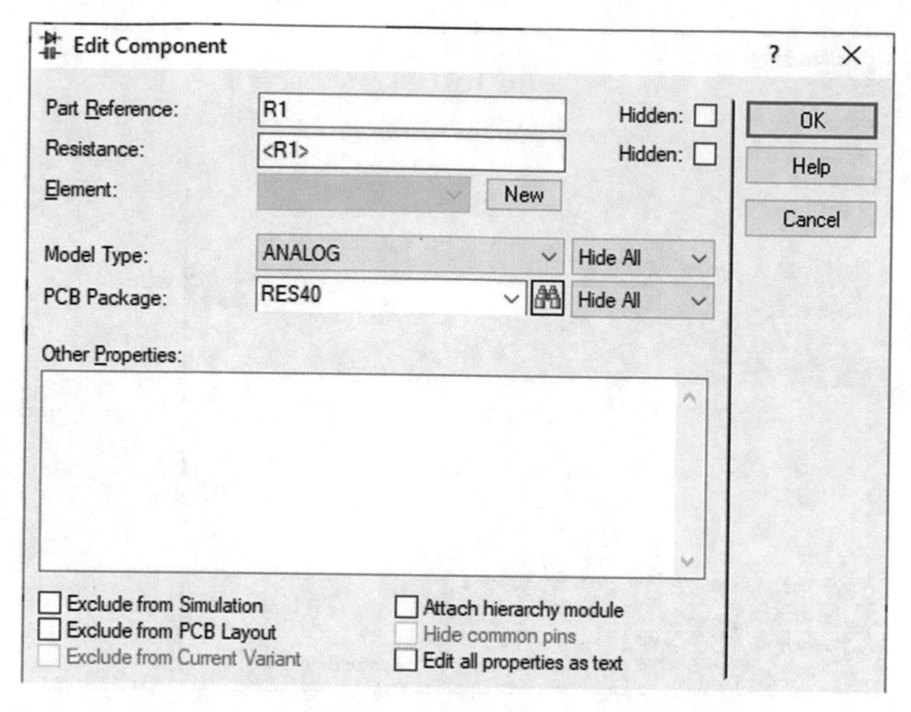

Fig. 1.69 Edit component window

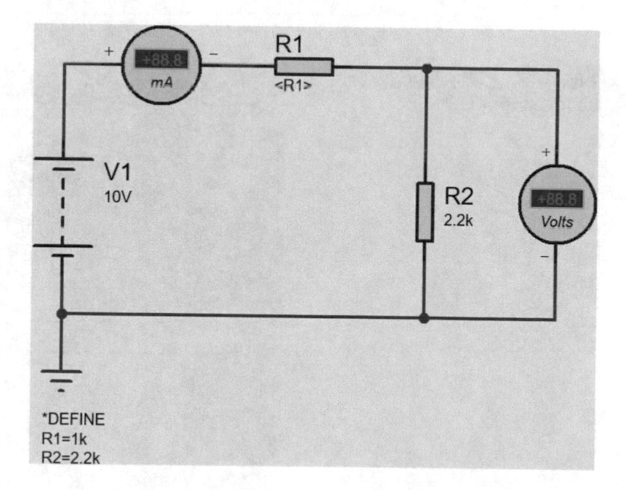

Fig. 1.70 Value of R_1 is changed to $\prec R_1 \succ$

Edit Component ? ×

Part Reference: R2 Hidden: ☐ OK
Resistance: <R2> Hidden: ☐ Help
Element: New Cancel
Model Type: ANALOG Hide All
PCB Package: RES40 Hide All
Other Properties:

☐ Exclude from Simulation ☐ Attach hierarchy module
☐ Exclude from PCB Layout ☐ Hide common pins
☐ Exclude from Current Variant ☐ Edit all properties as text

Fig. 1.71 Edit component window

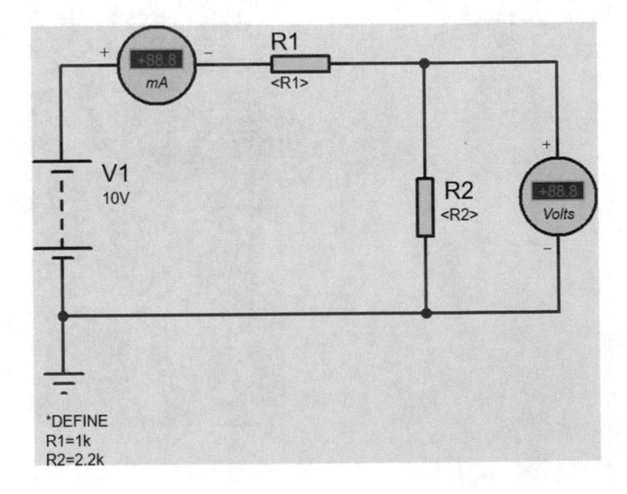

Fig. 1.72 Value of R_2 is changed to $\langle R_2 \rangle$

Run the simulation. The simulation result is shown in Fig. 1.73. The obtained result is the same as the result of Example 1.

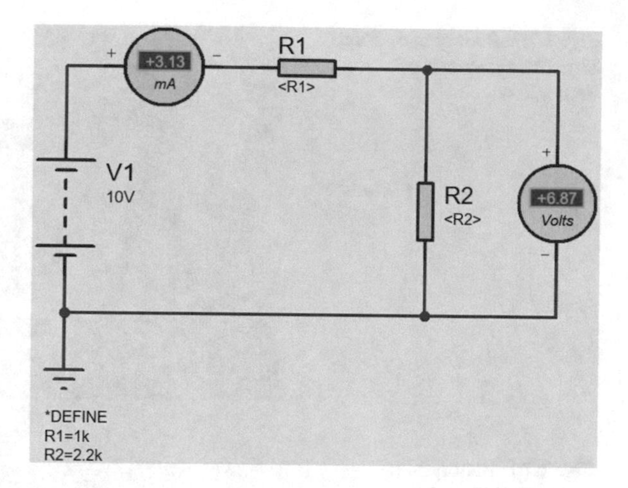

Fig. 1.73 Simulation result

Using variables is a simple way for apply changes to the schematic. For instance, consider the balanced load shown in Fig. 1.74. You need to double click the three resistors and enter their new values manually when you don't use the variables. When you define a variable, all the three resistor values are updated as soon as the variable value changes.

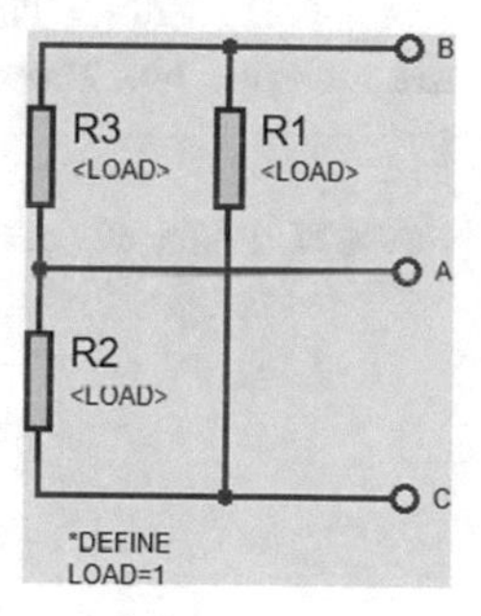

Fig. 1.74 Balanced delta connected load

1.6 Example 5: Removing the Unused Components from the Component List

In this example, we see how to remove unused components from the component list. Reopen the schematic of Example 1 and remove the battery (Fig. 1.75). The battery block is not used in the schematic however it is available in the component list (Fig. 1.76).

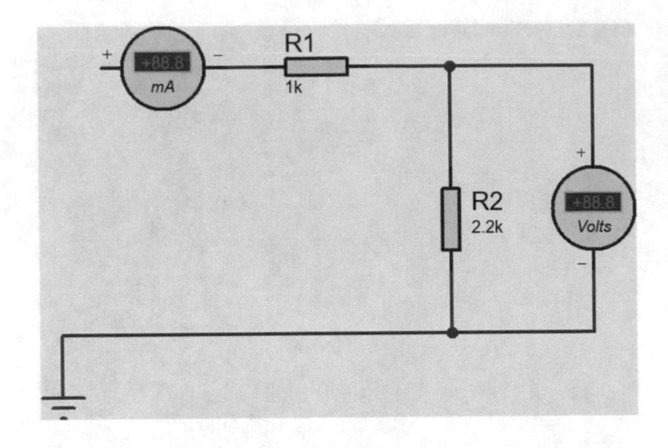

Fig. 1.75 Battery is removed from the schematic of Example 4

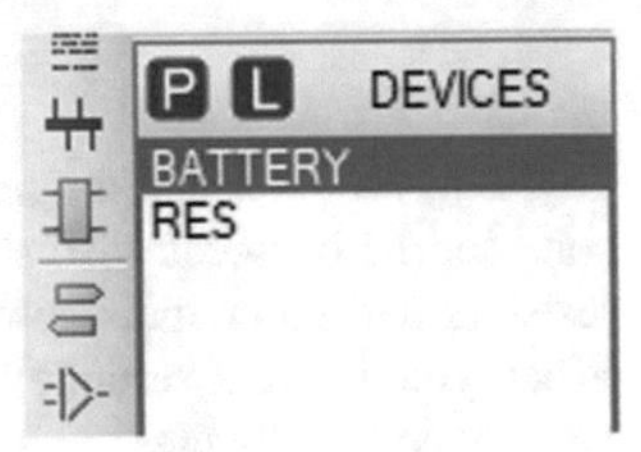

Fig. 1.76 Battery block is still available in the component list

Click the Edit > Tidy Design (Fig. 1.77). After clicking the Edit > Tidy Design, the message box shown in Fig. 1.78 appears on the screen. Click the OK button.

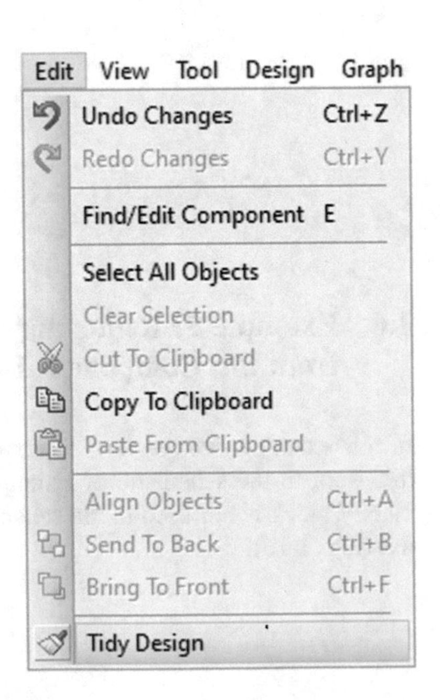

Fig. 1.77 Edit > tidy design

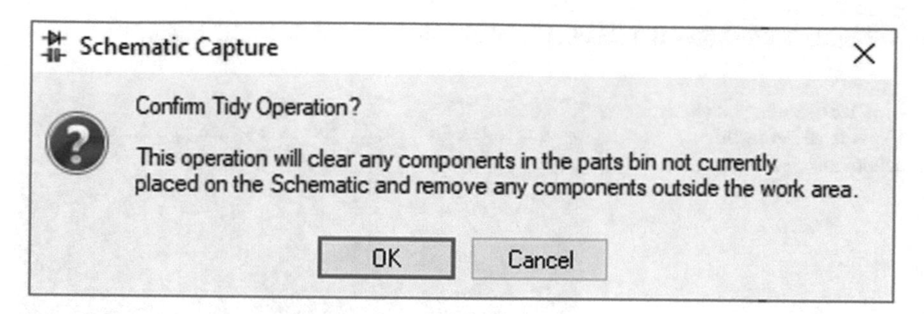

Fig. 1.78 Schematic capture window

After clicking the OK button, the unused components are removed from the component list (Fig. 1.79).

Fig. 1.79 Battery block is removed from the component list

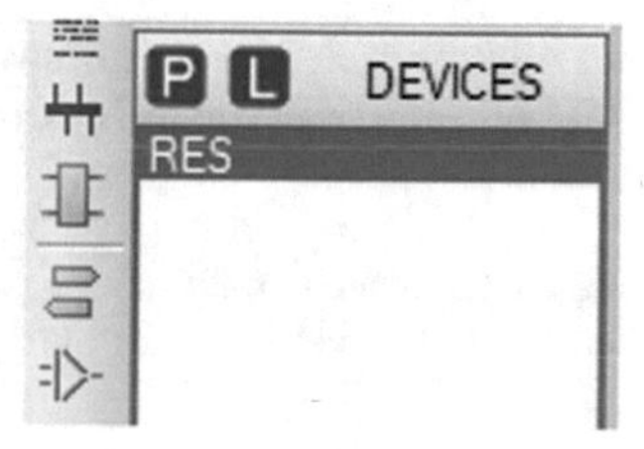

1.7 Example 6: V_{source} Block

In the Example 1, we used a "battery" block to simulate a DC voltage source. Proteus has another block called "v_{source}" (Fig. 1.80) which can be used to simulate a DC voltage source.

Fig. 1.80 Searching for a voltage source block

Figure 1.81 shows the schematic of Example 1 with a v_{source} block.

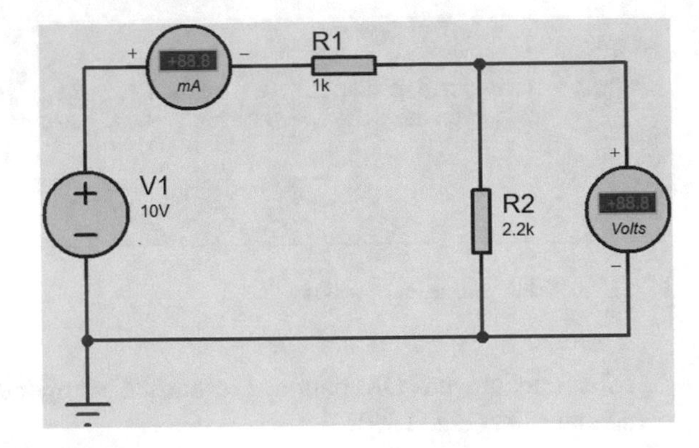

Fig. 1.81 Voltage source block is added to the schematic

1.8 Example 7: C_{source} Block

In this example, we learn how to simulate a constant DC current source. A constant DC current source can be simulated with the aid of "c_{source}" block (Fig. 1.82).

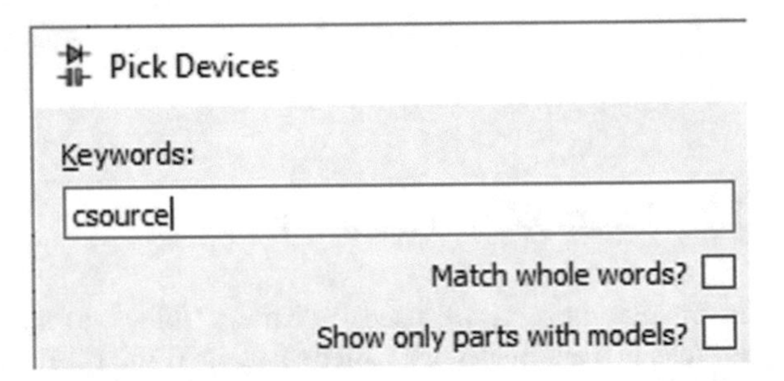

Fig. 1.82 Searching for current source block

Consider the schematic shown in Fig. 1.83. The voltage drop across the resistor is 2.2 V which is the correct value since $V_{R_1} = I_1 \times R_1 = 1\,\text{mA} \times 2.2\,\text{k}\Omega = 2.2\,\text{V}$.

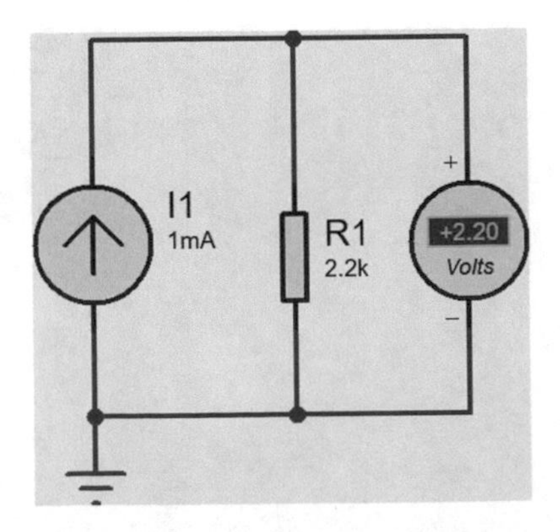

Fig. 1.83 Simulation result

1.9 Example 8: V_{sine} Block

Sinusoidal voltage sources can be simulated with the aid of v_{sine} block (Fig. 1.84). This block can be used to produce the voltage waveforms with equation $f(t) = A + B\sin(2\pi \times f \times t - \varphi_0) \times e^{-D\left(t-\frac{\varphi_0}{2\pi \times f}\right)} \times H\left(t - \frac{\varphi_0}{2\pi \times f}\right)$ where $0 \le \varphi_0 < 2\pi$ and H (.) shows the Heaviside step function $\left(\text{i.e., } H(t - t_0) = \begin{cases} 1 & t > t_0 \\ 0 & t < t_0 \end{cases}\right)$.

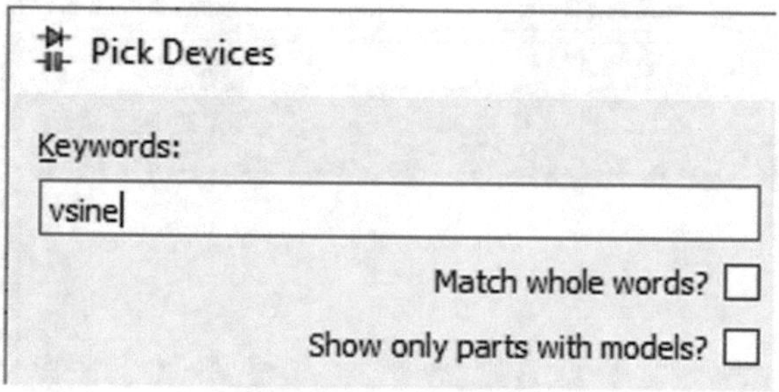

Fig. 1.84 Sinusoidal voltage source block

If you double click the v_{sine} block, the window shown in Fig. 1.85 appears on the screen. Definition of these parameters are shown in Fig. 1.86. Figure 1.86 is taken from Proteus help.

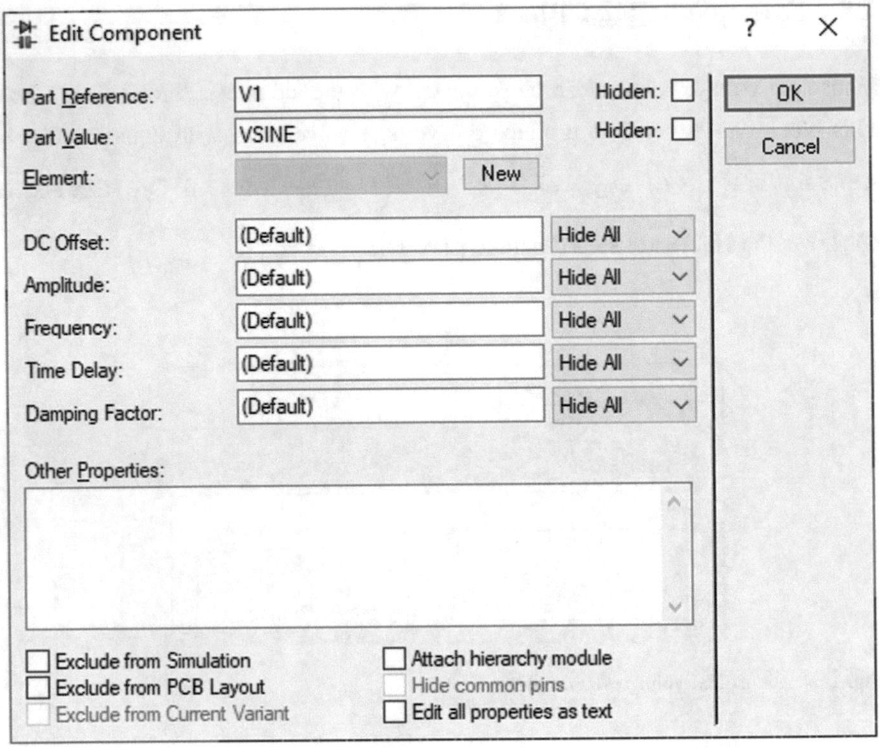

Edit Component ? ×

Part Reference: V1 Hidden: ☐ OK

Part Value: VSINE Hidden: ☐ Cancel

Element: New

DC Offset: (Default) Hide All

Amplitude: (Default) Hide All

Frequency: (Default) Hide All

Time Delay: (Default) Hide All

Damping Factor: (Default) Hide All

Other Properties:

☐ Exclude from Simulation ☐ Attach hierarchy module

☐ Exclude from PCB Layout ☐ Hide common pins

☐ Exclude from Current Variant ☐ Edit all properties as text

Fig. 1.85 Edit component window

Sine

This is used to produce continuous sinusoidal waves at a fixed frequency.

The output level is specified as a peak amplitude (VA) with optional DC offset (VO). The amplitude may also be specified in term of RMS or peak-peak values.

The frequency of oscillation can be given in Hz (FREQ), or as a period (PER), or in terms of the number of cycles over the entire graph.

A phase shift can be specified in either degrees (PHASE) or as a time delay (TD). In the latter case, oscillation does not start until the specified time.

Exponential decay of the waveform after the start of oscillations is specified by the damping factor, THETA.

Mathematically, the output is given by:

$$V = VO + VAe^{-(t-TD)THETA}\sin(2\pi FREQ(t-TD))$$

for t >= TD. For t < TD the output is simply the offset voltage, VO.

Fig. 1.86 Sinusoidal voltage source parameters

Assume that you want to generate the $V(t) = 10 + 20\sin(2\pi \times 50 \times t - 60°)$. First of all we need to write the waveform as $V(t) = 10 + 20\sin\left(2\pi \times 50 \times t - \frac{\pi}{3}\right) = 10 + 20\sin(2\pi \times 50 \times (t - 3.333m))$. Let's enter the values to v_{sine} block (Fig. 1.87).

Fig. 1.87 Edit component window

The settings shown in Fig. 1.87 generate the waveform shown in Fig. 1.88. The waveform equals to the value of DC offset (in this example 10 V) for [0, TD] interval. TD shows the value entered to the Time Delay box. So, the equation of the waveform in Fig. 1.88 can be written as $\begin{cases} 10 & 0 < t < 3.333\text{ ms} \\ 10 + 20\sin\left(2\pi \times 50 \times t - \frac{\pi}{3}\right) & t > 3.333\text{ ms} \end{cases}$.

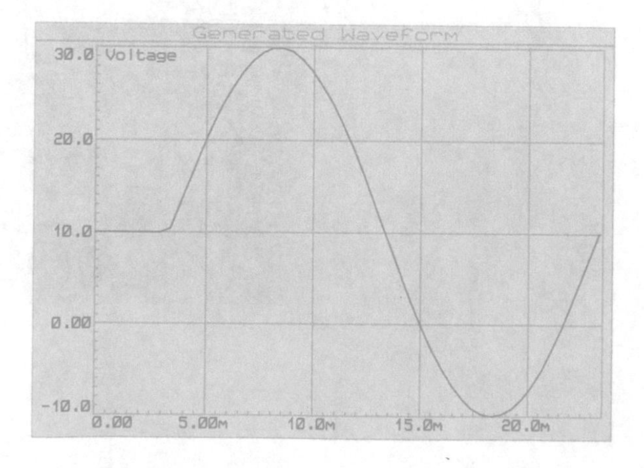

Fig. 1.88 Waveform generated with settings shown in Fig. 1.87

If you enter 100 to the Damping Factor box (Fig. 1.89), the waveform changes to what shown in Fig. 1.90. The equation of the waveform in Fig. 1.90 can be written as $\begin{cases} 10 & 0<t<3.333\,\text{ms} \\ \left(10+20\sin\left(2\pi\times 50\times t-\frac{\pi}{3}\right)\right)\times e^{-100(t-3.333m)} & t>3.333\,\text{ms} \end{cases}$

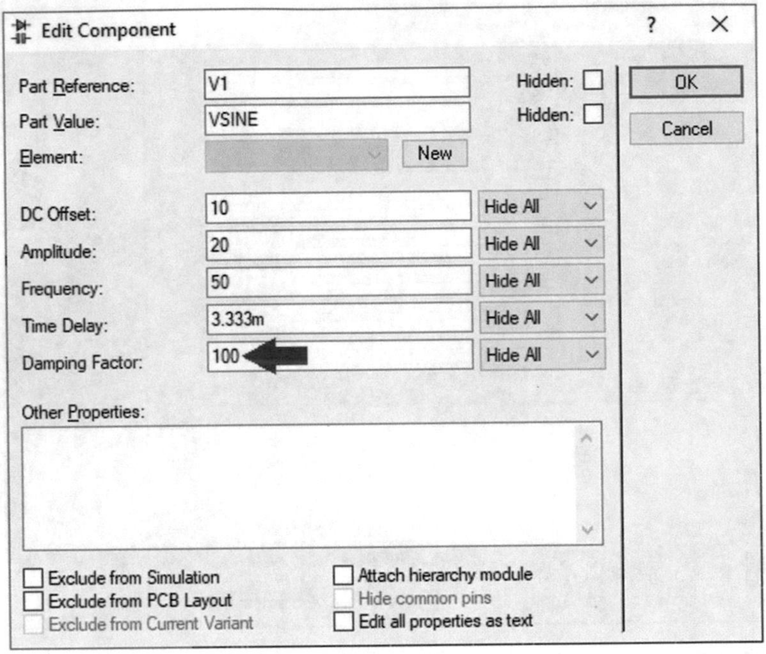

Fig. 1.89 Edit component window

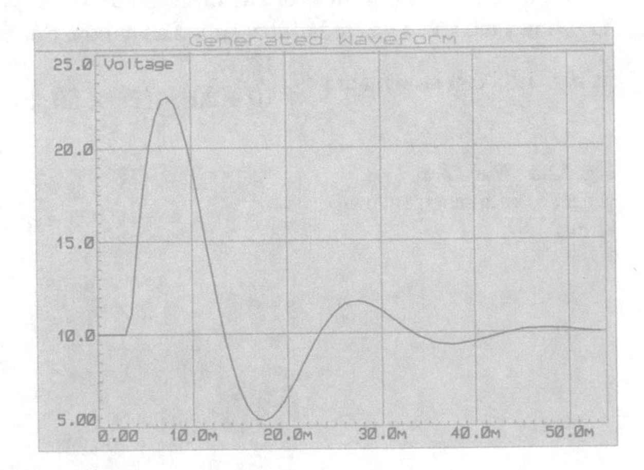

Fig. 1.90 Waveform generated with settings shown in Fig. 1.89

Assume that you want to generate $V(t) = 10 + 20\sin(2\pi \times 50 \times t + 60°)$. In this case we write $V(t)$ as $V(t) = 10 + 20\sin\left(2\pi \times 50 \times t + \frac{\pi}{3}\right) = 10 + 20\sin\left(2\pi \times 50 \times t + \frac{\pi}{3} - 2\pi\right) = 10 + 20\sin\left(2\pi \times 50 \times t - \frac{5\pi}{3}\right) = 10 + 20\sin(2\pi \times 50 \times (t - 16.667\,\text{ms}))$. Let's enter the values to v_{sine} block (Fig. 1.91).

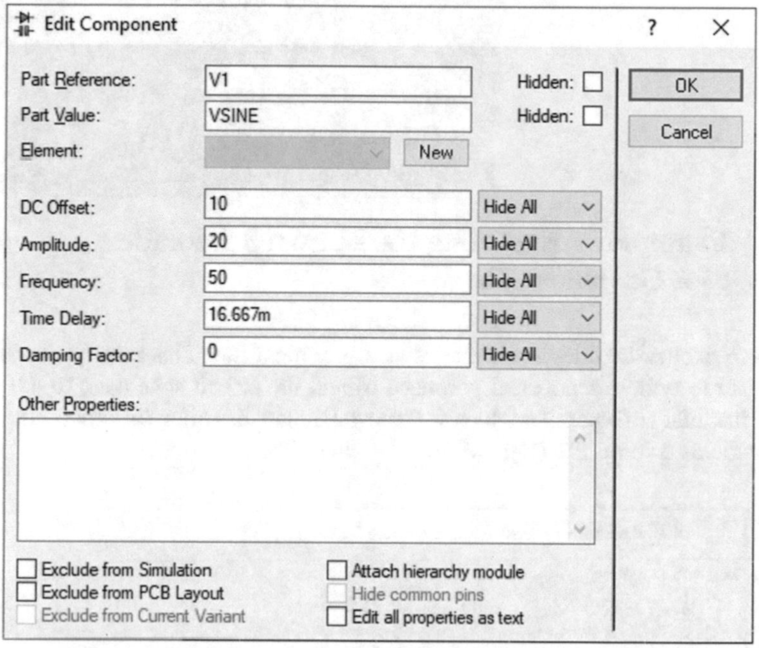

Fig. 1.91 Edit component block

The settings shown in Fig. 1.91 generate the waveform shown in Fig. 1.92. The waveform equals to the value of DC offset (in this example 10 V) for [0, TD] interval. TD shows the value entered to the Time Delay box. So, the equation of the waveform in Fig. 1.92 can be written as $\begin{cases} 10 & 0 < t < 16.667\,\text{ms} \\ 10 + 20\sin\left(2\pi \times 50 \times t + \frac{\pi}{3}\right) & t > 16.667\,\text{ms} \end{cases}$.

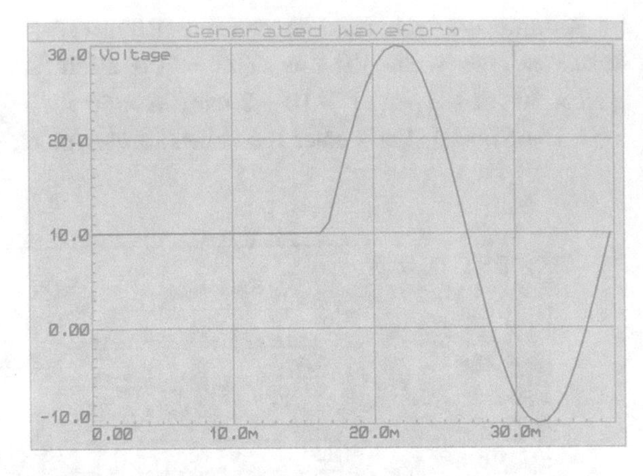

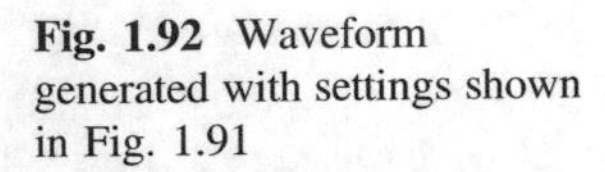

Fig. 1.92 Waveform generated with settings shown in Fig. 1.91

1.10 Example 9: Exporting the Drawn Schematic as a Graphical File

You can export the drawn schematics as a graphical file. This is very useful when you want to write a report and you need to add the circuit schematic to the report. Click the File > Export Graphics > Export Bitmap in order to export the drawn schematic as a .bmp file (Fig. 1.93).

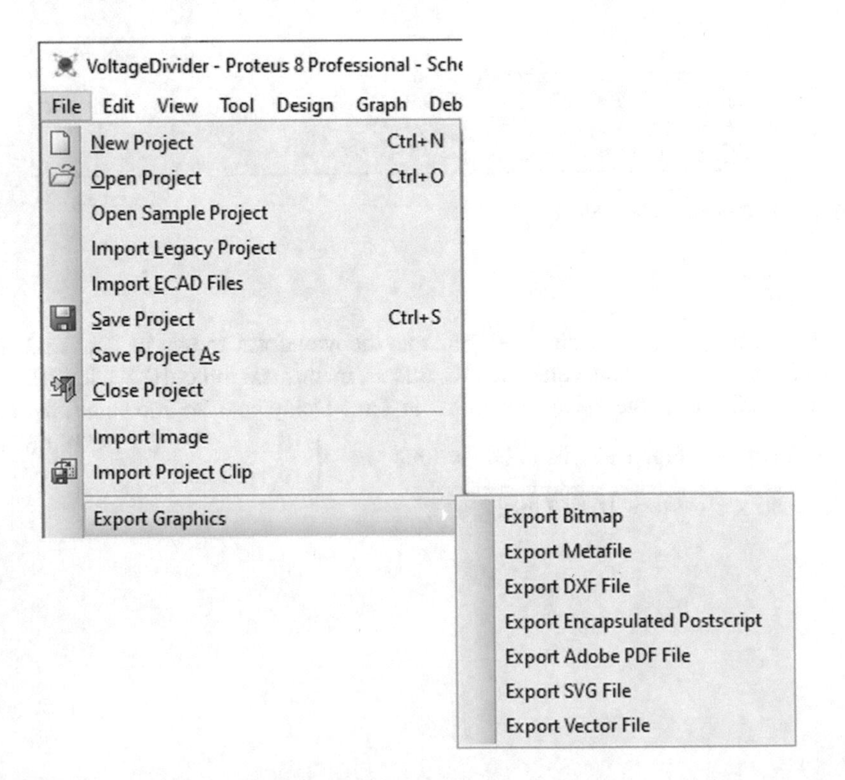

Fig. 1.93 File > export graphics

After clicking the Export Bitmap, the window shown in Fig. 1.94 appears on the screen. Use the Filename button to determine the file name and path. After clicking the OK button, the graphic file will be generated.

Fig. 1.94 Export bitmap window

You can export the drawn schematic as a pdf file as well. Click the File > Export Graphics > Export Adobe PDF File (Fig. 1.95) to export the drawn schematic as a pdf file.

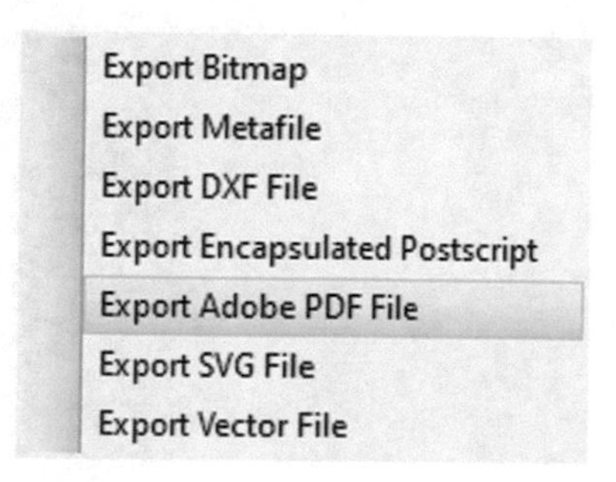

Fig. 1.95 File > export graphics > export adobe PDF file

After clicking the Export Adobe PDF File, the window shown in Fig. 1.96 appears on the screen. Use the Filename button to determine the file name and path. After clicking the OK button, the pdf file will be generated.

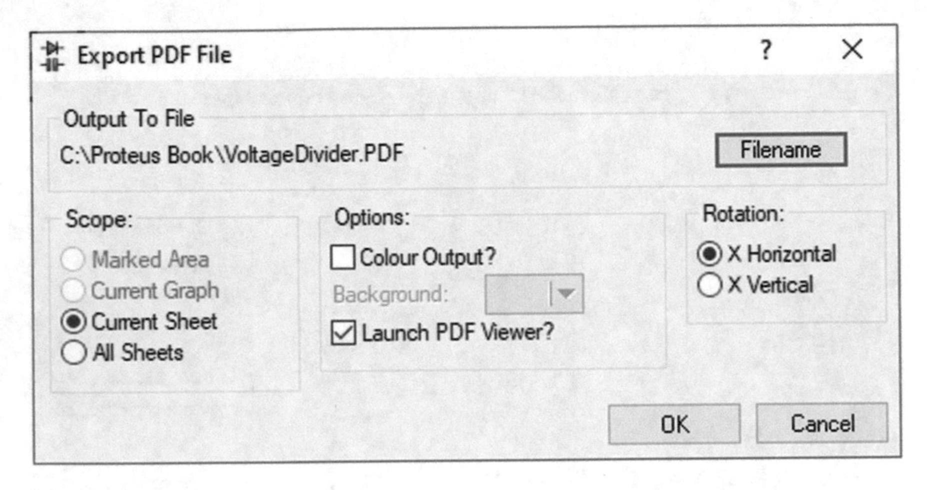

Fig. 1.96 Export PDF file window

1.11 Example 10: Measurement with Probes (I)

In Example 1, we used a DC voltmeter and ammeter to measure the voltage of resistor R_2 and circuit current. You can measure voltage and current with the aid of probes as well. Let's study a simple example. Consider the schematic shown in Fig. 1.97.

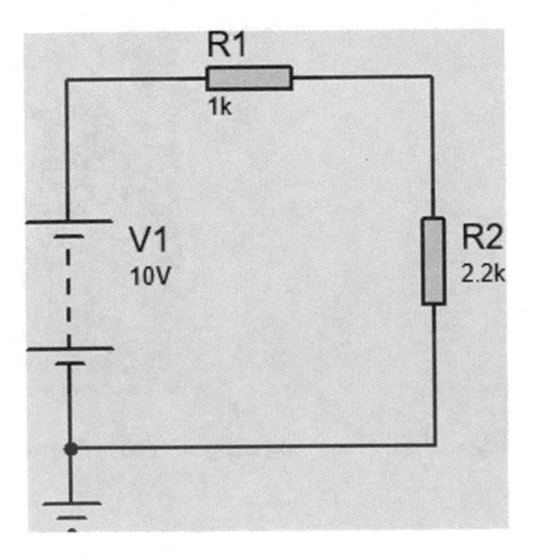

Fig. 1.97 Schematic of Example 10

Add two voltage probes (Fig. 1.98) to the circuit (Fig. 1.99).

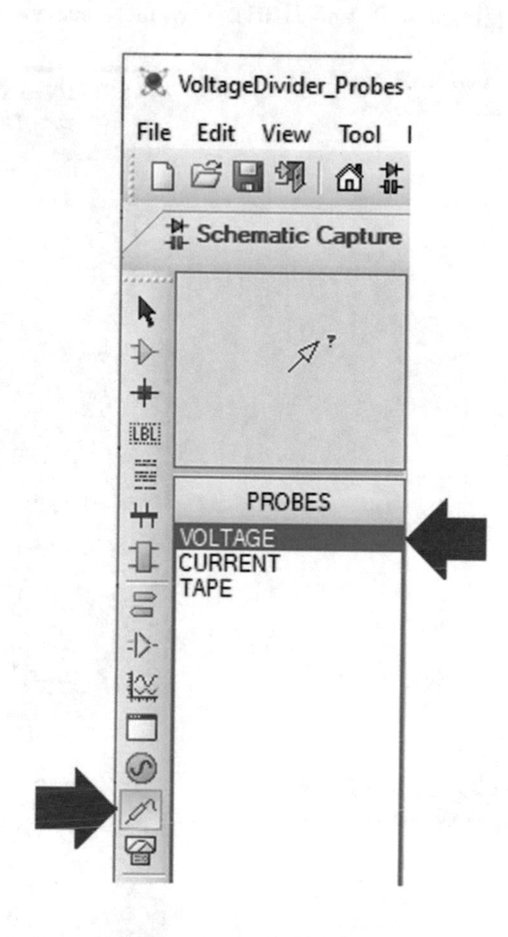

Fig. 1.98 Voltage probes

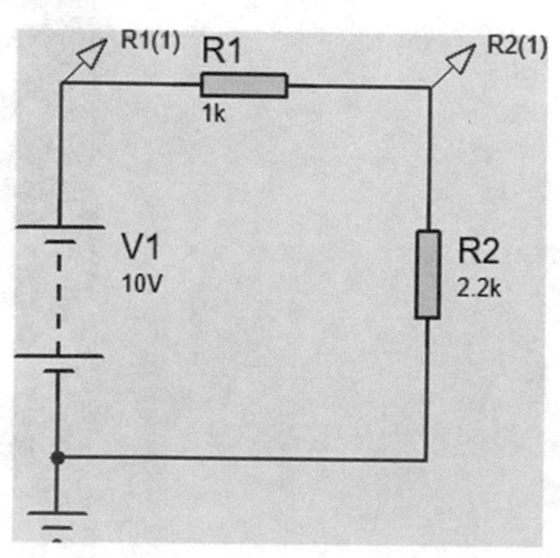

Fig. 1.99 Voltage probes are added to the schematic

Double click on the voltage probes and change their name to V_{in} and V_{out} (Figs. 1.100 and 1.101). Now, the schematic looks like Fig. 1.102.

Fig. 1.100 Edit voltage probe window

Fig. 1.101 Edit voltage probe window

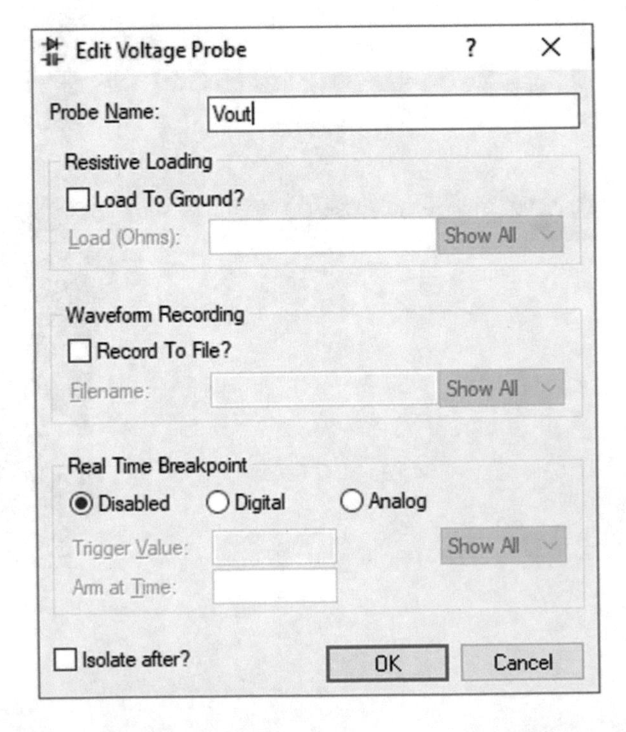

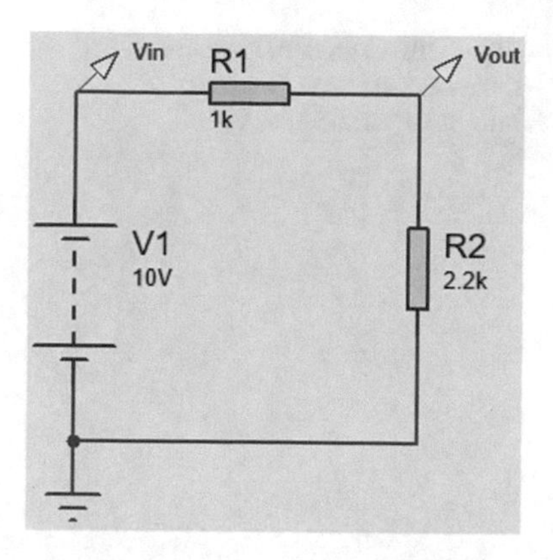

Fig. 1.102 Name of voltage probes are updated

Run the simulation. The simulation result is shown in Fig. 1.103. The probes show the voltage of nodes with respect to ground.

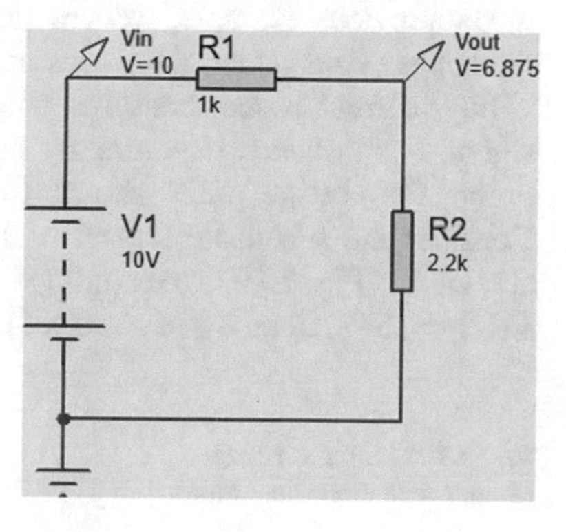

Fig. 1.103 Values measured by voltage probes are shown on them

You can measure the circuit current with a current probe as well. The current probe measures the current in the direction which is shown on its arrow. For instance, the current probe in Fig. 1.104 measures + 0.003125 A and the current probe in Fig. 1.105 measures − 0.003125 A. You can rotate a current probe with the aid of + or – key of numeric section of your keyboard.

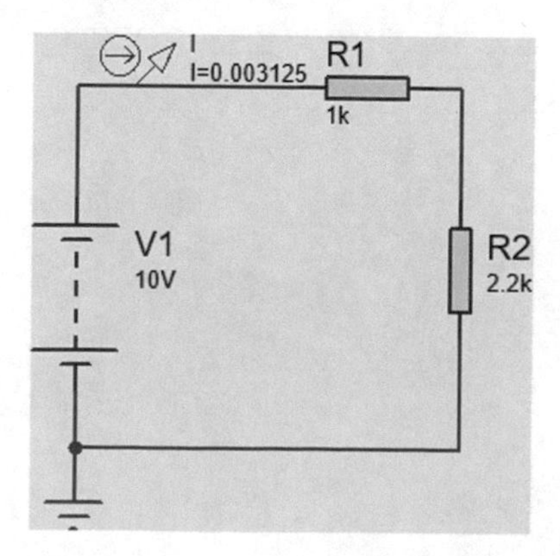

Fig. 1.104 Values measured by current probe are shown on it

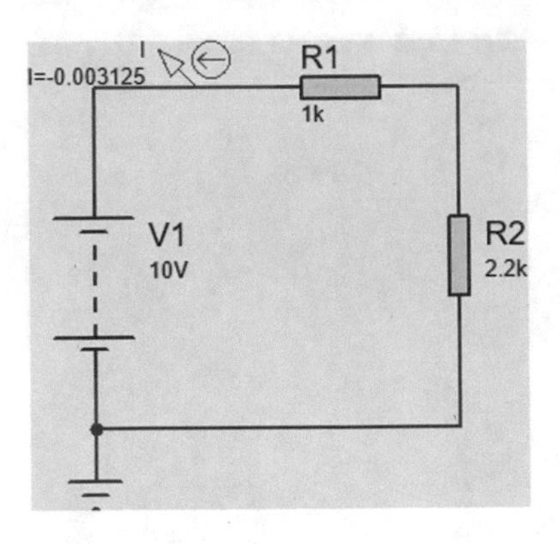

Fig. 1.105 The current probe measure the current entered into the + terminal of V_1

1.12 Example 11: Measurement with Probes (II)

In the previous example, we saw how probes can be used to measure purely DC quantities. Voltage/current probes can be used in AC circuits as well. Note that the voltage/current probe measures only the average value (DC component) of the signal. For instance, if you apply $v(t) = 10 + 20\sin(2 \times \pi \times 50 \times t)$ to a voltage probe, the voltage probe shows 10 V. Let's see this with a simple simulation. Consider the schematic shown in Fig. 1.106. Settings of voltage source V_1 are shown in Fig. 1.107. According to Fig. 1.107, the input voltage source is $V_1(t) = 33 + 10\sin(2 \times \pi \times 50 \times t)$.

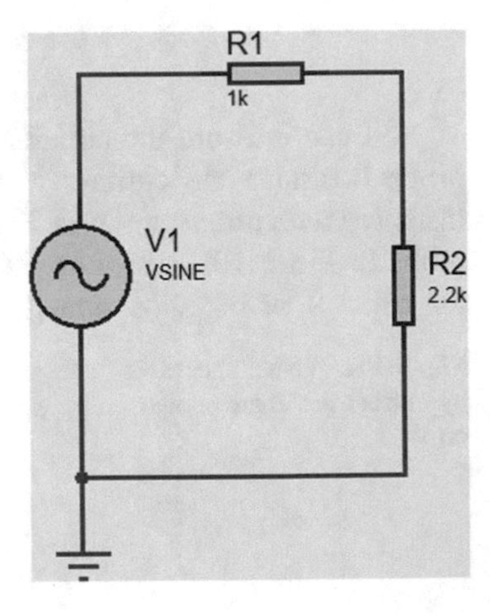

Fig. 1.106 Simple voltage divider with sinusoidal input

Edit Component

Part Reference: V1 Hidden:
Part Value: VSINE Hidden:
Element: New
OK
Cancel

DC Offset: 33 Hide All
Amplitude: 10 Hide All
Frequency: 50 Hide All
Time Delay: 0 Hide All
Damping Factor: 0 Hide All

Other Properties:

Exclude from Simulation
Exclude from PCB Layout
Exclude from Current Variant
Attach hierarchy module
Hide common pins
Edit all properties as text

Fig. 1.107 Edit component window

Add two voltage probes and a current probe to the schematic (Fig. 1.108).

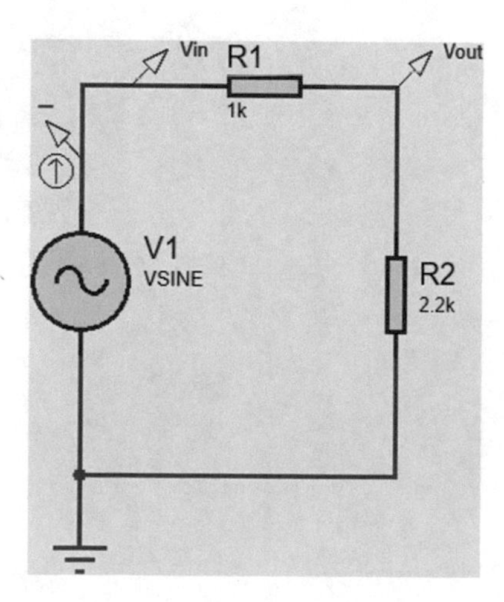

Fig. 1.108 Addition of probes to the schematic

Run the simulation. The simulation result is shown in Fig. 1.109. Note that the probes showed the DC component of signals.

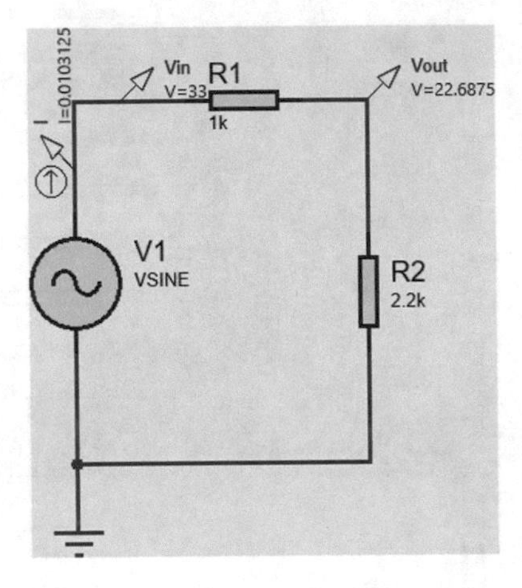

Fig. 1.109 Simulation result

Let's check the obtained result. The calculations shown in Fig. 1.110 show that Proteus result is correct.

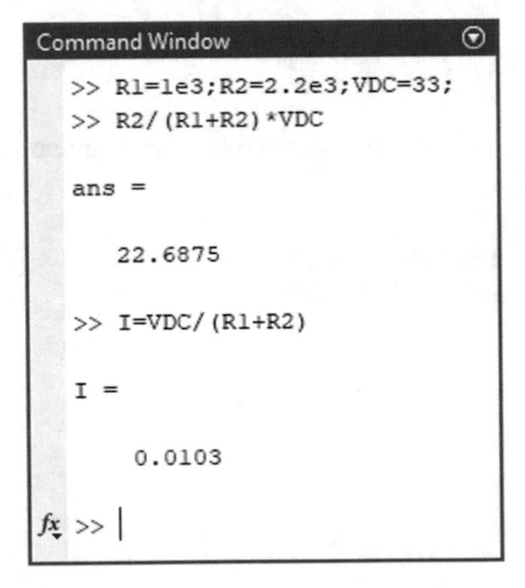

Fig. 1.110 MATLAB calculations

1.13 Example 12: Measurement with Probes (III)

In the previous two examples, we learned the details of measurement with voltage/ current probe. In this example, we learn more details voltage probes.

Open the schematic of previous example and double click the V_{out} voltage probe (Fig. 1.111).

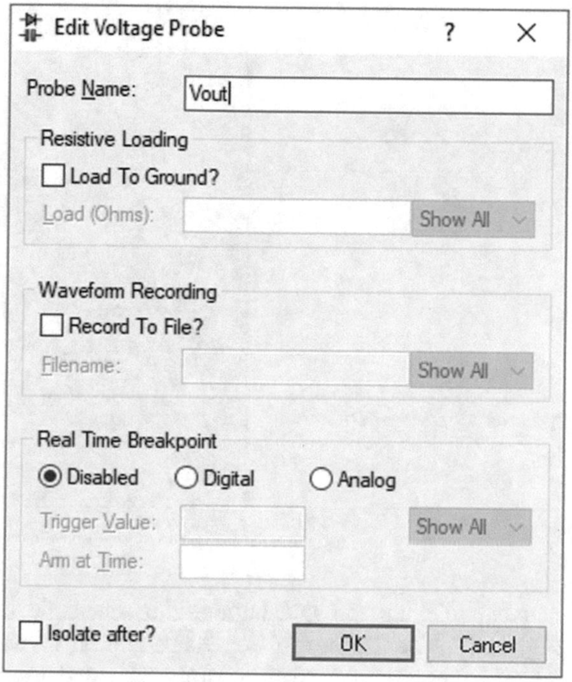

Fig. 1.111 Edit voltage probe window

Check the Load To Ground and enter 1000 to Load (Ω) box (Fig. 1.112). Then click the OK button.

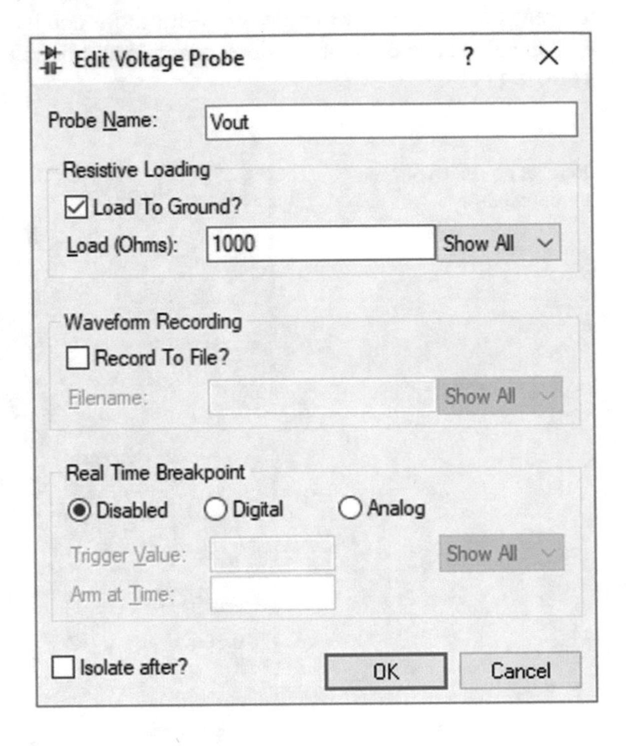

Fig. 1.112 Edit voltage probe window

After clicking the OK button, the schematic changes to what is shown in Fig. 1.113. The settings in Fig. 1.112 add an invisible resistor between the node which is connected to voltage probe V_{out} and ground. The equivalent circuit of Fig. 1.113 is shown in Fig. 1.114.

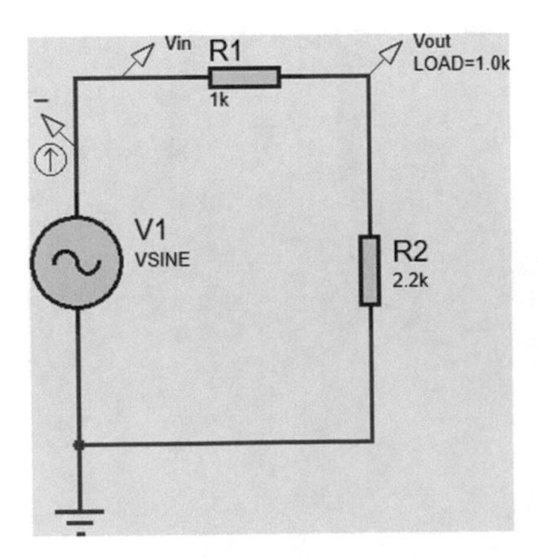

Fig. 1.113 Schematic with settings shown in Fig. 1.112

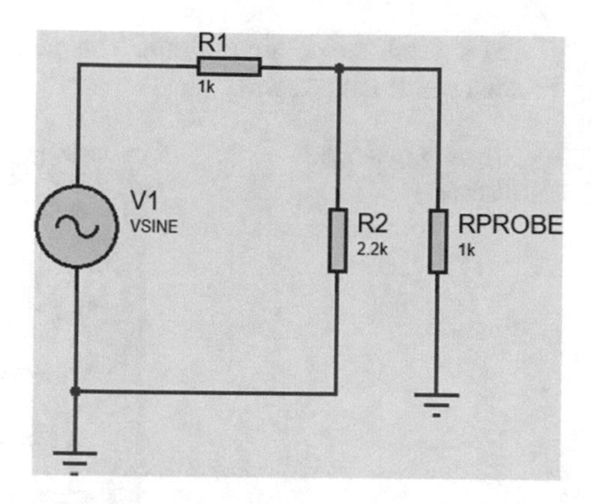

Fig. 1.114 Equivalent circuit for Fig. 1.113

Run the simulation. The simulation result is shown in Fig. 1.115 (Compare it with Fig. 1.109).

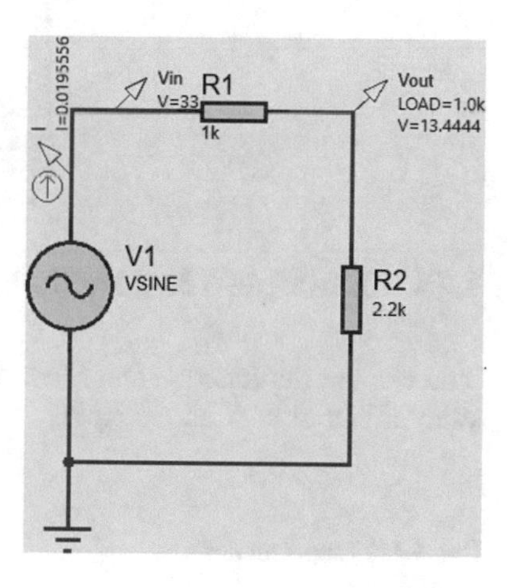

Fig. 1.115 Simulation result

Let's check the obtained result. The calculations shown in Fig. 1.116 show that Proteus result is correct.

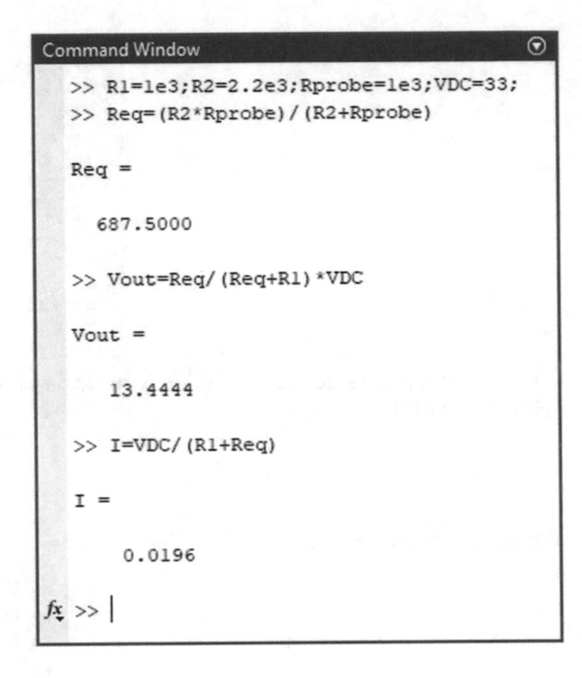

Fig. 1.116 MATLAB calculations

1.14 Example 13: Junction Dot Mode

You can use the Junction Dot Mode icon (Fig. 1.117) to make new junctions. Let's study this tool with an example.

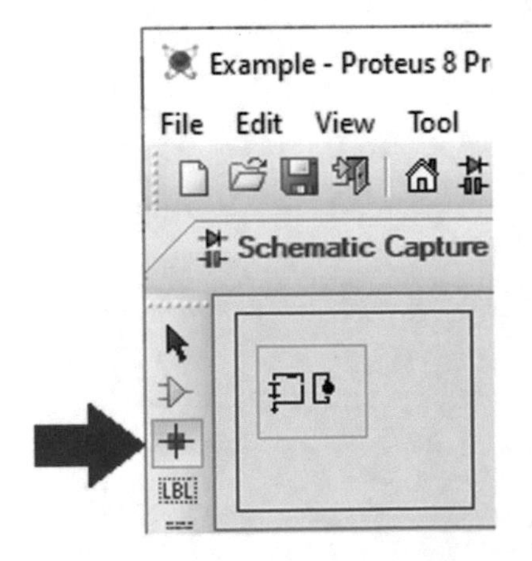

Fig. 1.117 Junction dot mode icon

Consider the circuit shown in Fig. 1.118. Upper terminal of R_2 must be connected to upper terminal of R_3 and lower terminal of R_2 must be connected to lower terminal of R_3. In order to do this, click the Junction Dot Mode icon and after that click the upper and lower terminals of R_2. This adds two new junctions across the resistor R_2 (Fig. 1.119). Now you can connect the resistors together (Fig. 1.120).

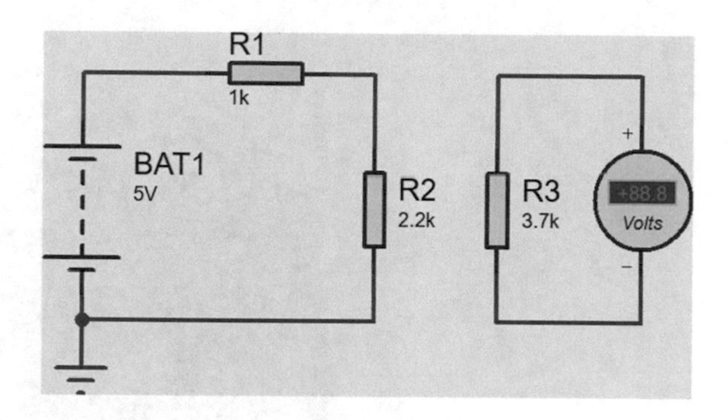

Fig. 1.118 Sample schematic

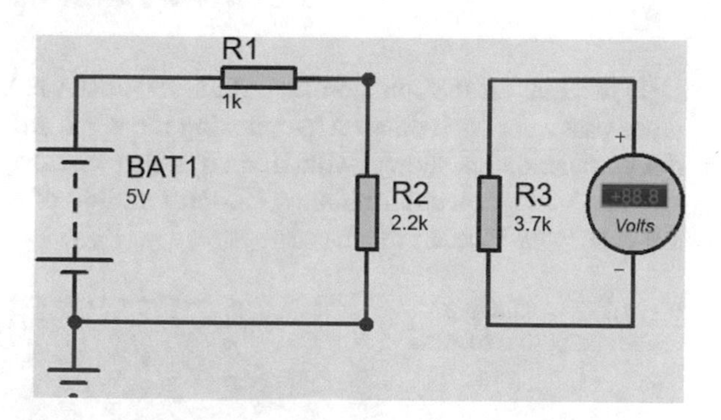

Fig. 1.119 Two junctions are added to resistor R_2

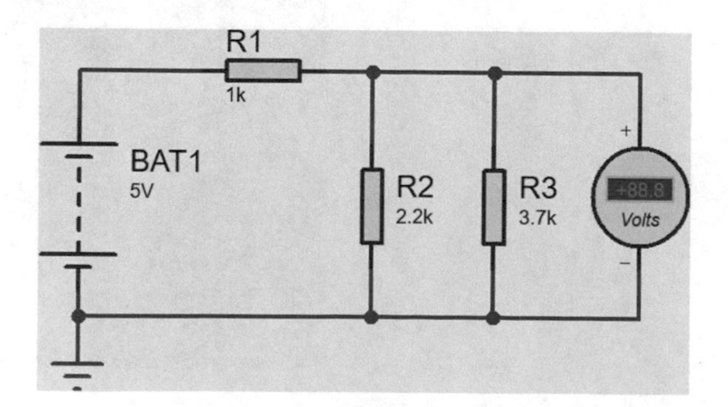

Fig. 1.120 R_2 is connected to R_3 with the aid of junctions added to R_2 in Fig. 1.119

Let's study another example. Consider the schematic shown in Fig. 1.121. There is no connection between the wires that connect the resistor R_2–R_4 and the resistor R_1–R_3.

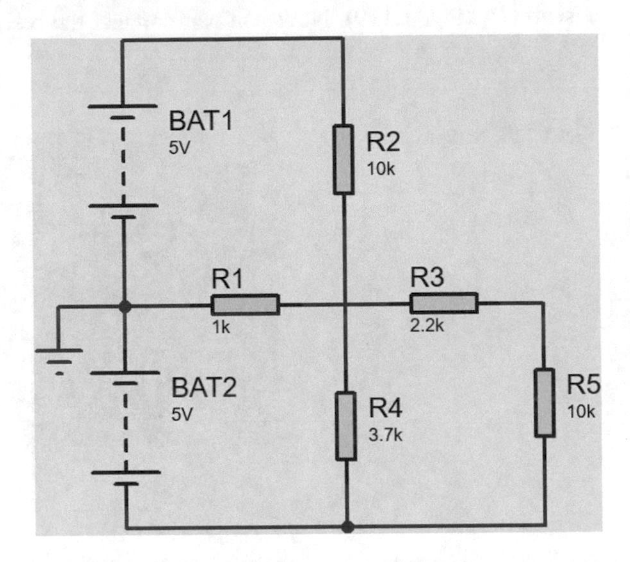

Fig. 1.121 Sample schematic

Now click on the Junction Dot Mode icon and click the point at which the two wires pass over each other. After clicking the wires are connected to each other and the connection is shown with a small filled circle (Fig. 1.122). Now the right terminal of R_1, lower terminal of R_2, left terminal of R_3, and upper terminal of R_4 are connected to each other.

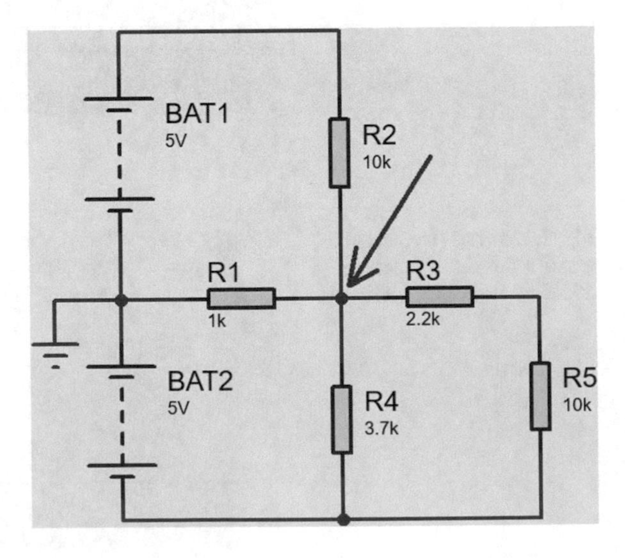

Fig. 1.122 Addition of contact point to schematic of Fig. 1.121

1.15 Example 14: Excluding a Component from Simulation

You can exclude a component from simulation without removing the component from the schematic. This example shows how to exclude a component from simulation without removing it from the schematic.

Consider the schematic shown in Fig. 1.123.

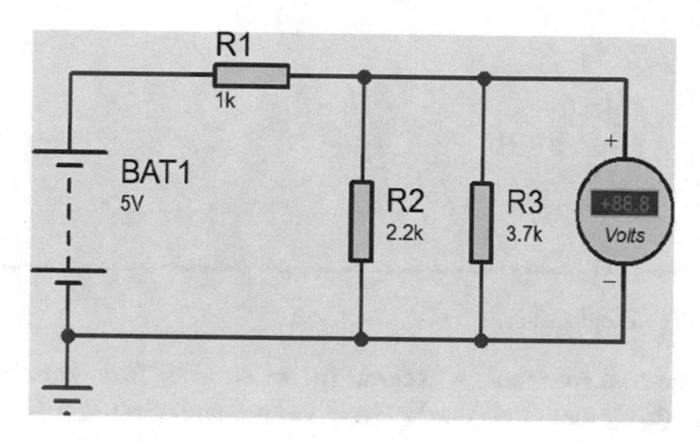

Fig. 1.123 Schematic of Example 14

Run the simulation. The simulation result is shown in Fig. 1.124.

Fig. 1.124 Simulation result

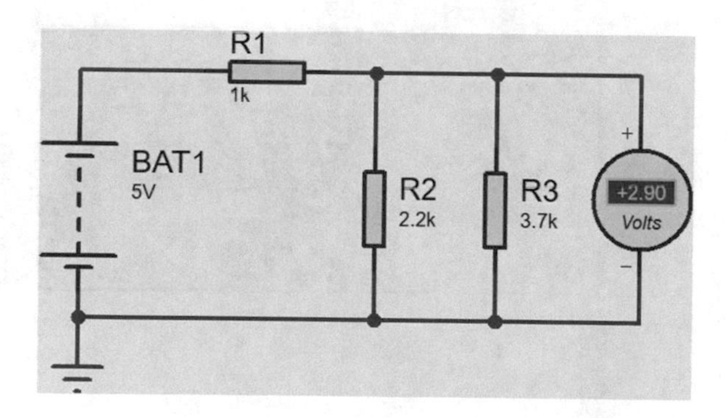

Let's check the obtained result. The calculation shown in Fig. 1.125 shows that the Proteus result is correct.

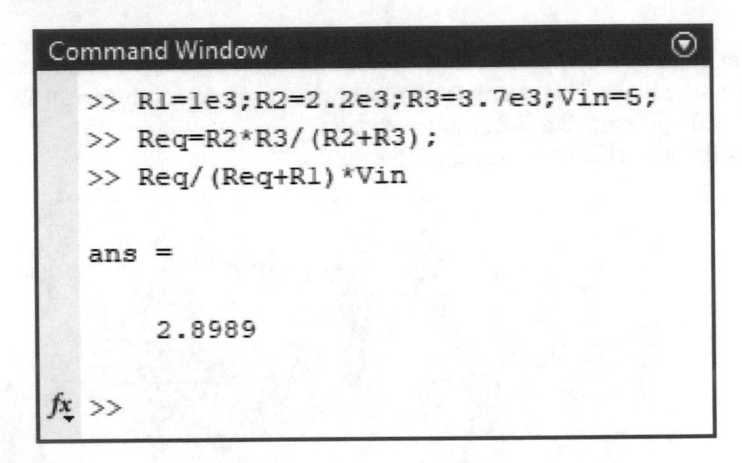

Fig. 1.125 MATLAB calculations

Now assume that we want to exclude the resistor R_3 from simulation. Double click the resistor R_3 and click the Exclude from Simulation box (Fig. 1.126) and click the OK button. Note that the schematic does not change after clicking the OK button (Fig. 1.127).

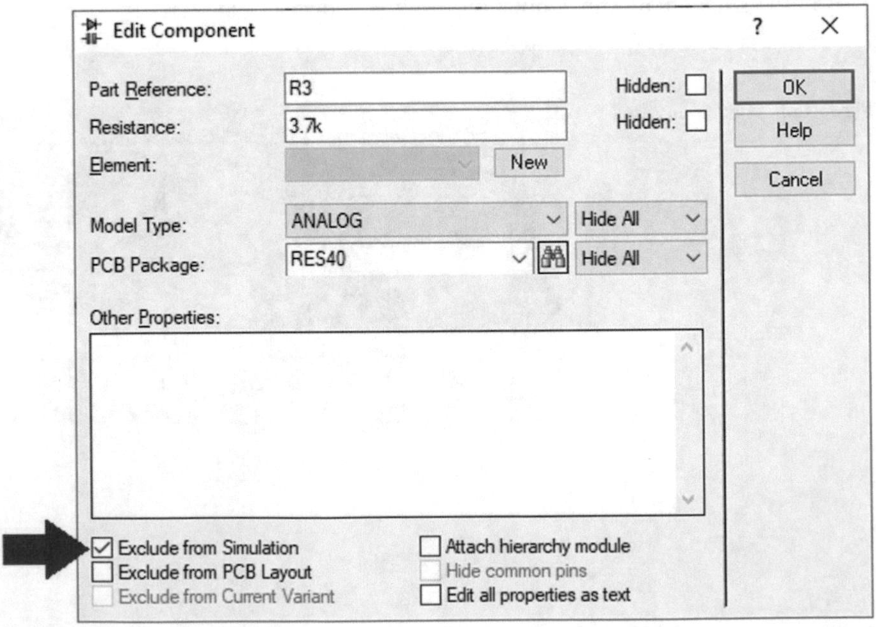

Fig. 1.126 Edit component window

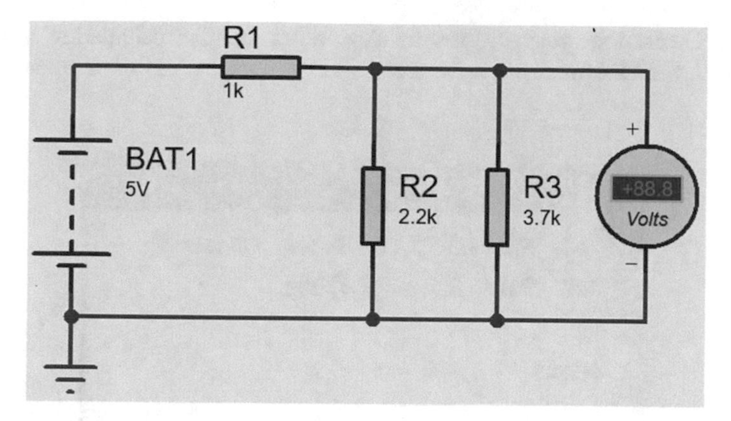

Fig. 1.127 Schematic is not changed

Run the simulation. The simulation result is shown in Fig. 1.128. Note that R_3 is excluded from the simulation and R_3 has no effect on the simulation result.

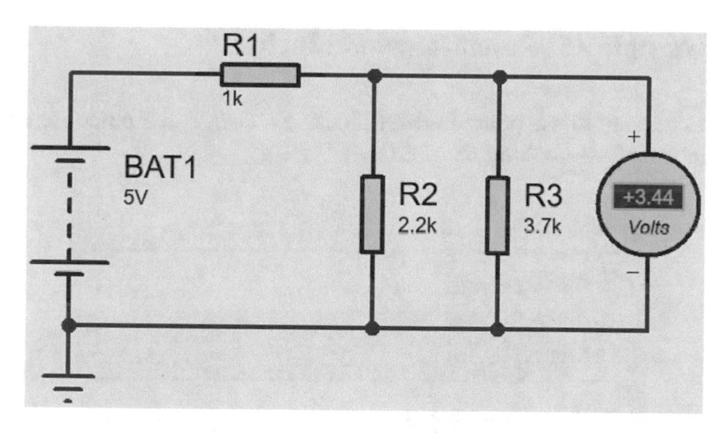

Fig. 1.128 Simulation result

Let's check the obtained result. According to the calculation shown in Fig. 1.129, the Proteus result is correct.

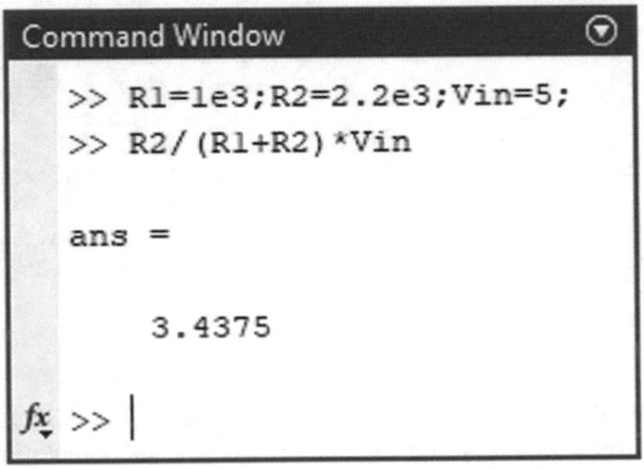

```
Command Window
>> R1=1e3;R2=2.2e3;Vin=5;
>> R2/(R1+R2)*Vin

ans =

    3.4375

fx >>
```

Fig. 1.129 MATLAB calculations

1.16 Example 15: Potentiometer Block

Proteus has a ready to use potentiometer block. You can add a potentiometer block to the schematic by searching for pot (Fig. 1.130).

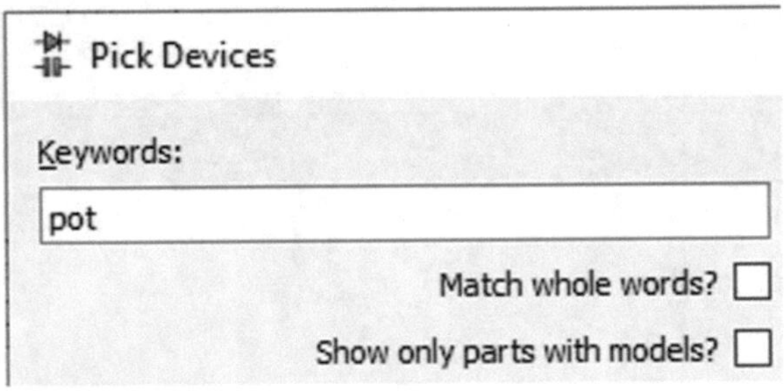

Pick Devices

Keywords:

pot

Match whole words? ☐

Show only parts with models? ☐

Fig. 1.130 Searching for a potentiometer block

Let's study a simple example. Consider the simple circuit shown in Fig. 1.131.

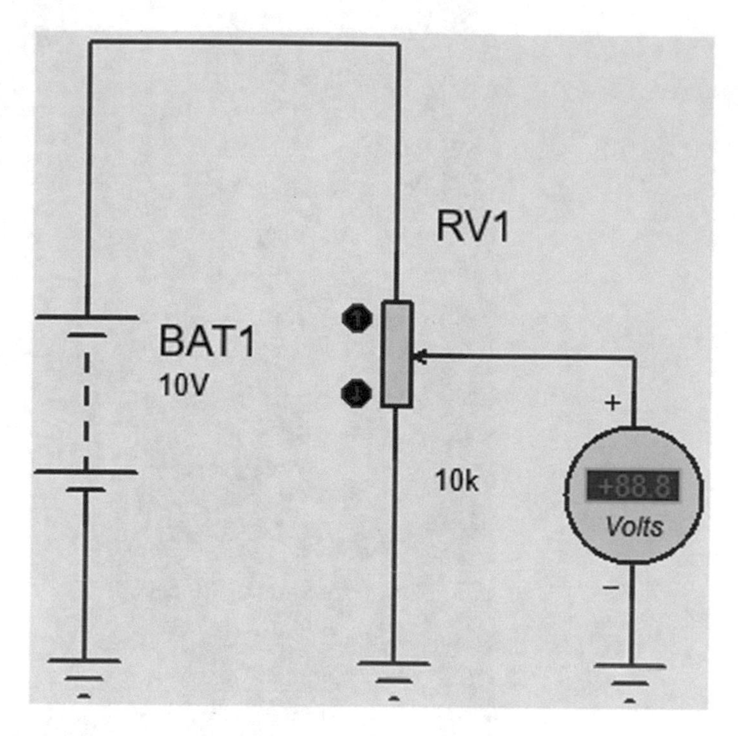

Fig. 1.131 Schematic of Example 15

Right click on the potentiometer block and click the Edit Properties (Fig. 1.132). This opens the window shown in Fig. 1.133. Thc potentiometer value is entered to the Resistance box. The Law drop down list determines the type of potentiometer (i.e., linear or logarithmic).

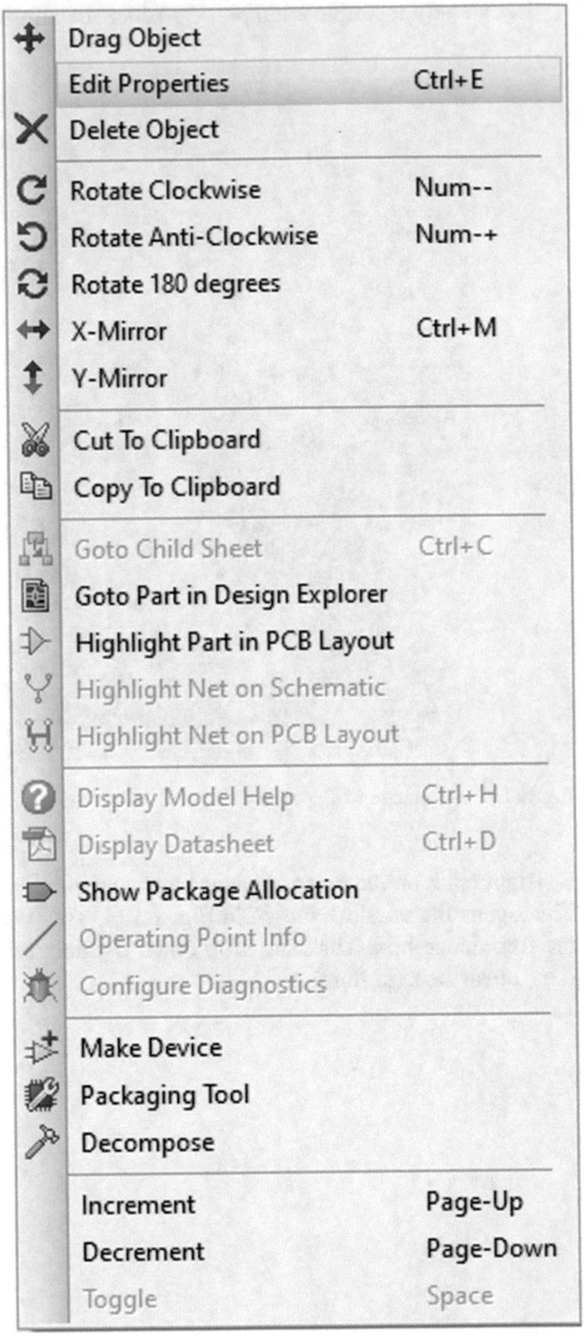

Fig. 1.132 Edit properties command

Drag Object
Edit Properties Ctrl+E
Delete Object
Rotate Clockwise Num--
Rotate Anti-Clockwise Num-+
Rotate 180 degrees
X-Mirror Ctrl+M
Y-Mirror
Cut To Clipboard
Copy To Clipboard
Goto Child Sheet Ctrl+C
Goto Part in Design Explorer
Highlight Part in PCB Layout
Highlight Net on Schematic
Highlight Net on PCB Layout
Display Model Help Ctrl+H
Display Datasheet Ctrl+D
Show Package Allocation
Operating Point Info
Configure Diagnostics
Make Device
Packaging Tool
Decompose
Increment Page-Up
Decrement Page-Down
Toggle Space

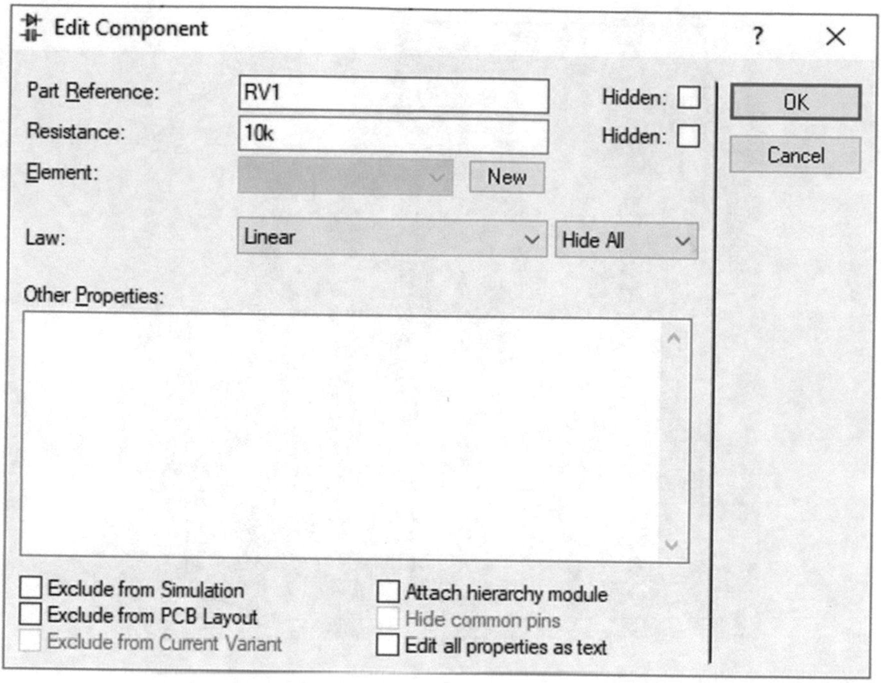

Fig. 1.133 Edit component window

Run the simulation. Use the arrows behind the potentiometer (Fig. 1.134) to change the position of the movable arm of potentiometer. When the resistance between center terminal and bottom terminal (ground) increases from minimum toward maximum, the voltmeter in Fig. 1.134 reads 0.01, 1.00, 2.00, 3.00, 4.00, 5.00, 6.00, 7.00, 8.00, 9.00, and 9.99 V. Note that the resistance between center terminal and bottom terminal changed linearly.

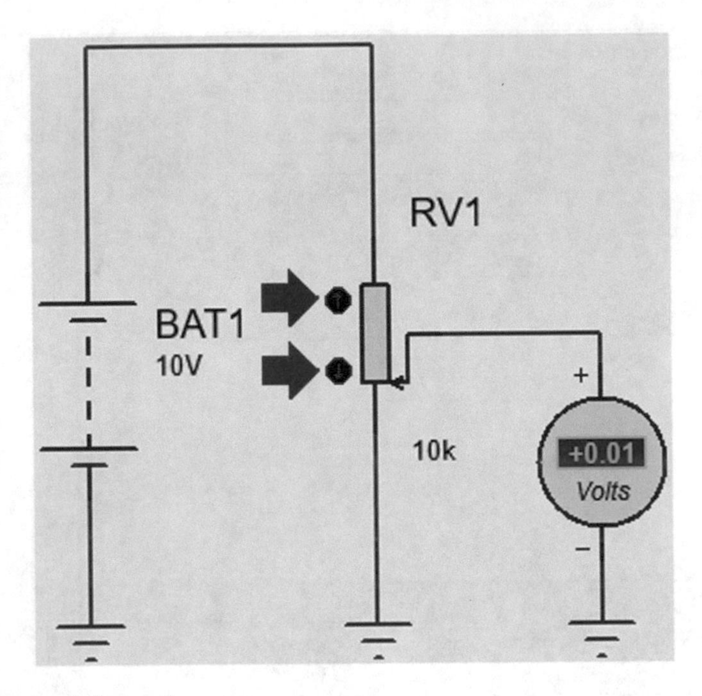

Fig. 1.134 Simulation result

Now right click on the potentiometer and click the Edit Properties. Then select the Log for “Law” drop down list (Fig. 1.135). In this case when the resistance between center terminal and bottom terminal (ground) increases from minimum toward maximum, the voltmeter in Fig. 1.134 reads 0.01, 0.23, 0.49, 0.86, 1.11, 1.51, 1.99, 2.61, 3.50, 5.00, 9.99 V. Note that the increase pattern is not linear.

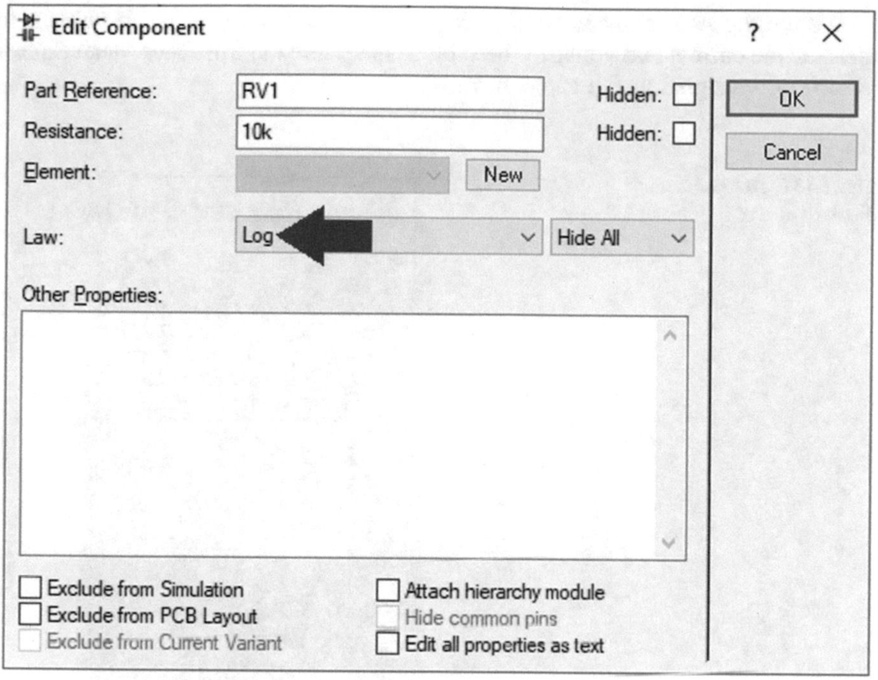

Fig. 1.135 Edit component window

1.17 Example 16: Measurement with AC Voltmeter/Ammeter (I)

In this example, we will analyze the circuit shown in Fig. 1.136. We want to measure the current drawn from voltage source V_1 and the voltage of resistor R_2. The input voltage source is $V_1 = 311 \times \sin(2 \times \pi \times 50 \times t + 30°)$.

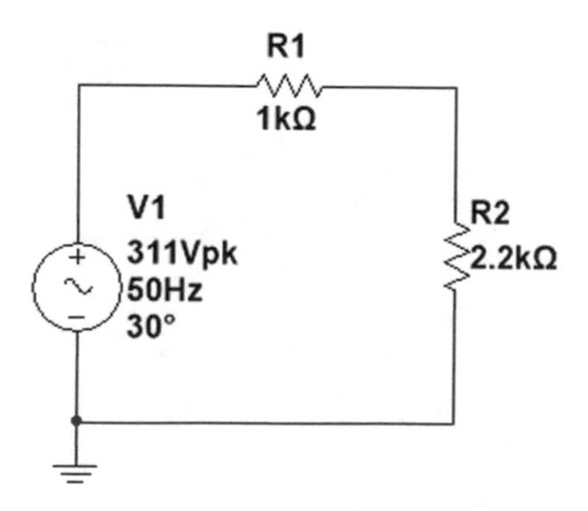

Fig. 1.136 Circuit for Example 16

We use the AC voltmeter (Fig. 1.137) and AC ammeter (Fig. 1.138) blocks to measure the current and voltage. These blocks measure the true Root Mean Square (RMS) of the signal which enters them.

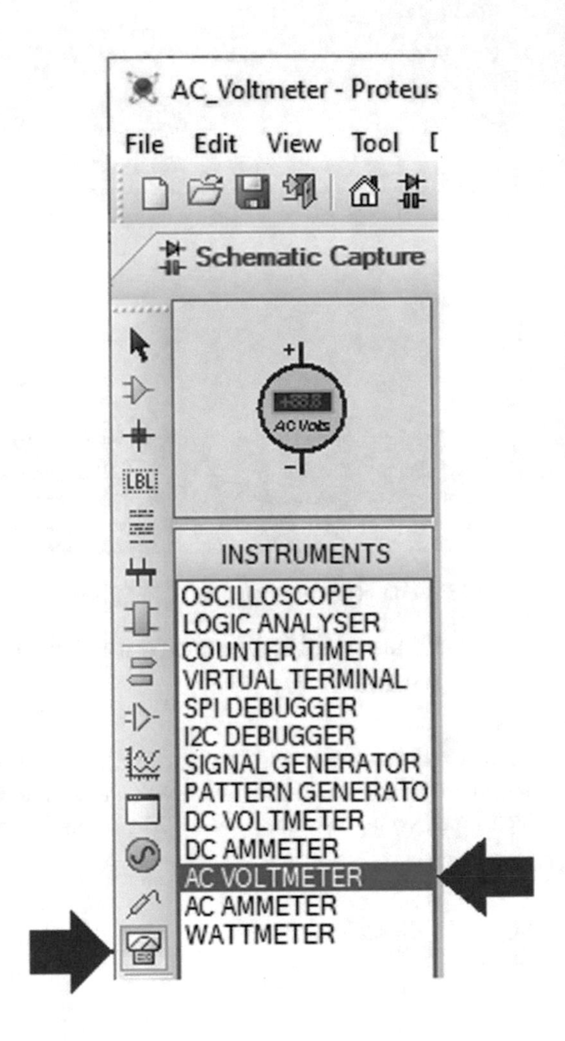

Fig. 1.137 AC voltmeter block

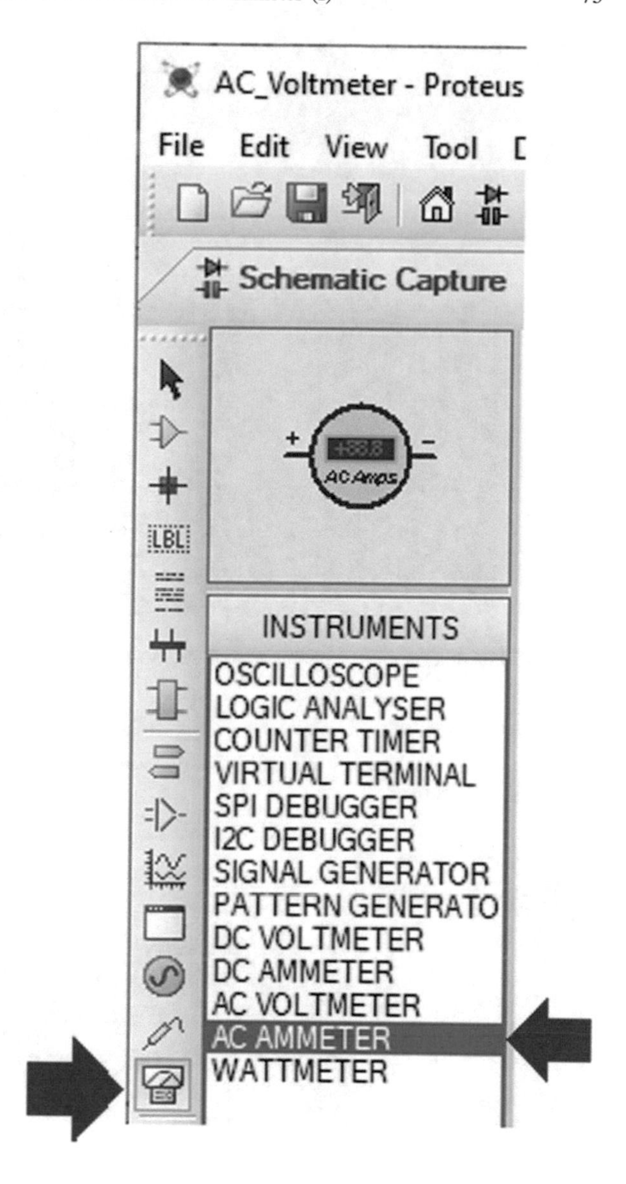

Fig. 1.138 AC ammeter block

Draw the schematic shown in Fig. 1.139.

Settings of AC ammeter and AC voltmeter are shown in Figs. 1.140 and 1.141, respectively. According to Fig. 1.141, the internal resistance of the AC voltmeter is 100 MΩ. The Time Constant box is filled with 20 ms which means 20 ms. The AC voltmeters and ammeters display true RMS values integrated over a user definable time constant.

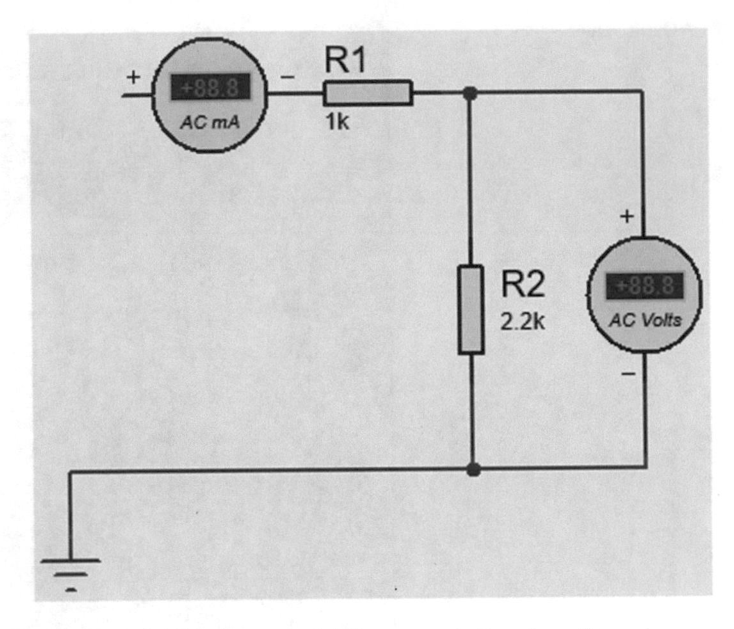

Fig. 1.139 Addition of AC voltmeter and AC ammeter to the schematic

Edit Component

Part Reference: Hidden:

Part Value: Hidden:

Element: New

Display Range: Milliamps Hide All

Time Constant: 20ms Hide All

Other Properties:

Exclude from Simulation
Exclude from PCB Layout
Exclude from Current Variant
Attach hierarchy module
Hide common pins
Edit all properties as text

OK
Help
Cancel

Fig. 1.140 Edit component window

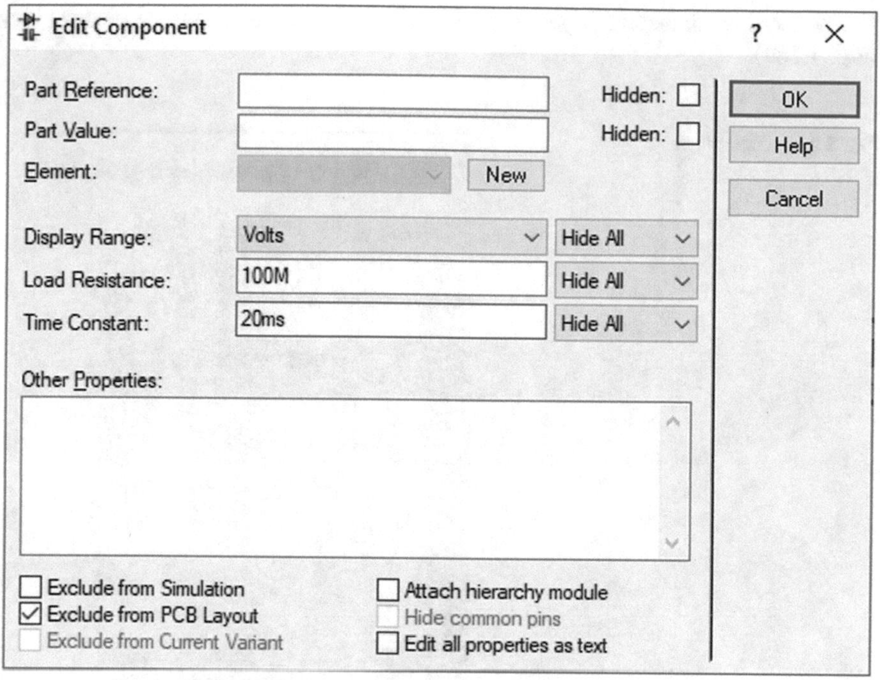

Edit Component

Part Reference: Hidden:
Part Value: Hidden:
Element: New
Display Range: Volts Hide All
Load Resistance: 100M Hide All
Time Constant: 20ms Hide All
Other Properties:
Exclude from Simulation
Exclude from PCB Layout
Exclude from Current Variant
Attach hierarchy module
Hide common pins
Edit all properties as text
OK
Help
Cancel

Fig. 1.141 Edit component window

The required sinusoidal input voltage can be generated by a sine generator block (Fig. 1.142).

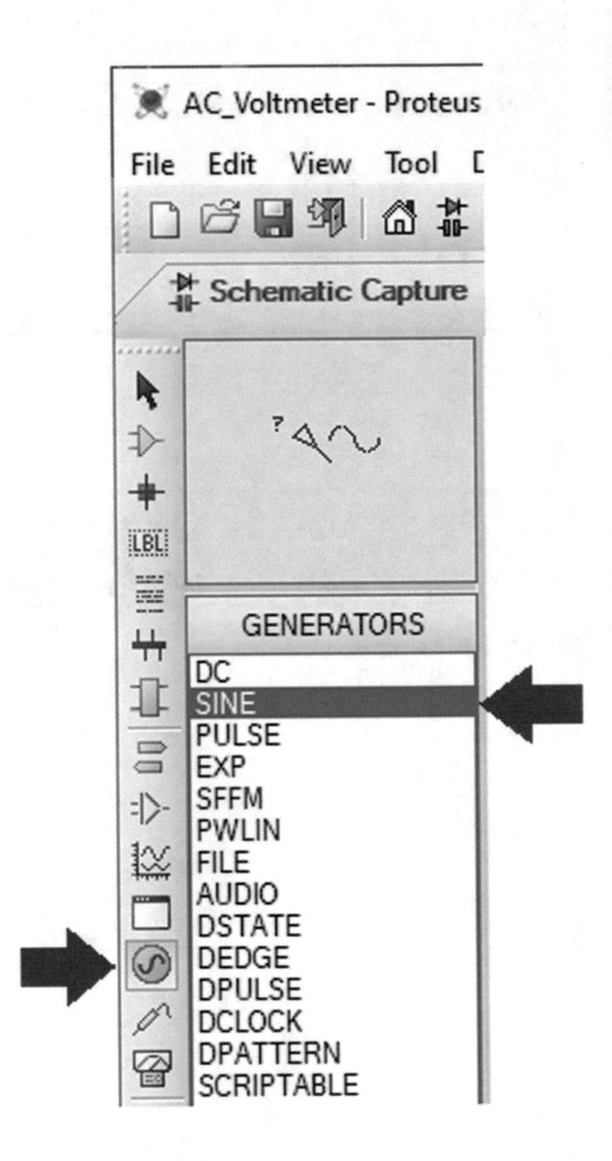

Fig. 1.142 Sine block

Add the sine generator block to the schematic (Fig. 1.143). Settings of the sine generator block are shown in Fig. 1.144.

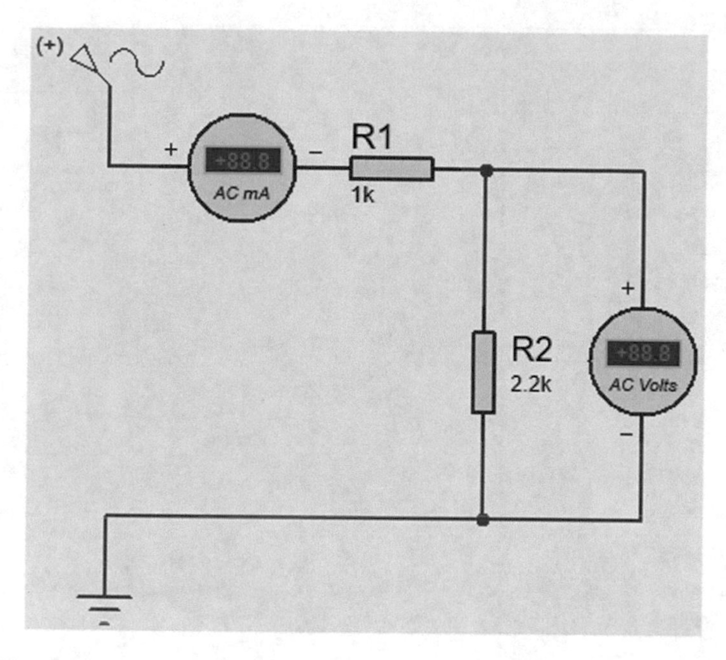

Fig. 1.143 Addition of a sine block to the schematic

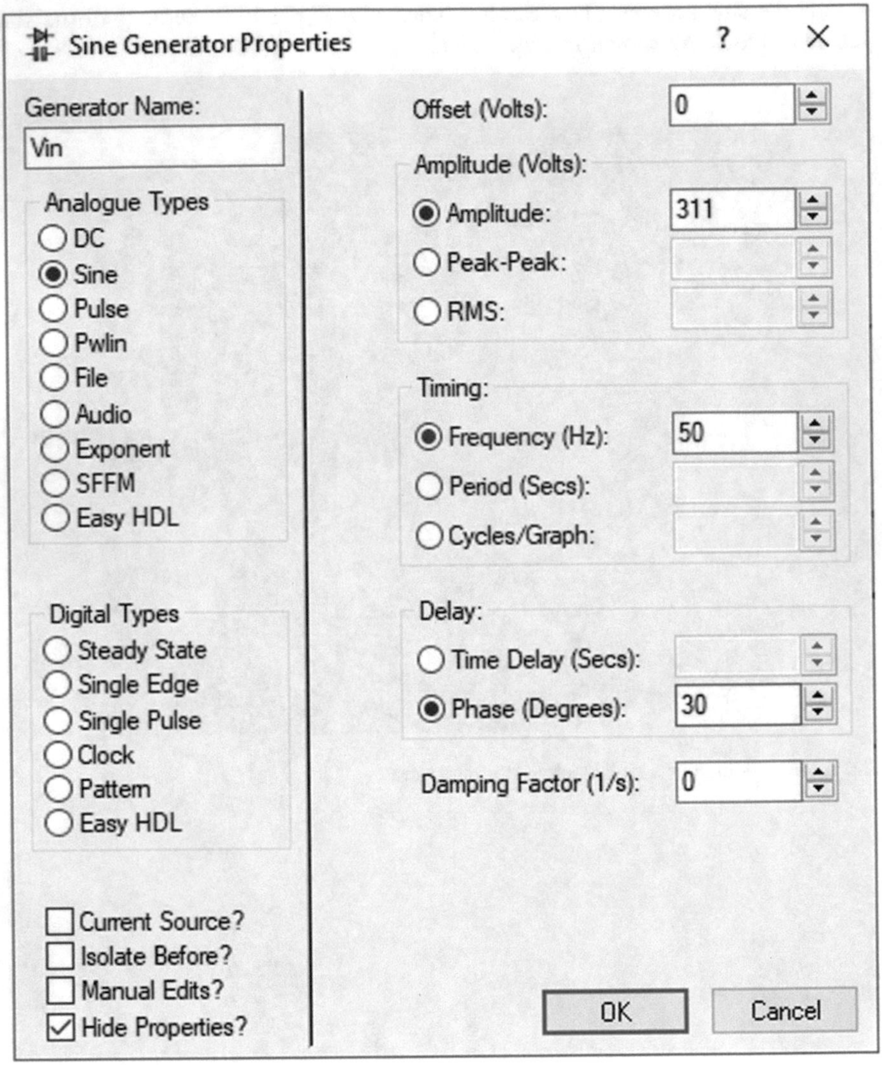

Fig. 1.144 Sine generator properties window

Run the simulation. Simulation result is shown in Fig. 1.145.

Let's check the result. According to the calculations voltage of resistor R_2 must be 151.1883 V_{RMS} and circuit current must be 68.7 mA. Proteus result is smaller than the expected values obtained in Fig. 1.146.

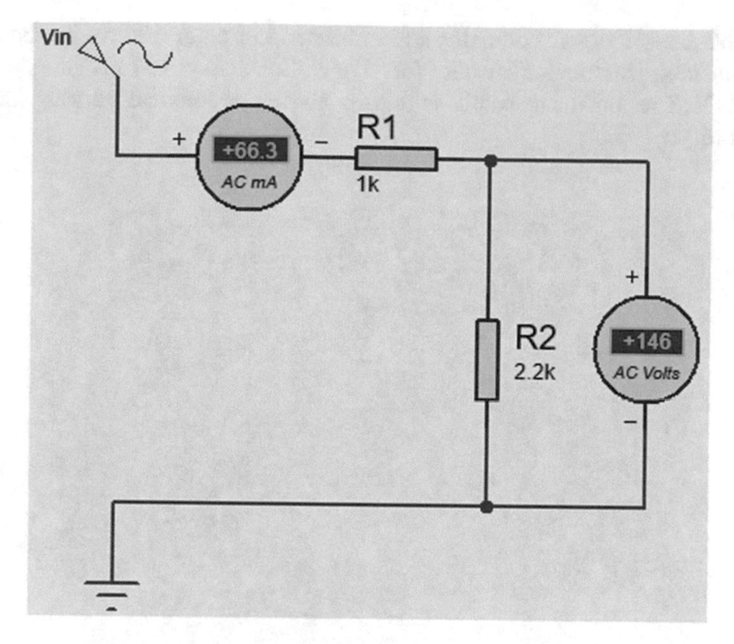

Fig. 1.145 Simulation result

Fig. 1.146 MATLAB calculations

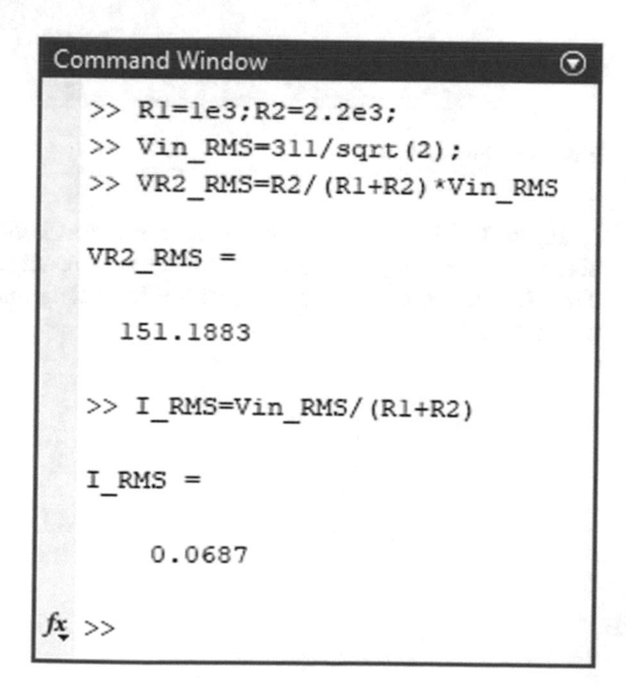

Double click the AC voltmeter and ammeter and enter 100 ms to their Time Constant box. Simulation result for Time Constant = 100 ms is shown in Fig. 1.147. The obtained result is closer to the theoretical results found in Fig. 1.146.

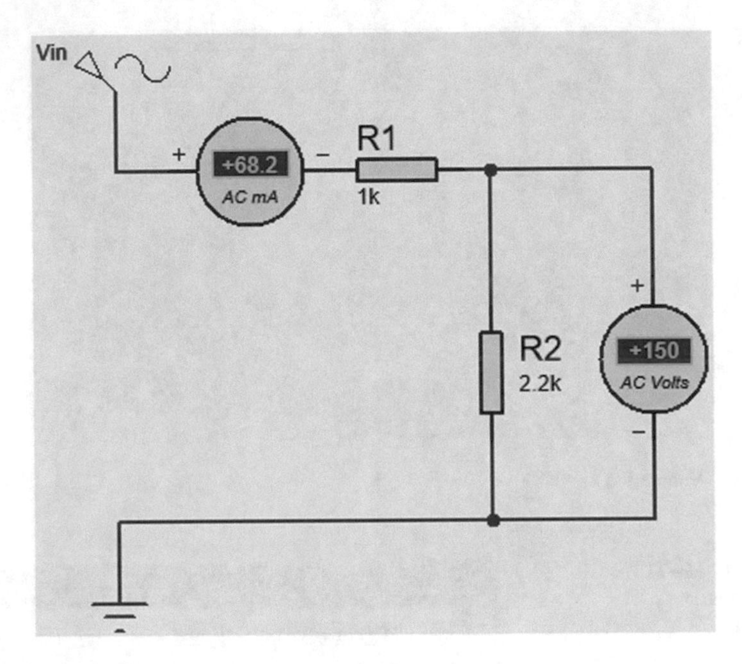

Fig. 1.147 Simulation result

Figure 1.148 shows the simulation result for Time Constant = 200 ms. The simulation result is quite close to the theoretical result found in Fig. 1.146. So, the Time Constant box must be filled with a big enough number to have an accurate measurement.

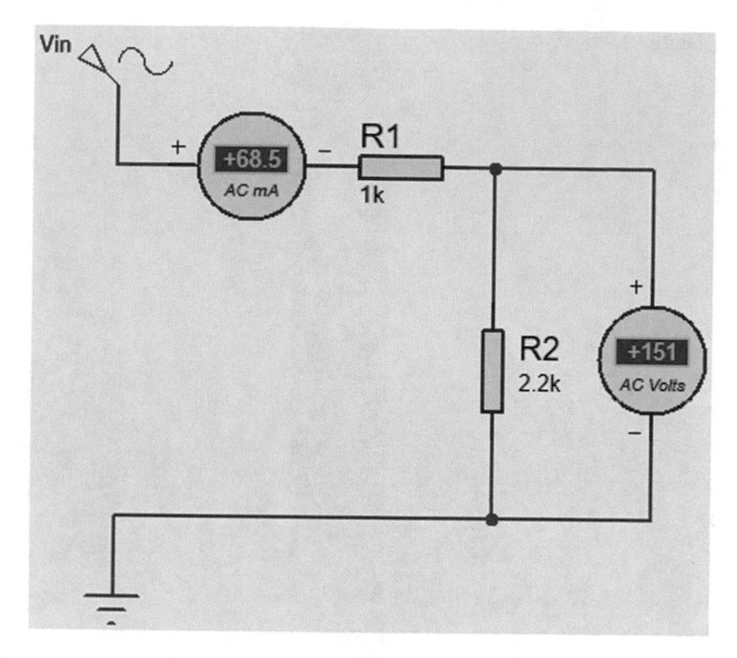

Fig. 1.148 Simulation result

Sometimes the number which is shown on the AC voltmeter or ammeter changes rapidly and it is difficult to read it. In such cases you can pause the simulation (Fig. 1.149) and read the numbers easily.

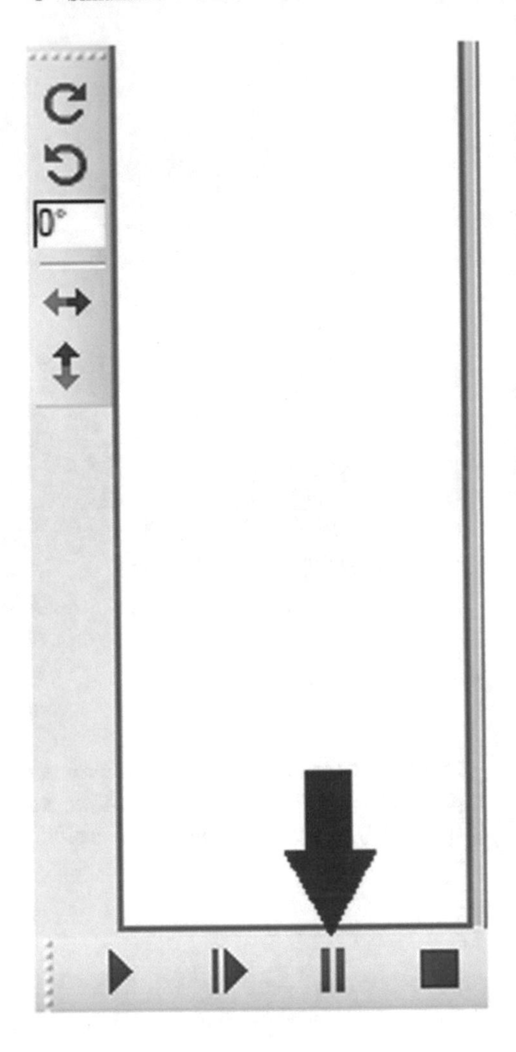

Fig. 1.149 Pause button

1.18 Example 17: Measurement with AC Voltmeter/ Ammeter (II)

The AC voltmeter/ammeter measures the true RMS of the applied signal. So, the AC voltmeter/ammeter block can measure the RMS of non-sinusoidal signals correctly. Let's test these block with a simple circuit.

Consider the schematic shown in Fig. 1.150. Settings of V_1, V_2, and V_3 are shown in Figs. 1.151, 1.152, and 1.153. The Time Constant box of AC voltmeter/ ammeter is filled with 1000 m.

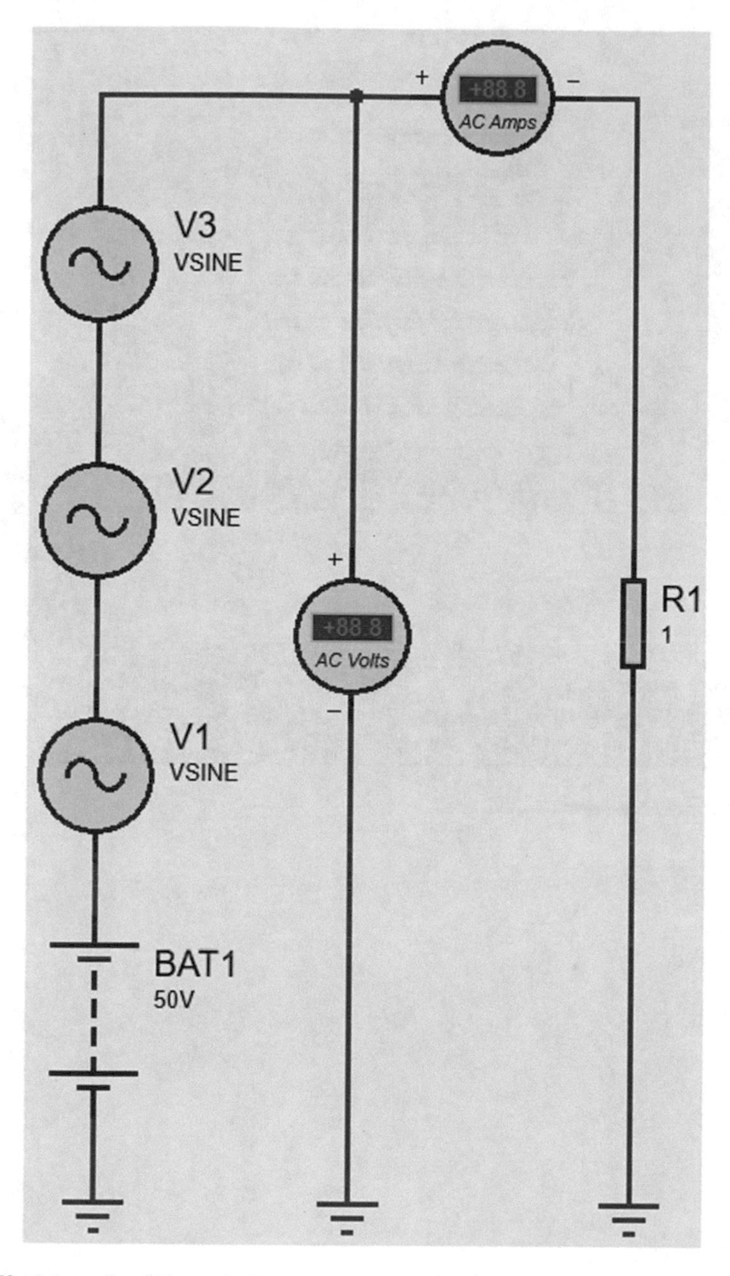

Fig. 1.150 Schematic of Example 17

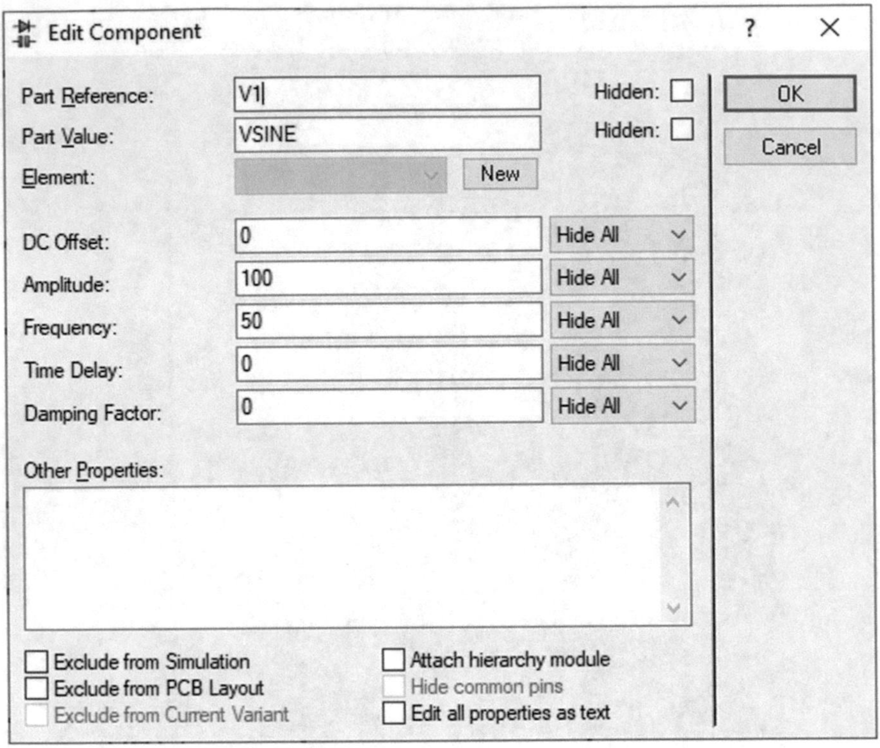

Fig. 1.151 Edit component window

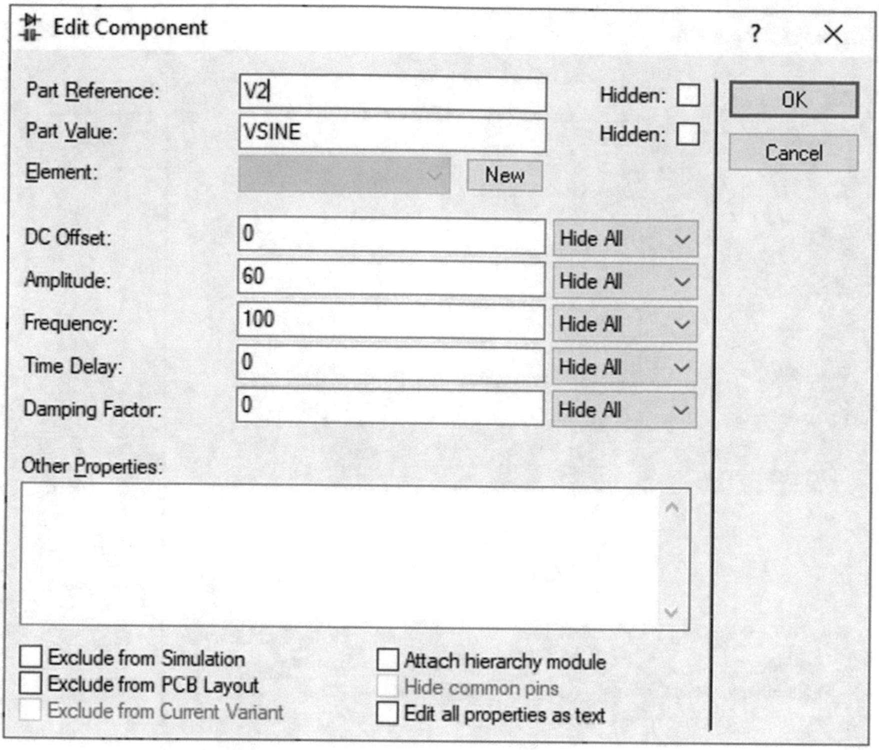

Fig. 1.152 Edit component window

Edit Component ? ×

Part Reference:	V3	Hidden: ☐	OK
Part Value:	VSINE	Hidden: ☐	Cancel
Element:		New	
DC Offset:	0	Hide All	
Amplitude:	30	Hide All	
Frequency:	150	Hide All	
Time Delay:	0	Hide All	
Damping Factor:	0	Hide All	

Other Properties:

☐ Exclude from Simulation ☐ Attach hierarchy module
☐ Exclude from PCB Layout ☐ Hide common pins
☐ Exclude from Current Variant ☐ Edit all properties as text

Fig. 1.153 Edit component window

According to Fig. 1.150, the resistor voltage is $v_R(t) = 50 + 100\sin(2 \times \pi \times 50 \times t) + 60\sin(2 \times \pi \times 100 \times t) + 30\sin(2 \times \pi \times 150 \times t)$. According to Ohm's law, $i_R(t) = \frac{v_R(t)}{R} = \frac{v_R(t)}{1} = v_R(t)$. So, the reading of voltmeter and ammeter must be the same. RMS of $v_R(t)$ is calculated in Fig. 1.154.

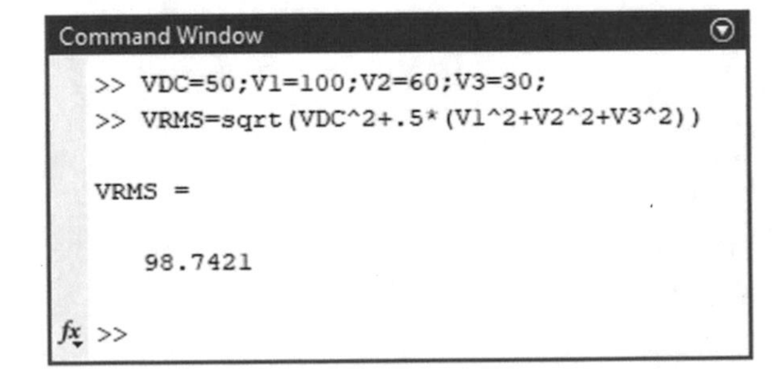

Fig. 1.154 MATLAB calculations

Run the simulation. Simulation result is shown in Fig. 1.155. The obtained result is quite close to the result shown in Fig. 1.154. So, the voltmeter/ammeter measures the RMS correctly even in presence of harmonics.

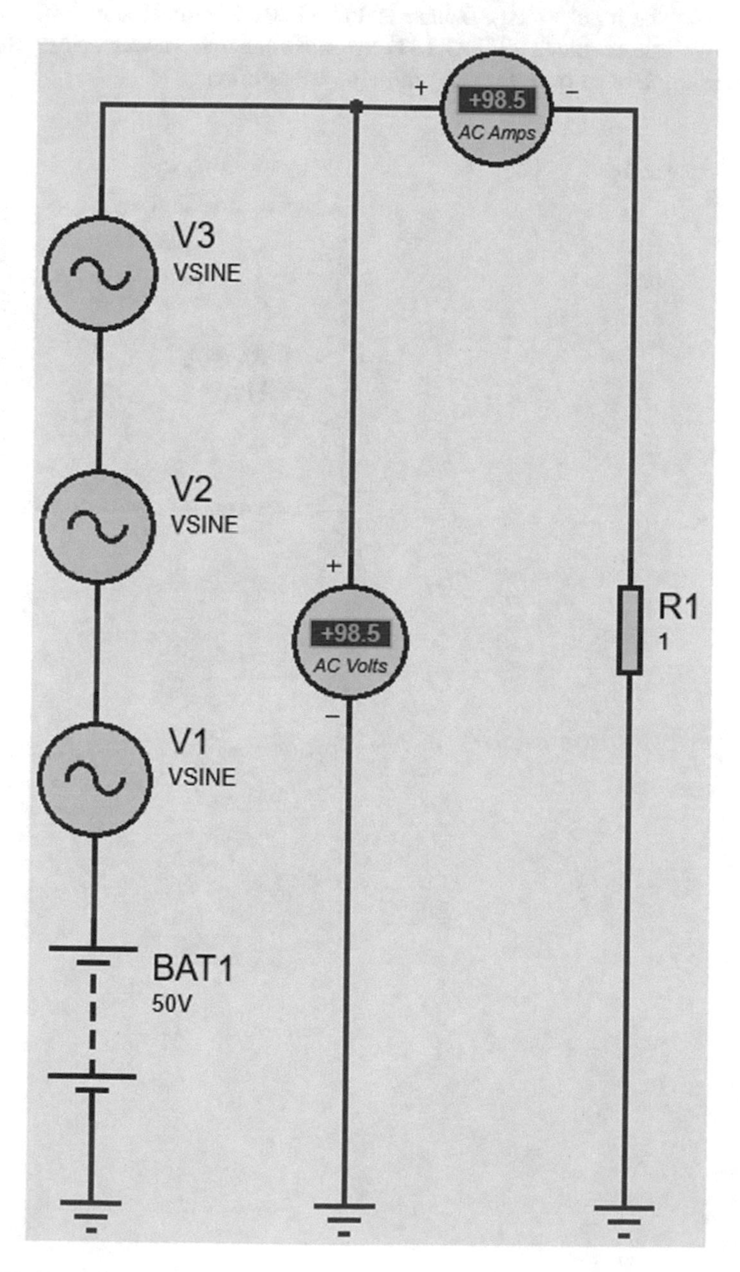

Fig. 1.155 Simulation result

1.19 Example 18: Wattmeter Block (I)

In this example, we want to measure the power drawn from voltage source V_1 (Fig. 1.156). The input voltage source is $V_1 = 220\sqrt{2} \times \sin(2 \times \pi \times 50 \times t)$. We use the Wattmeter block (Fig. 1.157) to measure the power. Note that the Wattmeter is intended only for use with sine waveforms.

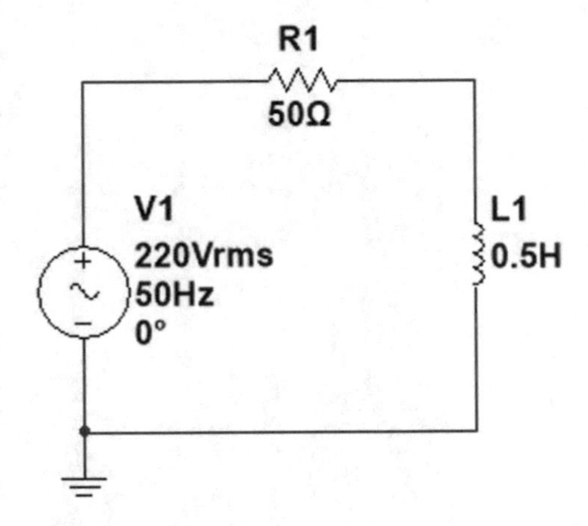

Fig. 1.156 Circuit for Example 18

Fig. 1.157 Wattmeter block

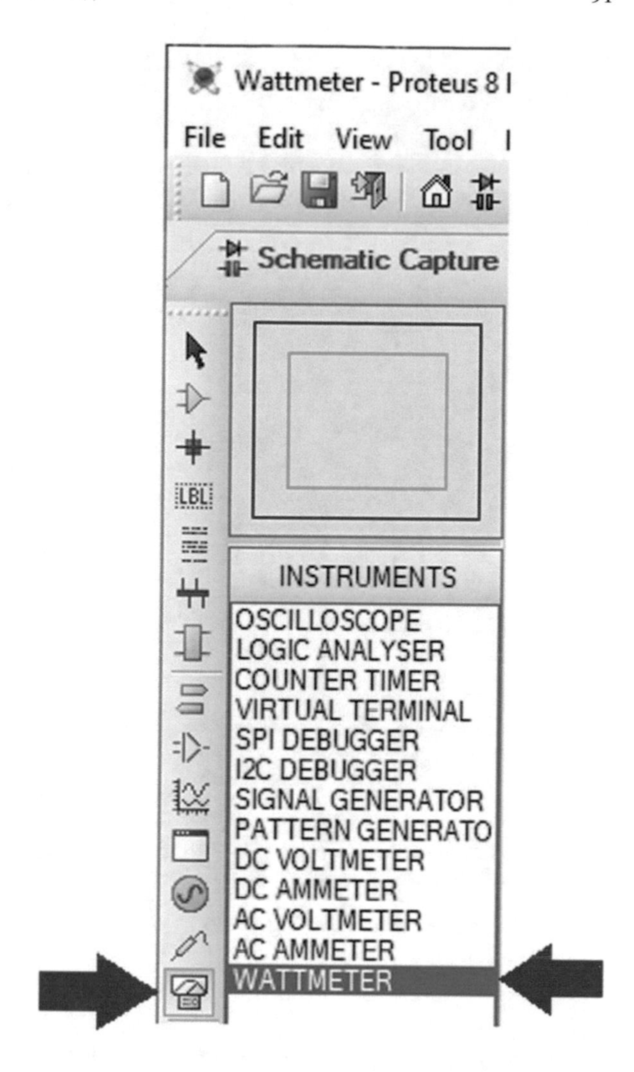

Draw the schematic shown in Fig. 1.158. You can add an inductor to the schematic by searching for "inductor" in the Pick Devices window (Fig. 1.159). Settings of V_{in} are shown in Fig. 1.160.

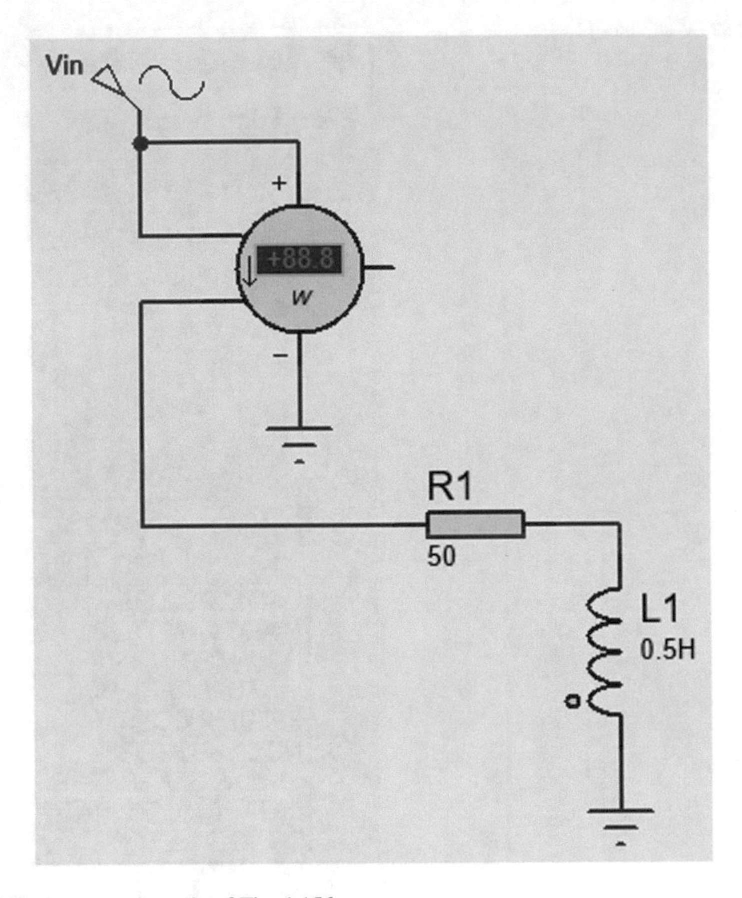

Fig. 1.158 Proteus schematic of Fig. 1.156

Fig. 1.159 Searching for inductor block

Pick Devices

Keywords:

inductor

Match whole words?

Show only parts with models?

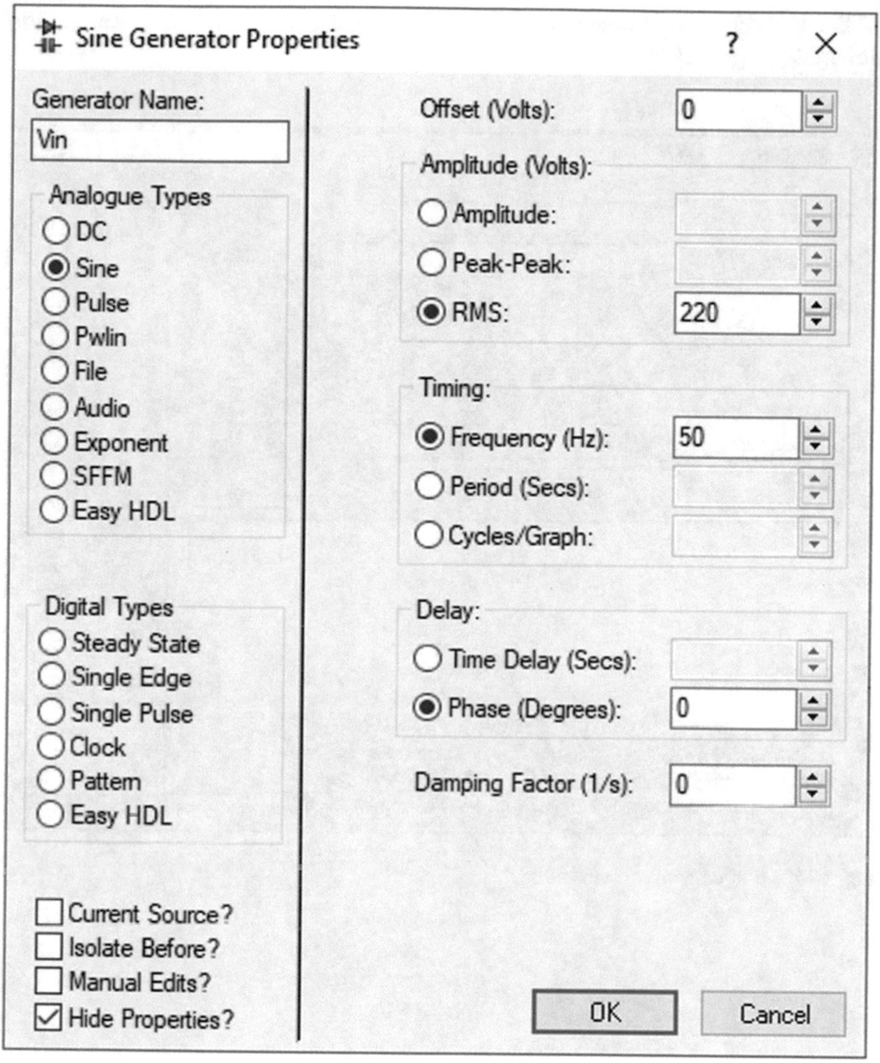

Fig. 1.160 Sine generator properties window

Double click the Wattmeter block and select the W for Display Range drop down list (Fig. 1.161).

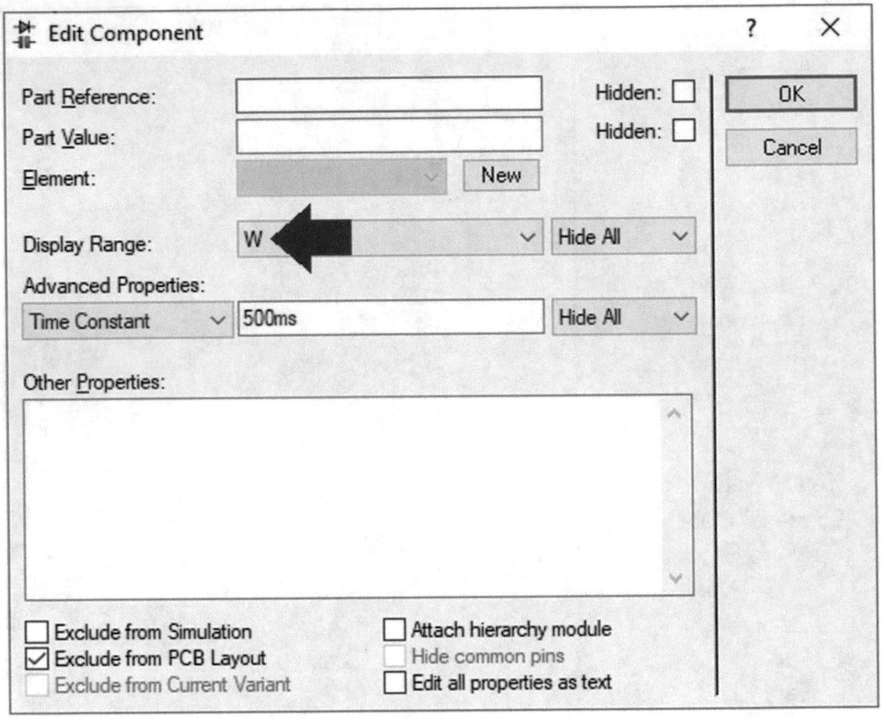

Fig. 1.161 Edit component window

Run the simulation. Simulation result is shown in Fig. 1.162. According to Fig. 1.162, the average power drawn from the input voltage source is 89 W.

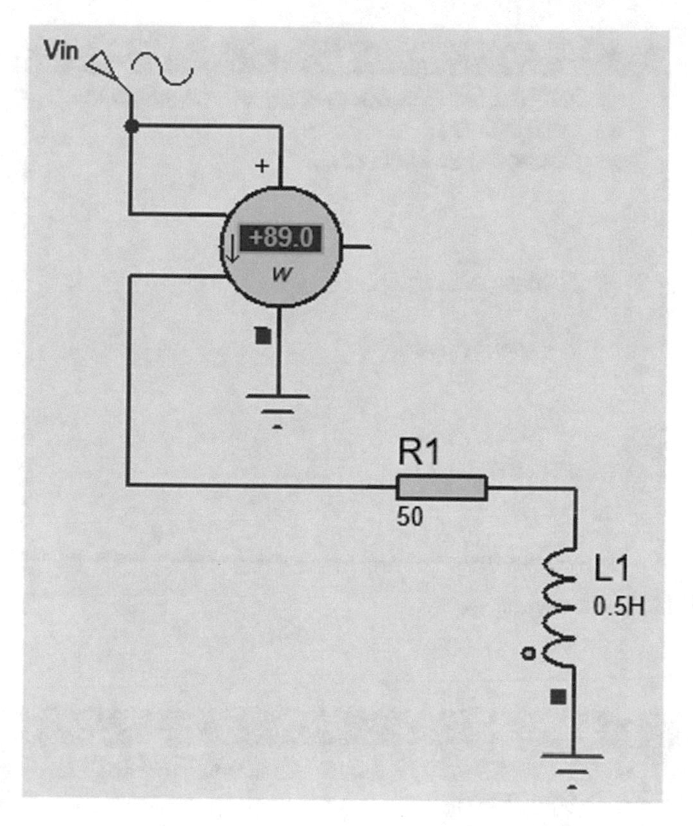

Fig. 1.162 Simulation result

Let's check the Proteus result. The calculations shown in Figs. 1.163 or 1.164 show that Proteus result is correct.

```
Command Window
>> R1=50;L1=0.5;VinRMS=220;f=50;w=2*pi*f;
>> ZL1=j*L1*w;
>> IRMS=VinRMS/(R1+ZL1)

IRMS =

   0.4048 - 1.2717i

>> P=R1*abs(IRMS)^2

P =

   89.0557

fx >>
```

Fig. 1.163 MATLAB calculations

```
Command Window
>> R1=50;L1=0.5;VinRMS=220;f=50;w=2*pi*f;
>> ZL1=j*L1*w;
>> IRMS=VinRMS/(R1+ZL1);
>> S=VinRMS*conj(IRMS)

S =

   8.9056e+01 + 2.7978e+02i

>> P=real(S)

P =

   89.0557

fx >>
```

Fig. 1.164 MATLAB calculations

Now double click the Wattmeter block and select the VA for Display Range drop down list (Fig. 1.165).

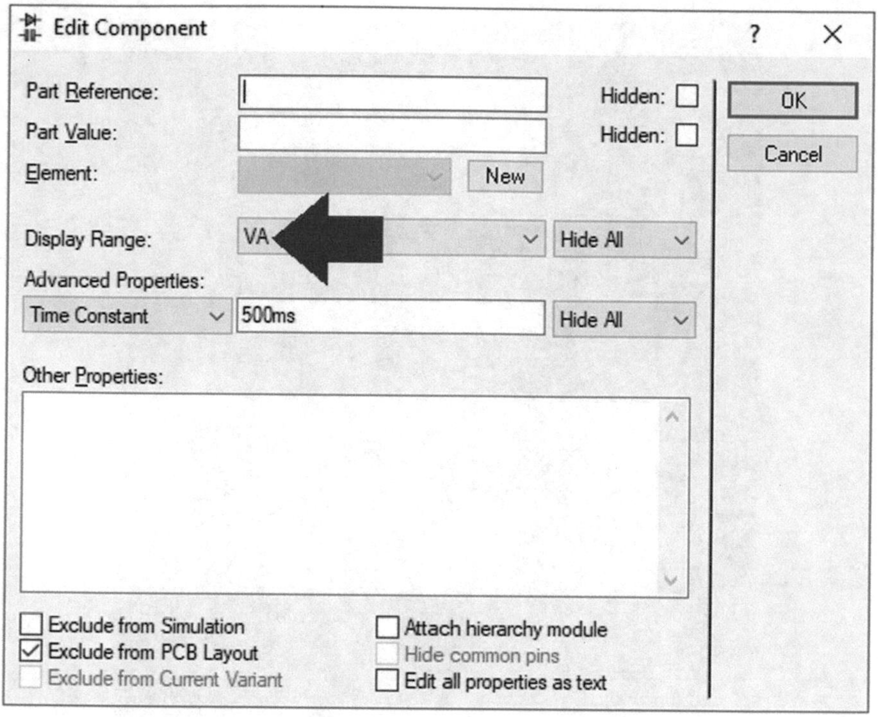

Fig. 1.165 Edit component window

Run the simulation. Simulation result is shown in Fig. 1.166. According to Fig. 1.166, the apparent power drawn from the input voltage source is 294 VA.

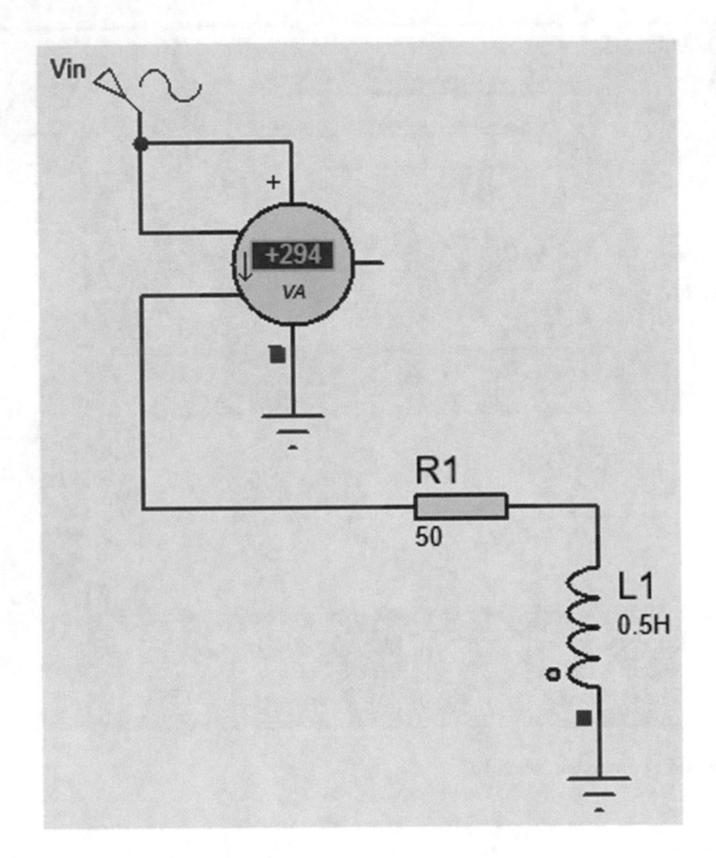

Fig. 1.166 Simulation result

Let's check the Proteus result. The calculations shown in Fig. 1.167 show that Proteus result is correct.

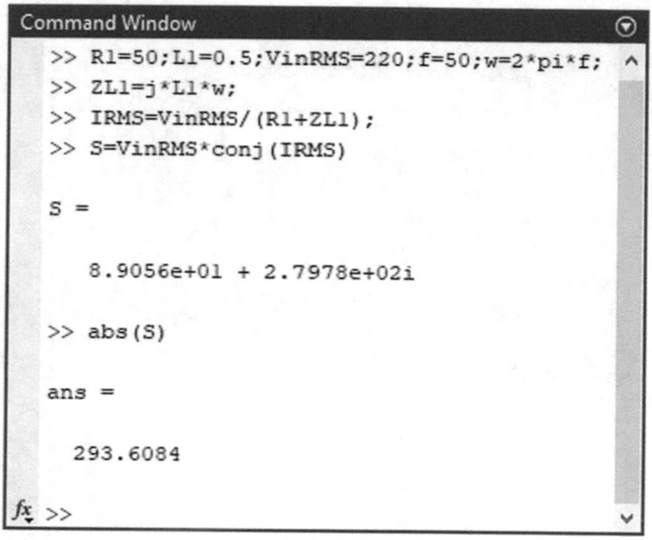

```
Command Window
>> R1=50;L1=0.5;VinRMS=220;f=50;w=2*pi*f;
>> ZL1=j*L1*w;
>> IRMS=VinRMS/(R1+ZL1);
>> S=VinRMS*conj(IRMS)

S =

   8.9056e+01 + 2.7978e+02i

>> abs(S)

ans =

  293.6084

fx >>
```

Fig. 1.167 MATLAB calculations

Now double click the Wattmeter block and select the VAR for Display Range drop down list (Fig. 1.168).

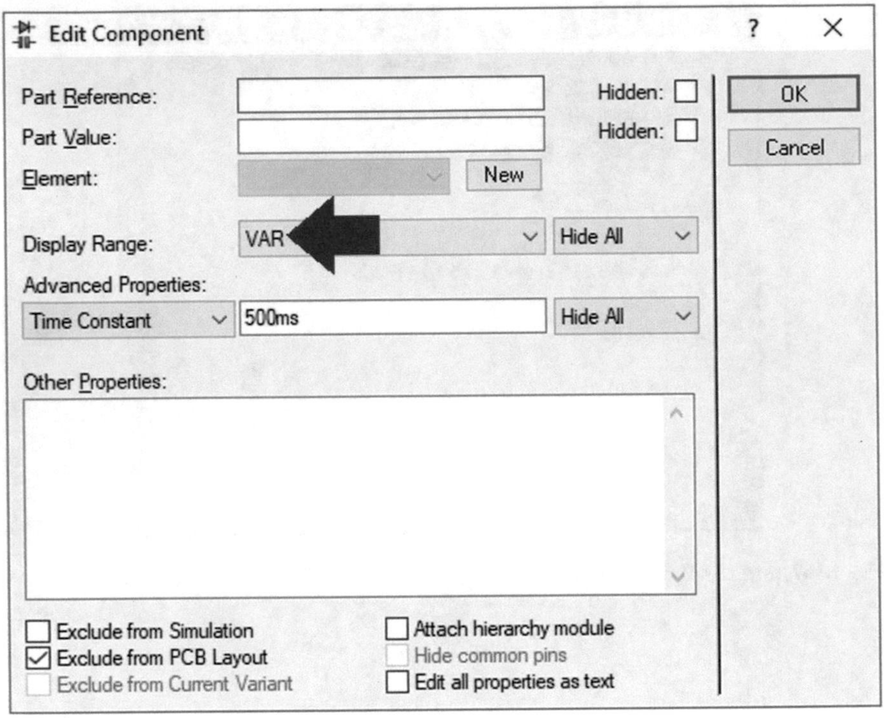

Fig. 1.168 Edit component window

Run the simulation. Simulation result is shown in Fig. 1.169. According to Fig. 1.169, the reactive power drawn from the input voltage source is + 280 VAR.

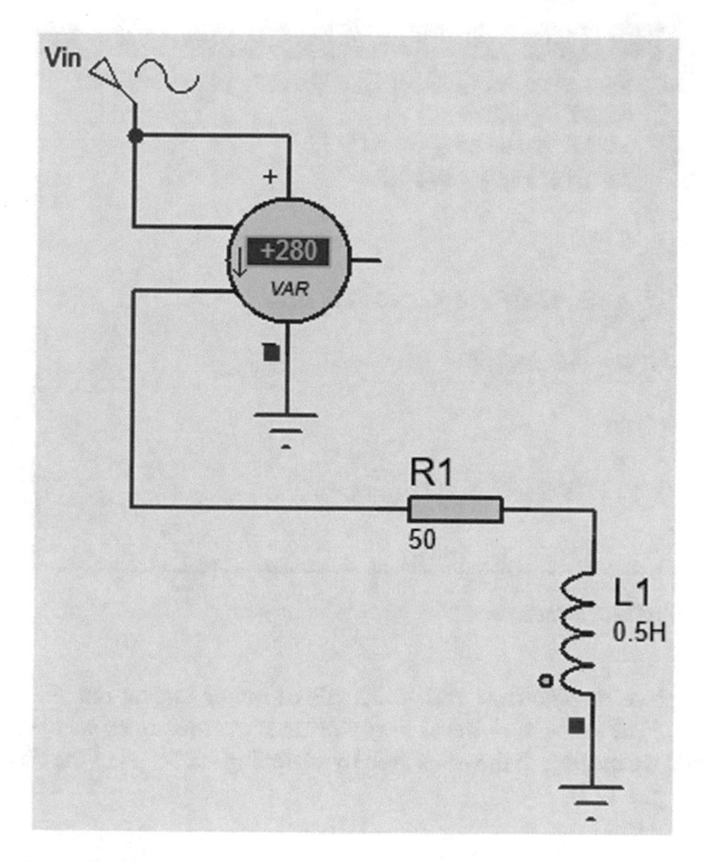

Fig. 1.169 Simulation result

Let's check the Proteus result. The calculations shown in Fig. 1.170 show that Proteus result is correct.

```
Command Window
>> R1=50;L1=0.5;VinRMS=220;f=50;w=2*pi*f;
>> ZL1=j*L1*w;
>> IRMS=VinRMS/(R1+ZL1);
>> S=VinRMS*conj(IRMS)

S =

   8.9056e+01 + 2.7978e+02i

>> Q=imag(S)

Q =

  279.7767

fx >>
```

Fig. 1.170 MATLAB calculations

Now replace the inductor with a 22 μF capacitor and re-run the simulation (Fig. 1.171). The reactive power is negative since the load is capacitive. Note that you can add a capacitor to the schematic by searching for "cap" in the Pick Devices window (Fig. 1.172).

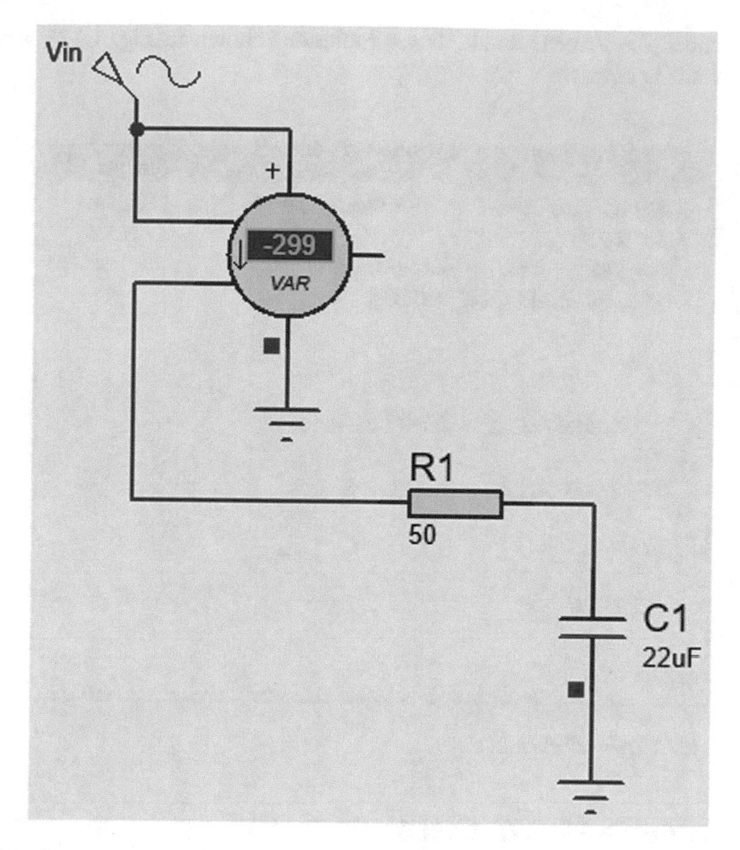

Fig. 1.171 Simulation result

Fig. 1.172 Searching for a capacitor block

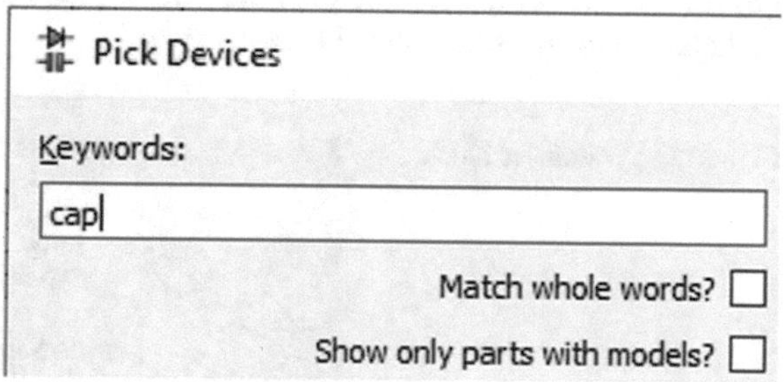

Pick Devices

Keywords:

cap

Match whole words?

Show only parts with models?

Let's check the Proteus result. The calculations shown in Fig. 1.173 show that Proteus result is correct.

```
Command Window
>> R1=50;C1=22e-6;VinRMS=220;f=50;w=2*pi*f;
>> ZC1=-j/(w*C1);
>> IRMS=VinRMS/(R1+ZC1);
>> S=VinRMS*conj(IRMS)

S =

   1.0327e+02 - 2.9883e+02i

>> Q=imag(S)

Q =

 -298.8299

fx >>
```

Fig. 1.173 MATLAB calculations

1.20 Example 19: Wattmeter Block (II)

In the previous example, we learned how to use Wattmeter block to measure different types of power. In this example, we use the right terminal of Wattmeter block (Fig. 1.174) to measure the phase difference between the voltage and current that enter into the Wattmeter block.

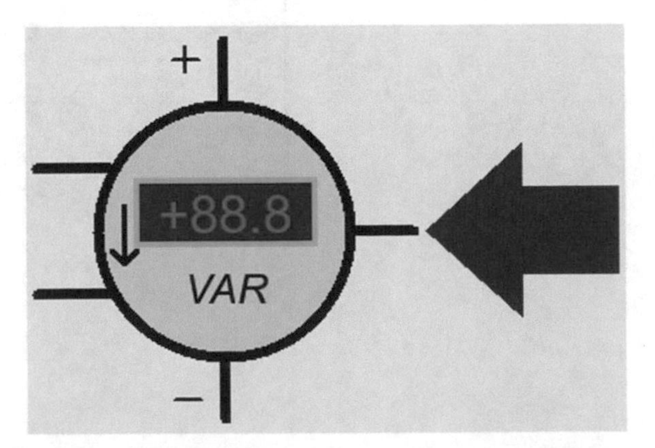

Fig. 1.174 Wattmeter block

Draw the schematic shown in Fig. 1.175.

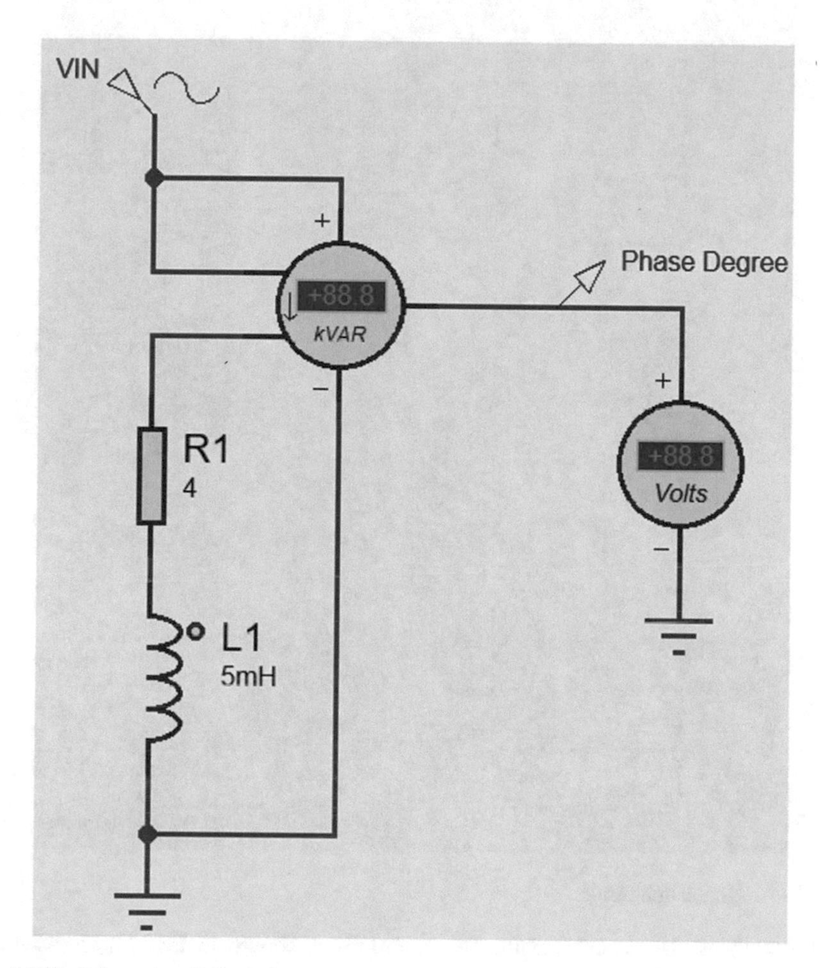

Fig. 1.175 Schematic of Example 19

Run the simulation. The simulation result is shown in Fig. 1.176. According to Fig. 1.176, the phase difference between the voltage and current that enters to the Wattmeter block is 21.457°.

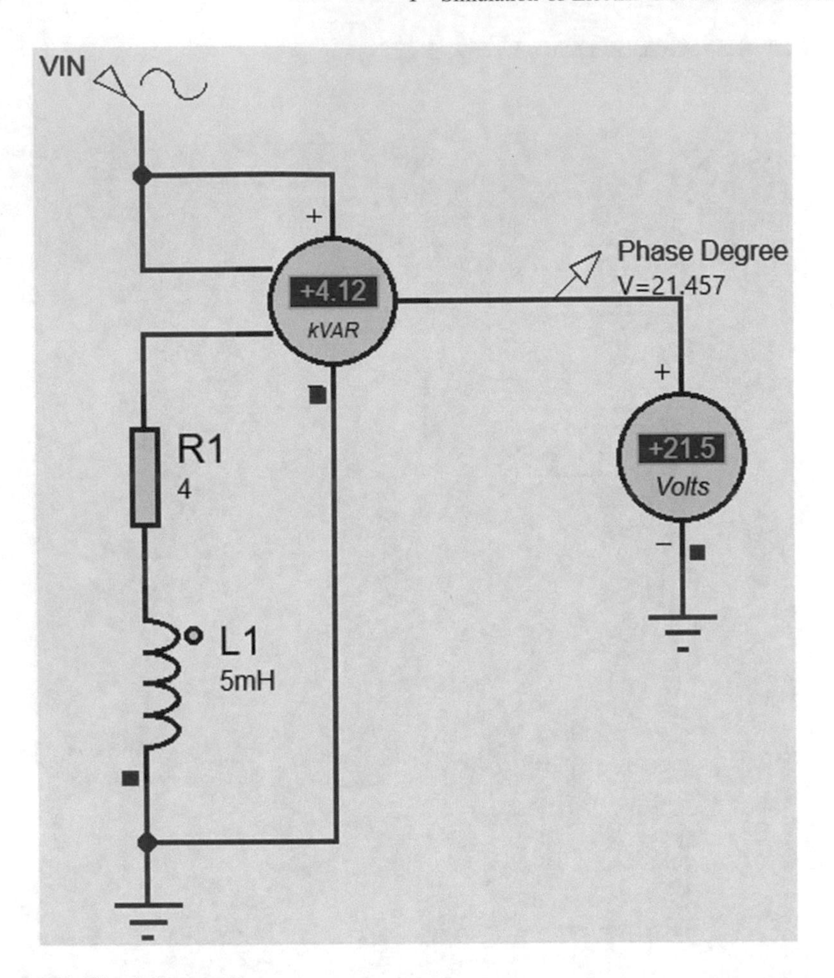

Fig. 1.176 Simulation result

Let's check the Proteus result. The calculations shown in Fig. 1.177 show that Proteus result is correct.

Fig. 1.177 MATLAB calculations

```
Command Window
>> R1=4;L1=5e-3;f=50;w=2*pi*f;
>> atand(L1*w/R1)

ans =

   21.4399

>>
```

Now replace the inductor with a 1000 μF capacitor and run the simulation. According to Fig. 1.178, the phase difference between the voltage and current that enters to the Wattmeter block is – 38.4838°.

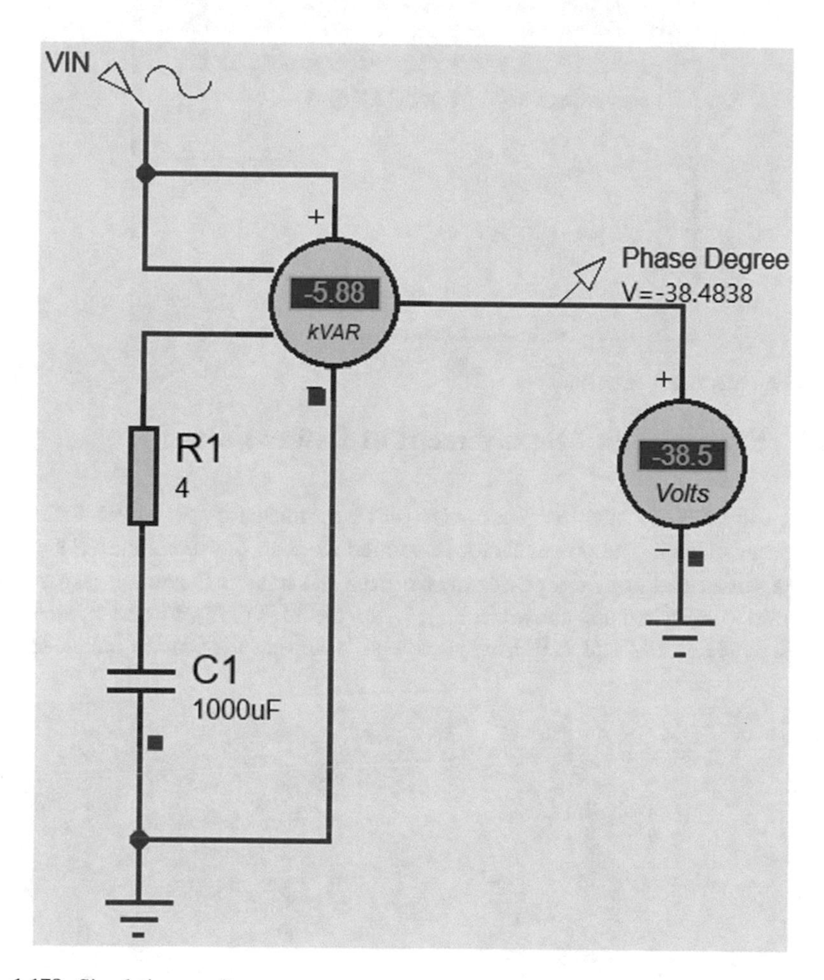

Fig. 1.178 Simulation result

Let's check the Proteus result. The calculations shown in Fig. 1.179 show that Proteus result is correct.

```
Command Window
>> R1=4;C1=1e-3;f=50;w=2*pi*f;
>> atand(-1/(w*C1*R1))

ans =

  -38.5119

fx >>
```

Fig. 1.179 MATLAB calculations

1.21 Example 20: Measurement of Power Factor

In this example, we use the Wattmeter block to measure the power factor of a three-phase circuit. The power factor is defined as $pf = \frac{P}{S}$ where P and S show the average power and apparent power drawn from the input AC source, respectively.

Consider the schematic shown in Fig. 1.180. Settings of V_a, V_b and V_c are shown in Figs. 1.181, 1.182 and 1.183, respectively. Note that the load is balanced.

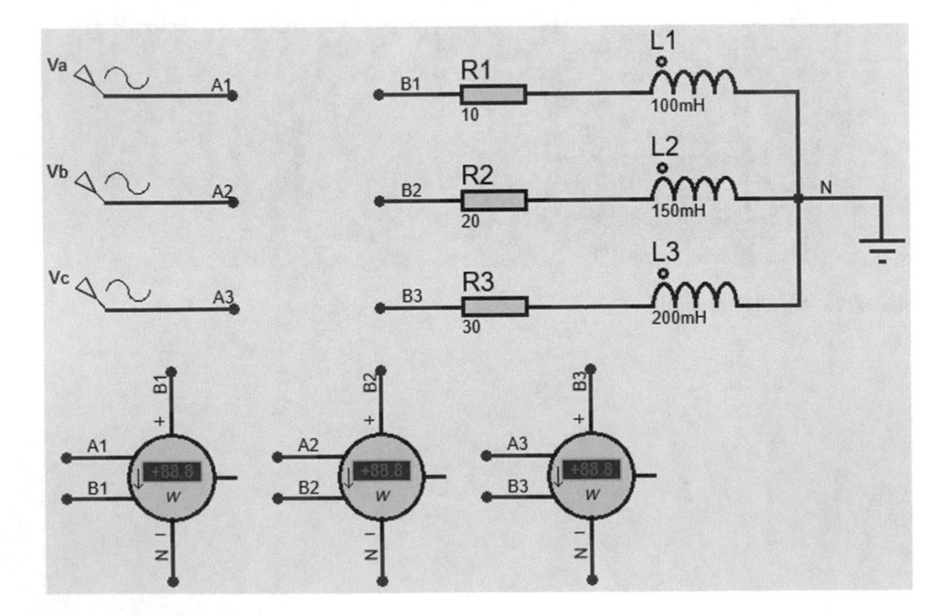

Fig. 1.180 Schematic of Example 20

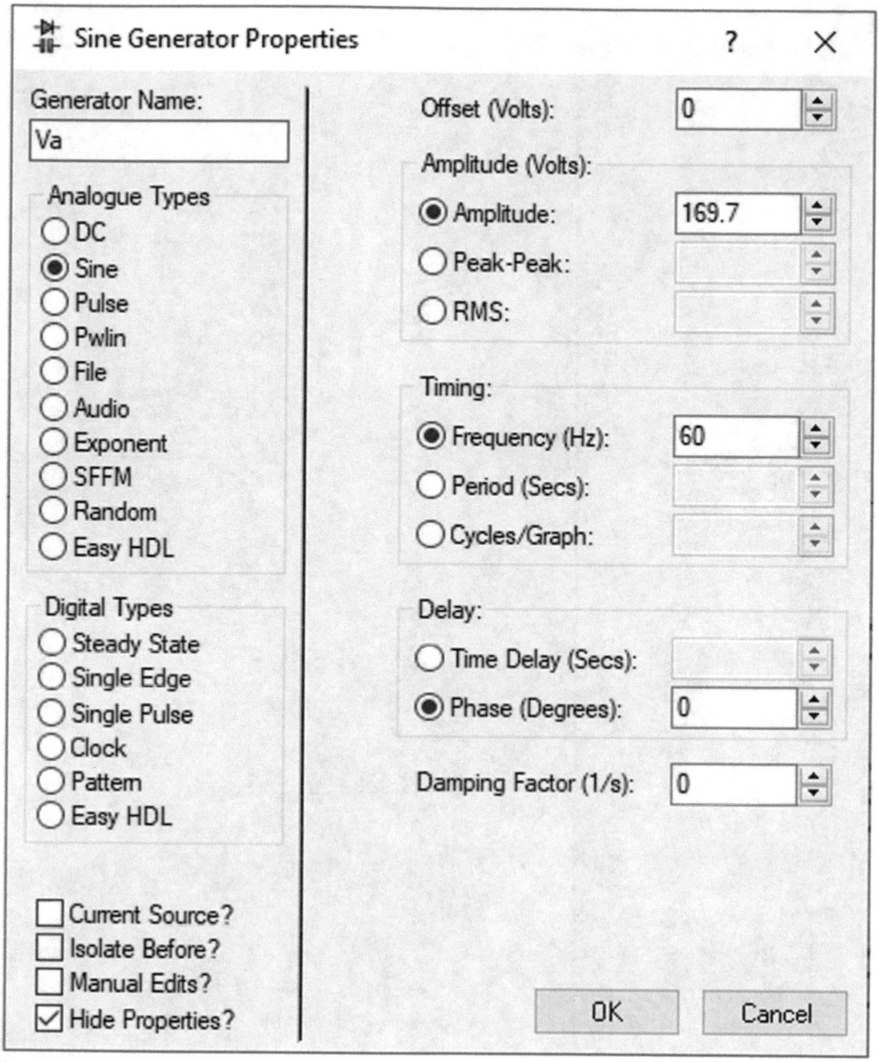

Fig. 1.181 Sine generator properties window

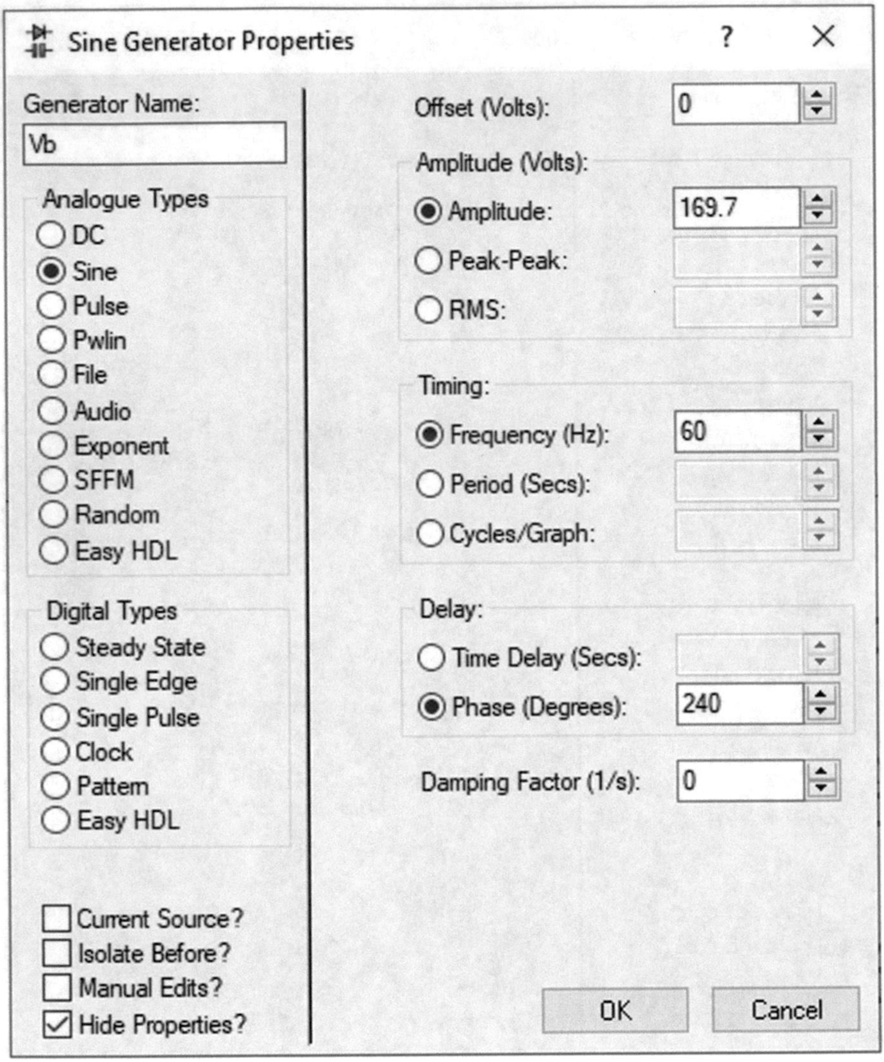

Sine Generator Properties ? ×

Generator Name:
Vb

Analogue Types
DC
Sine
Pulse
Pwlin
File
Audio
Exponent
SFFM
Random
Easy HDL

Digital Types
Steady State
Single Edge
Single Pulse
Clock
Pattern
Easy HDL

Current Source?
Isolate Before?
Manual Edits?
Hide Properties?

Offset (Volts): 0

Amplitude (Volts):
Amplitude: 169.7
Peak-Peak:
RMS:

Timing:
Frequency (Hz): 60
Period (Secs):
Cycles/Graph:

Delay:
Time Delay (Secs):
Phase (Degrees): 240

Damping Factor (1/s): 0

OK Cancel

Fig. 1.182 Sine generator properties window

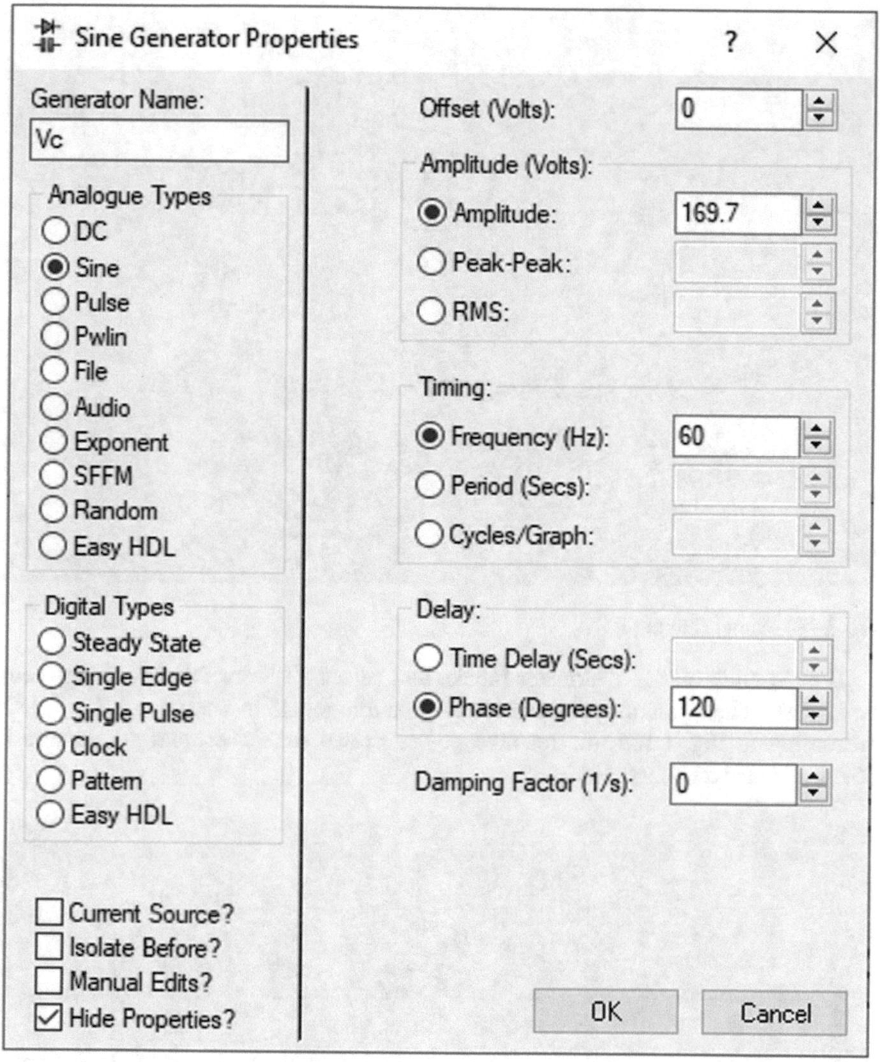

Fig. 1.183 Sine generator properties window

Run the simulation. The simulation result is shown in Fig. 1.184. According to Fig. 1.184, the average power drawn from the input AC source is 94.5 + 79.9 + 65.5 = 239.9 W.

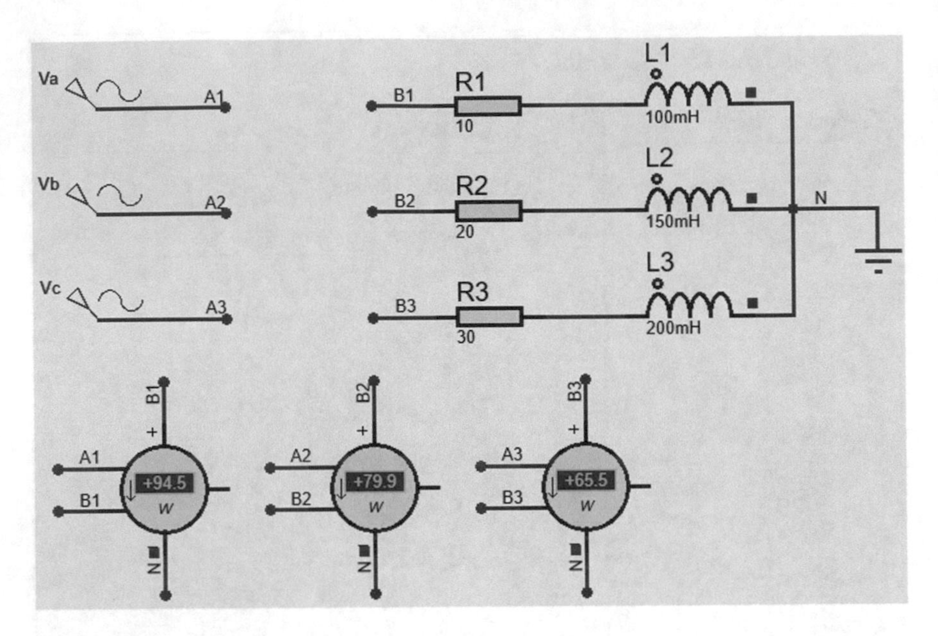

Fig. 1.184 Simulation result

Double click on the Wattmeter blocks and select VAR for Display Range drop down list. Then run the simulation. Simulation result is shown in Fig. 1.185. According to Fig. 1.185, the reactive power drawn from the input AC source is 357 + 226 + 165 = 748 VAR.

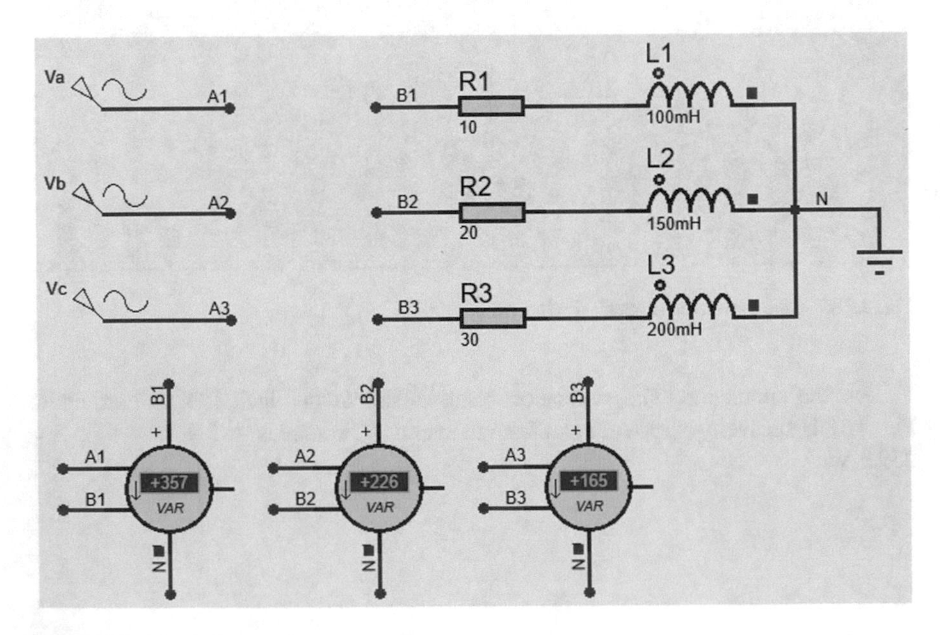

Fig. 1.185 Simulation result

So, the power drawn from the input AC source is $239.9 + j748$. The MATLAB commands shown in Fig. 1.186 calculate the power factor.

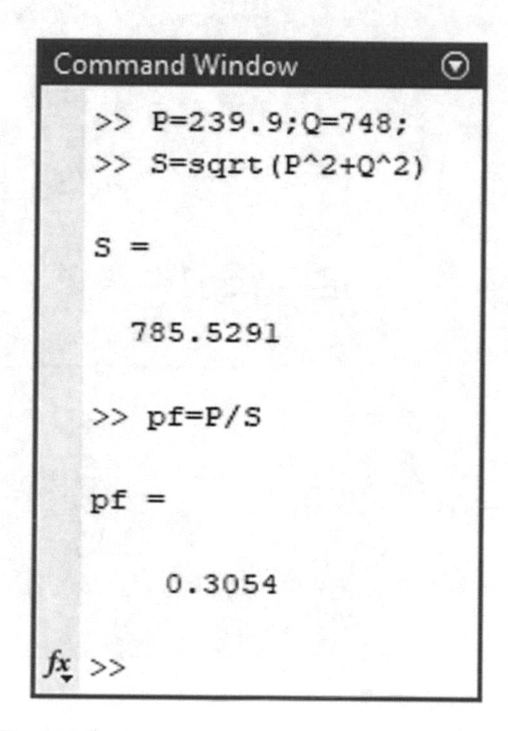

Fig. 1.186 MATLAB calculations

1.22 Example 21: Power Factor Correction

In this example, we want to compensate the circuit of Example 20 in order to increase its power factor. Add three capacitors in parallel to the load (Fig. 1.187).

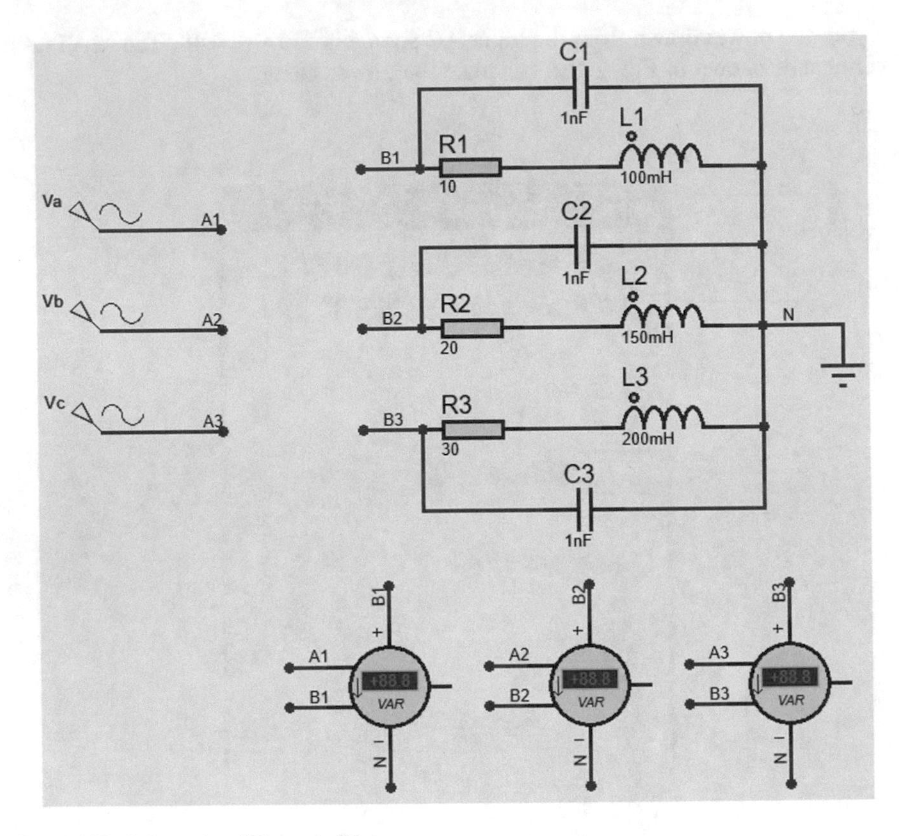

Fig. 1.187 Schematic of Example 21

Next step is to determine the capacitance of capacitors. Values of capacitors are determined with the aid of reactive powers measured by Wattmeters (Fig. 1.188). For instance, the load on phase A consumes 357 VAR. So, C_1 must generate 357 VAR. The MATLAB commands shown in Fig. 1.189 calculate the capacitor which generates 357 VAR.

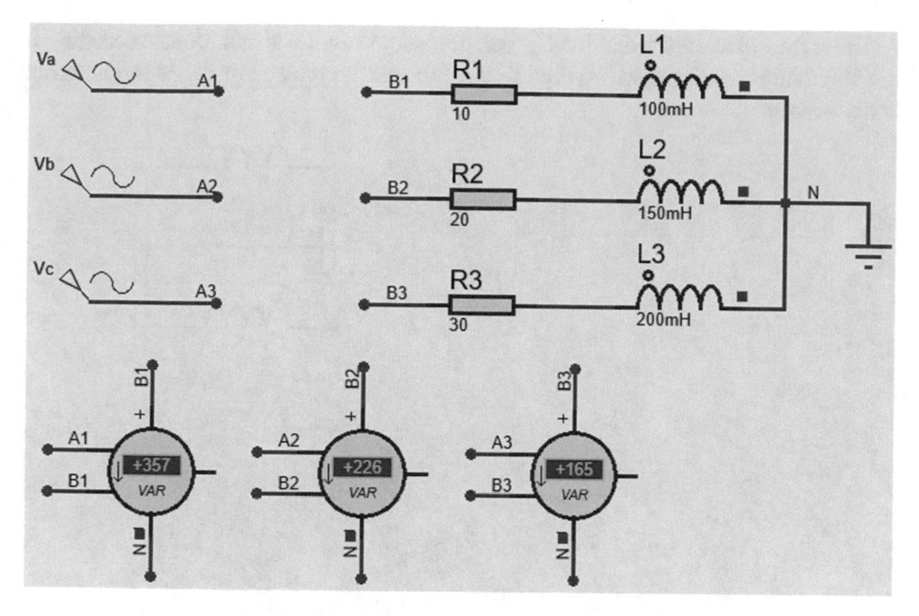

Fig. 1.188 Simulation result

```
Command Window
>> Vm=169.7;VRMS=Vm/sqrt(2);f=60;w=2*pi*f;
>> XC1=VRMS^2/357;
>> C1=1/(w*XC1)

C1 =

   6.5766e-05

fx >>
```

Fig. 1.189 MATLAB calculations

Enter the calculated value for C_1 to Proteus and run the simulation. According to the simulation result shown in Fig. 1.190, the reactive power of phase A decreased considerably.

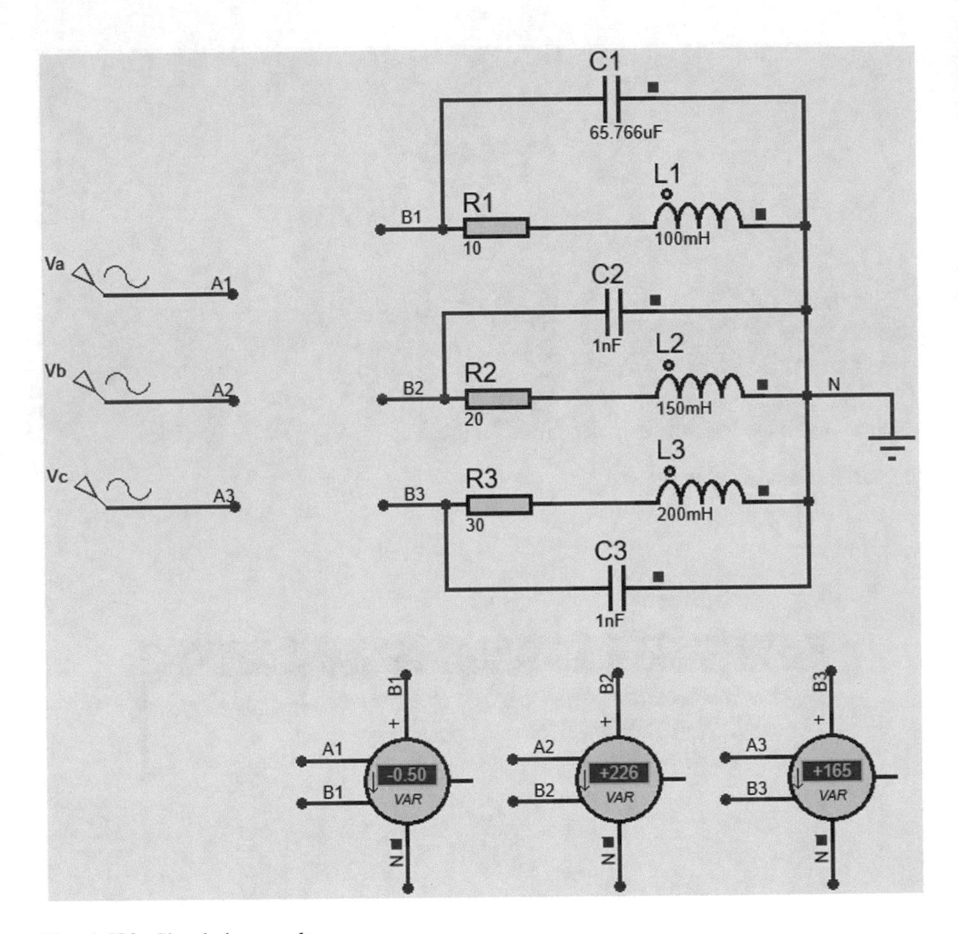

Fig. 1.190 Simulation result

Capacitors C_2 and C_3 must generate 226 VAR and 165 VAR, respectively. MATLAB commands shown in Fig. 1.191 calculate the suitable values for C_2 and C_3.

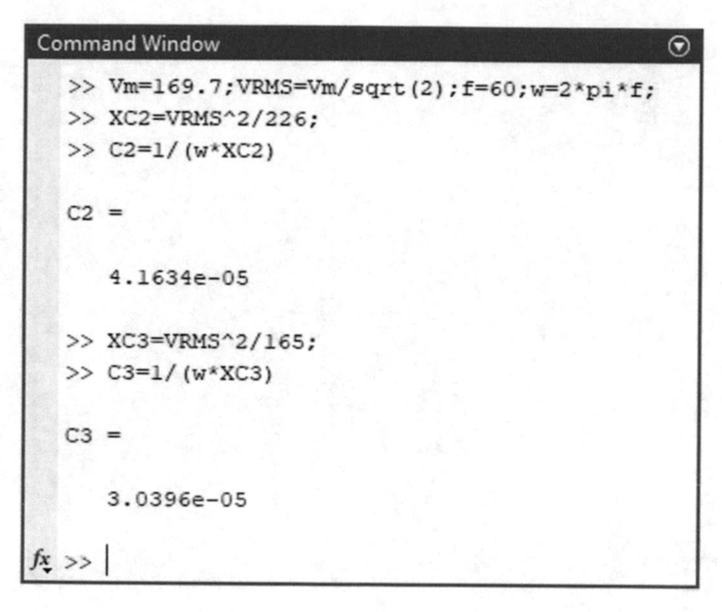

Fig. 1.191 MATLAB calculations

Enter the calculated values for C_2 and C_3 to Proteus and run the simulation. According to the simulation result shown in Fig. 1.192, the reactive power of input source decreased considerably.

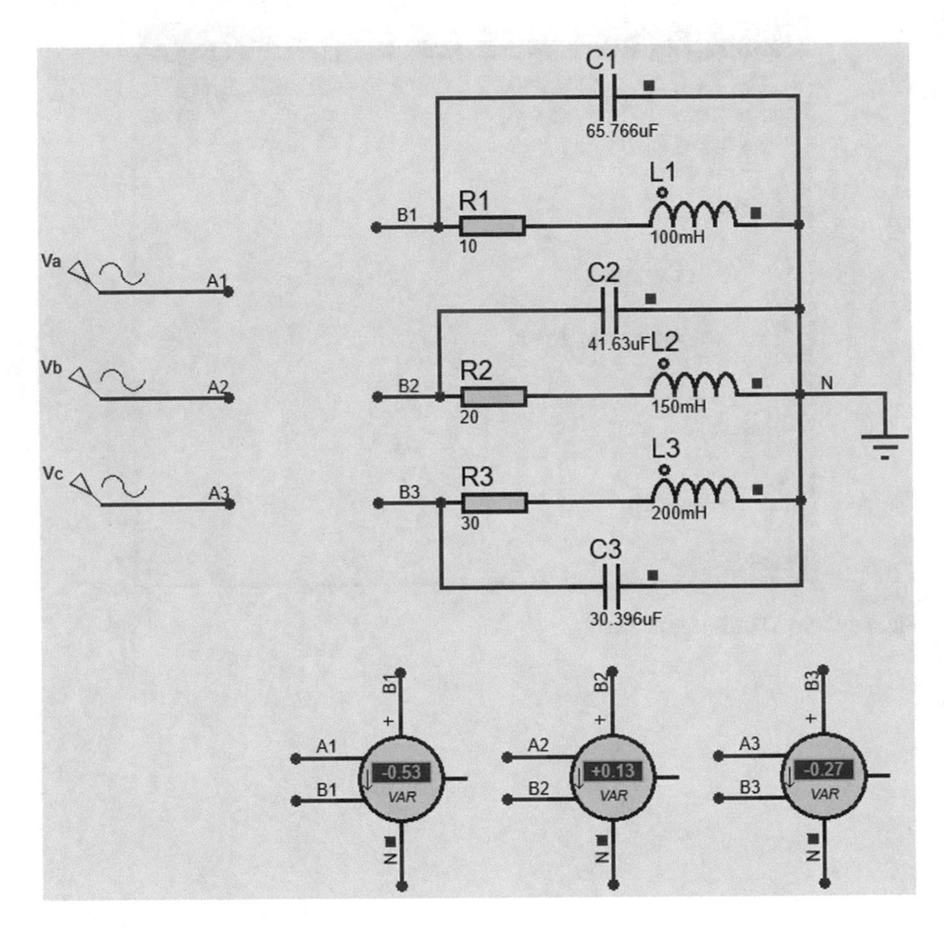

Fig. 1.192 Simulation result

1.23 Example 22: Measurement of Phase Difference

In this example, we want to measure the phase difference between point B and A in Fig. 1.193. The input voltage has frequency of 60 Hz and peak value of 1 V. From basic circuit theory, $V_B = \frac{j\times L\times\omega}{R+j\times L\times\omega}V_A = \frac{j\times L\times 2\pi f}{R+j\times L\times 2\pi f}V_A = \frac{j\times 5m\times 377}{4+j\times 5m\times 377}V_A = \frac{1.885j}{4+1.885j}V_A = 0.426e^{j64.76^\circ}V_A$. V_A and V_B show the phasor of voltage of node A and B, respectively. So, the phase difference between point B and A is 64.76°.

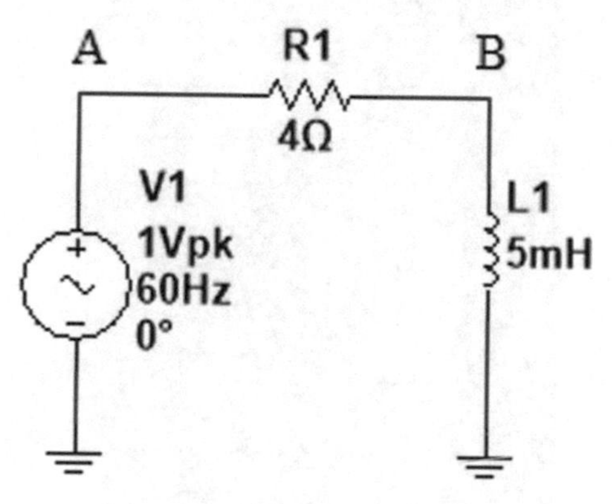

Fig. 1.193 Schematic of Example 22

We use an oscilloscope (Fig. 1.194) to measure the phase difference. Draw the schematic shown in Fig. 1.195. Settings of V_{in} are shown in Fig. 1.196.

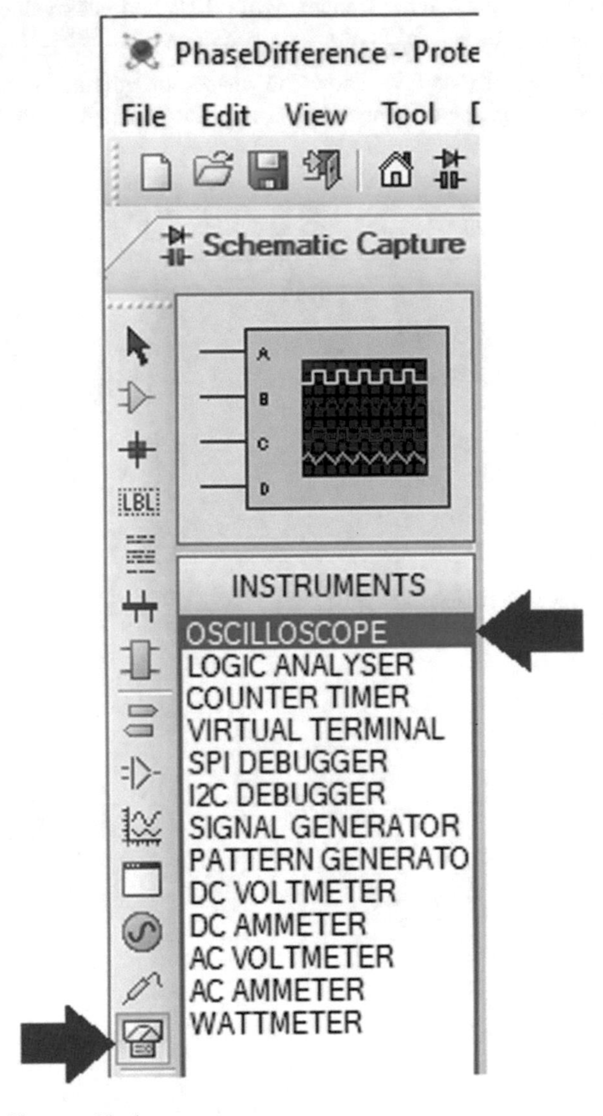

Fig. 1.194 Oscilloscope block

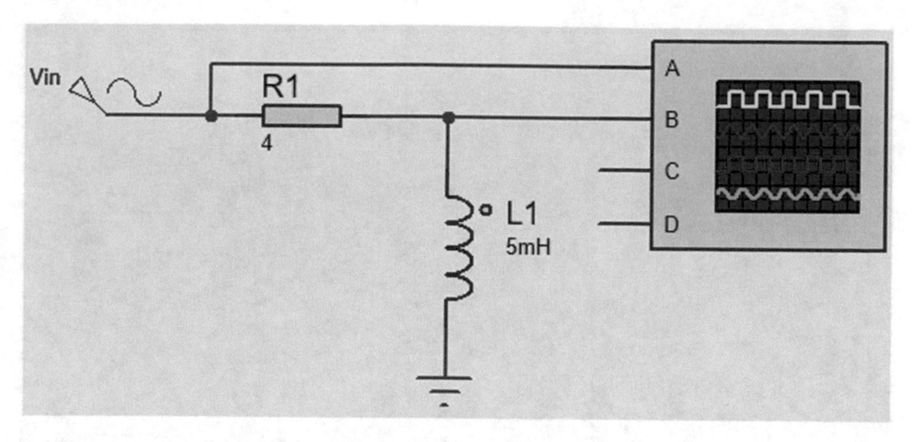

Fig. 1.195 Proteus equivalent of Fig. 1.193

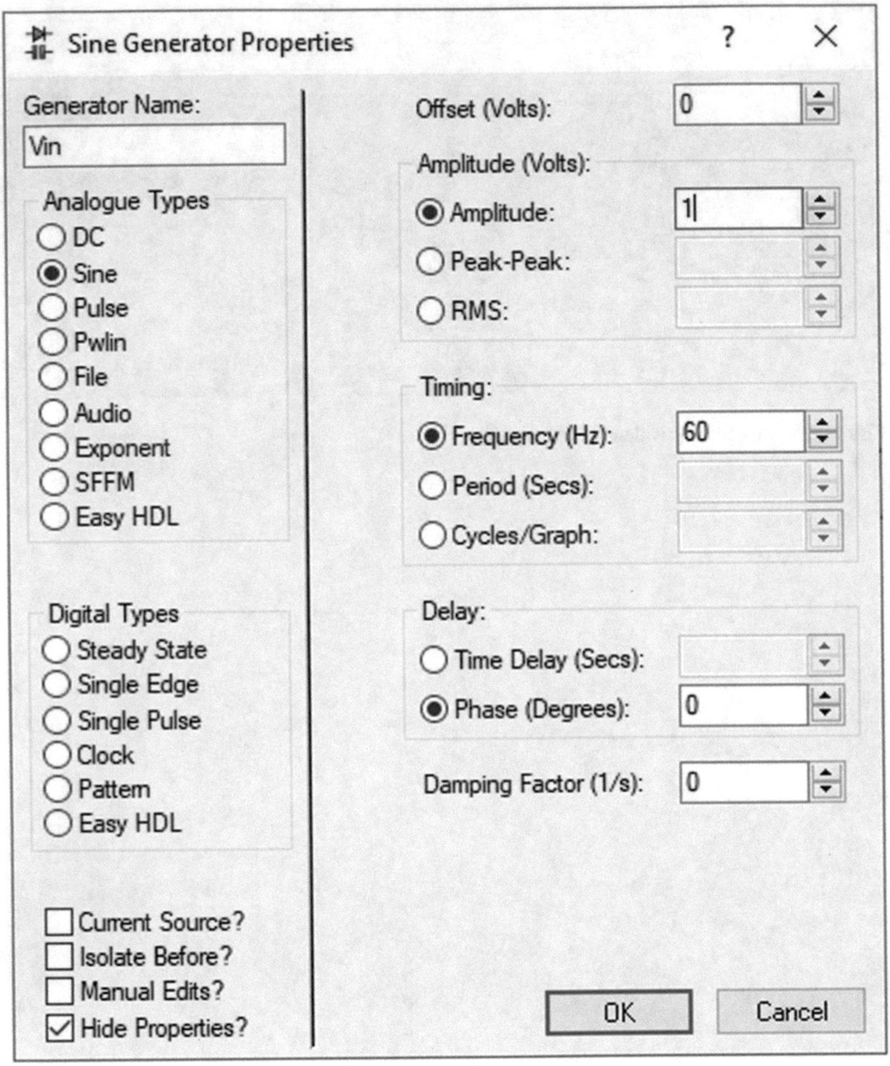

Fig. 1.196 Sine generator properties window

Run the simulation. The oscilloscope screen opens automatically after running the simulation (Fig. 1.197). Channel *A* waveform is shown with yellow color, Channel *B* waveform is shown with blue color, Channel *C* waveform is shown with magenta color and Channel *D* waveform is shown with green color.

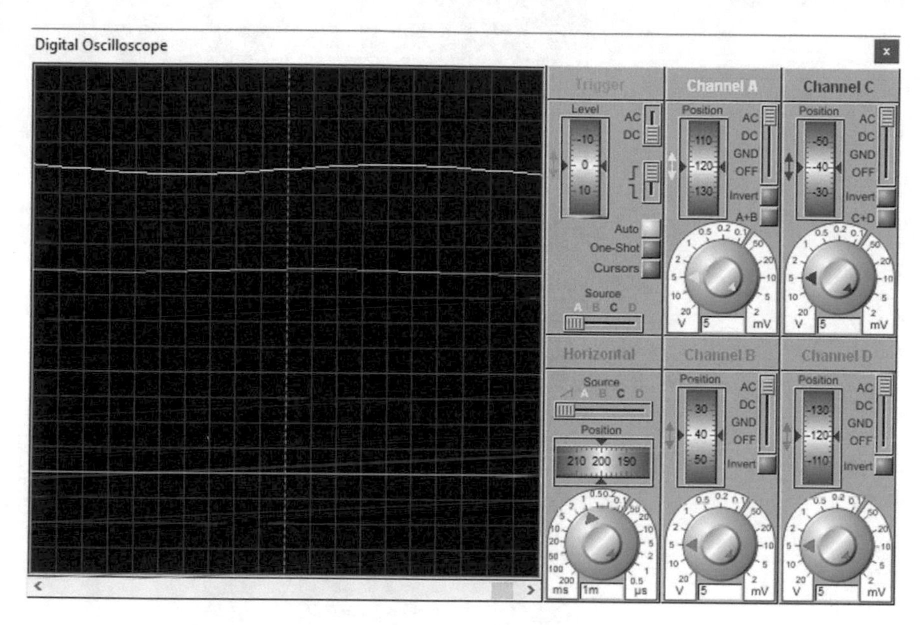

Fig. 1.197 Simulation result

If the oscilloscope screen didn't open automatically, click the Debug > Digital Oscilloscope (Fig. 1.198). This opens the oscilloscope screen for you.

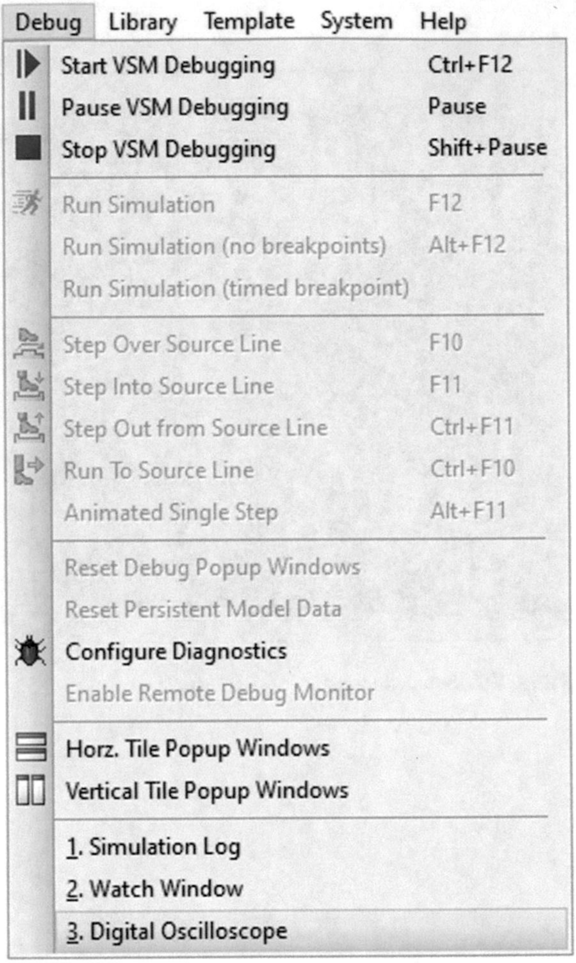

Fig. 1.198 Debug > 3. Digital oscilloscope

We didn't use the channel C and D of oscilloscope. Let's turn off the unused channels (Fig. 1.199).

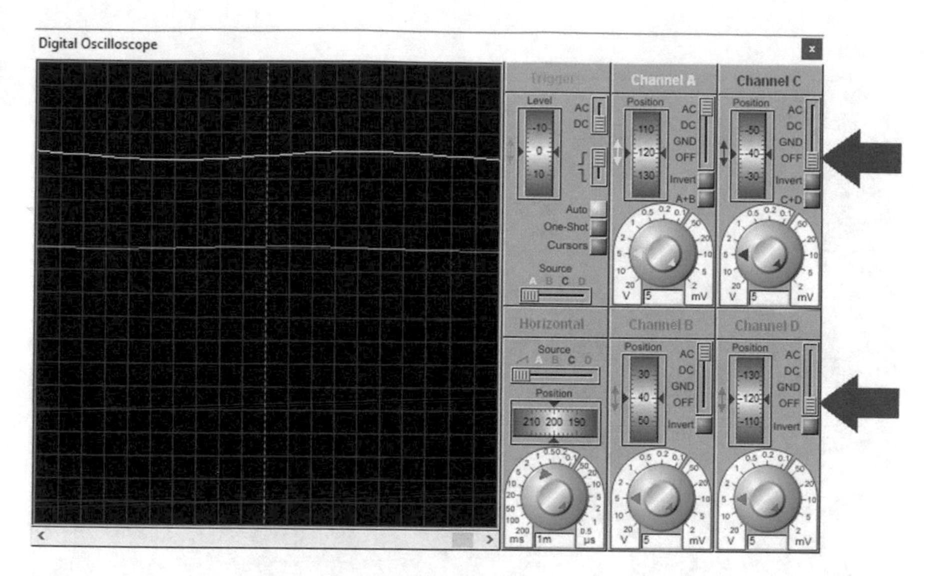

Fig. 1.199 Channel C and D are turned off

Default Volt/division of oscilloscope is 5 V/Div. Decrease the Volt/division to see the waveforms better (Fig. 1.200).

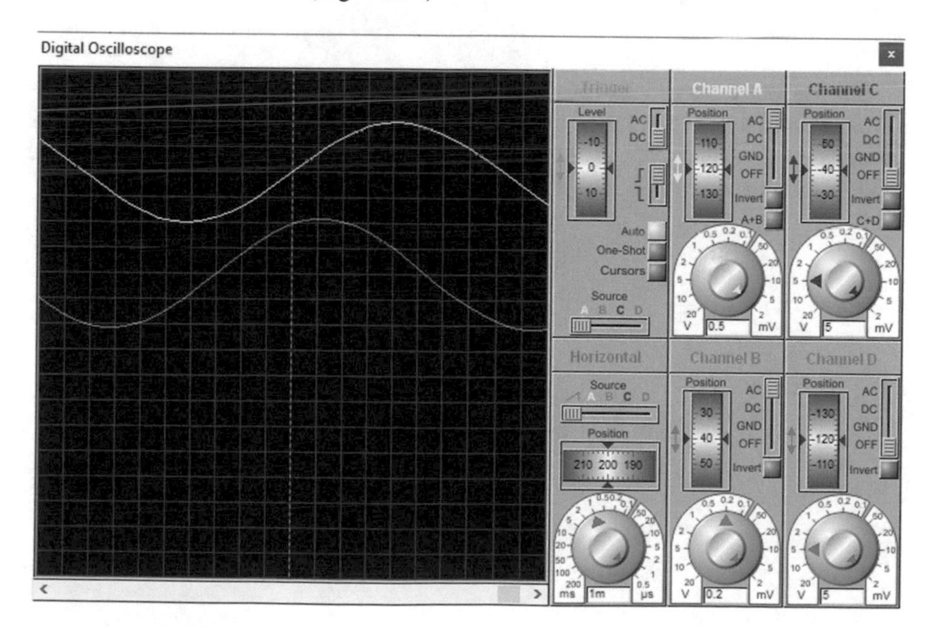

Fig. 1.200 Zoomed waveforms

Double click on the Position controls of channel *A* and *B* to make them zero (Fig. 1.201). Now the two waveforms are drawn on the same horizontal (time) axis.

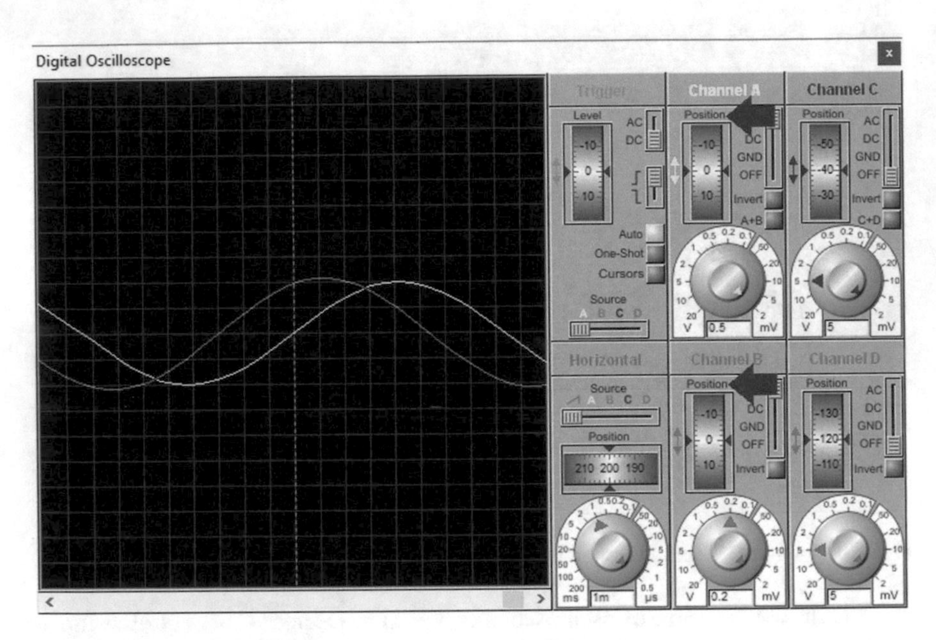

Fig. 1.201 Position controls are zero

Turn on the Cursors and measure the time difference between the two waveforms. According to Fig. 1.202, the time difference between the two waveforms is 3 ms. The calculations in Fig. 1.203 convert the 3 ms into degrees. The obtained value is quite close to the theoretical value.

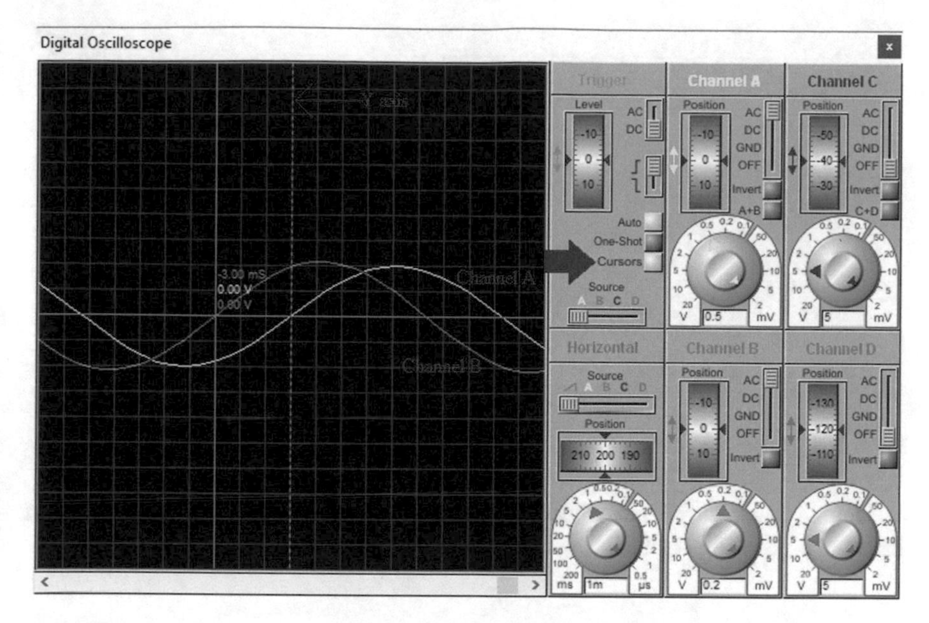

Fig. 1.202 Measurement of time difference between the waveforms

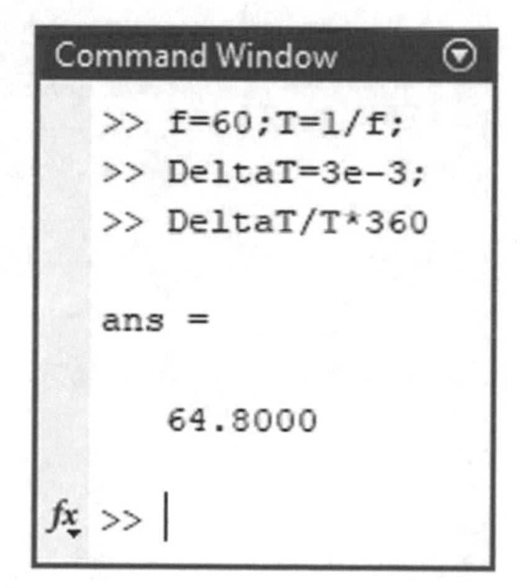

Fig. 1.203 MATLAB calculations

1.24 Example 23: Giving a Name to Oscilloscope Blocks

Sometimes your schematic contains more than one oscilloscope. In such a case, after running the simulation, oscilloscope screens are opened and you need a way to determine which waveform belongs to which oscilloscope. Fortunately, Proteus permits you to add a descriptive text to the oscilloscope window. Such a descriptive text permits you to easily determine the waveform belongs to which oscilloscope. Let's see how we can add such a text to the oscilloscopes.

Consider the schematic of previous example (Fig. 1.204).

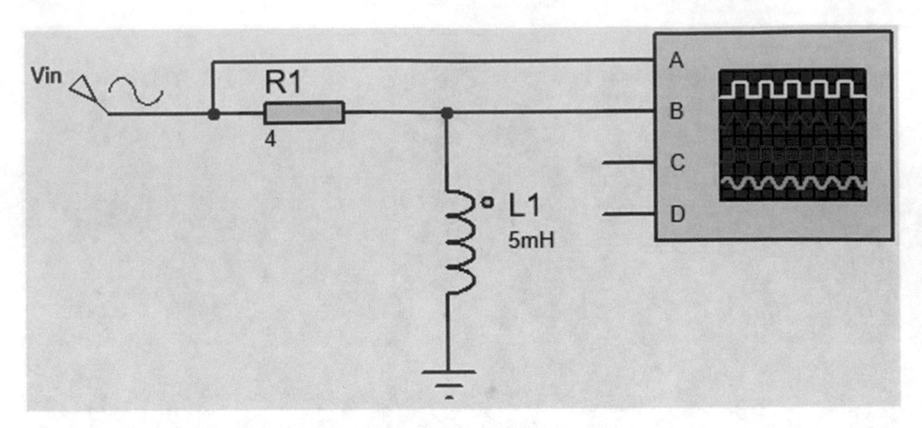

Fig. 1.204 Schematic of Example 23

Double click on the oscilloscope block and enter the desired text to Part Reference box (Fig. 1.205). Then click the OK button.

Edit Component

Part Reference: INPUTVOLTAGE_OUTPUTVOLTA Hidden:

Part Value: Hidden:

Element: New

VSM Model: DSO.DLL Hide All

TRIGCURSORS: TRUE Hide All

Other Properties:

{TRIGONESHOT=FALSE}

Exclude from Simulation

Exclude from PCB Layout

Exclude from Current Variant

Attach hierarchy module

Hide common pins

Edit all properties as text

OK

Help

Cancel

Fig. 1.205 Edit component window

After clicking the OK button, the schematic changes to what is shown in Fig. 1.206. The entered text is added to the schematic.

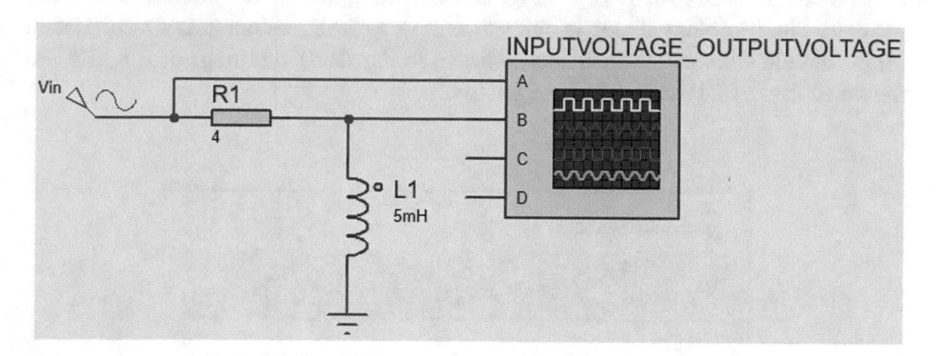

Fig. 1.206 Entered name is shown on the oscilloscope block

Run the simulation. The entered text appeared on the oscilloscope window (Fig. 1.207).

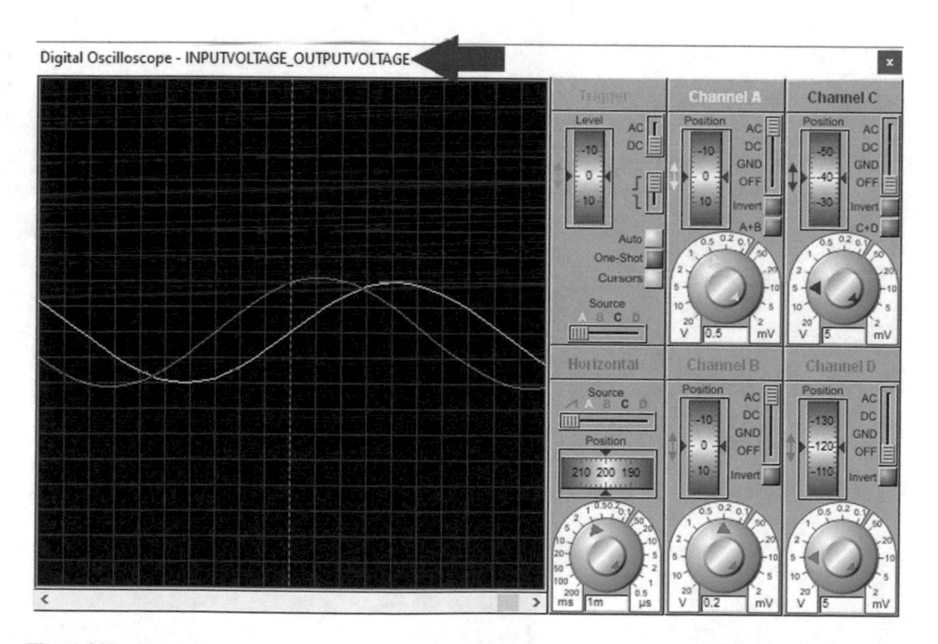

Fig. 1.207 Entered name is shown on the oscilloscope

1.25 Example 24: I_{sine} Block

Proteus has a block called "i_{sine}" (Fig. 1.208) which is able to generate sinusoidal currents. The i_{sine} block dialog is quite similar to v_{sine} block dialog. Let's simulate a simple circuit. Consider the schematic shown in Fig. 1.209. Settings of I_1 and I_2 are shown in Figs. 1.210 and 1.211, respectively.

Pick Devices

Keywords:

isine

Match whole words?

Show only parts with models?

Fig. 1.208 Search for i_{sine} block

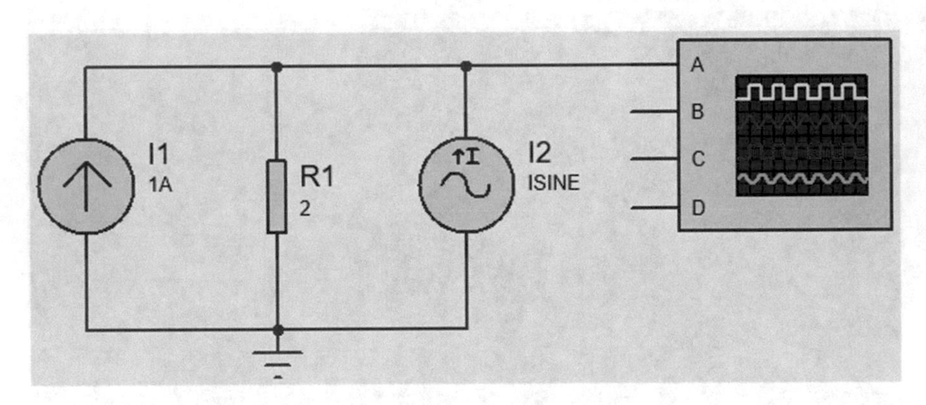

Fig. 1.209 Schematic for Example 24

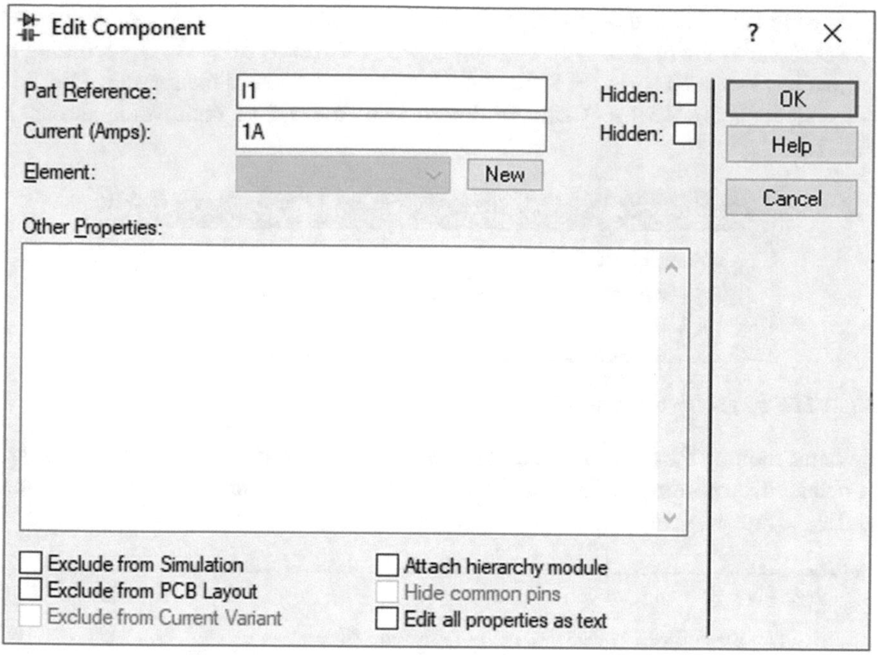

Fig. 1.210 Edit component window

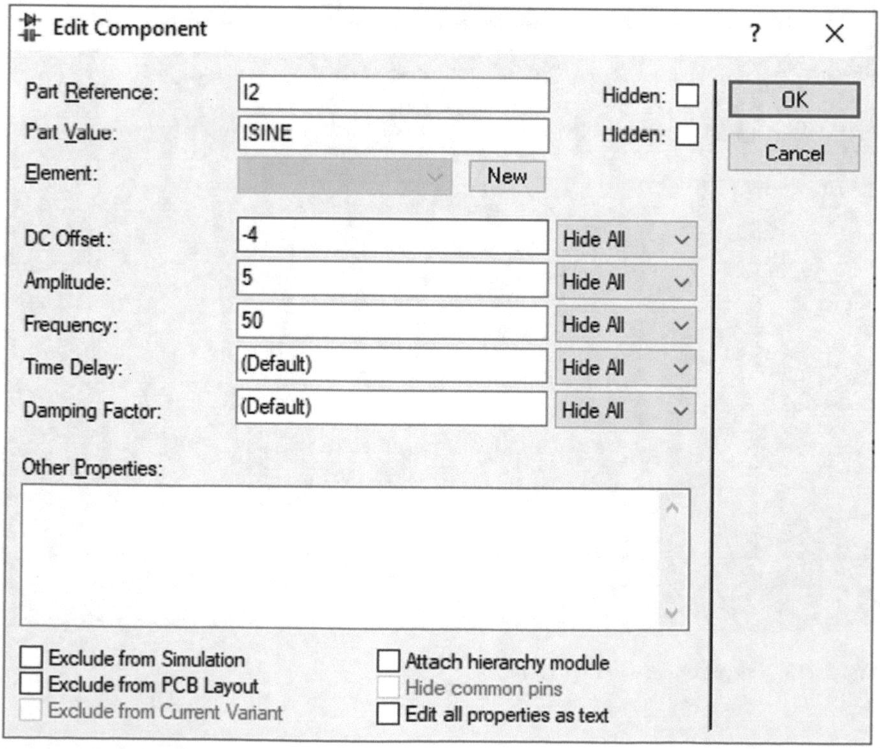

Fig. 1.211 Edit component window

I_1 generates 1 A and I_2 generates $-4 + 5 \times \sin(2\pi \times 50 \times t)$ A. So, the current that enters into the resistor is $I_1 + I_2 = -3 + 5 \times \sin(2\pi \times 50 \times t)$ A. According to Ohm's law $-6 + 10 \times \sin(2\pi \times 50 \times t)$ V is generated across the resistor. Graph of $-6 + 10 \times \sin(2\pi \times 50 \times t)$ can be drawn with the aid of commands shown in Fig. 1.212.

```
Command Window
>> syms t
>> ezplot(-6+10*sin(2*pi*50*t),[0 0.1])
fx >>
```

Fig. 1.212 MATLAB commands

After running the commands shown in Fig. 1.212, the result shown in Fig. 1.213 is obtained. According to Fig. 1.213 the maximum and the minimum of graph are 4 and − 16, respectively.

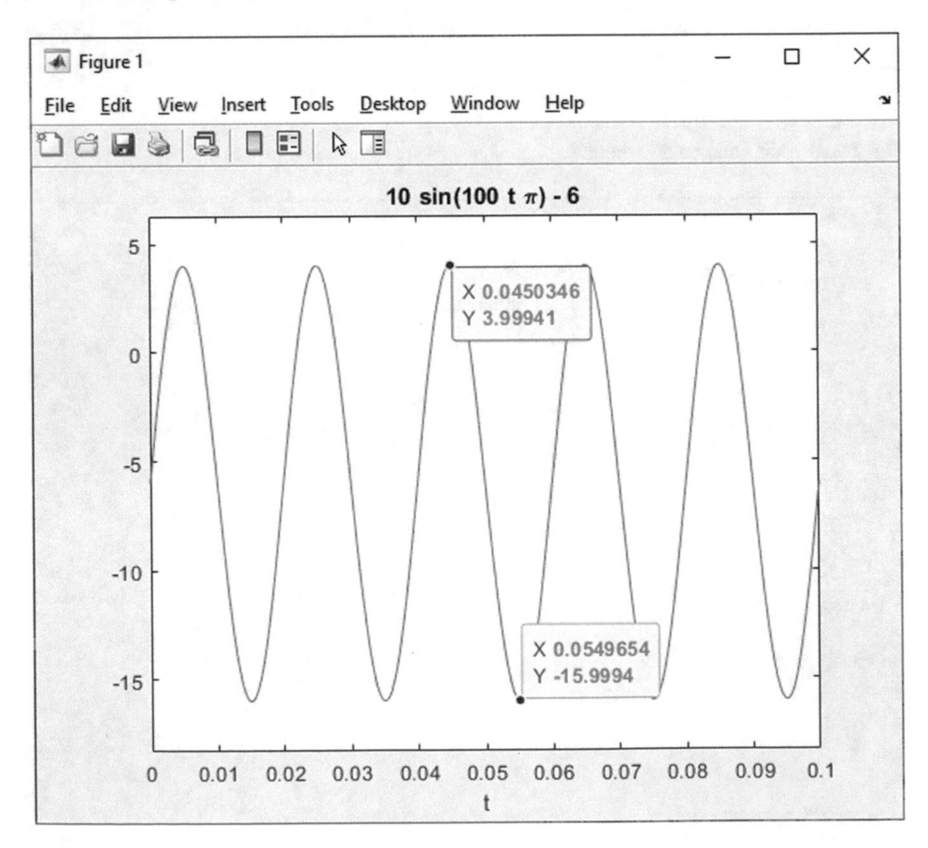

Fig. 1.213 Output of code in Fig. 1.212

Run the simulation. The simulation result is shown in Fig. 1.214. Let's measure the maximum and minimum of the obtained waveform. According to Fig. 1.215, the maximum and minimum of the obtained waveform are + 4 V and − 16 V, respectively.

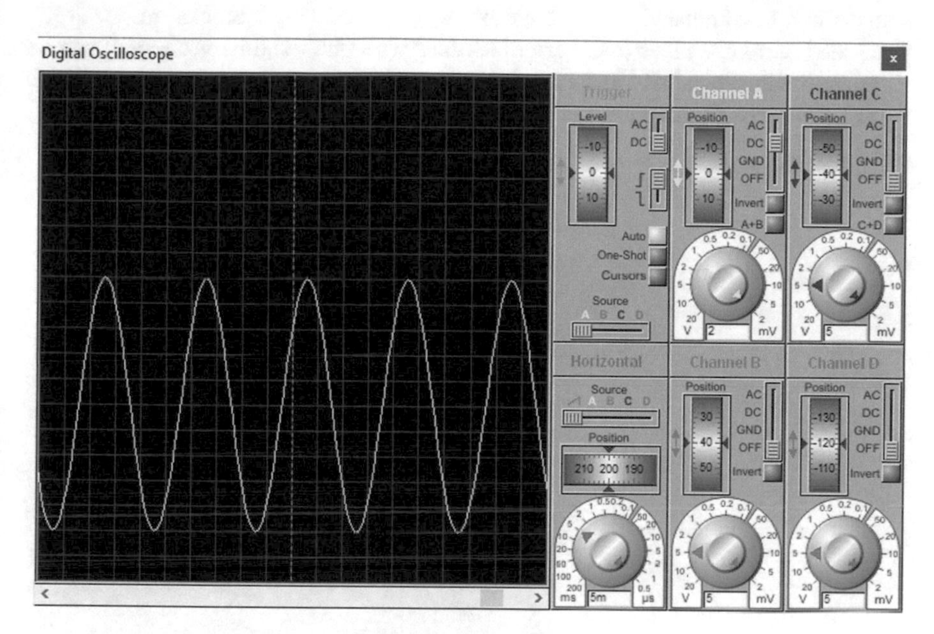

Fig. 1.214 Simulation result

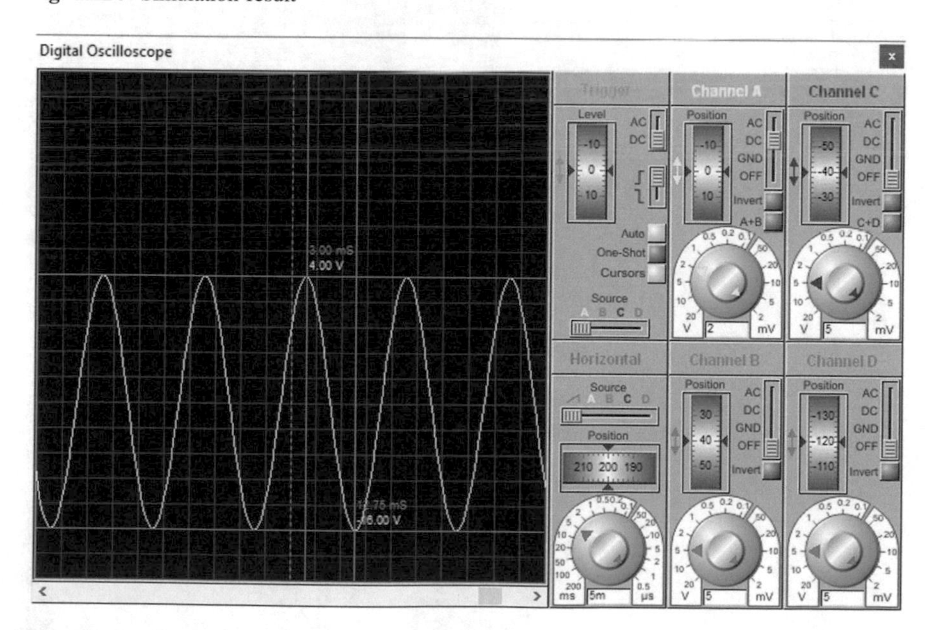

Fig. 1.215 Measurement with the aid of cursors

1.26 Example 25: Grounded Current Sources

In the previous examples, we saw that DC current sources and sinusoidal current sources can be simulated with the aid of "c_{source}" and "i_{sine}" blocks, respectively. Grounded current sources (i.e., current sources with one terminal grounded) can be simulated with the aid of blocks available in the generators section (Fig. 1.216) as well. Let's see an example.

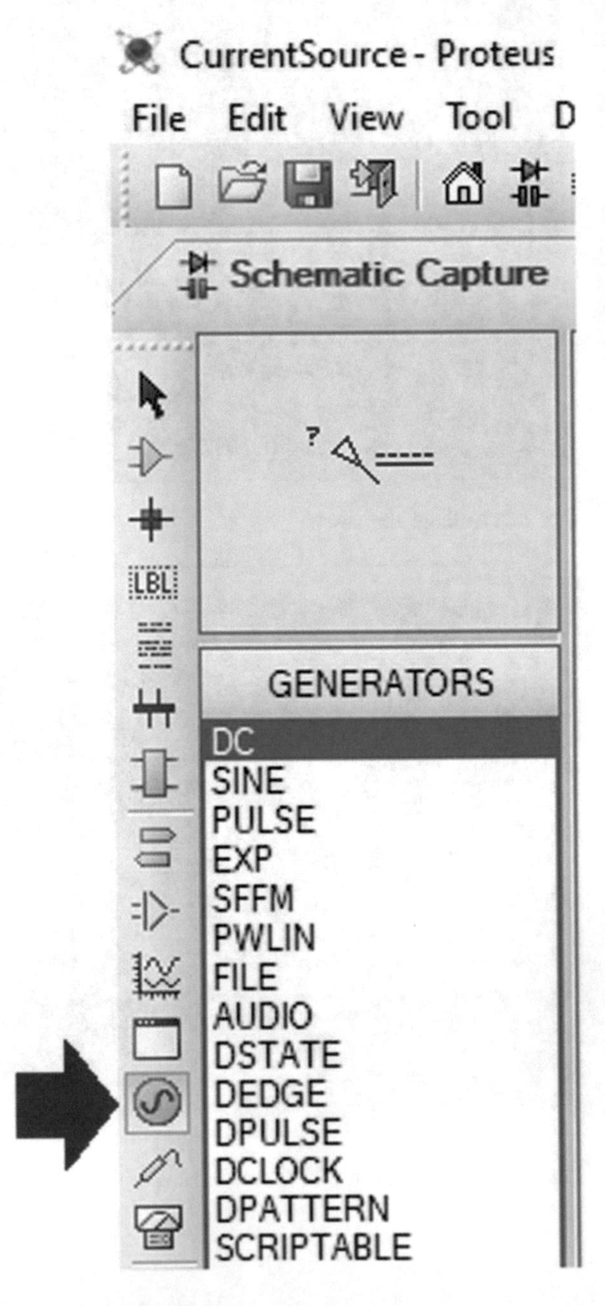

Fig. 1.216 DC block

In this example, we want to simulate the previous example with the aid of blocks available in the generators section. Open the schematic of previous example (Fig. 1.217) and remove the current source blocks I_1 and I_2 from it.

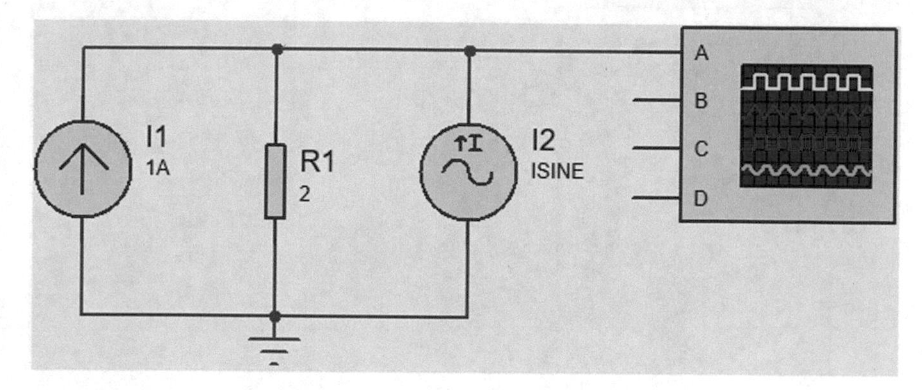

Fig. 1.217 Schematic for Example 24

Add a DC and SINE generator to the schematic (Fig. 1.218). Settings of I_1 and I_2 are shown in Figs. 1.219 and 1.220, respectively. If you run the simulation, you obtain the result shown in Fig. 1.214.

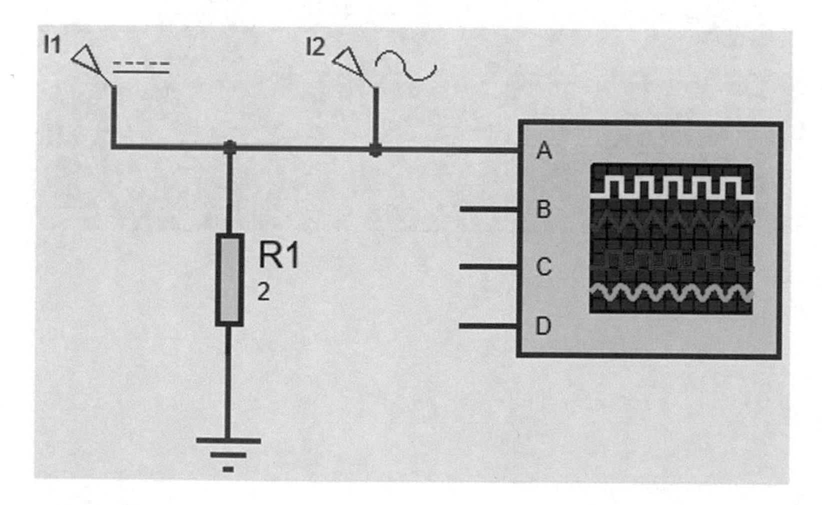

Fig. 1.218 Schematic for Example 25

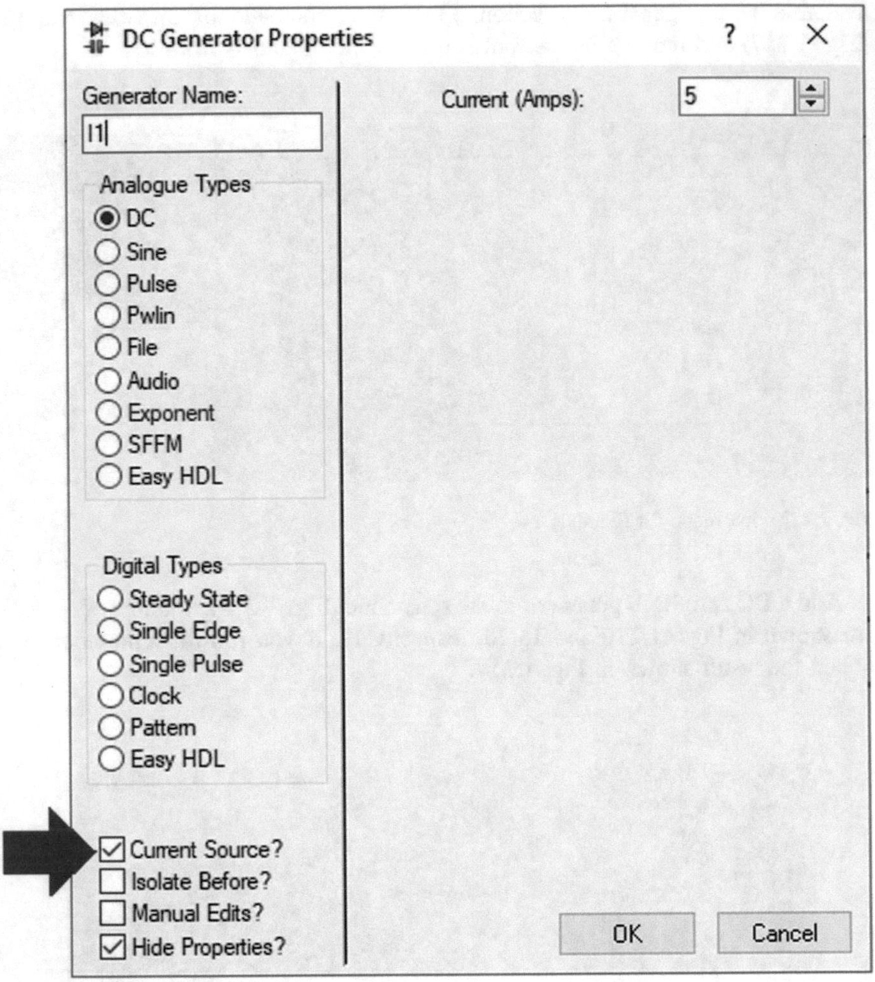

DC Generator Properties

Generator Name: I1

Current (Amps): 5

Analogue Types: DC, Sine, Pulse, Pwlin, File, Audio, Exponent, SFFM, Easy HDL

Digital Types: Steady State, Single Edge, Single Pulse, Clock, Pattern, Easy HDL

Current Source?
Isolate Before?
Manual Edits?
Hide Properties?

OK Cancel

Fig. 1.219 Settings of I_1

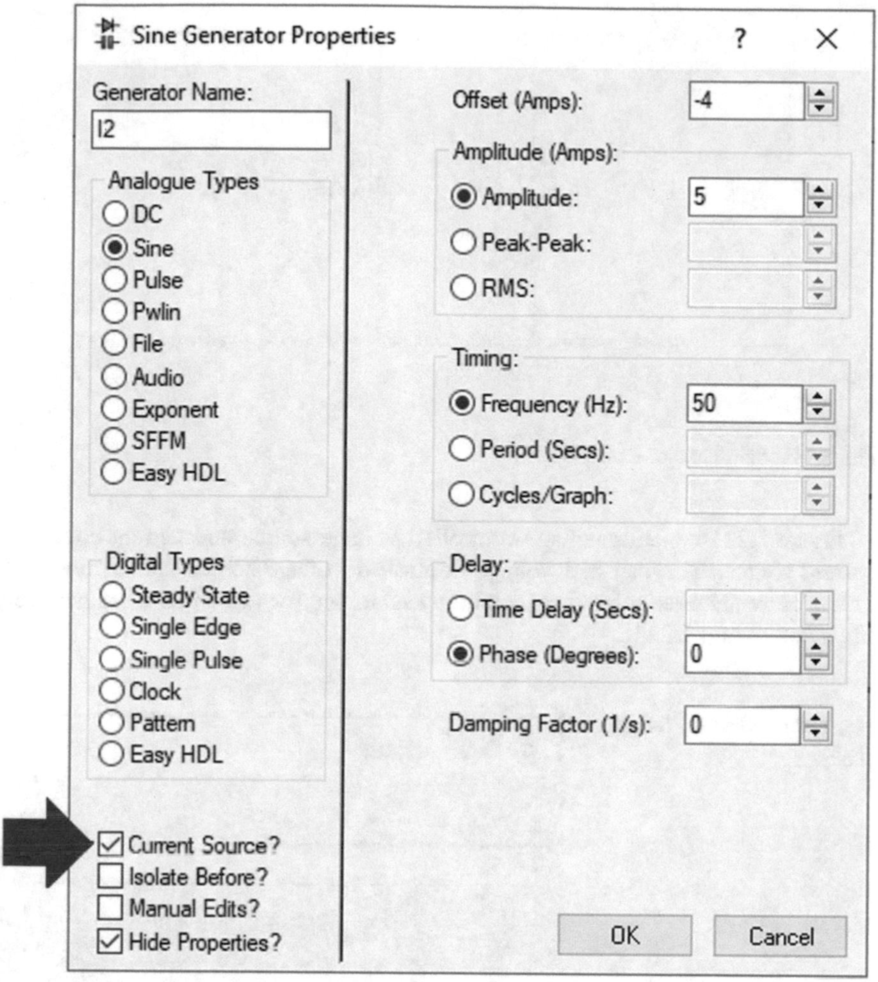

Fig. 1.220 Settings of I_2

1.27 Example 26: Thevenin Equivalent Circuit

In this example, we want to find the Thevenin equivalent circuit with respect to the terminals "*a*" and "*b*" for the circuit shown in Fig. 1.221.

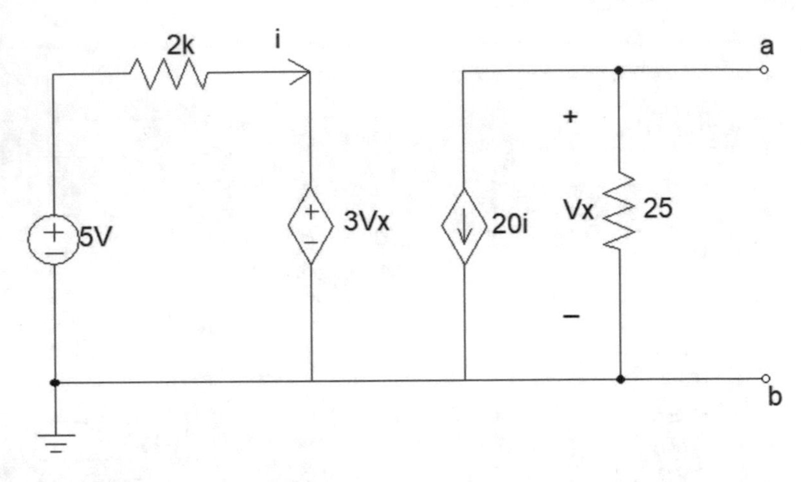

Fig. 1.221 Schematic of Example 26

Figure 1.221 contains voltage controlled voltage source and current controlled current source. You can add voltage controlled voltage source and current controlled current source to your schematic by searching for v_{cvs} and c_{ccs}, respectively (Figs. 1.222 and 1.223).

Fig. 1.222 Searching for v_{cvs} block

Pick Devices

Keywords:

vcvs

Match whole words?

Show only parts with models?

Fig. 1.223 Searching for c_{ccs} block

Pick Devices

Keywords:

cccs

Match whole words?

Show only parts with models?

Draw the schematic shown in Fig. 1.224. Settings of E_1 and F_1 are shown in Figs. 1.225 and 1.226, respectively.

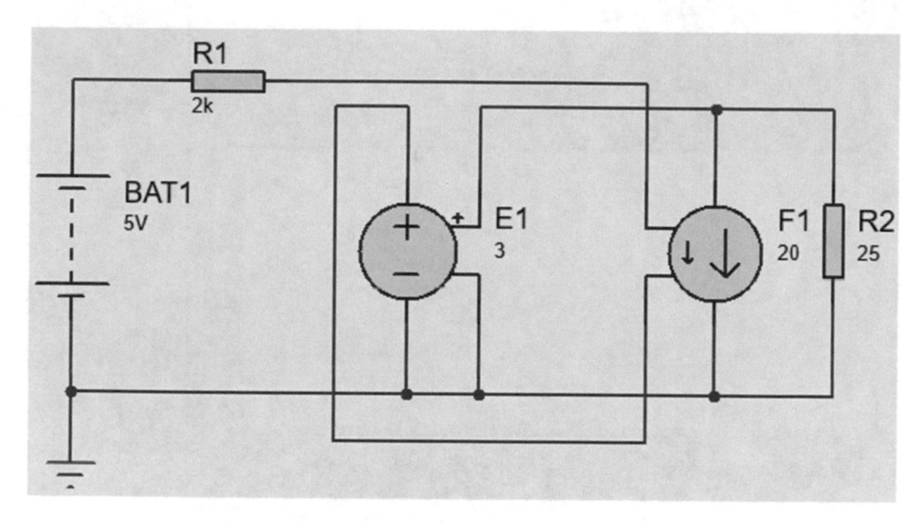

Fig. 1.224 Proteus equivalent of Fig. 1.221

Edit Component ? X

Part Reference: E1 Hidden: ☐ OK

Voltage Gain: 3 Hidden: ☐ Help

Element: New Cancel

Other Properties:

☐ Exclude from Simulation ☐ Attach hierarchy module

☐ Exclude from PCB Layout ☐ Hide common pins

☐ Exclude from Current Variant ☐ Edit all properties as text

Fig. 1.225 Settings of E_1

Edit Component

Part Reference: F1 Hidden: ☐ OK

Current Gain: 20 Hidden: ☐ Help

Element: New Cancel

Other Properties:

☐ Exclude from Simulation ☐ Attach hierarchy module

☐ Exclude from PCB Layout ☐ Hide common pins

☐ Exclude from Current Variant ☐ Edit all properties as text

Fig. 1.226 Settings of F_1

According to Fig. 1.227 the Thevenin voltage is -5 V.

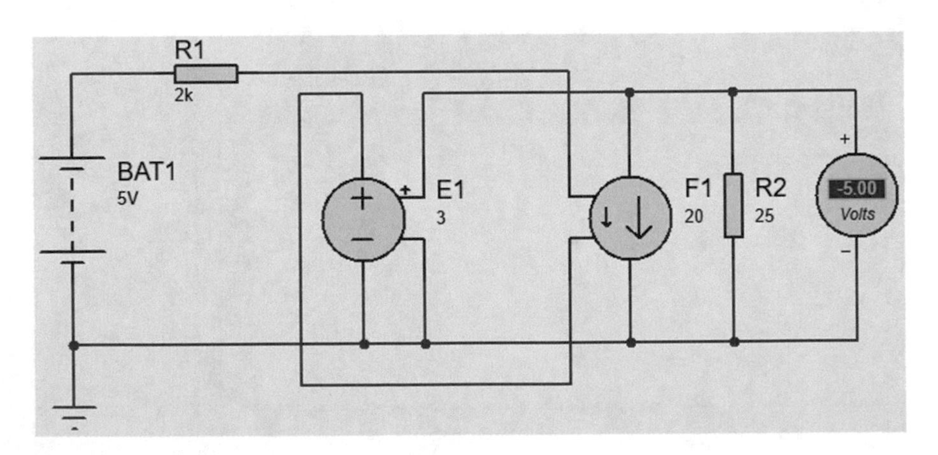

Fig. 1.227 Simulation result

According to Fig. 1.228, the short circuit current is − 50 mA. So, the Thevenin resistor is $R_{TH} = \frac{V_{OC}}{I_{SC}} = \frac{-5\,\text{V}}{-50\,\text{mA}} = 100\,\Omega$. Thevenin equivalent circuit of Fig. 1.221 is shown in Fig. 1.229.

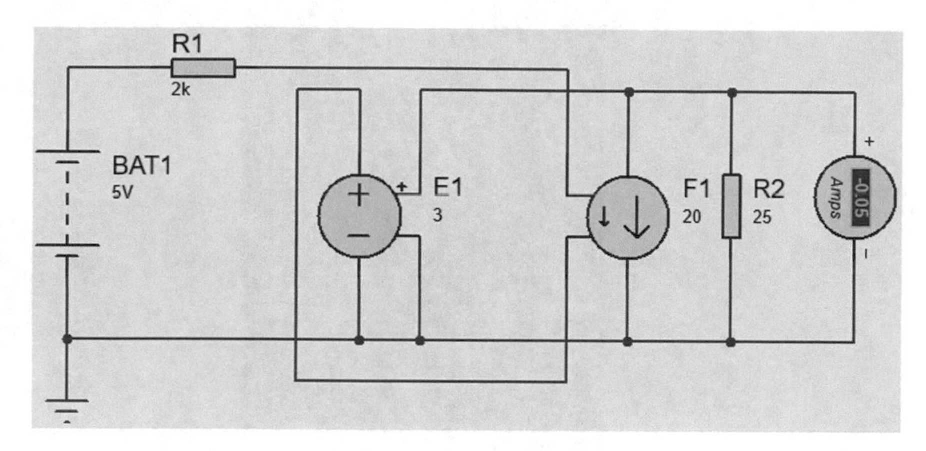

Fig. 1.228 Simulation result

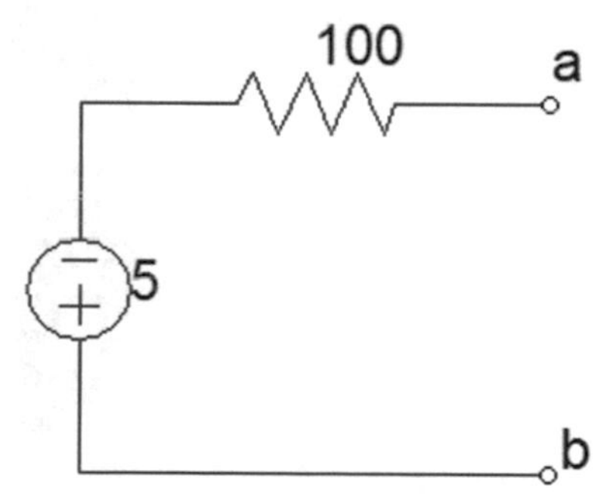

Fig. 1.229 Thevenin equivalent circuit for Fig. 1.221

1.28 Example 27: Making Connections Without Using the Wire (I)

The schematic of previous example is a little bit crowded. Connecting the components with wires is not the only way to make connection. You can make connections with the aid of terminal blocks (Fig. 1.230) as well. In this technique, no wire is drawn on the schematic. So, it does not make your schematic crowded.

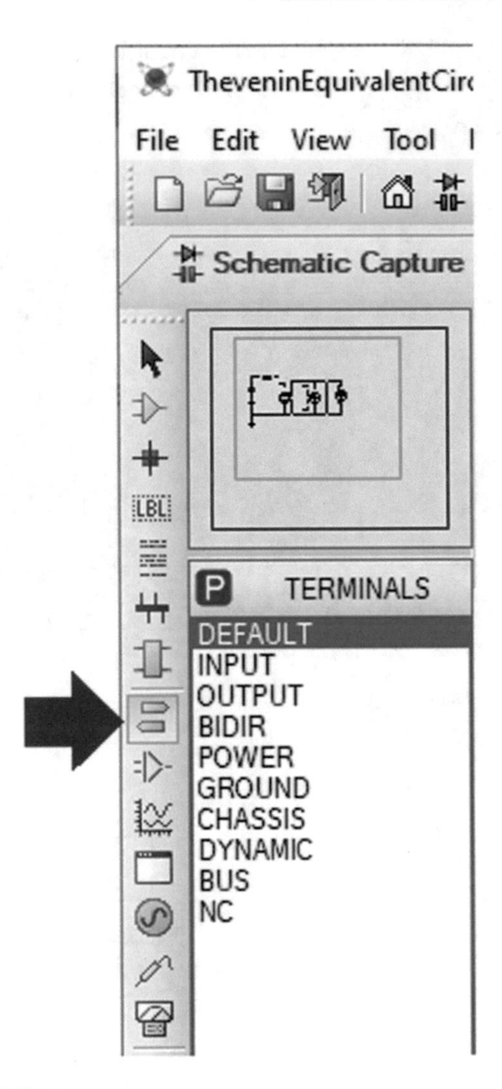

Fig. 1.230 Terminal block

Open the schematic of previous example and remove the wires (Fig. 1.231).

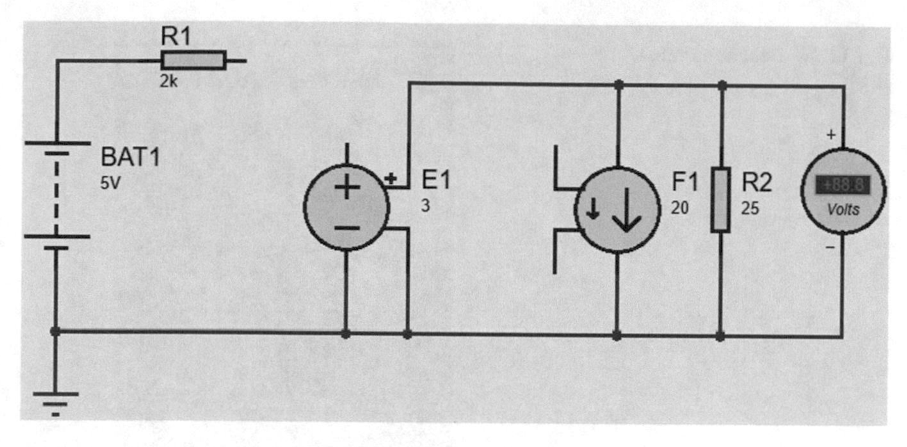

Fig. 1.231 Some wires are removed from schematic of Example 26

Connect four DEFAULT terminals (Fig. 1.232) to the schematic (Fig. 1.233).

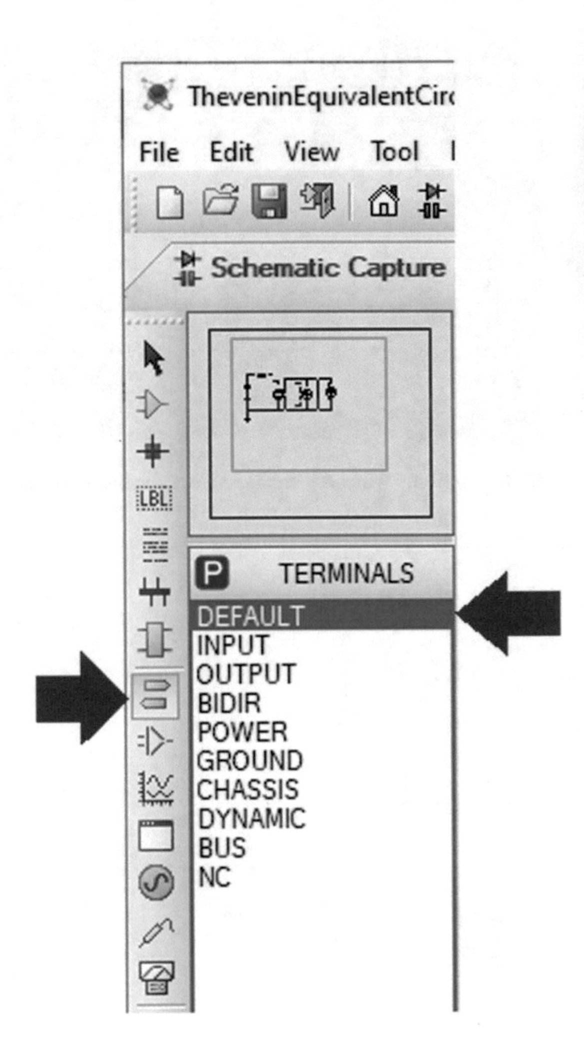

Fig. 1.232 Default terminal block

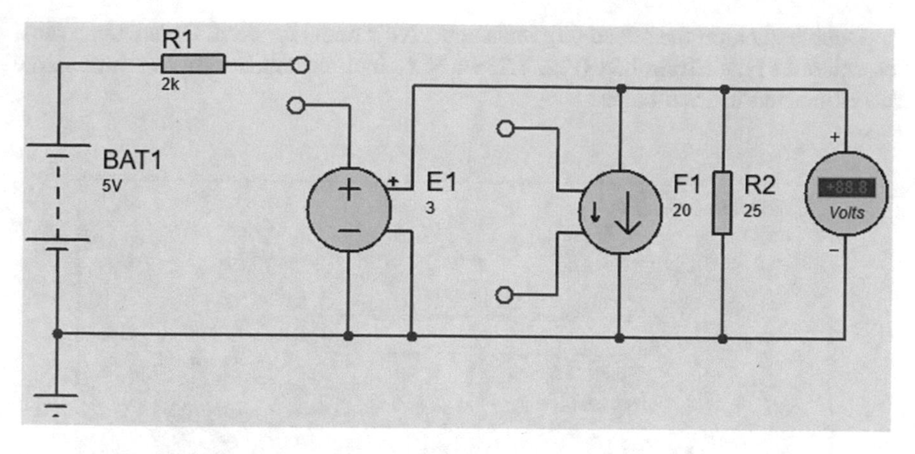

Fig. 1.233 Four default terminals are connected to the schematic

Double click on the added terminals and give a name to them. Given name must be entered to the String box (Fig. 1.234). Note that terminals with the same name are connected to each other.

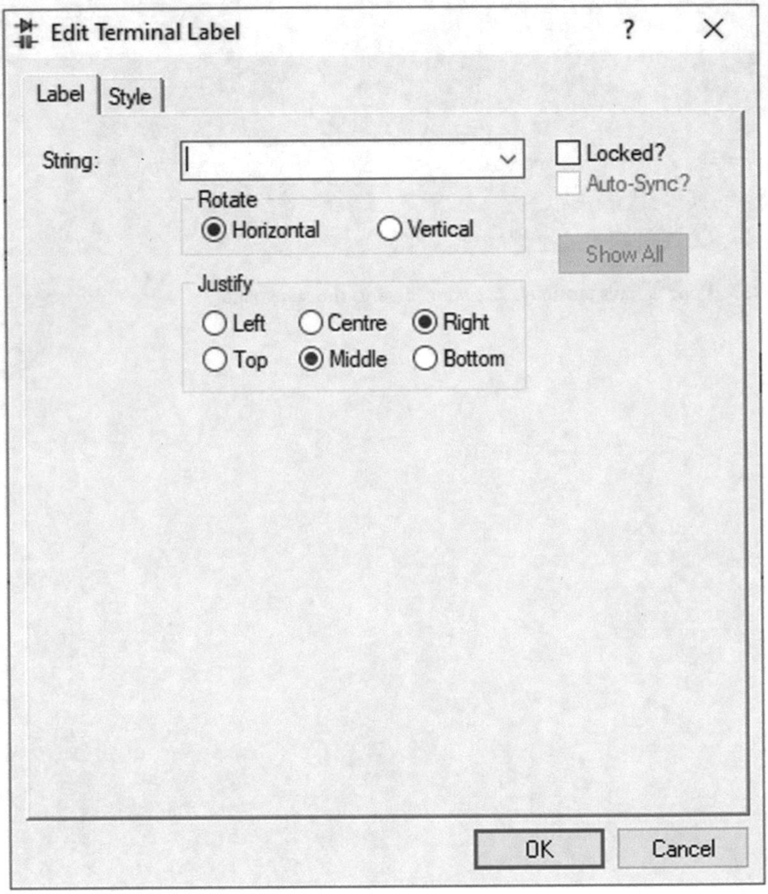

Edit Terminal Label ? ×

Label | Style

String: | Locked? | Auto-Sync?

Rotate: Horizontal | Vertical

Show All

Justify: Left | Centre | Right | Top | Middle | Bottom

OK | Cancel

Fig. 1.234 Edit terminal label window

Schematic shown in Fig. 1.235 is equivalent to the schematic shown in Fig. 1.224. However, the schematic shown in Fig. 1.235 is easier to understand for the user.

If you run the simulation, the same results as Example 26 are obtained.

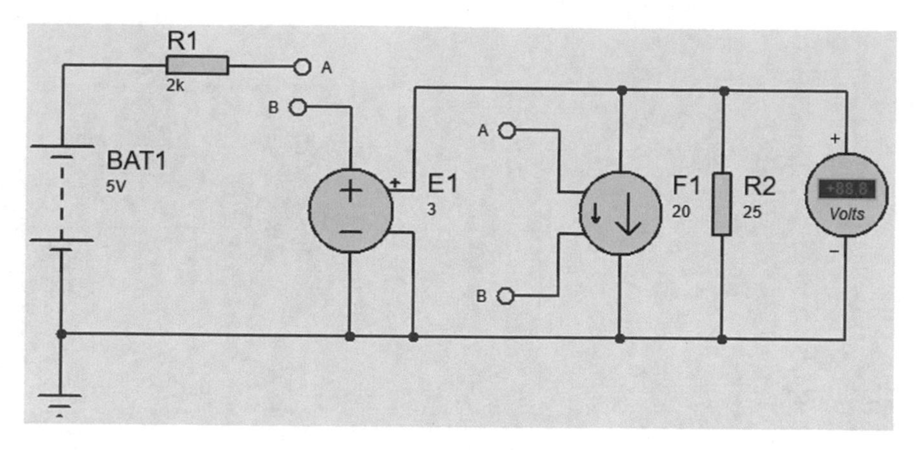

Fig. 1.235 This schematic is equivalent to Fig. 1.224

1.29 Example 28: Making Connections Without Using the Wire (II)

In the previous example, we learned one way to make connection without using wires. In this example, we learn another method to make connections without using wires. Open the schematic of Example 26 and remove the wires (Fig. 1.236).

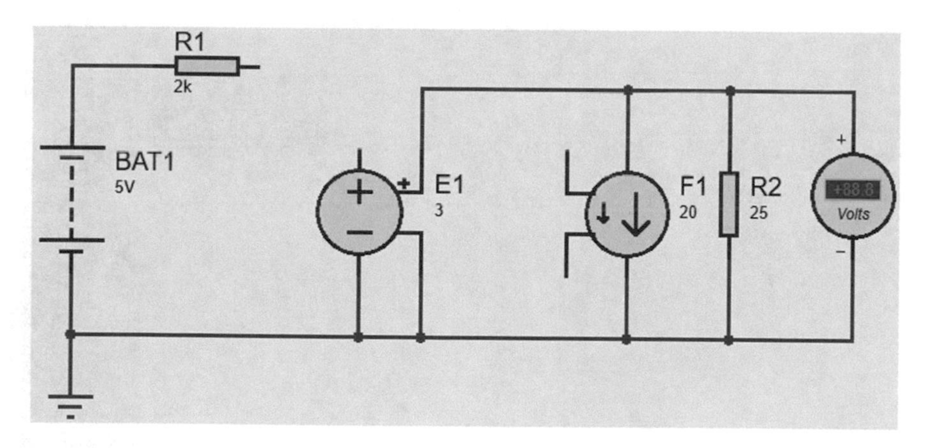

Fig. 1.236 Some wires are removed from schematic of Example 26

Add a piece of wire to the terminals of the components (Fig. 1.237). Use the Wire Label Mode icon (Fig. 1.238) to assign a name to each piece of wire (Fig. 1.239). After clicking the Wire Label Mode icon, click on the wires and enter the desired name to the String box of the opened window (Fig. 1.240). Note that the nodes with the same name are connected to each other.

If you run the simulation, the same results as Example 26 are obtained.

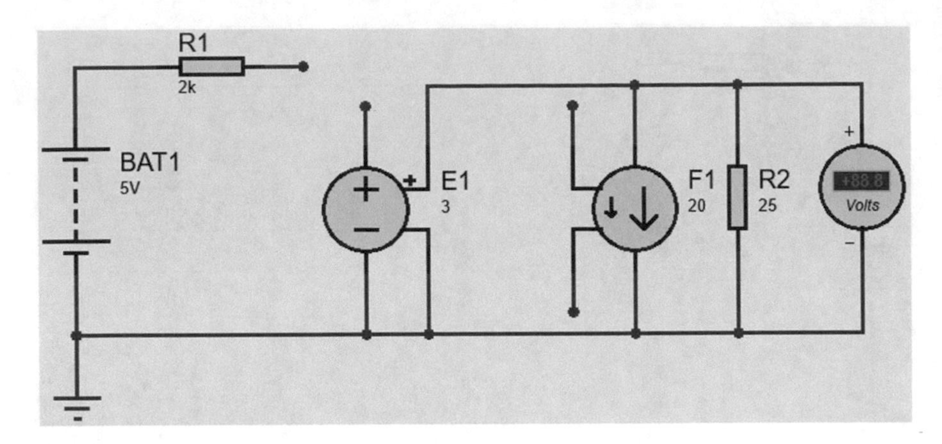

Fig. 1.237 Piece of wire is connected to terminals of the components

Fig. 1.238 Wire label mode icon

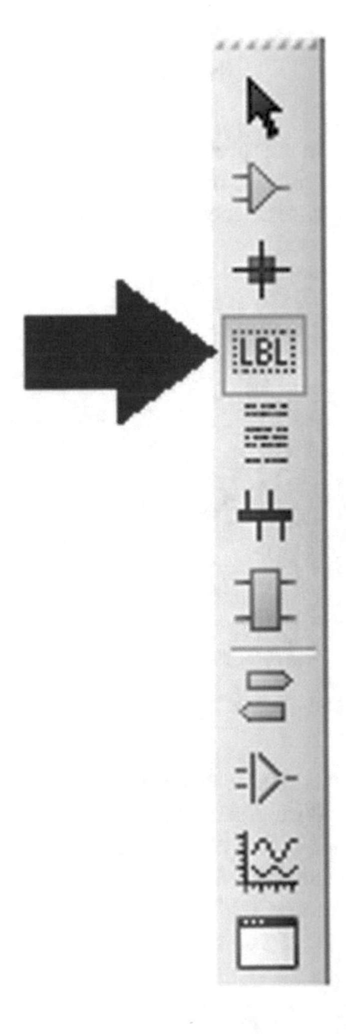

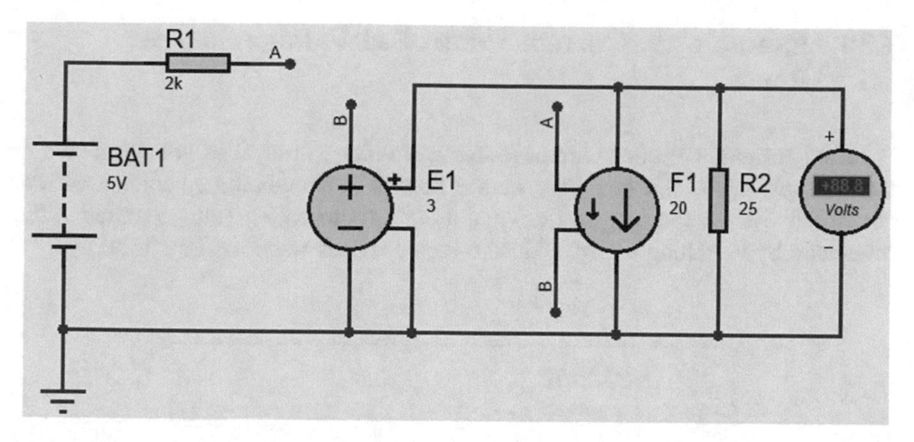

Fig. 1.239 Wires are labeled

Edit Wire Label
Label | Style | Net Class | Differential Pair
String:
Locked?
Auto-Sync?
Rotate
Horizontal
Vertical
Show All
Justify
Left
Centre
Right
Top
Middle
Bottom
OK
Cancel

Fig. 1.240 Edit wire label window

1.30 Example 29: Current Controlled Voltage Source Block

We used current controlled current source and voltage controlled voltage source in the Example 26. In this example, we see how to simulate circuits contain current controlled voltage sources. You can add a current controlled voltage source to the schematic by searching for "c_{cvs}" in the Pick Devices window (Fig. 1.241).

Pick Devices
Keywords:
ccvs
Match whole words?
Show only parts with models?

Fig. 1.241 Searching for c_{cvs} block

Consider the circuit shown in Fig. 1.242. We want to use Proteus to measure the voltage of node "out". From basic circuit theory we expect the voltage of node "out" to be 10 V.

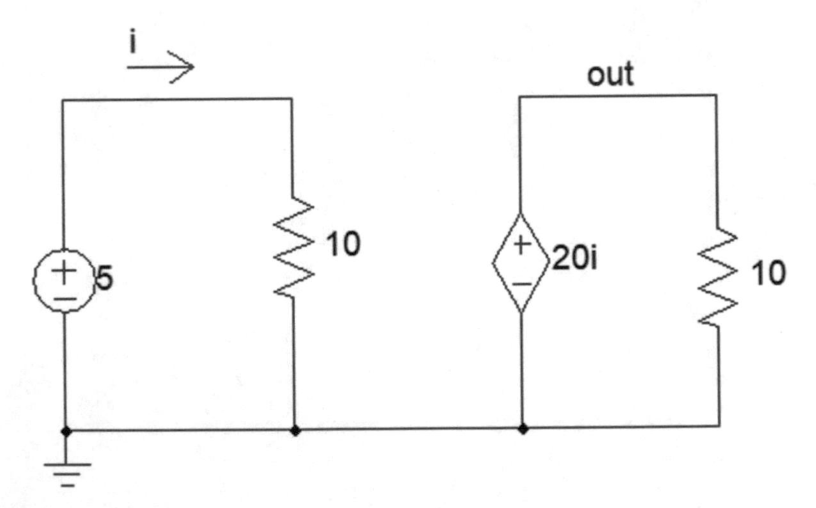

Fig. 1.242 Schematic of Example 29

Draw the schematic shown in Fig. 1.243. Settings of H_1 are shown in Fig. 1.244.

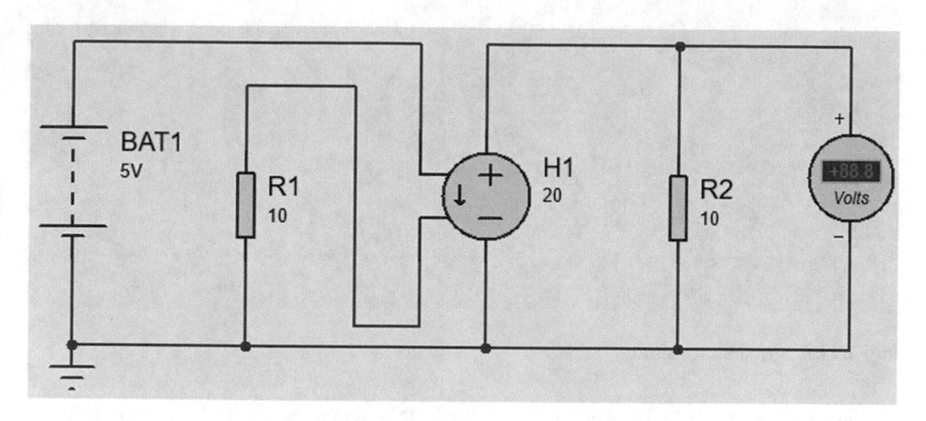

Fig. 1.243 Proteus equivalent of Fig. 1.242

Edit Component ? ×

Part Reference: H1 Hidden: ☐ OK

Transresistance (Volts/ 20 Hidden: ☐ Help

Element: New Cancel

Other Properties:

☐ Exclude from Simulation ☐ Attach hierarchy module

☐ Exclude from PCB Layout ☐ Hide common pins

☐ Exclude from Current Variant ☐ Edit all properties as text

Fig. 1.244 Settings of H_1

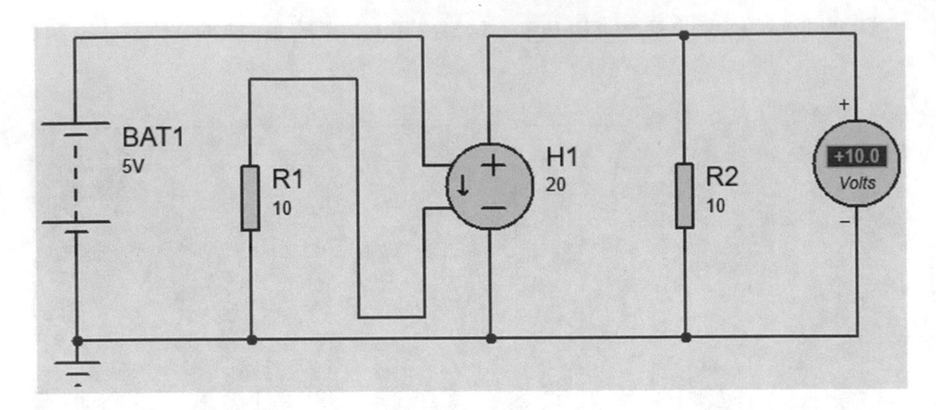

Fig. 1.245 Simulation result

Simulation result is shown in Fig. 1.245. The voltage is 10 V as expected.

1.31 Example 30: Current Sensor

Current controlled voltage source block can be used as current sensor as well. In this example, we use the current controlled voltage source block as current sensor.

Consider the schematic shown in Fig. 1.246. Settings of V_{in} are shown in Fig. 1.247. We want to observe the current waveform of this circuit on the oscilloscope screen. So, we need to convert the current waveform into the voltage waveform.

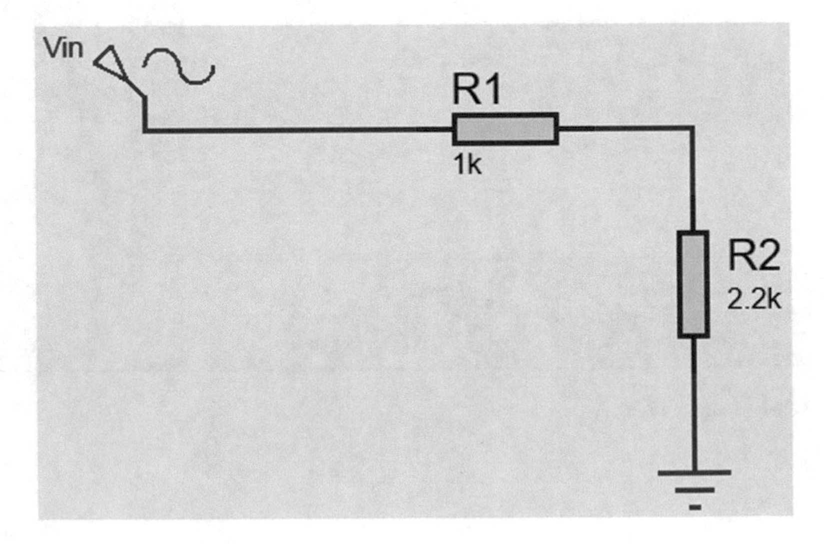

Fig. 1.246 Schematic of Example 30

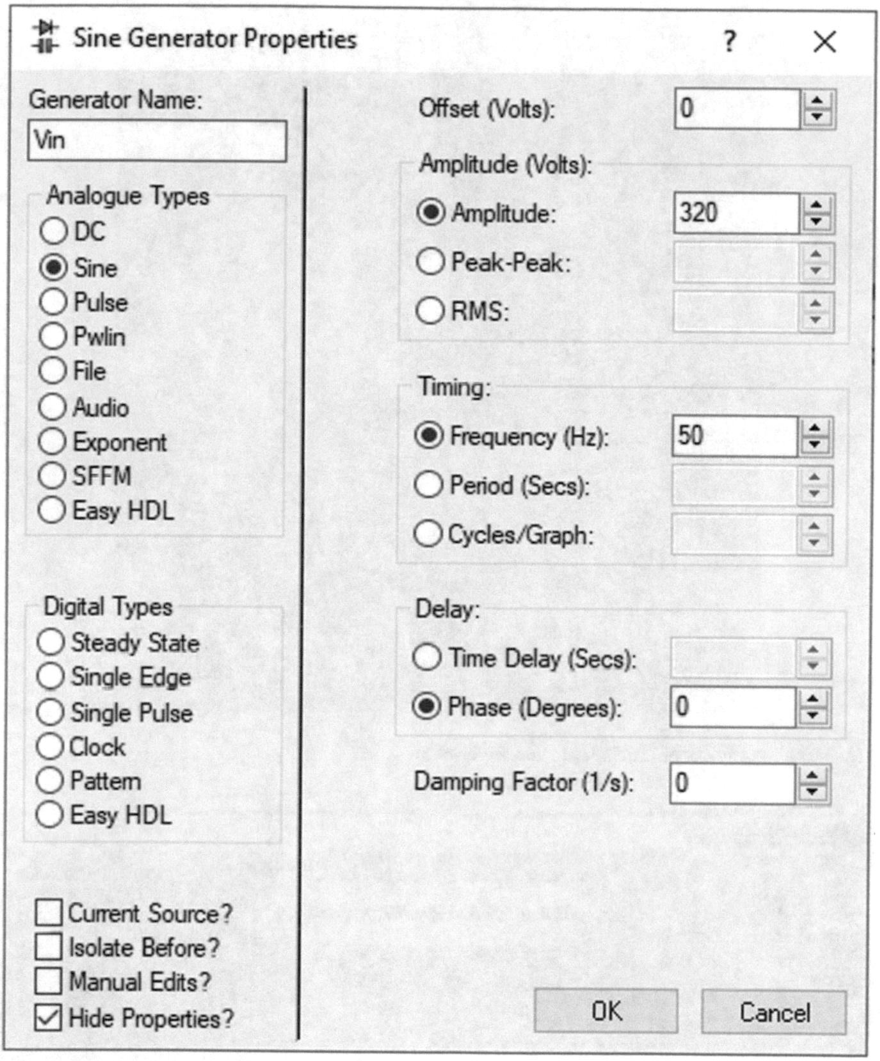

Sine Generator Properties ? ×

Generator Name:
Vin

Analogue Types
- ○ DC
- ◉ Sine
- ○ Pulse
- ○ Pwlin
- ○ File
- ○ Audio
- ○ Exponent
- ○ SFFM
- ○ Easy HDL

Digital Types
- ○ Steady State
- ○ Single Edge
- ○ Single Pulse
- ○ Clock
- ○ Pattern
- ○ Easy HDL

- ☐ Current Source?
- ☐ Isolate Before?
- ☐ Manual Edits?
- ☑ Hide Properties?

Offset (Volts): 0

Amplitude (Volts):
- ◉ Amplitude: 320
- ○ Peak-Peak:
- ○ RMS:

Timing:
- ◉ Frequency (Hz): 50
- ○ Period (Secs):
- ○ Cycles/Graph:

Delay:
- ○ Time Delay (Secs):
- ◉ Phase (Degrees): 0

Damping Factor (1/s): 0

OK Cancel

Fig. 1.247 Settings of V_{in}

Remove the wire between the V_{in} and resistor R_1. Then add a current dependent voltage source between V_{in} and resistor R_1 (Fig. 1.248). Settings of H_1 are shown in Fig. 1.249. According to Fig. 1.249, the trans impedance of the current dependent voltage source H_1 is 1 Ω. So, the current of 1 A generates 1 V of output.

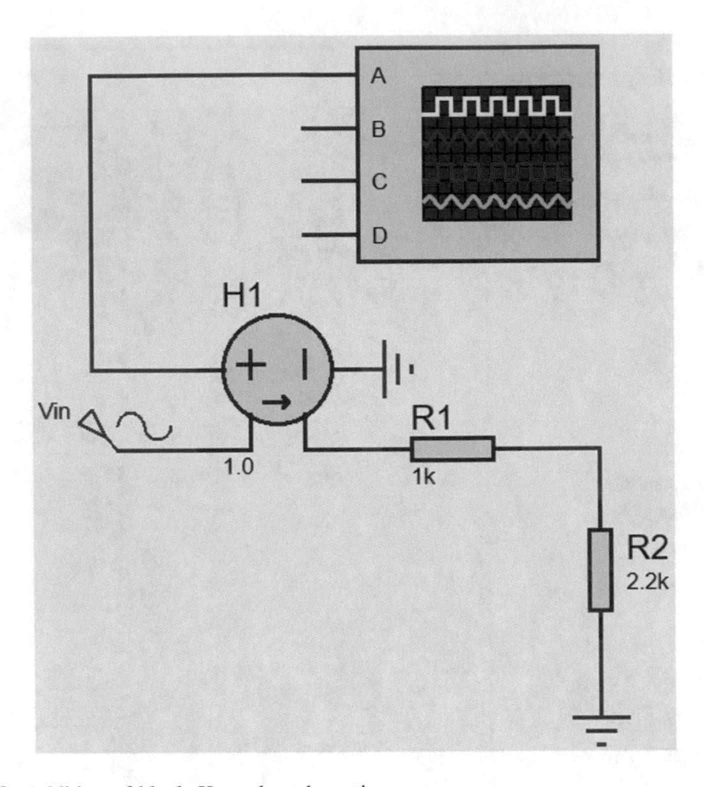

Fig. 1.248 Addition of block H_1 to the schematic

Edit Component ? ×

Part Reference: H1 Hidden: ☐ OK

Transresistance (Volts/A) 1.0 Hidden: ☐ Help

Element: New Cancel

Other Properties:

☐ Exclude from Simulation ☐ Attach hierarchy module
☐ Exclude from PCB Layout ☐ Hide common pins
☐ Exclude from Current Variant ☐ Edit all properties as text

Fig. 1.249 Settings of H_1

Run the simulation. The simulation result is shown in Fig. 1.250. The peak of the waveform on screen is about 100 mV. So, the peak of current waveform is 100 mA.

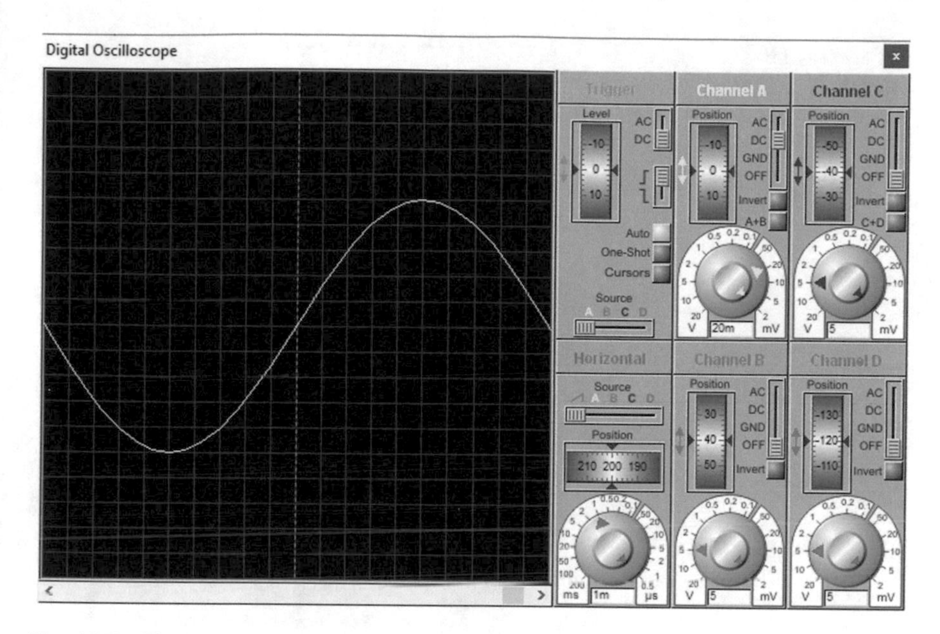

Fig. 1.250 Simulation result

Let's check the Proteus result. According to Ohm's law, peak of circuit current is $I = \frac{320}{1k + 2.2k} = 100\,\text{mA}$. This shows that Proteus result is correct.

1.32 Example 31: Voltage Controlled Current Source Block

In this example, we see how to simulate circuits that contain voltage controlled current sources. You can add a voltage controlled current source to the schematic by searching for "v_{ccs}" in the Pick Devices window (Fig. 1.251).

Fig. 1.251 Searching for v_{ccs}

Pick Devices

Keywords:

vccs

Match whole words?

Show only parts with models?

Consider the schematic shown in Fig. 1.252. This circuit contains a voltage controlled current source. We want to measure the value of V_x. From basic circuit theory, $V_x + 2V_x - 0.3V_x = 5$ or $V_x = 1.8518\,\text{V}$.

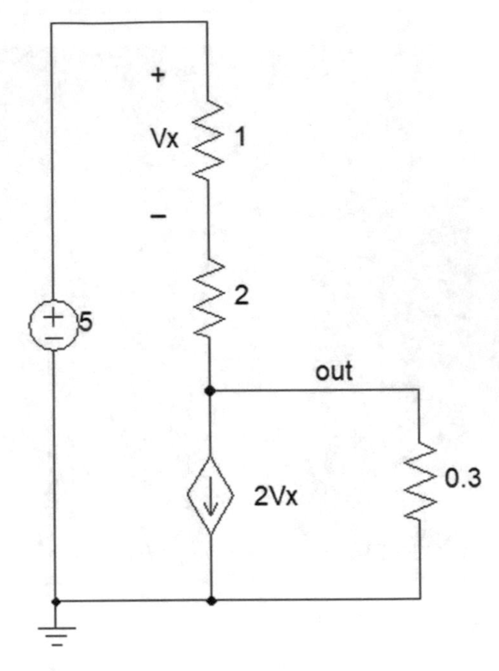

Fig. 1.252 Circuit for Example 31

Draw the schematic shown in Fig. 1.253. Settings of G_1 are shown in Fig. 1.254.

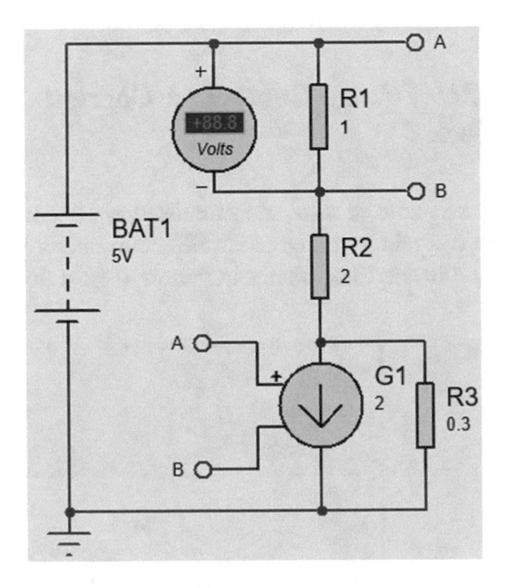

Fig. 1.253 Proteus equivalent of Fig. 1.252

Edit Component

Part Reference: G1 Hidden:

Transconductance (Am 2 Hidden:

Element: New

Other Properties:

OK

Help

Cancel

Exclude from Simulation

Exclude from PCB Layout

Exclude from Current Variant

Attach hierarchy module

Hide common pins

Edit all properties as text

Fig. 1.254 Settings of G_1

Run the simulation. The simulation result is shown in Fig. 1.255. Obtained result is the same as the value predicted by theory.

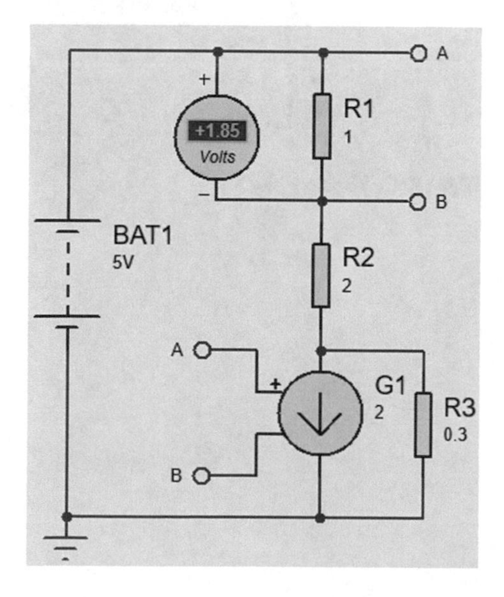

Fig. 1.255 Simulation result

1.33 Example 32: Three-Phase Voltage Source Block

Proteus has a ready to use *Y*-connected three-phase source. You can add a *Y*-connected three-phase source to the schematic by searching for "v3phase" in the Pick Devices window (Fig. 1.256). Equations of generated voltages are shown in Fig. 1.257.

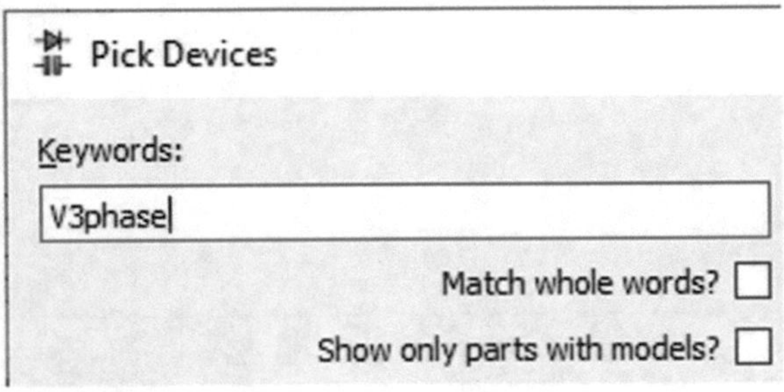

Fig. 1.256 Searching for V3phase block

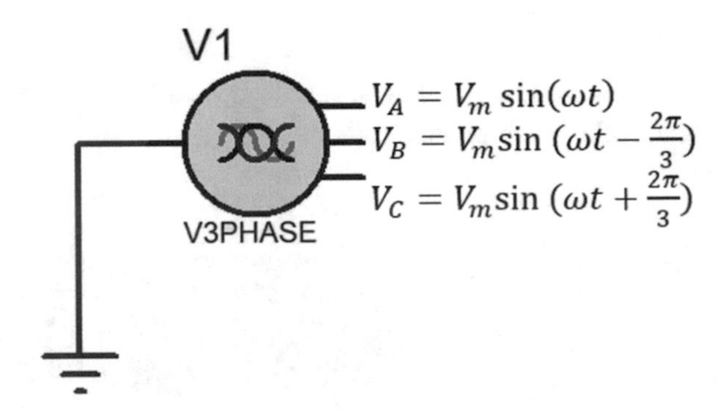

Fig. 1.257 V3phase block

Draw the schematic shown in Fig. 1.258. Settings of V_1 are shown in Fig. 1.259.

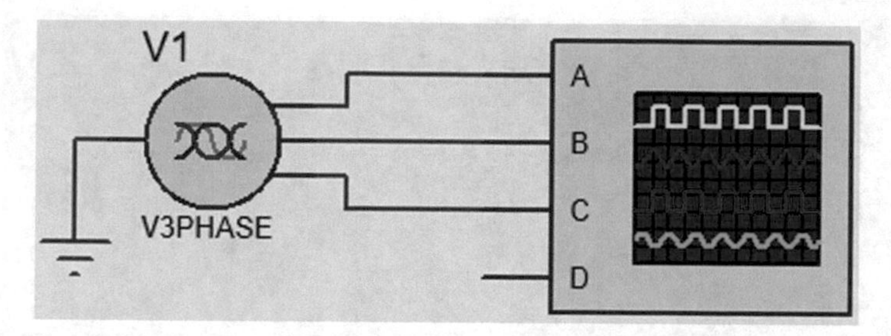

Fig. 1.258 Schematic of Example 32

Edit Component ? ×

Part Reference:	V1	Hidden: ☐	OK
Part Value:	V3PHASE	Hidden: ☐	Cancel
Element:		New	
Amplitude (Volts)	1	Hide All	
Amplitude mode:	Peak	Hide All	
Frequency (Hz)	50	Hide All	
Time Delay (sec)	0	Hide All	
Damping Factor:	0	Hide All	
LISA Model File:	V3PHASE	Hide All	
Advanced Properties:			
Line resistance (Ohms)	1m	Hide All	

Other Properties:

☐ Exclude from Simulation ☐ Attach hierarchy module
☐ Exclude from PCB Layout ☐ Hide common pins
☐ Exclude from Current Variant ☐ Edit all properties as text

Fig. 1.259 Settings of V_1

Run the simulation. Simulation result is shown in Fig. 1.260.

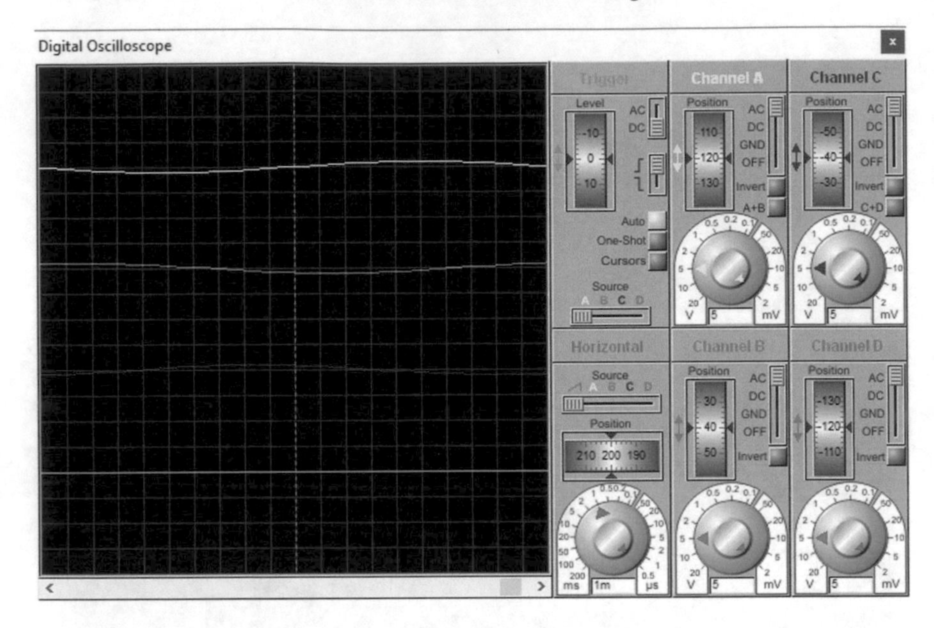

Fig. 1.260 Simulation result

Turn off channel *D* and double click the Position control of channel *A*, *B* and *C*. This makes the position controls zero and three waveforms are shown on the same horizontal (time) axis. Decrease the Volt/Div. to obtain a better view (Fig. 1.261).

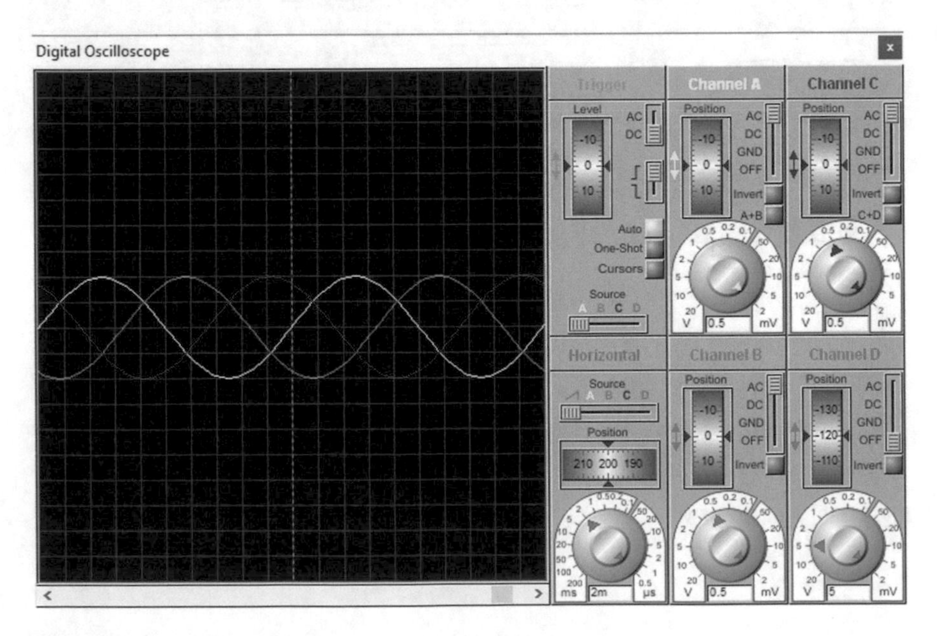

Fig. 1.261 Zoomed waveforms

1.34 Example 33: Voltage Difference Measurement (I)

You can use the oscilloscope to measure the voltage difference between two nodes. For instance, assume that we want to measure the voltage difference between phase *A* and phase *B* (Fig. 1.262). Settings of V_1 are shown in Fig. 1.259.

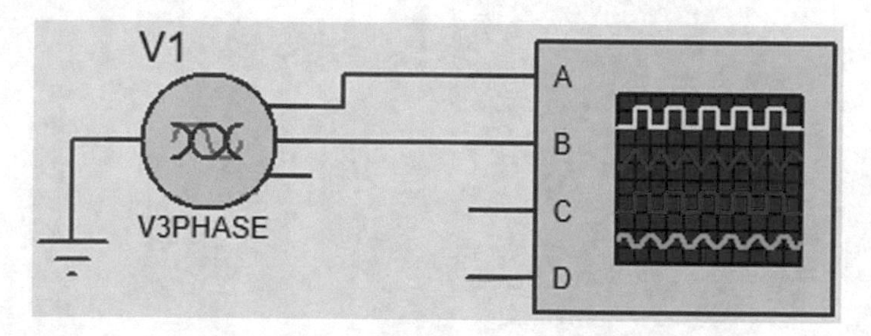

Fig. 1.262 Schematic of Example 33

Run the simulation. Click the Invert button of channel *B* and click the *A* + *B* of channel *A* (Fig. 1.263). What shown on the oscilloscope screen is $V_A - V_B$.

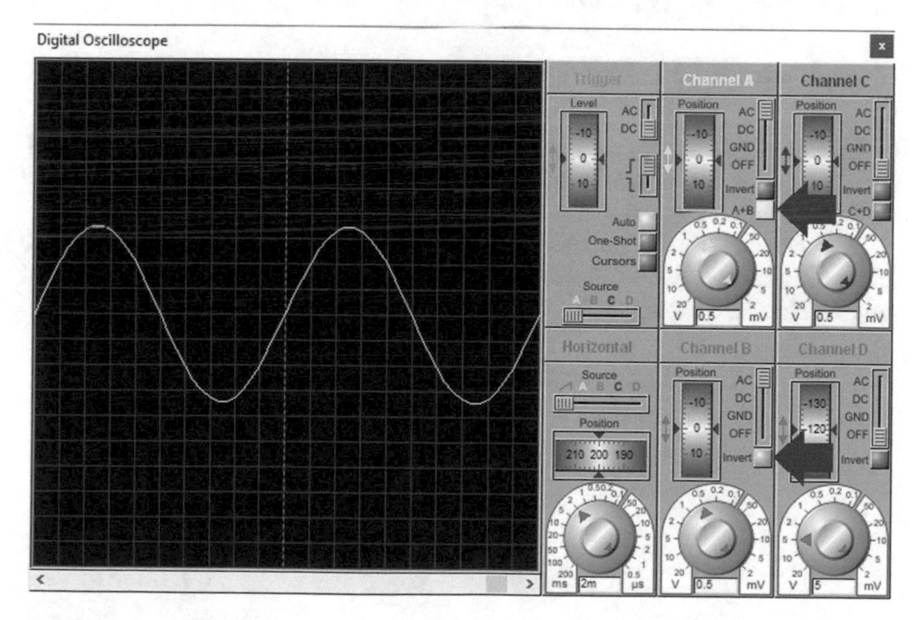

Fig. 1.263 Simulation result

According to Fig. 1.264 peak of voltage $V_A - V_B$ is 1.73 V.

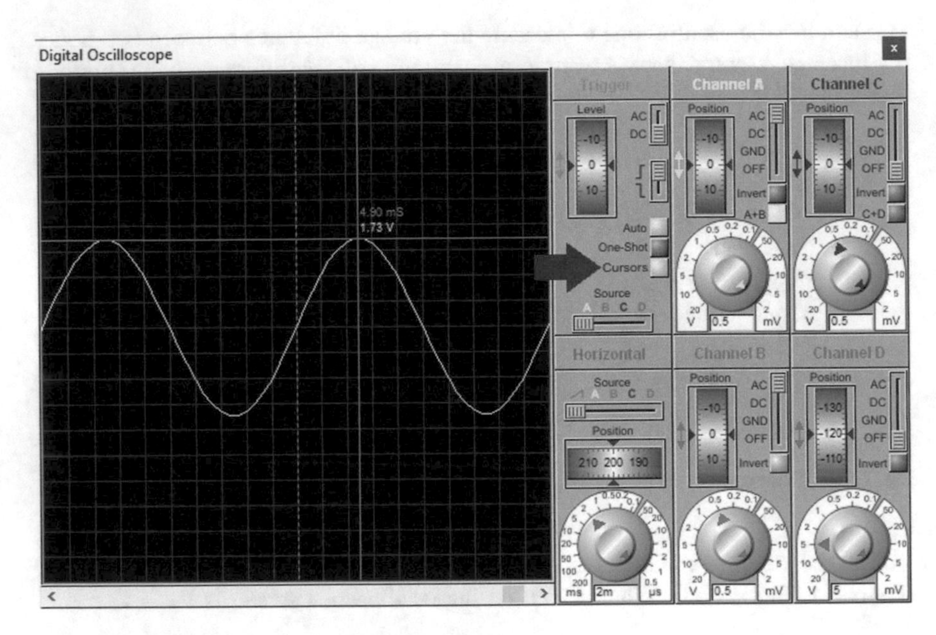

Fig. 1.264 Peak of $\mathbf{V_A} - \mathbf{V_B}$ is 1.73 V

1.35 Example 34: Voltage Difference Measurement (II)

In the previous example, we learned one way to use to observe the voltage difference between two nodes. In this example, we learn three more ways to observe the voltage difference between two nodes.

First method uses the summer block (Fig. 1.265). Add a summer block to the schematic of Example 33 and change it to what is shown in Fig. 1.266. Settings of summer block are shown in Fig. 1.267.

Pick Devices

Keywords:

summer

Match whole words?

Show only parts with models?

Fig. 1.265 Searching for summer block

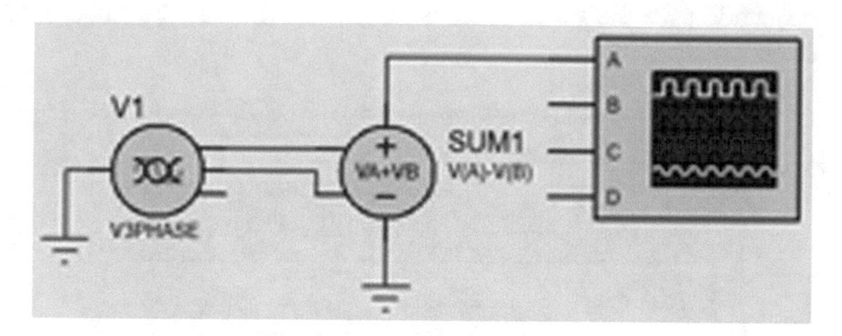

Fig. 1.266 A summer block is added to the schematic

Edit Component ? ×

Part Reference: SUM1 Hidden: ☐ OK

Transfer Function: V(A)-V(B) Hidden: ☐ Cancel

Element: New

Other Properties:

☐ Exclude from Simulation ☐ Attach hierarchy module

☐ Exclude from PCB Layout ☐ Hide common pins

☐ Exclude from Current Variant ☐ Edit all properties as text

Fig. 1.267 Settings of SUM1

Second method uses the subtract block (Fig. 1.268). Draw the schematic shown in Fig. 1.269.

Pick Devices

Keywords:

subtract

Match whole words?

Show only parts with models?

Fig. 1.268 Searching for subtract block

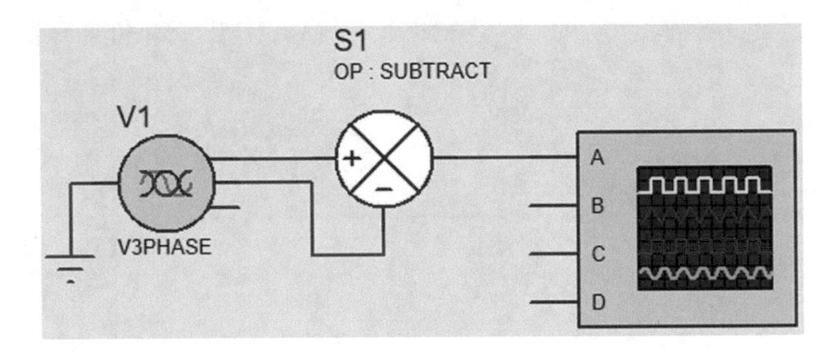

Fig. 1.269 A subtract block is added to the schematic

Third method uses the voltage controlled voltage source block (Fig. 1.270). Settings of voltage controlled voltage source block are shown in Fig. 1.271.

All of the aforementioned methods lead to the same result as the one obtained in Example 33.

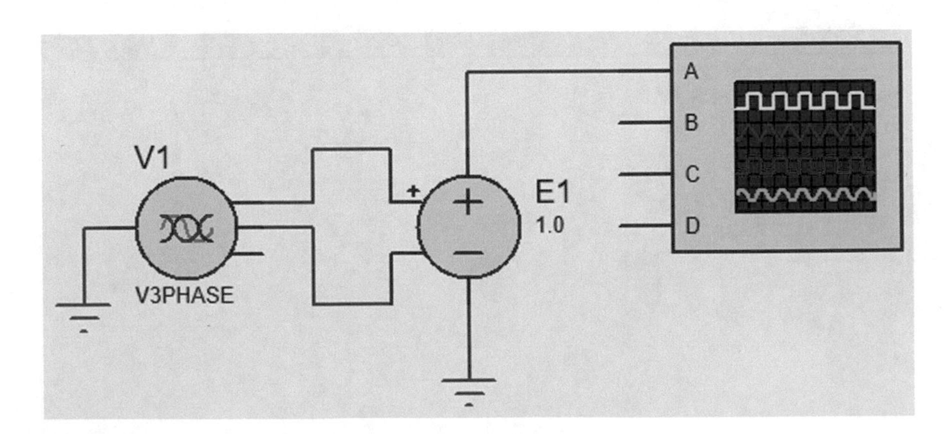

Fig. 1.270 A voltage controlled voltage source block is added to the schematic

Edit Component ? ×

Part Reference: E1 Hidden: ☐ OK

Voltage Gain: 1.0 Hidden: ☐ Help

Element: New Cancel

Other Properties:

☐ Exclude from Simulation ☐ Attach hierarchy module

☐ Exclude from PCB Layout ☐ Hide common pins

☐ Exclude from Current Variant ☐ Edit all properties as text

Fig. 1.271 Settings of E_1 block

1.36 Example 35: Transient Analysis (I)

Transient analysis permits you to study the behavior of a circuit during the time interval that you want. In this example, we want to study the behavior of the circuit shown in Fig. 1.272 during the [0, 100 ms] time interval. Settings of V_a, V_b and V_c are shown in Figs. 1.273, 1.274 and 1.275, respectively.

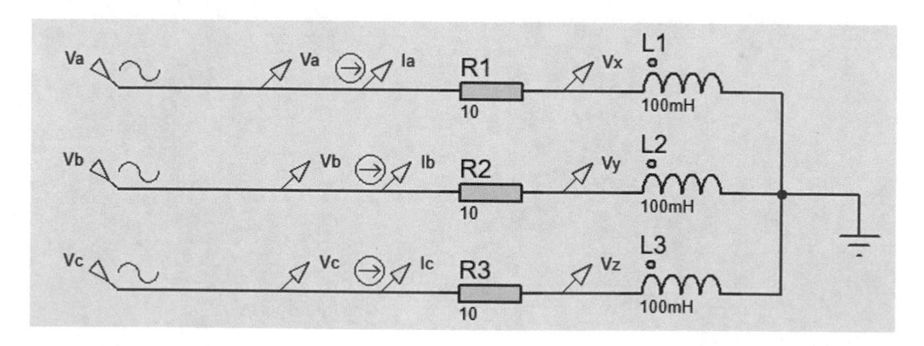

Fig. 1.272 Schematic of Example 35

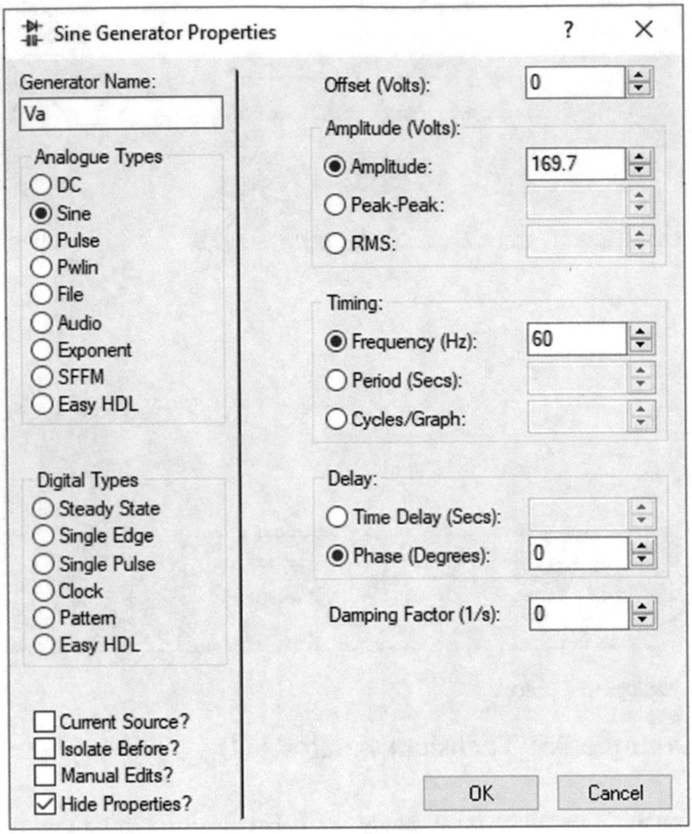

Fig. 1.273 Settings of V_a

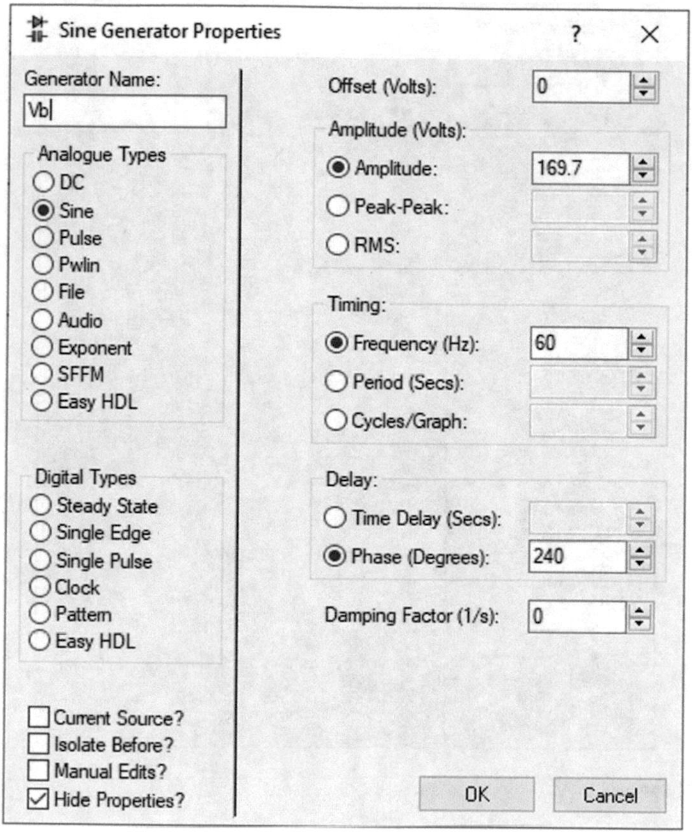

Sine Generator Properties

Generator Name: Vb

Analogue Types: DC, Sine (selected), Pulse, Pwlin, File, Audio, Exponent, SFFM, Easy HDL

Digital Types: Steady State, Single Edge, Single Pulse, Clock, Pattern, Easy HDL

Current Source? Isolate Before? Manual Edits? Hide Properties? (checked)

Offset (Volts): 0

Amplitude (Volts): Amplitude: 169.7; Peak-Peak; RMS

Timing: Frequency (Hz): 60; Period (Secs); Cycles/Graph

Delay: Time Delay (Secs); Phase (Degrees): 240

Damping Factor (1/s): 0

OK Cancel

Fig. 1.274 Settings of V_b

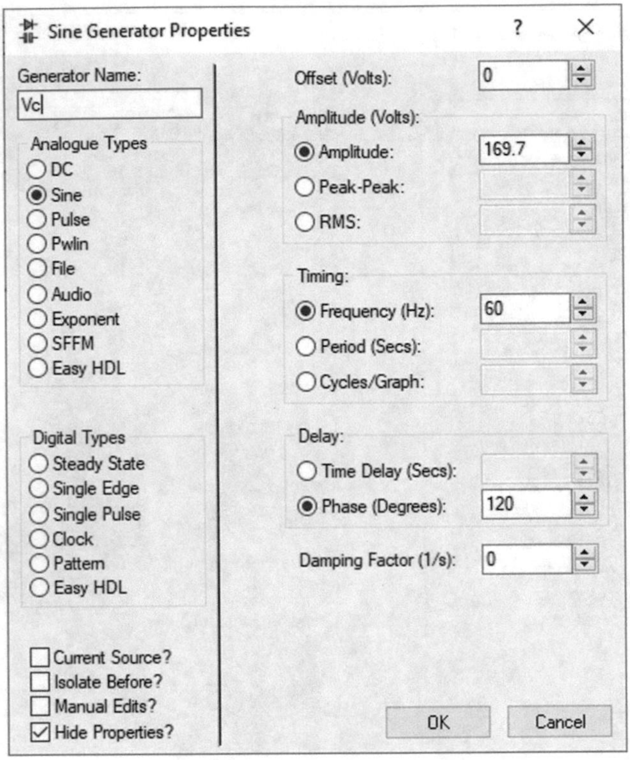

Fig. 1.275 Settings of V_c

Click the analogue graph (Fig. 1.276). After clicking the analogue graph, the mouse pointer changes to a pencil. Click on the schematic and draw a rectangle. This adds an analogue graph to the schematic (Fig. 1.277).

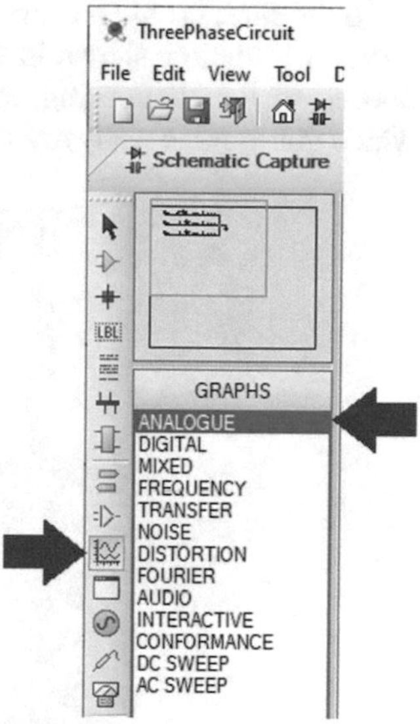

Fig. 1.276 Analogue graph block

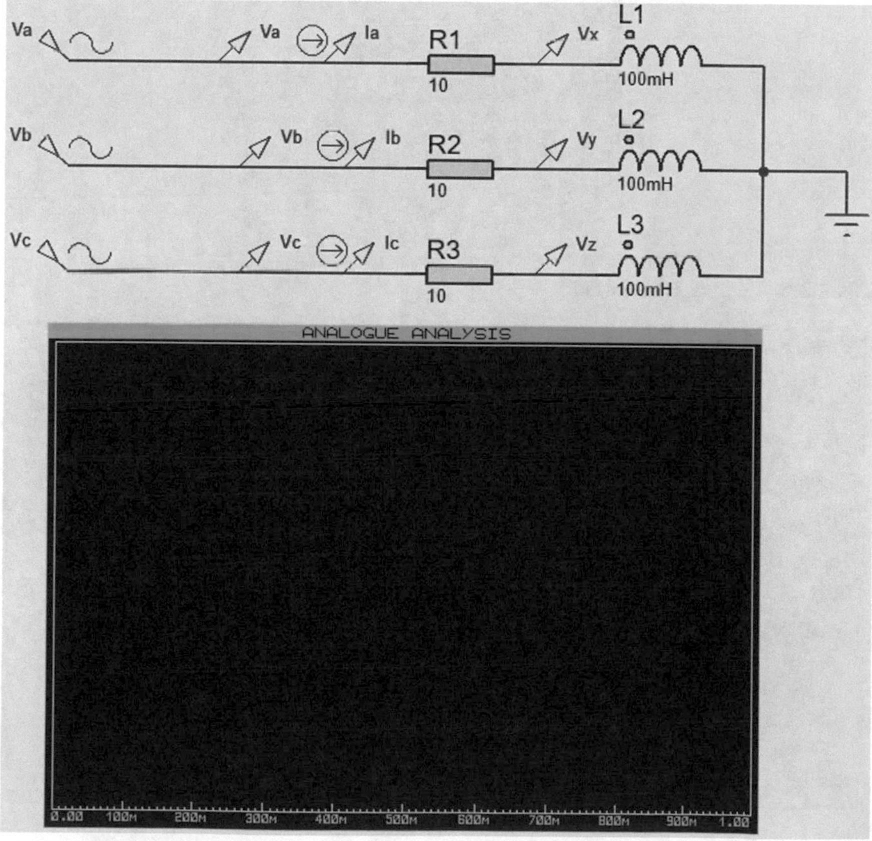

Fig. 1.277 An analogue graph is added to the schematic

Right click on the analogue graph and click the Edit Graph (Fig. 1.278). This opens the window shown in Fig. 1.279. The window shown in Fig. 1.279 can be opened by double clicking on the analogue graph as well. Enter the time interval which you want to study to Start time and Stop time boxes and click the OK button.

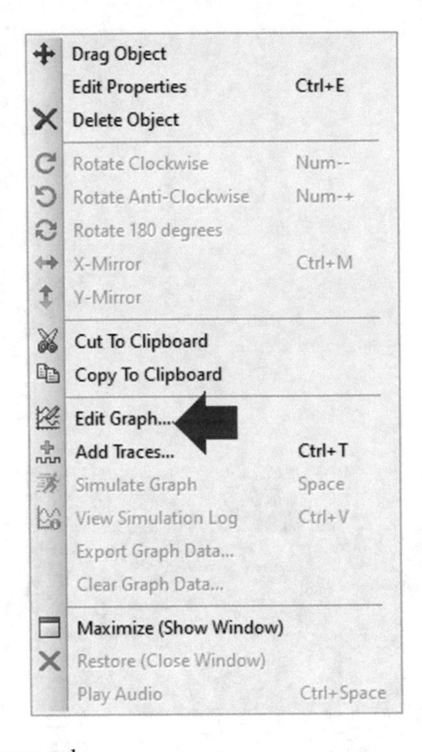

Fig. 1.278 Edit graph command

Edit Transient Graph

Graph title: ANALOGUE ANALYSIS

Start time: 0

Stop time: 100m

Left Axis Label:

Right Axis Label:

Options

Initial DC solution:

Always simulate:

Log netlist(s):

SPICE Options

Set Y-Scales

User defined properties:

OK Cancel

Fig. 1.279 Edit transient graph window

Right click on the analogue graph and click the Add Traces (Fig. 1.280). This opens the window shown in Fig. 1.281.

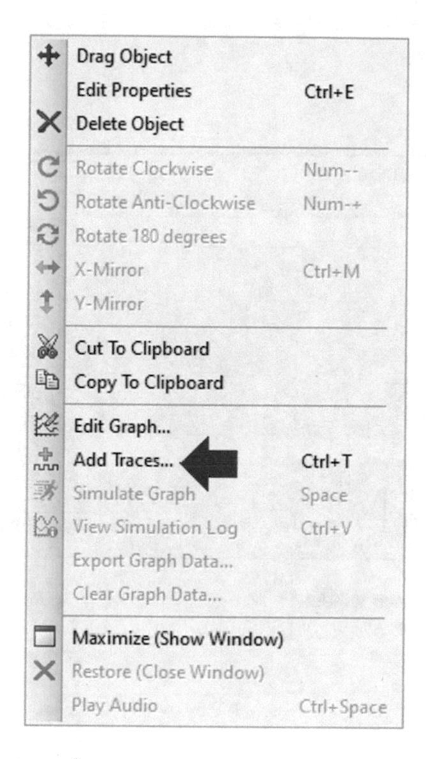

Fig. 1.280 Add traces command

Fig. 1.281 Add transient trace window

Assume that we want to see the current I_a. In order to do this, select the ROOT_I_a for Probe P_1 drop down list (Fig. 1.282) and click the OK button. After clicking the OK button, the schematic changes to what is shown in Fig. 1.283. Note that trace I_a is added to the analogue graph. Added traces can be removed by right clicking on them.

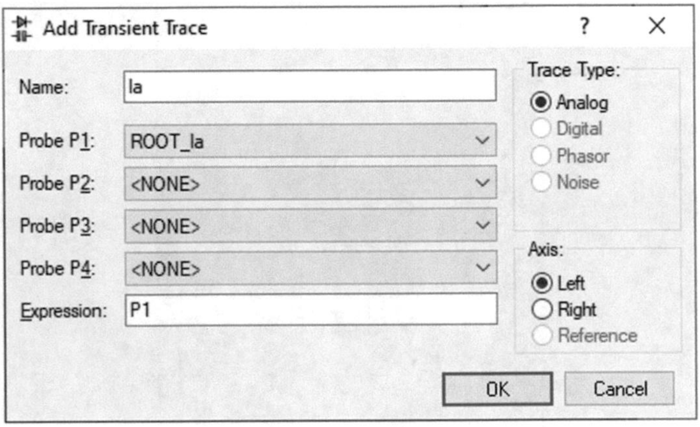

Fig. 1.282 Add transient trace window

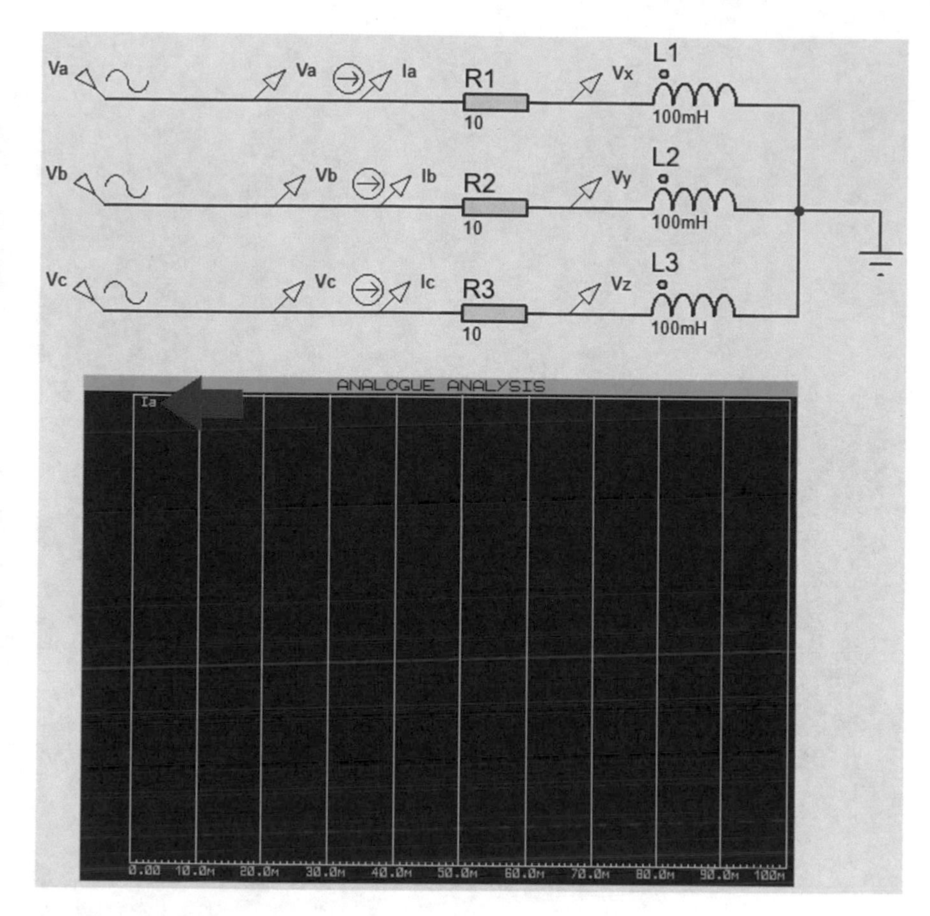

Fig. 1.283 I_a is added to the analogue analysis graph

Right click on the analogue graph again and click the Add Traces (Fig. 1.280). Then select the ROOT_Ib for Probe P_1 (Fig. 1.284) and click the OK button. After clicking the OK button, the schematic changes to what is shown in Fig. 1.285.

Add Transient Trace ? ×

Name: Ib

Probe P1: ROOT_Ib

Probe P2: <NONE>

Probe P3: <NONE>

Probe P4: <NONE>

Expression: P1

Trace Type: Analog, Digital, Phasor, Noise

Axis: Left, Right, Reference

OK Cancel

Fig. 1.284 Add transient trace window

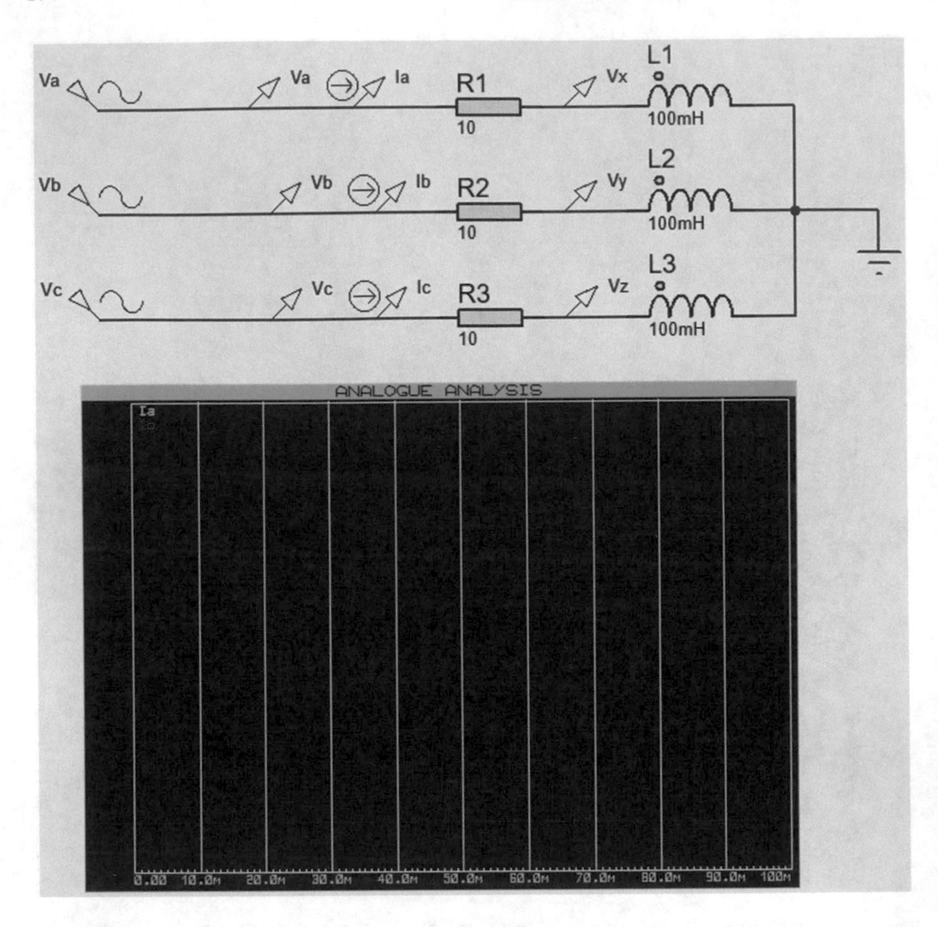

Fig. 1.285 I_b is added to the analogue analysis graph

Right click on the analogue graph again and click the Add Traces (Fig. 1.280). Then select the ROOT_Ic for Probe P_1 (Fig. 1.286) and click the OK button. After clicking the OK button, the schematic changes to what is shown in Fig. 1.287.

Add Transient Trace ? ×

Name: Ic

Probe P1: ROOT_Ic

Probe P2: <NONE>

Probe P3: <NONE>

Probe P4: <NONE>

Expression: P1

Trace Type: Analog, Digital, Phasor, Noise

Axis: Left, Right, Reference

OK Cancel

Fig. 1.286 Add transient trace window

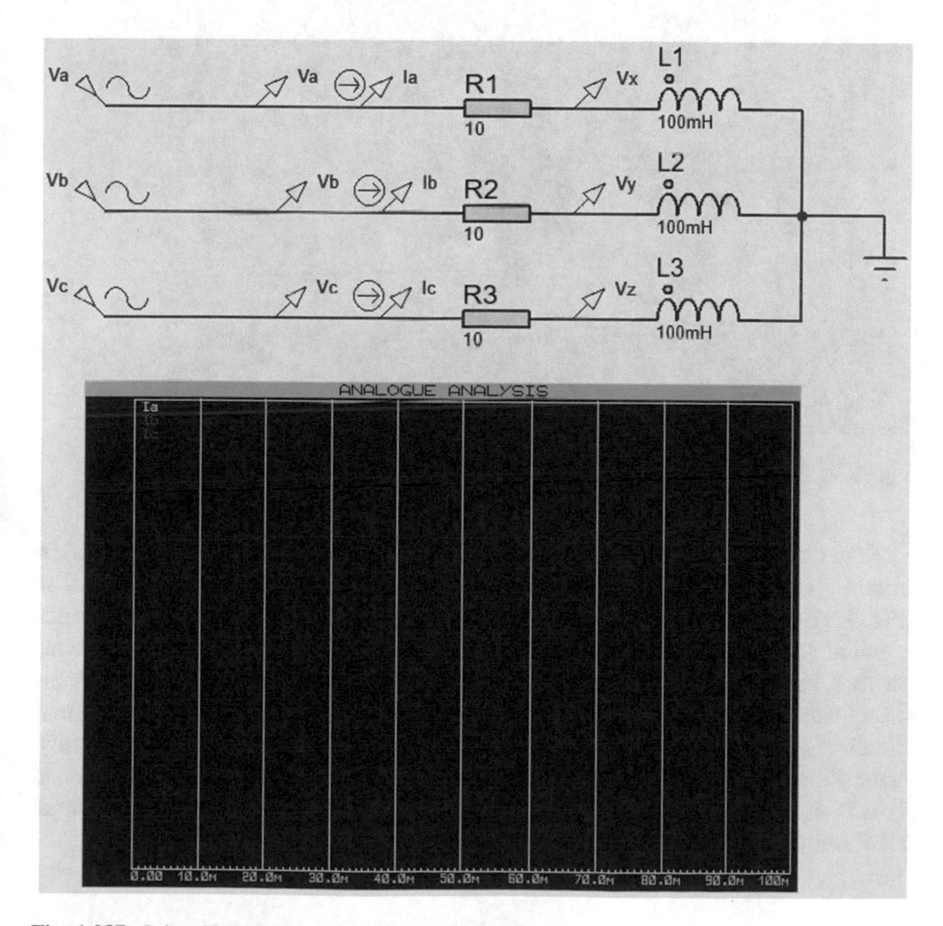

Fig. 1.287 I_c is added to the analogue analysis graph

Press the space bar key of your keyboard to run the simulation. The simulation result is shown in Fig. 1.288.

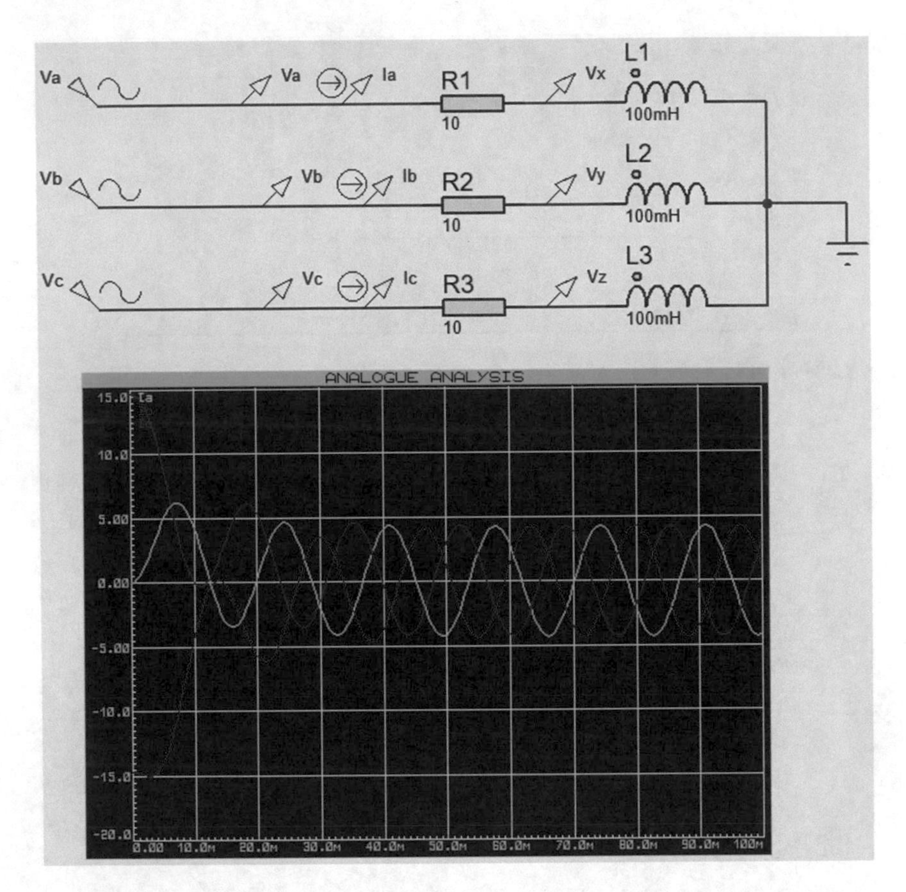

Fig. 1.288 Simulation result

Note that I_b and I_c are not started from zero. I_b starts from − 14.69 A and I_c starts from + 14.69 A. Let's see why? When Initial DC solution box is checked (Fig. 1.279), Proteus does a steady-state DC analysis and uses the obtained results as initial condition of the circuit (Fig. 1.289). In DC steady-state analysis, inductors act as a short circuit and capacitors act as open circuit. Figure 1.290 shows the steady-state DC equivalent circuit. Note that inductors are replaced with short circuit. V_{a0}, V_{b0} and V_{c0} show the value of V_a, V_b and V_c at $t = 0$, respectively. According to the calculations shown in Fig. 1.291, the current through R_1, R_2 and R_3 is 0 A, − 14.69 A and 14.69 A, respectively. These values are used as initial condition of the circuit.

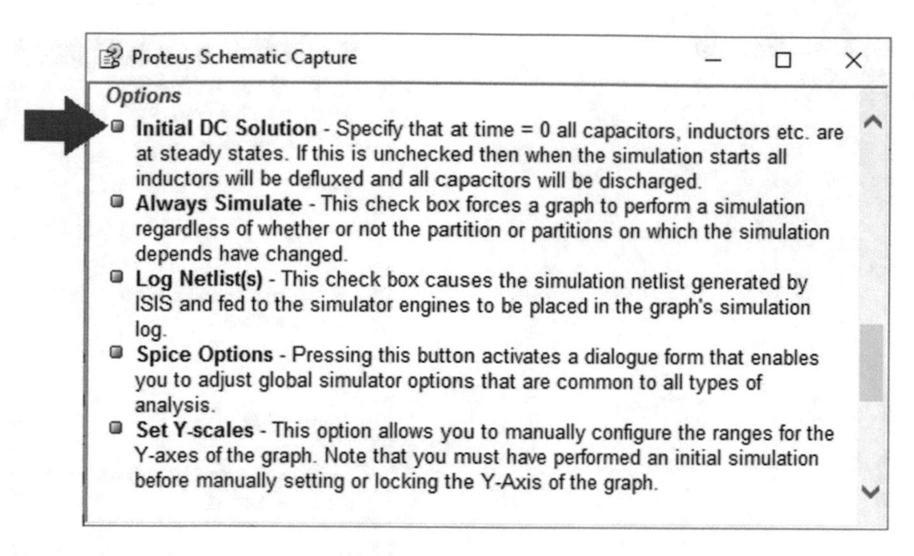

Fig. 1.289 Proteus calculates the DC operating point in the beginning of the simulation

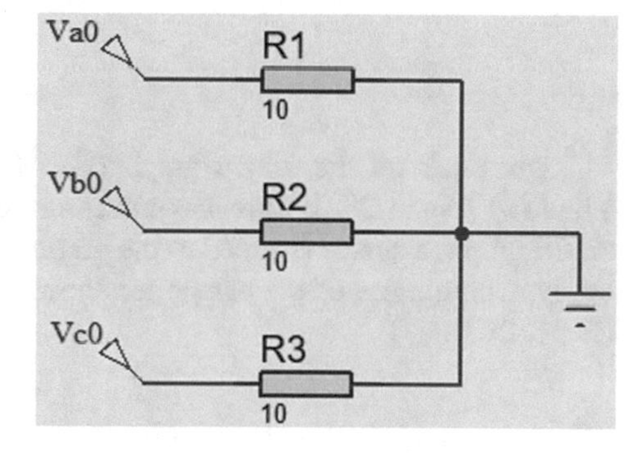

Fig. 1.290 Steady-state DC equivalent circuit of Fig. 1.272

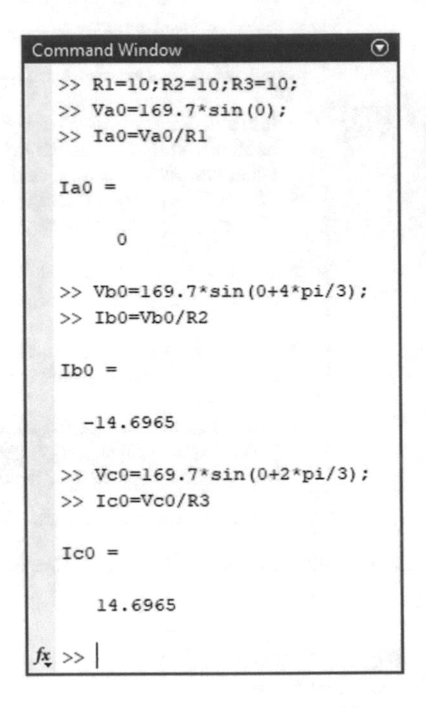

Fig. 1.291 MATLAB calculations

Right click on the drawn graph (Fig. 1.288) and click the Maximize (Show Window) (Fig. 1.292). This maximizes the drawn graph (Fig. 1.293). If you click on the graph, a cursor is added to the graph. This cursor helps you to read the graph easily. Coordinates of the cursor are shown in the left and right side of the screen (Fig. 1.293).

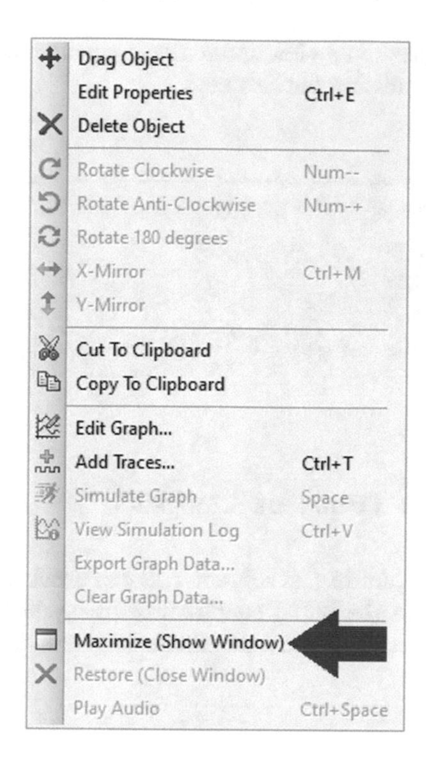

Fig. 1.292 Maximize command

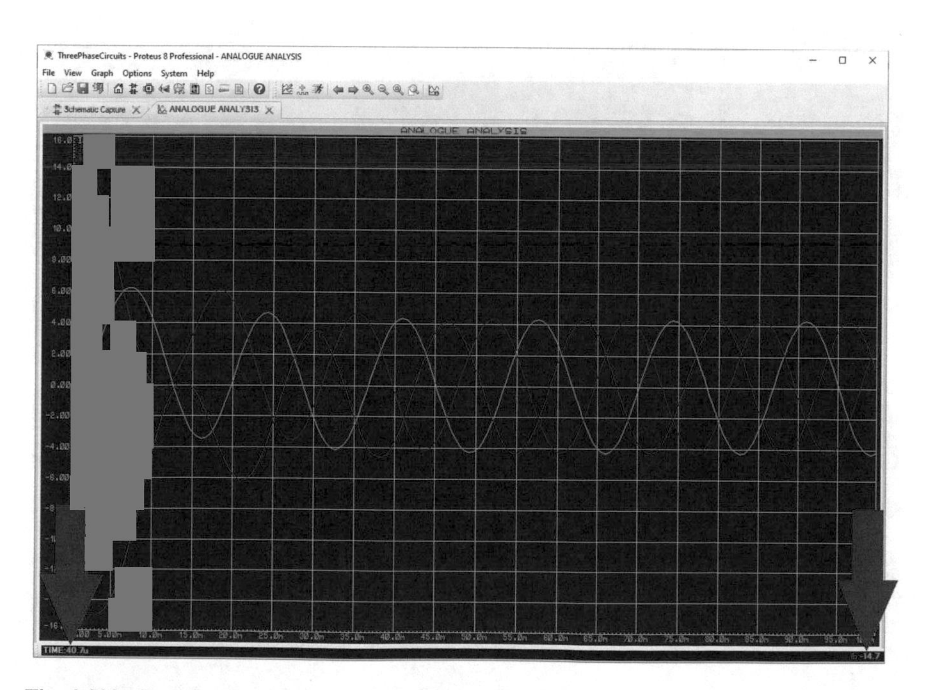

Fig. 1.293 Coordinates of the cursor are shown in the sides

You can close the maximum view and return to small view by clicking the close button (Fig. 1.294) or pressing the Esc key.

Fig. 1.294 Close button

1.37 Example 36: Transient Analysis (II)

In the previous example, Initial DC solution box was checked and Proteus used the result of DC analysis as the initial condition of the system. If you uncheck the Initial DC solution box (Fig. 1.295), the simulation starts from zero initial conditions (Fig. 1.296).

Edit Transient Graph

Graph title: ANALOGUE ANALYSIS

Start time: 0

Stop time: 100m

Left Axis Label:

Right Axis Label:

Options

Initial DC solution:

Always simulate:

Log netlist(s):

SPICE Options

Set Y-Scales

User defined properties:

OK Cancel

Fig. 1.295 Edit transient graph window

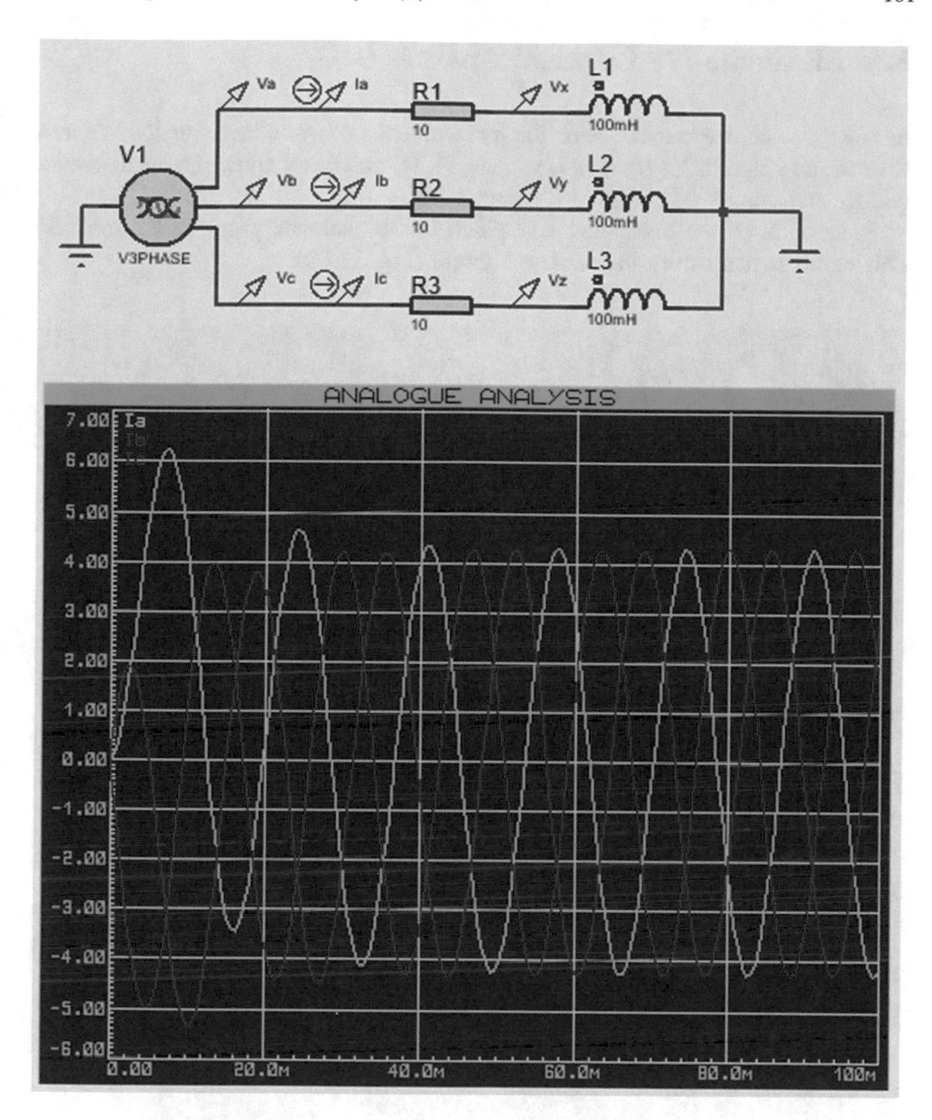

Fig. 1.296 Simulation result

1.38 Example 37: Transient Analysis (III)

In this example, we want to draw the instantaneous power of resistor R_1. We need to draw the graph of $V_{R_1}(t) \times I_{R_1}(t)$ where $V_{R_1}(t)$ and $I_{R_1}(t)$ show the instantaneous voltage and current of resistor R_1, respectively.

Right click on the traces that you added to the analogue graph in Example 36. This removes them from the analogue graph (Fig. 1.297).

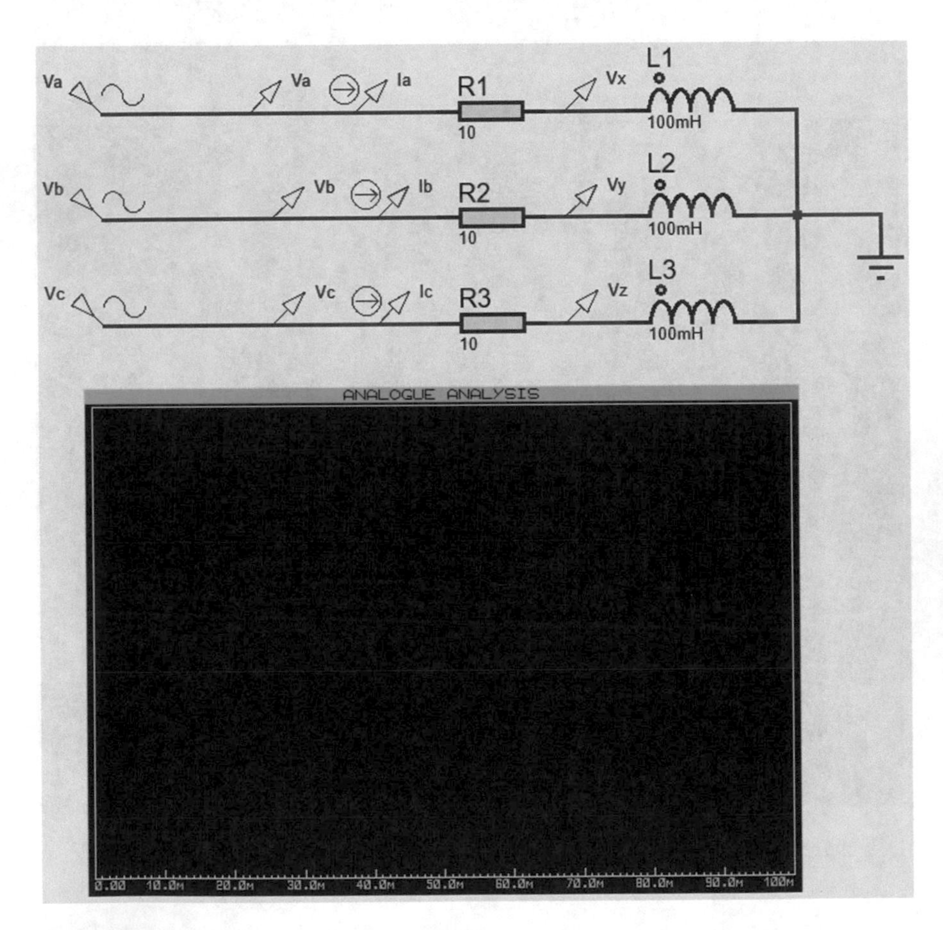

Fig. 1.297 Analog graph contains no waveform

Right click on the analogue graph and click the Add Traces. Then select ROOT_V_a for Probe P_1, ROOT_V_x for Probe P_2 and ROOT_Ia for Probe P_3. Enter $(P_1 - P_2) * P_3$ to the Expression box (Fig. 1.298) and click the OK button. The entered expression ($(P_1 - P_2) * P_3$) draws the graph of $(V_a - V_x) \times I_a$ which is the instantaneous power of resistor. V_a, V_x and I_a show the voltage of probe V_a, voltage of probe V_b and current of probe I_a, respectively.

Add Transient Trace ? ×
Name: R1 power
Probe P1: ROOT_Va
Probe P2: ROOT_Vx
Probe P3: ROOT_Ia
Probe P4: <NONE>
Expression: (P1-P2)*P3
Trace Type: Analog, Digital, Phasor, Noise
Axis: Left, Right, Reference
OK Cancel

Fig. 1.298 Add transient trace window

Run the simulation by pressing the space bar key. The simulation result is shown in Fig. 1.299. The graph shown on the screen is the instantaneous power of resistor R_1.

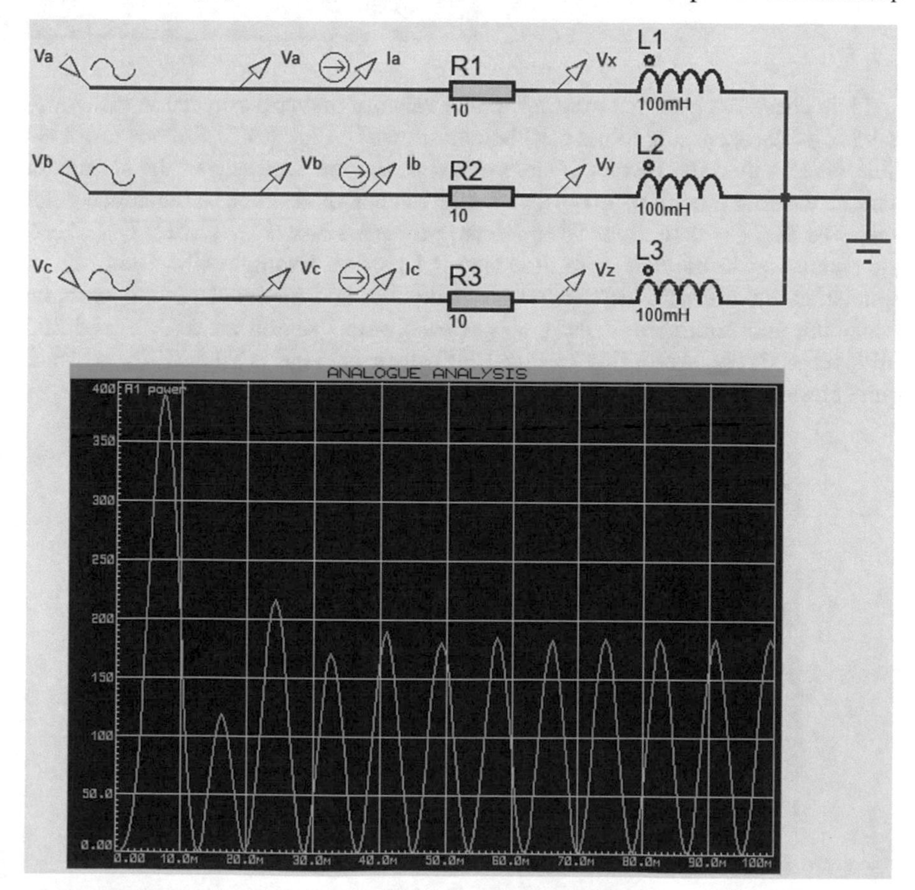

Fig. 1.299 Simulation result

Maximize the obtained graph and use the cursors to read the steady-state values of instantaneous power. The maximum and minimum of the graph at steady-state region are 184 W and 1.72 W, respectively. The average power can be calculated by averaging the maximum and minimum value of steady-state region. According to Fig. 1.300, the average power is 92.86 W.

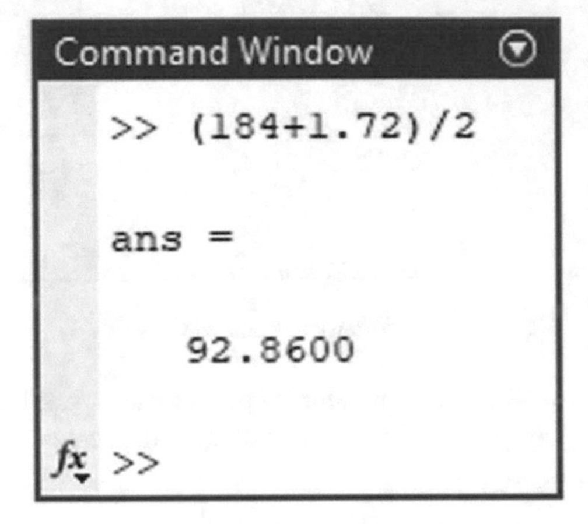

Fig. 1.300 MATLAB calculation

Let's check the obtained result. Average value of power dissipated in resistor R_1 is 94.6544 W according to the calculations shown in Fig. 1.301. So, our result is a little bit less than the correct value. Let's increase the accuracy of the simulation and see whether it helps us to obtain a better value. Double click the analogue graph and enter T_{MAX} = 0.1u to the User defined properties box (Fig. 1.302). T_{MAX} force the simulation to be done with time step 0.1 μs (see Example 40). Then run the simulation and measure the steady-state maximum and minimum. In this case, the maximum and minimum of the graph at steady-state region are 189 W and 20.3 mW, respectively. According to Fig. 1.303, their average is 94.5101 W which is quite close to the correct value (Fig. 1.301).

```
Command Window
>> Vm=169.7;R1=10;L1=100e-3;f=60;w=2*pi*f;
>> VRMS=Vm/sqrt(2);
>> IRMS=abs(VRMS/(R1+j*L1*w));
>> PR1=R1*IRMS^2

PR1 =

    94.6544

fx >>
```

Fig. 1.301 MATLAB calculation

Edit Transient Graph

Graph title: ANALOGUE ANALYSIS

Start time: 0

Stop time: 100m

Left Axis Label:

Right Axis Label:

Options

Initial DC solution: ☑

Always simulate: ☑

Log netlist(s): ☐

SPICE Options

Set Y-Scales

User defined properties:

TMAX=0.1u

OK Cancel

Fig. 1.302 Edit transient graph window

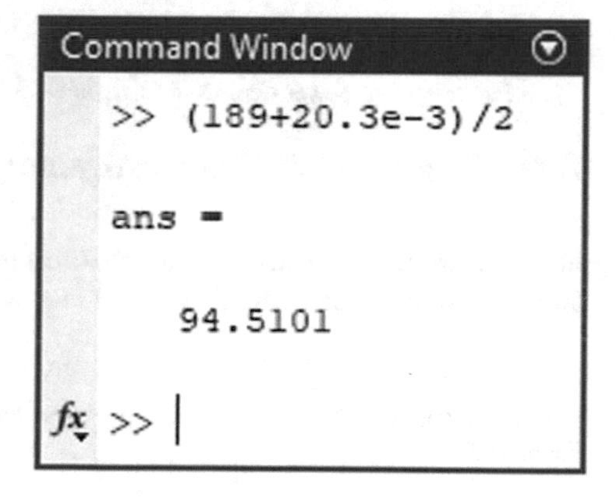

Fig. 1.303 MATLAB calculation

The average value of resistor R_1 power can be measured with the aid of Wattmeter block as well (Fig. 1.304).

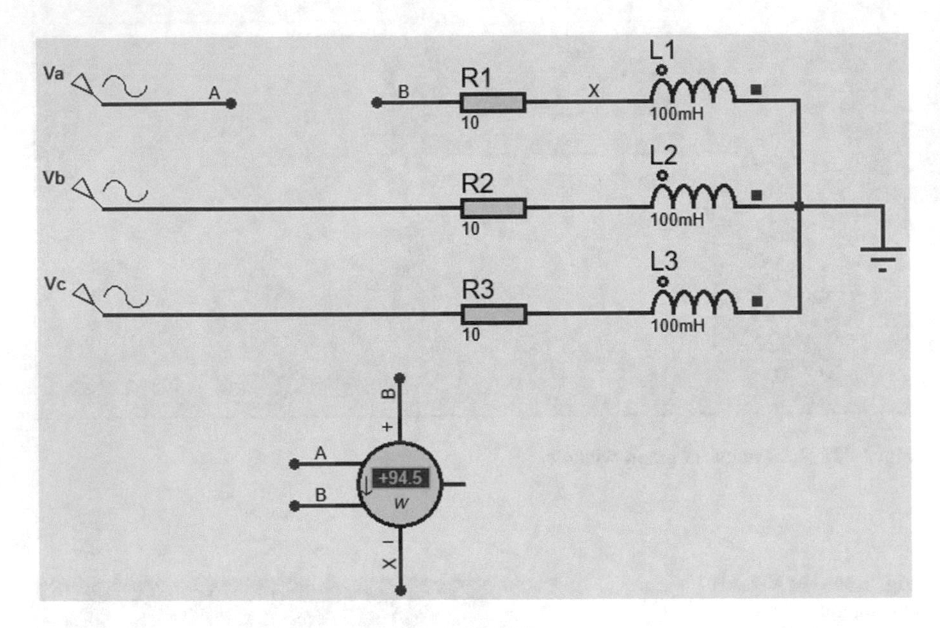

Fig. 1.304 Average power dissipated in R_1 is 94.5 W

1.39 Example 38: Transient Analysis (IV)

In this example, we want to simulate the circuit shown in Fig. 1.305. Initial condition of inductor and capacitor is 0.1 A and 2 V, respectively.

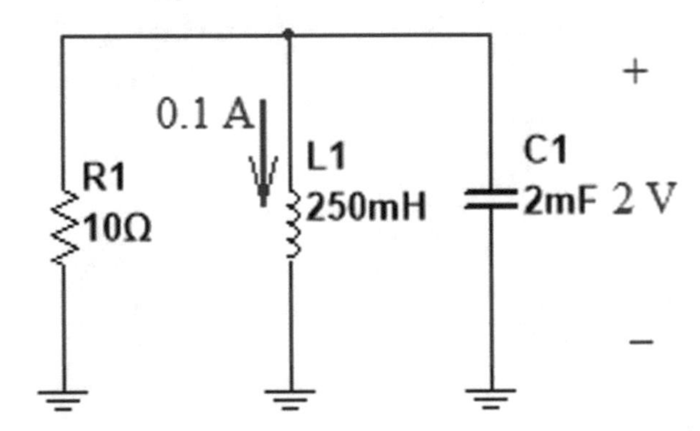

Fig. 1.305 Circuit for Example 38

Draw the schematic shown in Fig. 1.306.

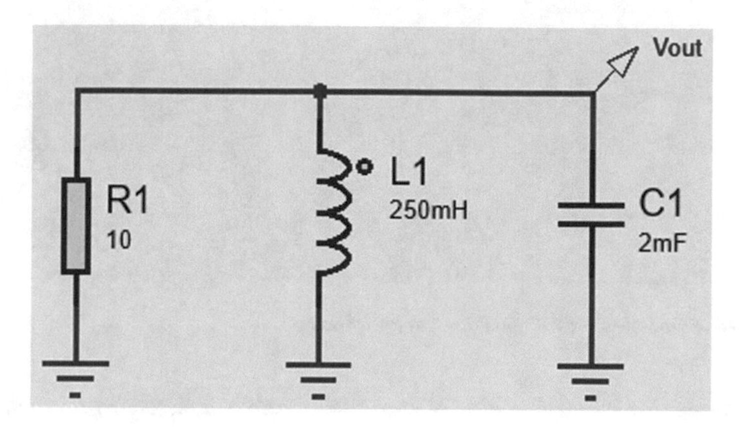

Fig. 1.306 Proteus equivalent of Fig. 1.305

Let's enter the initial condition of inductor. Double click the inductor and enter PREFLUX = 0.1 to the Other Properties box (Fig. 1.307) and click the OK button. After clicking the OK button, the schematic changes to what shown in Fig. 1.308. The current that enters into the terminal with dot on it is assumed to be positive. The PREFLUX = 0.1 means that initial current of the inductor is 0.1 A and this current enters into the terminal with dot on it.

Edit Component ? ×
Part Reference: L1 Hidden:
Inductance (Henrys): 250mH Hidden:
Element: New
OK
Help
Cancel
Other Properties:
PREFLUX=0.1
Exclude from Simulation
Exclude from PCB Layout
Exclude from Current Variant
Attach hierarchy module
Hide common pins
Edit all properties as text

Fig. 1.307 Settings of L_1

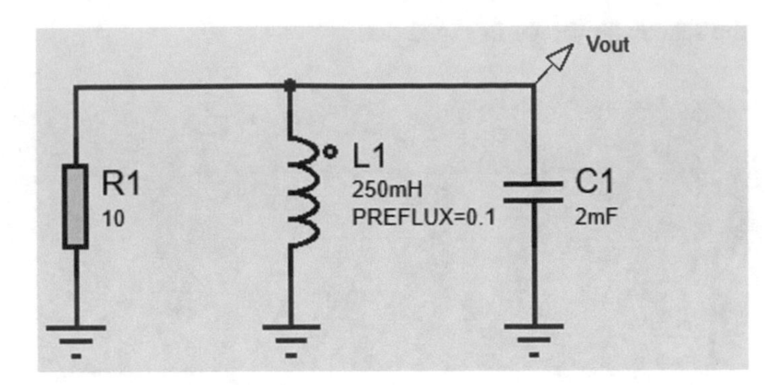

Fig. 1.308 Addition of initial condition the inductor L_1

Let's enter the initial condition of capacitor. Double click the capacitor and enter PRECHARGE = 2 to the Other Properties box (Fig. 1.309) and click the OK button. After clicking the OK button, the schematic changes to what shown in Fig. 1.310. The PRECHARGE = 2 means that initial voltage of the capacitor is 2 V.

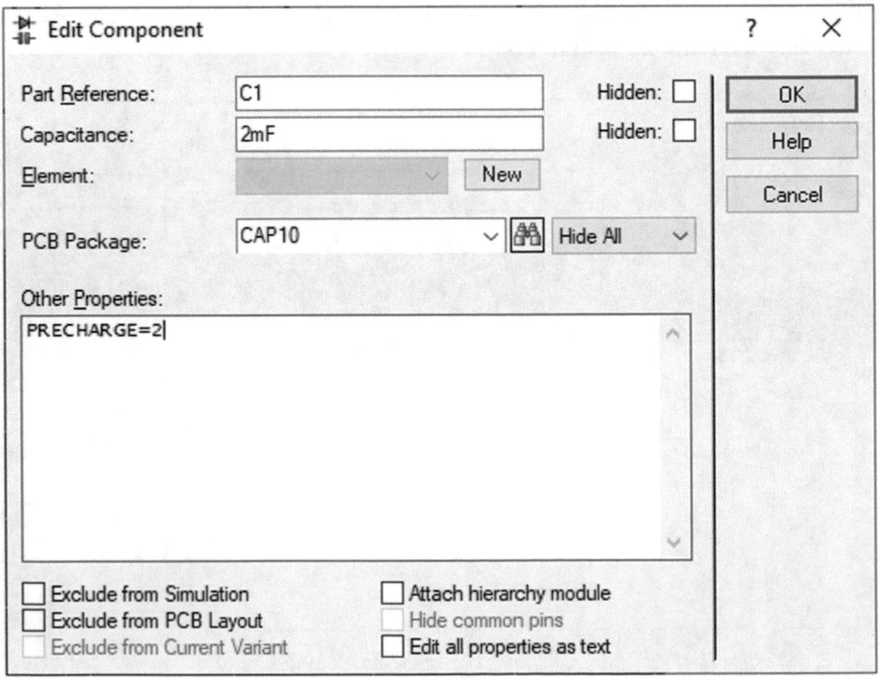

Fig. 1.309 Settings of C_1

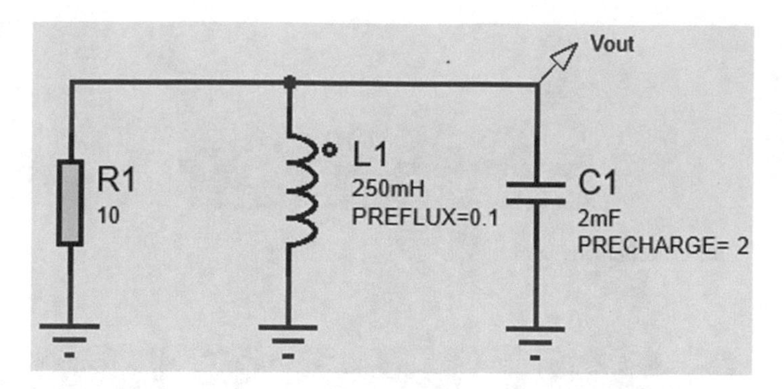

Fig. 1.310 Initial conditions are shown on the schematic

The initial voltage of capacitor C_1 in Fig. 1.310 is 2 V. However, we don't know whether $V_{\text{uppert terminal}} - V_{\text{lower terminal}} = 2\text{ V}$ or $V_{\text{lower terminal}} - V_{\text{uppert terminal}} = 2\text{ V}$. Let's use a DC voltmeter to measure the capacitor voltage. Convert the schematic to what shown in Fig. 1.311 and click the run icon (Fig. 1.312). Simulation result is shown in Fig. 1.313. The DC voltmeter reads 2 V so we can deduce that $V_{\text{uppert terminal}} - V_{\text{lower terminal}} = 2\text{V}$. According to Fig. 1.313 this is the initial condition that we need.

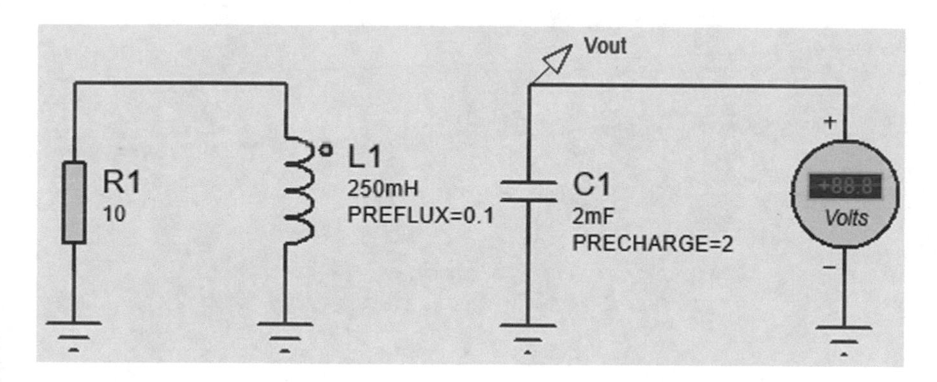

Fig. 1.311 Wire between inductor and capacitor is removed

Fig. 1.312 Run the schematic icon

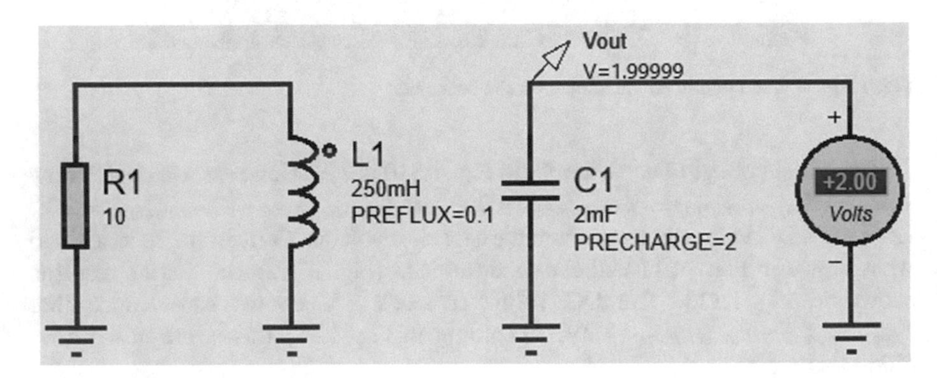

Fig. 1.313 Simulation result

Remove the DC voltmeter, connect the capacitor to the inductor and add an analogue graph to the schematic (Fig. 1.314). Settings of analogue graph is shown in Fig. 1.315.

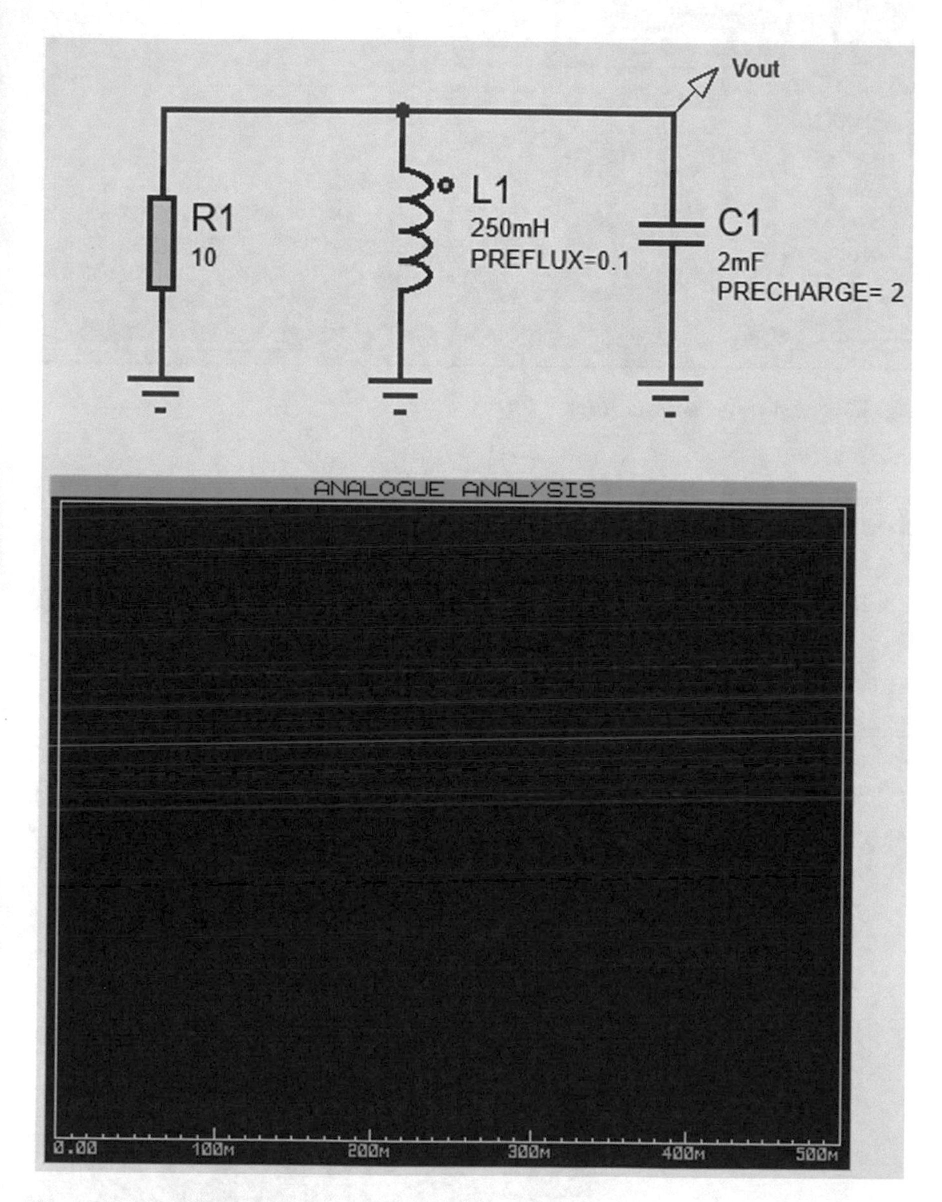

Fig. 1.314 Completed schematic of Example 38

Edit Transient Graph ? ×

Graph title: ANALOGUE ANALYSIS
Start time: 0
Stop time: 500m
Left Axis Label:
Right Axis Label:
Options
Initial DC solution: ☑
Always simulate: ☑
Log netlist(s): ☐
SPICE Options
Set Y-Scales
User defined properties:
OK Cancel

Fig. 1.315 Settings of analogue graph

Press the space bar key of keyboard to run the simulation. Simulation result is shown in Fig. 1.316.

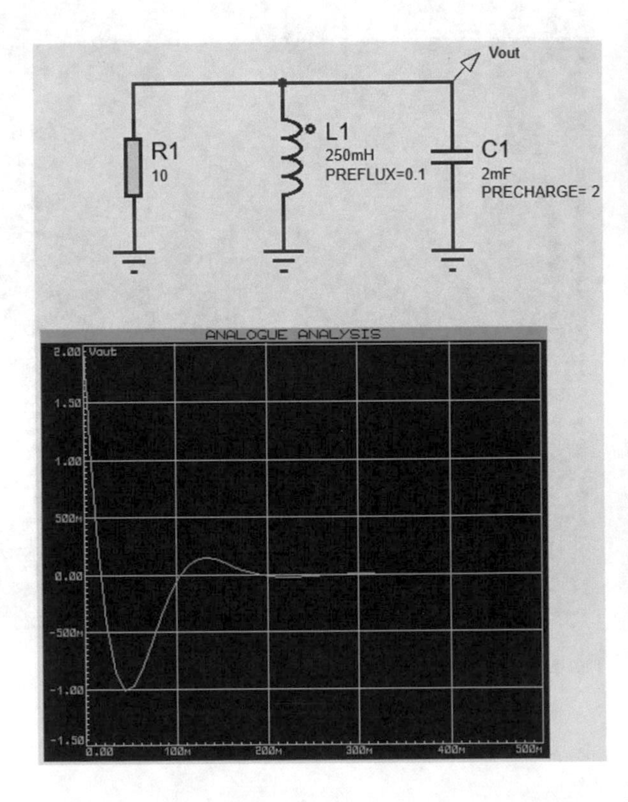

Fig. 1.316 Simulation result

1.40 Example 39: Transient Analysis (V)

In the previous example, we used a voltmeter to determine the polarity of initial voltage of capacitor. There is no need to use such a test if you use an electrolytic capacitor. An electrolytic capacitor can be added to the schematic by searching for "cap-elec" in the Pick Devices window (Fig. 1.317).

Fig. 1.317 Searching for electric capacitor

The polarity of voltage for an electrolytic capacitor is shown in Fig. 1.318. For instance, when you write PRECHARGE = 2 in the Other Properties box of an electrolytic capacitor, it means that $V_{\text{Terminal without hatch}} - V_{\text{Terminal with hatch}} = 2\,\text{V}$.

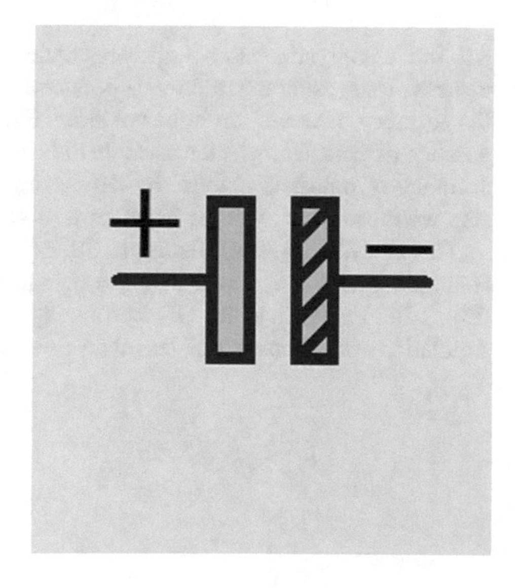

Fig. 1.318 Electric capacitor symbol

Figure 1.319 shows the schematic of Example 38 with an electrolytic capacitor. If you run this schematic, the result obtained in Fig. 1.316 is obtained.

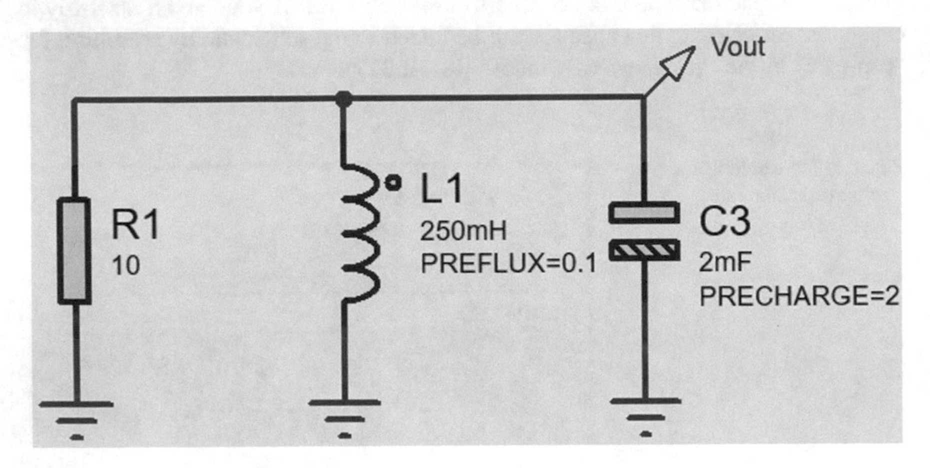

Fig. 1.319 Schematic of Example 38 with an electrolytic capacitor

1.41 Example 40: Increasing the Accuracy of Transient Analysis Graph

All the circuit simulation softwares calculate the waveforms in discrete time instants. Decreasing the difference between two consecutive time instants increases the accuracy of waveform however simulation requires more time to be done since number of calculations increases. In this example we learn how to increase the accuracy of transient analysis by decreasing the time step of simulation. By time step we mean the difference between two consecutive instant in the output graph.

The analysis result of Example 38 is shown in Fig. 1.320. Double click the analogue graph and enter $T_{MAX} = 0.1$ μ to the User defined properties box (Fig. 1.321) and click the OK button. $T_{MAX} = 0.1$ μ tells the Proteus to do the simulation with time step of less than or equal to 0.1 μs.

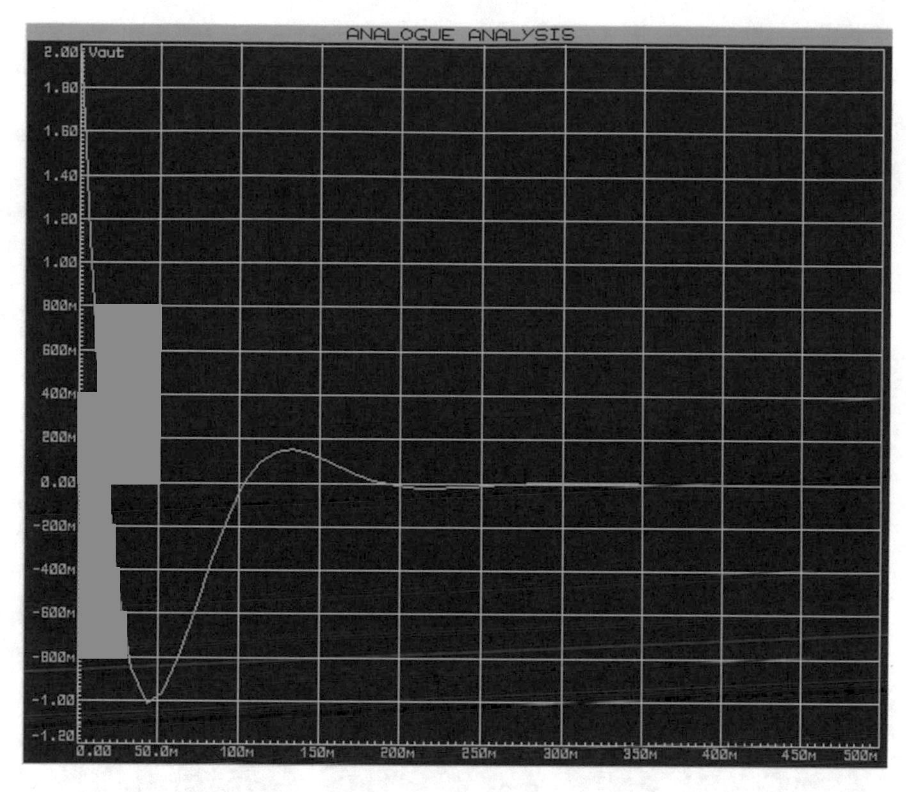

Fig. 1.320 Simulation result

Edit Transient Graph
Graph title: ANALOGUE ANALYSIS
Start time: 0
Stop time: 500m
Left Axis Label:
Right Axis Label:
Options
Initial DC solution:
Always simulate:
Log netlist(s):
SPICE Options
Set Y-Scales
User defined properties:
TMAX=0.1u
OK
Cancel

Fig. 1.321 Defining the maximum step size of simulation

Run the simulation by pressing the space bar key of your keyboard. The simulation result is shown in Fig. 1.322. If you compare Figs. 1.320 with 1.322, you understand that Fig. 1.322 is smoother in comparison to Fig. 1.320.

Decreasing the time step is one of the solutions when you are not happy with the smoothness of drawn graph.

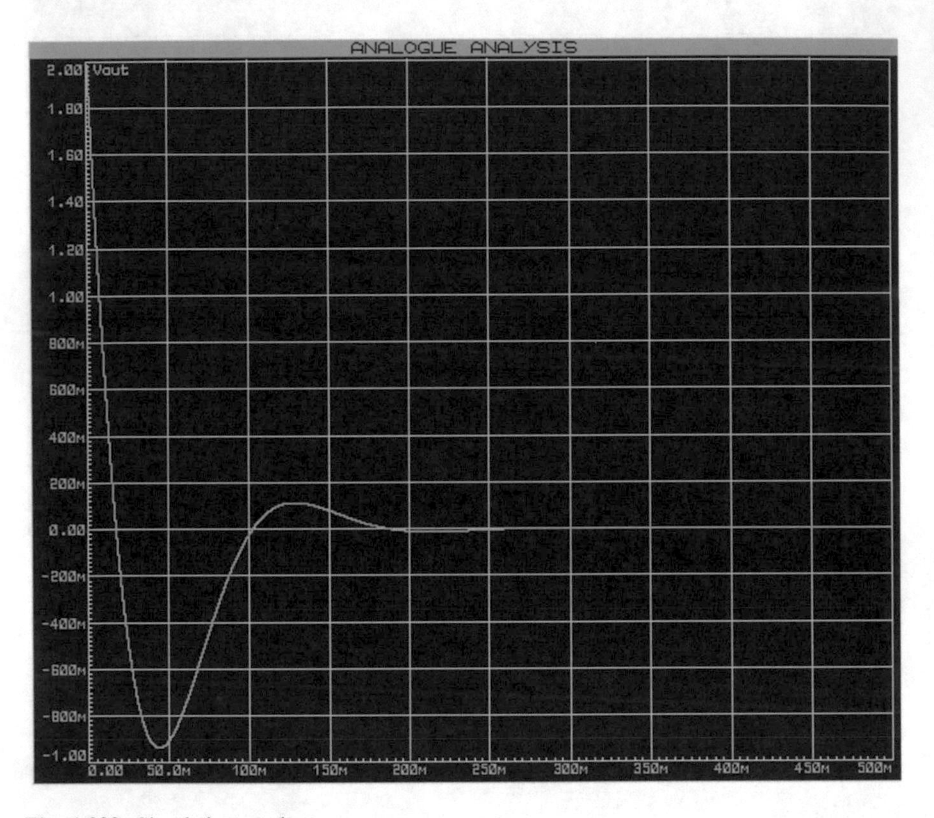

Fig. 1.322 Simulation result

1.42 Example 41: Copying the Waveform Graph into the Clipboard Memory

You can copy the waveform that you see in the analogue graph into clipboard and paste it into other softwares. This is very useful when you want to prepare a report/presentation and you need to show the waveforms.

In order to copy the waveforms into clipboard, right click on the analogue graph and click the Copy To Clipboard (Fig. 1.323). You can paste the copied waveform into other softwares by pressing the Ctrl + V.

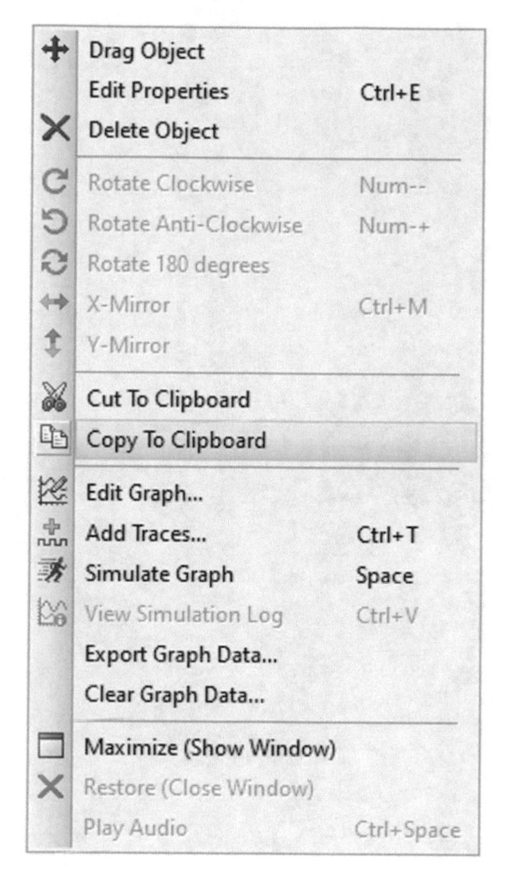

Fig. 1.323 Copy to clipboard command

1.43 Example 42: Exporting the Waveforms into MATLAB®

The waveforms shown in analogue graph can be exported into MATLAB® environment. Exporting to MATLAB permits further process of the waveforms (for instance see Example 52). In this example we learn how to import the waveform data into MATLAB environment.

Open the schematic of Example 40 and run the simulation. Then right click on the drawn graph and click the Export Graph Data (Fig. 1.324). After clicking the Export Graph Data, Graph Data window appears on the screen and permit you to determine the name and path of the file which contains the graph data.

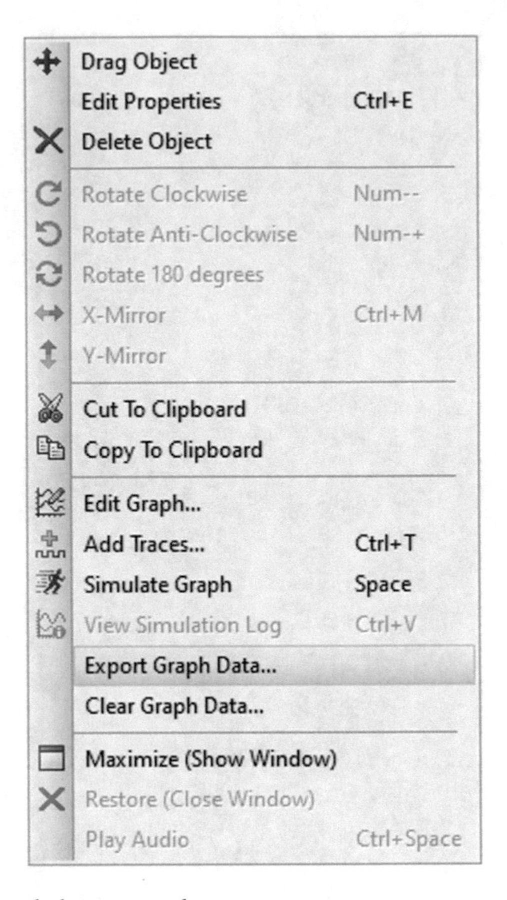

Fig. 1.324 Export graph data command

You can open the saved file with Notepad (Fig. 1.325). In this example, the saved file contains two columns. First column is the time column and the second column is the output voltage values.

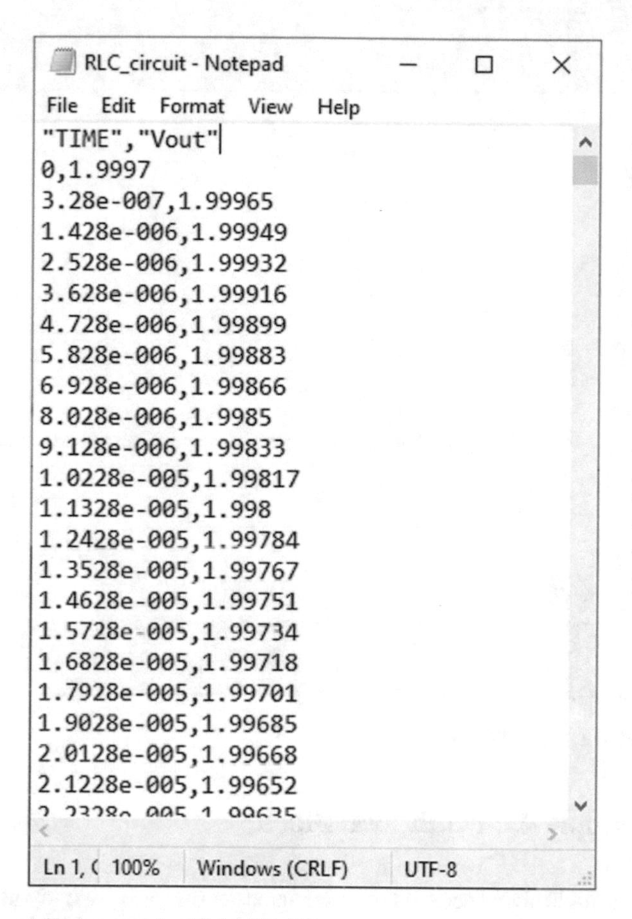

Fig. 1.325 Saved file is opened with MATLAB

The MATLAB commands shown in Fig. 1.326 read the saved file and draws it graph. Output of this code is shown in Fig. 1.327.

Command Window

```
>> F=importdata('C:\Proteus Book\RLC_circuit.dat');
>> time=F.data(:,1);
>> Vout=F.data(:,2);
>> plot(time,Vout)
fx >>
```

Fig. 1.326 MATLAB commands

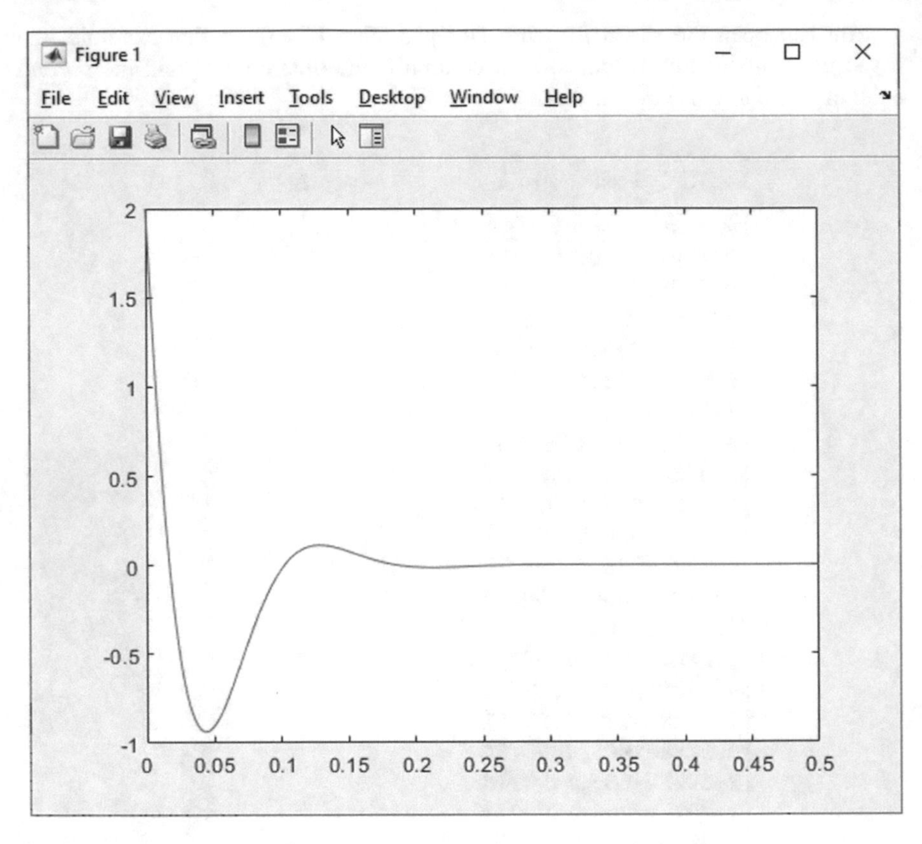

Fig. 1.327 Output of MATLAB code in Fig. 1.326

1.44 Example 43: Multiplier Block

Proteus has a multiplier block which can used to multiply two signals. The multiplier block can be added to the schematic by searching for "multiplier" in the Pick Devices window (Fig. 1.328).

Fig. 1.328 Searching for multiplier block

Consider the schematic shown in Fig. 1.329. In this schematic, the multiplier block generates the instantaneous power waveform of load of phase *A*. V_a, V_b and V_c simulate a three-phase input with line-line RMS value of $120\sqrt{3} = 207.85$ V and frequency of 60 Hz.

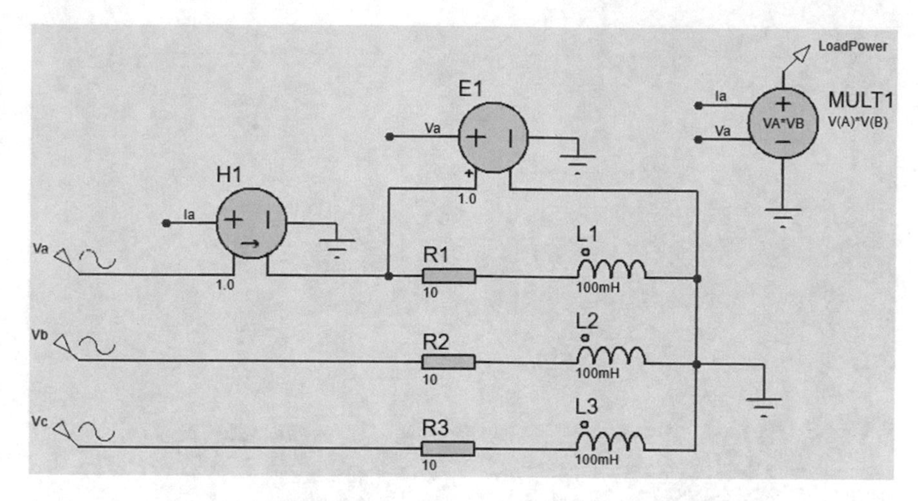

Fig. 1.329 Schematic of Example 43

Let's add an analogue graph to the schematic and simulate the circuit for [0, 100 ms] time interval. Simulation result is shown in Fig. 1.330.

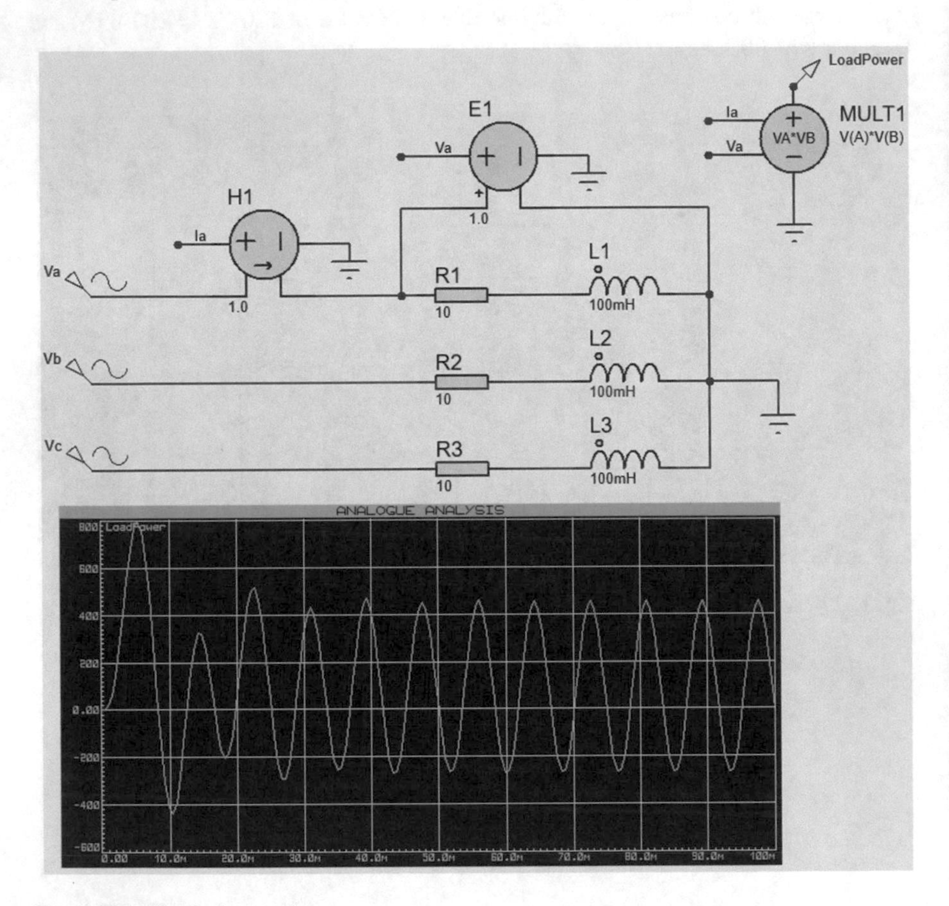

Fig. 1.330 Simulation result

Let's check the obtained result. The following MATLAB code draws the graph of instantaneous power of phase *A* load for [0, 100 ms] time interval.

```
clc

clear all

R1=10;L1=100e-3;f=60;T=1/f;w=2*pi*f;Vm=169.7;

syms i(t) V1(t)

Va=Vm*sin(w*t);

ode=L1*diff(i,t)+R1*i==Va;

cond=i(0)==0;

Ia(t)=dsolve(ode,cond);

p=Va*Ia;

ezplot(p,[0,0.1])
```

Output of the code is shown in Fig. 1.331. You can compare different points of the two graphs (Figs. 1.330 and 1.331) to ensure that Proteus result is quite close to theoretical result.

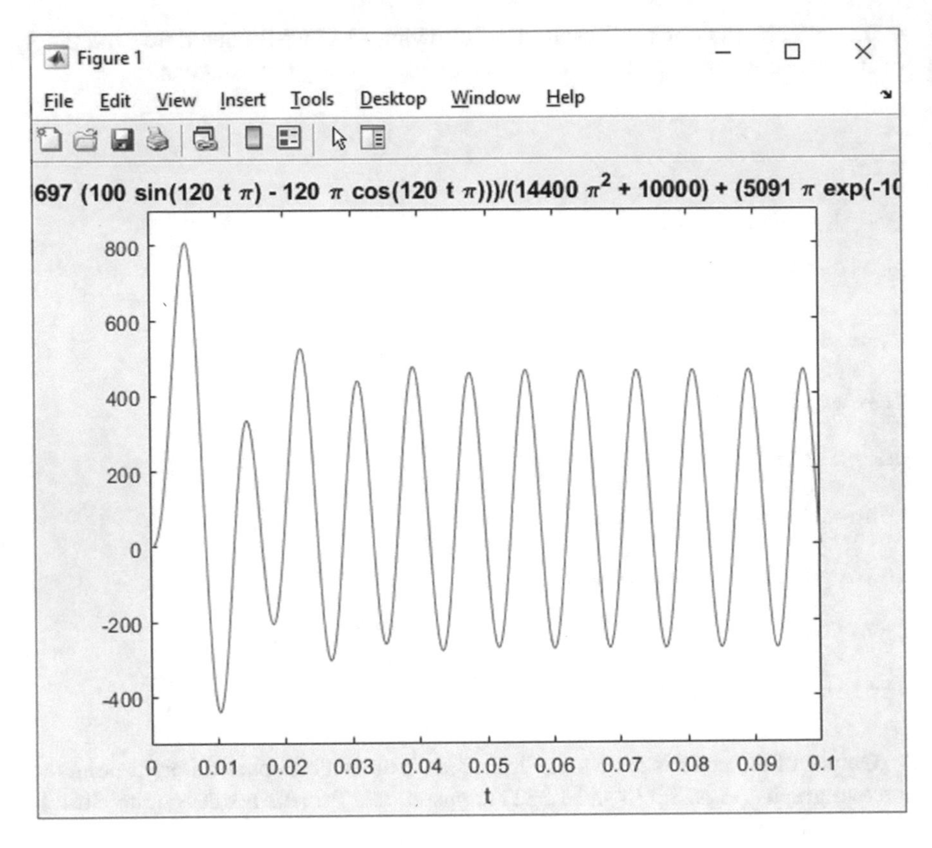

Fig. 1.331 Output of MATLAB code

The output of multiplier block can be connected directly to the oscilloscope block as well (Fig. 1.332).

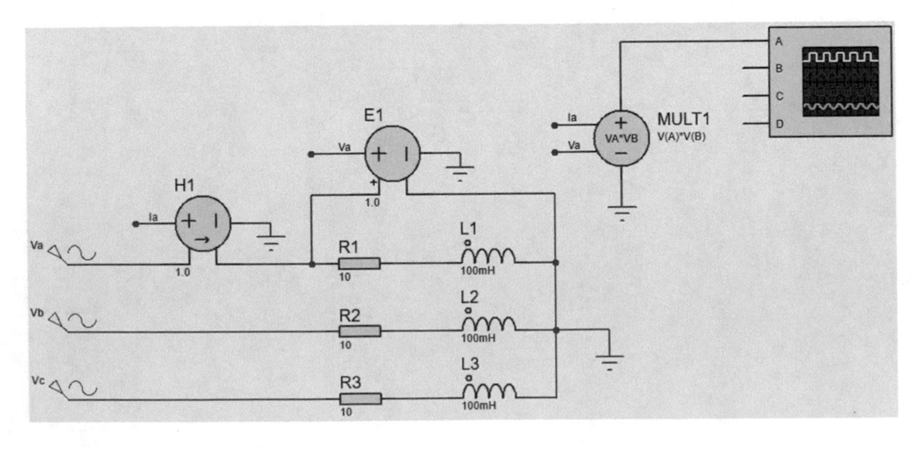

Fig. 1.332 Output of multiplier block is connected to the oscilloscope block

1.45 Example 44: Gain Block

Proteus has a gain block which can used to multiply a signal by a constant value. For instance, assume that you want to see a big voltage with an oscilloscope block. In this case, if you put a gain block with value of less than 1 between signal source and the oscilloscope block (Fig. 1.333), the signal that enters the oscilloscope block decreases and you can see it better.

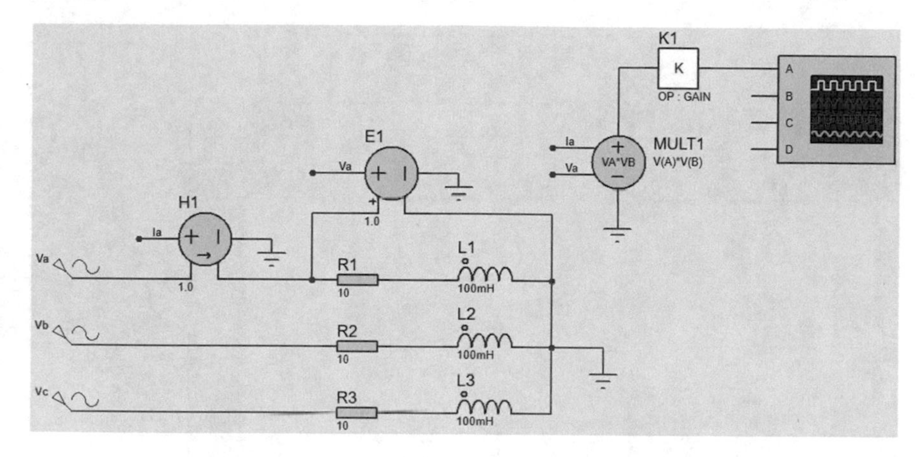

Fig. 1.333 Schematic of Example 44

The gain block can be added to the schematic by searching for "op: gain" in the Pick Devices window (Fig. 1.334).

Fig. 1.334 Searching for gain block

Pick Devices

Keywords:

op : gain

Match whole words?

Show only parts with models?

If you double click on the gain block, the window shown in Fig. 1.335 appears on the screen. Required gain must be entered into the Amplification coefficient box (Fig. 1.335).

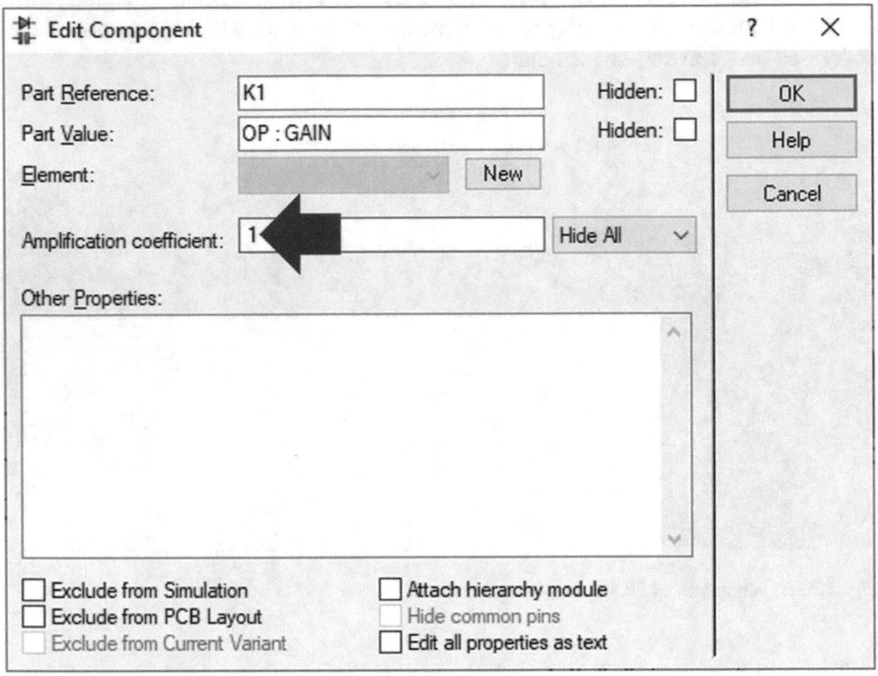

Fig. 1.335 Settings of K_1 block

1.46 Example 45: Coupled Inductors (I)

In this example we want to simulate coupled inductors. Assume that we have three inductors L_1, L_2 and L_3. Self-inductance of L_1, L_2 and L_3 are 10 mH, 20 mH and 30 mH, respectively.

The mutual inductance between L_1 and L_2 is $M_{12} = 10$ mH. So, the coupling coefficient between L_1 and L_2 is $\frac{M_{12}}{\sqrt{L_1 L_2}} = 0.707$. The mutual inductance between L_1 and L_3 is $M_{13} = 16$ mH. So, the coupling coefficient between L_1 and L_3 is $\frac{M_{13}}{\sqrt{L_1 L_3}} = 0.924$. The mutual inductance between L_2 and L_3 is $M_{23} = 12$ mH. So, the coupling coefficient between L_2 and L_3 is $\frac{M_{23}}{\sqrt{L_2 L_3}} = 0.490$. We want to model these inductors in Proteus. Put three inductors on the schematic and set their values (Fig. 1.336).

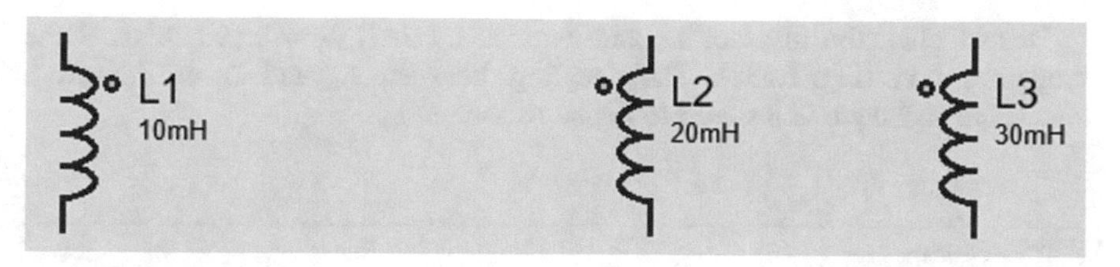

Fig. 1.336 Three inductors are placed on the schematic

Double click the inductor L_1 and enter the MUTUAL_L_2 = 0.707 and MUTUAL_L_3 = 0.924 to the Other Properties box (Fig. 1.337). These two lines sets the coupling factor between inductor L_1 and inductors L_2 and L_3.

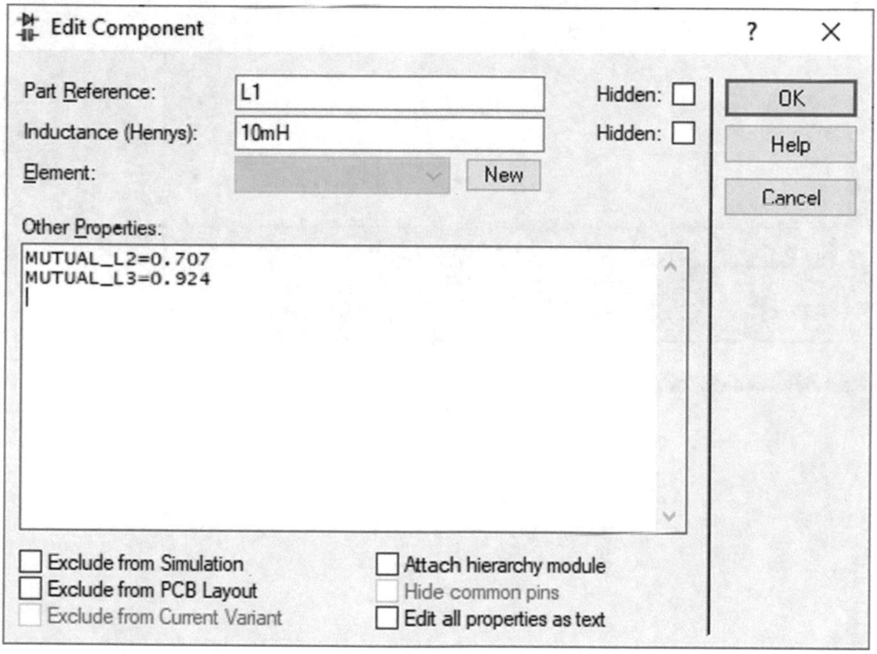

Fig. 1.337 Settings of L_1

Double click the inductor L_2 and enter MUTUAL_L_3 = 0.490 to the Other Properties box (Fig. 1.338). The coupling between L_2 and L_1 is defined in Fig. 1.337 and there is no need to define it once more.

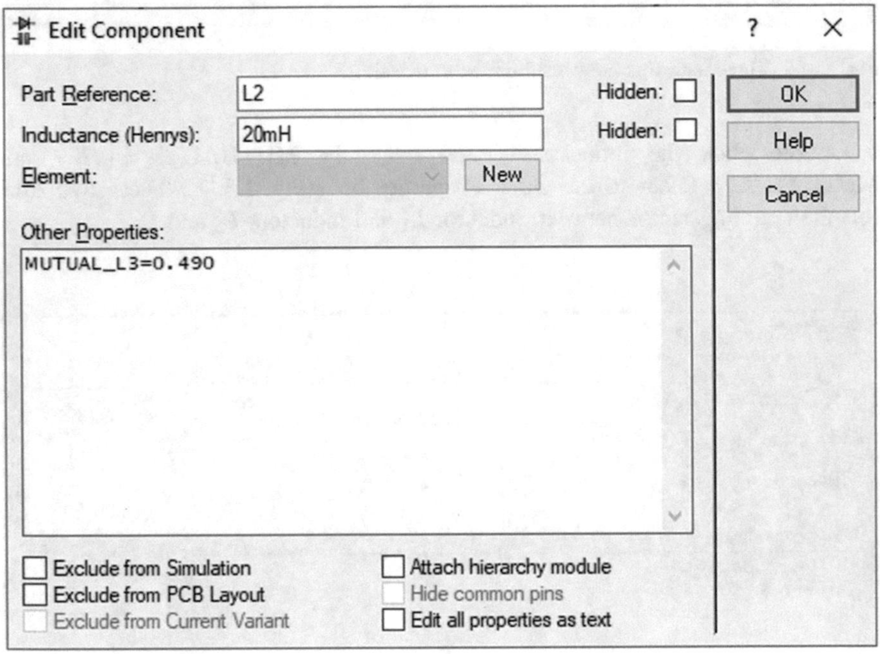

Fig. 1.338 Settings of L_2

Settings of inductor L_3 is shown in Fig. 1.339. The coupling between inductor L_3 and inductors L_1 and L_2 are determined in Figs. 1.337 and 1.338 and there is no need to define it once more. That is why the Other Properties box of inductor L_3 is empty.

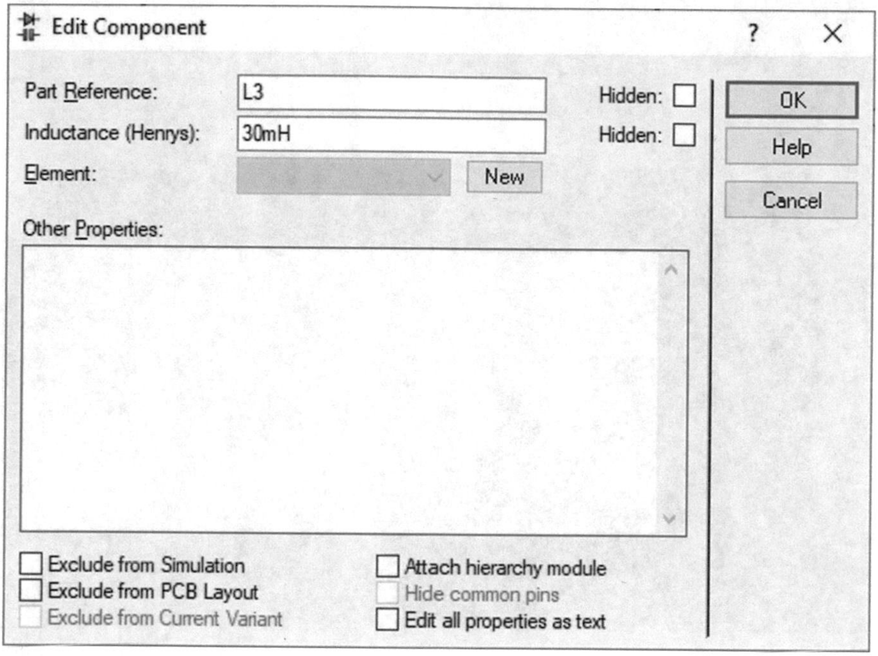

Edit Component ? X
Part Reference: L3 Hidden: OK
Inductance (Henrys): 30mH Hidden: Help
Element: New Cancel
Other Properties:
Exclude from Simulation
Exclude from PCB Layout
Exclude from Current Variant
Attach hierarchy module
Hide common pins
Edit all properties as text

Fig. 1.339 Settings of L_3

Now the schematic looks like Fig. 1.340. Let's test the inductors with a simple circuit. Draw the schematic shown in Fig. 1.341. Settings of V_{in} and analogue graph are shown in Figs. 1.342 and 1.343, respectively.

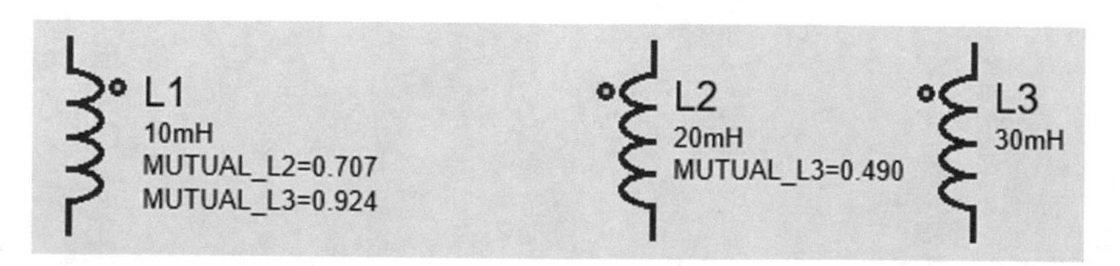

Fig. 1.340 Entered commands are shown on the schematic

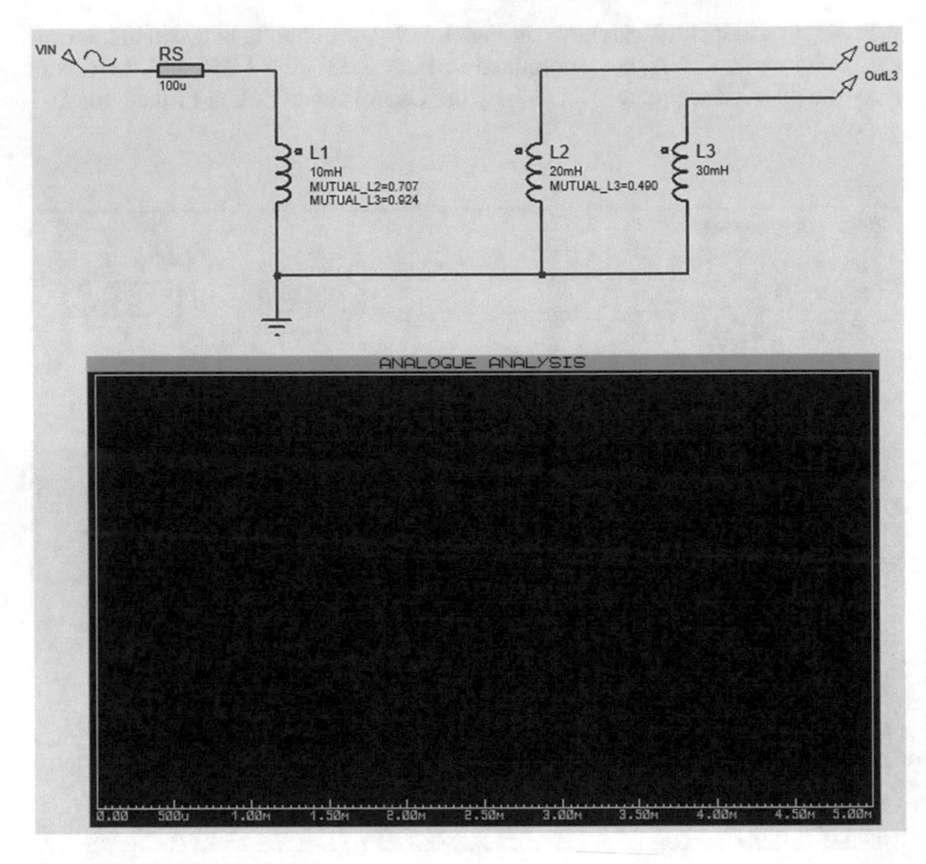

Fig. 1.341 Completed schematic

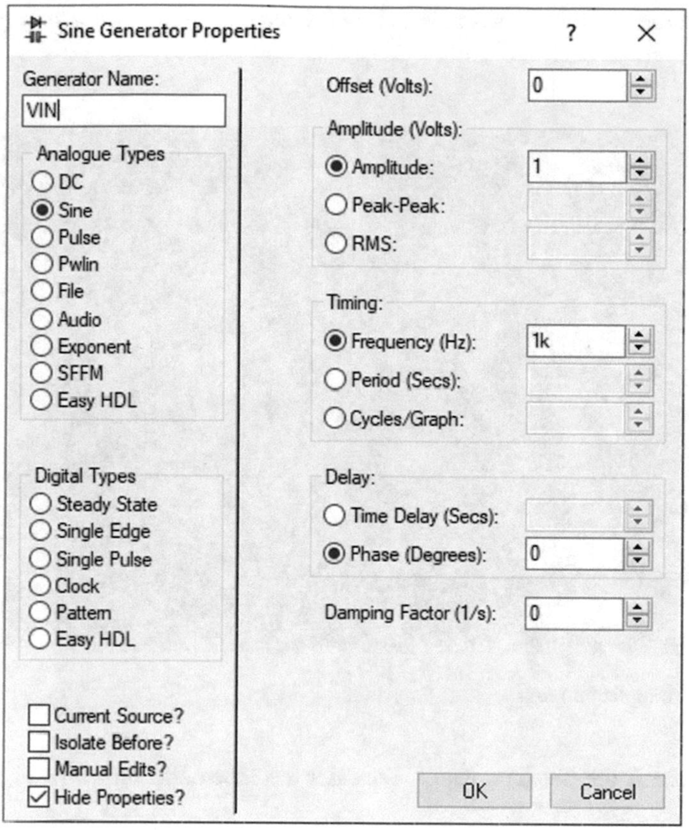

Sine Generator Properties ? X

Generator Name:
VIN

Analogue Types
DC
Sine
Pulse
Pwlin
File
Audio
Exponent
SFFM
Easy HDL

Digital Types
Steady State
Single Edge
Single Pulse
Clock
Pattern
Easy HDL

Current Source?
Isolate Before?
Manual Edits?
Hide Properties?

Offset (Volts): 0
Amplitude (Volts):
Amplitude: 1
Peak-Peak:
RMS:

Timing:
Frequency (Hz): 1k
Period (Secs):
Cycles/Graph:

Delay:
Time Delay (Secs):
Phase (Degrees): 0

Damping Factor (1/s): 0

OK Cancel

Fig. 1.342 Settings of V_{IN}

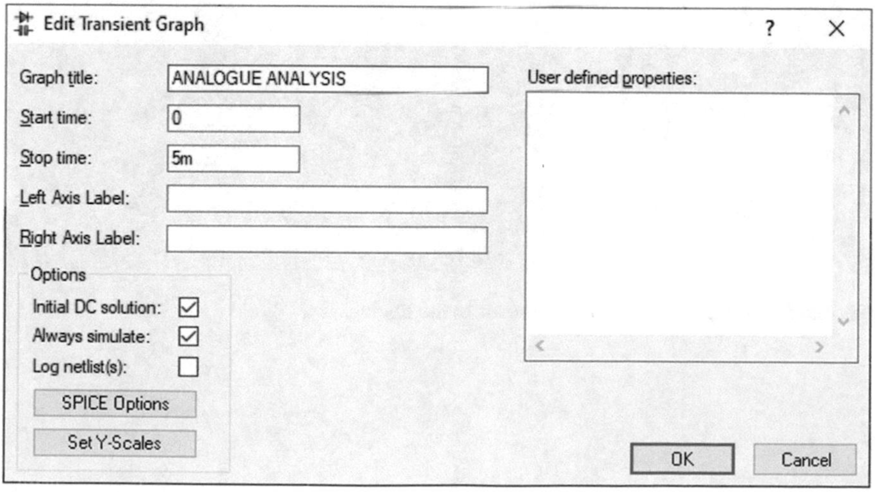

Edit Transient Graph ? X

Graph title: ANALOGUE ANALYSIS
Start time: 0
Stop time: 5m
Left Axis Label:
Right Axis Label:

Options
Initial DC solution: ☑
Always simulate: ☑
Log netlist(s): ☐
SPICE Options
Set Y-Scales

User defined properties:

OK Cancel

Fig. 1.343 Settings of analog analysis

Simulation result is shown in Fig. 1.344. Note that L_1 induced a voltage into the L_2 and L_3.

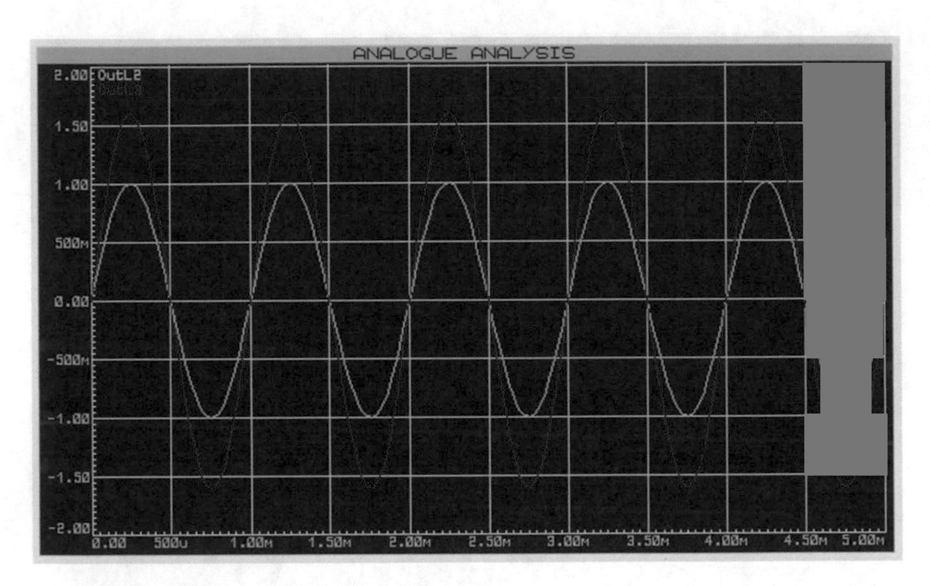

Fig. 1.344 Simulation result

Let's check the obtained result. Consider the schematic shown in Fig. 1.345.

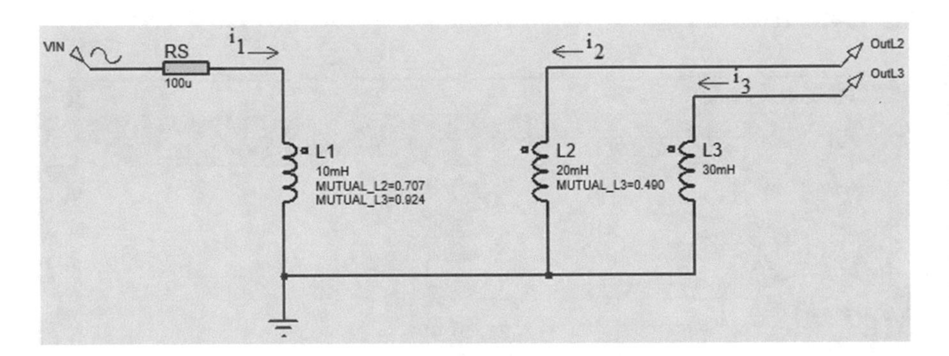

Fig. 1.345 Currents i_1, i_2 and i_3 are shown in the figure

Differential equations of Fig. 1.345 are (note that $M_{ij} = M_{ji}$ and $k_{ij} = k_{ji}$):

$$L_1 \frac{di_1}{dt} + M_{12} \frac{di_2}{dt} + M_{13} \frac{di_3}{dt} = V_{IN} \Rightarrow L_1 \frac{di_1}{dt}$$
$$+ M_{12} \times 0 + M_{13} \times 0 = V_{IN} \Rightarrow L_1 \frac{di_1}{dt} = V_{IN}$$
$$L_2 \frac{di_2}{dt} + M_{21} \frac{di_1}{dt} + M_{31} \frac{di_3}{dt} = V_{L_2} \Rightarrow L_2 \times 0$$
$$+ M_{21} \frac{di_1}{dt} + M_{31} \times 0 = V_{L_2} \Rightarrow M_{21} \frac{di_1}{dt} = V_{L_2}$$
$$L_3 \frac{di_3}{dt} + M_{31} \frac{di_1}{dt} + M_{32} \frac{di_2}{dt} = V_{L_3} \Rightarrow L_3 \times 0$$
$$+ M_{31} \frac{di_1}{dt} + M_{32} \times 0 = V_{L_3} \Rightarrow M_{31} \frac{di_1}{dt} = V_{L_3}$$

So,

$$V_{L_2} = \frac{M_{21}}{L_1} V_{IN} = \frac{k_{21} \sqrt{L_1 \times L_2}}{L_1} = k_{21} \sqrt{\frac{L_2}{L_1}}$$
$$V_{I_3} = \frac{M_{31}}{L_1} V_{IN} = \frac{k_{31} \sqrt{L_1 \times L_3}}{L_1} = k_{31} \sqrt{\frac{L_3}{L_1}}$$

Amplitude of V_{L_2} and V_{L_3} are calculated in Figs. 1.346 and 1.347, respectively. Obtained values are quite close to Proteus result (Fig. 1.344).

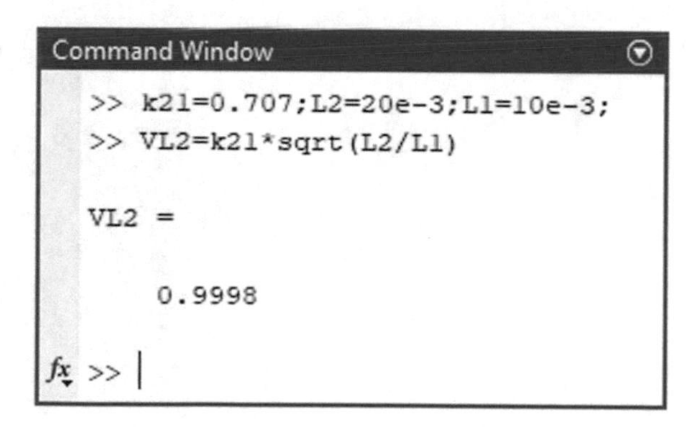

Fig. 1.346 MATLAB calculations

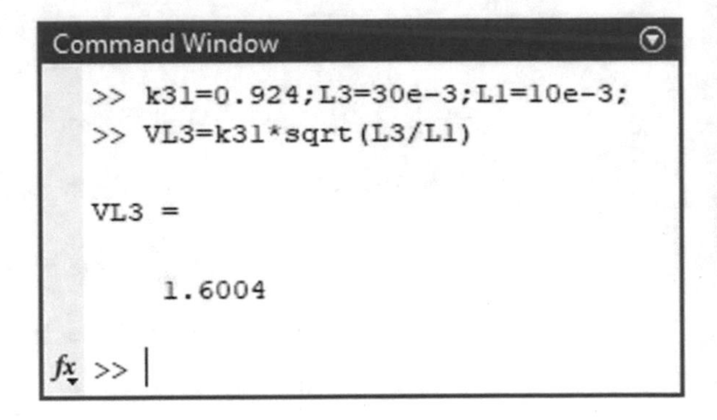

Fig. 1.347 MATLAB calculations

1.47 Example 46: Coupled Inductors (II)

The general way of modeling coupled inductors is studied in the previous example. We study another way to model coupled inductors in this example. In this example, we use a transformer with one primary winding and one secondary winding (Fig. 1.348). Method of this example can be used to model only two coupled inductors.

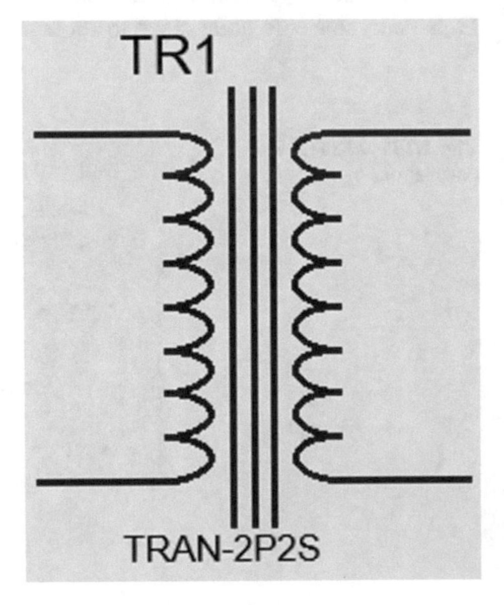

Fig. 1.348 Transformer with one primary and one secondary

The Proteus does not show the dots of “TRAN-2P2S” on it, however the location of dots is similar to what shown in Fig. 1.349.

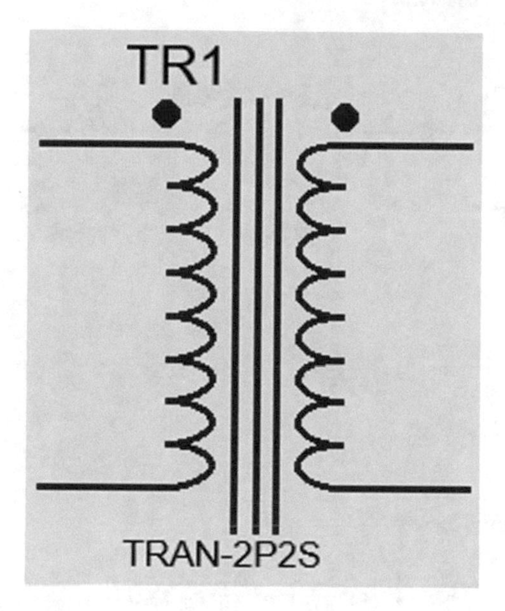

Fig. 1.349 Location of dots

The transformer shown in Fig. 1.348 can be added to the schematic by searching for “tran-2p2s” in the Pick Devices window (Fig. 1.350).

Pick Devices

Keywords:

TRAN-2P2S

Match whole words?

Show only parts with models?

Fig. 1.350 Searching for a transformer with one primary and one secondary

Let's simulate the circuit shown in Fig. 1.351. V_{in} is a step voltage and M is the mutual inductance between L_1 and L_2. The coupling coefficient between the two coils is $k = \frac{M}{\sqrt{L_1 L_2}} = \frac{0.9m}{\sqrt{1m \times 1.1m}} = 0.8581$.

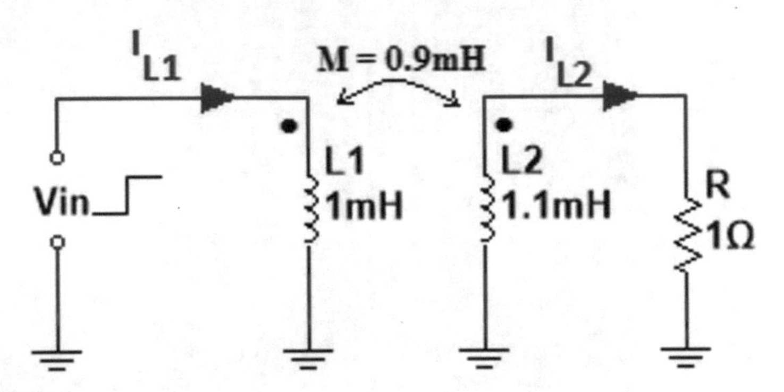

Fig. 1.351 Sample circuit

From basic circuit theory,

$$\begin{cases} L_1 \frac{di_{L_1}}{dt} - M \frac{di_{L_2}}{dt} = V_{in}(t) \\ Ri_{L_2} + L_2 \frac{di_{L_2}}{dt} - M \frac{di_{L_1}}{dt} = 0 \end{cases}$$

Take the Laplace transform of both side:

$$\begin{bmatrix} L_1 s & -Ms \\ -Ms & R + L_2 s \end{bmatrix} \times \begin{bmatrix} I_{L_1}(s) \\ I_{L_2}(s) \end{bmatrix} = \begin{bmatrix} V_{in}(s) \\ 0 \end{bmatrix}$$

So,

$$\begin{bmatrix} I_{L_1}(s) \\ I_{L_2}(s) \end{bmatrix} = \begin{bmatrix} L_1 s & -Ms \\ -Ms & R + L_2 s \end{bmatrix}^{-1} \times \begin{bmatrix} V_{in}(s) \\ 0 \end{bmatrix}$$

$V_{in}(s) = \frac{1}{s}$, so

$$\begin{bmatrix} I_{L_1}(s) \\ I_{L_2}(s) \end{bmatrix} = \begin{bmatrix} \frac{(11s + 10000) \times 10000}{s^2 \times (29s + 100000)} \\ \frac{90000}{s(29s + 100000)} \end{bmatrix}$$

You can use the following MATLAB commands to see the time domain graph of I_{L_1} and I_{L_2}.

```
s=tf('s');

I1=(11*s+10000)*10000/s/(29*s+100000);

I2=90000/(29*s+100000);

figure(1)

step(I1,[0:0.06/100:0.06]),grid on

figure(2)

step(I2), grid on
```

Output of the code is shown in Figs. 1.352 and 1.353.

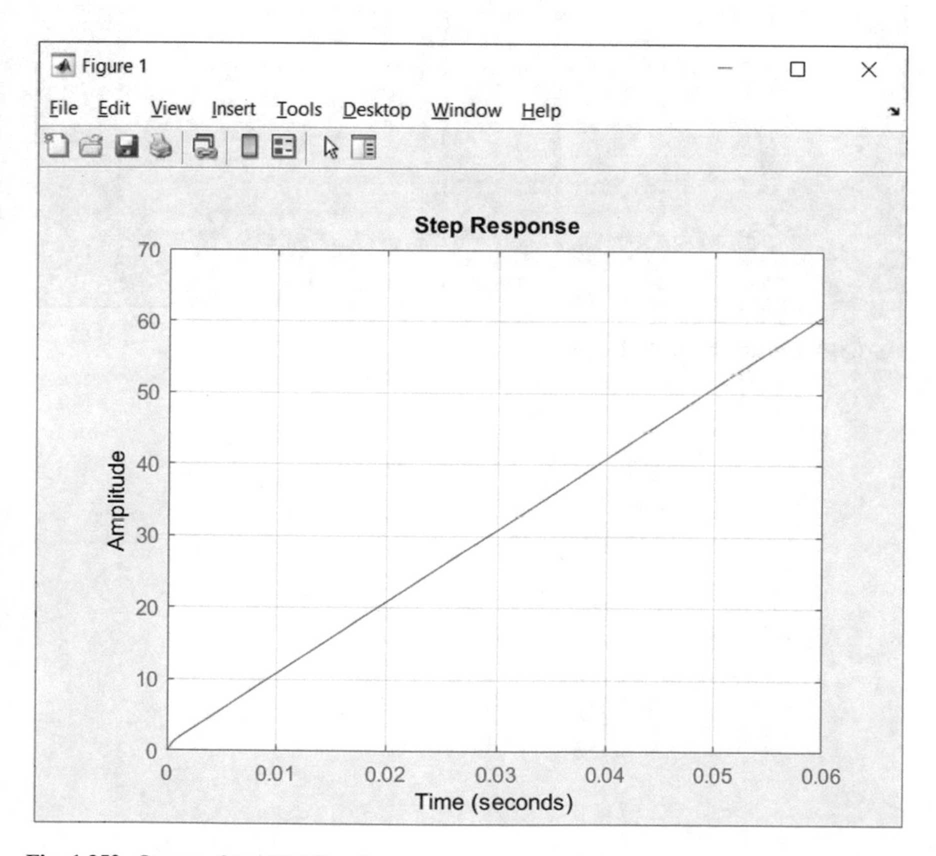

Fig. 1.352 Output of MATLAB code

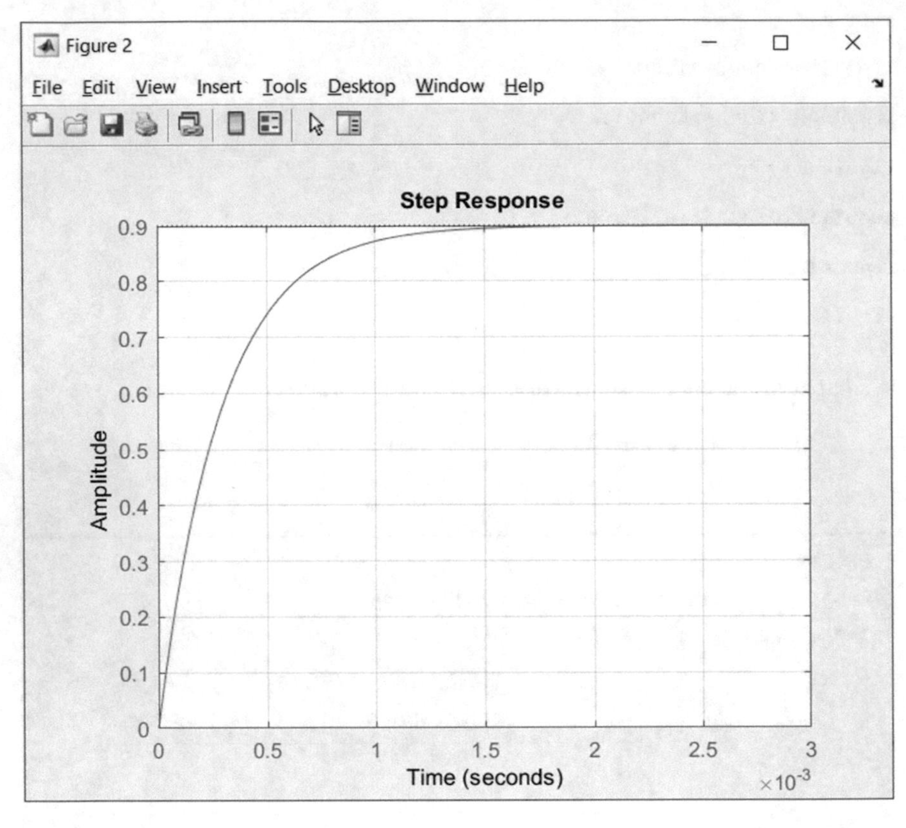

Fig. 1.353 Output of MATLAB code

Draw the schematic shown in Fig. 1.354. Settings of $TR1(P1)$ are shown in Fig. 1.355. The settings shown in Fig. 1.355 generate the voltage shown in Fig. 1.356. Settings of the transformer are shown in Fig. 1.357.

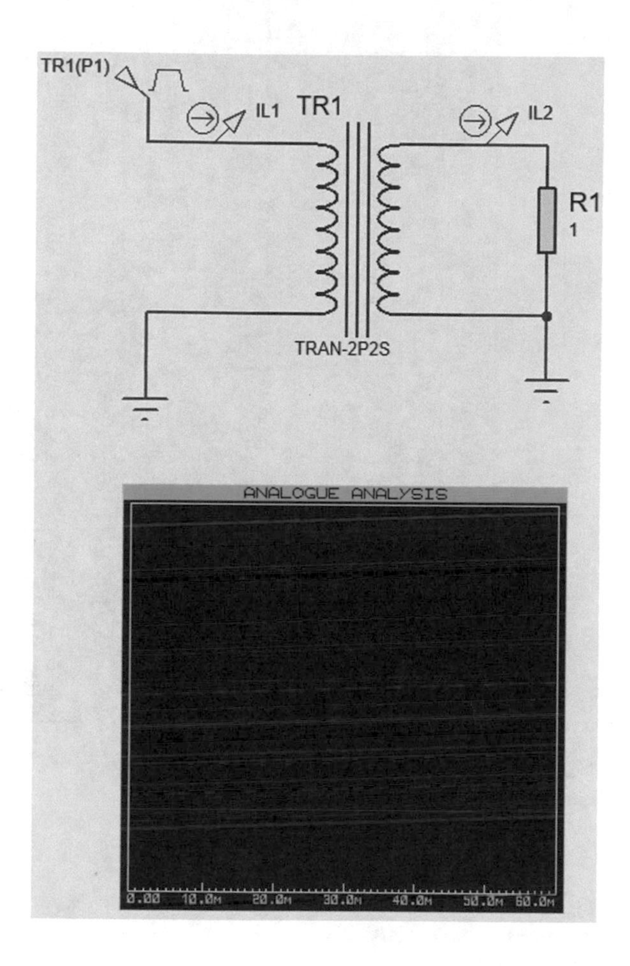

Fig. 1.354 Proteus equivalent of Fig. 1.351

Pulse Generator Properties ? ×

Generator Name: TR1(P1)

Analogue Types
DC
Sine
Pulse
Pwlin
File
Audio
Exponent
SFFM
Easy HDL

Digital Types
Steady State
Single Edge
Single Pulse
Clock
Pattern
Easy HDL

Current Source?
Isolate Before?
Manual Edits?
Hide Properties?

Initial (Low) Voltage: 0
Pulsed (High) Voltage: 1
Start (Secs): 0
Rise Time (Secs): 1n
Fall Time (Secs): 1n

Pulse Width:
Pulse Width (Secs): 1
Pulse Width (%):

Frequency/Period:
Frequency (Hz): 1
Period (Secs):
Cycles/Graph:

OK Cancel

Fig. 1.355 Settings of $TR1(P1)$

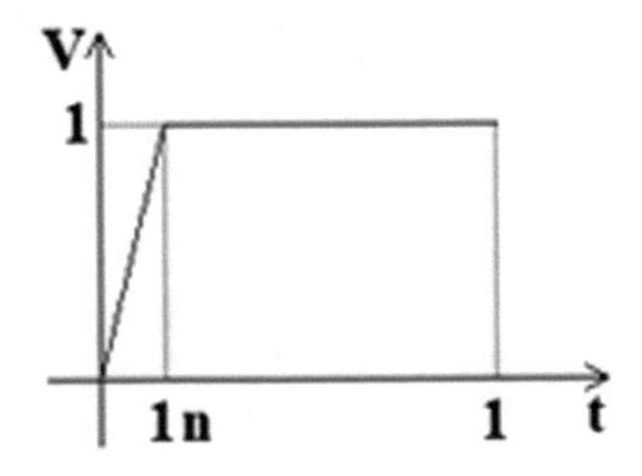

Fig. 1.356 Waveform generated with settings shown in Fig. 1.355

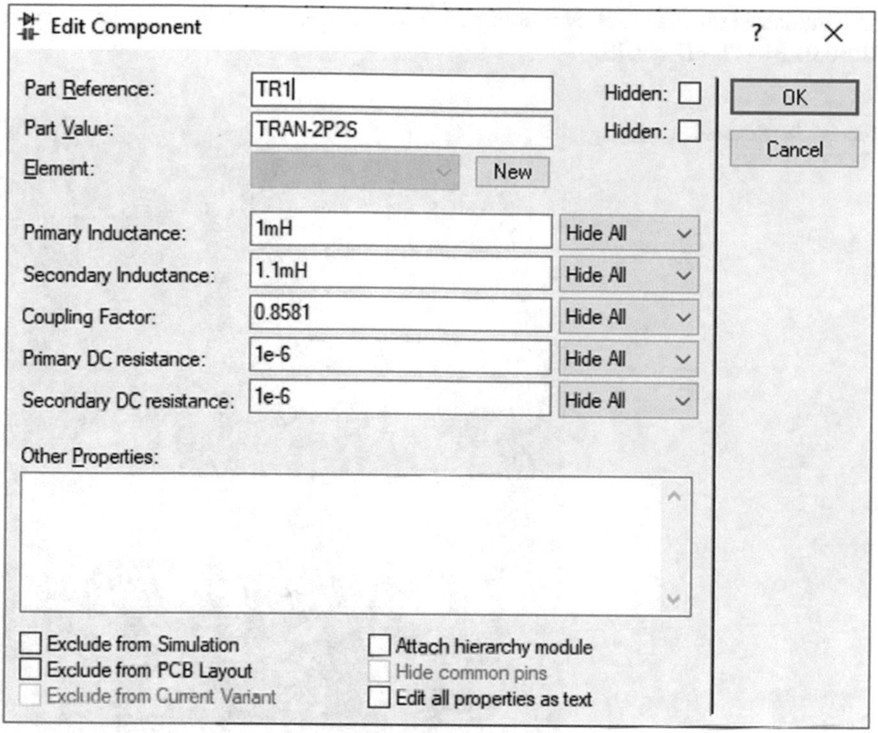

Edit Component ? ×

Part Reference:	TR1	Hidden: ☐	OK
Part Value:	TRAN-2P2S	Hidden: ☐	Cancel
Element:		New	
Primary Inductance:	1mH	Hide All	
Secondary Inductance:	1.1mH	Hide All	
Coupling Factor:	0.8581	Hide All	
Primary DC resistance:	1e-6	Hide All	
Secondary DC resistance:	1e-6	Hide All	

Other Properties:

☐ Exclude from Simulation
☐ Exclude from PCB Layout
☐ Exclude from Current Variant
☐ Attach hierarchy module
☐ Hide common pins
☐ Edit all properties as text

Fig. 1.357 Settings of transformer *TR*1

Simulation results are shown in Figs. 1.358 and 1.359. Obtained results are quite close to MATLAB results.

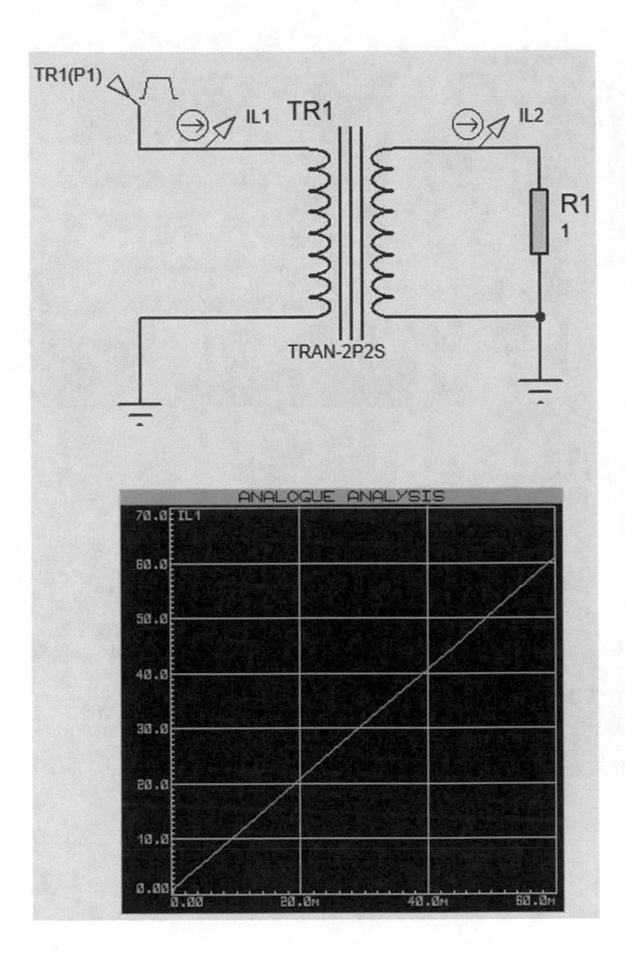

Fig. 1.358 Simulation result (I_{L_1})

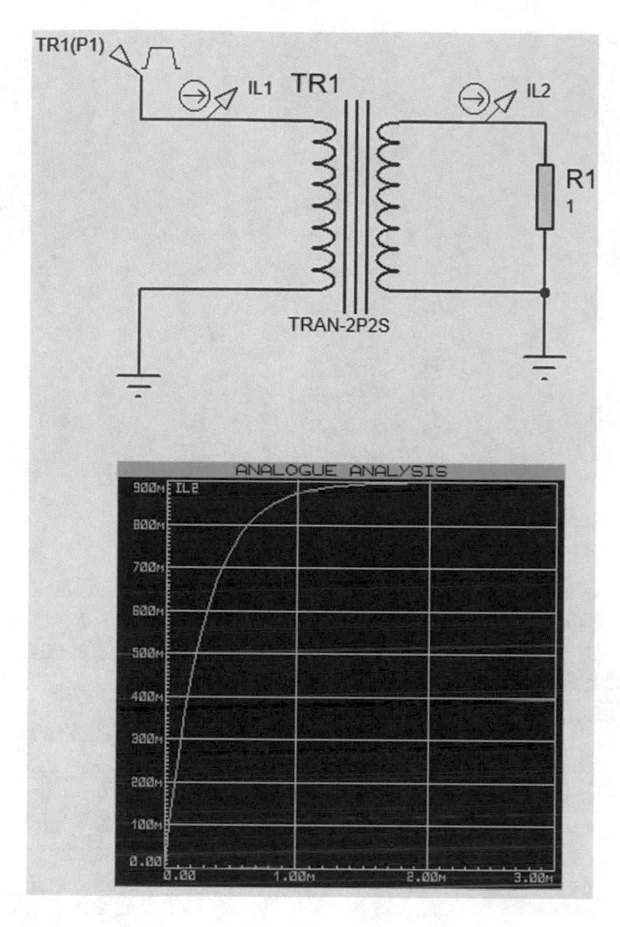

Fig. 1.359 Simulation result (I_{L_2})

1.48 Example 47: Single-Phase Transformer

In this example we want to simulate a single-phase transformer with one primary winding and one secondary winding. Draw the schematic shown in Fig. 1.360. Settings of V_{in} and transformer are shown in Figs. 1.361 and 1.362, respectively.

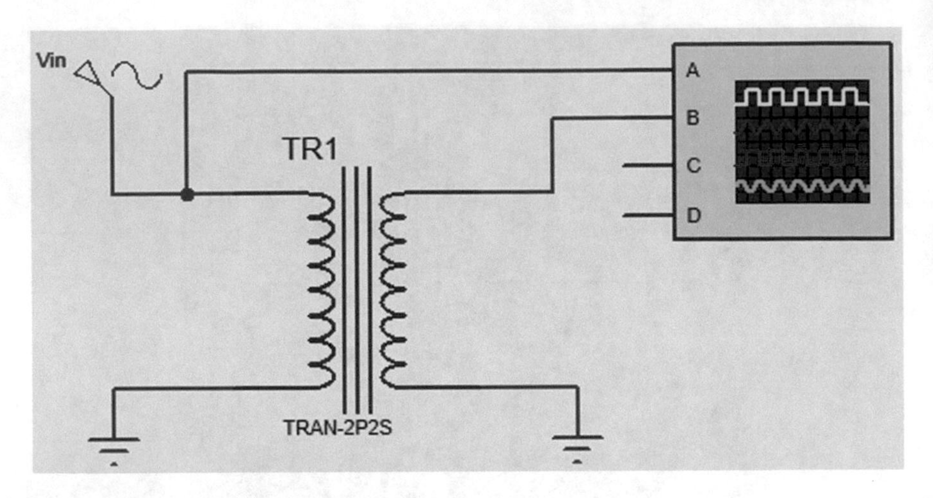

Fig. 1.360 Schematic of Example 47

Sine Generator Properties ? ×

Generator Name: Vin

Analogue Types: DC, Sine (selected), Pulse, Pwlin, File, Audio, Exponent, SFFM, Easy HDL

Digital Types: Steady State, Single Edge, Single Pulse, Clock, Pattern, Easy HDL

☐ Current Source?
☐ Isolate Before?
☐ Manual Edits?
☑ Hide Properties?

Offset (Volts): 0

Amplitude (Volts):
Amplitude (selected): 1
Peak-Peak:
RMS:

Timing:
Frequency (Hz) (selected): 50
Period (Secs):
Cycles/Graph:

Delay:
Time Delay (Secs):
Phase (Degrees) (selected): 0

Damping Factor (1/s): 0

OK Cancel

Fig. 1.361 Settings of V_{in}

The settings in Fig. 1.362 simulate a transformer with turn ratio of $\frac{N_p}{N_s} = \sqrt{\frac{L_p}{L_s}} = \sqrt{\frac{4H}{1H}} = 2$. Simulation result is shown in Fig. 1.363. Note that output voltage is half of input voltage since $\frac{V_p}{V_s} = \frac{N_p}{N_s} = 2$.

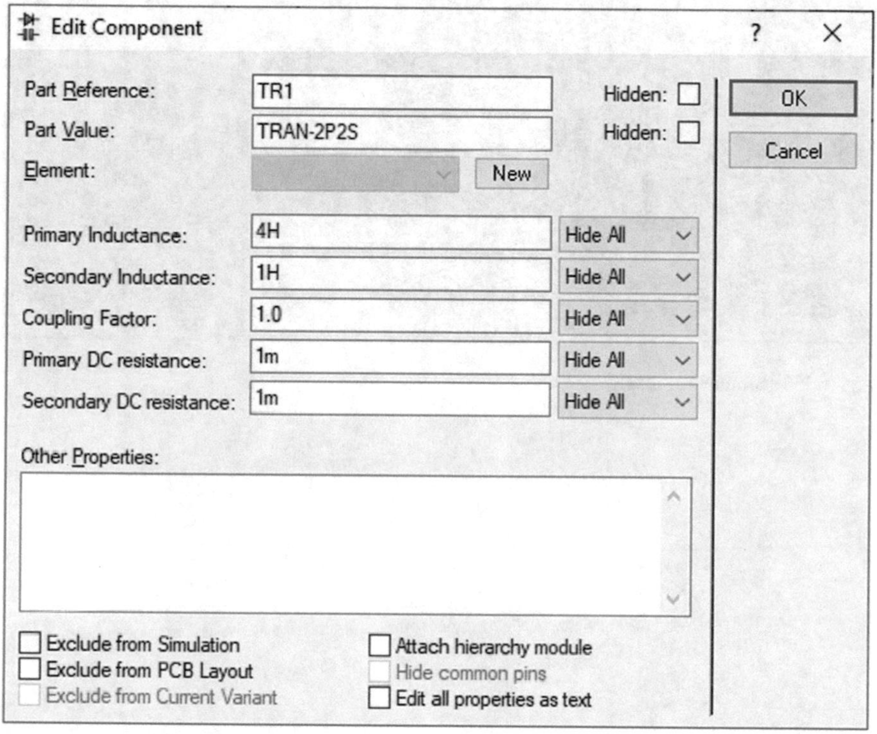

Fig. 1.362 Settings of transformer *TR*1

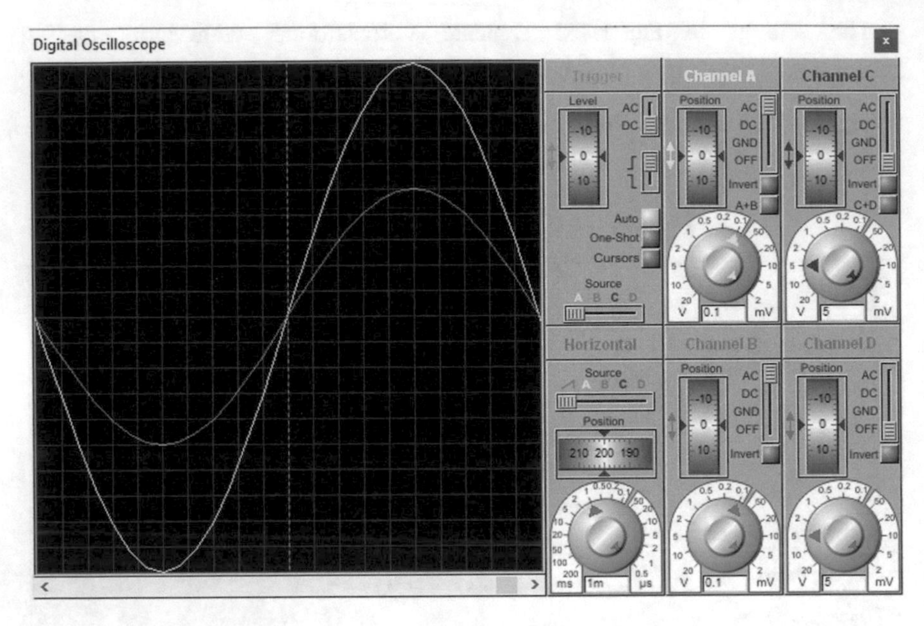

Fig. 1.363 Simulation result

The settings shown in Fig. 1.364 can be used to simulate a transformer with turn ratio of $\frac{N_p}{N_s} = 2$ as well. However, when the value primary and secondary inductances increase, the transformer become closer to the ideal case. Remember that primary and secondary inductances of an ideal transformer are infinity.

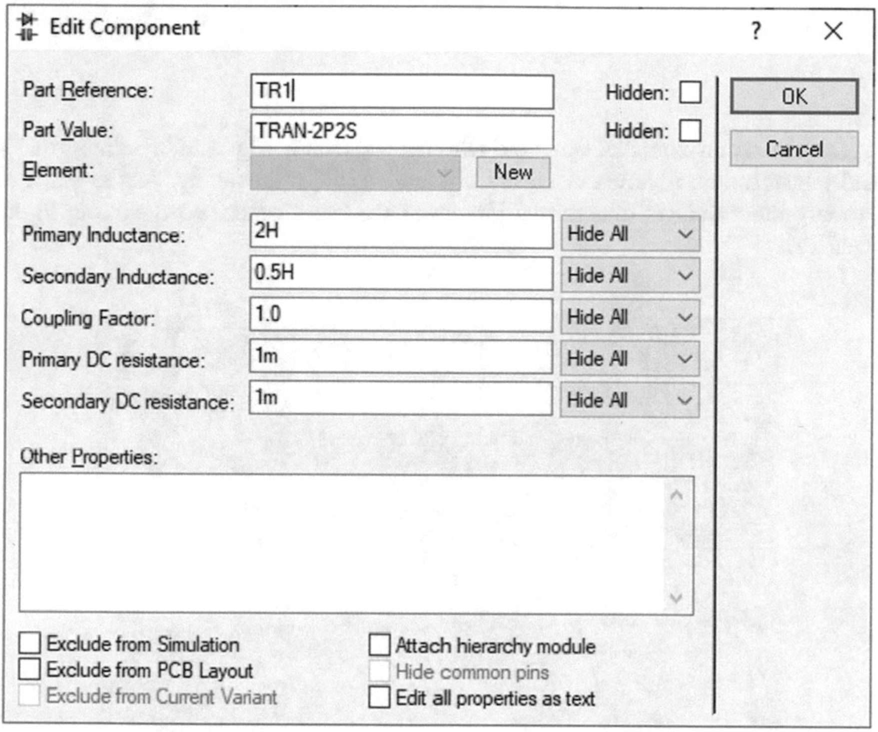

Fig. 1.364 Settings of transformer *TR*1

1.49 Example 48: Single-Phase Transformer with Two Outputs

In this example, we want to simulate a single-phase transformer with one primary winding and two secondary windings. A transformer with one primary winding and two secondary windings can be added to the schematic by searching for "tran-1p2s" in the Pick Devices window (Fig. 1.365).

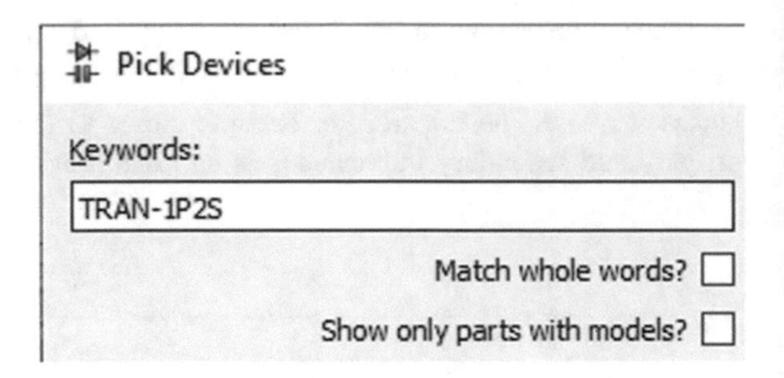

Fig. 1.365 Searching for a transformer with 1 primary and 2 secondary

Let's study an example. Draw the schematic shown in Fig. 1.366. Settings of V_{in} and transformer are shown in Figs. 1.367 and 1.368, respectively. As the value of Primary Inductance of transformer increases, the transformer become closer to the ideal case.

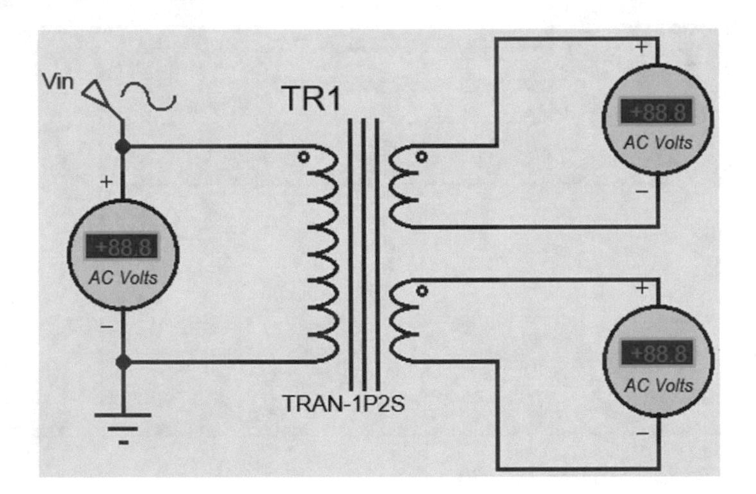

Fig. 1.366 Schematic for Example 48

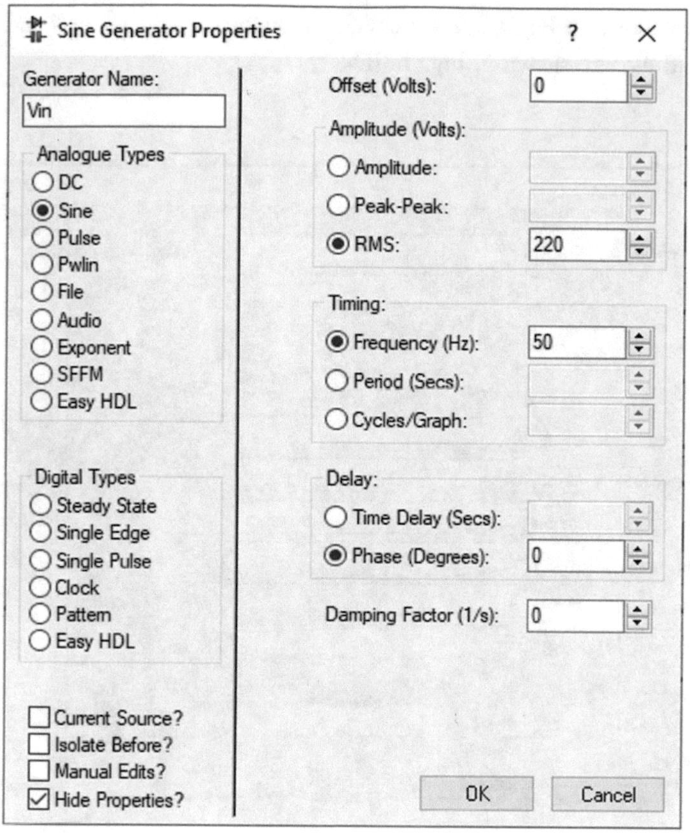

Sine Generator Properties ? ×

Generator Name: Vin

Analogue Types
- DC
- Sine (selected)
- Pulse
- Pwlin
- File
- Audio
- Exponent
- SFFM
- Easy HDL

Digital Types
- Steady State
- Single Edge
- Single Pulse
- Clock
- Pattern
- Easy HDL

- [ ] Current Source?
- [ ] Isolate Before?
- [ ] Manual Edits?
- [x] Hide Properties?

Offset (Volts): 0

Amplitude (Volts):
- Amplitude:
- Peak-Peak:
- RMS (selected): 220

Timing:
- Frequency (Hz) (selected): 50
- Period (Secs):
- Cycles/Graph:

Delay:
- Time Delay (Secs):
- Phase (Degrees) (selected): 0

Damping Factor (1/s): 0

OK Cancel

Fig. 1.367 Settings of V_{in}

Settings shown in Fig. 1.368 simulates a transformer with $\frac{N_p}{N_{s1}} = 5$ and $\frac{N_P}{N_{s2}} = 10$. N_p, N_{s1} and N_{s2} are shown in Fig. 1.369.

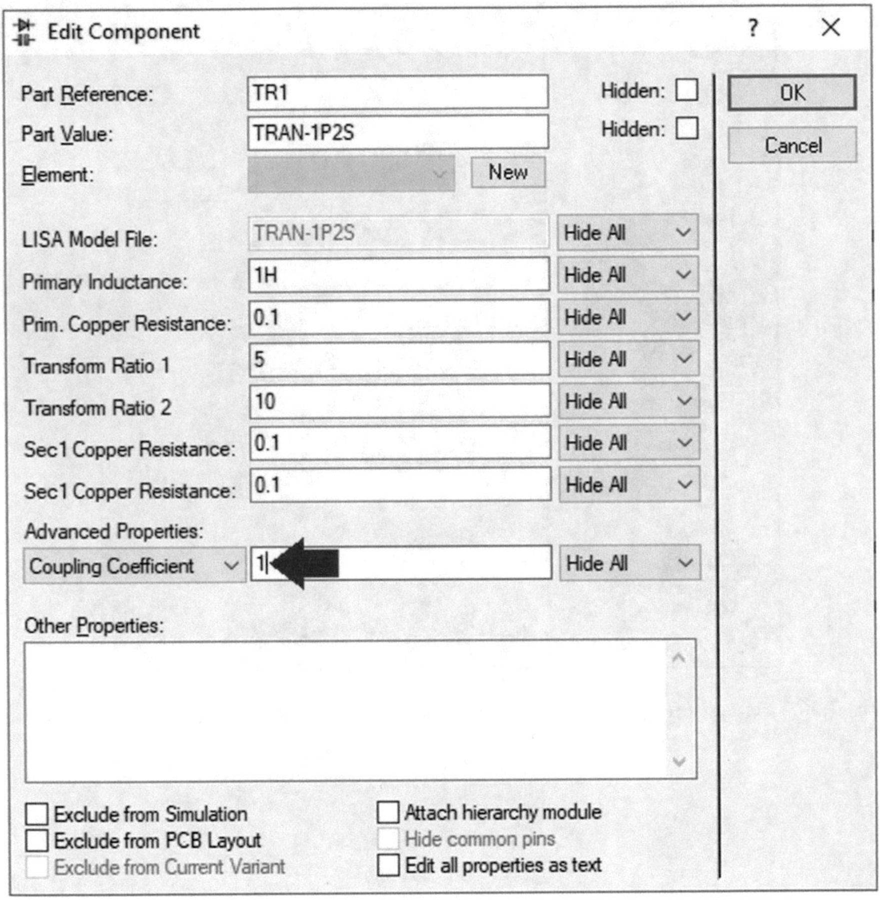

Fig. 1.368 Settings of *TR*1

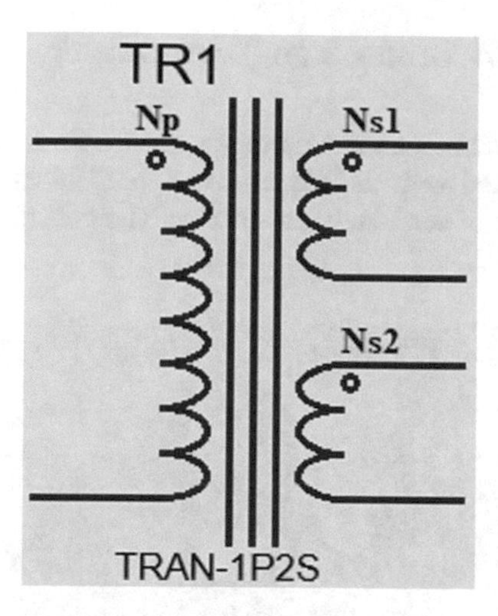

Fig. 1.369 Turns ratio for 1P2S transformer

Run the simulation. Simulation result is shown in Fig. 1.370. Note that $\frac{V_p}{V_{s1}} = \frac{N_p}{N_{s1}} = \frac{220}{44} = 5$ and $\frac{V_p}{V_{s2}} = \frac{N_p}{N_{s2}} = \frac{220}{22} = 10$.

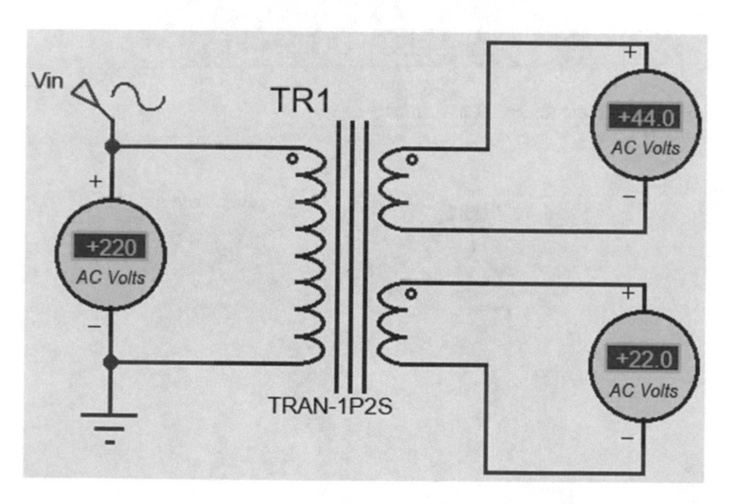

Fig. 1.370 Simulation result

1.50 Example 49: Center Tap Transformer

In this example, we want to simulate a center tap transformer. A center tap transformer can be simulated with the aid of "tran-2p3s" block (Fig. 1.371). Proteus does not show the dots of center tap transformer. The dots of center tap transformer are shown in Fig. 1.372.

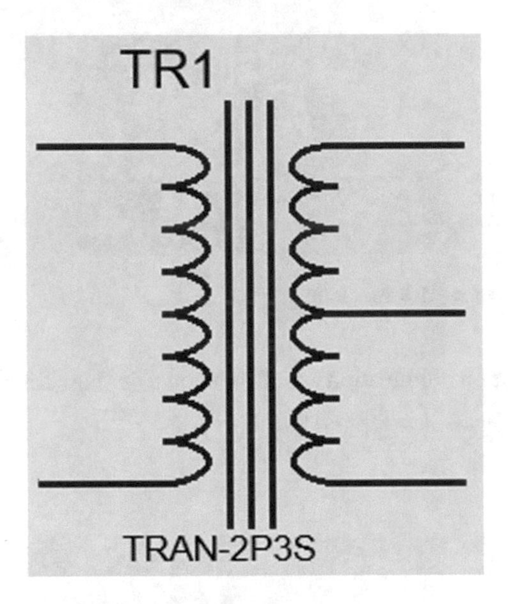

Fig. 1.371 Symbol for center tap transformer

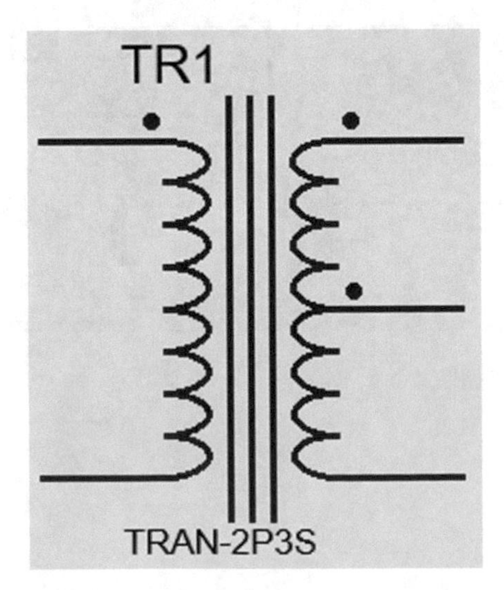

Fig. 1.372 Location of dots

Center tap transformer can be added to the schematic by searching for "tran-2p3s" in the Pick Devices window (Fig. 1.373).

Pick Devices

Keywords:

TRAN-2P3S

Match whole words?

Show only parts with models?

Fig. 1.373 Searching for 2P3S transformer

Let's study an example. Draw the schematic shown in Fig. 1.374. Settings of V_{in} and the transformer are shown in Figs. 1.375 and 1.376, respectively.

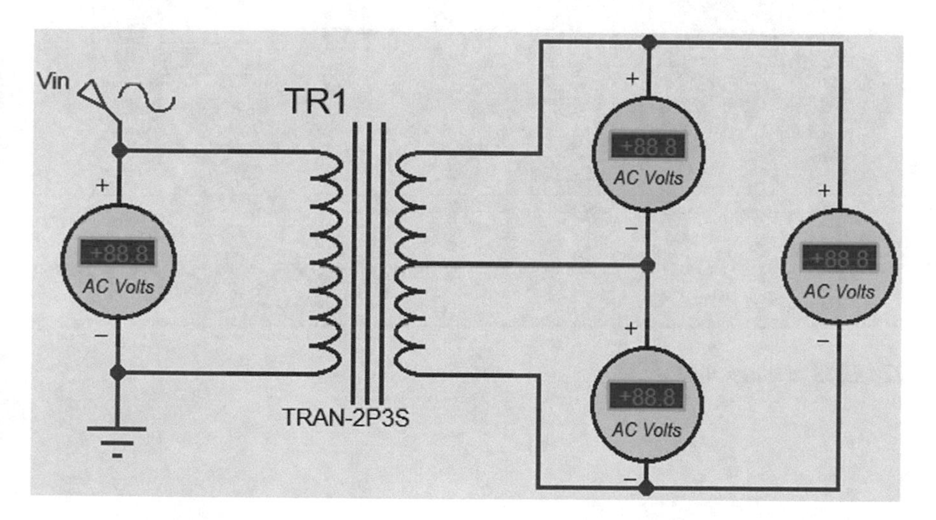

Fig. 1.374 Schematic for Example 49

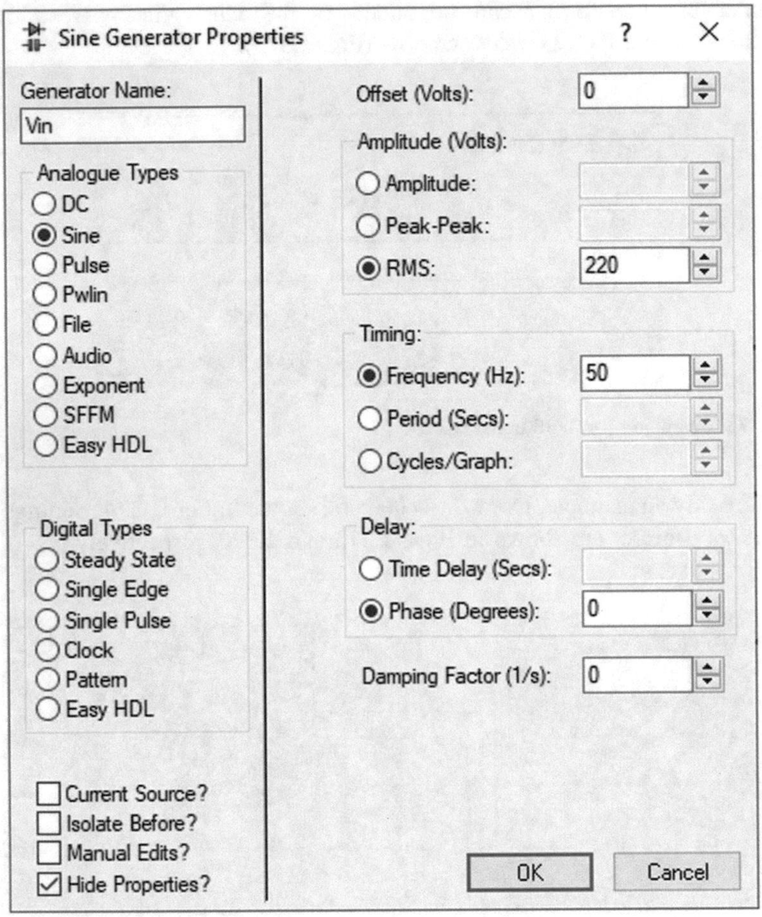

Fig. 1.375 Settings for V_{in}

Edit Component

Part Reference:	TR1	Hidden: ☐
Part Value:	TRAN-2P3S	Hidden: ☐
Element:		New
Primary Inductance:	1H	Hide All
Total Secondary Inductance:	.006H	Hide All
Coupling Factor:	1	Hide All
Primary DC resistance:	1m	Hide All
Secondary DC resistance:	1m	Hide All

OK
Cancel

Other Properties:

☐ Exclude from Simulation
☐ Exclude from PCB Layout
☐ Exclude from Current Variant
☐ Attach hierarchy module
☐ Hide common pins
☐ Edit all properties as text

Fig. 1.376 Settings for transformer *TR*1

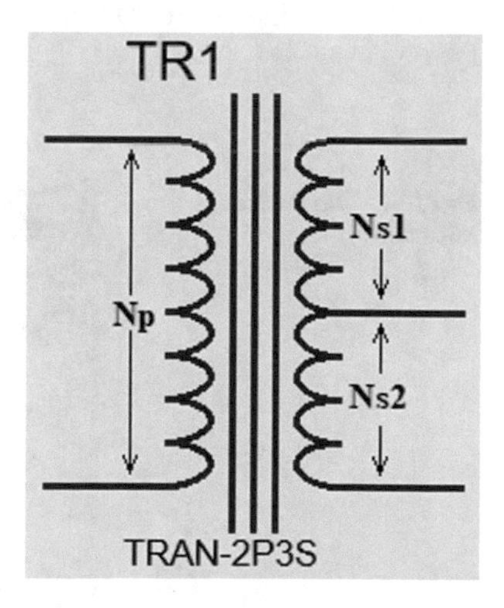

Fig. 1.377 Turns ratio for 2P3S transformer

The settings shown in Fig. 1.376 simulates a transformer with turns ration of $\frac{N_p}{N_{s1}} = \frac{N_p}{N_{s2}} = \sqrt{\frac{L_p}{\frac{L_s}{2}}} = \sqrt{\frac{1}{\frac{0.006}{2}}} = \sqrt{333.333} = 18.257$. Note that $N_{s1} = N_{s2}$. N_p, N_{s1} and N_{s2} are shown in Fig. 1.377.

Simulation result is shown in Fig. 1.378.

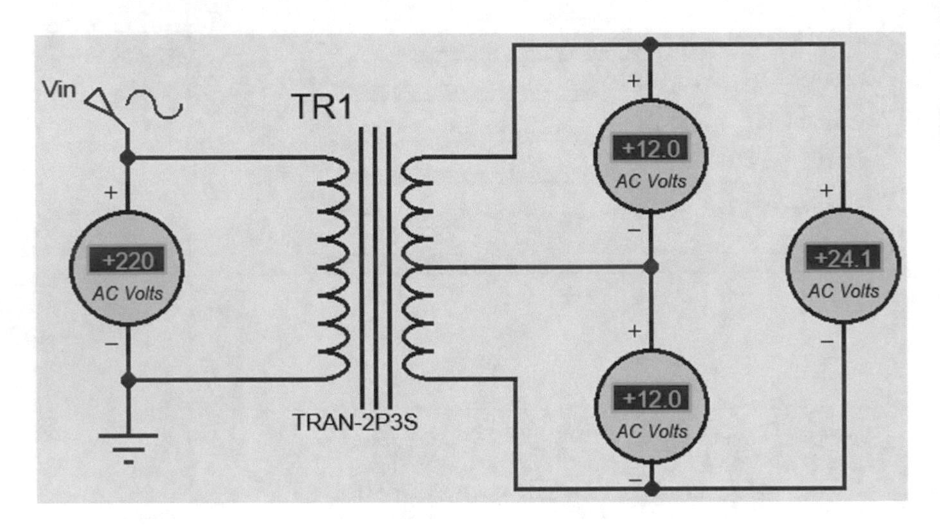

Fig. 1.378 Simulation result

Let's check the obtained result. The calculations shown in Fig. 1.379 shows that Proteus result is correct.

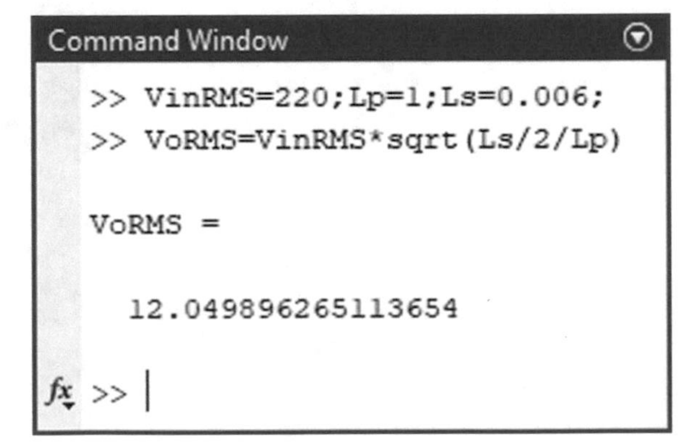

Fig. 1.379 MATLAB calculations

1.51 Example 50: Three-Phase Transformer

In this example we want to simulate a three-phase transformer. Three-phase transformers can be simulated with the aid of “tran-3phase” block (Fig. 1.380). The three windings of primary and secondary can be connected in start (*Y*) or delta (Δ) configurations.

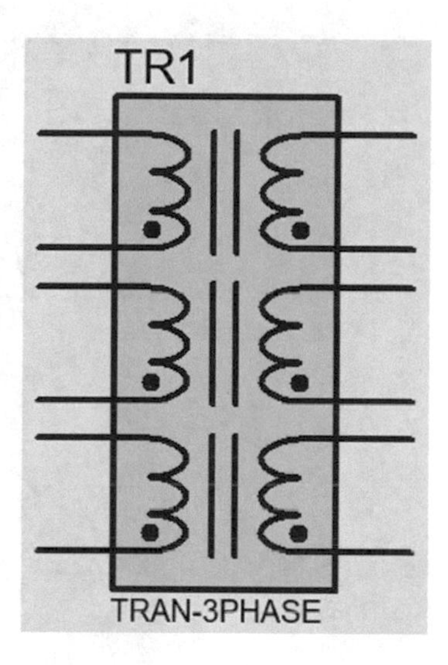

Fig. 1.380 Proteus symbol for three-phase transformer

A three-phase transformer can be added to the schematic by searching for “tran-3phase” in the Pick Devices window (Fig. 1.381).

Pick Devices

Keywords:

TRAN-3PHASE

Match whole words?

Show only parts with models?

Fig. 1.381 Searching for three-phase transformer

Let's study an example. Draw the schematic shown in Fig. 1.382. Settings of V_1 and the transformer are shown in Figs. 1.383 and 1.384, respectively. As the value of Primary Inductance of transformer increases, the transformer become closer to the ideal case.

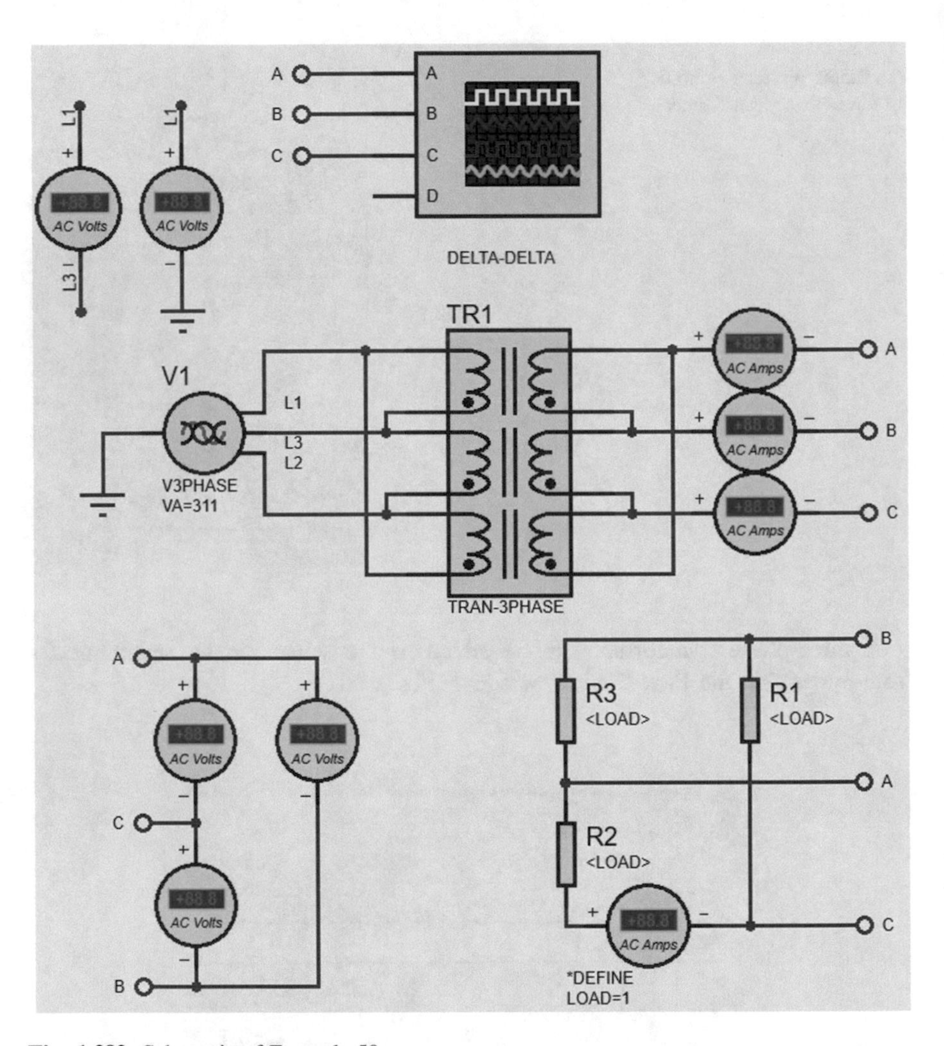

Fig. 1.382 Schematic of Example 50

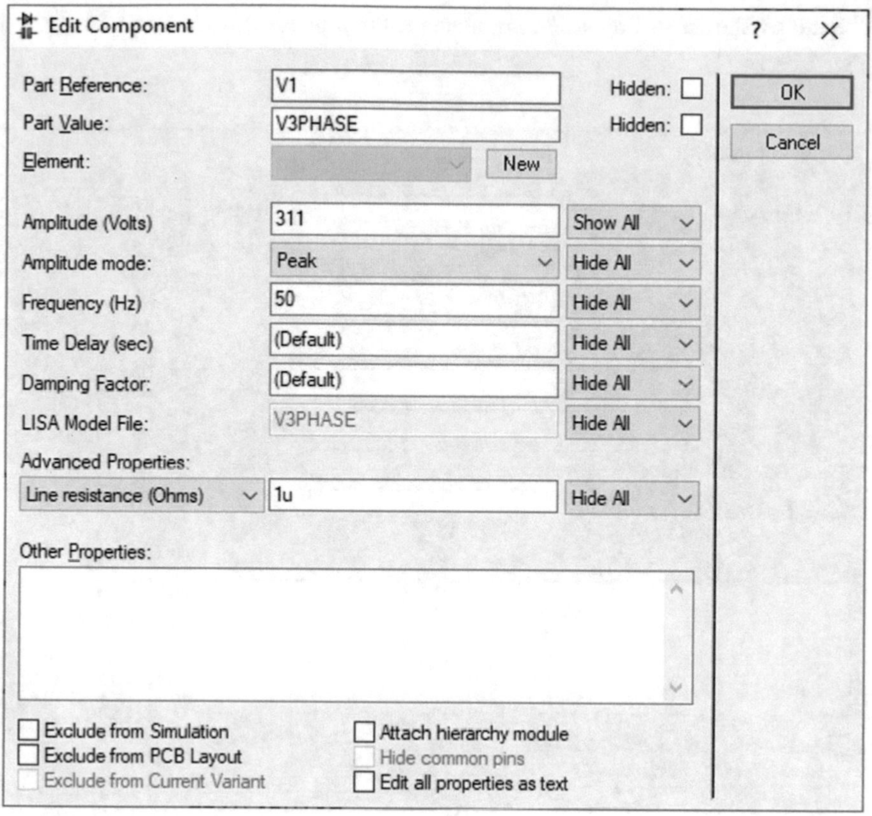

Edit Component

Part Reference: V1 Hidden:
Part Value: V3PHASE Hidden:
Element: New
OK
Cancel
Amplitude (Volts) 311 Show All
Amplitude mode: Peak Hide All
Frequency (Hz) 50 Hide All
Time Delay (sec) (Default) Hide All
Damping Factor: (Default) Hide All
LISA Model File: V3PHASE Hide All
Advanced Properties:
Line resistance (Ohms) 1u Hide All
Other Properties:
Exclude from Simulation
Exclude from PCB Layout
Exclude from Current Variant
Attach hierarchy module
Hide common pins
Edit all properties as text

Fig. 1.383 Setting of three-phase source V_1

Settings shown in Fig. 1.384 simulates a three-phase transformer with $\frac{N_p}{N_s} = \frac{10}{1}$ (Fig. 1.385).

Edit Component ? ×

Part Reference: TR1 Hidden: ☐ OK

Part Value: TRAN-3PHASE Hidden: ☐ Device Notes

Element: New Cancel

LISA Model File: TRAN-3PHASE Hide All

Primary Inductance (H) 5H Hide All

Turns Ratio (N:1) 10 Hide All

Primary Resistance (Ohms) 1m Hide All

Secondary Resistance (Ohms) 0.01m Hide All

Other Properties:

☐ Exclude from Simulation ☐ Attach hierarchy module

☐ Exclude from PCB Layout Hide common pins

Exclude from Current Variant ☐ Edit all properties as text

Fig. 1.384 Settings of transformer *TR*1

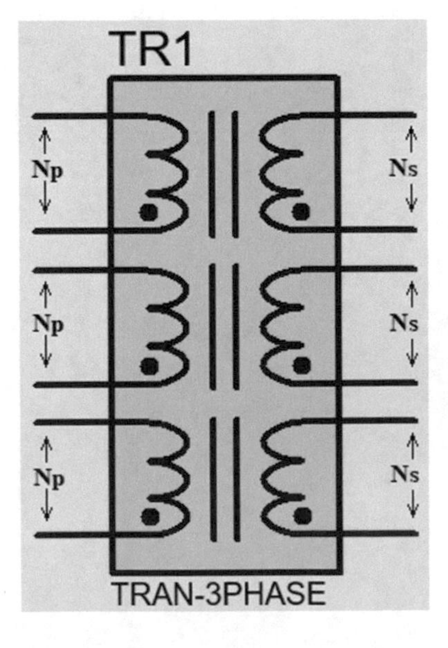

Fig. 1.385 Turns ratio for three-phase transformer

Run the simulation. The simulation result is shown in Fig. 1.386. Note that $\frac{V_p}{V_s}$ is around 10 as expected.

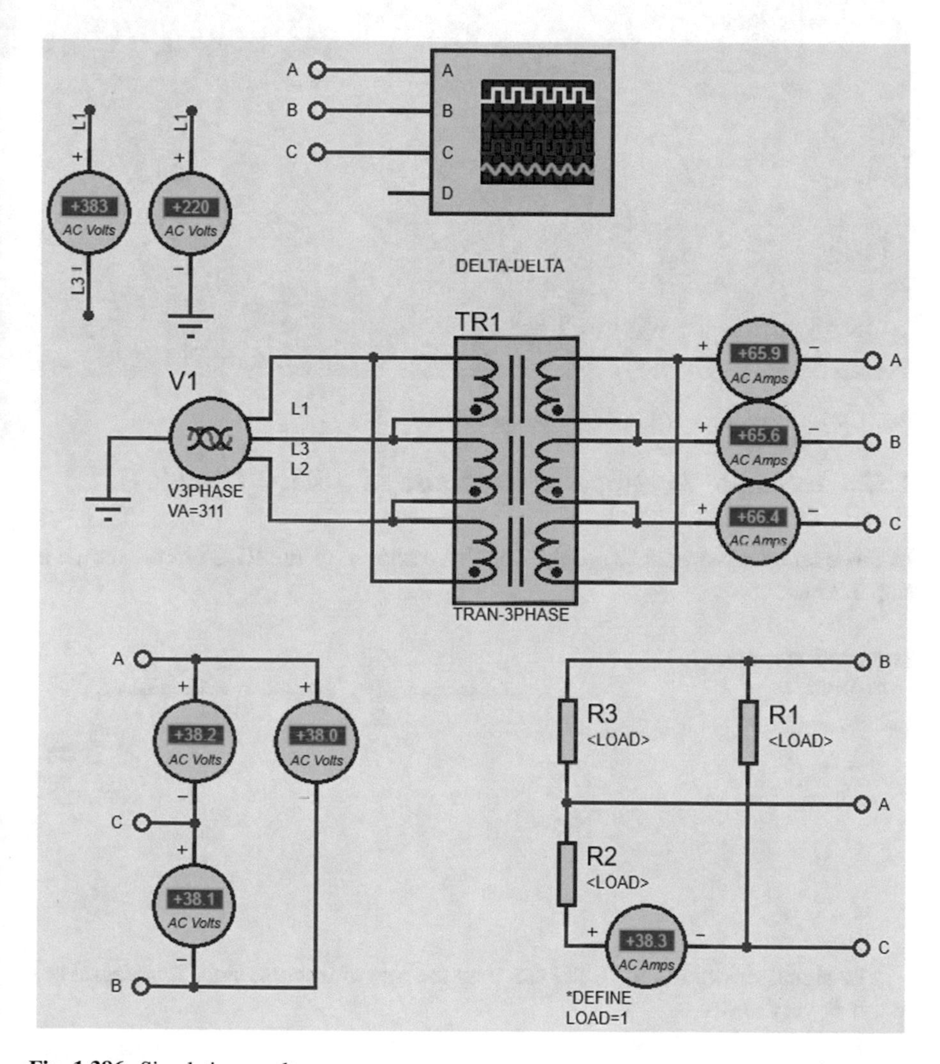

Fig. 1.386 Simulation result

Schematic shown in Fig. 1.387 shows a three-phase transformer with delta-*Y* connections.

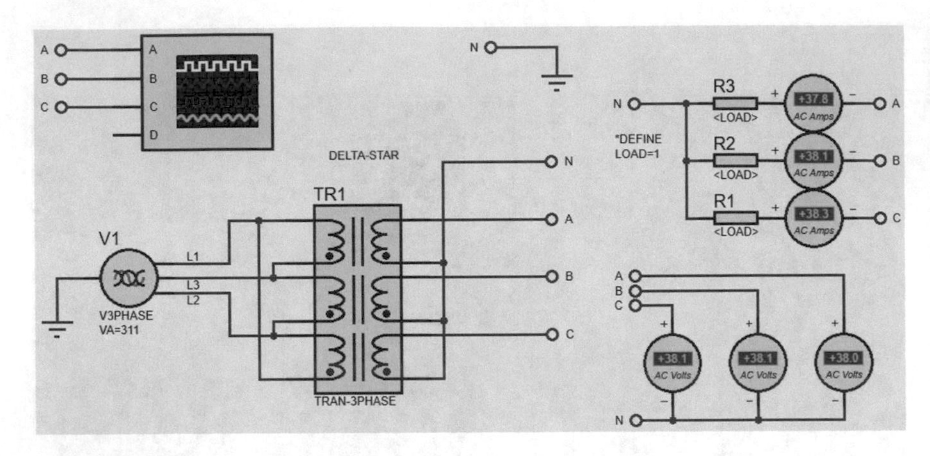

Fig. 1.387 Transformer *TR*1 is delta-*Y* connected

1.52 Example 51: Impulse Response of a RLC Circuit (I)

In this example, we want to see the impulse response of the RLC circuit shown in Fig. 1.388.

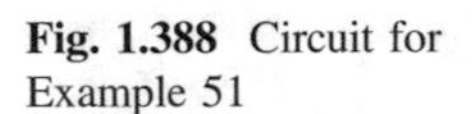
Fig. 1.388 Circuit for Example 51

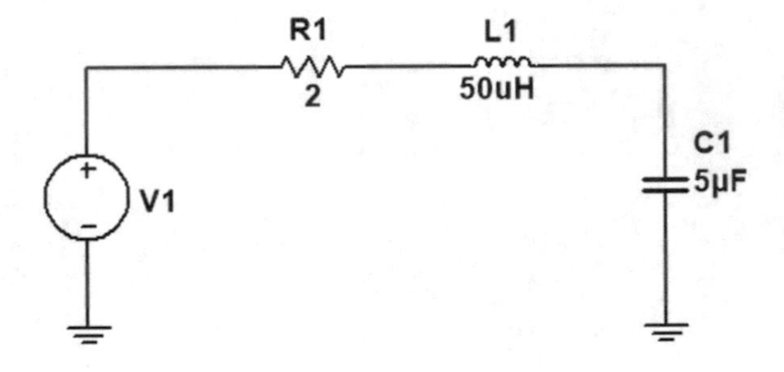

The signal shown in Fig. 1.389 can play the role of impulse since its integral is 1 and it is very short.

Fig. 1.389 Impulse input. Area under the pulse equals to 1

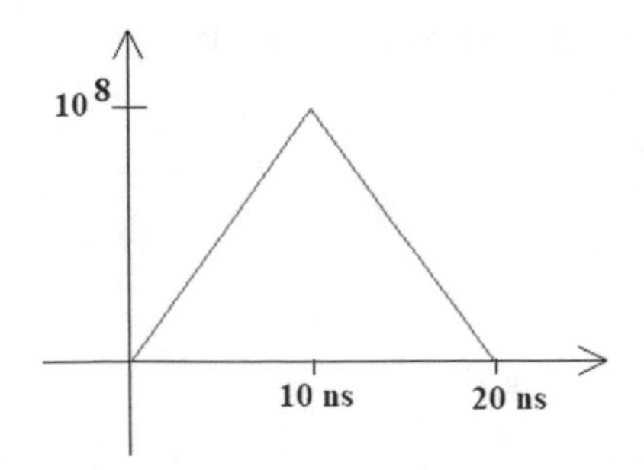

You can apply a signal with smaller amplitude to the circuit as well. For instance, you can apply the signal shown in Fig. 1.390 to the circuit and multiply the output response by 10^7 to obtain the unit impulse response (note that this circuit is linear and obeys the superposition principle).

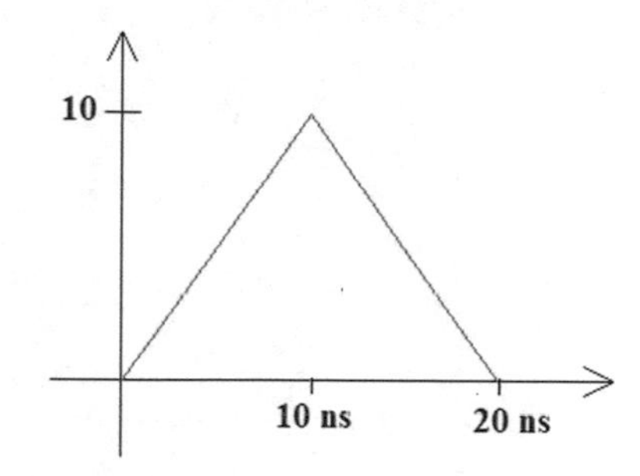

Fig. 1.390 Area under the pulse is not equal to 1

The signal shown in Fig. 1.389 can be generated with the aid of a PWLIN generator (Fig. 1.391).

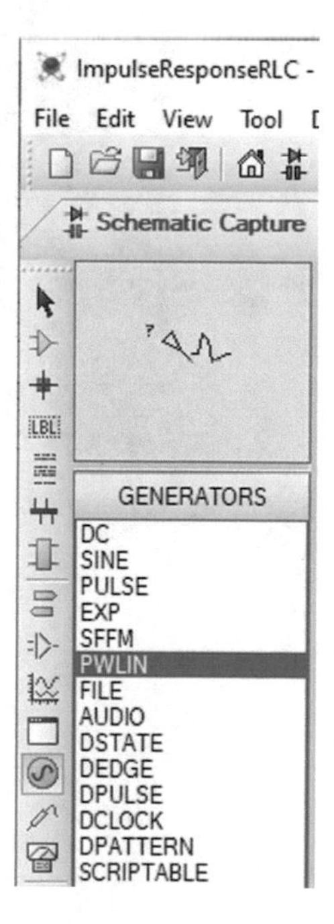

Fig. 1.391 PWLIN generator block

Draw the schematic shown in Fig. 1.392.

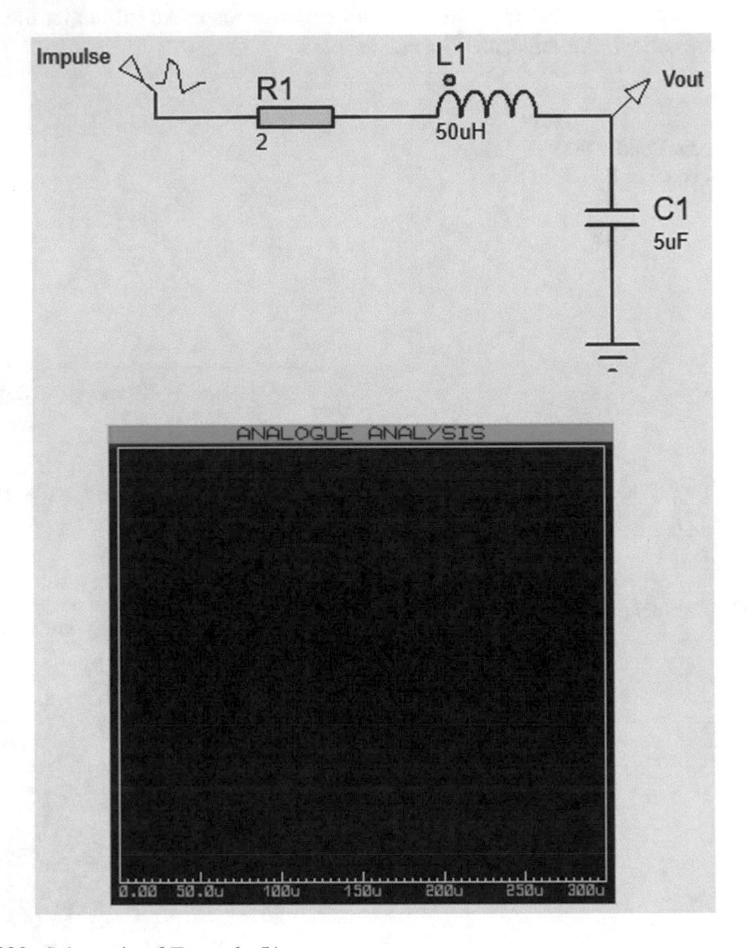

Fig. 1.392 Schematic of Example 51

Double click the Impulse source (Fig. 1.393). Use the mouse to draw the required signal (Fig. 1.394).

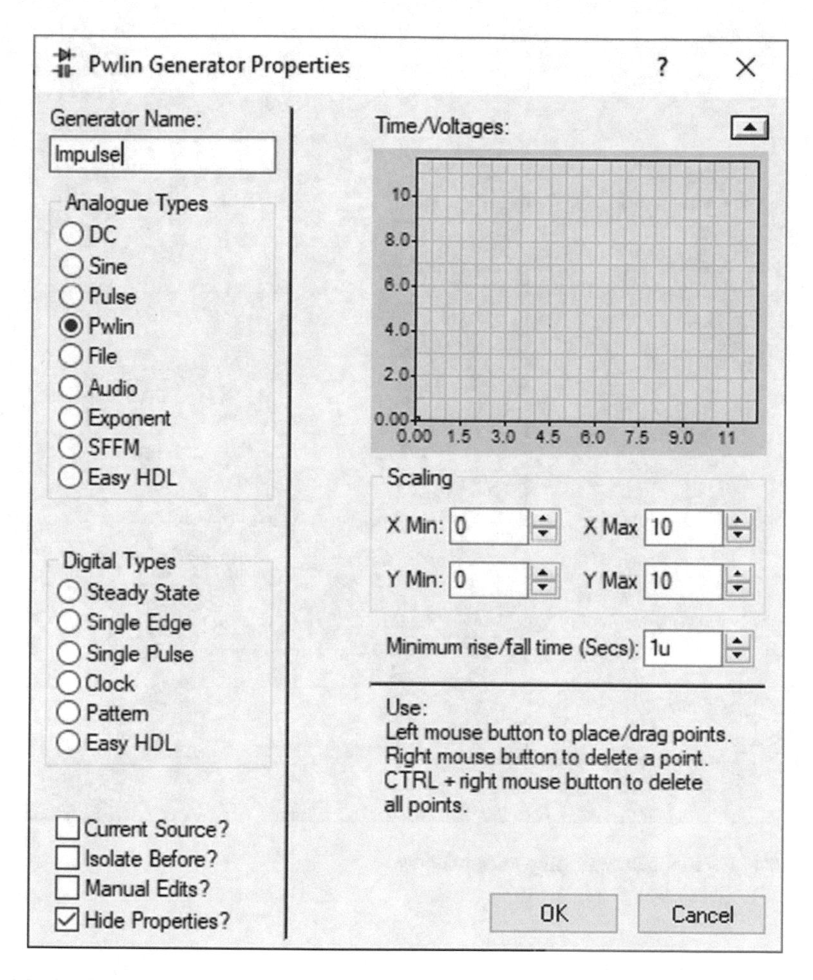

Fig. 1.393 PWLIN generator properties window

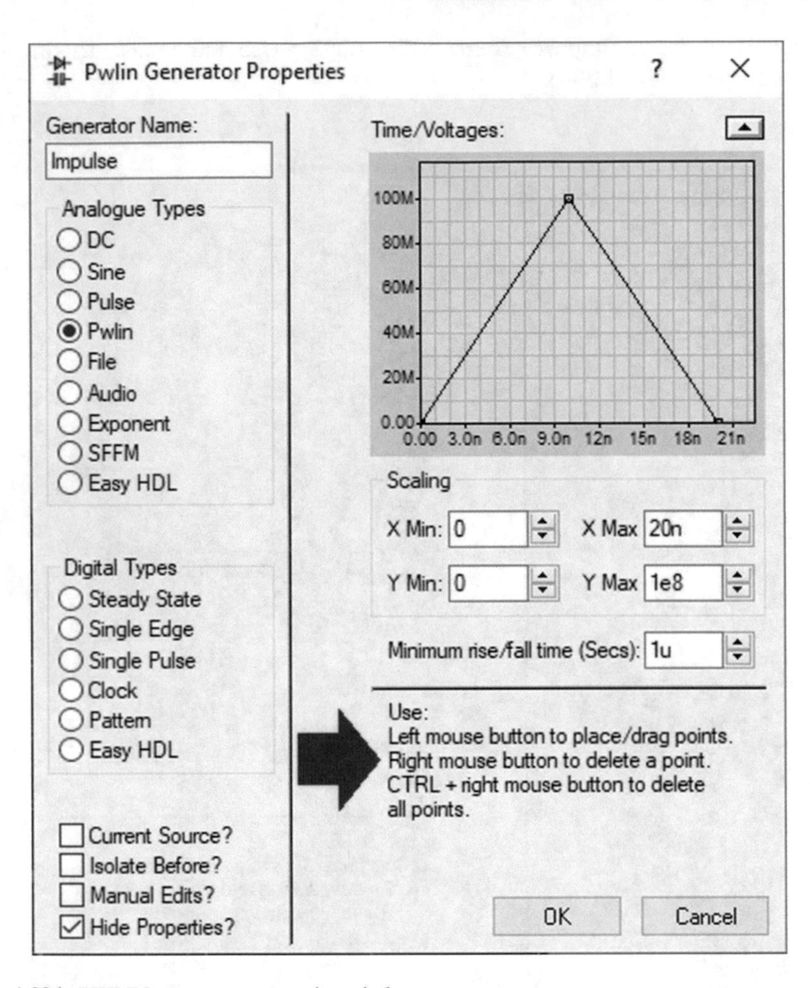

Fig. 1.394 PWLIN generator properties window

Run the simulation. The simulation result is shown in Fig. 1.395.

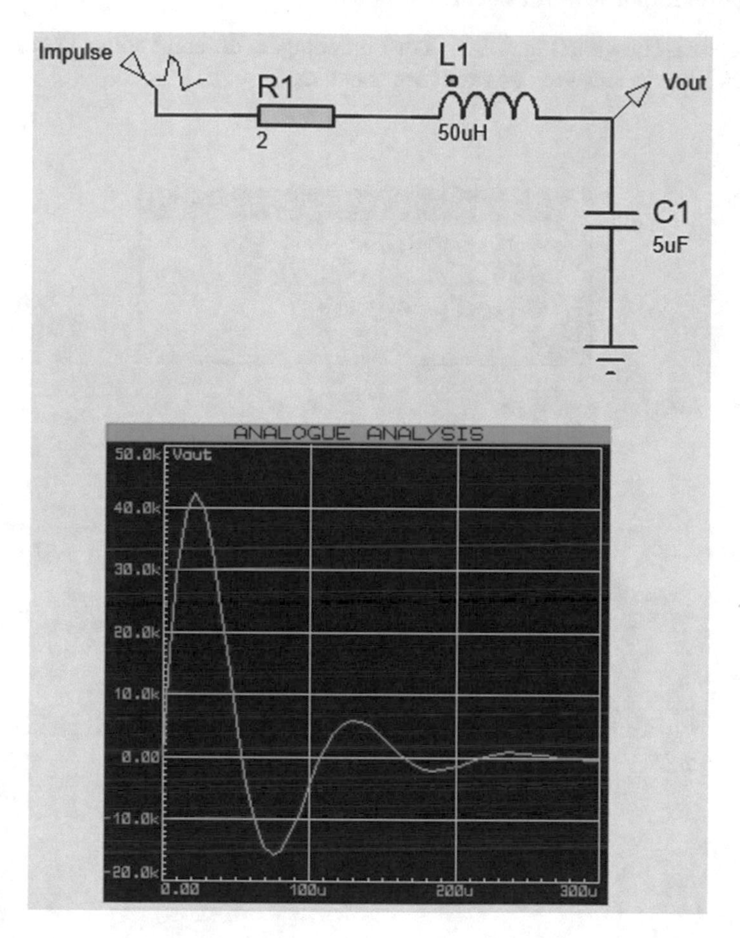

Fig. 1.395 Simulation result

Let's check the obtained result. The MATLAB code shown in Fig. 1.396 draws the impulse response of the circuit $\left(\frac{V_{C_1}(s)}{V_1(s)} = \frac{\frac{1}{C_1 s}}{R_1 + L_1 s + \frac{1}{C_1 s}} = \frac{1}{L_1 C_1 s^2 + R_1 C_1 s + 1}\right)$. Output of this code is shown in Fig. 1.397. You can compare different points in order to see MATLAB result and Proteus result are the same.

```
Command Window
>> R1=2;L1=50e-6;C1=5e-6;
>> H=tf([1],[L1*C1 R1*C1 1]);
>> impulse(H,0.5e-3)
fx >>
```

Fig. 1.396 MATLAB commands

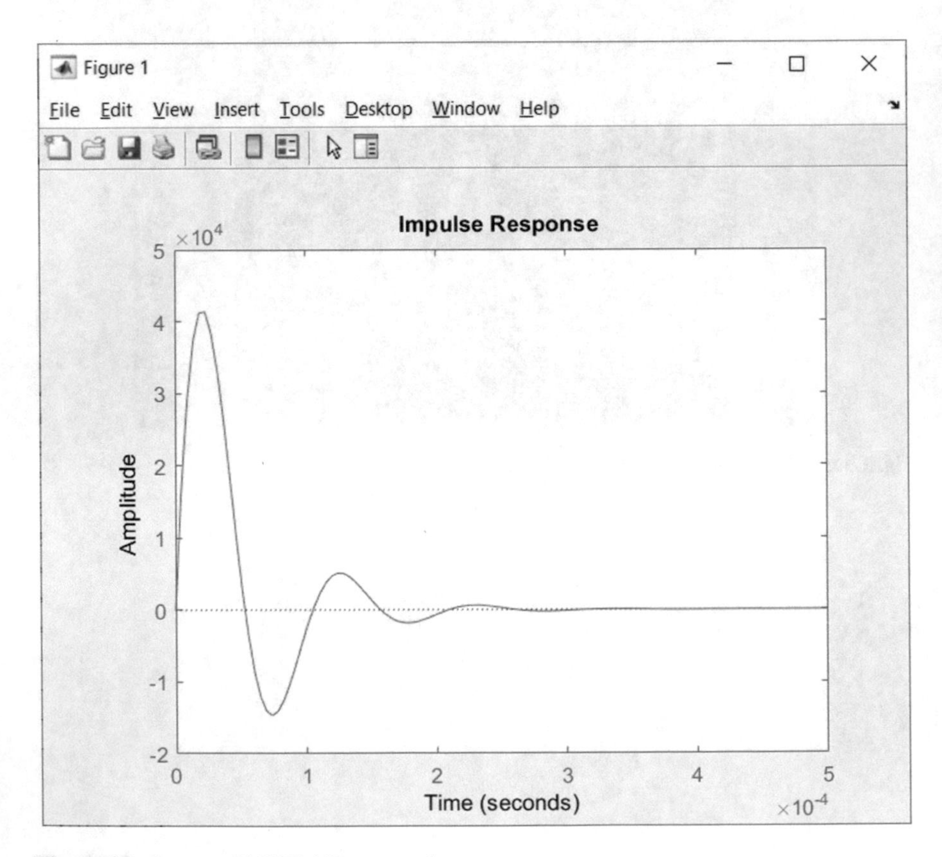

Fig. 1.397 Output of MATLAB commands

1.53 Example 52: Impulse Response of an RLC Circuit (II)

We observed the unit impulse response of a RLC circuit in previous example. In this example we want to obtain a mathematical equation for the graph that obtained in the previous example. We use MATLAB to solve this problem.

The waveform shown in Fig. 1.395 is imported into the MATLAB environment (Figs. 1.398 and 1.399).

```
Command Window
>> F=importdata('C:\Proteus Book\ImpulseResponse.dat');
>> time=F.data(:,1);
>> Vout=F.data(:,2);
>> plot(time,Vout)
fx >>
```

Fig. 1.398 MATLAB commands

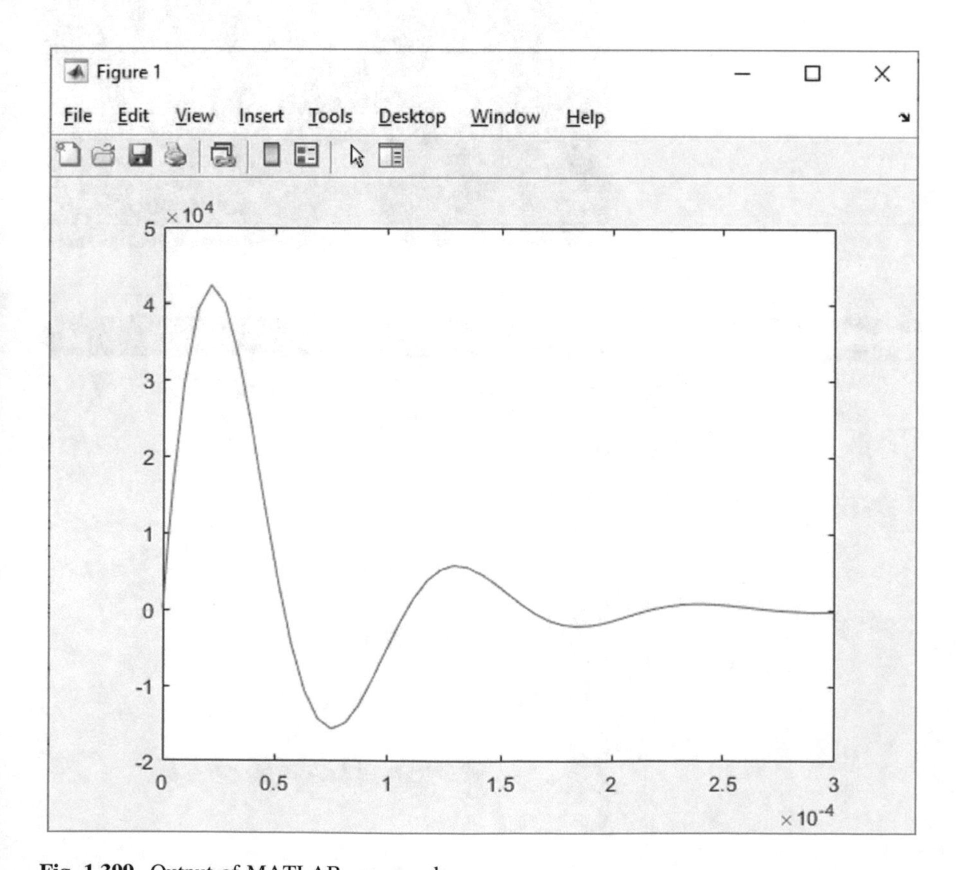

Fig. 1.399 Output of MATLAB commands

Let's obtain the mathematical equation of the graph shown in Fig. 1.400. From basic circuit theory we know that such a response belongs to function $f(t) = ae^{-bt}\sin(\omega t)$. Value of ω can be found easily. Just maximize the graph shown in Fig. 1.400 and use a cursor to measure the frequency of the waveform. According to Fig. 1.401, $\omega = 5.7727 \times 10^4$. So, $f(t) = ae^{-bt}\sin(5.7727 \times 10^4 t)$. Now we need to find the values of a and b.

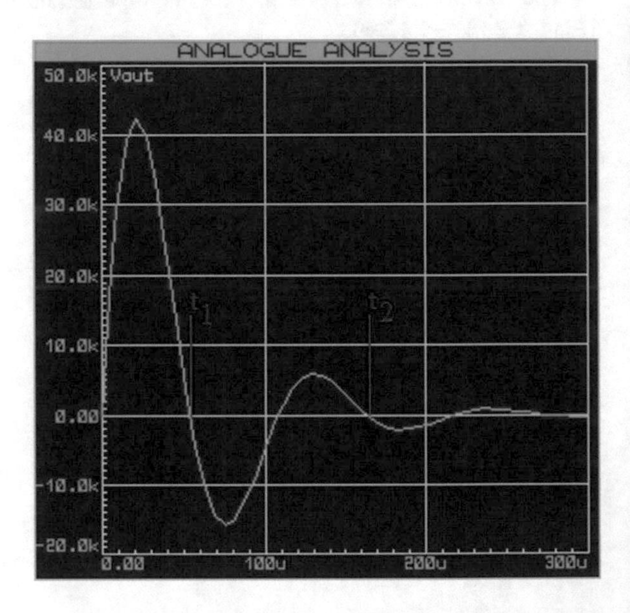

Fig. 1.400 Measurement of time difference between two roots

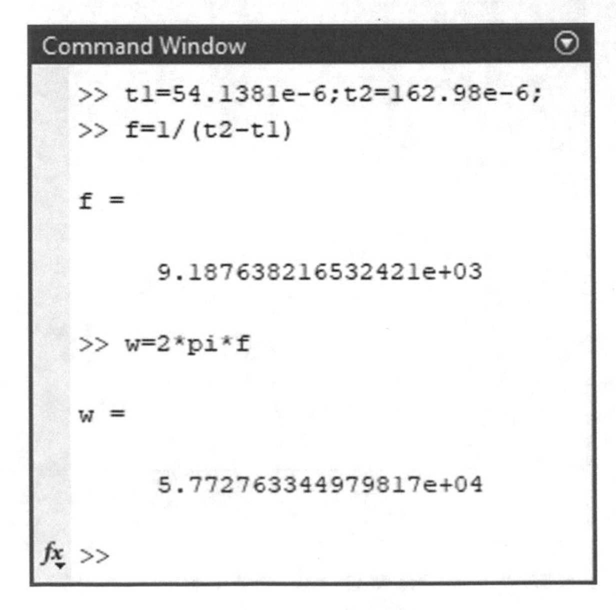

Fig. 1.401 MATLAB calculations

Let's use Curve Fitting Toolbox® to find the best values for a and b. Run the Curve Fitting Toolbox with the aid of c_{ftool} command (Fig. 1.402) and do the settings similar to Fig. 1.403. According to Fig. 1.403, $a = 6.668 \times 10^4$ and $b = 1.852 \times 10^4$. So, mathematical equation of impulse response of the circuit is $6.668 \times 10^4 e^{-1.852\times10^4 t}\sin(5.7727 \times 10^4 t)$.

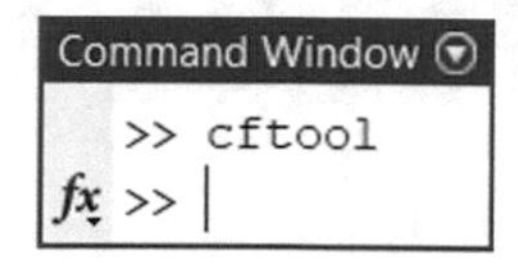

Fig. 1.402 Running the curve fitting toolbox

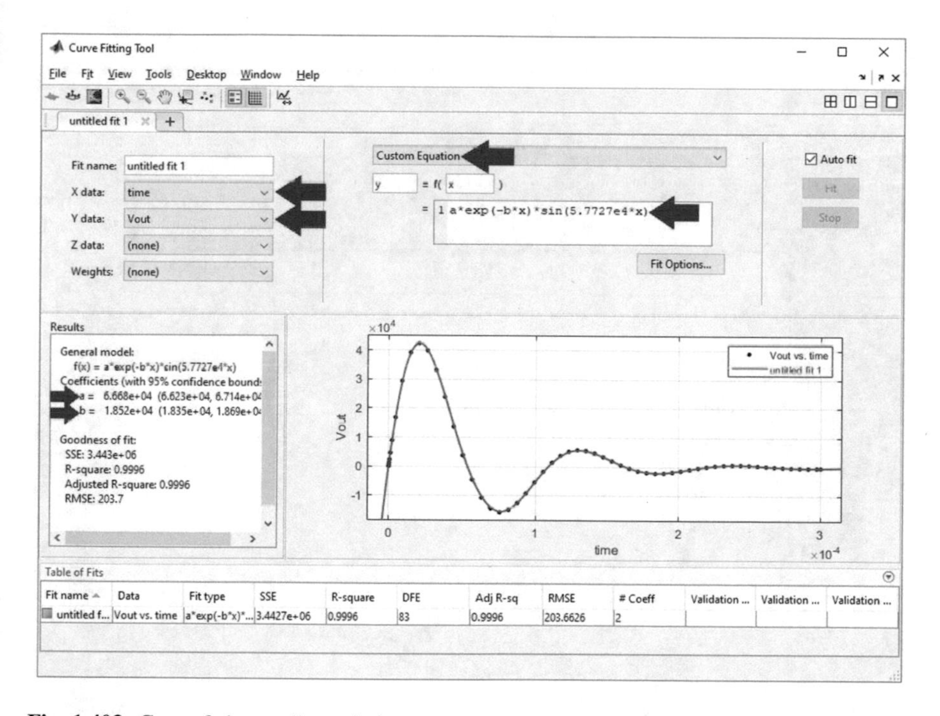

Fig. 1.403 Curve fitting toolbox window

The commands shown in Fig. 1.404 show the Proteus result and graph of $f(t) = 6.668 \times 10^4 e^{-1.852\times10^4 t}\sin(5.7727 \times 10^4 t)$ simultaneously. Output of this code is shown in Fig. 1.405. The two graphs overlap. This shows that $f(t) = 6.668 \times 10^4 e^{-1.852\times10^4 t}\sin(5.7727 \times 10^4 t)$ is a good function to represent our data.

```
Command Window
>> F=importdata('C:\Proteus Book\ImpulseResponse.dat');
>> time=F.data(:,1);
>> Vout=F.data(:,2);
>> plot(time,Vout),hold on
>> plot(time,6.668e4*exp(-1.852e4*time).*sin(5.7727e4*time))
fx >>
```

Fig. 1.404 MATLAB commands

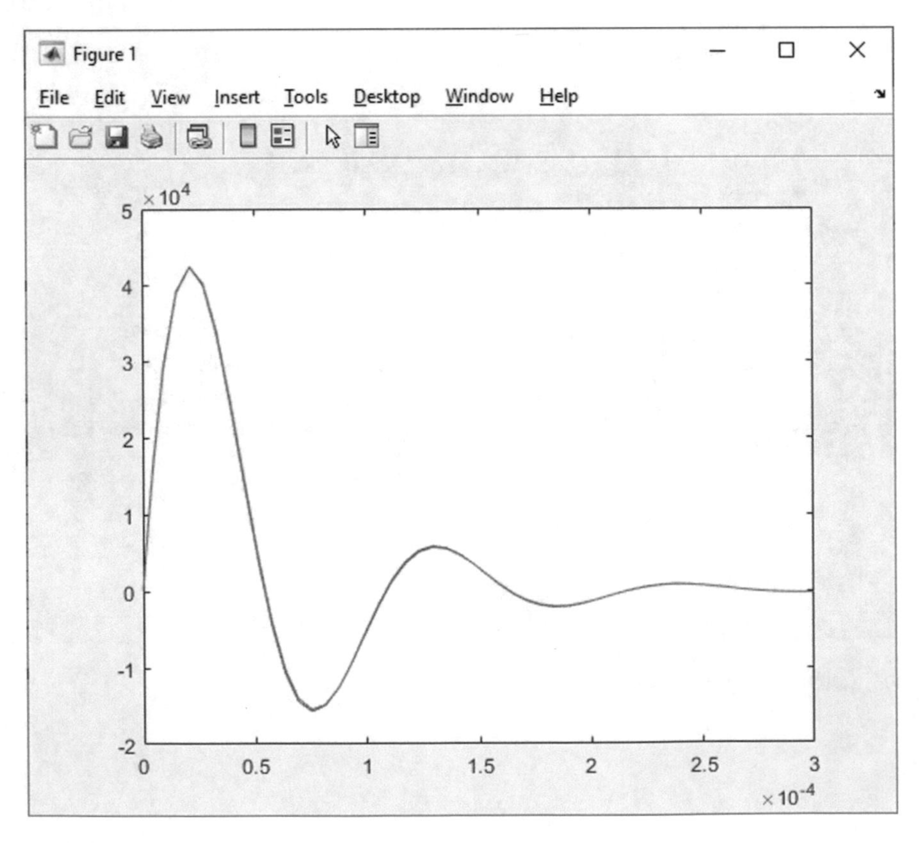

Fig. 1.405 Output of MATLAB commands

The MATLAB code shown in Fig. 1.406 draws the difference between Proteus result and $f(t) = 6.668 \times 10^4 e^{-1.852 \times 10^4 t} \sin(5.7727 \times 10^4 t)$ function. The output of this code is shown in Fig. 1.407. According to Fig. 1.407, the maximum value of error is about 517.

```
Command Window
>> F=importdata('C:\Proteus Book\ImpulseResponse.dat');
>> time=F.data(:,1);
>> Vout=F.data(:,2);
>> plot(time,Vout-6.668e4*exp(-1.852e4*time).*sin(5.7727e4*time))
fx >> |
```

Fig. 1.406 MATLAB commands

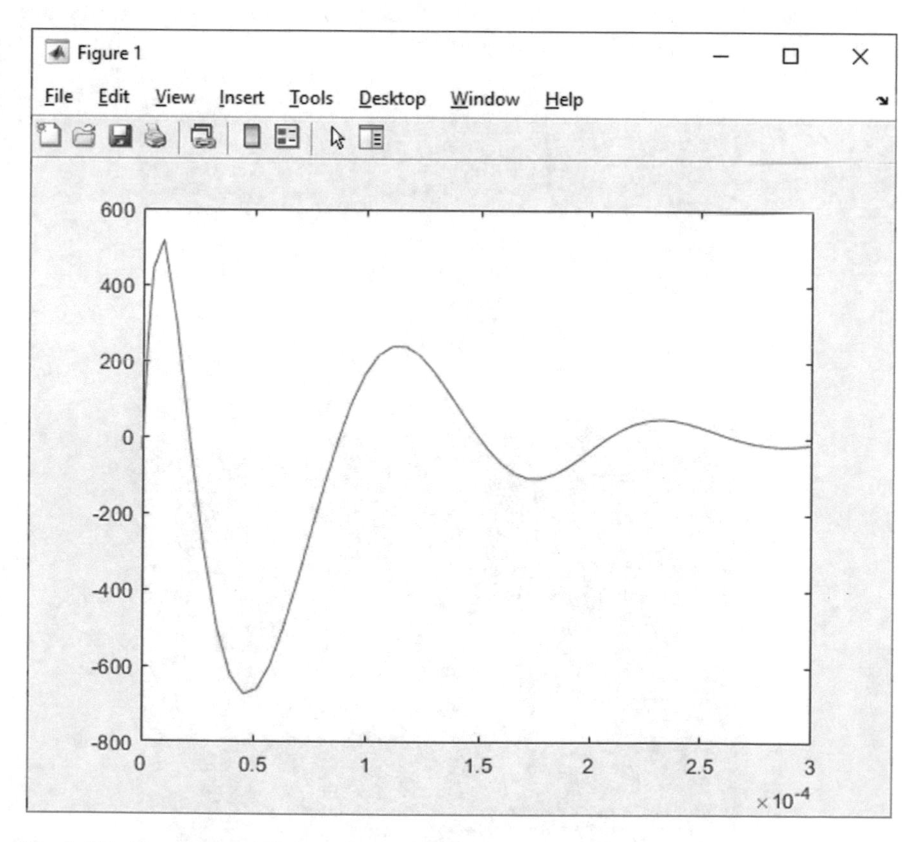

Fig. 1.407 Output of MATLAB commands

1.54 Example 53: Step Response of a RC Circuit

In this example, we want to observe the step response of a RC circuit shown in Fig. 1.408. It is quite easy to show that the transfer function this circuit is $\frac{V_{out}(s)}{V_{in}(s)} = \frac{1}{(RC)^2 s^2 + 3RCs + 1}$.

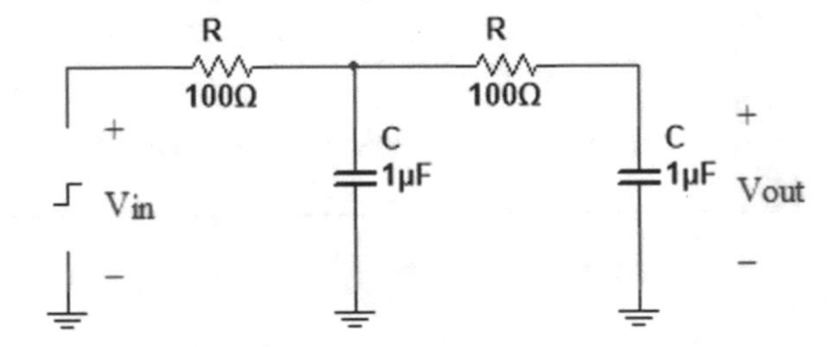

Fig. 1.408 Circuit for Example 53

Draw the schematic shown in Fig. 1.409. Settings of analogue graph and V_{in} are shown in Figs. 1.410 and 1.411, respectively.

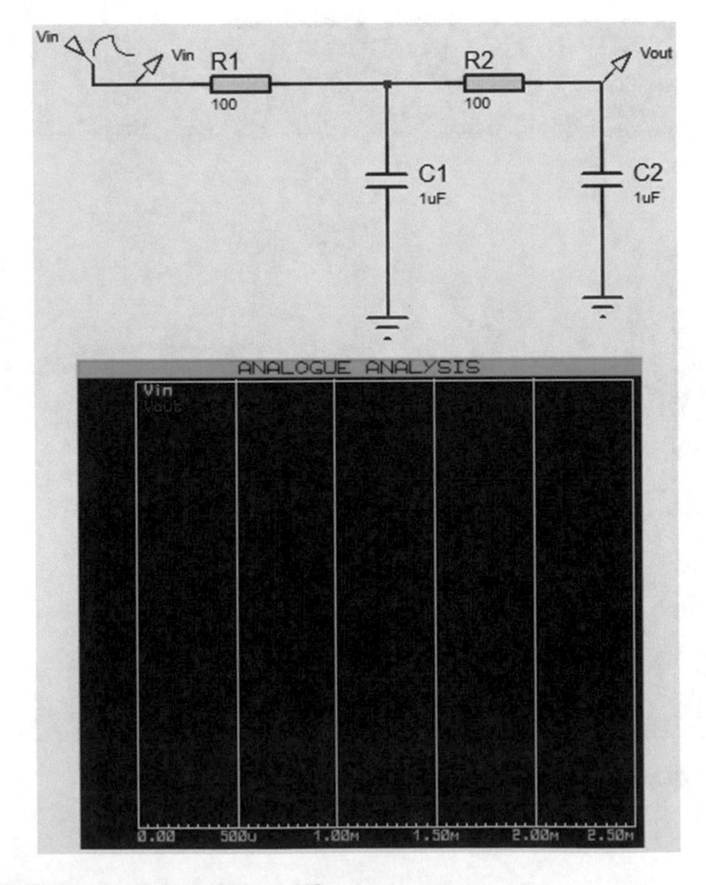

Fig. 1.409 Proteus equivalent of Fig. 1.408

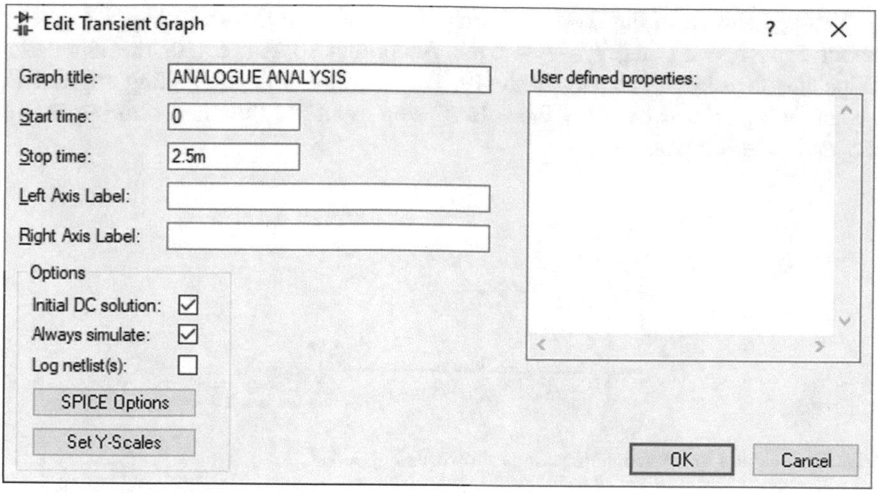

Edit Transient Graph

Graph title: ANALOGUE ANALYSIS

Start time: 0

Stop time: 2.5m

Left Axis Label:

Right Axis Label:

User defined properties:

Options

Initial DC solution: ☑

Always simulate: ☑

Log netlist(s): ☐

SPICE Options

Set Y-Scales

OK Cancel

Fig. 1.410 Edit transient graph window

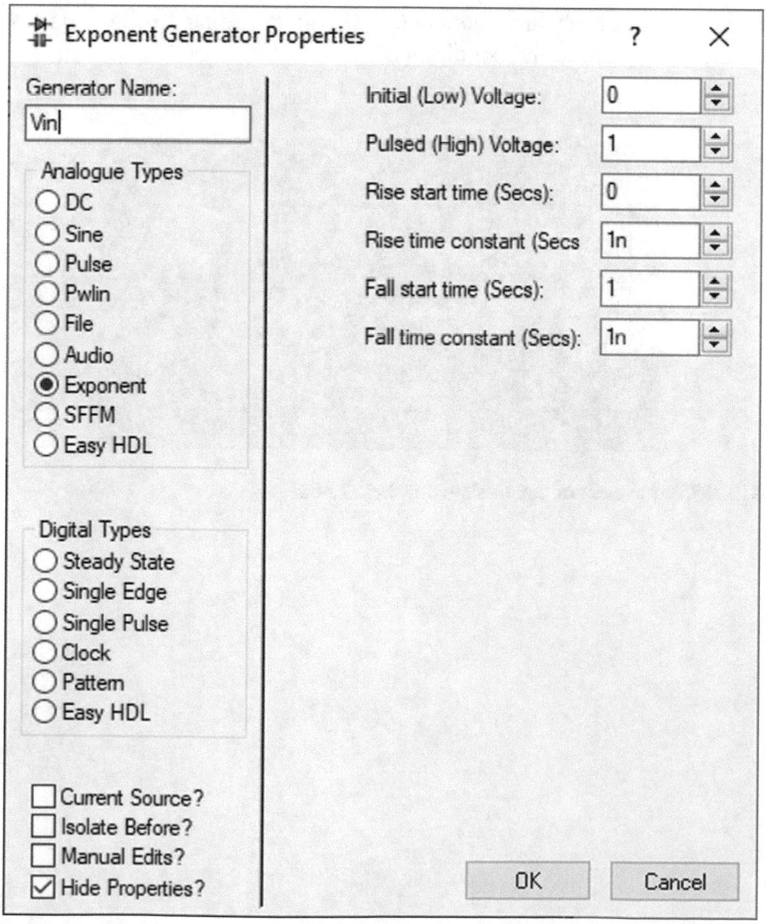

Exponent Generator Properties

Generator Name: Vin

Analogue Types: DC, Sine, Pulse, Pwlin, File, Audio, Exponent (selected), SFFM, Easy HDL

Digital Types: Steady State, Single Edge, Single Pulse, Clock, Pattern, Easy HDL

☐ Current Source?
☐ Isolate Before?
☐ Manual Edits?
☑ Hide Properties?

Initial (Low) Voltage: 0

Pulsed (High) Voltage: 1

Rise start time (Secs): 0

Rise time constant (Secs: 1n

Fall start time (Secs): 1

Fall time constant (Secs): 1n

OK Cancel

Fig. 1.411 Exponent generator properties window

Settings shown in Fig. 1.411 generate the waveform shown in Fig. 1.412. t_1 is about 5 ns, $t_2 = 1$ s and $t_3 - t_2 \approx 5$ ns. According to Fig. 1.410, the simulation calculates the circuit behavior in the [0, 2.5 ms] time interval. Falling edge of V_{in} occurs at 1 s. So, it can play the role of step input for the circuit during the [0, 2.5 ms] time interval.

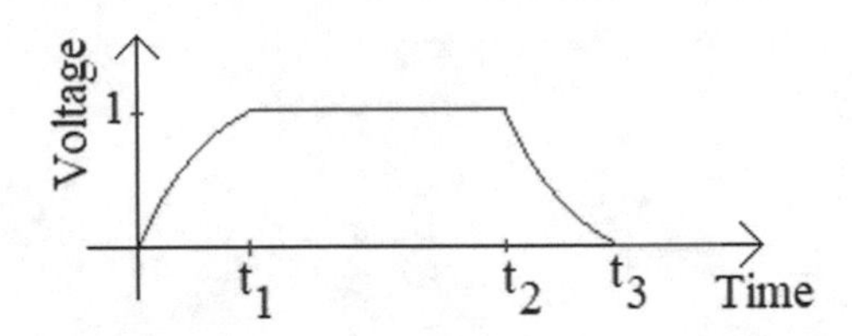

Fig. 1.412 Pulse generated with settings shown in Fig. 1.411

Initial DC solution box is checked in Fig. 1.410. So, a DC steady-state analysis is done at $t = 0$ and its results are used as initial condition. The V_{in} is zero at $t = 0$. Therefore, the DC steady-state equivalent circuit is similar to Fig. 1.413 and the circuit starts from zero initial conditions.

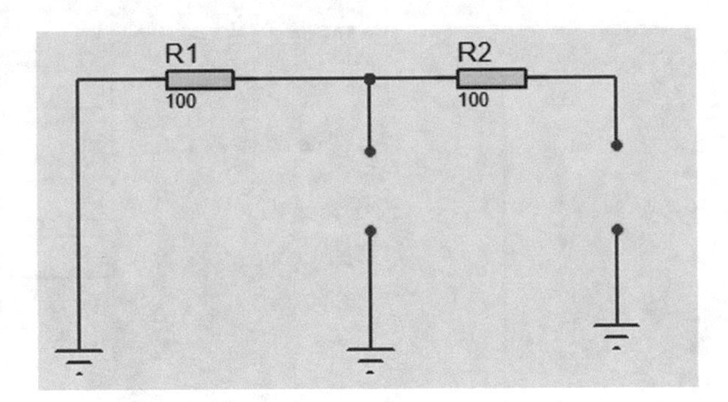

Fig. 1.413 DC equivalent of circuit shown in Fig. 1.408

Run the simulation. Simulation result is shown in Fig. 1.414.

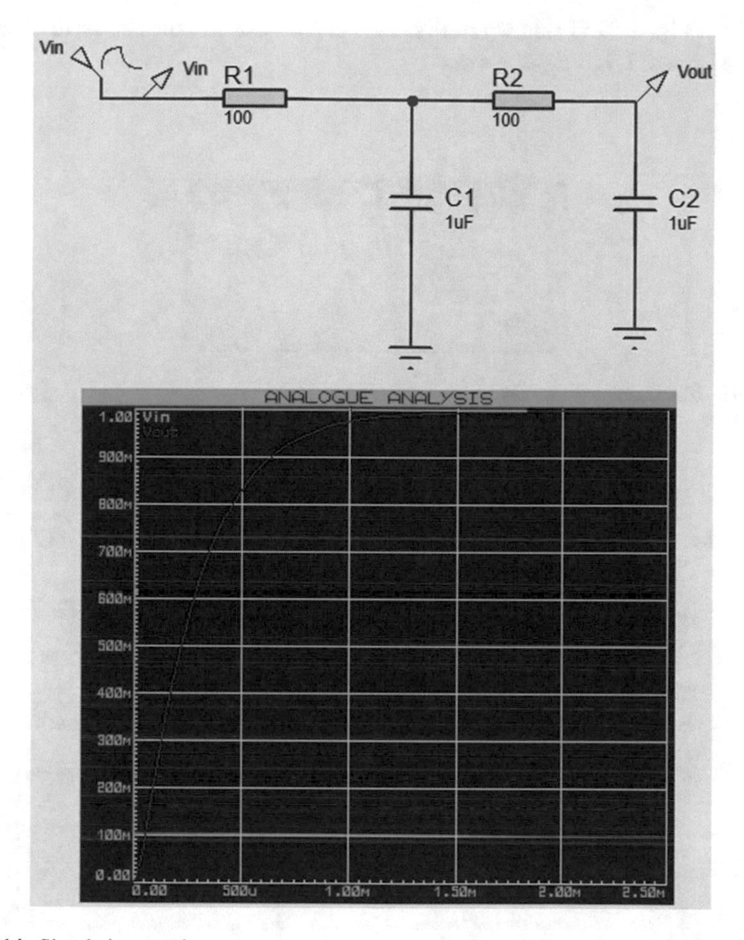

Fig. 1.414 Simulation result

Let's check the obtained result. The code shown in Fig. 1.415 draws the step response of the circuit. Note that in this circuit $\frac{V_{out}(s)}{V_{in}(s)} = \frac{1}{(RC)^2 s^2 + 3RCs + 1}$. Output of the code in Fig. 1.415 is shown in Fig. 1.416. You can use the cursor to ensure that Figs. 1.414 and 1.416 are the same.

```
Command Window
>> R=100;C=1e-6;
>> H=tf(1,[(R*C)^2 3*R*C 1]);
>> step(H)
>> grid on
fx >>
```

Fig. 1.415 MATLAB commands

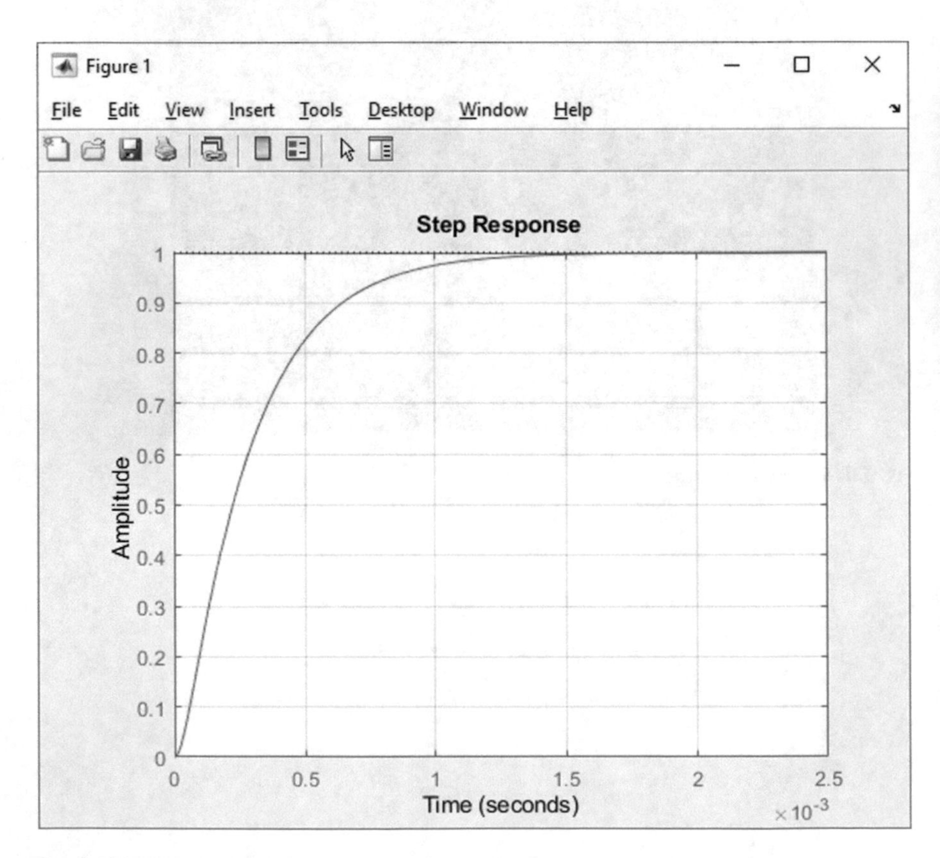

Fig. 1.416 Output of MATLAB commands

1.55 Example 54: Pulse Response of a RC Circuit

In this example, we want to observe the output of Example 53 for a pulse with duration of 1 ms and amplitude of 1 V (Fig. 1.417).

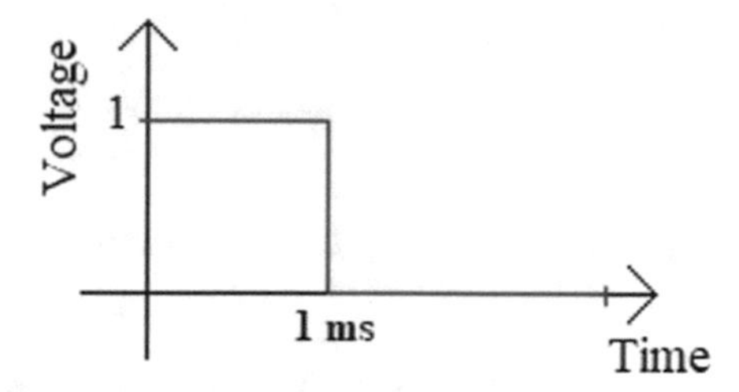

Fig. 1.417 A pulse with duration of 1 ms

Open the schematic of Example 53 and double click the V_{in}. Then change its settings to what shown in Fig. 1.418.

Exponent Generator Properties ? X

Generator Name:
Vin

Analogue Types
DC
Sine
Pulse
Pwlin
File
Audio
Exponent
SFFM
Easy HDL

Digital Types
Steady State
Single Edge
Single Pulse
Clock
Pattern
Easy HDL

Current Source?
Isolate Before?
Manual Edits?
Hide Properties?

Initial (Low) Voltage: 0
Pulsed (High) Voltage: 1
Rise start time (Secs): 0
Rise time constant (Secs 1n
Fall start time (Secs): 1m
Fall time constant (Secs): 1n

OK Cancel

Fig. 1.418 Settings of V_{in}

Settings shown in Fig. 1.418 generate the waveform shown in Fig. 1.419. t_1 is about 5 ns, $t_2 = 1$ ms and $t_3 - t_2 \approx 5$ ns.

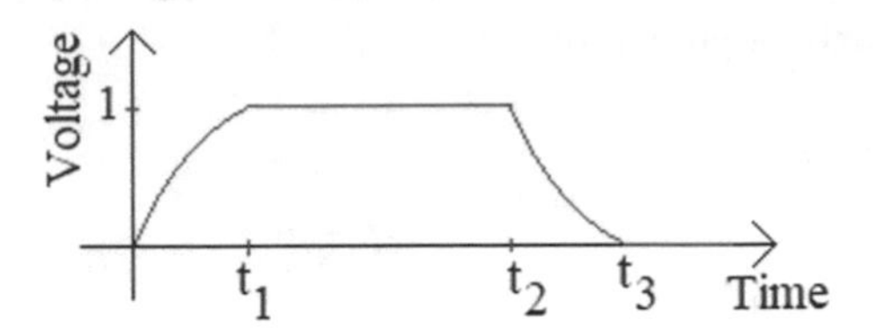

Fig. 1.419 Waveform generated with settings shown in Fig. 1.418

Run the simulation. The simulation result is shown in Fig. 1.420.

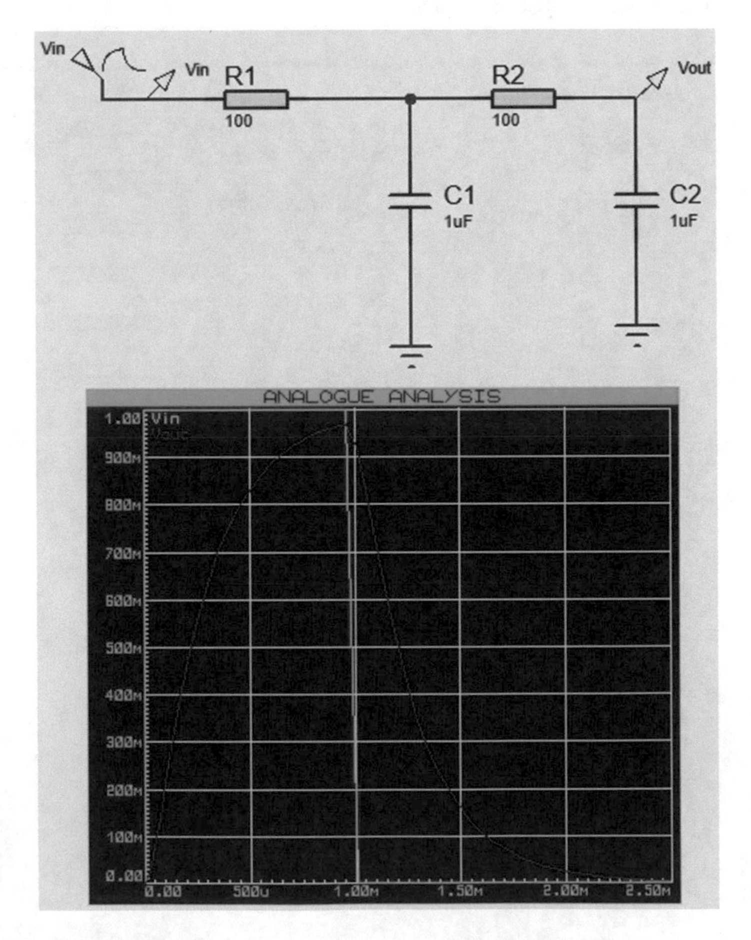

Fig. 1.420 Simulation result

1.56 Example 55: Frequency Response of Electric Circuits (I)

In this example, we want to observe the frequency response ($\frac{V_{\text{out}}(j\omega)}{V_{\text{in}}(j\omega)}$) of the circuit shown in Fig. 1.421.

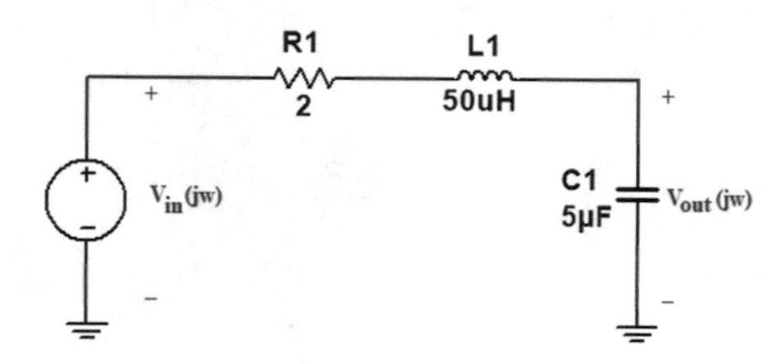

Fig. 1.421 Circuit for Example 55

Draw the schematic shown in Fig. 1.422. "in" is a sine generator block (Fig. 1.423).

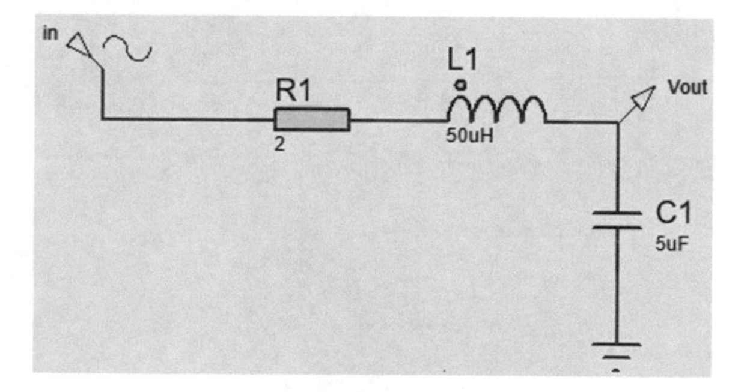

Fig. 1.422 Proteus equivalent of Fig. 1.421

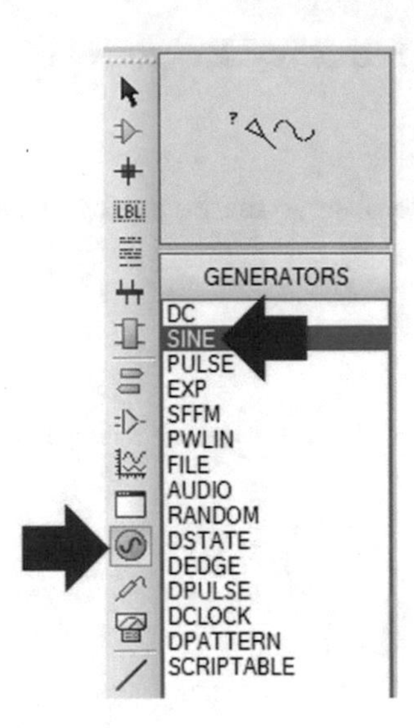

Fig. 1.423 Sine generator block

Settings of "in" are shown in Fig. 1.424. Note that these settings have no effect on the frequency simulation.

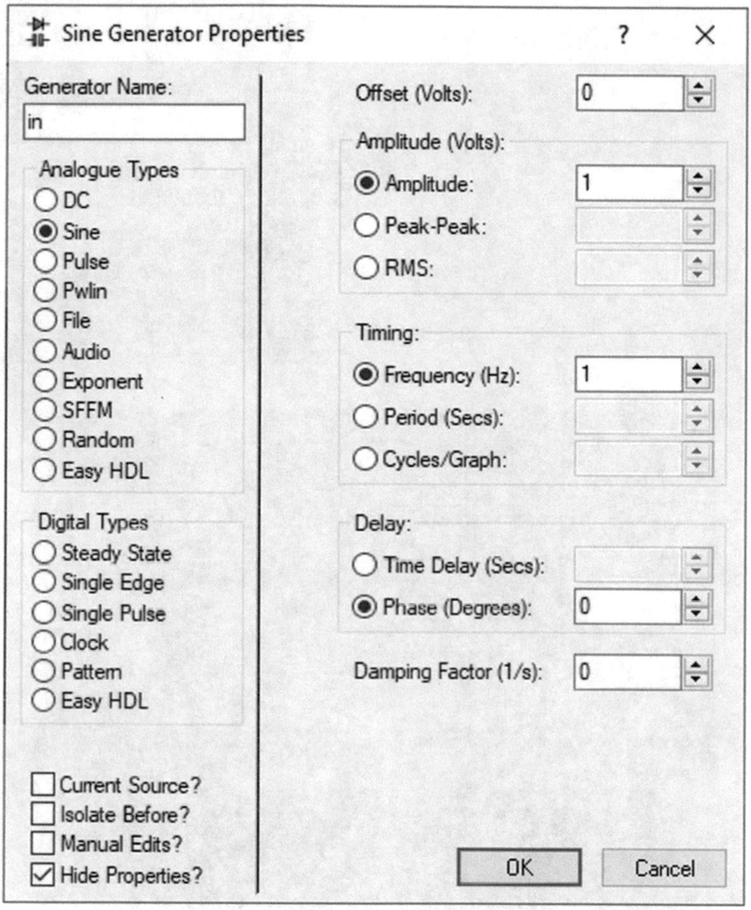

Sine Generator Properties

Generator Name: in

Analogue Types: DC, Sine (selected), Pulse, Pwlin, File, Audio, Exponent, SFFM, Random, Easy HDL

Digital Types: Steady State, Single Edge, Single Pulse, Clock, Pattern, Easy HDL

Current Source? Isolate Before? Manual Edits? Hide Properties? (checked)

Offset (Volts): 0

Amplitude (Volts): Amplitude: 1; Peak-Peak; RMS

Timing: Frequency (Hz): 1; Period (Secs); Cycles/Graph

Delay: Time Delay (Secs); Phase (Degrees): 0

Damping Factor (1/s): 0

OK Cancel

Fig. 1.424 Setting of sine generator block

Add a frequency graph (Fig. 1.425) to the schematic (Fig. 1.426).

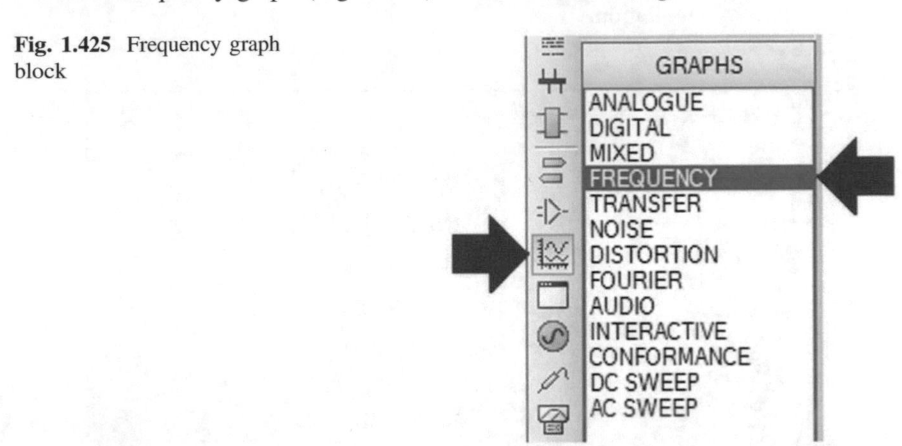

Fig. 1.425 Frequency graph block

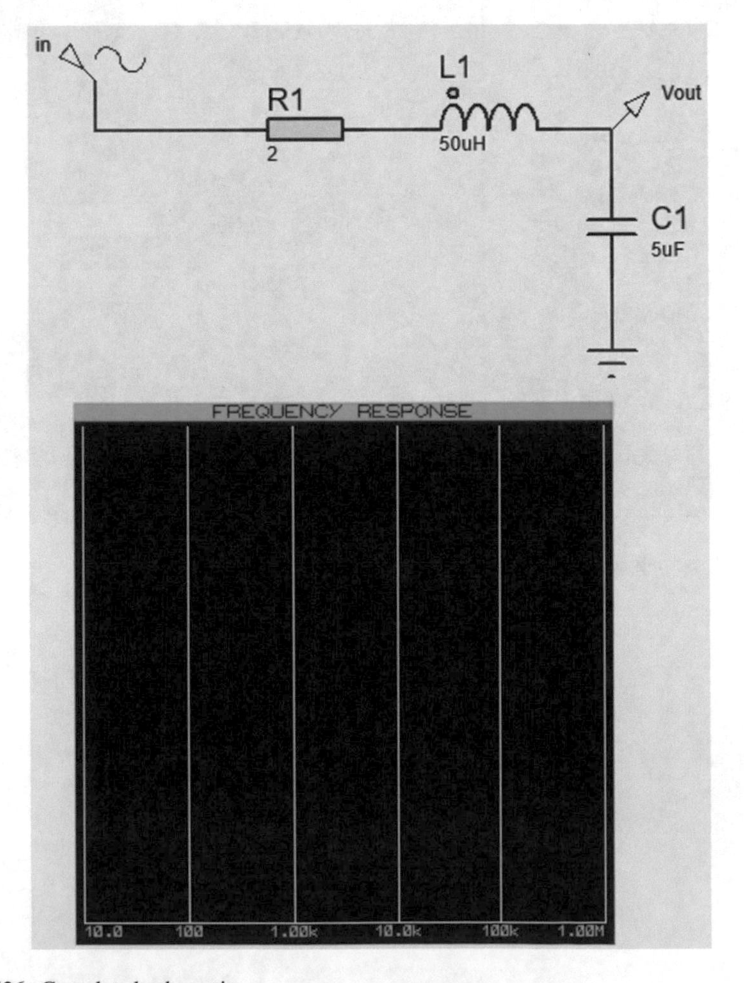

Fig. 1.426 Completed schematic

Double click the frequency graph and do the settings similar to Fig. 1.427. Then click the OK button. The Reference box determines the input point of the circuit. The magnitude of the input is always 1 and its phase is always 0°. Settings of the input source (Fig. 1.424) have no effect on the frequency simulations.

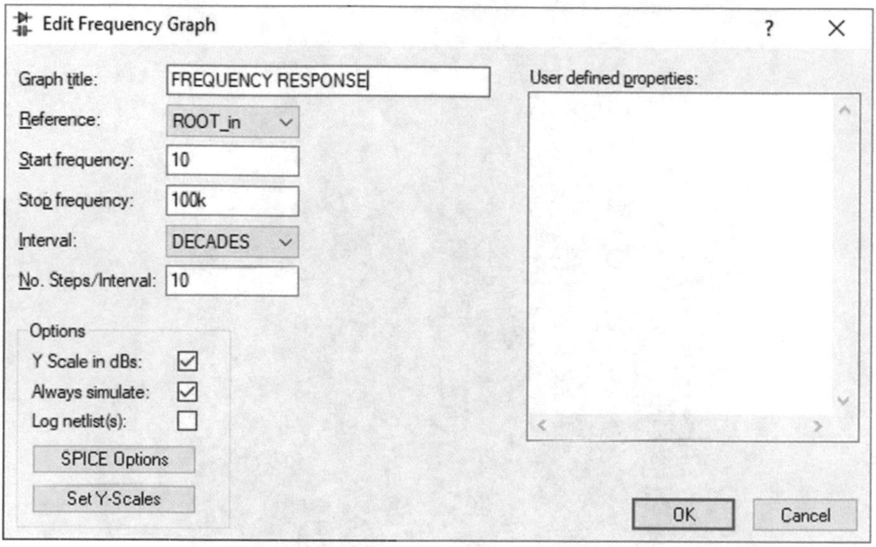

Fig. 1.427 Settings of frequency response graph

Start frequency and Stop frequency boxes determines the lower and upper band of analysis. Settings in Fig. 1.427 draw the frequency response for [10 Hz, 100 kHz] interval.

Generally, DECADES is selected for Interval box. No. Steps/Interval box determines the accuracy (smoothness) of the output curve. Entering a bigger number increases the accuracy however simulation takes more time to be done. When Y Scale in dBs box is checked, the vertical axis is shown in dB, i.e., $20\log_{10}(\text{gain})$.

Now drag and drop the V_{out} voltage probe to the left top corner of frequency graph (Fig. 1.428). This adds the V_{out} to the GAIN section of frequency graph. So, the output of simulation shows the gain graph $\left(\text{magnitude of } \frac{V_{\text{out}(j\omega)}}{V_{\text{in}}(j\omega)}, \text{ i.e., } \left|\frac{V_{\text{out}(j\omega)}}{V_{\text{in}}(j\omega)}\right|\right)$. Proteus draws the $20\log_{10}\left(\left|\frac{V_{\text{out}(j\omega)}}{V_{\text{in}}(j\omega)}\right|\right)$ since Y Scale in dBs box is checked.

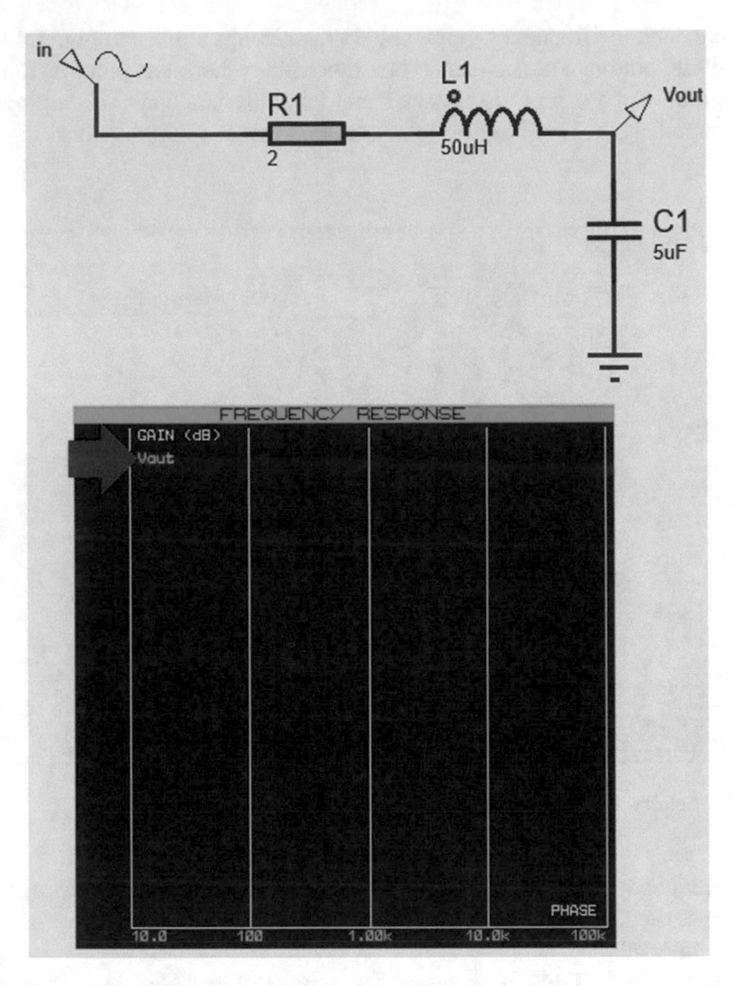

Fig. 1.428 V_{out} is dropped on the left top corner of the graph

Drag and drop the V_{out} probe to the right bottom corner of frequency graph (Fig. 1.429). The V_{out} is added to the PHASE section of frequency graph. So, the output of simulation shows the phase graph (i.e., $\measuredangle \frac{V_{out(j\omega)}}{V_{in}(j\omega)} = \measuredangle V_{out}(j\omega) - \measuredangle V_{in}(j\omega)$) as well.

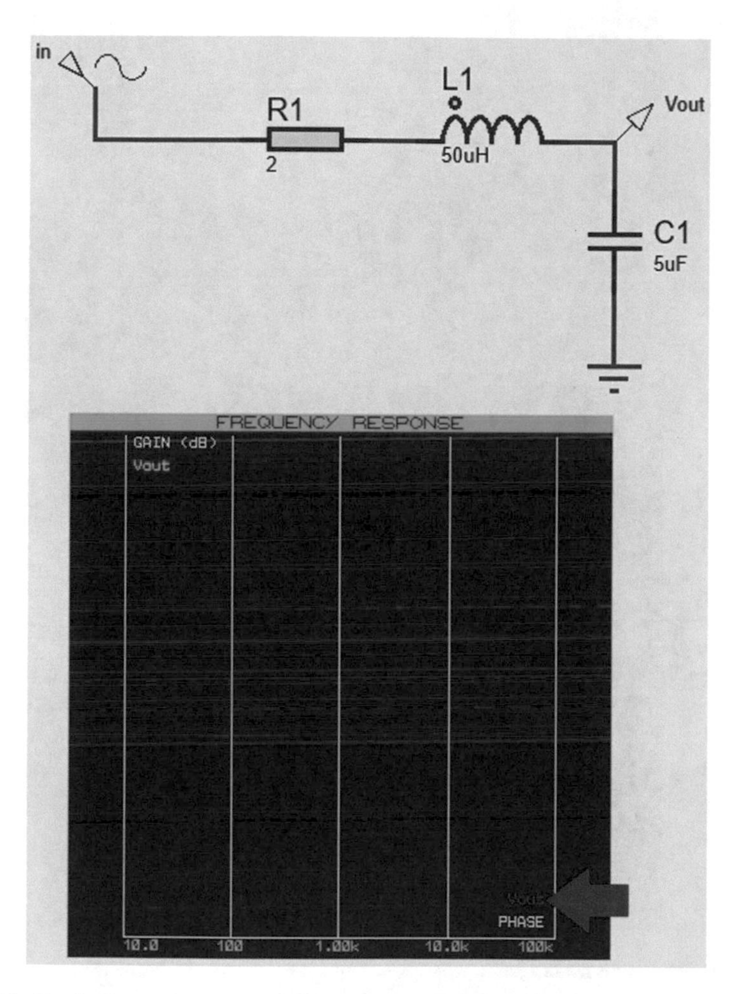

Fig. 1.429 V_{out} is dropped on the right bottom corner of the graph

Press the space bar key of keyboard to run the simulation. The simulation result is shown in Fig. 1.430. Magnitude and phase graphs are shown with green and red color, respectively.

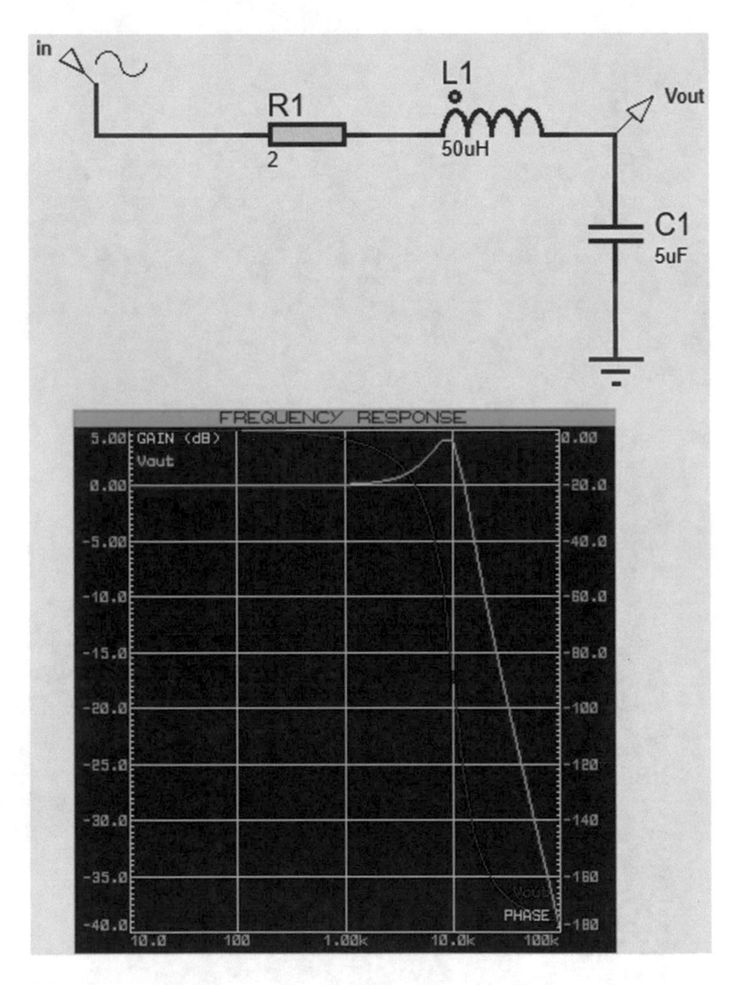

Fig. 1.430 Simulation result

You can increase the accuracy (smoothness) of graph by double clicking the frequency graph and enter a bigger number into No. Steps/Interval box (Fig. 1.431). Figure 1.432 shows the simulation result with settings shown in Fig. 1.431.

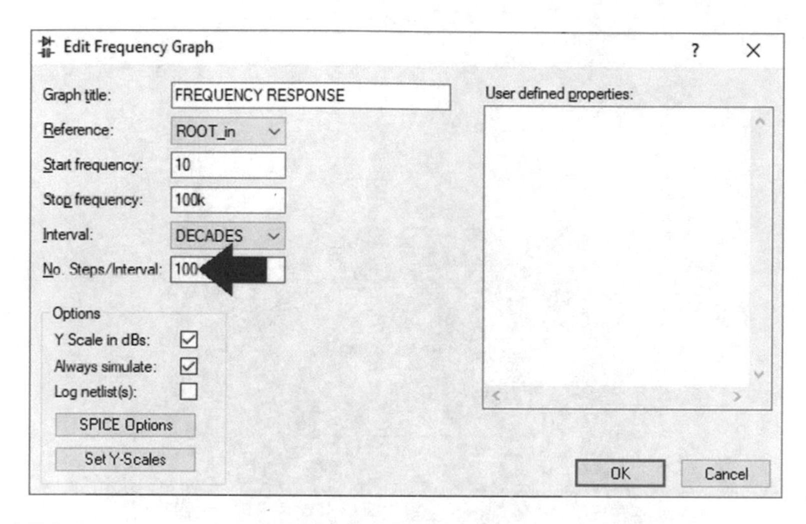

Fig. 1.431 Frequency response graph settings

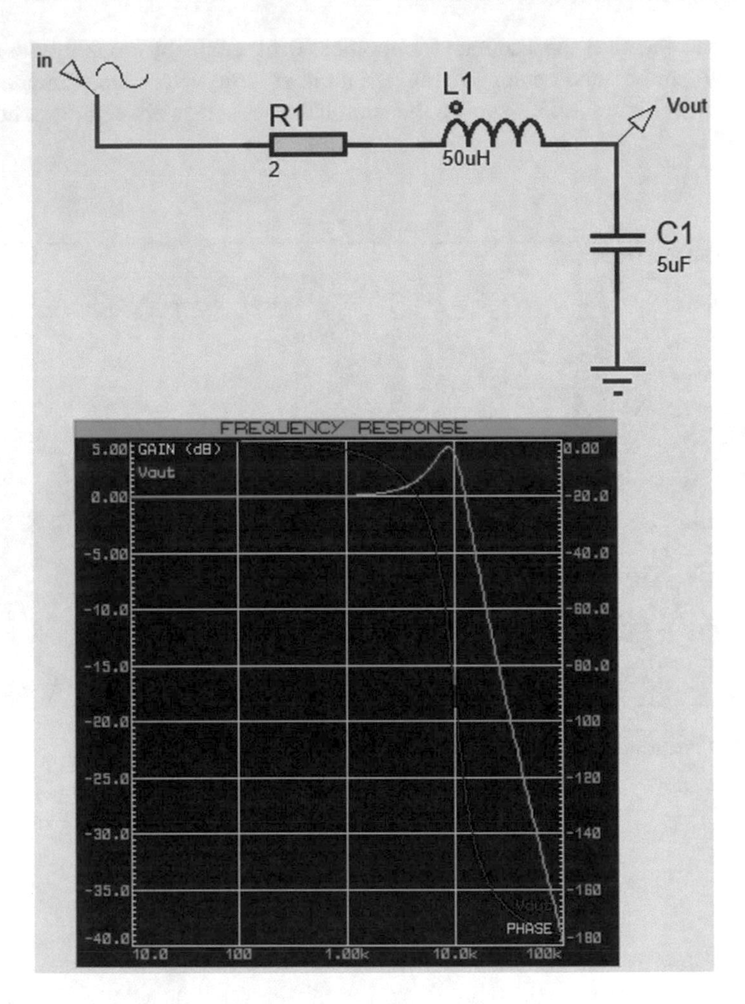

Fig. 1.432 Simulation result

Let's check the obtained result. The commands shown in Fig. 1.433 draws the frequency response of the circuit. Output of this code is shown in Fig. 1.434. Note that the horizontal axis has the unit of rad/s. Let's change it to kHz.

```
Command Window
>> R1=2;L1=50e-6;C1=5e-6;
>> H=tf([1],[L1*C1 R1*C1 1]);
>> wmin=2*pi*10;wmax=2*pi*1e5;
>> w=logspace(log10(wmin),log10(wmax));
>> bode(H,w)
fx >>
```

Fig. 1.433 MATLAB commands

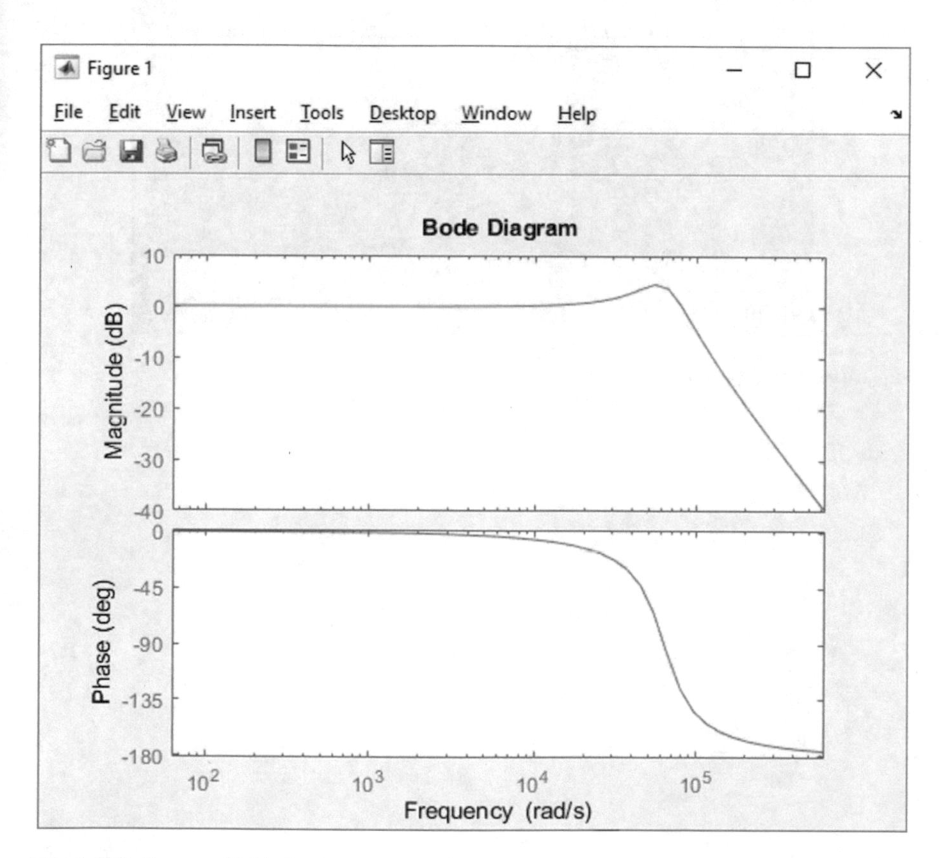

Fig. 1.434 Output of MATLAB commands

Double click on Fig. 1.434. This opens the Property Editor: Bode Diagram window (Fig. 1.435).

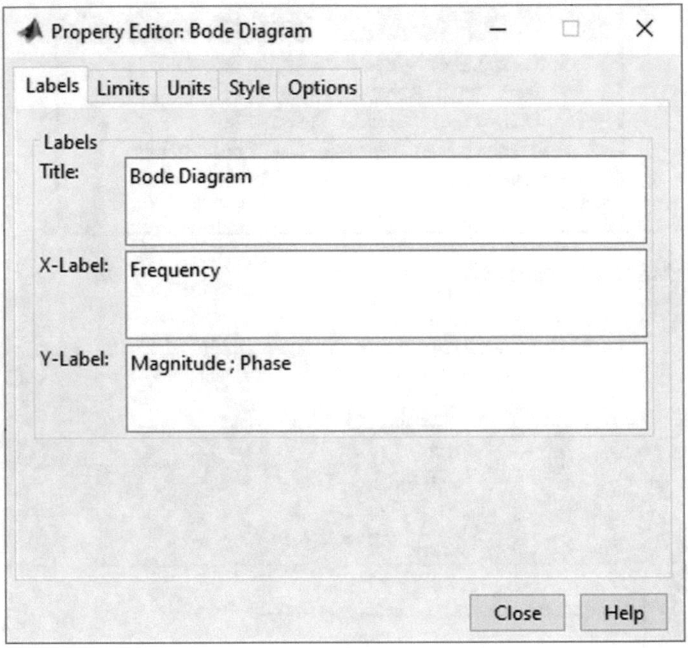

Fig. 1.435 Property editor window

Go to the Units tab and select the kHz for Frequency drop down list (Fig. 1.436). Then click the Close.

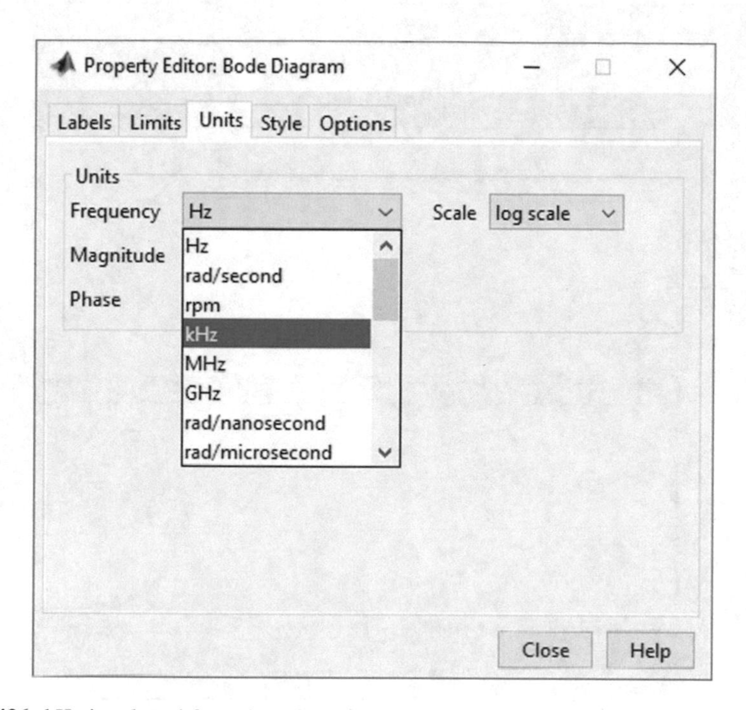

Fig. 1.436 kHz is selected from drop down list

After clicking the Close button, the graph changes to what shown in Fig. 1.437. Note that the horizontal axis has the unit of kHz. You can read different points of this graph easily. Simply click on the point that you want to know its coordinate and MATLAB shows the coordinate for you. Compare different points of this graph with the Proteus graph (Fig. 1.432) in order to ensure that Proteus result is correct.

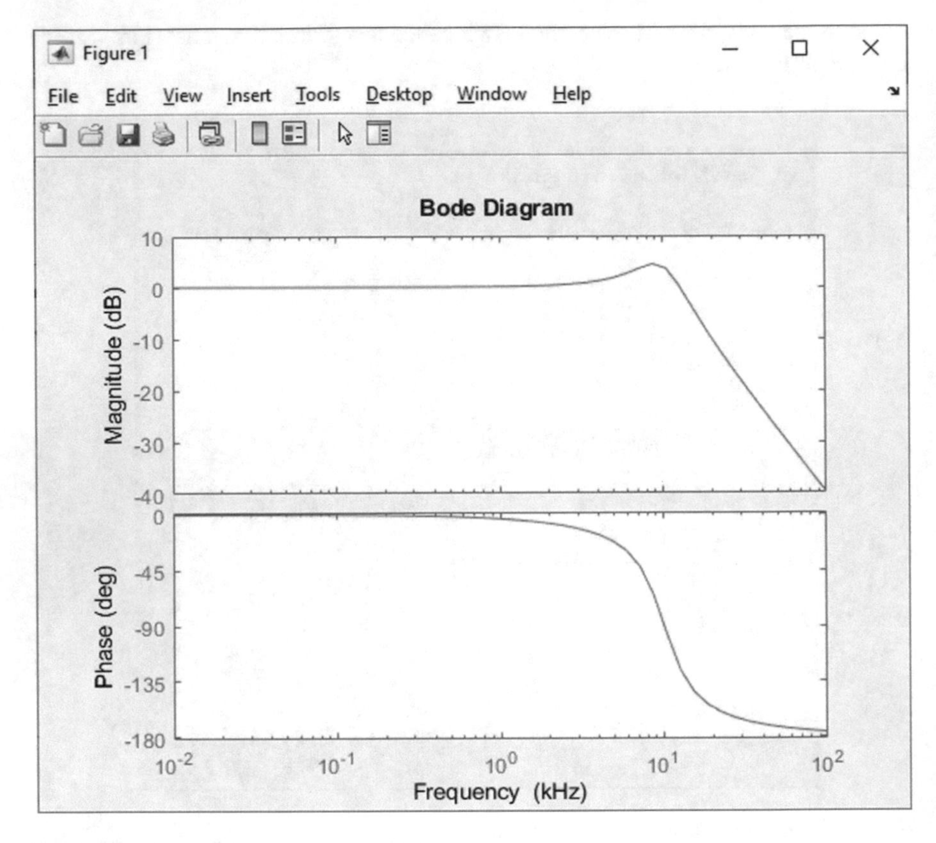

Fig. 1.437 Horizontal axis has unit of kHz

1.57 Example 56: Frequency Response of Electric Circuits (II)

In this example, we want to observe the magnitude of $\frac{i(j\omega)}{V_{\text{in}}(j\omega)}$ for the circuit shown in Fig. 1.438.

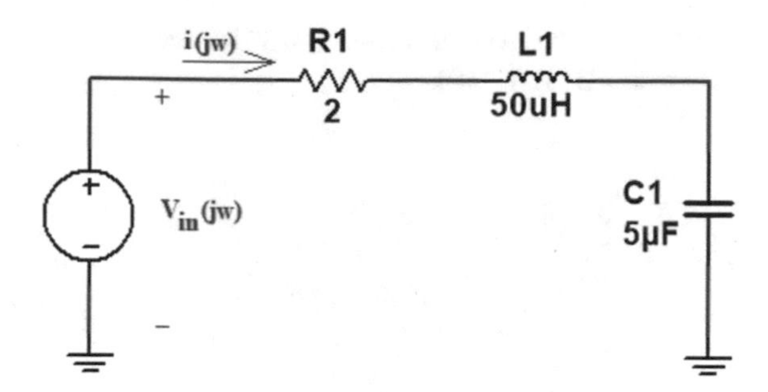

Fig. 1.438 Circuit for Example 56

Open the schematic of Example 55 and right click on the traces that you added to the frequency graph. This removes them from the graph (Fig. 1.439).

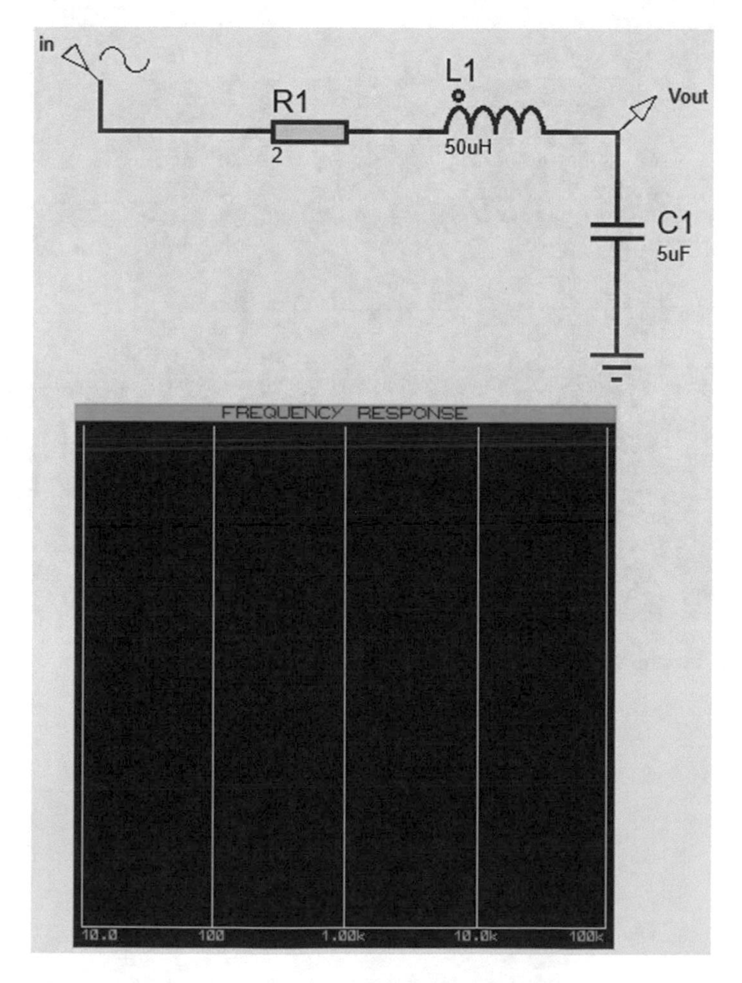

Fig. 1.439 Schematic of Example 55

Remove the voltage probe form the schematic and add a current probe to the schematic (Fig. 1.440).

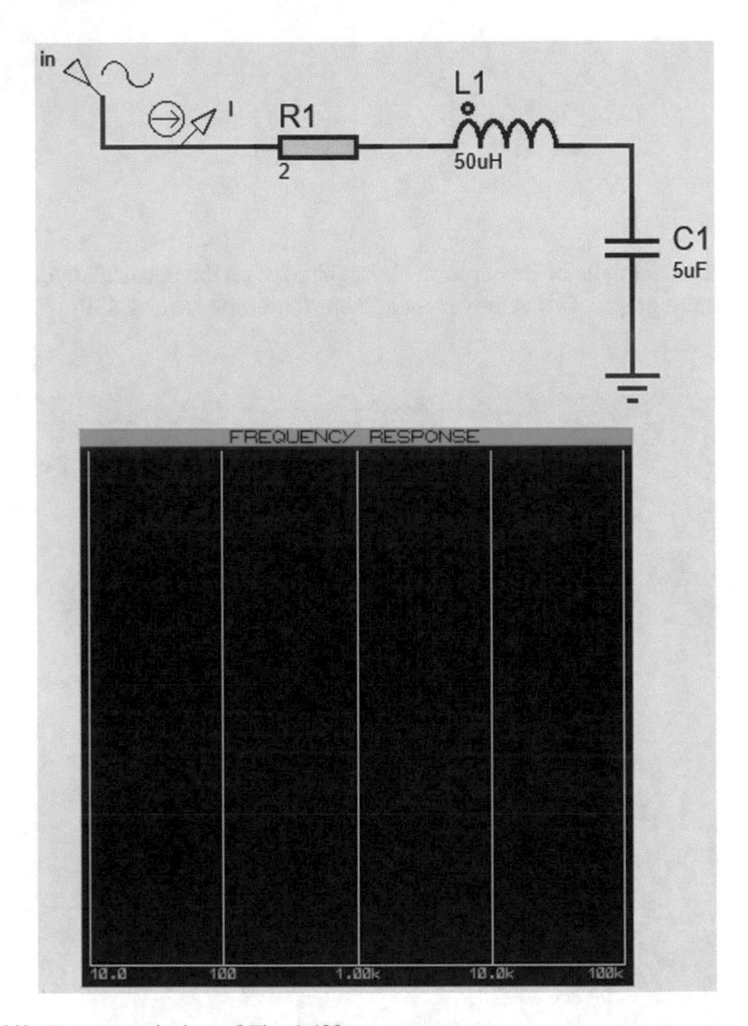

Fig. 1.440 Proteus equivalent of Fig. 1.438

Drag and drop the current probe to left top corner of frequency graph (Fig. 1.441).

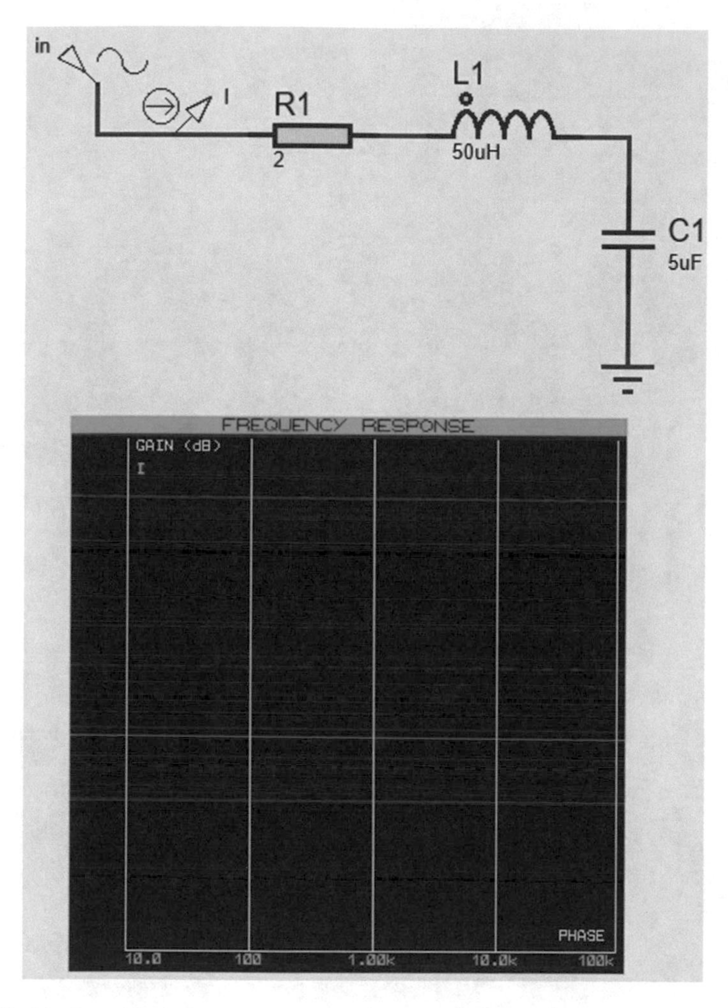

Fig. 1.441 I is dropped on the left top corner of the graph

Run the simulation by pressing the space bar key of keyboard. Simulation result is shown in Fig. 1.442.

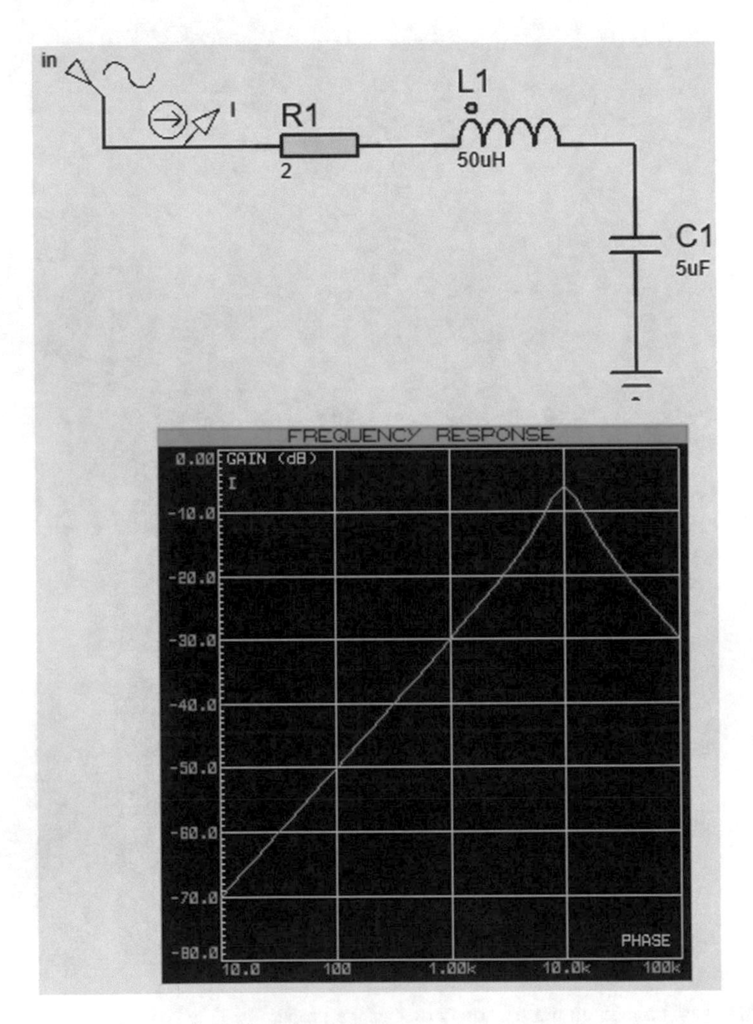

Fig. 1.442 Simulation result

Let's check the obtained result. The commands shown in Fig. 1.443 draw the magnitude graph of $\frac{i(j\omega)}{V_{\text{in}}(j\omega)}$. Output of these commands is shown in Fig. 1.444. Change the horizontal axis unit into kHz (Fig. 1.445). Now, you can compare different points of the MATLAB graph with Proteus graph to ensure that Proteus result is correct.

```
Command Window
>> R1=2;L1=50e-6;C1=5e-6;
>> H=tf([C1 0],[L1*C1 R1*C1 1]);
>> wmin=2*pi*10;wmax=2*pi*1e5;
>> w=logspace(log10(wmin),log10(wmax));
>> bodemag(H,w)
fx >>
```

Fig. 1.443 MATLAB commands

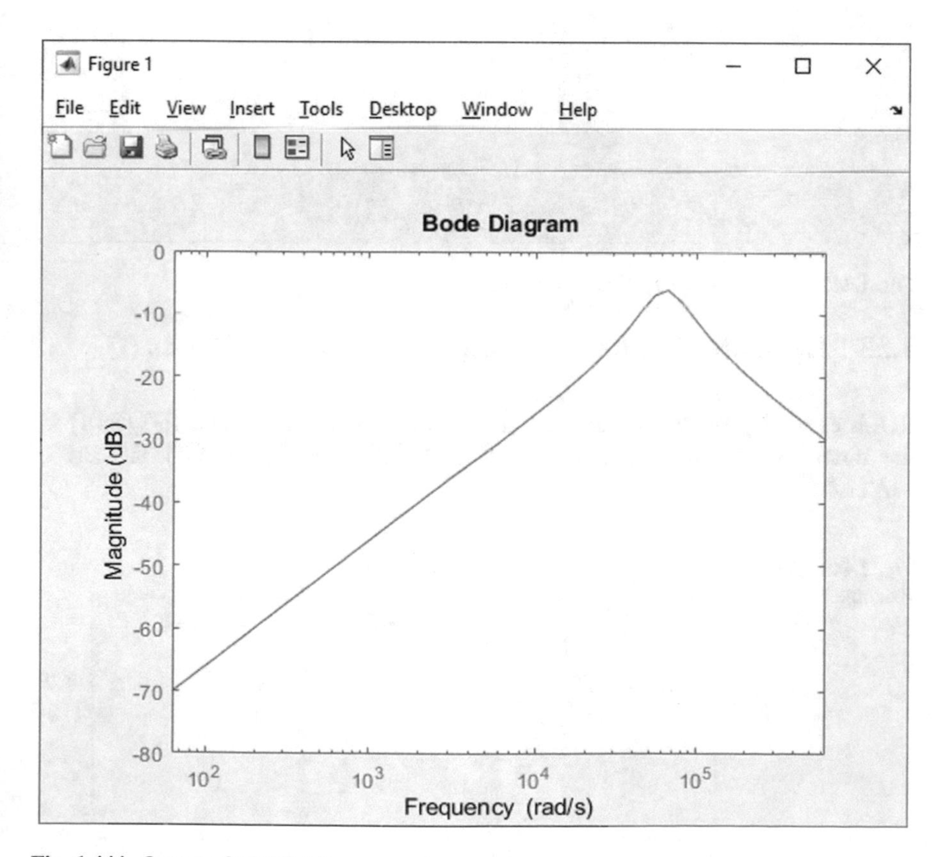

Fig. 1.444 Output of MATLAB commands

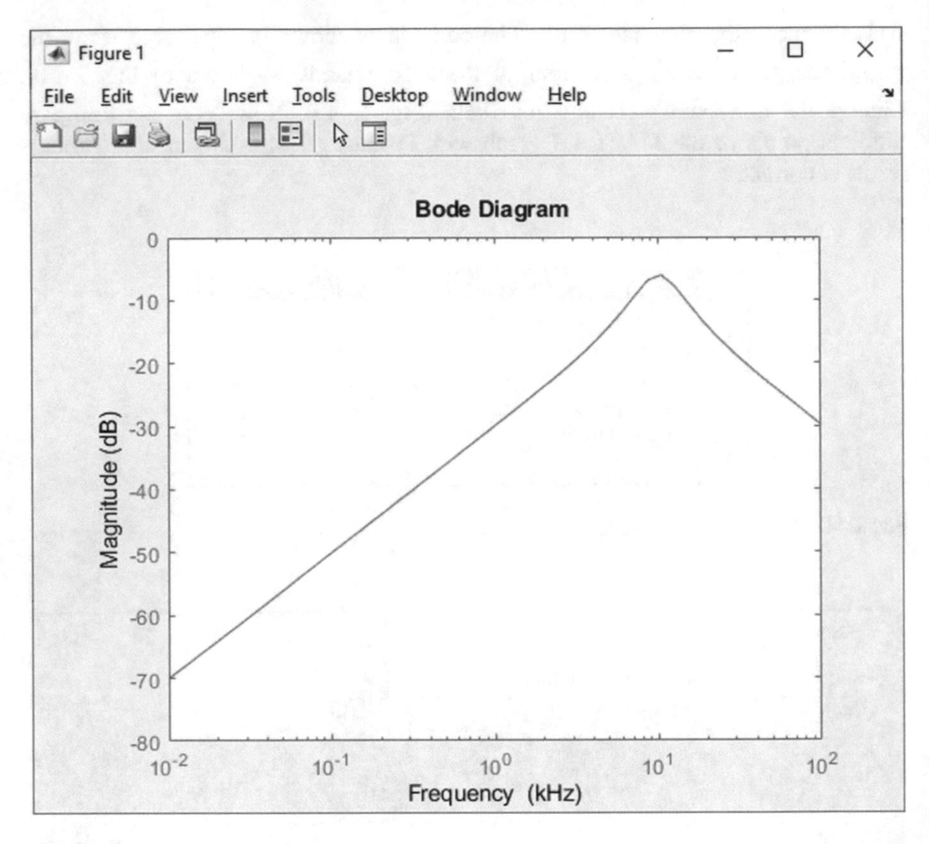

Fig. 1.445 Horizontal axis has the unit of kHz

1.58 Example 57: Input Impedance of Electric Circuits (I)

In this example, we want to draw the magnitude of input impedance ($|Z_{\text{in}}(j\omega)|$) for the circuit shown in Fig. 1.446. Input impedance is drawn with the aid of MATLAB.

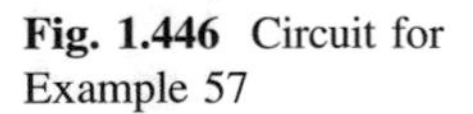

Fig. 1.446 Circuit for Example 57

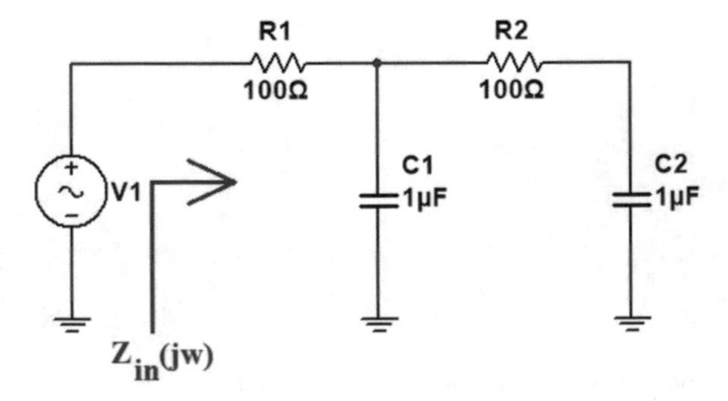

The commands shown in Fig. 1.447 draw the magnitude of input impedance of the circuit. Output of these commands are shown in Fig. 1.448.

```
Command Window
>> R=100;C=1e-6;f1=100;f2=1e3;
>> s=tf('s');
>> Zin=R+(R+1/(C*s))/(C*s*(R+2/(C*s)))

Zin =

  1e-14 s^3 + 3e-10 s^2 + 1e-06 s
  -------------------------------
       1e-16 s^3 + 2e-12 s^2

Continuous-time transfer function.

>> w=logspace(log10(2*pi*f1),log10(2*pi*f2));
>> bodemag(Zin,w)
>> grid on
fx >>
```

Fig. 1.447 MATLAB commands

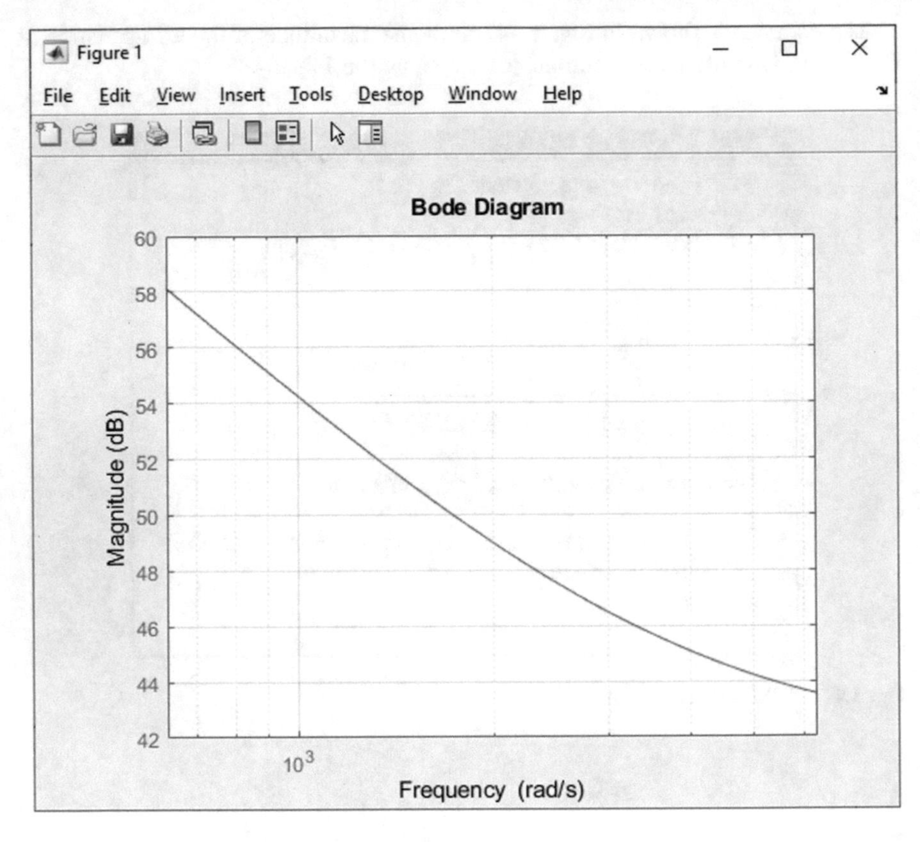

Fig. 1.448 Output of MATLAB commands

The horizontal axis of Fig. 1.448 is in rad/s. Change it into Hz (Fig. 1.449).

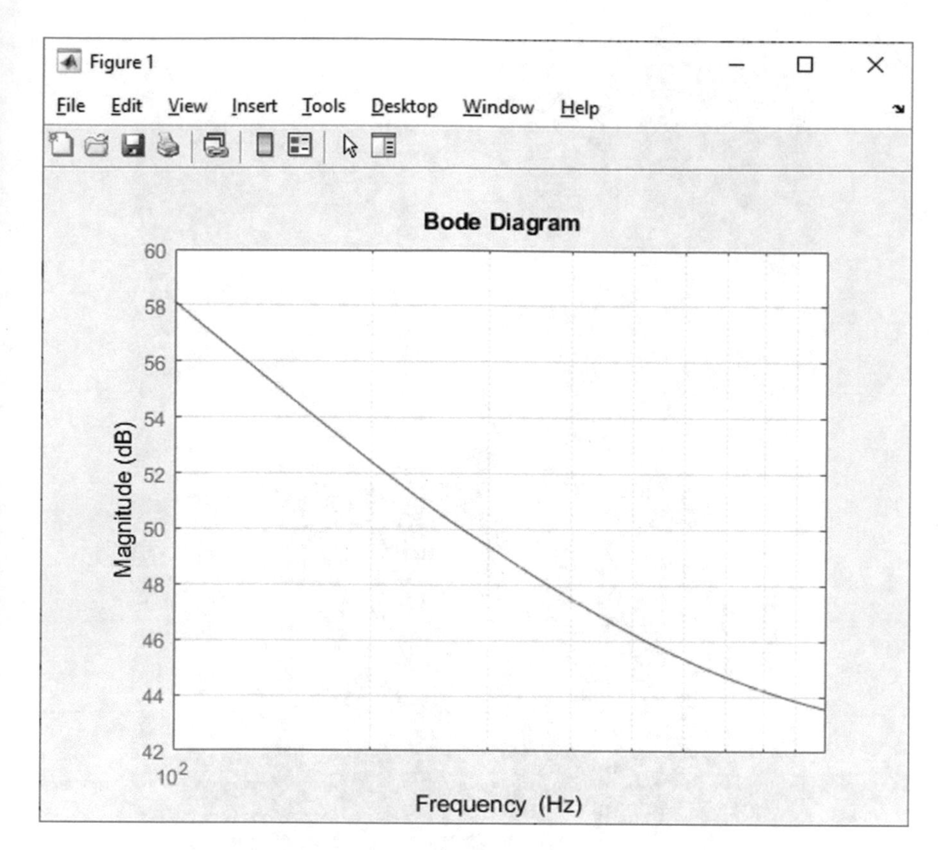

Fig. 1.449 Horizontal axis has the unit of kHz

Let's use Proteus to generate the result obtained in Fig. 1.449. Draw the schematic shown in Fig. 1.450.

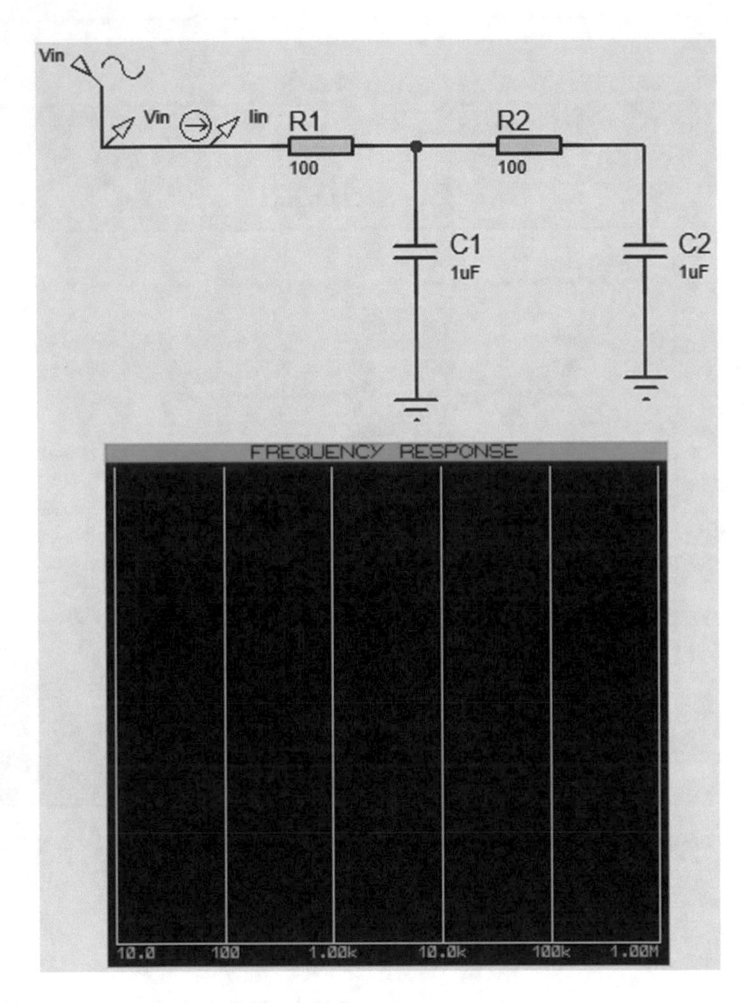

Fig. 1.450 Proteus equivalent of Fig. 1.446

Drag and drop the voltage probe Vin and the current probe Iin to the left top corner of frequency graph (Fig. 1.451).

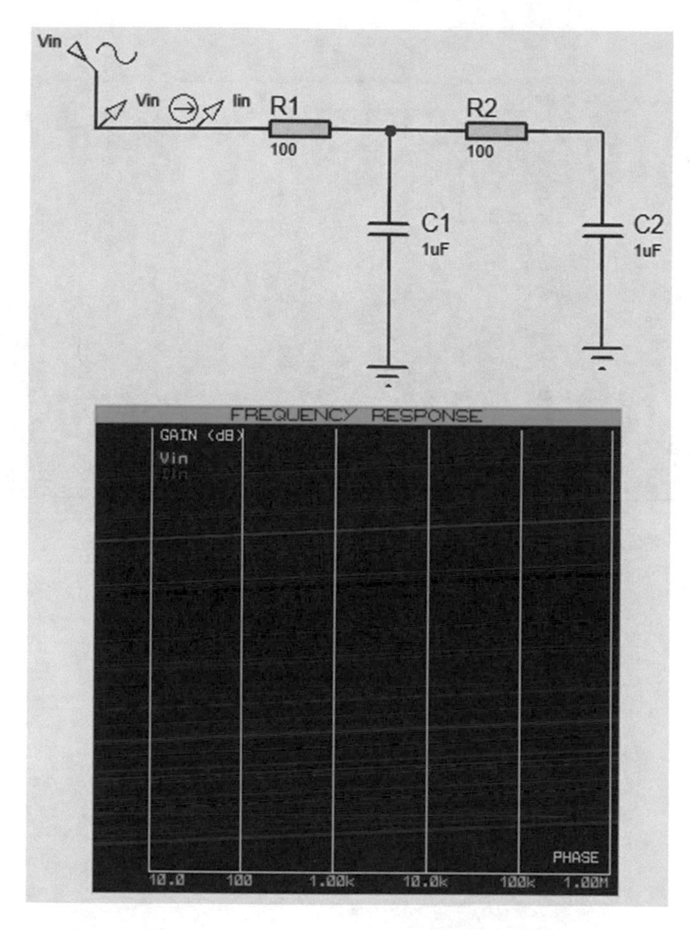

Fig. 1.451 V_{in} and I_{in} are dropped on the left top corner of the graph

We want to analyze the circuit on the [100 Hz, 1 kHz] range. Double click the frequency graph and do the settings similar to Fig. 1.452. Note that *Y* Scale in dBs is checked. So, the vertical axis is in dB.

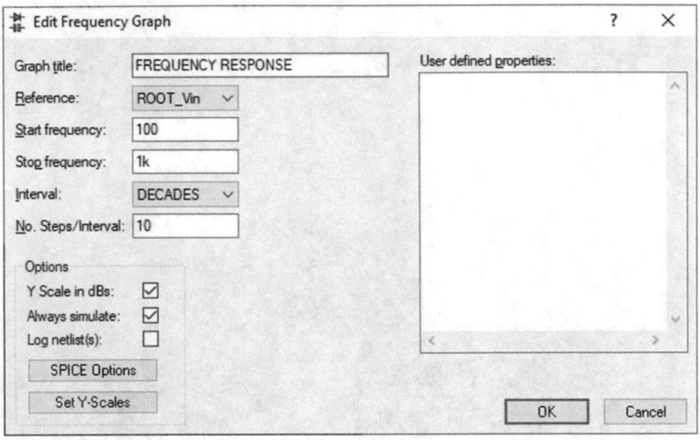

Edit Frequency Graph
Graph title: FREQUENCY RESPONSE
Reference: ROOT_Vin
Start frequency: 100
Stop frequency: 1k
Interval: DECADES
No. Steps/Interval: 10
Options
Y Scale in dBs: ☑
Always simulate: ☑
Log netlist(s): ☐
SPICE Options
Set Y-Scales
User defined properties:
OK
Cancel

Fig. 1.452 Settings of frequency response graph

Run the simulation by pressing the space bar key of your keyboard. Simulation result is shown in Fig. 1.453.

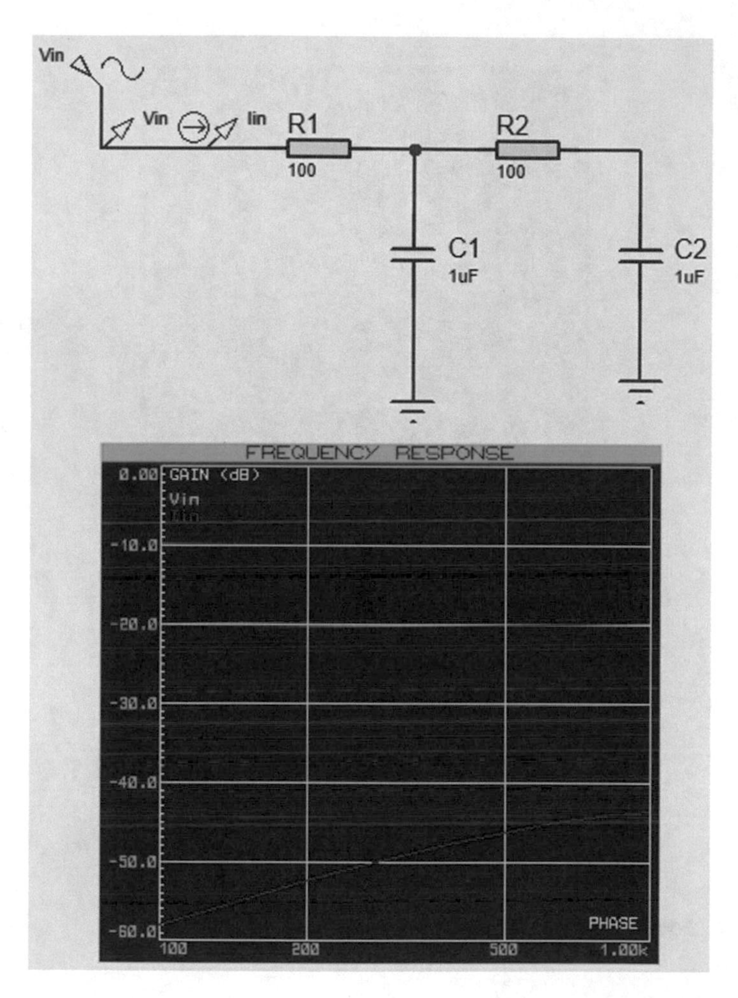

Fig. 1.453 Simulation result

Right click the on the obtained graph and click the Export Graph Data (Fig. 1.454). Then save the obtained result with the desired name in the desired path.

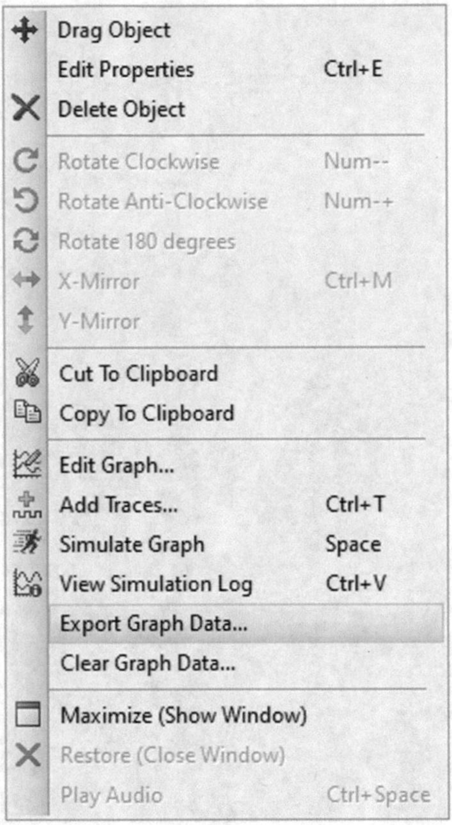

Fig. 1.454 Export graph data command

Remember that input impedance is defined as $Z_{\text{in}}(j\omega) = \frac{V_{\text{in}}(j\omega)}{I_{\text{in}}(j\omega)}$. If we use the dB we have $Z_{\text{in,dB}}(j\omega) = 20\log_{10}\left(\frac{V_{\text{in}}(j\omega)}{I_{\text{in}}(j\omega)}\right) = 20\log_{10}(V_{\text{in}}(j\omega)) - 20\log_{10}(I_{\text{in}}(j\omega)) = V_{\text{in,dB}}(j\omega) - I_{\text{in,dB}}(j\omega)$.

The commands shown in Fig. 1.455 draw the input impedance of the circuit. Note that the data comes from Proteus is in dB. So, we need to calculate the difference between voltage and current to obtain the input impedance. Output of the code in Fig. 1.455 is shown in Fig. 1.456. Compare this figure with Fig. 1.449. The results are the same. So, we obtained the input frequency correctly.

```
Command Window
>> F=importdata('C:\Proteus Book\InputImpData.dat');
>> semilogx(F.data(:,1),F.data(:,2)-F.data(:,3))
>> grid on
>> xlabel('Freq(Hz.)')
>> ylabel('Input impedance(dB)')
fx >>
```

Fig. 1.455 MATLAB commands

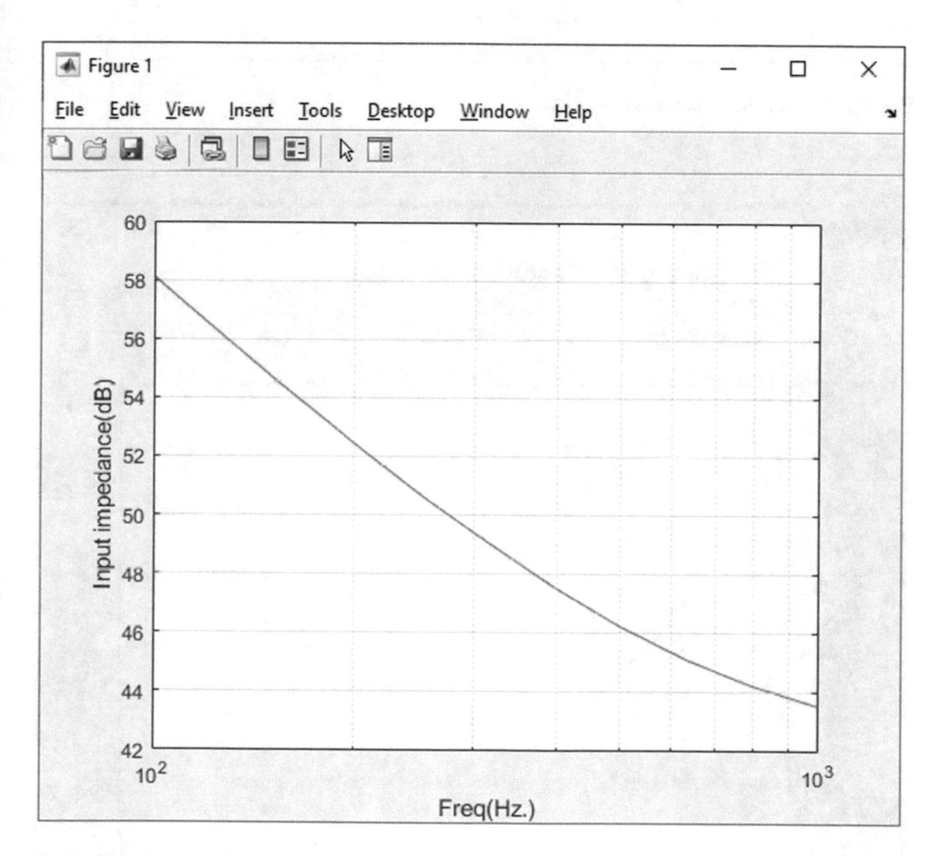

Fig. 1.456 Output of MATLAB commands

Vertical axis of Fig. 1.456 is in dB. You can change the code to have the vertical axis in Ohms. The commands shown in Fig. 1.457 draw the input impedance of the circuit with vertical axis in Ohms. Output is shown in Fig. 1.458.

```
Command Window
>> F=importdata('C:\Proteus Book\InputImpData.dat');
>> semilogx(F.data(:,1),10.^((F.data(:,2)-F.data(:,3))/20))
>> grid on
>> xlabel('Freq(Hz.)')
>> ylabel('Input impedance(Ohm)')
fx >>
```

Fig. 1.457 MATLAB commands

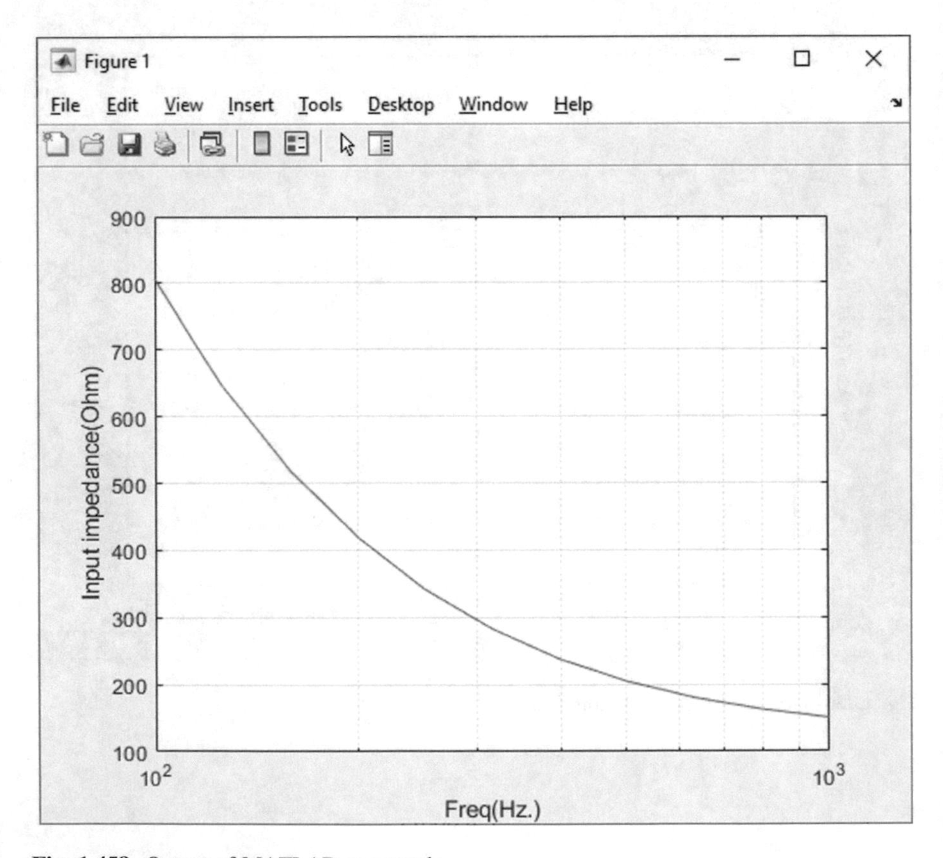

Fig. 1.458 Output of MATLAB commands

1.59 Example 58: Input Impedance of Electric Circuits (II)

In the previous example we imported the data of voltage probe V_{in} and current probe I_{in} to MATLAB and used them to draw to the input impedance. However, importing the data of voltage probe V_{in} is not necessary. The reason is that the frequency response simulation assumes that input source has amplitude of 1 and phase of 0°. So, V_{in} is 0 dB for all frequencies (Fig. 1.459) and the input impedance can be drawn with the aid current probe I_{in} only using the relationship $Z_{in,dB}(j\omega) = 20\log_{10}\left(\frac{V_{in}(j\omega)}{I_{in}(j\omega)}\right) = 20\log_{10}(V_{in}(j\omega)) - 20\log_{10}(I_{in}(j\omega)) = 0 - I_{in,dB}(j\omega) = -I_{in,dB}(j\omega)$. In this example we want to draw the input impedance of Example 57 without using the voltage probe data.

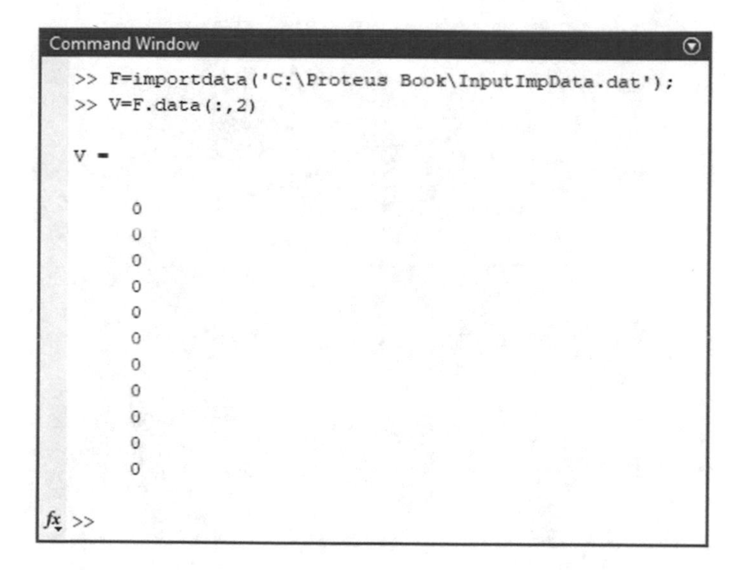

Fig. 1.459 Magnitude of V_{in} is 0 dB

Open the schematic of Example 57 and remove the voltage probe from it (Fig. 1.460).

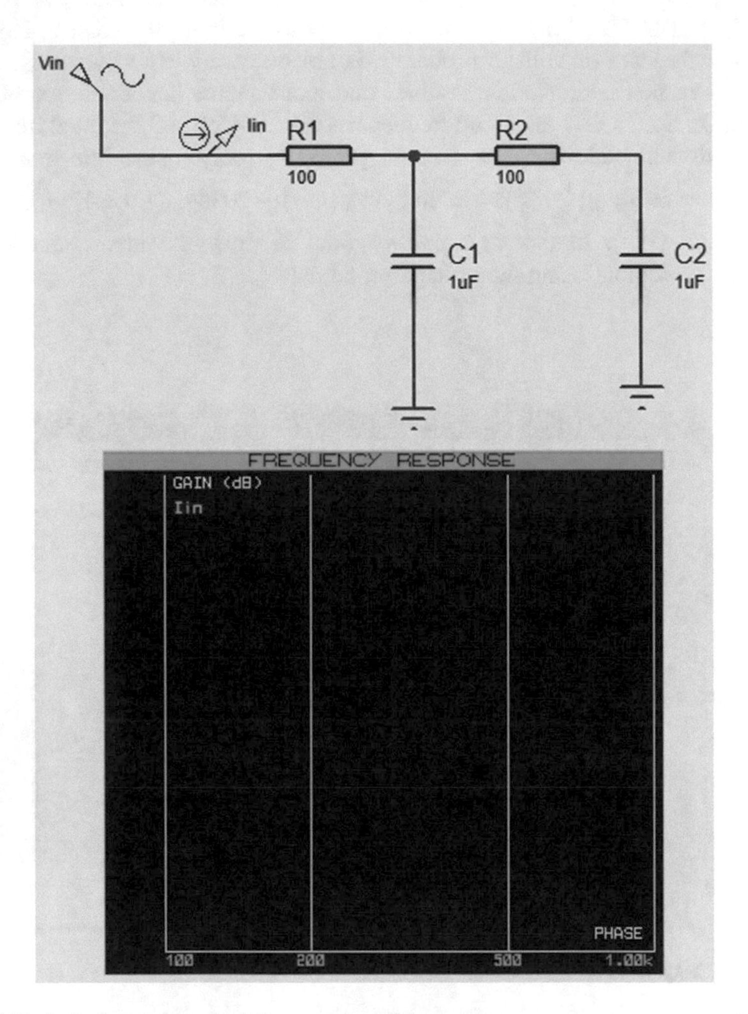

Fig. 1.460 I_{in} is dropped on the left top corner of the graph

Run the simulation. Simulation result is shown in Fig. 1.461. Right click on the graph and export it.

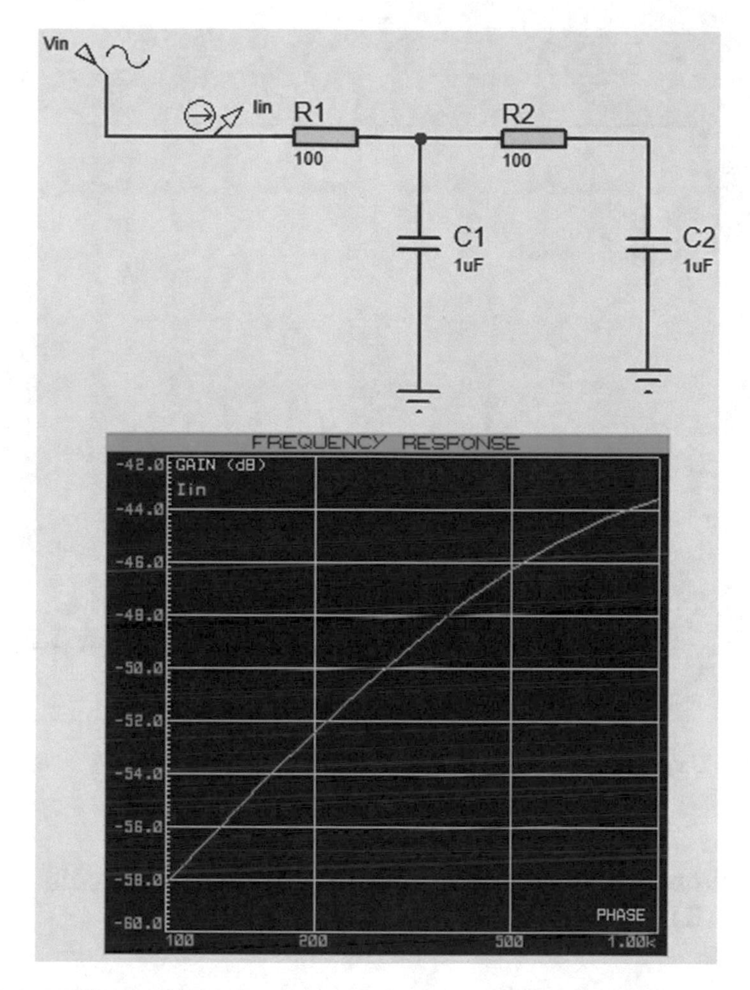

Fig. 1.461 Simulation result

The commands shown in Fig. 1.462 draw the input impedance of the circuit. Output of this code is shown in Fig. 1.463. The output is the same as Fig. 1.458.

```
Command Window
>> F=importdata('C:\Proteus Book\InputImpData2.dat');
>> semilogx(F.data(:,1),10.^(-F.data(:,2)/20))
>> grid on
>> xlabel('Freq(Hz.)')
>> ylabel('Input impedance(Ohm)')
fx >> |
```

Fig. 1.462 MATLAB command

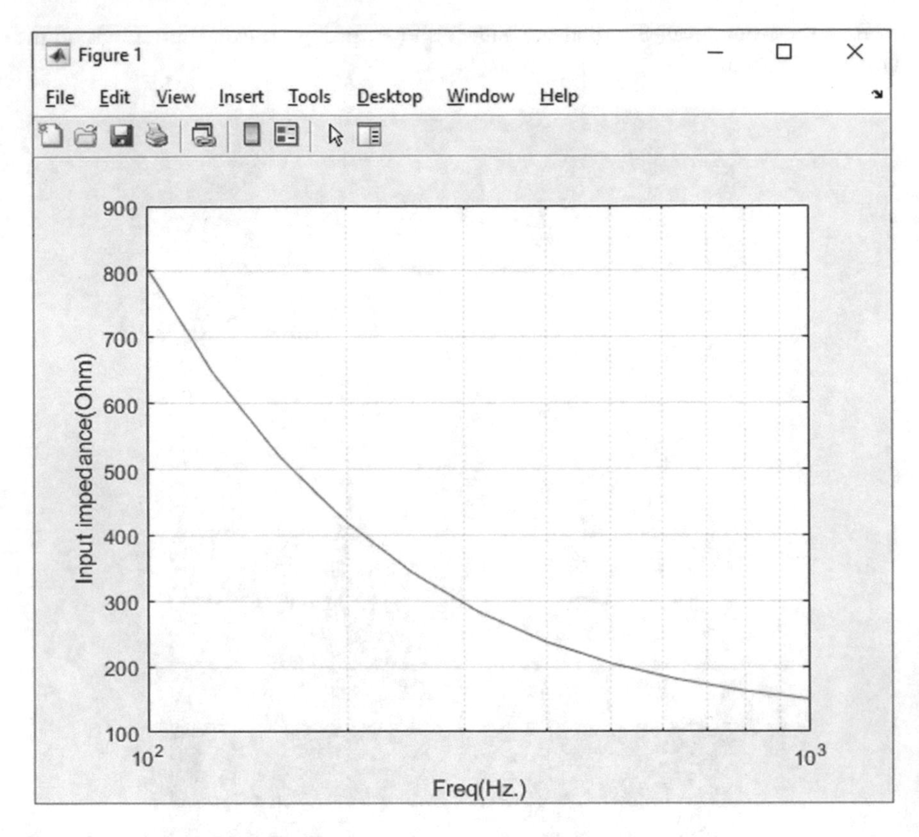

Fig. 1.463 Output of MATLAB command

1.60 Example 59: Input Impedance of Electric Circuits (III)

In the previous two examples we used MATLAB in order to extract the input impedance of a circuit. In this example we extract the input impedance of the circuit shown in Fig. 1.464 without using MATLAB.

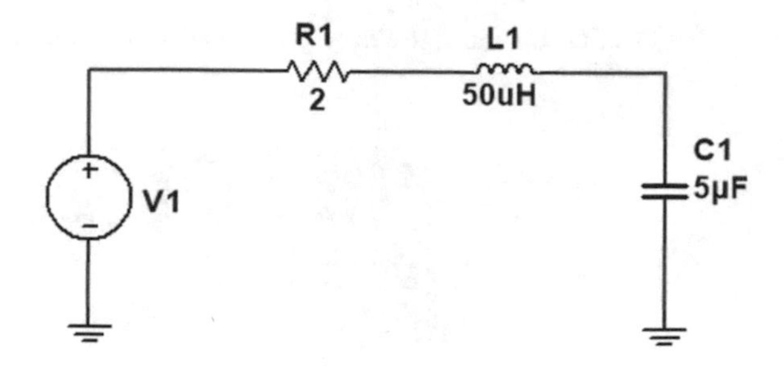

Fig. 1.464 Circuit for Example 59

Draw the schematic shown in Fig. 1.465.

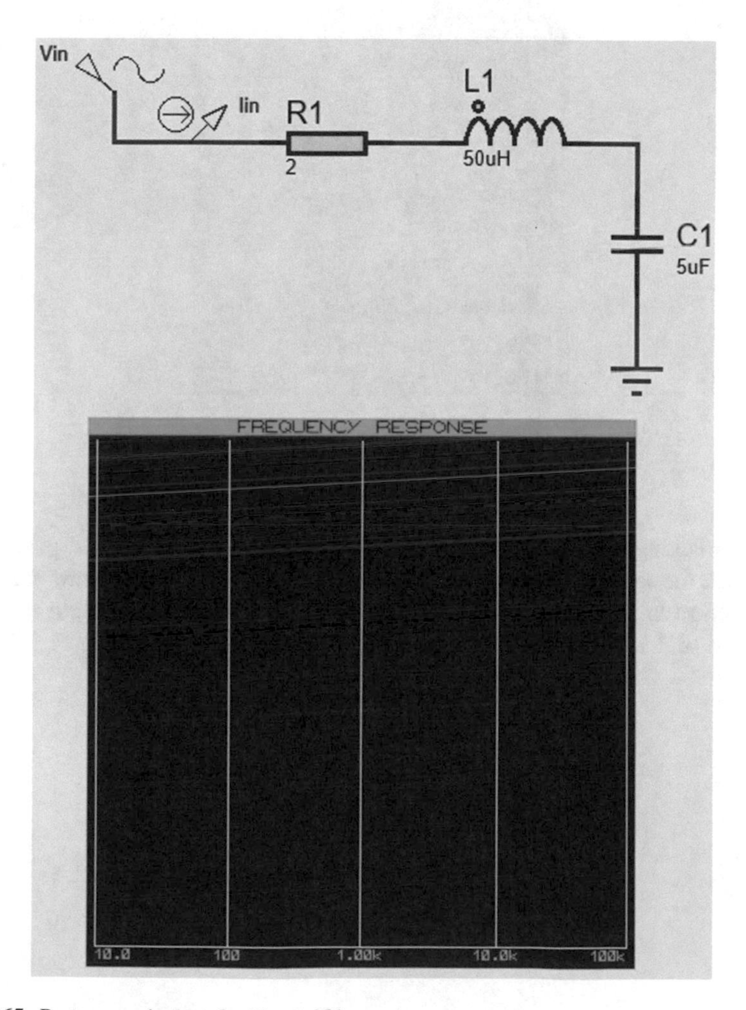

Fig. 1.465 Proteus equivalent for Fig. 1.464

Right click on the frequency graph and click the Add Traces (Fig. 1.466).

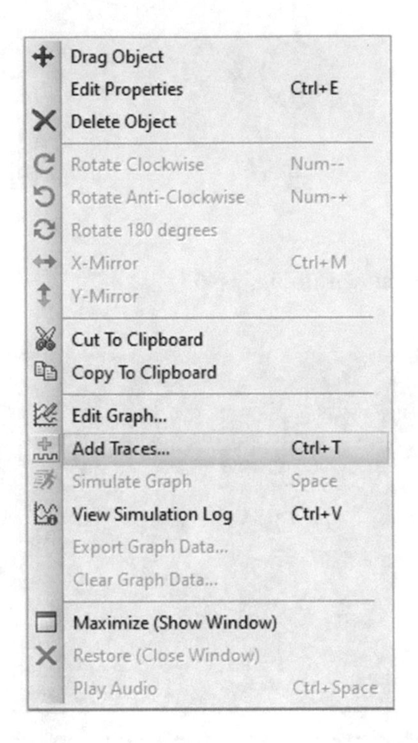

Fig. 1.466 Add Traces command

After clicking the Add Traces, the window shown in Fig. 1.467 appears on the screen. Do the settings similar to Fig. 1.468. Settings in Fig. 1.468 draw the ratio of input voltage to input current which is the input impedance. Since input source has amplitude of 1 and phase of 0°, you can use the settings shown in Fig. 1.469 as well.

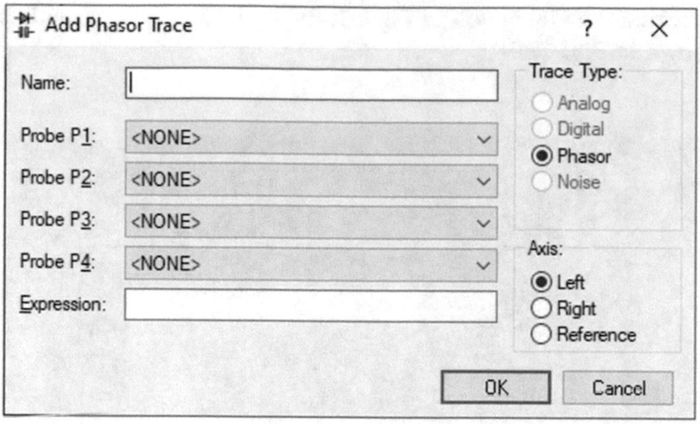

Fig. 1.467 Add phasor trace window

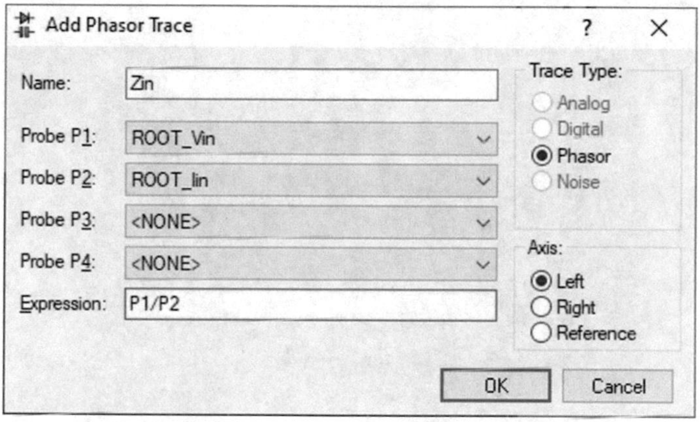

Fig. 1.468 Add phasor trace window

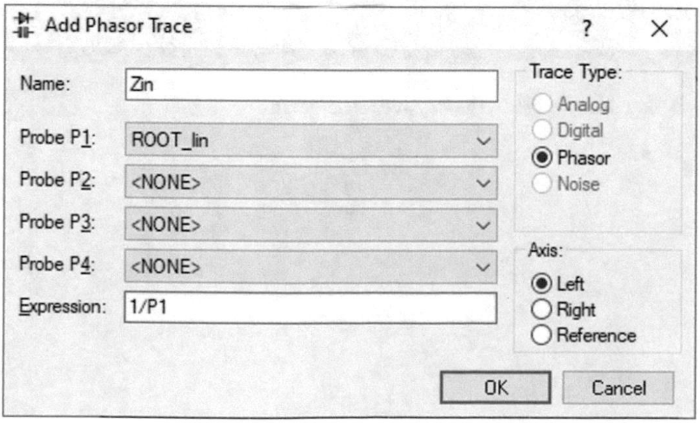

Fig. 1.469 Add phasor trace window

After clicking the OK button in Figs. 1.468 or 1.469, the schematic changes to what is shown in Fig. 1.470.

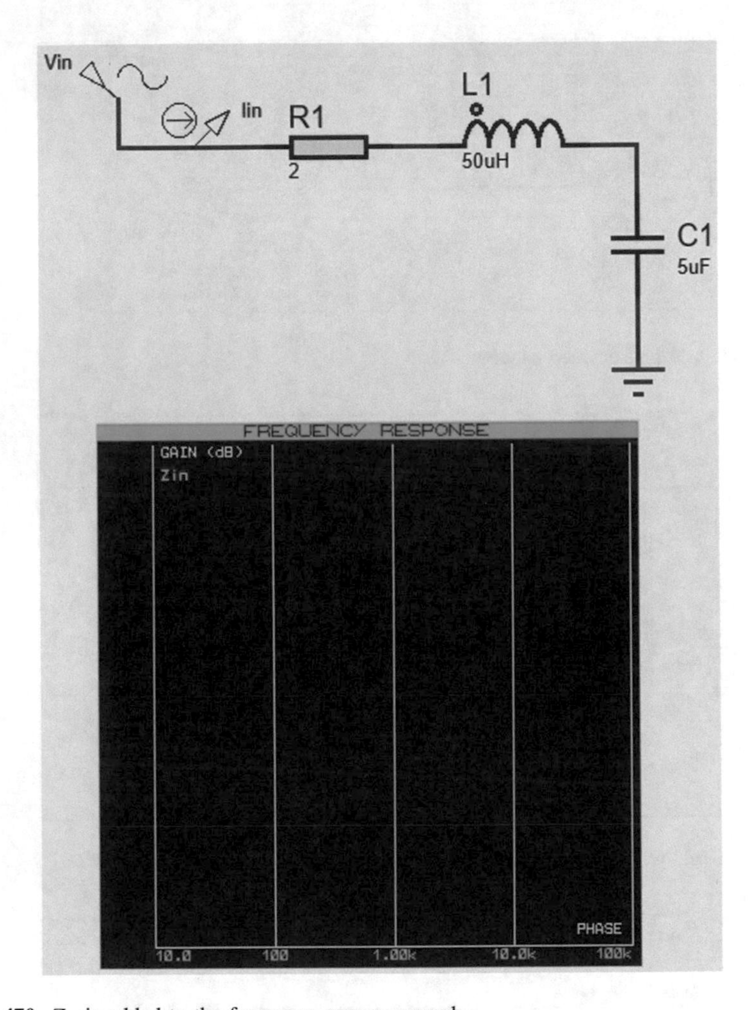

Fig. 1.470 Z_{in} is added to the frequency response graph

Run the simulation by pressing the spacebar key of keyboard. Simulation result is shown in Fig. 1.471. This graph shows the magnitude of input impedance.

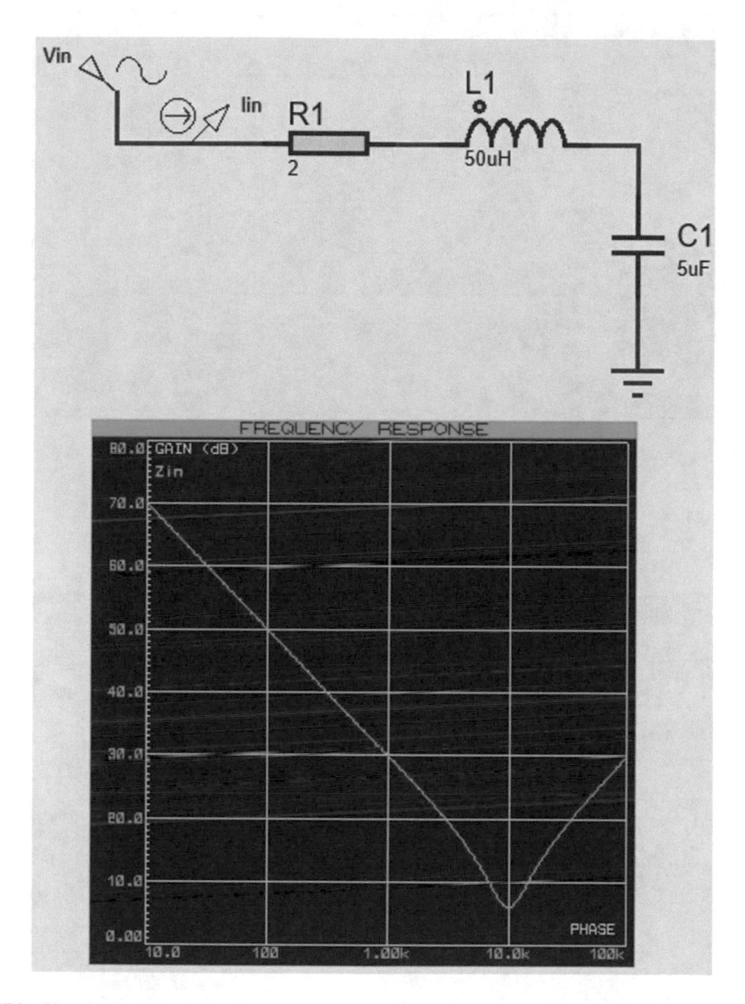

Fig. 1.471 Simulation result

The phase graph can be drawn easily as well. Simply right click on the graph, click the Add Traces and do the settings similar to Figs. 1.472 or 1.473.

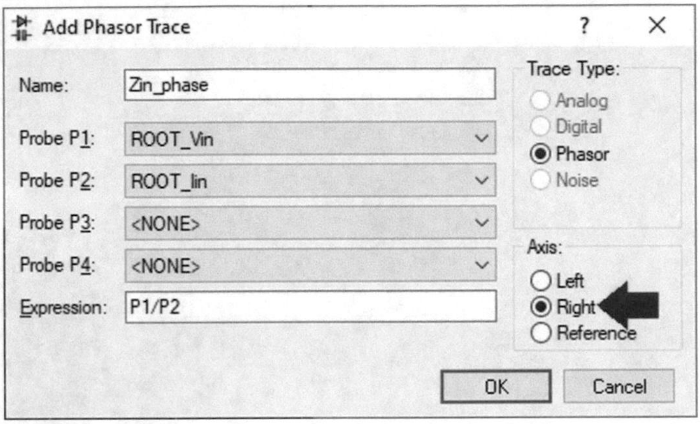

Fig. 1.472 Add phasor trace window

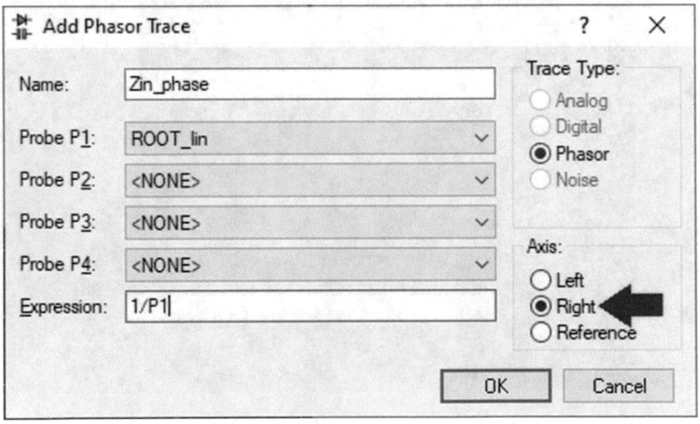

Fig. 1.473 Add phasor trace window

Run the simulation by pressing the spacebar key. Simulation result is shown in Fig. 1.474.

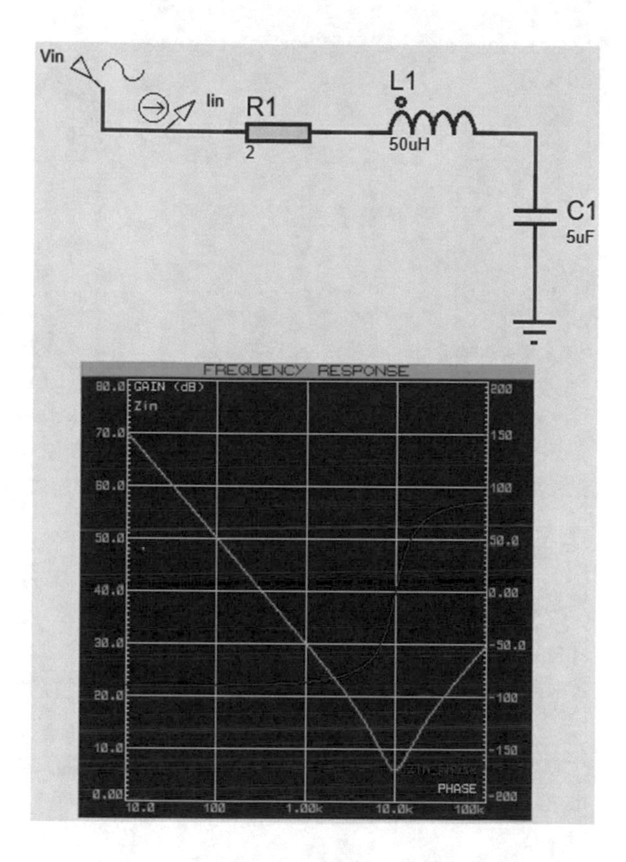

Fig. 1.474 Simulation result

Let's check the obtained result. The commands shown in Fig. 1.475 draw the input impedance of the circuit shown in Fig. 1.464. The output of this code is shown in Fig. 1.476. Convert the horizontal axis into Hz (Fig. 1.477). Now you can compare Fig. 1.477 with Proteus result (Fig. 1.474). Compare different points of the two graphs in order to ensure that Proteus result is correct.

Fig. 1.475 MATLAB commands

```
Command Window
>> R1=2;L1=50e-6;C1=5e-6;f1=10;f2=100e3;
>> s=tf('s');
>> Zin=R1+L1*s+1/C1/s;
>> w=logspace(log10(2*pi*f1),log10(2*pi*f2));
>> bode(Zin,w)
>> grid on
fx >>
```

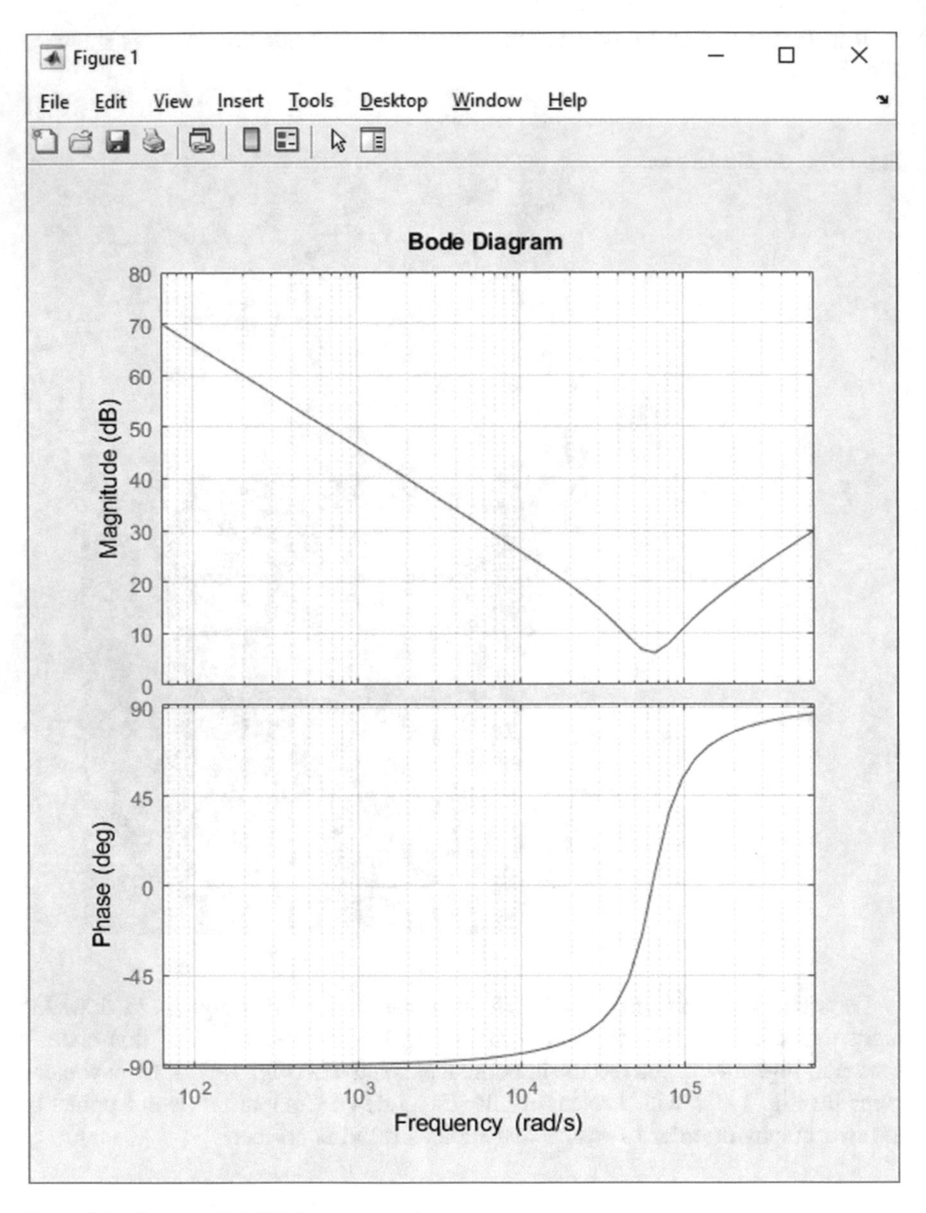

Fig. 1.476 Output of MATLAB commands

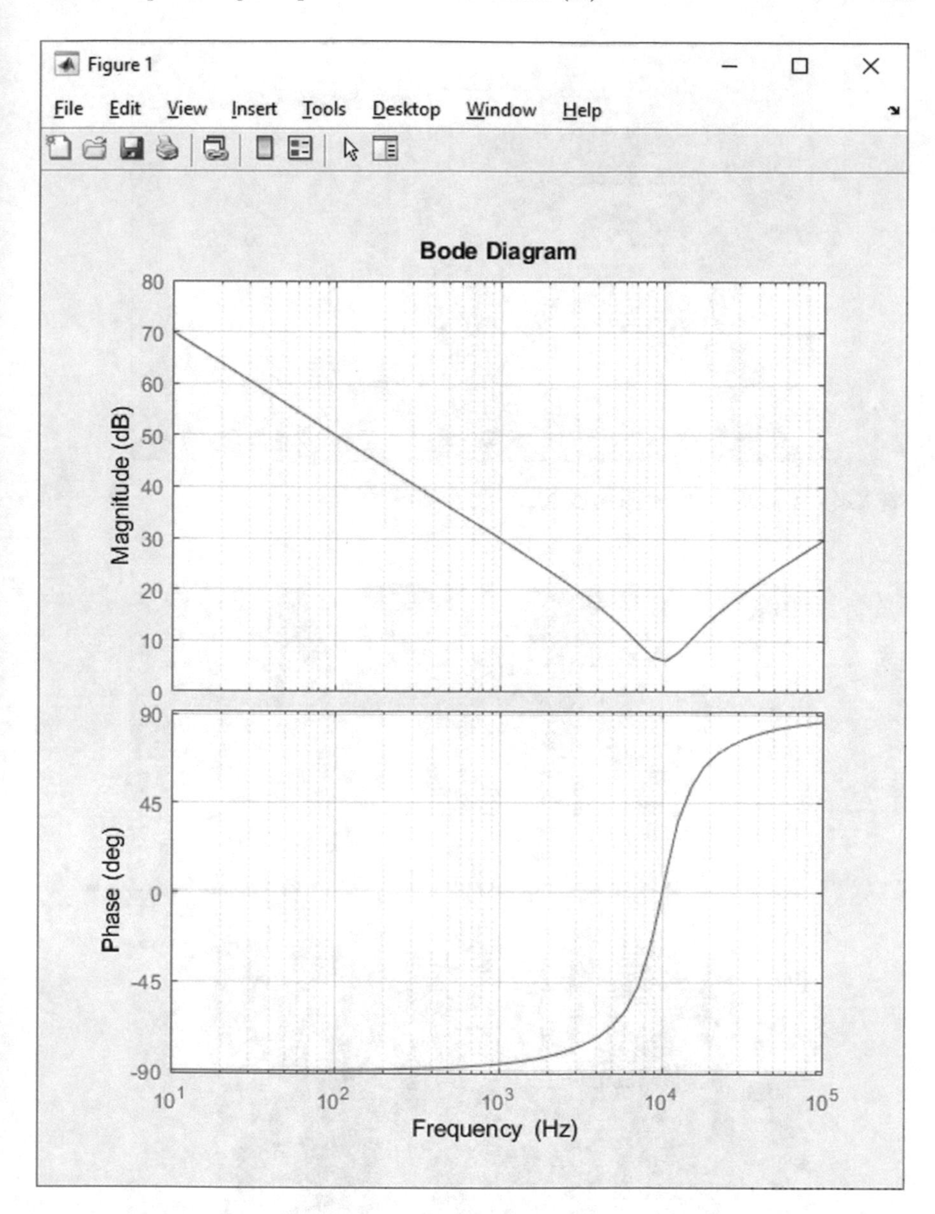

Fig. 1.477 Horizontal axis has the unit of kHz

You can convert the unit of vertical axis into Ohm's as well. In order to do this, double click on the graph and remove the *Y* Scale in dBs (Fig. 1.478). Simulation result is shown in Fig. 1.479. In Fig. 1.479 the vertical axis is in Ohm's.

Fig. 1.478 *Y* scale in dBs is unchecked

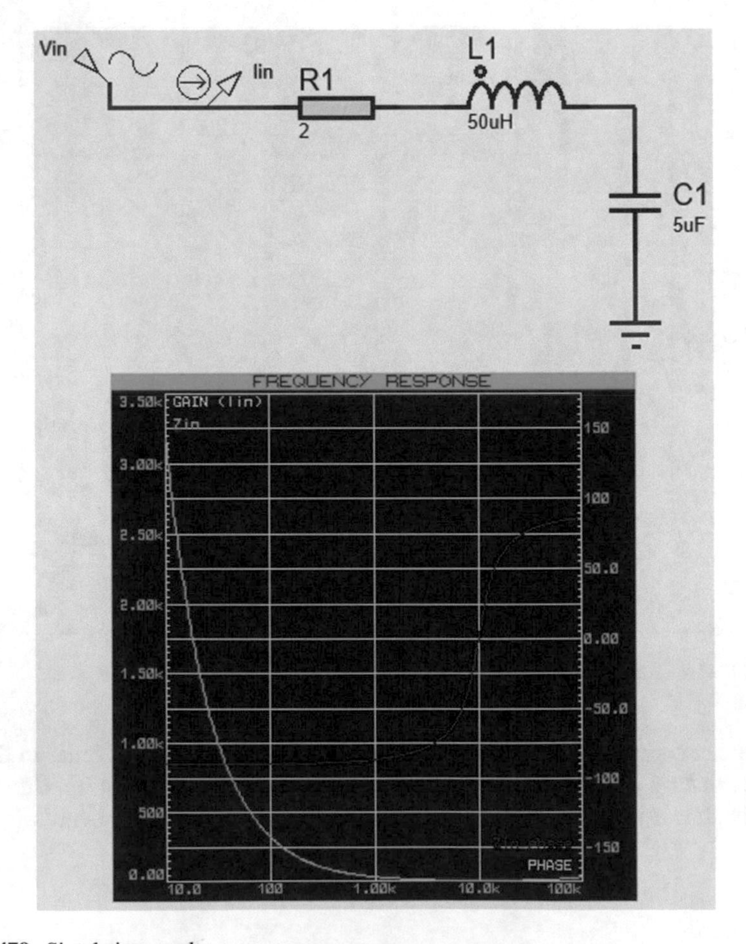

Fig. 1.479 Simulation result

1.61 Example 60: AC Sweep Analysis

You can use AC sweep analysis (Fig. 1.480) to change the value of a component and see its effect on the frequency response of the circuit.

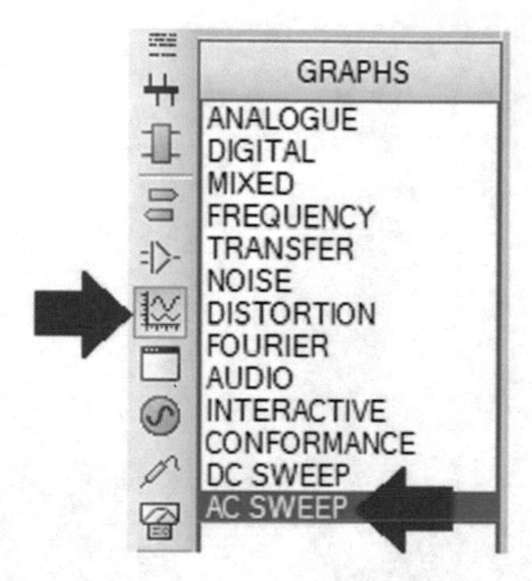

Fig. 1.480 AC sweep analysis block

Let's study an example. Draw the schematic shown in Fig. 1.481. We want to change the value of capacitor and see its effect on the frequency response of the circuit $\left(\frac{V_{out}(j\omega)}{V_{in}(j\omega)}\right)$.

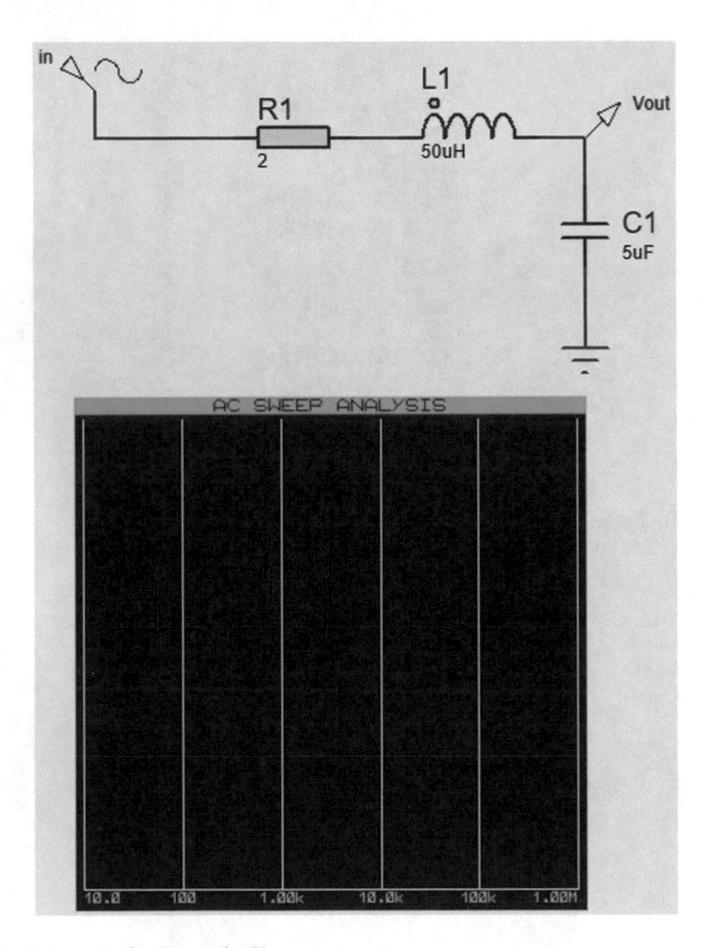

Fig. 1.481 Schematic for Example 60

Double click the capacitor and enter "*X*" to the Capacitance box (Fig. 1.482). Then click the OK button.

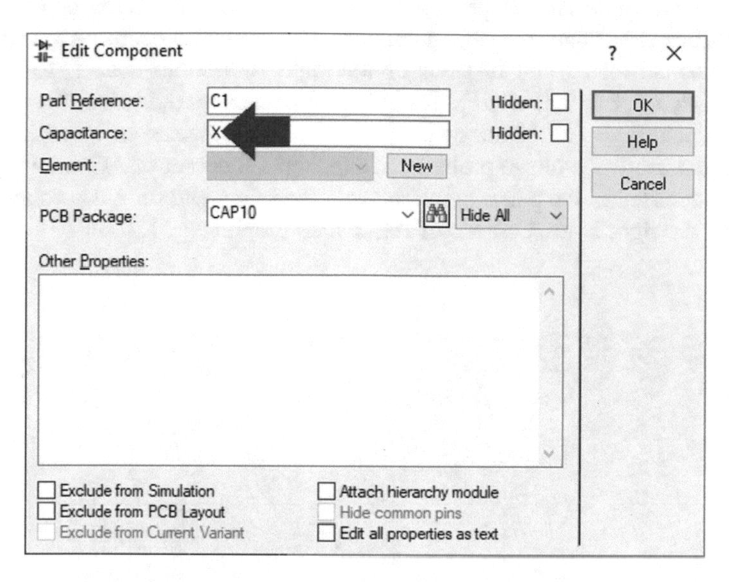

Fig. 1.482 Edit component window

Double click the AC sweep graph and change the settings to what shown in Fig. 1.483. Then click the OK button.

Edit AC Sweep Graph

Graph title: AC SWEEP ANALYSIS

User defined properties:

Reference: ROOT_in

Sweep var: X

Start frequency: 10

Start value: 1e-006

Stop frequency: 100k

Stop value: 1e-005

Interval: DECADES

Nom. value: 5e-006

No. Steps/Interval: 10

No. steps: 10

Options

Y Scale in dBs?

Always simulate?

Log netlist(s)?

SPICE Options

Set Y-Scales

OK

Cancel

Fig. 1.483 Edit AC sweep graph

Reference, Start frequency, Stop frequency, Interval and No. Steps/Interval boxes are studied in Example 55.

The Sweep var box determines the variable which is changed by AC sweep analysis. Start value and Stope value boxes determine the range of change and No. steps box determines the step of changes. For instance, according to Fig. 1.483, the distance between 1 and 10 μF is divided into 10 sections (i.e., 1, 1.9, 2.8, 3.7, 4.6, 5.5, 6.4, 7.3, 8.2, 9.1, 10 μF). The AC sweep analyzes the circuit for each value separately and draws all of the obtained frequency responses on a graph.

Drag and drop the voltage probe V_{out} to the top left corner of AC sweep graph. If you are interested in the phase graph as well, then drag and drop the voltage probe V_{out} onto the right bottom corner of AC sweep graph (Fig. 1.484).

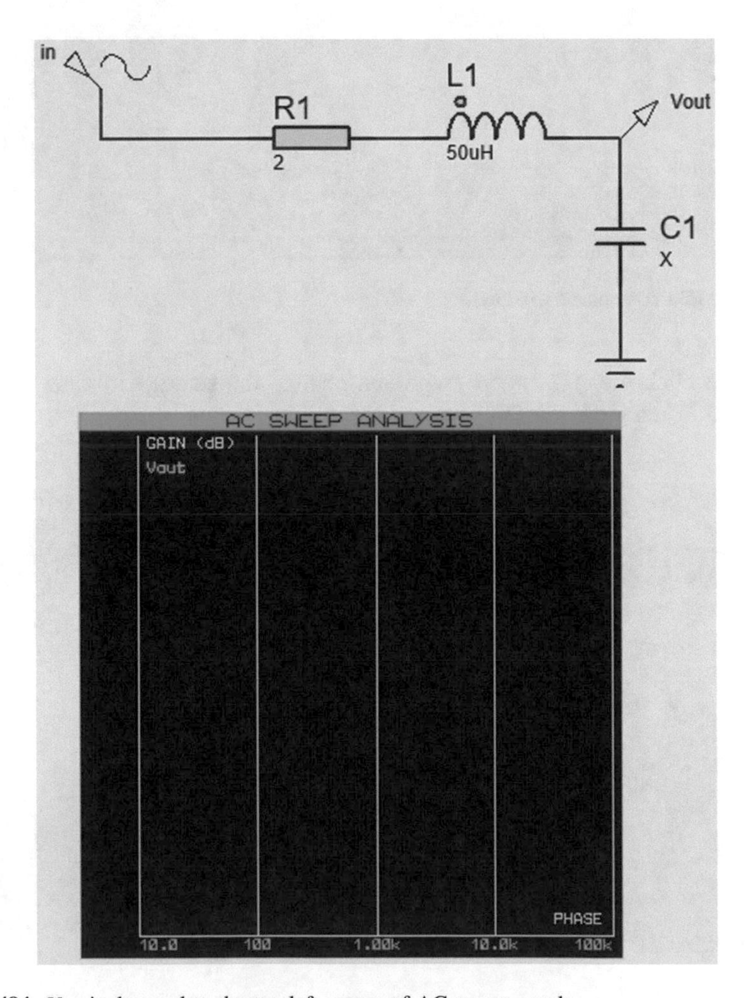

Fig. 1.484 V_{out} is dropped to the top left corner of AC sweep graph

Run the simulation by pressing the space bar key of your keyboard. Simulation result is shown in Fig. 1.485.

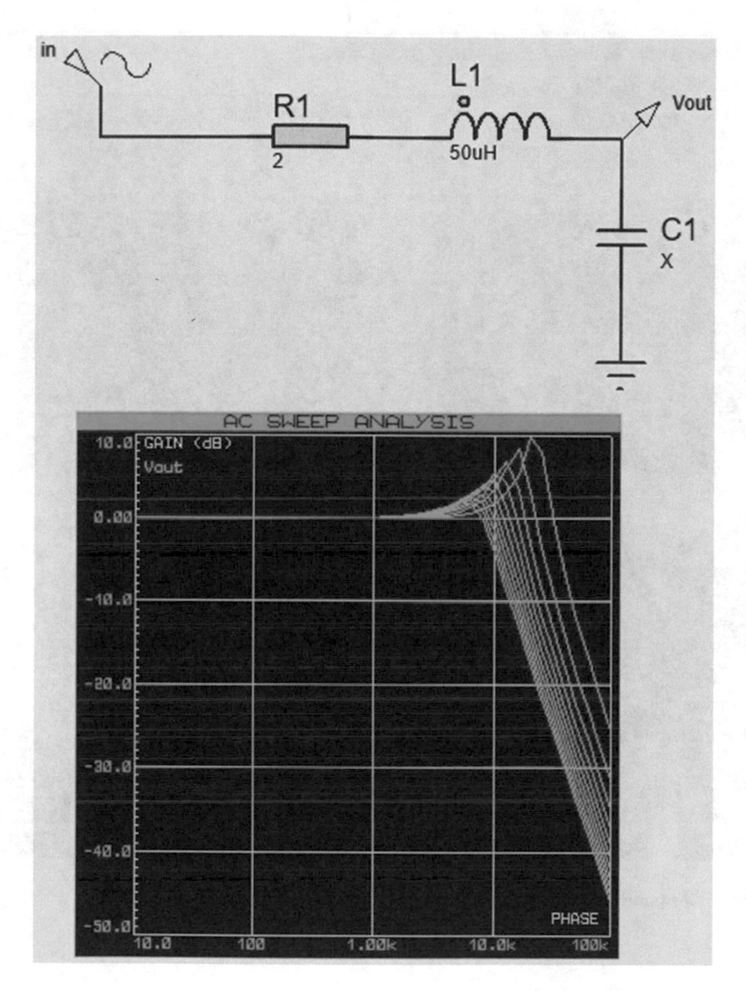

Fig. 1.485 Simulation result

Right click on the obtained graph and click the Maximize (Show Window) to maximize the graph (Fig. 1.486).

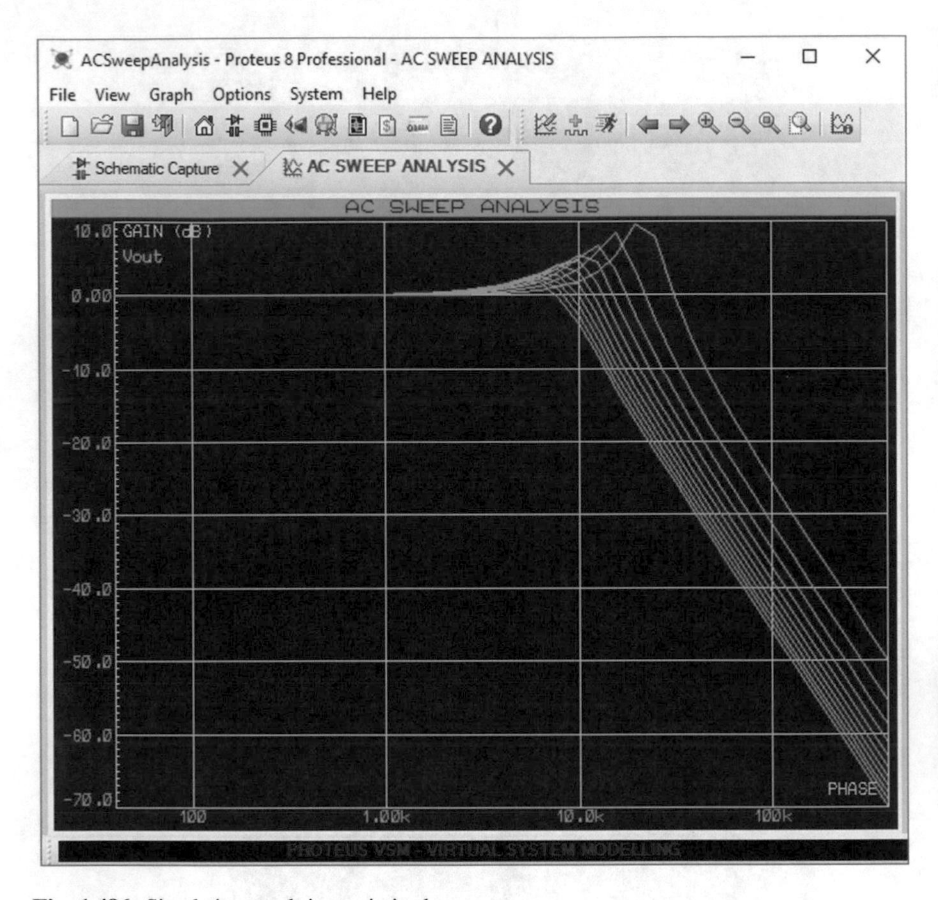

Fig. 1.486 Simulation result is maximized

Click the desired point of the graph to read its coordinate. The coordinate of the point which you clicked and the value of capacitor for that point are shown on the bottom left and right corners of the screen (Fig. 1.487).

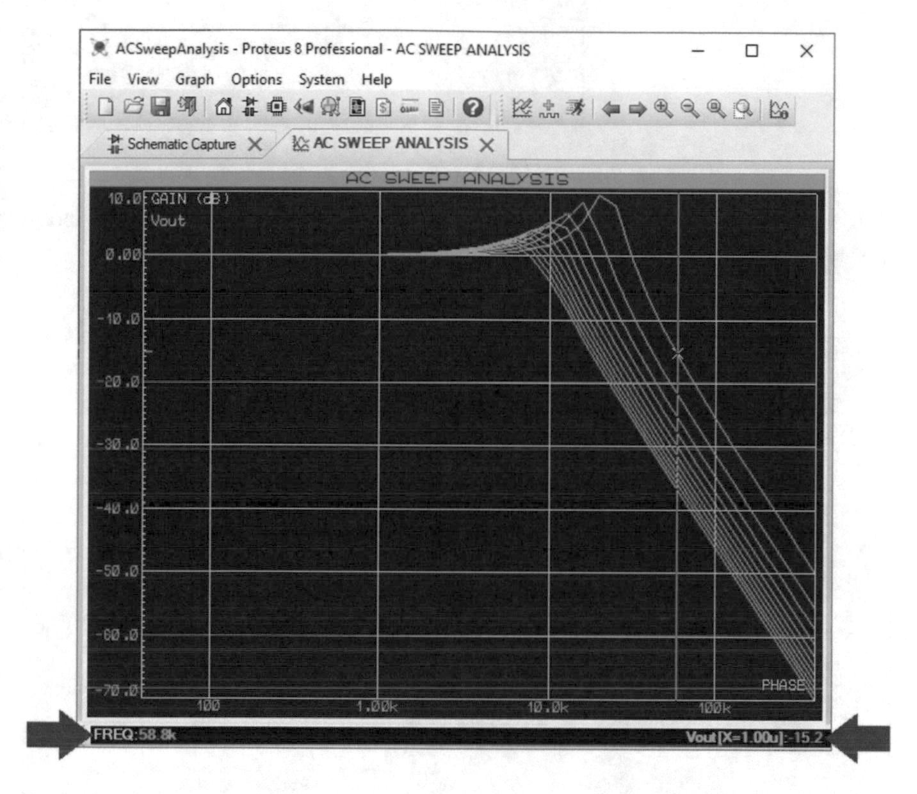

Fig. 1.487 Coordinate of clicked point is shown in the sides of screen

1.62 Example 61: Samples Simulations

Proteus has many ready to use sample simulations. Studying these samples are an effective way to extend your Proteus knowledge. Click the File > Open Sample Project (Fig. 1.488) to open the sample simulation files.

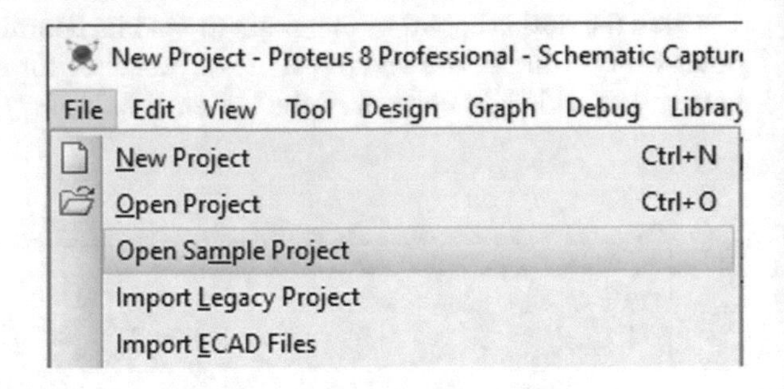

Fig. 1.488 File > open sample project

After clicking the Open Sample Project, the Sample Projects Browser window appears on the screen (Fig. 1.489).

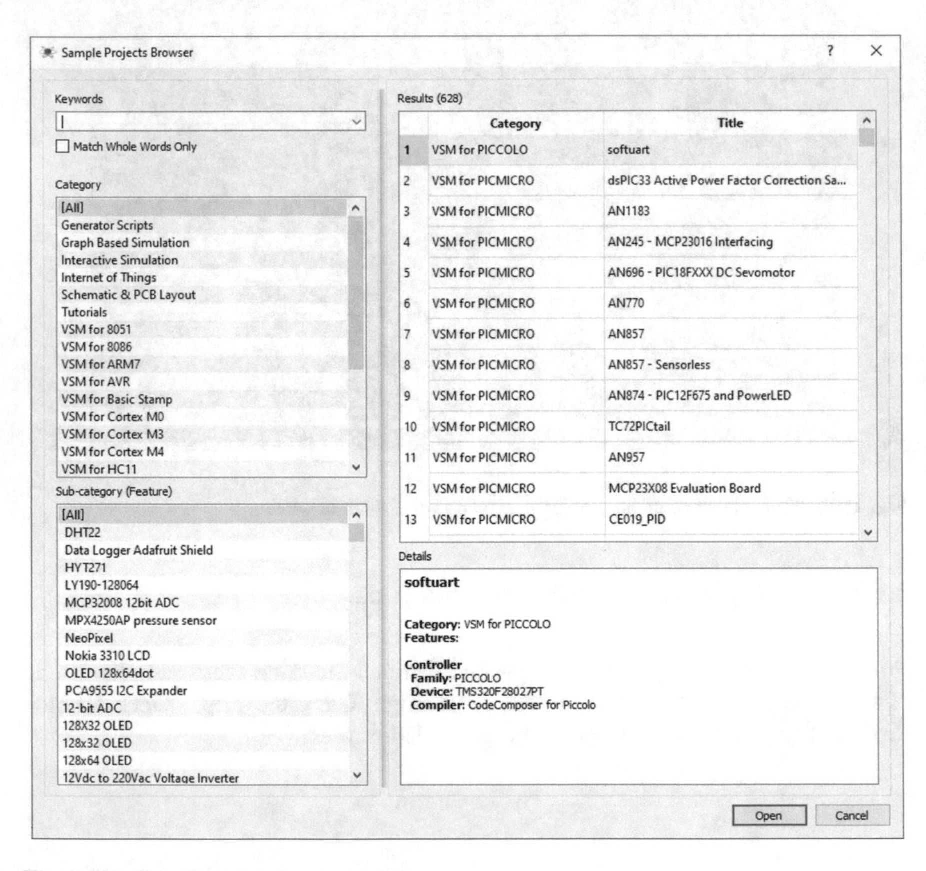

Fig. 1.489 Sample project browser window

Don't forget to see the Graph Based Simulation and Tutorials sections (Figs. 1.490 and 1.491).

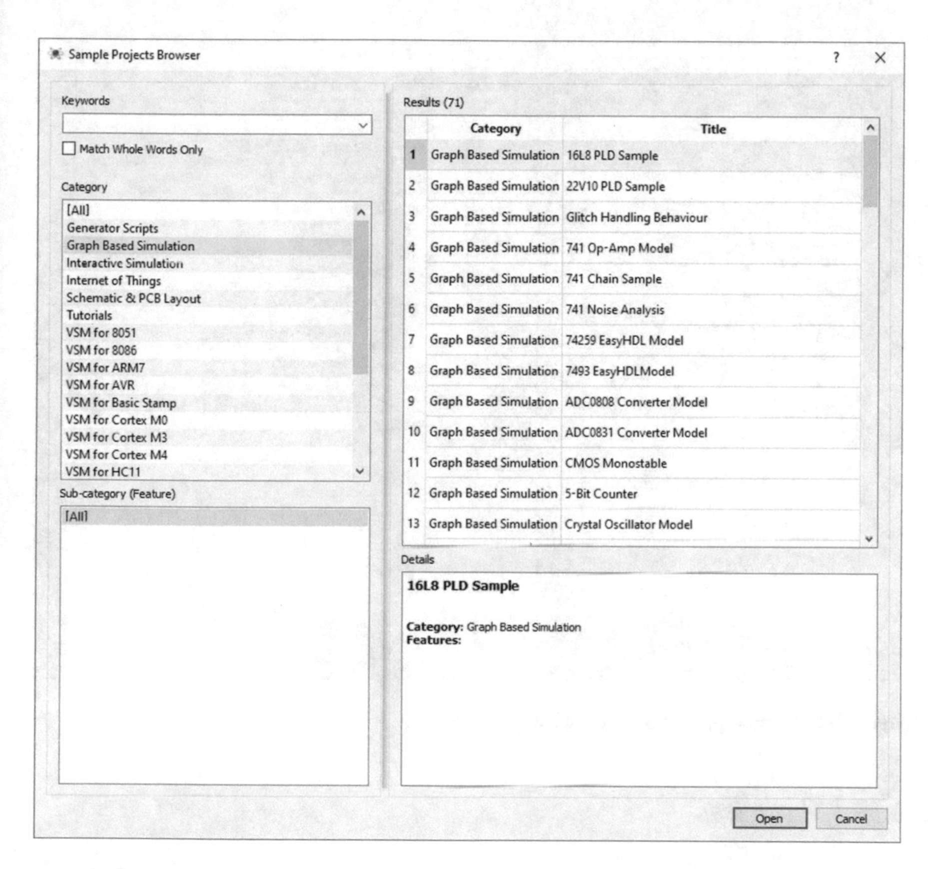

Fig. 1.490 Sample project browser window

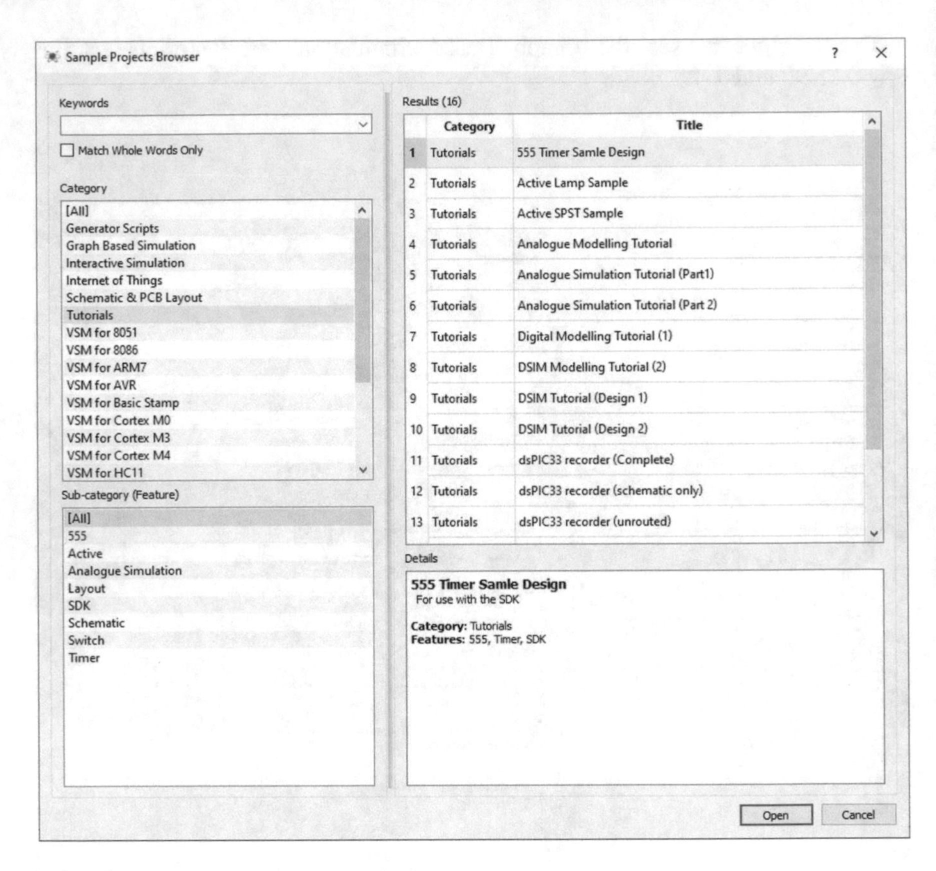

Fig. 1.491 Sample project browser window

1.63 Exercises

1. (a) Find the Thevenin equivalent circuit with respect to the terminals “a” and “b” for the circuit shown in Figs. 1.492 and 1.493.
   (b) Use Proteus to check the result of part (a).

Fig. 1.492 Circuit for exercise 1

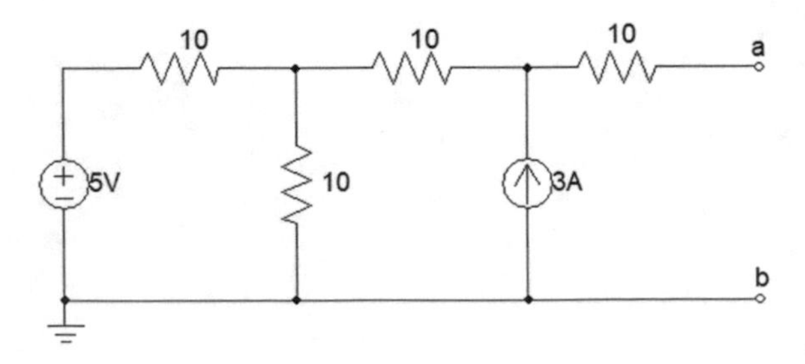

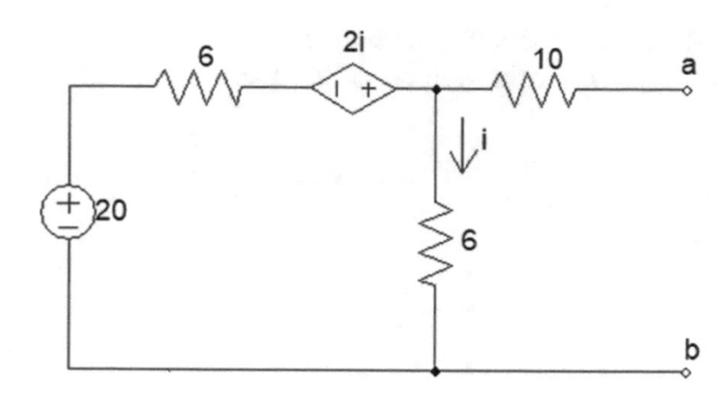

Fig. 1.493 Circuit for exercise 1

2. Simulate the circuit shown in Fig. 1.494 with Proteus. Initial conditions are shown on the figure.

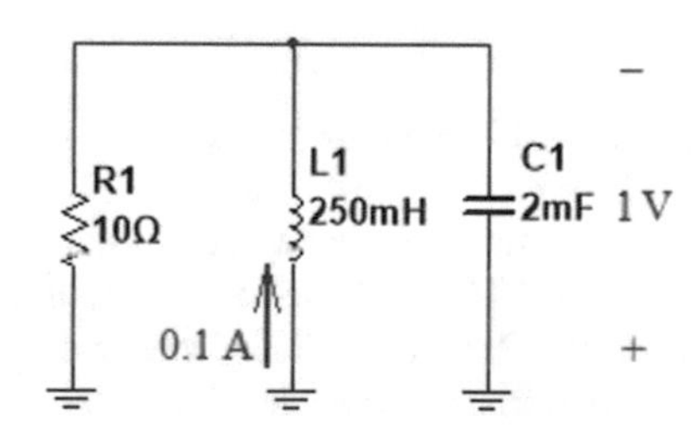

Fig. 1.494 Circuit for exercise 2

3. In the circuit shown in Fig. 1.495, $V_1 = 10 + 25\sin(2\pi \times 60t)$. Initial conditions are $V_{C,0} = 10\,\text{V}$ and $i_{L,0} = 0\,\text{A}$. Use Proteus to observe the circuit current.

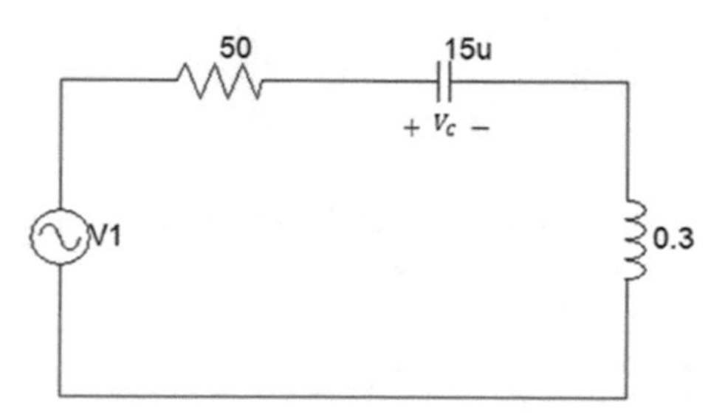

Fig. 1.495 Circuit for exercise 3

4. (a) Calculate the current i in the circuit of Fig. 1.496.
   (b) Use Proteus to check the result of part (a).

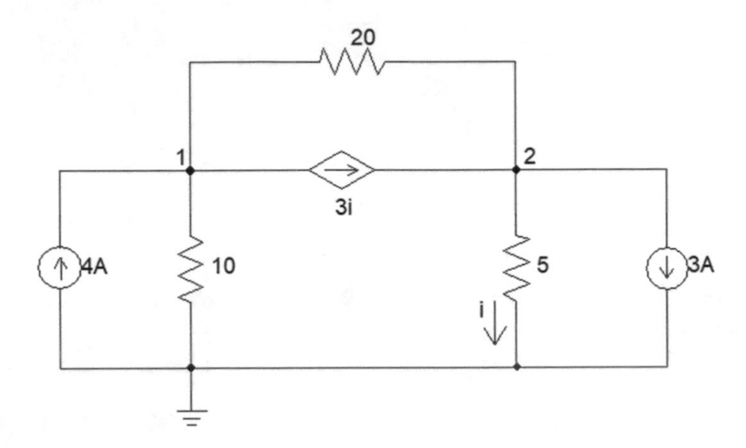

Fig. 1.496 Circuit for exercise 4

5. Set up a Proteus simulation to measure the RMS of a triangular wave. Use MATLAB or hand calculation to verify the obtained result.
6. Use Proteus to find the value of the load resistor R_{load} which consumes the maximum power in Fig. 1.497.

Fig. 1.497 Circuit for exercise 6

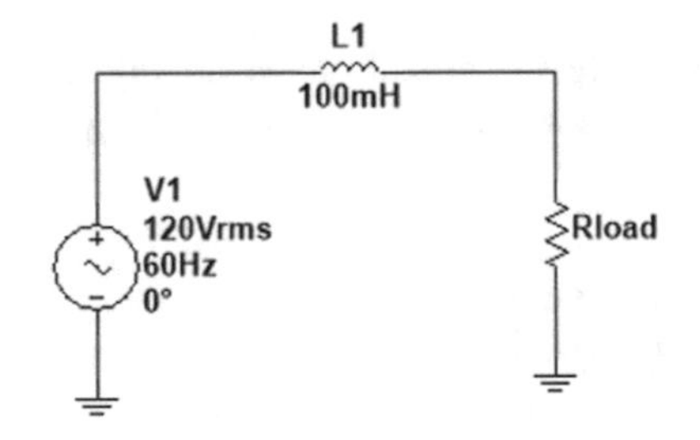

7. Calculate the impulse response of the circuit shown in Fig. 1.498 (Output is capacitor voltage).

Fig. 1.498 Circuit for exercise 7

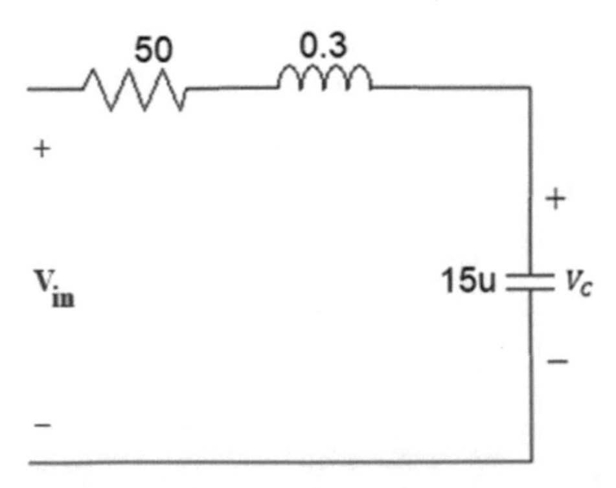

References for Further Study

Asadi F (2022) Electric circuit analysis with EasyEDA, Springer
Alexander C, Sadiku MNO (2016) Fundamentals of electric circuits, 6th edn. McGraw-Hill
Hayt W, Kemmerly J, Durbin S (2021) Engineering circuit analysis, 9th edn. McGraw-Hill
Nilsson J, Riedel S (2018) Electric circuits, 11th edn. Pearson
Thomas RE, Rosa AJ, Toussaint GJ (2020) The analysis and design of linear circuits, 9th edn. Wiley

Chapter 2
Simulation of Electronic Circuits with Proteus®

Abstract This chapter shows how an electronic circuit can be analyzed in Proteus. In this chapter you will learn how to do a DC sweep analysis, how to simulate a rectifier circuit and measure its parameters, how to simulate a transistor amplifier and measure its gain, input impedance and output impedance, how to simulate a differential pair and measure its CMRR and how to measure THD of an oscillator.

Keywords DC sweep · Diode · Zener diode · Rectifier · Half wave rectifier · Full wave rectifier · Ripple · Common mode rejection ratio · CMRR · Common emitter · Common base · Common collector · Input impedance · Output impedance · Frequency response of an amplifier · Optocoupler · Colpitts Oscillator · Relay

2.1 Introduction

In this chapter, you will learn how to analyze electronic circuits in Proteus. The theory behind the studied circuits can be found in any standard electronic/ microelectronic textbook (Razavi 2021; Rashid 2016; Sedra et al. 2019). Similar to previous chapter, doing some hand calculations for the given circuits and comparing the hand analysis results with Proteus results are recommended.

2.2 Example 1: DC Sweep Analysis

Consider the circuit shown in Fig. 2.1. In this example, we want to change the value of V_1 from 0 to 3 V with $\frac{3}{50} = 60$ mV steps and see its effect on the current drawn from the voltage source V_1. DC sweep analysis can be used to solve this problem.

F. Asadi, *Essential Circuit Analysis Using Proteus®*, Energy Systems in Electrical Engineering, https://doi.org/10.1007/978-981-19-4353-9_2

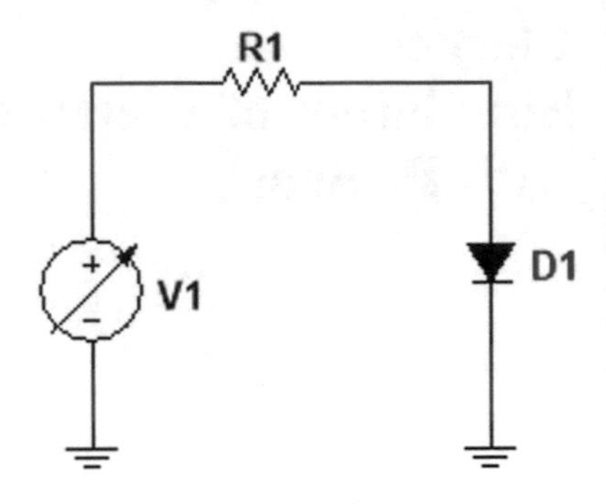

Fig. 2.1 Circuit for Example 1

Draw the schematic shown in Fig. 2.2. The diode 1N4007 can be added to the schematic by searching for 1N4007 in Pick Devices window (Fig. 2.3).

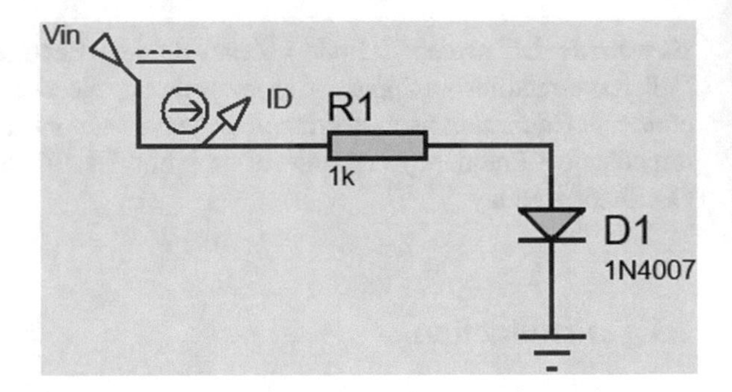

Fig. 2.2 Proteus equivalent of schematic in Fig. 2.1

Pick Devices
Keywords:
1n4007
Match whole words?
Show only parts with models?

Fig. 2.3 Searching for 1N4007

Add a DC sweep graph (Fig. 2.4) to the schematic (Fig. 2.5).

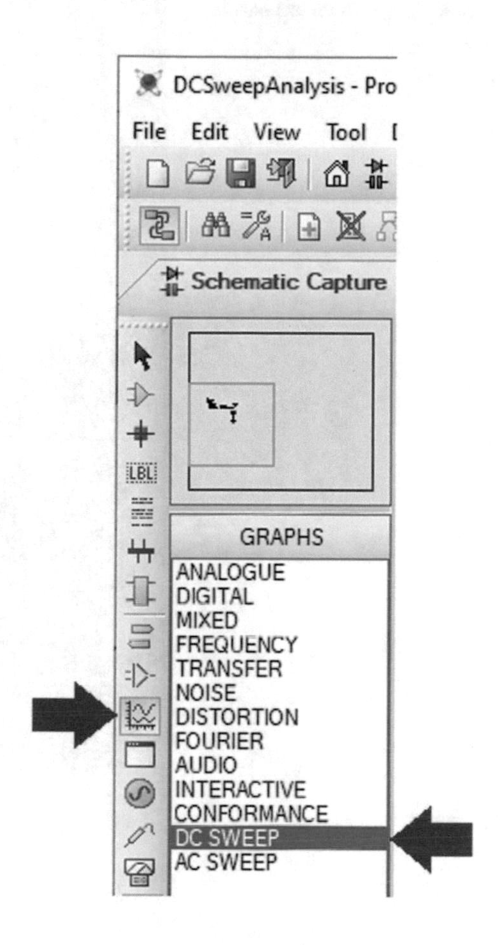

Fig. 2.4 DC sweep graph

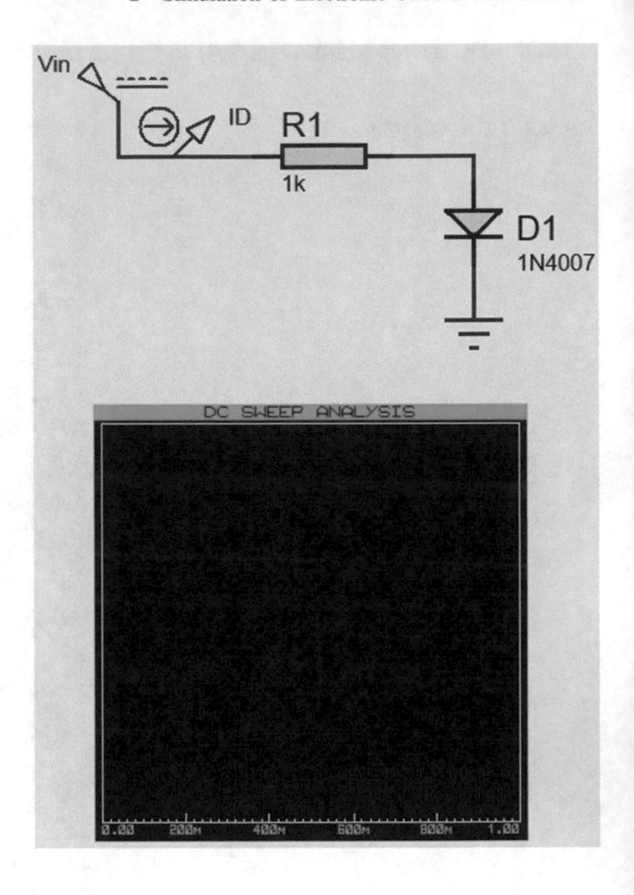

Fig. 2.5 Addition of DC sweep graph to the schematic

Double click the V_{in} and enter X to the Voltage (Volts) box (Fig. 2.6).

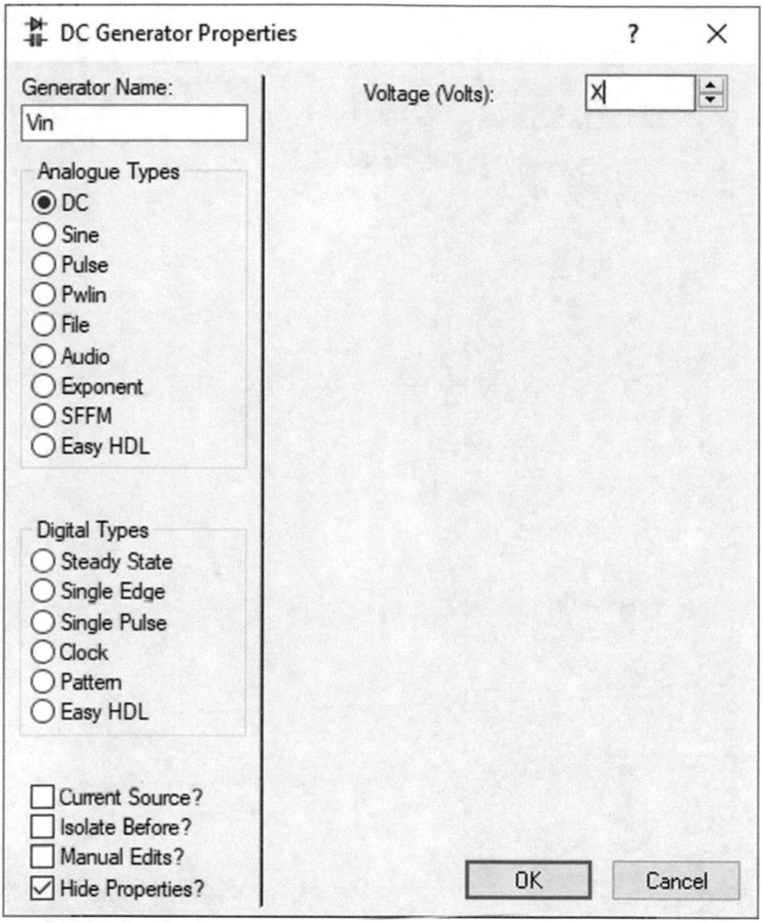

Fig. 2.6 Settings of V_{in}

Click the Manual Edits box (Fig. 2.7).

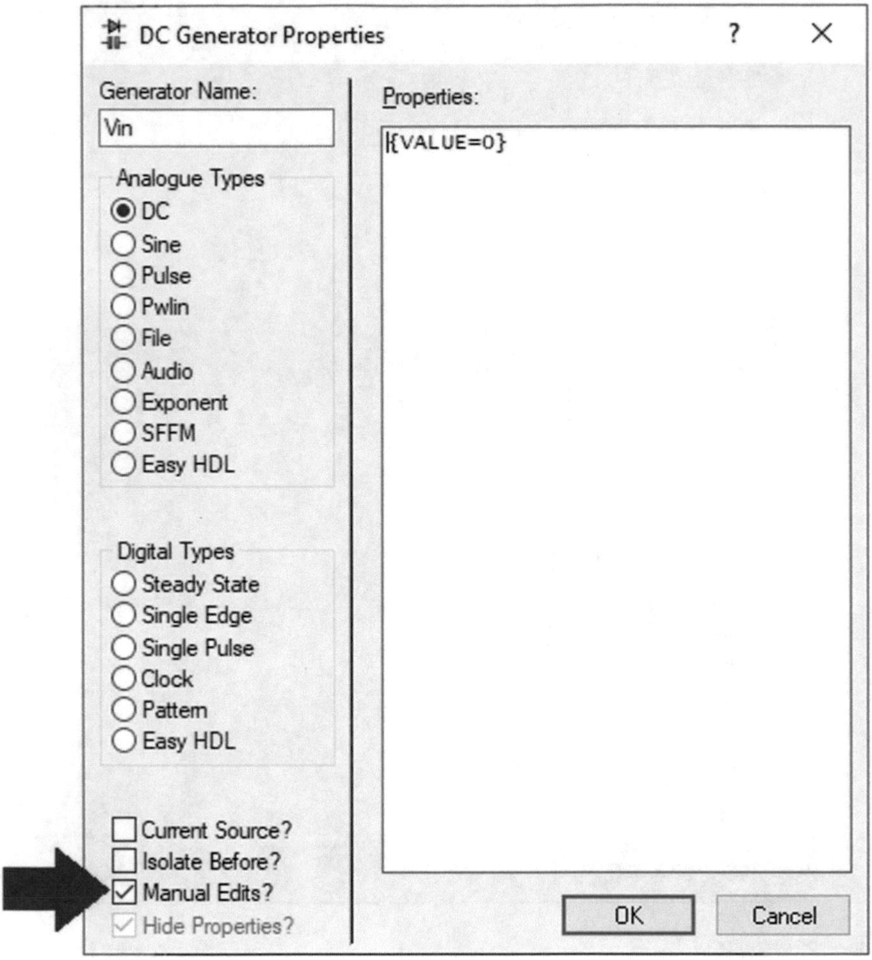

Fig. 2.7 Settings of V_{in}

Change the {VALUE = 0} to {VALUE = X} and click the OK button (Fig. 2.8).

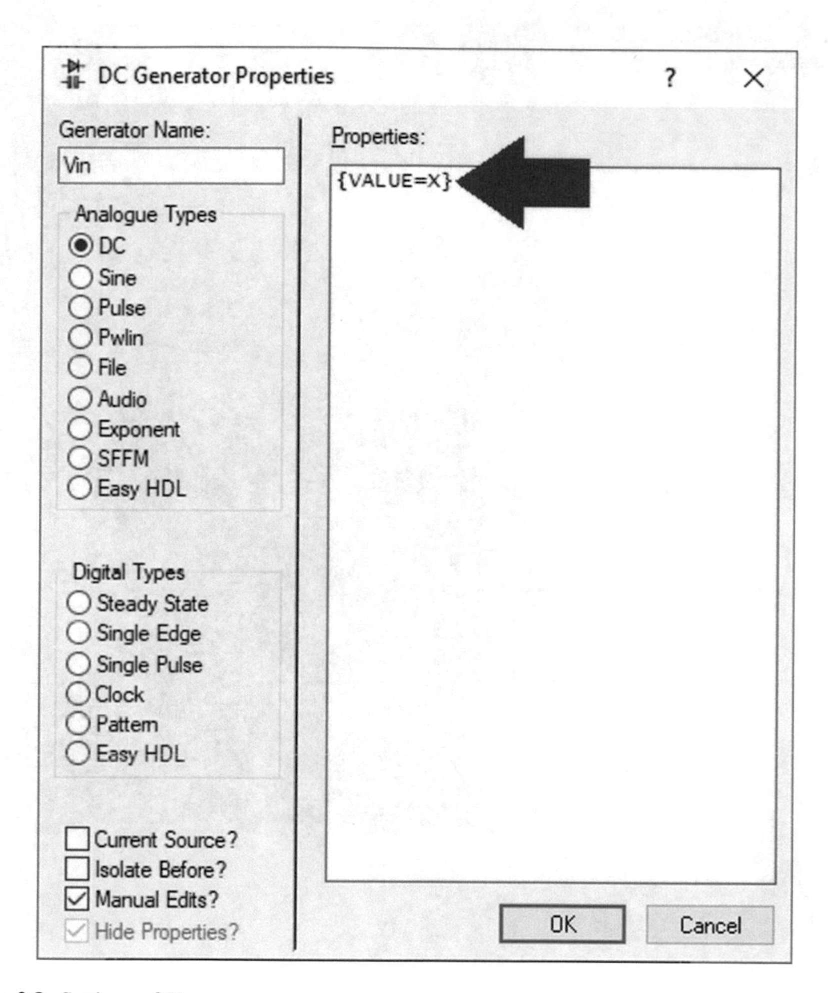

Fig. 2.8 Settings of V_{in}

Drag and drop the current probe "ID" onto the DC sweep graph (Fig. 2.9).

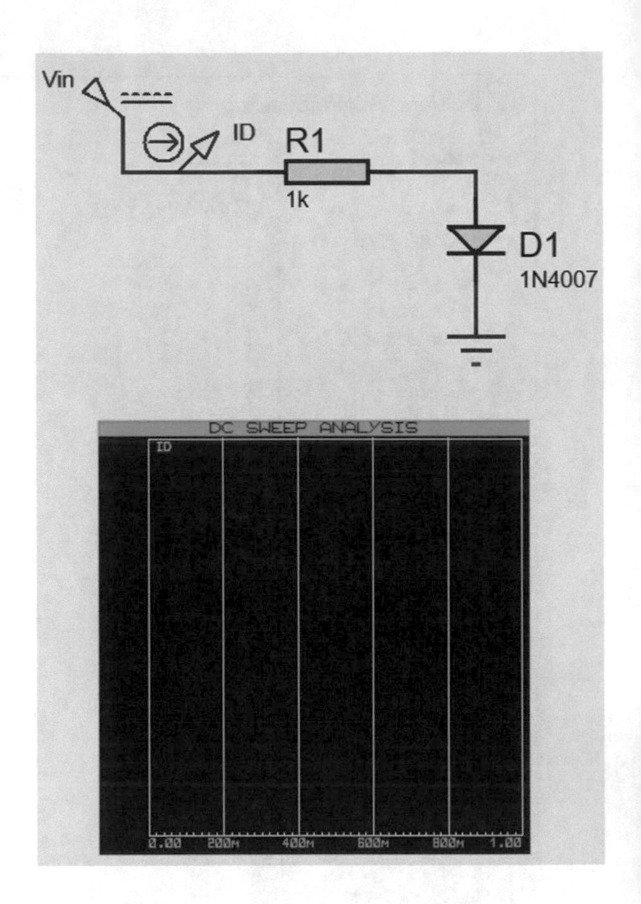

Fig. 2.9 ID is added to the DC sweep graph

Double click the DC sweep graph and change the settings to what is shown in Fig. 2.10. These settings change the value of variable X (voltage of V_{in}) from 0 to 3 V with $\frac{\text{Stop value}-\text{Start value}}{\text{No. steps}} = \frac{3-0}{50} = 60$ mV steps.

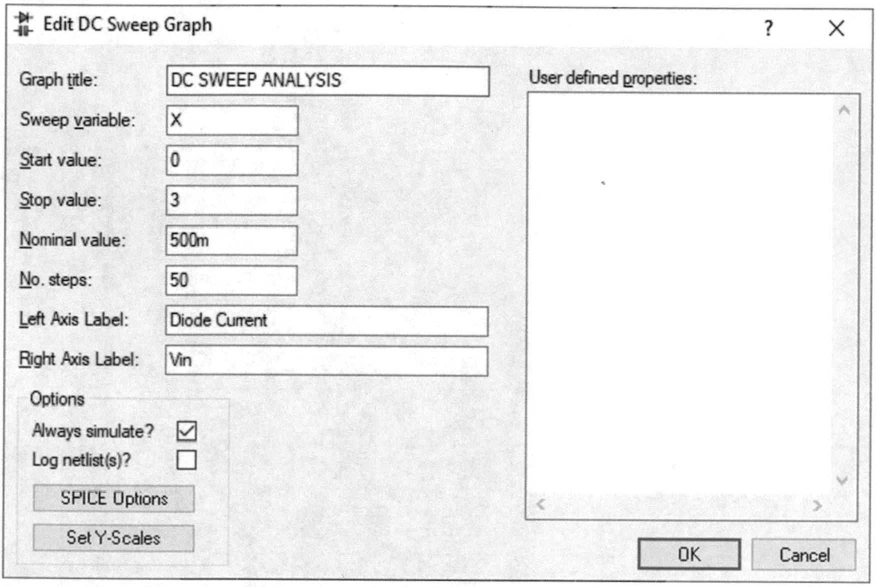

Fig. 2.10 DC sweep graph settings

Run the simulation. Simulation result is shown in Fig. 2.11. Value of variable X is shown on the horizontal axis, and value of current probe ID is shown on the vertical axis.

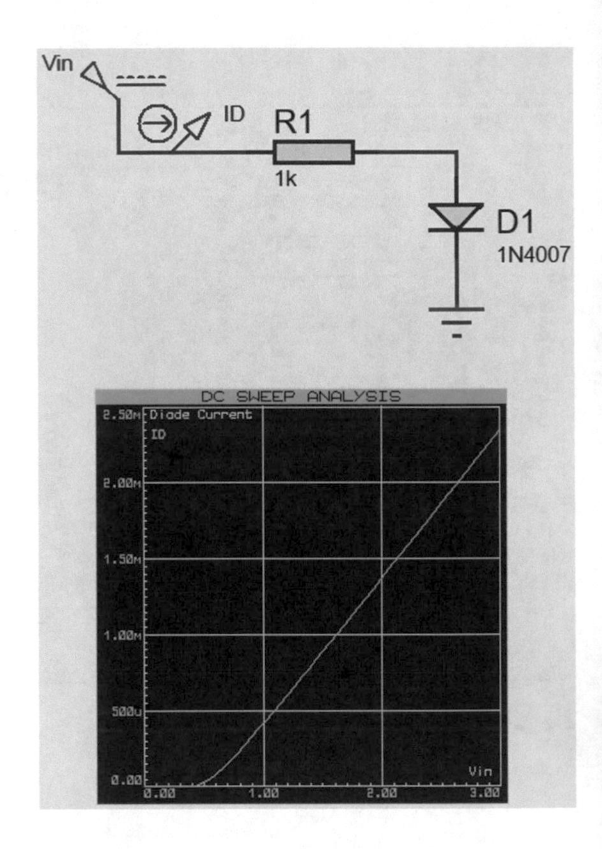

Fig. 2.11 Simulation result

Right click on the graph and click the Maximize (Show Window). This maximizes the simulation result. If you click on the maximized graph, a cursor is added and permits you to read the graph (Fig. 2.12).

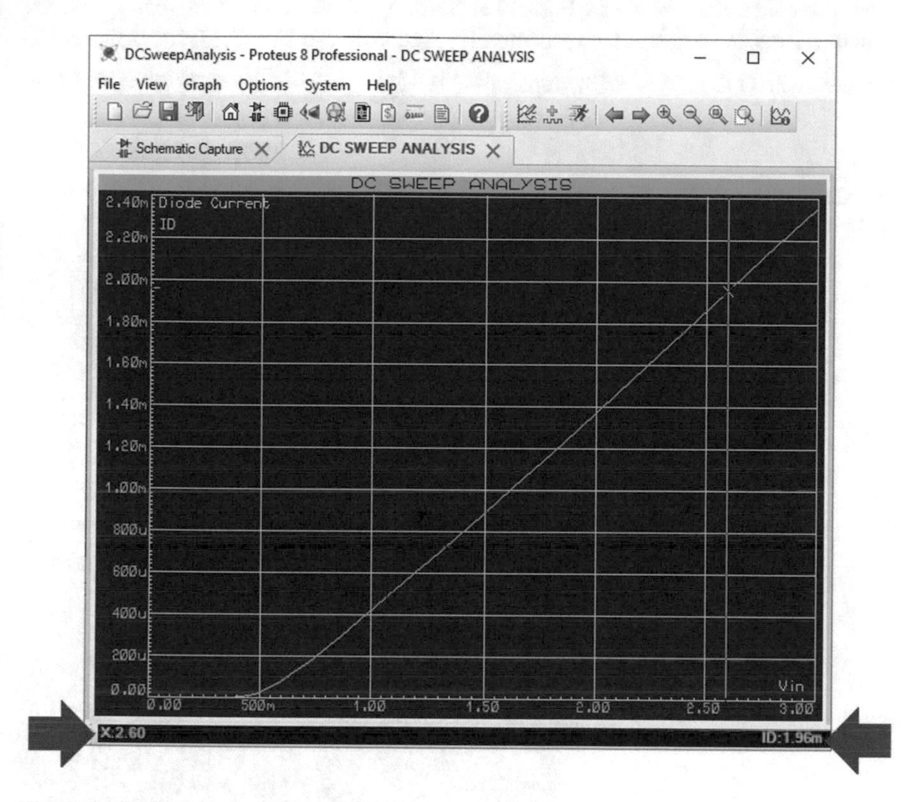

Fig. 2.12 Simulation result is maximized

2.3 Example 2: Diode IV Characteristic

In this example, we want to see the current voltage (IV) characteristic of a diode. Draw the schematic shown in Fig. 2.13. Settings of VD and DC sweep graph are shown in Figs. 2.14 and 2.15, respectively. According to Fig. 2.15, the voltage VD changes from 0 to 0.65 V with steps of $\frac{\text{Stop value} - \text{Start value}}{\text{No. steps}} = \frac{0.65-0}{100} = 6.5\ \text{mV}$.

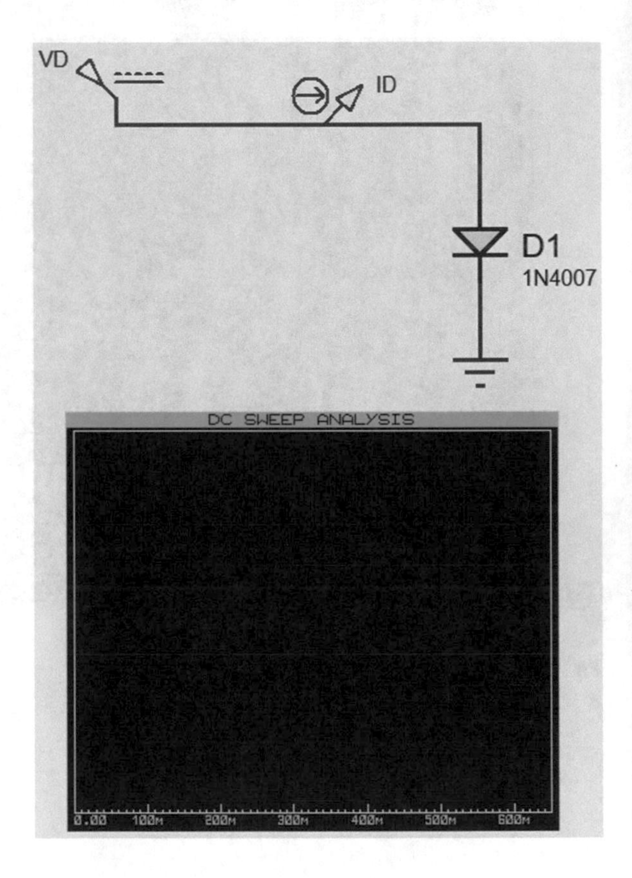

Fig. 2.13 Schematic of Example 2

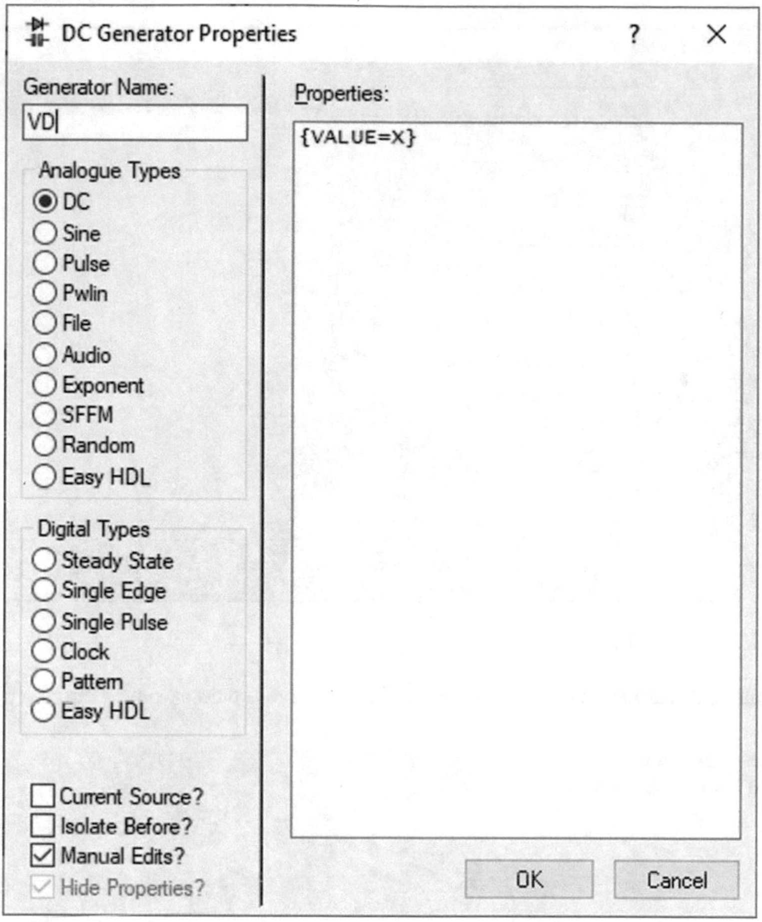

DC Generator Properties ? ×

Generator Name:
VD

Analogue Types
DC
Sine
Pulse
Pwlin
File
Audio
Exponent
SFFM
Random
Easy HDL

Digital Types
Steady State
Single Edge
Single Pulse
Clock
Pattern
Easy HDL

Current Source?
Isolate Before?
Manual Edits?
Hide Properties?

Properties:
{VALUE=X}

OK Cancel

Fig. 2.14 Settings of VD

Edit DC Sweep Graph

Graph title: DC SWEEP ANALYSIS
Sweep variable: X
Start value: 0
Stop value: 0.65
Nominal value: 500m
No. steps: 100
Left Axis Label: VD
Right Axis Label: ID
Options
Always simulate? ☑
Log netlist(s)? ☐
SPICE Options
Set Y-Scales
User defined properties:
OK Cancel

Fig. 2.15 Settings of DC sweep graph

Drag and drop the current probe ID onto the DC sweep graph (Fig. 2.16).

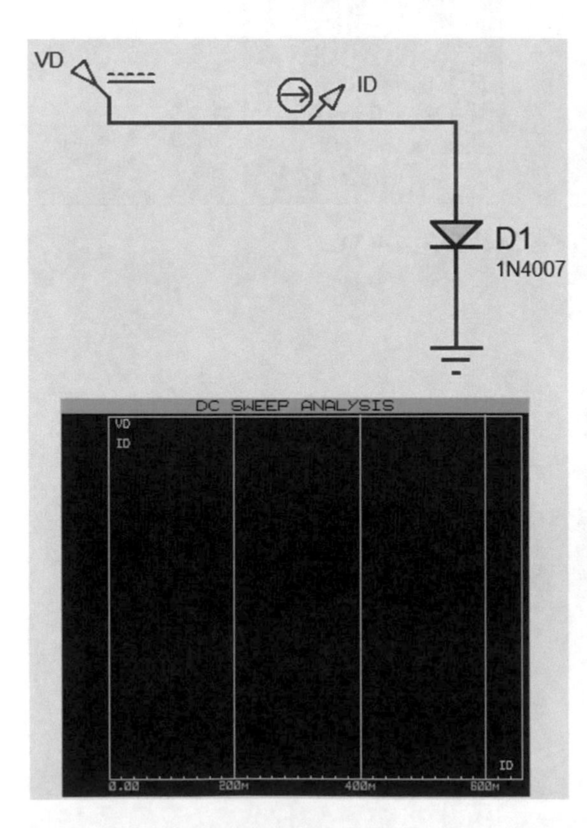

Fig. 2.16 Current probe ID is dropped onto the DC sweep graph

Run the simulation. Simulation result is shown in Fig. 2.17.

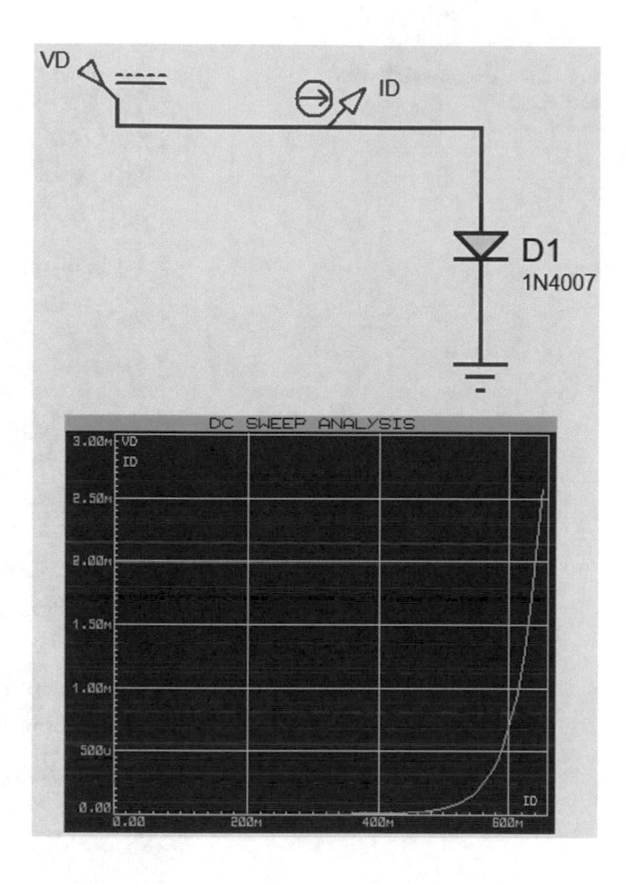

Fig. 2.17 Simulation result

Let's export the obtained result. Right click on the DC sweep graph and click the Export Graph Data (Fig. 2.18). Data are exported into "DiodeIV.dat" in drive C.

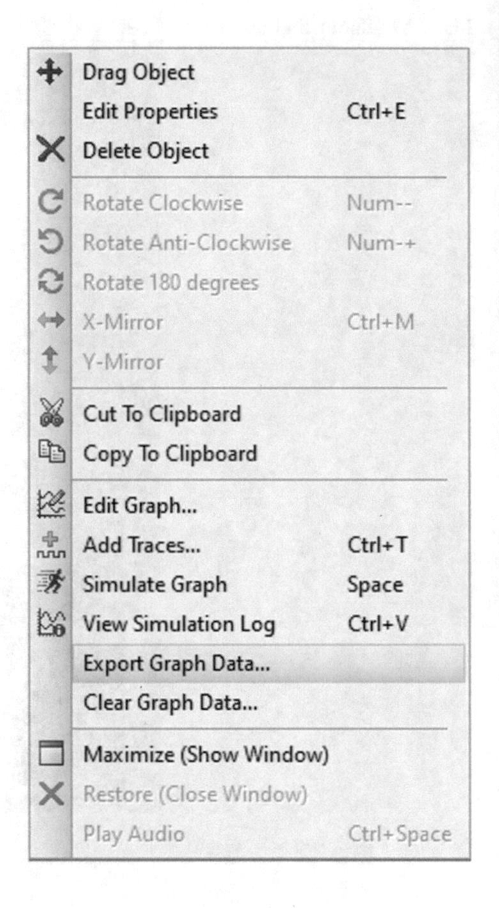

Fig. 2.18 Export graph data command

The MATLAB commands shown in Fig. 2.19 read the exported file (DiodeIV.dat) and draw its graph. Output of this code is shown in Fig. 2.20.

Fig. 2.19 MATLAB commands

Command Window

```
>> F1=importdata('C:\DiodeIV.dat');
>> Voltage1=F1.data(:,1);
>> Current1=F1.data(:,2);
>> plot(Voltage1,1000*Current1,'r')
>> xlabel('Diode Voltage (V)')
>> ylabel('Diode Current (mA)')
>> title('1N4007 IV curve')
fx >>
```

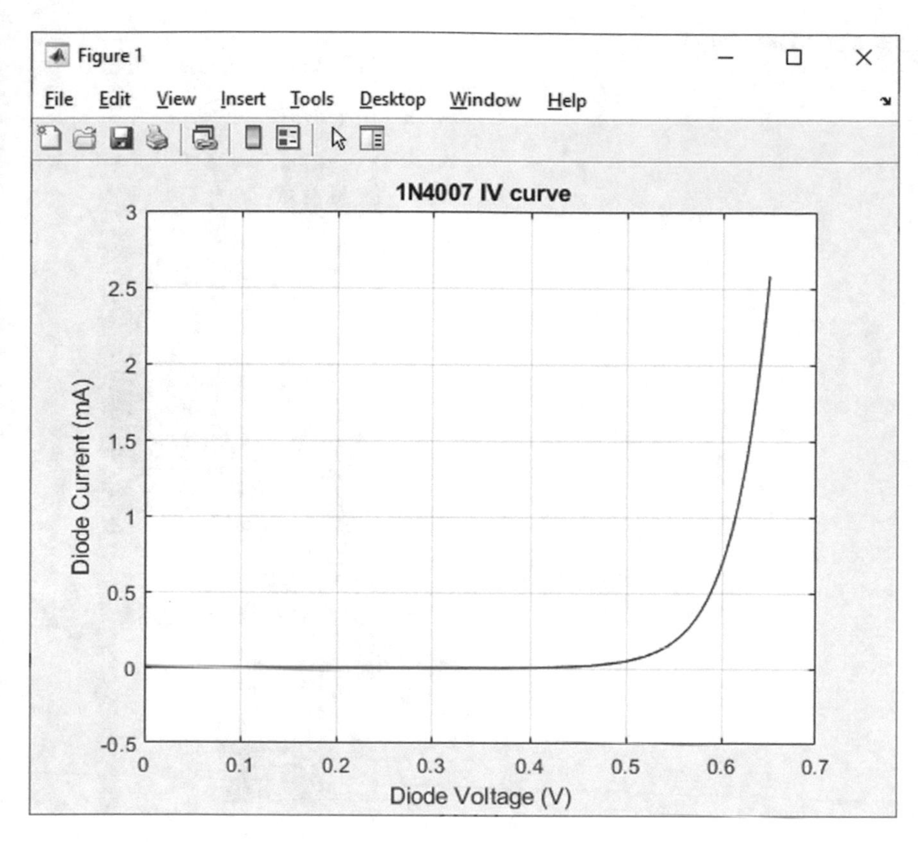

Fig. 2.20 Output of MATLAB commands

2.4 Example 3: DC Transfer Curve Analysis

In this example, we want to draw the graph of collector current versus collector emitter voltage for different values of base current. The graph of collector current versus collector emitter voltage for different values of base current can be obtained with the aid of transfer graph (Fig. 2.21).

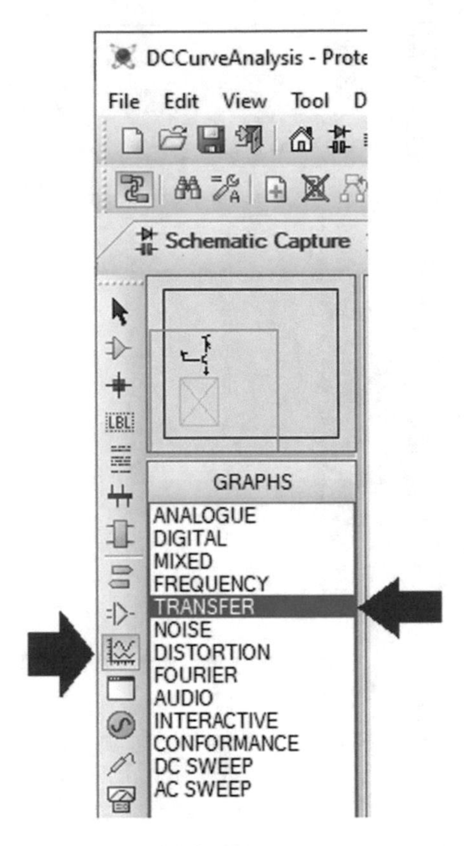

Fig. 2.21 Transfer block

Let's start. Draw the schematic shown in Fig. 2.22. Settings of IB and VCE are shown in Fig. 2.23 and 2.24, respectively.

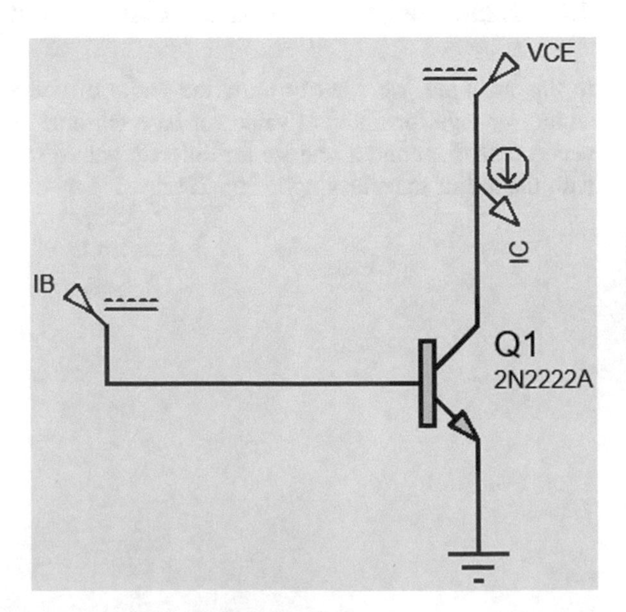

Fig. 2.22 Schematic of Example 3

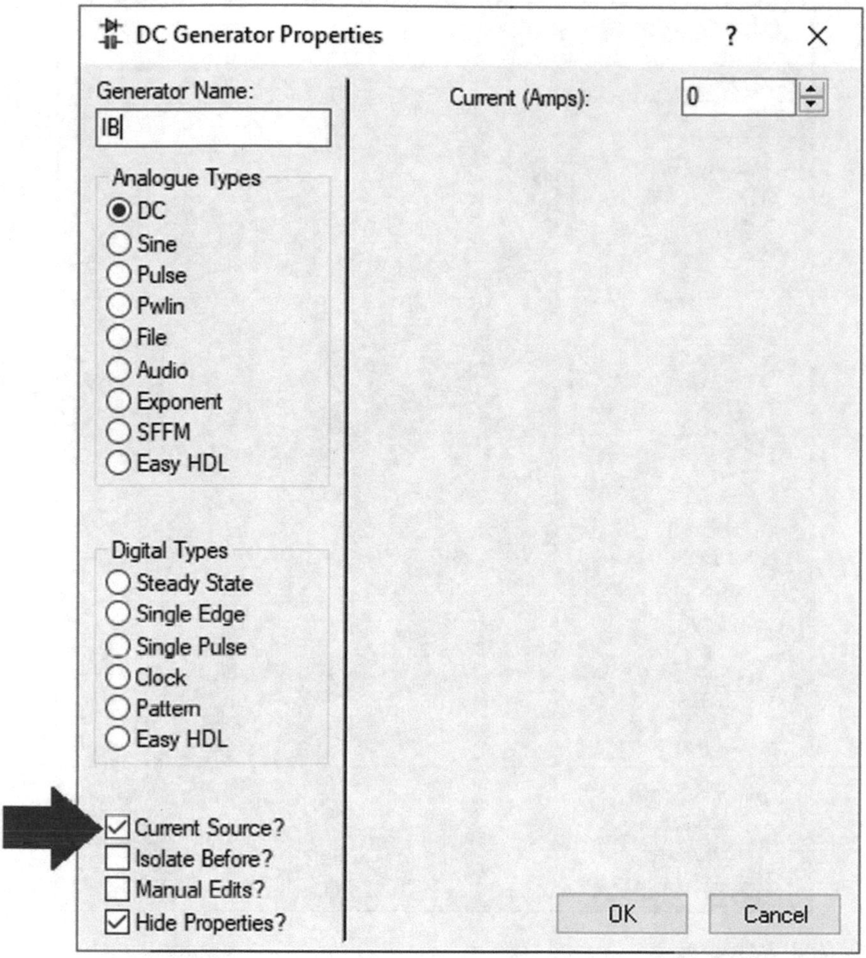

Fig. 2.23 IB settings

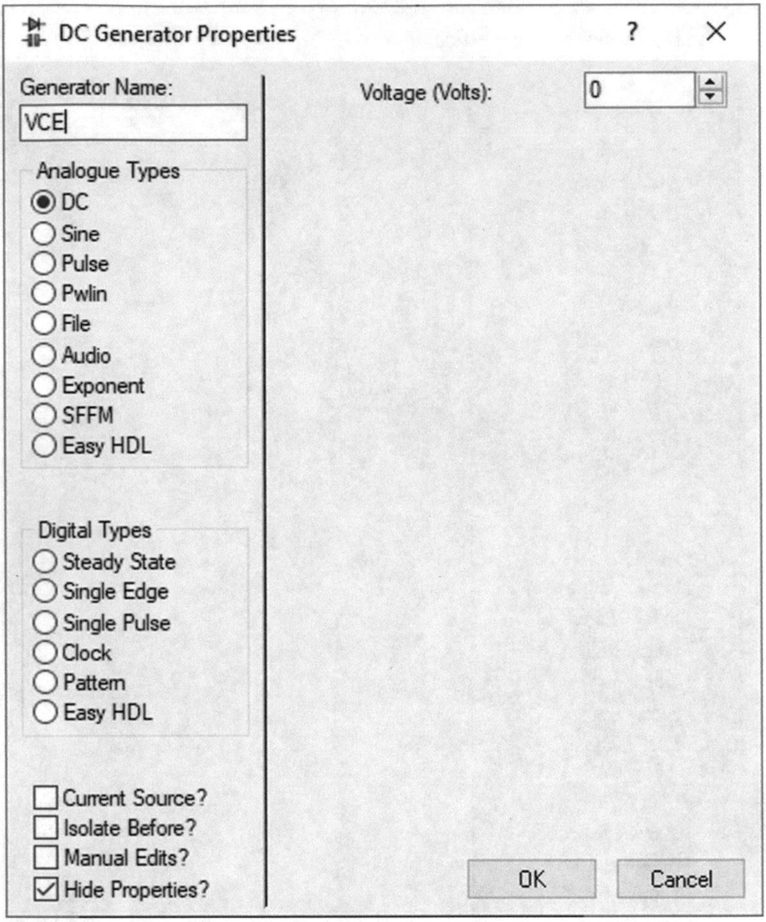

Fig. 2.24 VCE settings

Add a transfer graph to the schematic (Fig. 2.25).

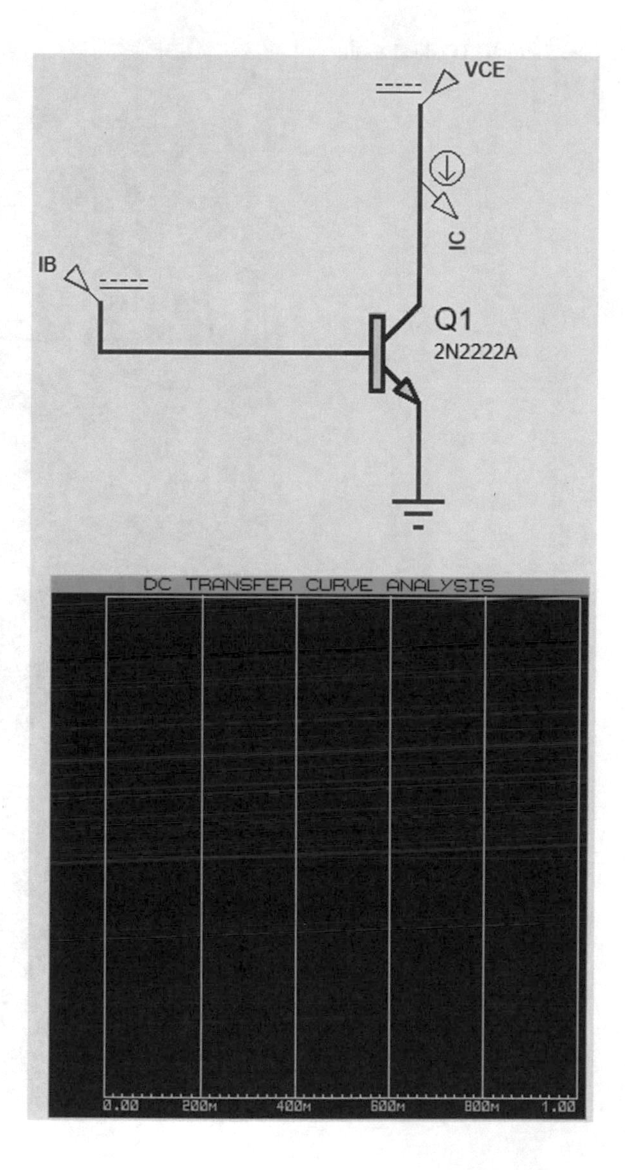

Fig. 2.25 DC transfer curve graph is added to the schematic

Drag and drop current probe "IC" onto the graph (Fig. 2.26).

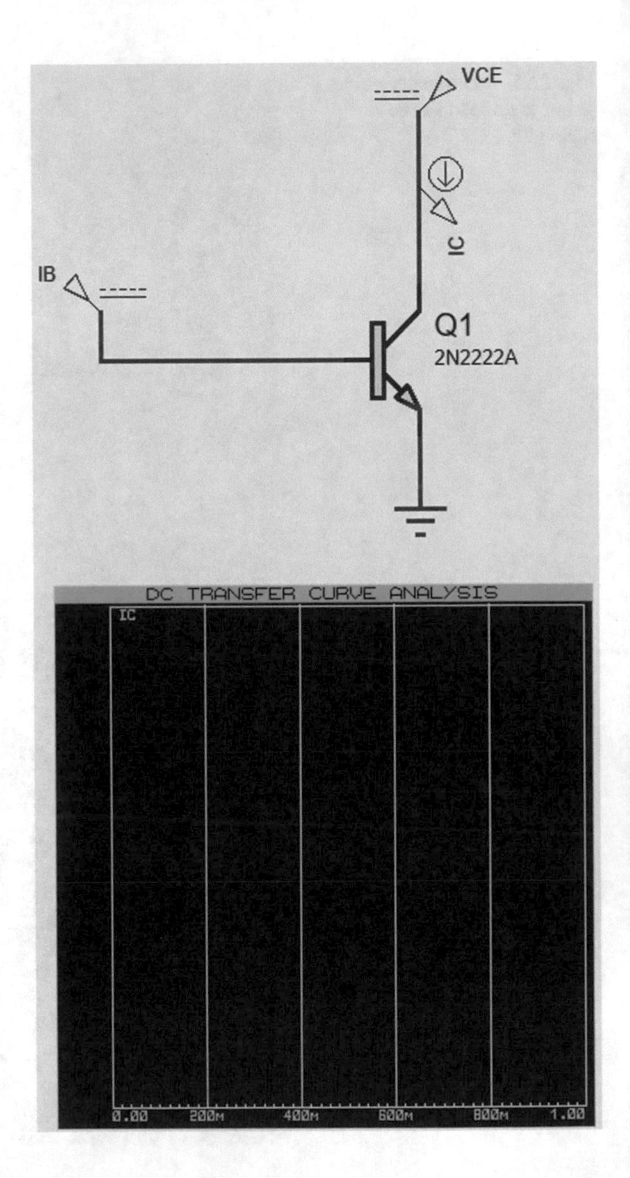

Fig. 2.26 IC is added to the graph

Double click on the graph and change the settings to what is shown in Fig. 2.27. Then click the OK button. According to Fig. 2.27, the VCE is changed from 0 to 2 V with 20 mV steps, and the base current is changed from 10 μA to 1 mA with 99 μA steps.

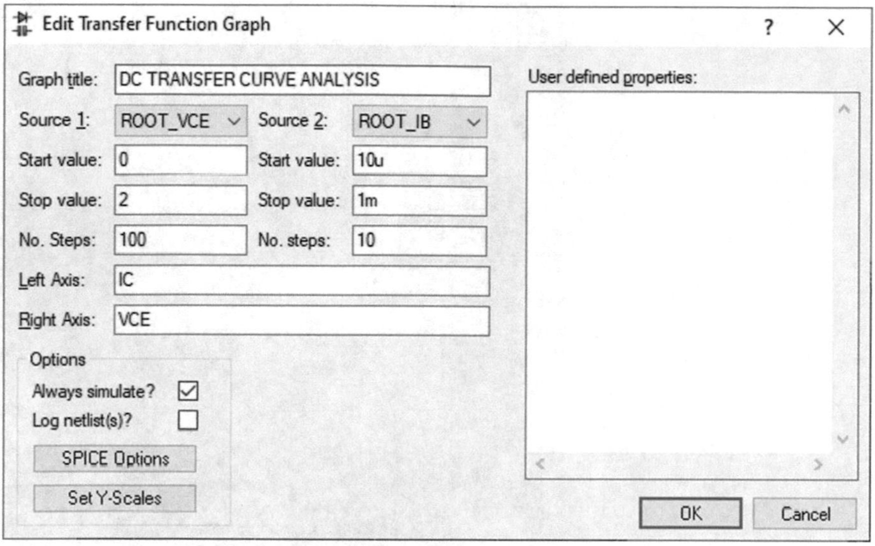

Fig. 2.27 Edit transfer function graph window

Run the simulation. Simulation result is shown in Fig. 2.28.

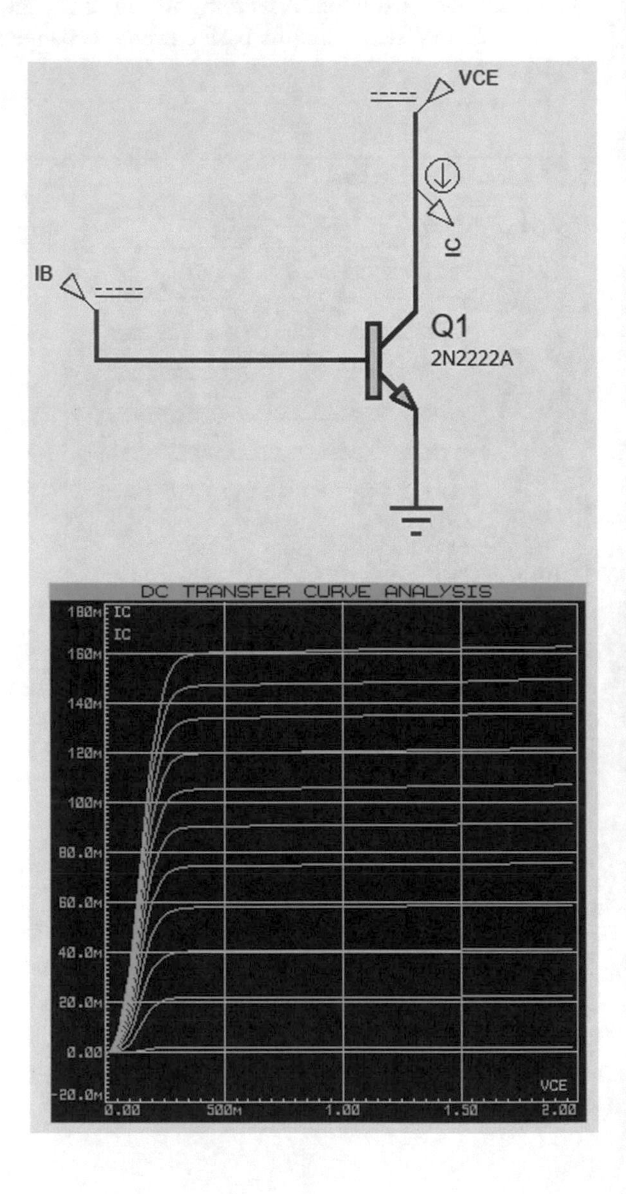

Fig. 2.28 Simulation result

Right click on the graph and click the Maximize (Show Window). This maximizes the simulation result. If you click on the maximized graph, a cursor is added and permits you to read the graph (Fig. 2.29).

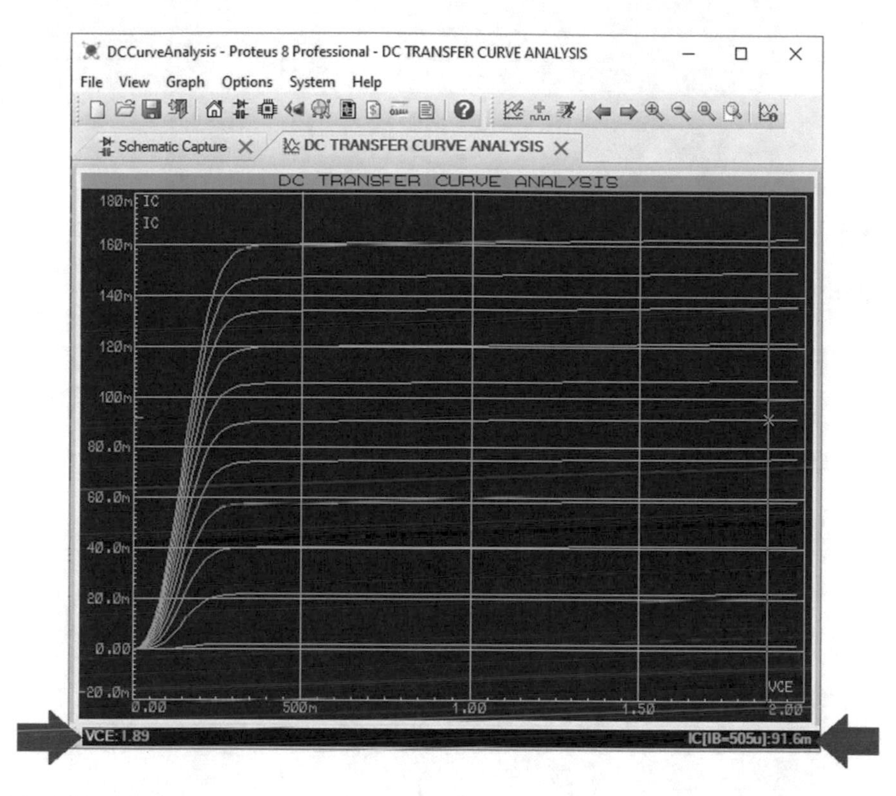

Fig. 2.29 Simulation result is maximized

2.5 Example 4: Small Signal Resistance of Diode

In this example, we want to measure the AC resistance of a diode. Draw the schematic shown in Fig. 2.30. Settings of input voltage source are shown in Fig. 2.31. Note that the RMS of the sinusoidal component is 0. So, a purely DC voltage is applied to the circuit. Current probe ID measures the average value of current passed from the circuit.

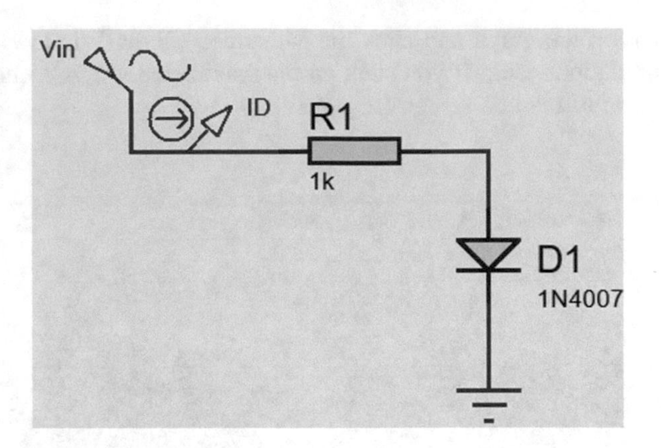

Fig. 2.30 Schematic of Example 4

Sine Generator Properties

Generator Name: Vin

Analogue Types: DC, Sine (selected), Pulse, Pwlin, File, Audio, Exponent, SFFM, Easy HDL

Digital Types: Steady State, Single Edge, Single Pulse, Clock, Pattern, Easy HDL

Current Source? / Isolate Before? / Manual Edits? / Hide Properties? (checked)

Offset (Volts): 1.65

Amplitude (Volts): Amplitude: / Peak-Peak: / RMS (selected): 0

Timing: Frequency (Hz) (selected): 1 / Period (Secs): / Cycles/Graph:

Delay: Time Delay (Secs): / Phase (Degrees) (selected): 0

Damping Factor (1/s): 0

OK Cancel

Fig. 2.31 V_{in} settings

Run the simulation. According to Fig. 2.32, 1.03436 mA is passed from the circuit.

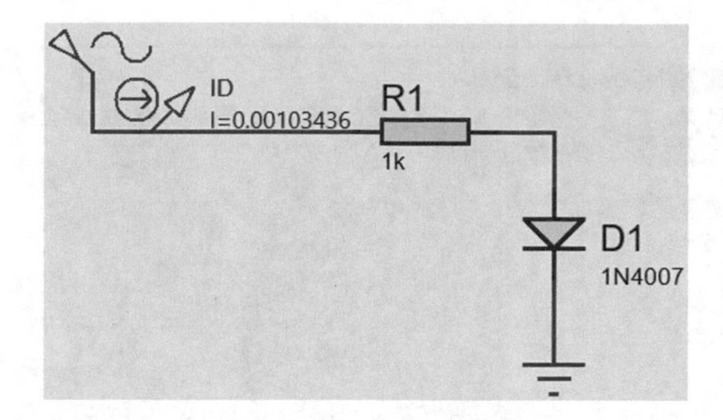

Fig. 2.32 Simulation result

Add an oscilloscope to the schematic (Fig. 2.33).

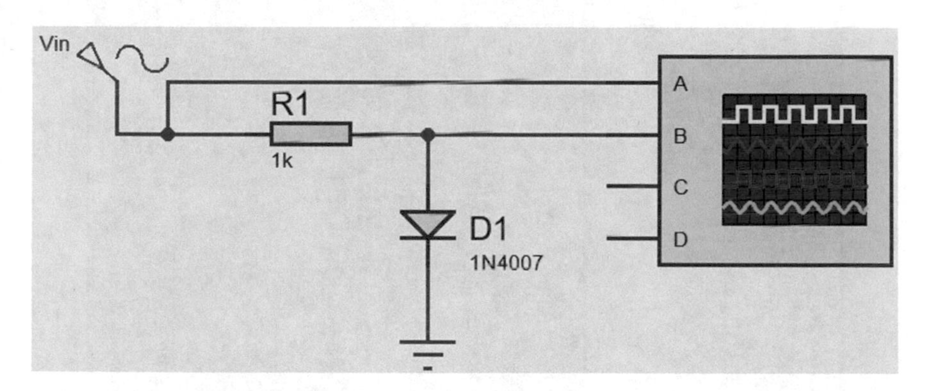

Fig. 2.33 An oscilloscope block is added to the schematic

Double click the V_{in} and add an AC component to it (Fig. 2.34).

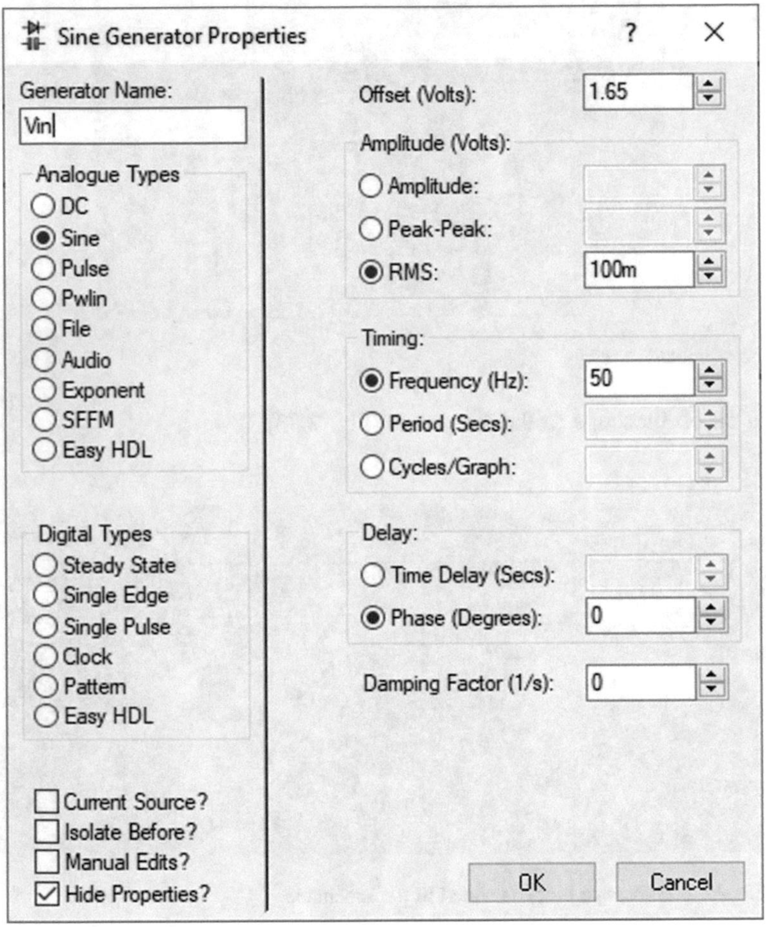

Fig. 2.34 Settings of V_{in}

Run the simulation. Simulation result is shown in Fig. 2.35. Note that both of channel A and B have AC coupling.

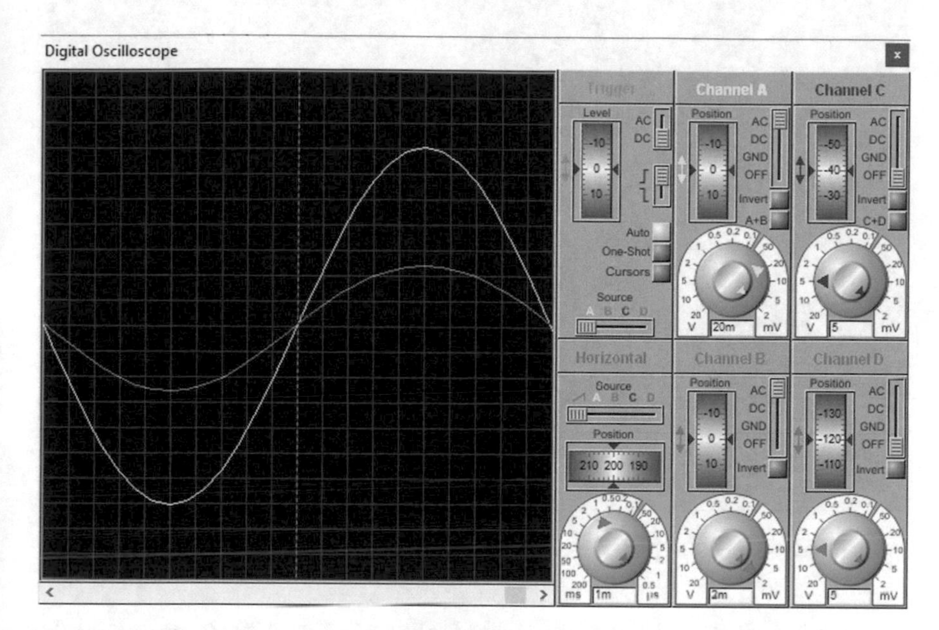

Fig. 2.35 Simulation result

Use cursors to measure the amplitude of input and output (Figs. 2.36 and 2.37). According to Figs. 2.36 and 2.37, amplitude of input voltage and amplitude of voltage across the diode are 141 mV and 4.8 mV, respectively.

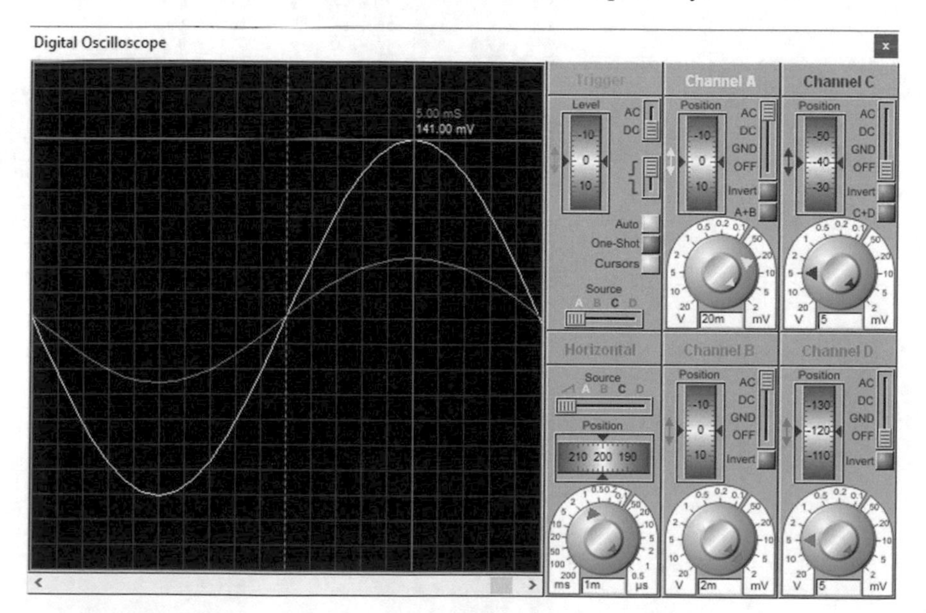

Fig. 2.36 Measurement of peak value of input

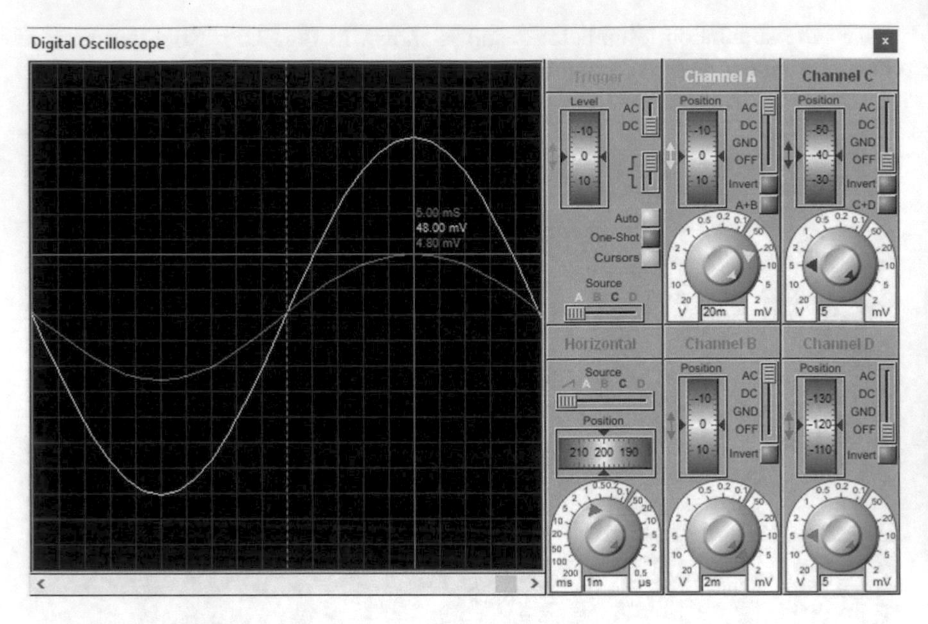

Fig. 2.37 Measurement of peak value of output

The AC equivalent circuit of the circuit is shown in Fig. 2.38. R_{ac} shows the AC resistance of the diode. R_{ac} is unknown and its value must be calculated.

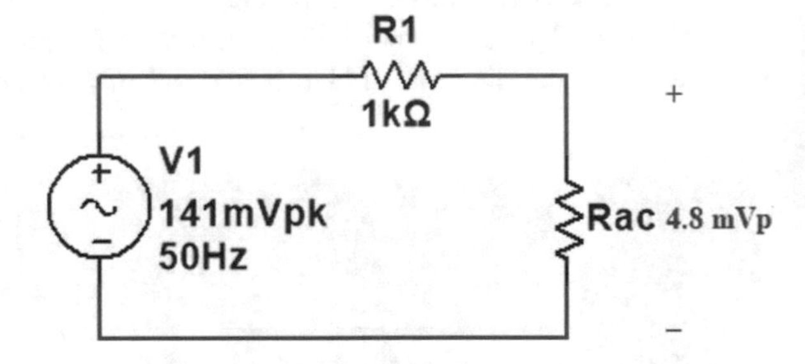

Fig. 2.38 Equivalent circuit for Fig. 2.30

The MATLAB commands shown in Fig. 2.39 calculate the value of R_{ac}. According to Fig. 2.39, the value of R_{ac} is 35.2423 Ω.

```
Command Window
>> R1=1e3;Vinac=141e-3;Voac=4.8e-3;
>> syms Rd
>> eval(solve(Rd/(Rd+R1)*Vinac==Voac))

ans =

   35.2423

fx >> |
```

Fig. 2.39 MATLAB commands

2.6 Example 5: Doing the Simulation at a Specific Temperature

Proteus does the simulation at 27 °C by default. In this example, we learn how to do the simulation at a specific temperature. Draw the schematic shown in Fig. 2.40. We want to measure the voltage drop of the diode at 80 °C.

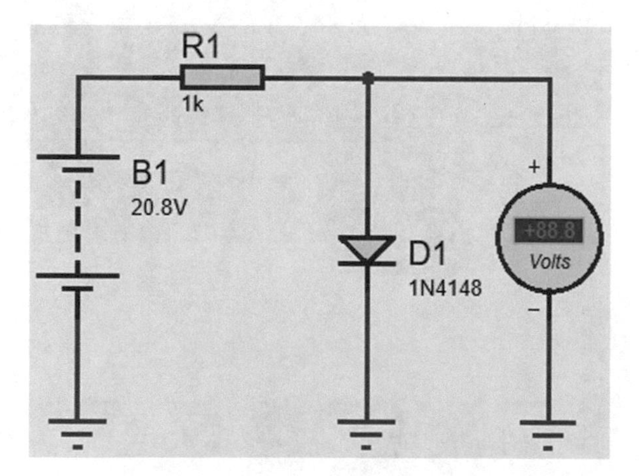

Fig. 2.40 Schematic for Example 5

Run the simulation. According to Fig. 2.41, the voltage drop of the diode is 0.78 V.

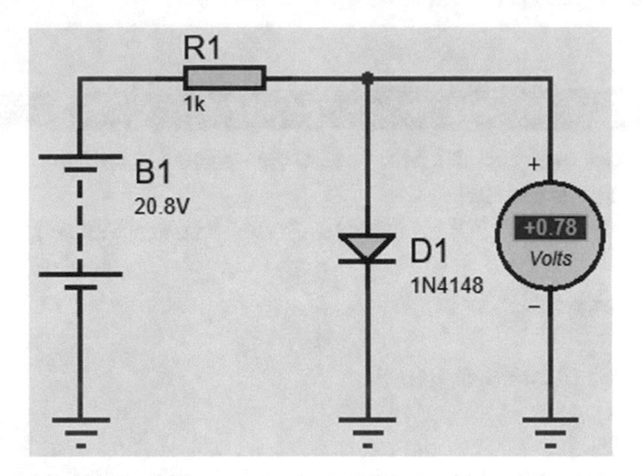

Fig. 2.41 Simulation result

Stop the simulation and double click the diode. Then write "TEMP = 80" in the Other Properties box and click the OK button (Fig. 2.42). After clicking the OK button, the schematic changes to what is shown in Fig. 2.43.

Edit Component
Part Reference: D1 Hidden:
Part Value: 1N4148 Hidden:
Element: New
SPICE Model: 1N4148 Hide All
SPICE Library: DIODESINC Hide All
PCB Package: DO35 Hide All
Other Properties:
TEMP=80
Exclude from Simulation
Exclude from PCB Layout
Exclude from Current Variant
Attach hierarchy module
Hide common pins
Edit all properties as text
OK
Cancel

Fig. 2.42 Settings of D_1

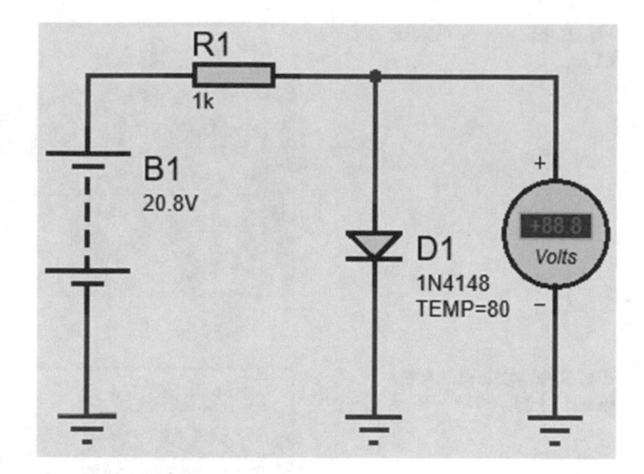

Fig. 2.43 TEMP = 80 is added to the schematic

Run the simulation. Simulation result is shown in Fig. 2.44. According to Fig. 2.44, the voltage drop of diode is 0.71 V at 80 °C.

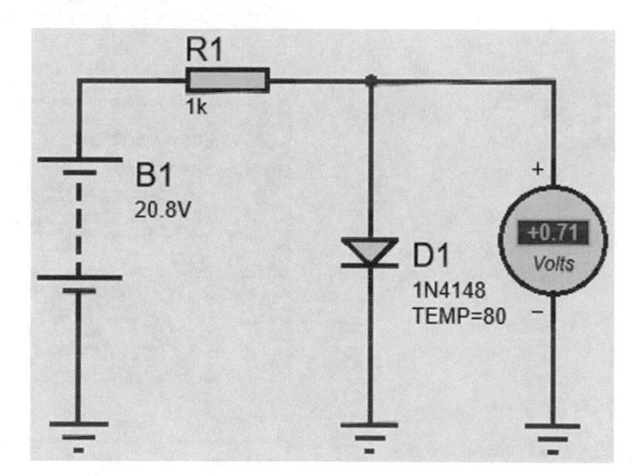

Fig. 2.44 Simulation result

2.7 Example 6: LED and Push Button Blocks

In this example, we want to study the LED and push button blocks. Red led can be added to the schematic by searching for "led-red" in the Pick Devices window (Fig. 2.45). Green and blue LEDs can be added to the schematic by searching for "led-green" and "led-blue", respectively (Figs. 2.46 and 2.47). Push button can be added to the schematic by searching for "button" (Fig. 2.48).

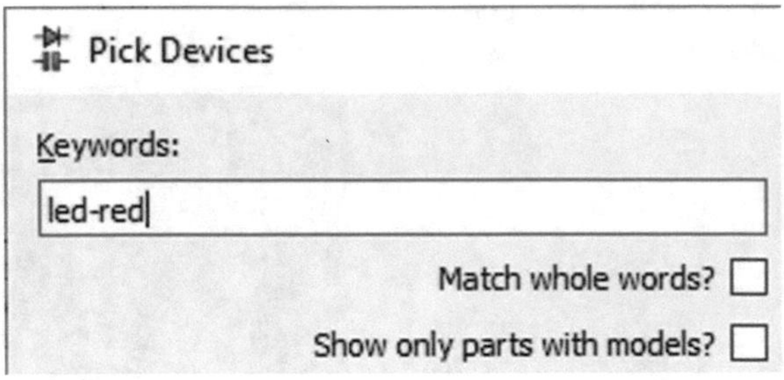

Fig. 2.45 Searching for red LED

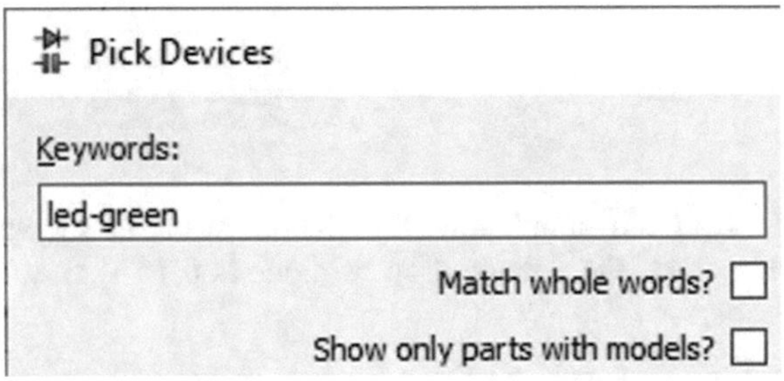

Fig. 2.46 Searching for green LED

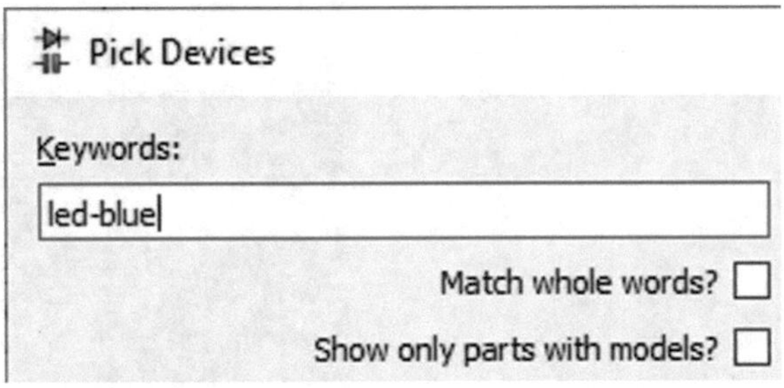

Fig. 2.47 Searching for blue LED

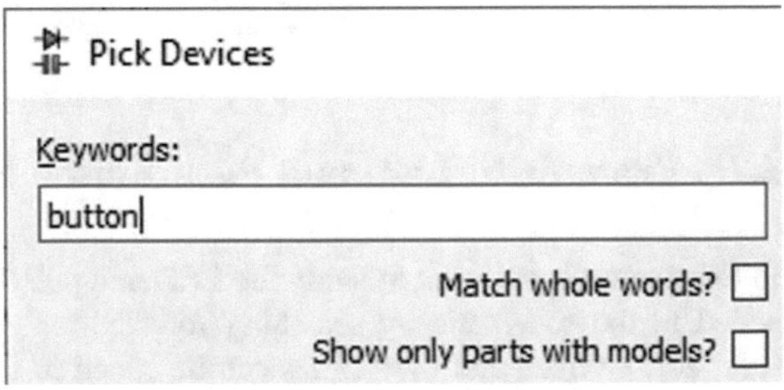

Fig. 2.48 Searching for button

Let's start. Draw the schematic shown in Fig. 2.49.

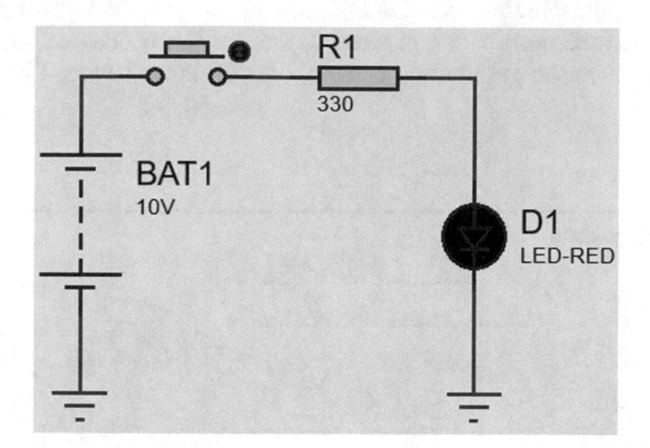

Fig. 2.49 Schematic of Example 6

Right click on the LED and click the Edit Properties. Then select the Analog for Model Type box (Fig. 2.50) and click the OK button.

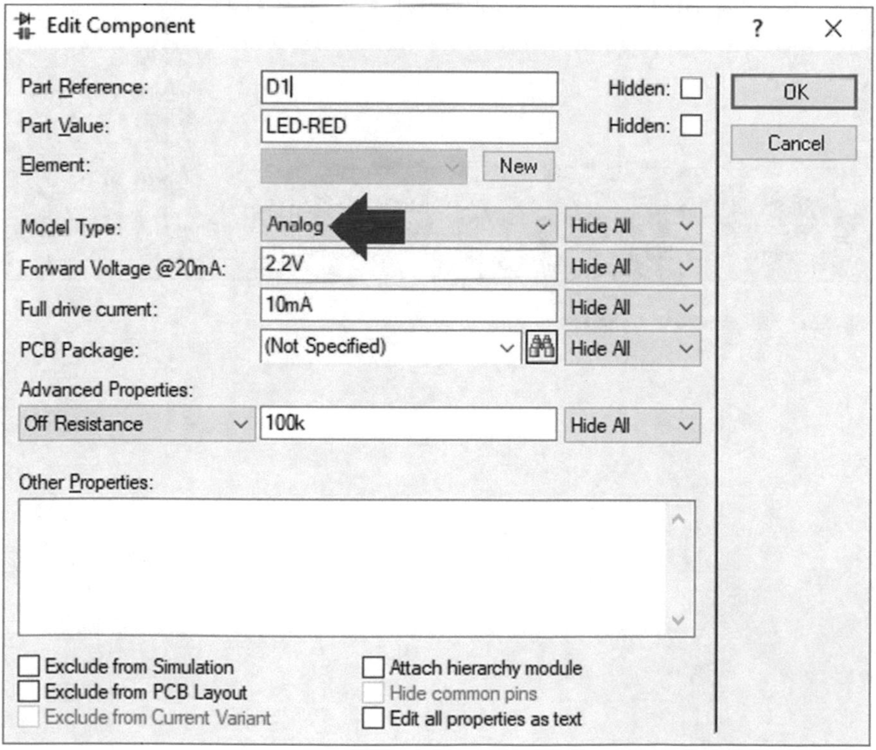

Fig. 2.50 Settings of D_1

Right click on the push button and click the Edit Properties. After clicking the Edit Properties, the window shown in Fig. 2.51 appears on the screen. According to Fig. 2.51, when the push button is pressed, the resistance between its terminals is 100 mΩ and when the push button is open, the resistance between its terminals is 100 MΩ.

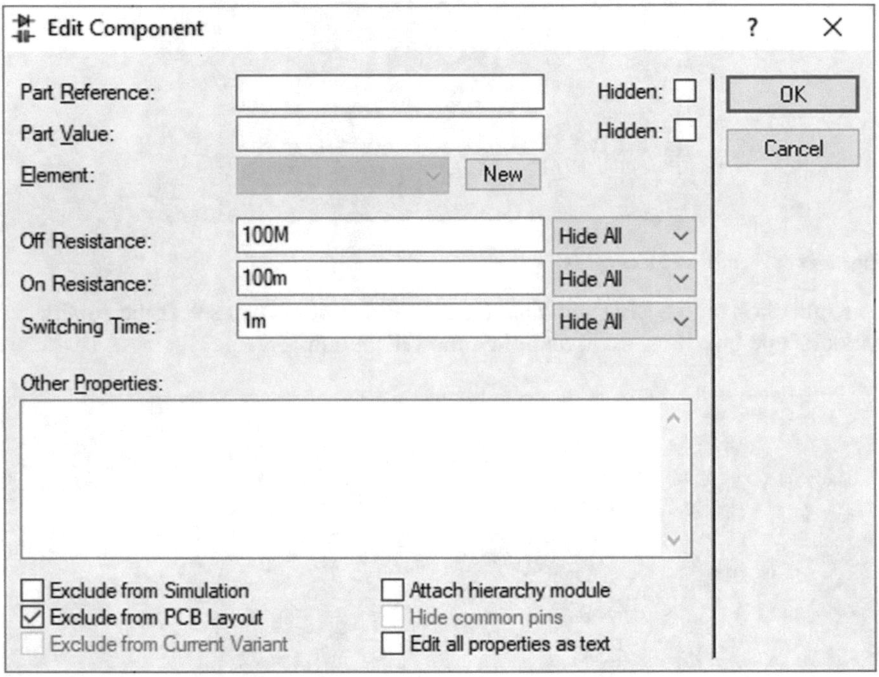

Fig. 2.51 Edit component window

Run the simulation. If you click on the region shown in Fig. 2.52, the push is closed and the led turns on. However, the push button is opened as soon as you release the mouse button.

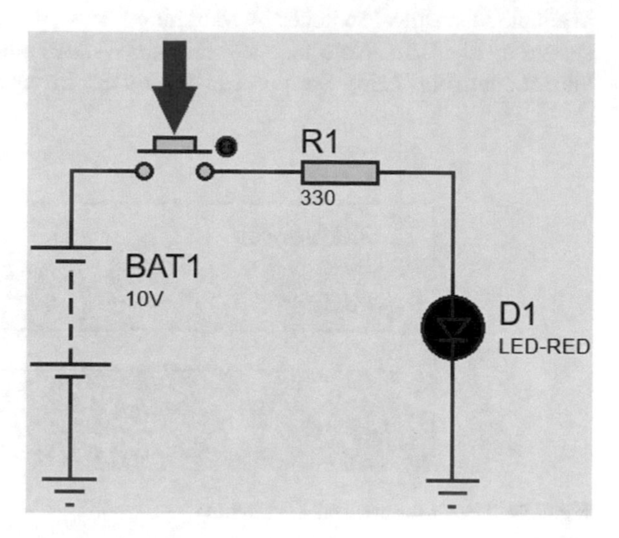

Fig. 2.52 Click on the push button to turn on the LED

If you click on the small arrow behind the push button (Fig. 2.53), the push is closed and it is not opened after releasing the mouse button. The push button is opened if you click the arrow behind the push button again.

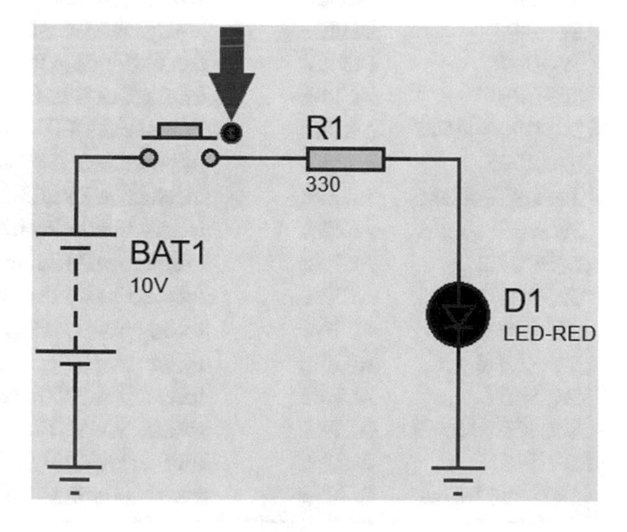

Fig. 2.53 You can click on the push button to turn on the LED

2.8 Example 7: Different Kinds of Mechanical Switches

Different kinds of mechanical switches are available in Proteus. You can see the available mechanical switches by searching for "sw-" (Fig. 2.54). Search results are shown in Fig. 2.55. Note that SW-DIP4, SW-DIP7 and SW-DIP8 have no simulator model (Fig. 2.56). So, you can't use them in your simulations.

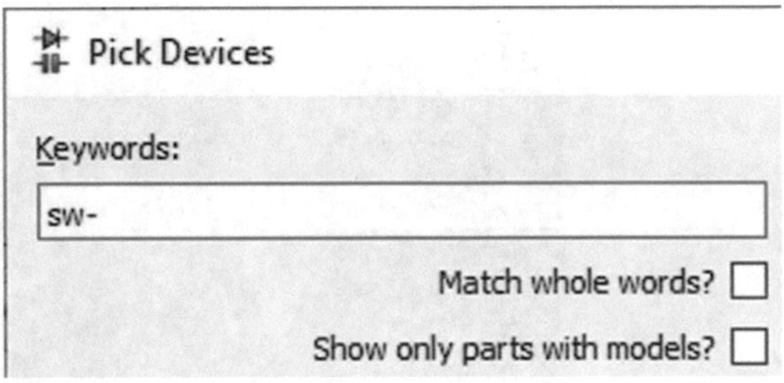

Pick Devices

Keywords:

sw-

Match whole words? ☐

Show only parts with models? ☐

Fig. 2.54 Search for mechanical switches

Showing local results: 16

Device	Library	Description
SW-DIP4	DEVICE	4 way DIP switch
SW-DIP7	DEVICE	7 way DIP switch
SW-DIP8	DEVICE	8 way DIP switch
SW-DPDT	ACTIVE	Interactive DPDT Switch (Latched Action)
SW-DPDT-MOM	ACTIVE	Interactive DPDT Switch (Momentary Action)
SW-DPST	ACTIVE	Interactive DPST Switch (Latched Action)
SW-DPST-MOM	ACTIVE	Interactive DPST Switch (Momentary Action)
SW-ROT-12	ACTIVE	Interactive 12 Position Rotary Switch
SW-ROT-3	ACTIVE	Interactive 3 Position Rotary Switch
SW-ROT-4	ACTIVE	Interactive 4 Position Rotary Switch
SW-ROT-5	ACTIVE	Interactive 5 Position Rotary Switch
SW-ROT-6	ACTIVE	Interactive 6 Position Rotary Switch
SW-SPDT	ACTIVE	Interactive SPDT Switch (Latched Action)
SW-SPDT-MOM	ACTIVE	Interactive SPDT Switch (Momentary Action)
SW-SPST	ACTIVE	Interactive SPST Switch (Latched Action)
SW-SPST-MOM	ACTIVE	Interactive SPST Switch (Momentary Action)

Fig. 2.55 Available mechanical switches

Showing local results: 16

Device	Library	Description
SW-DIP4	DEVICE	4 way DIP switch
SW-DIP7	DEVICE	7 way DIP switch
SW-DIP8	DEVICE	8 way DIP switch
SW-DPDT	ACTIVE	Interactive DPDT Switch (Latched Action)
SW-DPDT-MOM	ACTIVE	Interactive DPDT Switch (Momentary Action)
SW-DPST	ACTIVE	Interactive DPST Switch (Latched Action)
SW-DPST-MOM	ACTIVE	Interactive DPST Switch (Momentary Action)
SW-ROT-12	ACTIVE	Interactive 12 Position Rotary Switch
SW-ROT-3	ACTIVE	Interactive 3 Position Rotary Switch
SW-ROT-4	ACTIVE	Interactive 4 Position Rotary Switch
SW-ROT-5	ACTIVE	Interactive 5 Position Rotary Switch
SW-ROT-6	ACTIVE	Interactive 6 Position Rotary Switch
SW-SPDT	ACTIVE	Interactive SPDT Switch (Latched Action)
SW-SPDT-MOM	ACTIVE	Interactive SPDT Switch (Momentary Action)
SW-SPST	ACTIVE	Interactive SPST Switch (Latched Action)
SW-SPST-MOM	ACTIVE	Interactive SPST Switch (Momentary Action)

Preview

No Simulator Model

1 2 3 4 8 7 6 5

Fig. 2.56 Some of the switches can't be used in simulations

2.9 Example 8: Turning on and off a Lamp

In this example, we want to turn on and off a lamp with a switch. Draw the schematic shown in Fig. 2.57. The lamp and switch blocks can be added to the schematic by searching for "lamp" and "switch" in the Pick Devices window (Figs. 2.58 and 2.59).

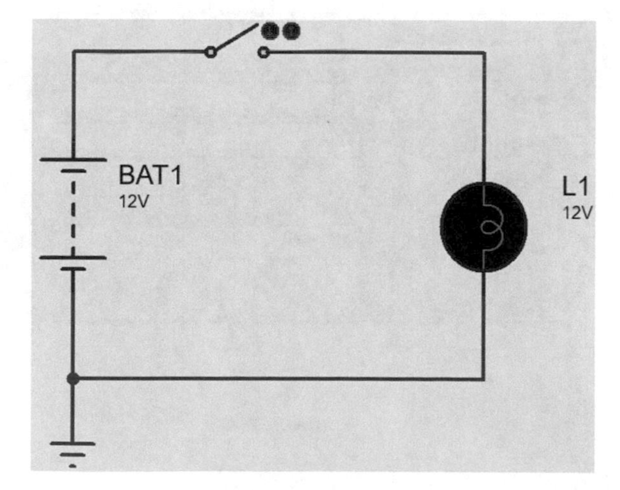

Fig. 2.57 Schematic for Example 8

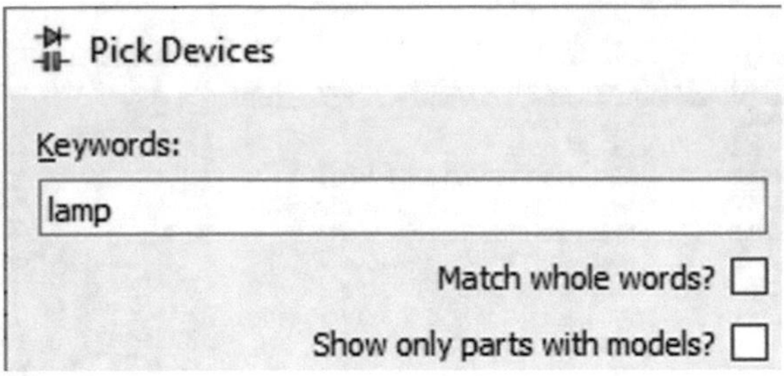

Fig. 2.58 Search for lamp

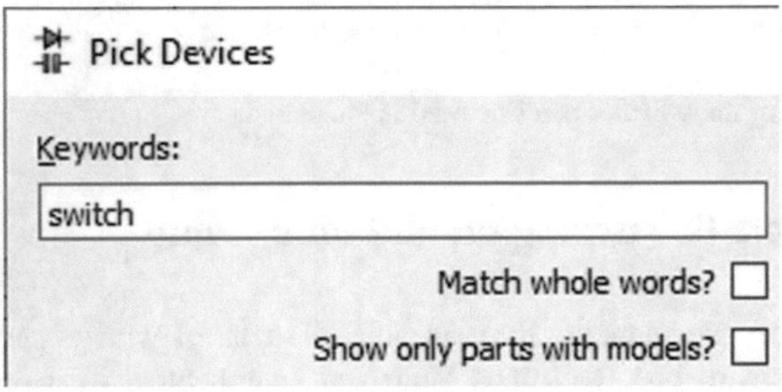

Fig. 2.59 Search for switch

Settings of the lamp are shown in Fig. 2.60.

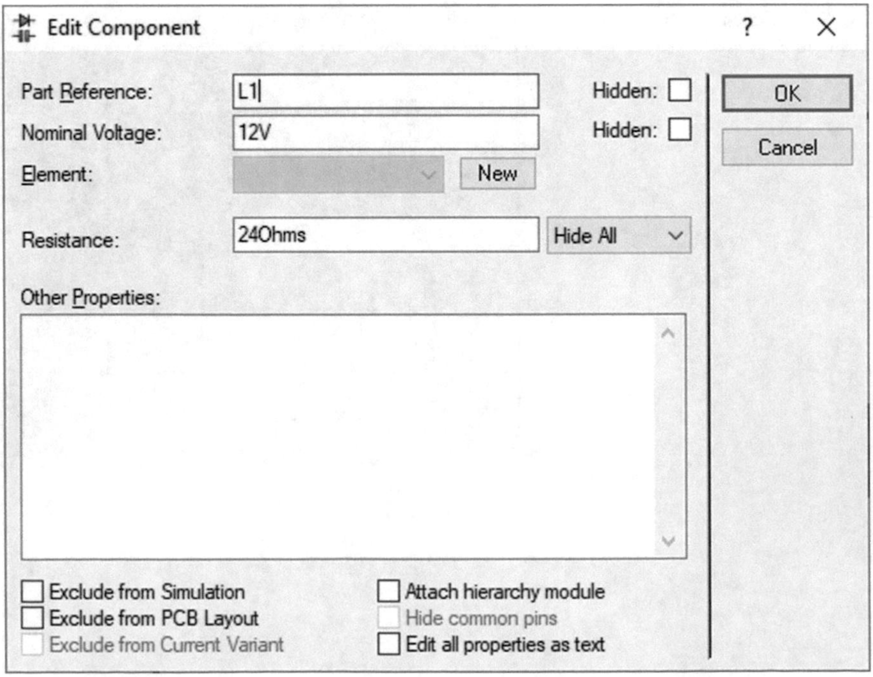

Fig. 2.60 Settings of lamp L_1

Run the simulation. You can turn on and off the lamp by clicking the arrows behind the switch (Fig. 2.61).

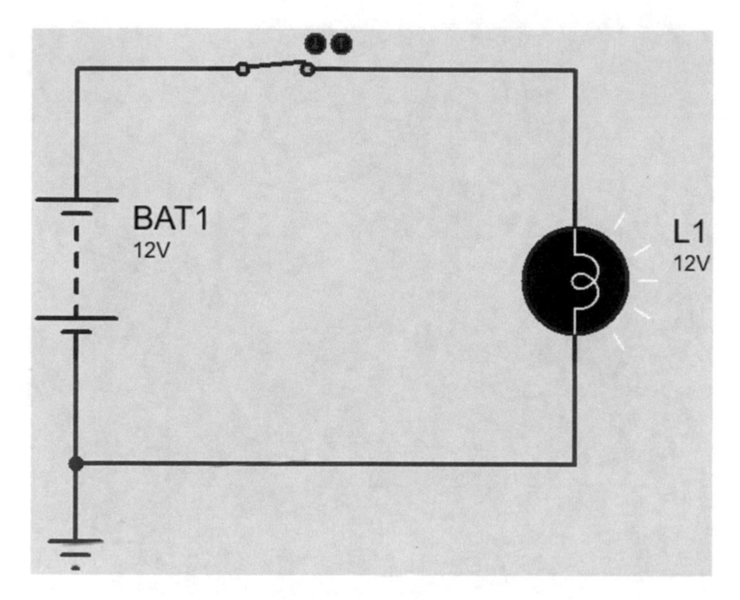

Fig. 2.61 Lamp turns on when the switch is pressed

2.10 Example 9: Turning on and off a Lamp from Two Different Points

You can use two Single Pole Double Through (SPDT) switches to control a lamp from two (physically) different points. SPDT switch can be added to the schematic by searching for "sw-spdt" in the Pick Devices window (Fig. 2.62).

Fig. 2.62 Searching for sw-spdt switch

Draw the schematic shown in Fig. 2.63.

Run the simulation. Note that both of the SW_1 and SW_2 can turn on and off the lamp.

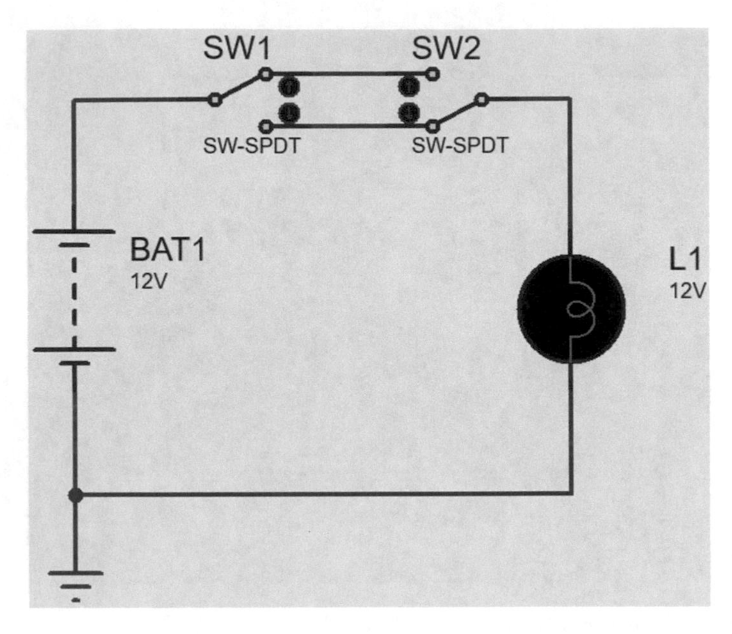

Fig. 2.63 Schematic for Example 9

2.11 Example 10: Measurement of Output Voltage Ripple for Half Wave Diode Rectifier

In this example, we want to measure the output voltage ripple of a half wave rectifier. Draw the schematic shown in Fig. 2.64. Settings of V_{in} are shown in Fig. 2.65.

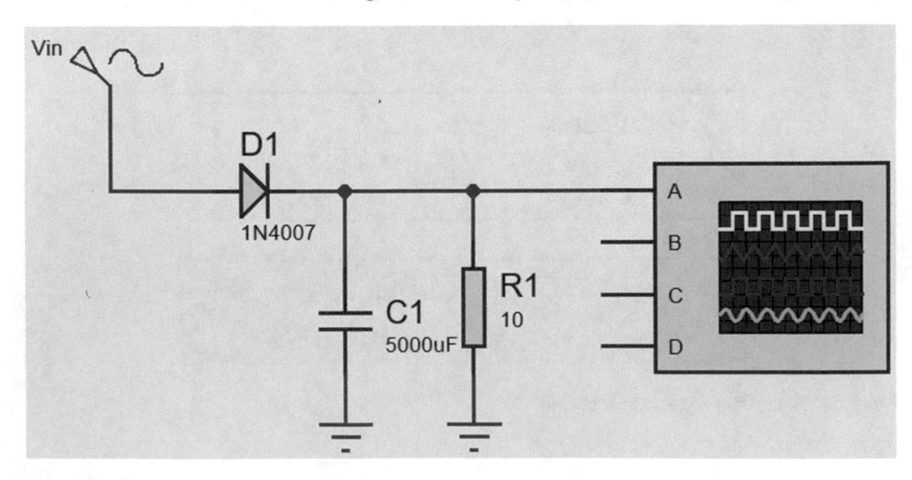

Fig. 2.64 Schematic for Example 10

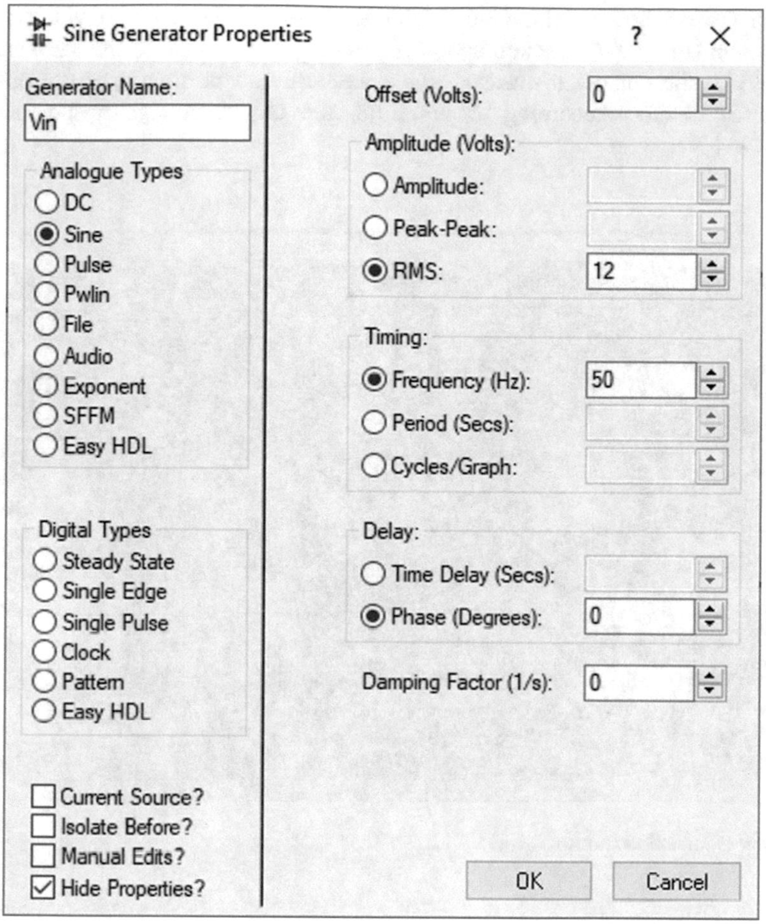

Sine Generator Properties ? ×

Generator Name:
Vin

Analogue Types
DC
Sine
Pulse
Pwlin
File
Audio
Exponent
SFFM
Easy HDL

Digital Types
Steady State
Single Edge
Single Pulse
Clock
Pattern
Easy HDL

Current Source?
Isolate Before?
Manual Edits?
Hide Properties?

Offset (Volts): 0

Amplitude (Volts):
Amplitude:
Peak-Peak:
RMS: 12

Timing:
Frequency (Hz): 50
Period (Secs):
Cycles/Graph:

Delay:
Time Delay (Secs):
Phase (Degrees): 0

Damping Factor (1/s): 0

OK Cancel

Fig. 2.65 Settings of V_{in}

Run the simulation. Simulation result is shown in Fig. 2.66. If you click the Pause icon (Fig. 2.67), the simulation is paused and you can do the measurement easily. Use the cursors to measure the maximum and minimum of the obtained graph (Fig. 2.68). According to Fig. 2.68, the output voltage ripple is 15.8–11.4 = 4.4 V.

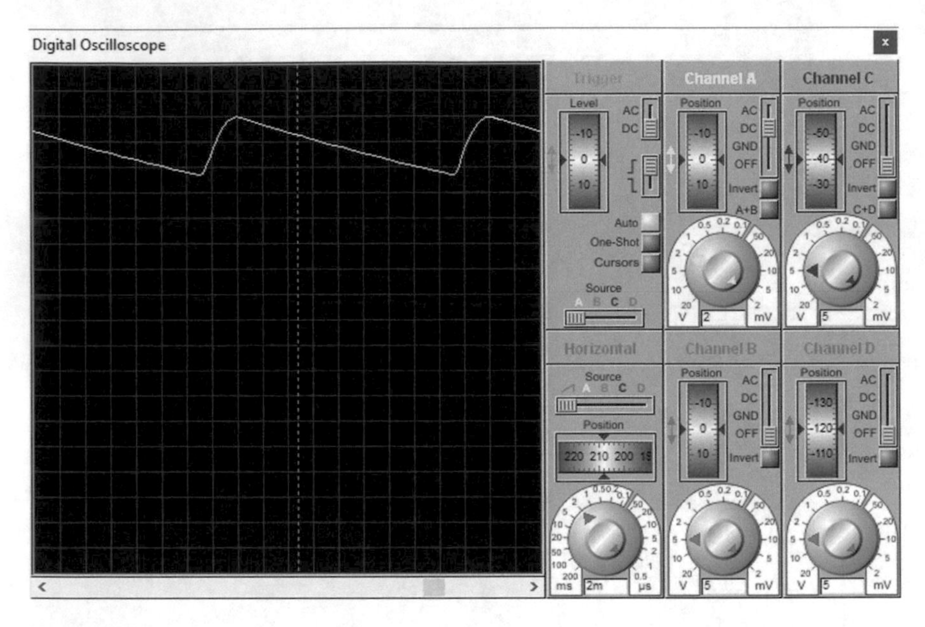

Fig. 2.66 Simulation result

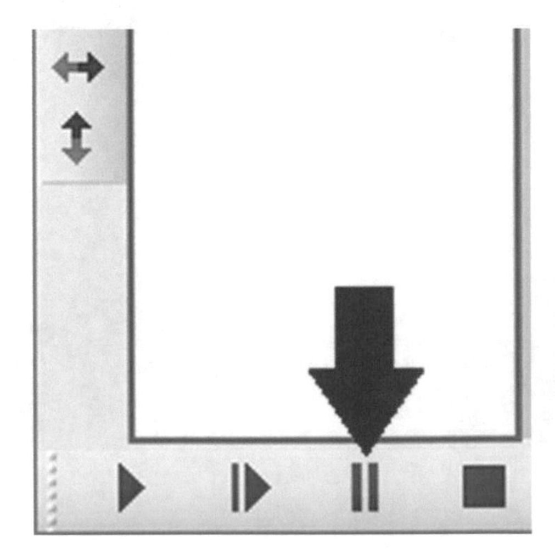

Fig. 2.67 The pause icon

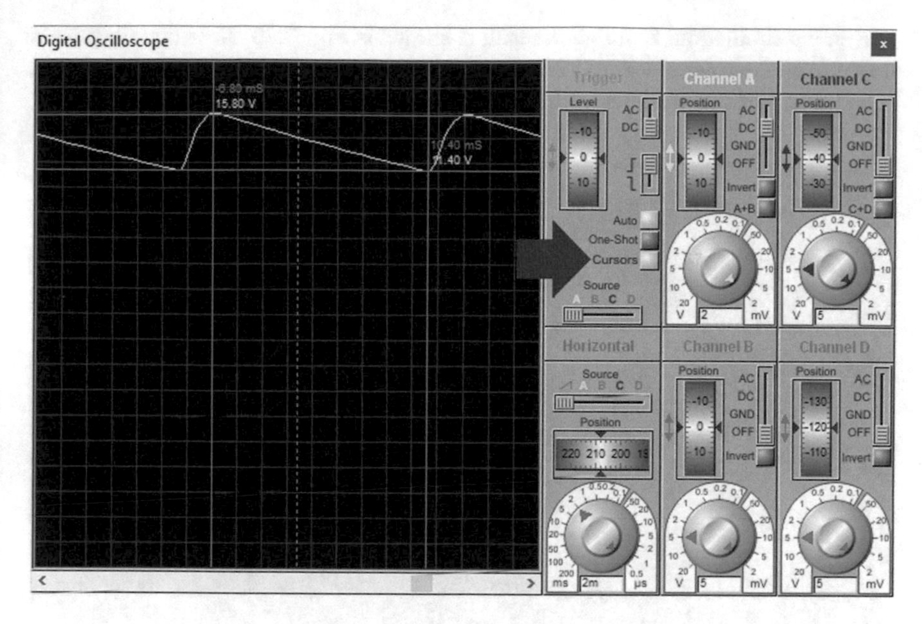

Fig. 2.68 Measurement of peak-peak values

2.12 Example 11: Input Current of Half Wave Rectifier

In this example, we want to observe the input current of rectifier of Example 10. We use a current-dependent voltage source to show the input current on the oscilloscope screen. Draw the schematic shown in Fig. 2.69.

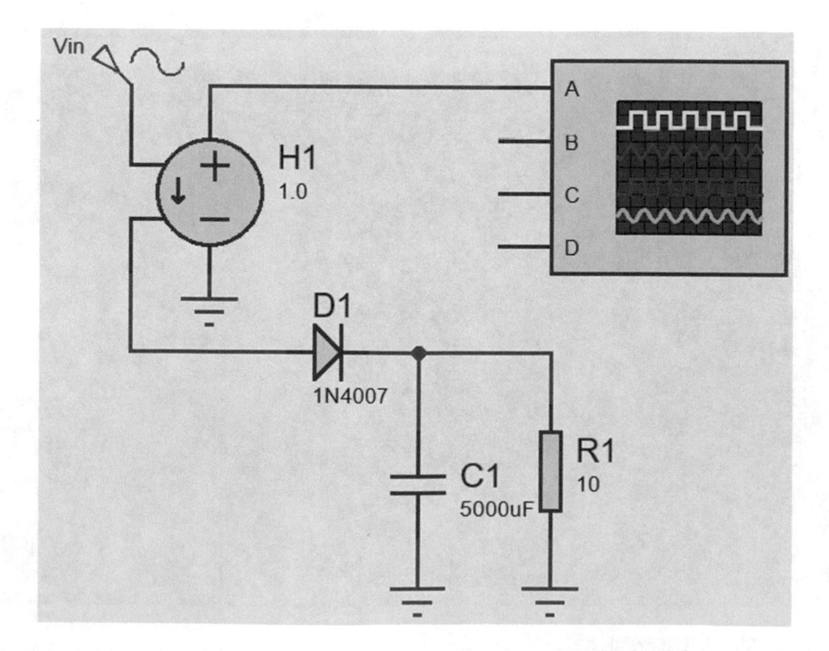

Fig. 2.69 Schematic of Example 11

Run the simulation. Simulation result is shown in Fig. 2.70. Note that the current drawn from the source is impulsive. So, the input current contains many harmonics. Remember that the presence of harmonics decreases the power factor.

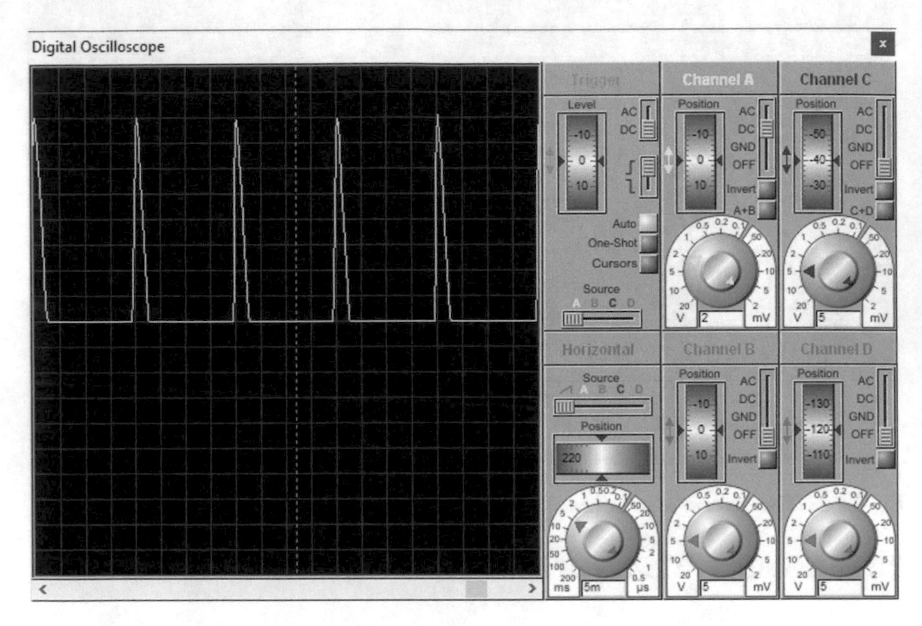

Fig. 2.70 Simulation result

If you decrease the capacitance C_1, the amplitude of impulses decreases (Fig. 2.71). However, decreasing the capacitance C_1 increases the output voltage ripple.

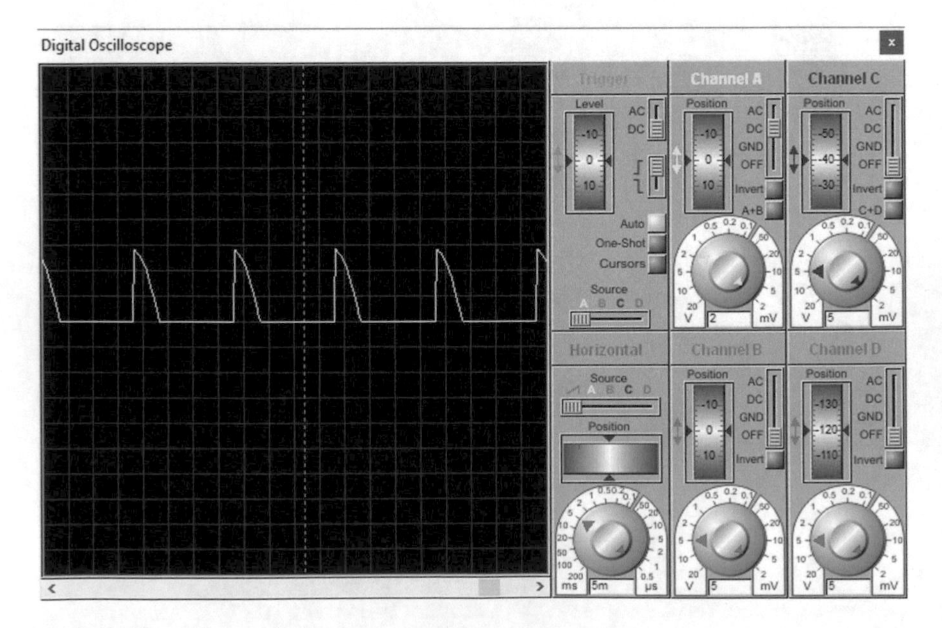

Fig. 2.71 Simulation result

2.13 Example 12: Full Wave Rectifier

In this example, we want to simulate a full wave rectifier. Draw the schematic shown in Fig. 2.72. Settings of input AC source V_1 and transformer TR1 are shown in Figs. 2.73 and 2.74.

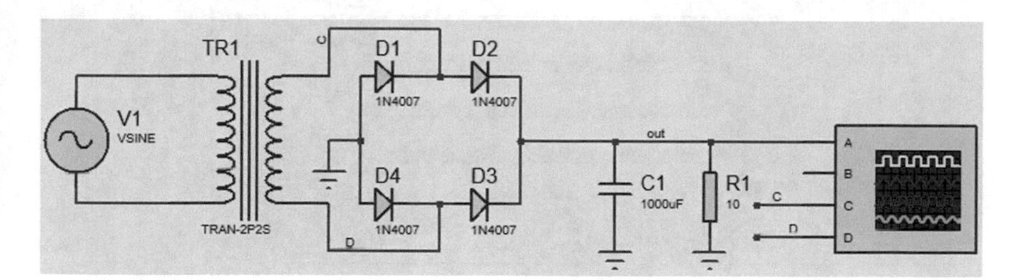

Fig. 2.72 Schematic of Example 12

Edit Component ? ×

Part Reference: V1 Hidden: ☐ OK
Part Value: VSINE Hidden: ☐ Cancel
Element: New

DC Offset: 0 Hide All
Amplitude: 169.7 Hide All
Frequency: 60 Hide All
Time Delay: 0 Hide All
Damping Factor: 0 Hide All

Other Properties:

☐ Exclude from Simulation ☐ Attach hierarchy module
☐ Exclude from PCB Layout ☐ Hide common pins
☐ Exclude from Current Variant ☐ Edit all properties as text

Fig. 2.73 Settings of V_1

Edit Component

Part Reference: TR1 Hidden: ☐
Part Value: TRAN-2P2S Hidden: ☐
Element: New

OK
Cancel

Primary Inductance: 1H Hide All
Secondary Inductance: 0.01H Hide All
Coupling Factor: 1.0 Hide All
Primary DC resistance: 1m Hide All
Secondary DC resistance: 1m Hide All

Other Properties:

☐ Exclude from Simulation ☐ Attach hierarchy module
☐ Exclude from PCB Layout ☐ Hide common pins
☐ Exclude from Current Variant ☐ Edit all properties as text

Fig. 2.74 Settings of TR1

Run the simulation. Simulation result is shown in Fig. 2.75.

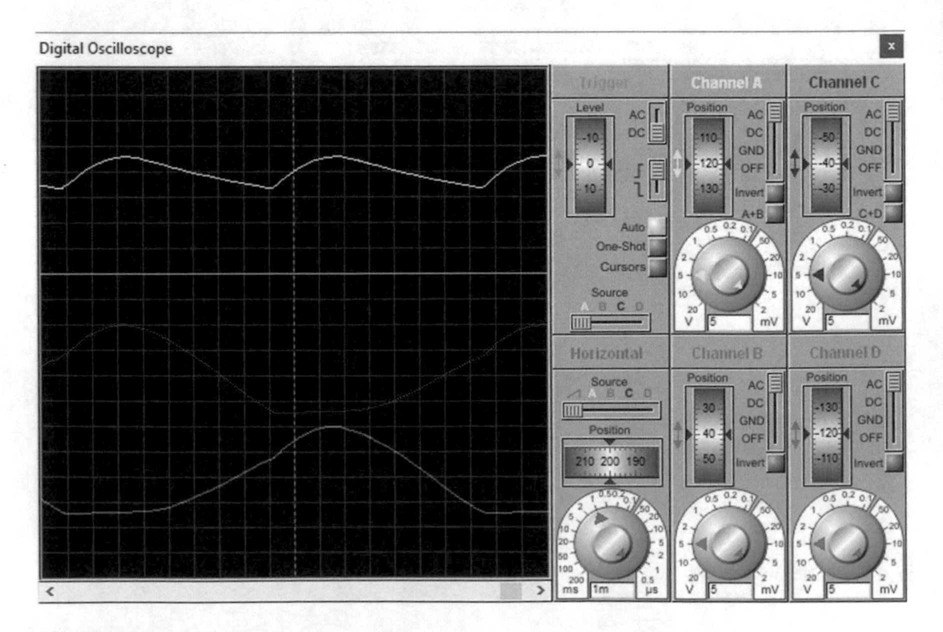

Fig. 2.75 Simulation result

Turn off the channel B, click the $C + D$ button of channel C and click the Invert button of channel D. Then double click on the Position control of channel A and C. Now you can see the resistor R_1 voltage and voltage of secondary of transformer TR1 (voltage difference between node C and D) simultaneously (Fig. 2.76). If you measure the frequency of resistor R_1 voltage, you see that its frequency is 120 Hz. Remember that in full wave rectifiers load voltage frequency is two times bigger than the input AC source frequency.

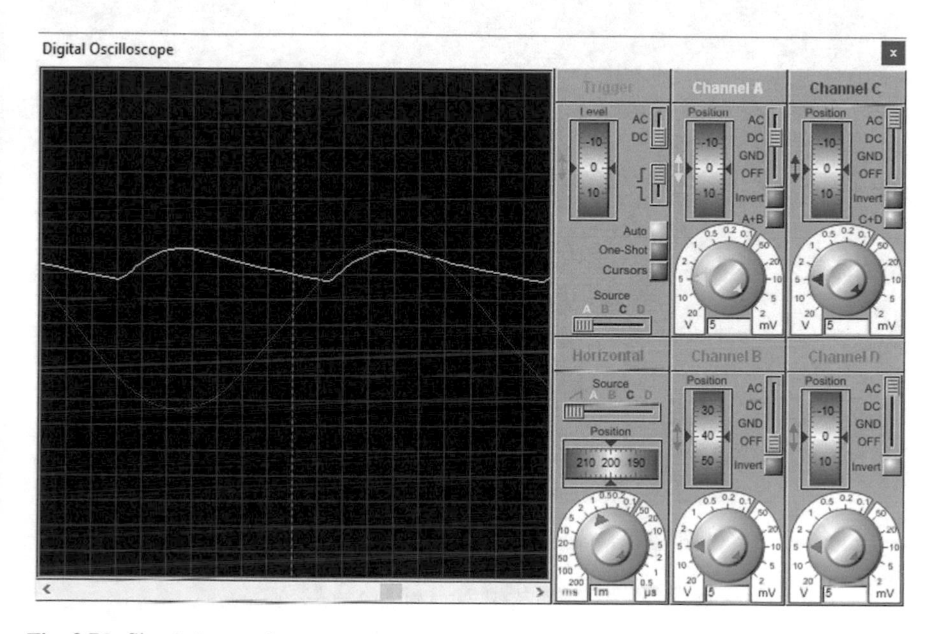

Fig. 2.76 Simulation result

Pause the simulation (Fig. 2.77) to do the measurement more easily. According to Fig. 2.78, the amplitude of output of transformer is around 17 V.

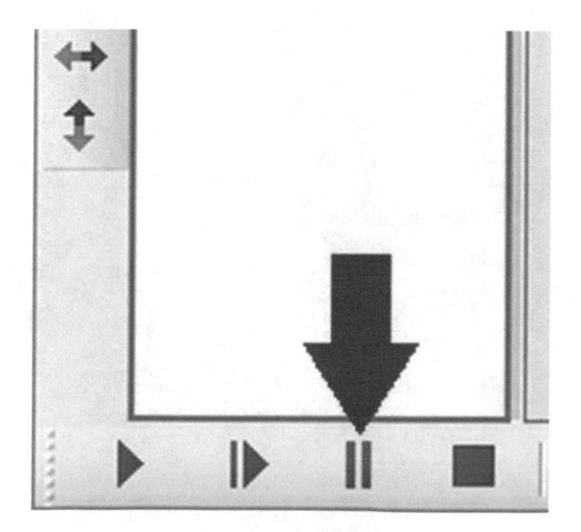

Fig. 2.77 Pause icon

Let's check the obtained result. According to the settings shown in Fig. 2.74, turn ratio of the transformer equals to $N = \frac{Np}{Ns} = \sqrt{\frac{Lp}{Ls}} = \sqrt{\frac{1H}{0.01H}} = 10$. So, RMS value of secondary winding voltage is $\frac{V_p}{N} = \frac{120}{10} = 12$ V. The amplitude of secondary winding voltage is $12\sqrt{2} = 16.97$ V. This verifies the result obtained in Fig. 2.78.

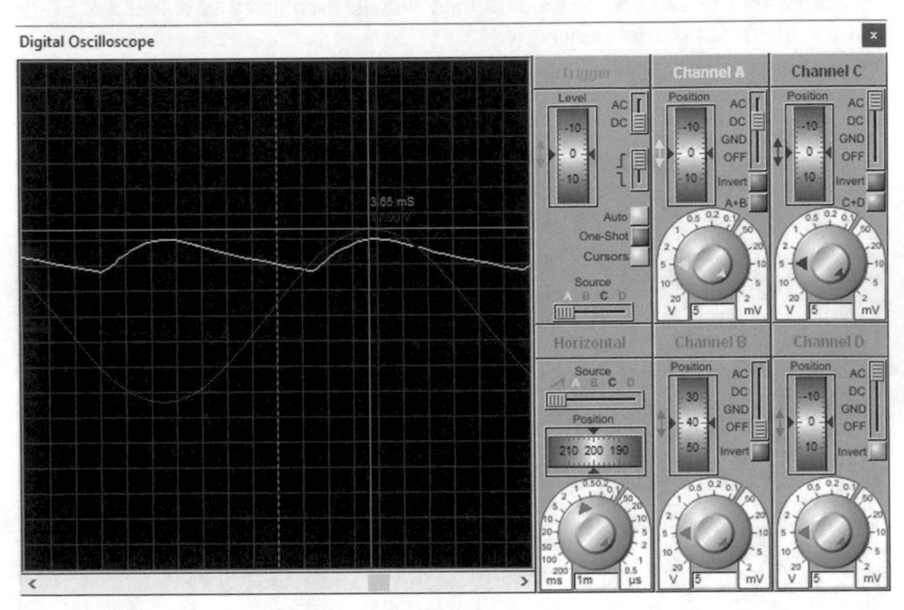

Fig. 2.78 Peak of voltage at transformer secondary is 17 V

According to Fig. 2.79, peak of load R_1 voltage is 15 V.

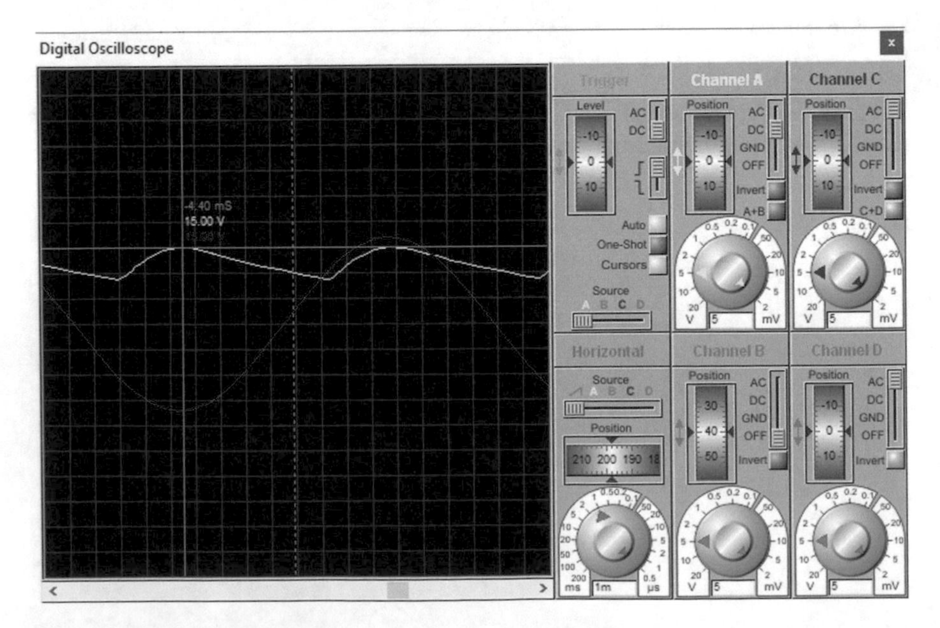

Fig. 2.79 Peak of load R_1 voltage is 15 V

Let's measure the output voltage ripple. According to Fig. 2.80, the output voltage ripple is 15–8.75 = 6.25 V.

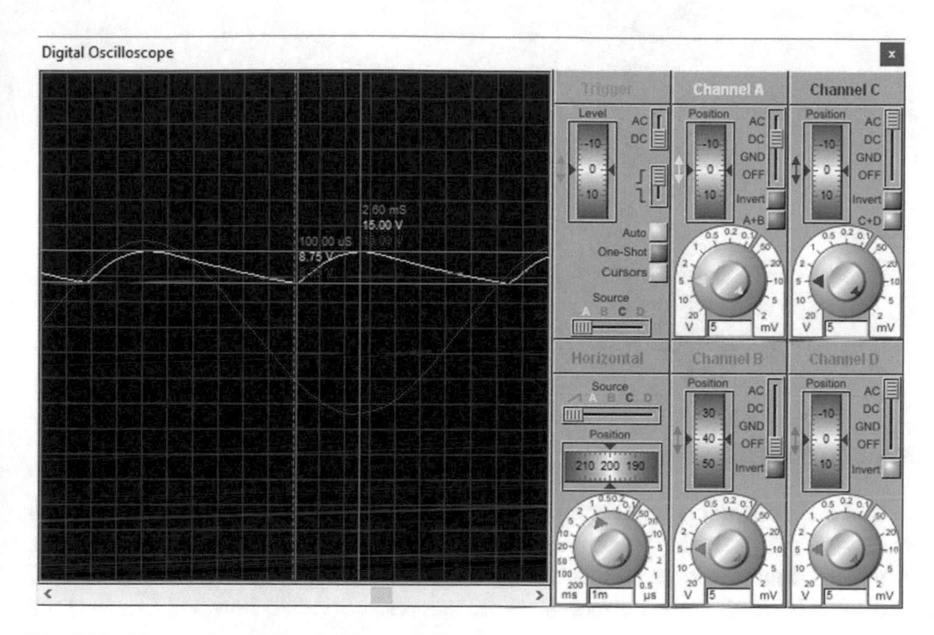

Fig. 2.80 Measurement of load voltage ripple

2.14 Example 13: Measurement of Average Value of Output Voltage for Full Wave Rectifier

In this example, we want to measure the average and RMS values of load voltage for circuit in the previous example. Connect a DC and an AC voltmeter to output. According to Fig. 2.81, reading of DC voltmeter and AC voltmeter is 10.4 V and 12.2 V, respectively.

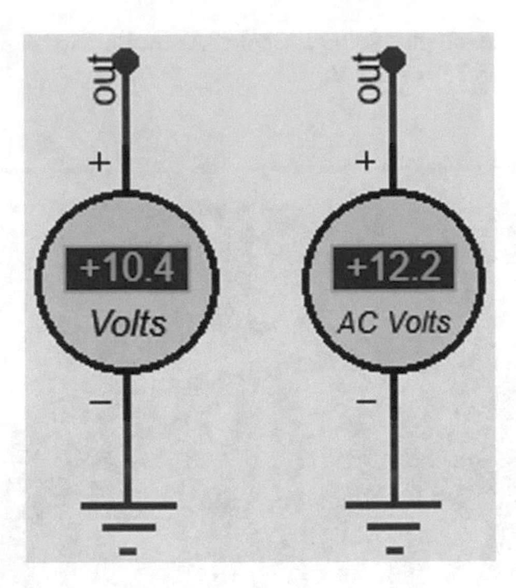

Fig. 2.81 Simulation result

Let's check the obtained results with another software. Simulation of the same circuit in NI Multisim® environment is shown in Fig. 2.82. According to the Multisim result, average and RMS values of load voltage are 12.1 V and 12.2 V, respectively. Both software measured the same RMS value; however, the average values are different.

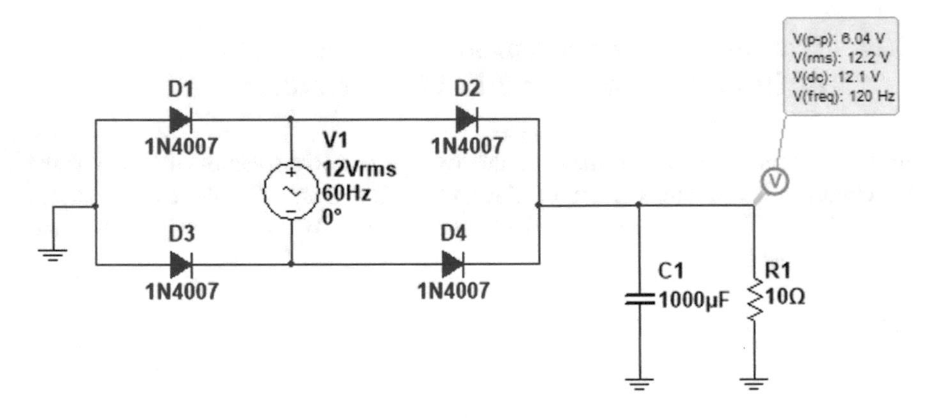

Fig. 2.82 NI Multisim simulation result for the circuit of Example 12

In Fig. 2.81, the DC voltmeter is connected to the output directly. Let's put a low pass filter block between the output and DC voltmeter (Fig. 2.83). Settings of the low pass filter are shown in Fig. 2.84. The low pass filter decreases the amplitude of the harmonics, which reach the DC voltmeter. You can add a low pass filter to the schematic by searching for "LP_F" in the Pick Devices window (Fig. 2.85).

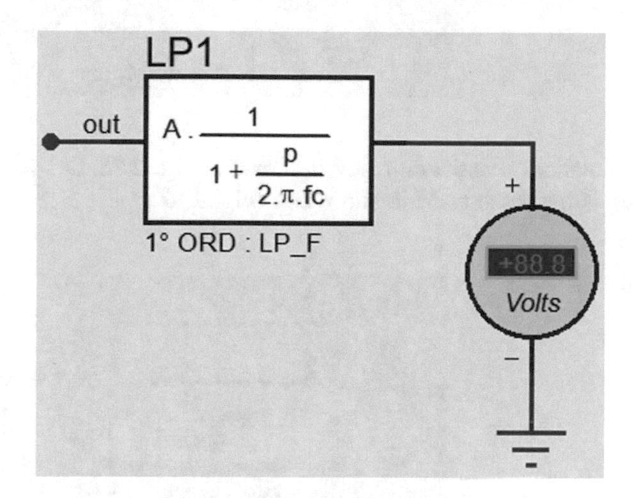

Fig. 2.83 Addition of low pass filter to the schematic

Edit Component ? X

Part Reference: LP1 Hidden: ☐ OK
Part Value: 1° ORD : LP_F Hidden: ☐ Help
Element: New Cancel
Static gain: 1.0 Hide All
Cut-off frequency [in Hz] 5 Hide All
Other Properties:

☐ Exclude from Simulation ☐ Attach hierarchy module
☐ Exclude from PCB Layout ☐ Hide common pins
☐ Exclude from Current Variant ☐ Edit all properties as text

Fig. 2.84 Setting of low pass filter LP_1

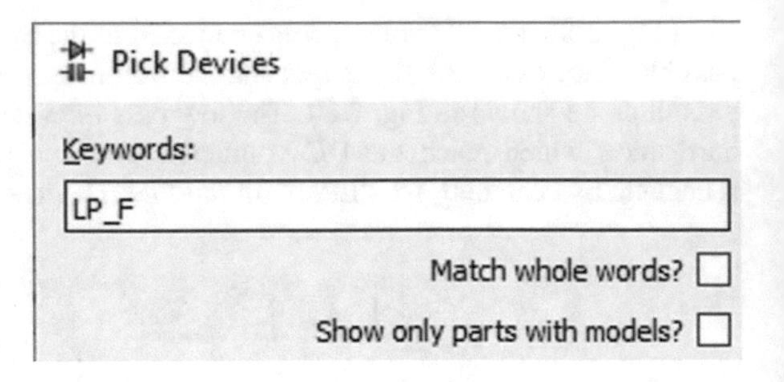

Fig. 2.85 Search for LP_F

Run the simulation. Simulation result is shown in Fig. 2.86. DC voltmeter reads 12.1 V, which is the same as Multisim result (Fig. 2.82).

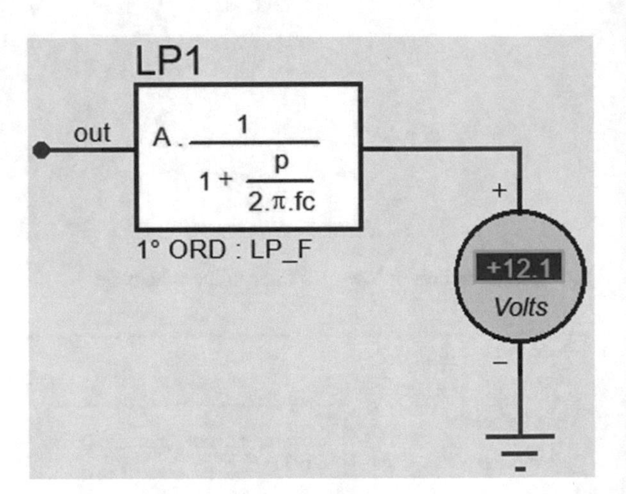

Fig. 2.86 Simulation result

You can use a simple RC filter instead of low pass filter block as well (Fig. 2.87).

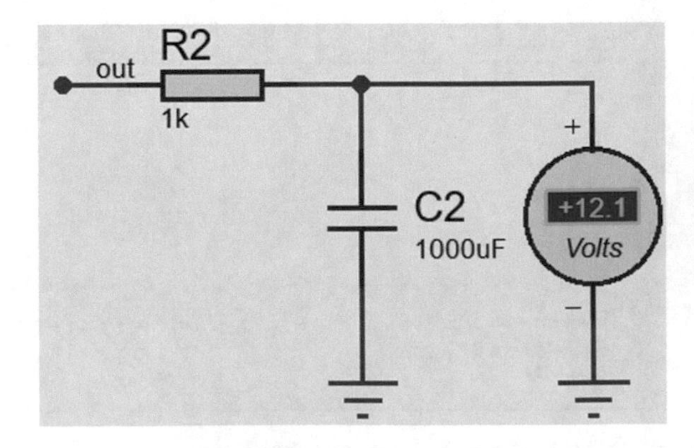

Fig. 2.87 RC filter can be used to remove the high-frequency components

2.15 Example 14: Current Passed from Rectifier Diodes

In this example, we want to measure the maximum and average current passed from the diodes of Example 12. Measurement of maximum and average values of current passed from the devices help us to select a suitable diode for the circuit.

Let's start. Draw the schematic shown in Fig. 2.88.

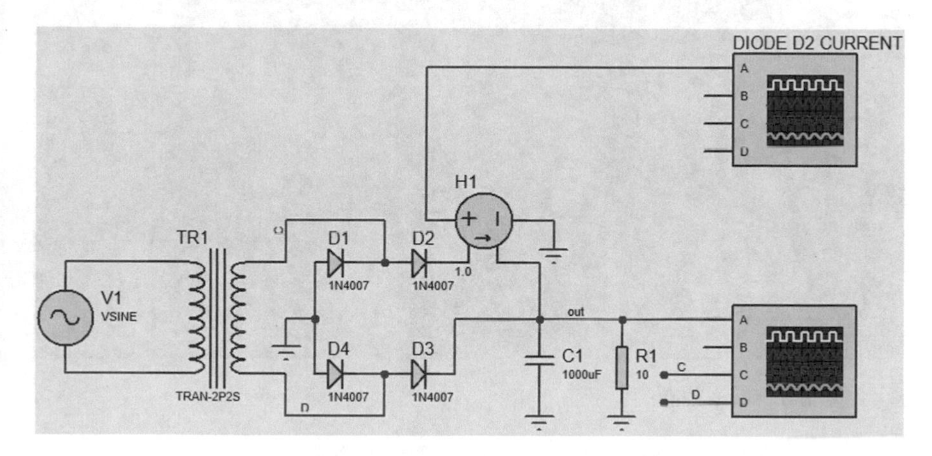

Fig. 2.88 Schematic of Example 14

Run the simulation. Simulation result is shown in Fig. 2.89. According to Fig. 2.90, peak value of the waveform is about 5.45 A.

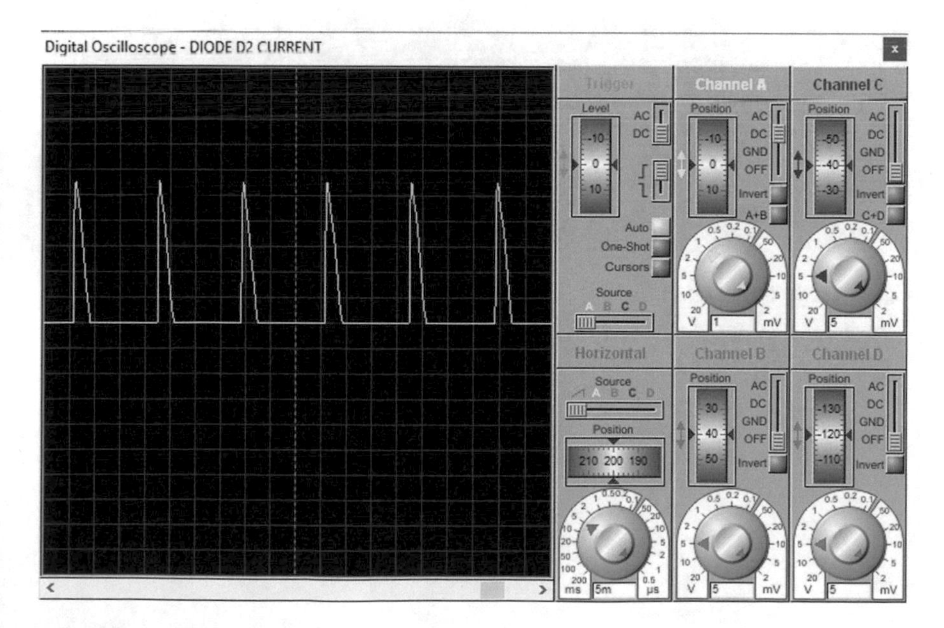

Fig. 2.89 Simulation result

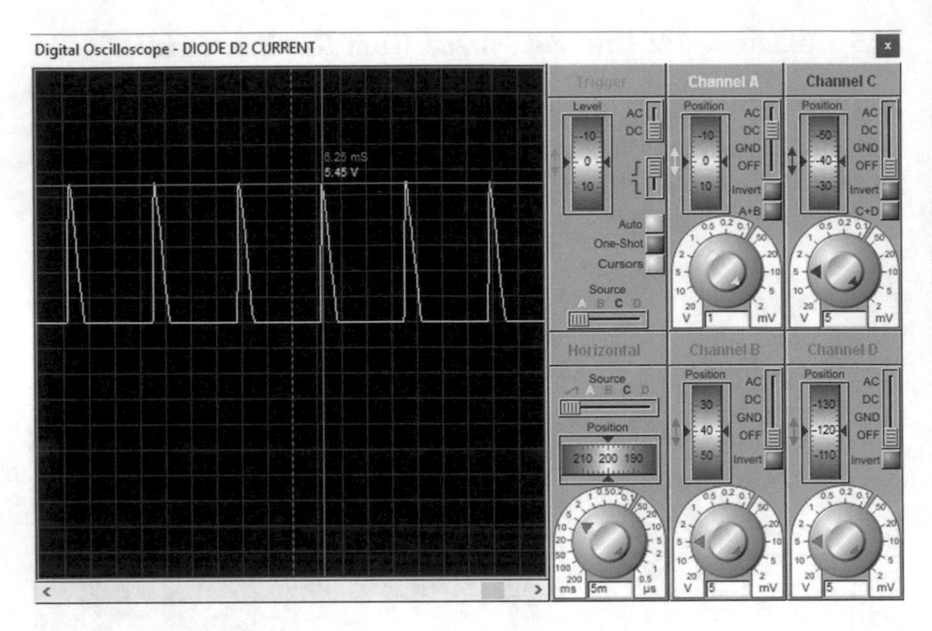

Fig. 2.90 Simulation result

Let's measure the average current passed from diode D_2. Put a DC ammeter in series with the diode D_2. According to Fig. 2.91, the DC ammeter reads 0 mA.

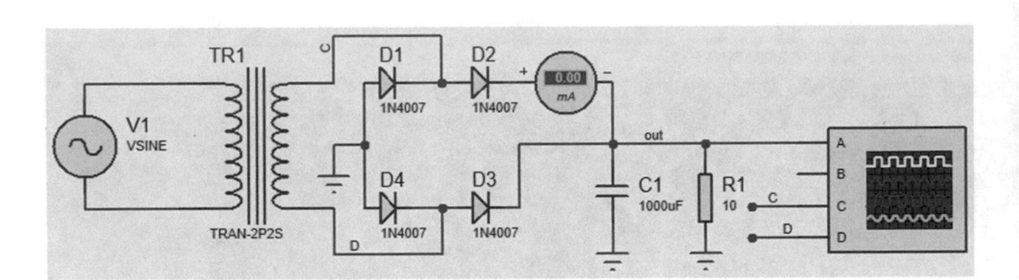

Fig. 2.91 Simulation result

Replace the DC ammeter with an AC ammeter. According to Fig. 2.92, the RMS value of current passed from diode D_2 is 1.55 A.

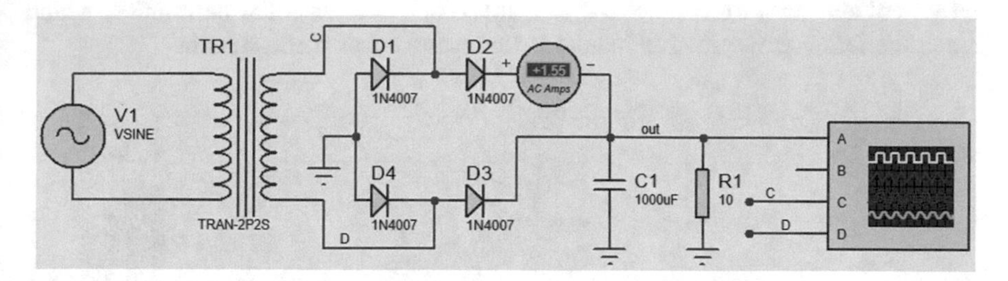

Fig. 2.92 Simulation result

Let's check the obtained results with another software. Simulation of the same circuit in NI Multisim environment is shown in Fig. 2.93. According to the Multisim result, average and RMS values of diode D_2 current are 607 mA and 1.6 A, respectively. Measured RMS values are quite close; however, the average values are different.

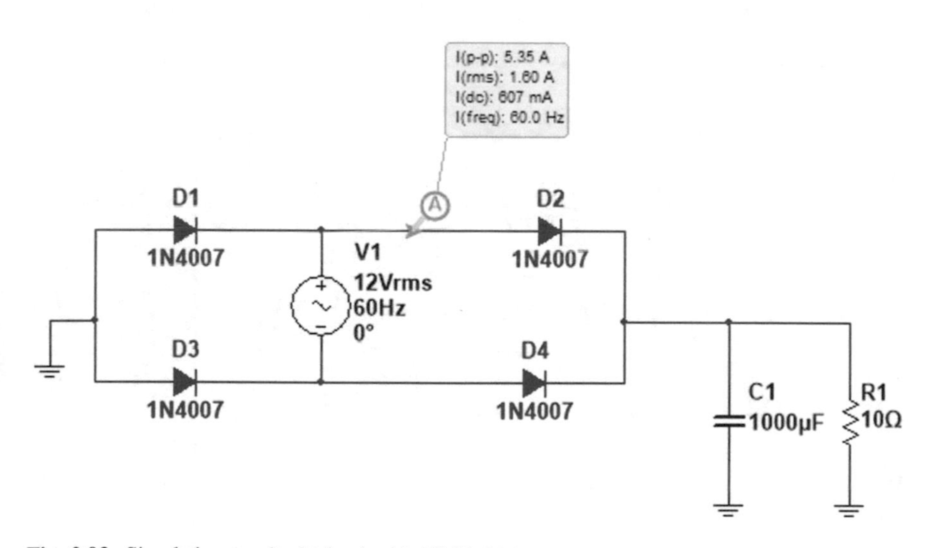

Fig. 2.93 Simulation result obtained with NI Multisim

Let's measure the average value of current passed from diode D_2 with the aid of schematic shown in Fig. 2.94. Settings of the low pass filter are shown in Fig. 2.95. The current-dependent voltage source H_1 converts the current passed from diode D_2 into a voltage waveform. The low pass filter LP_1 decreases the harmonics, which reach the DC voltmeter. This increases the accuracy of measurement.

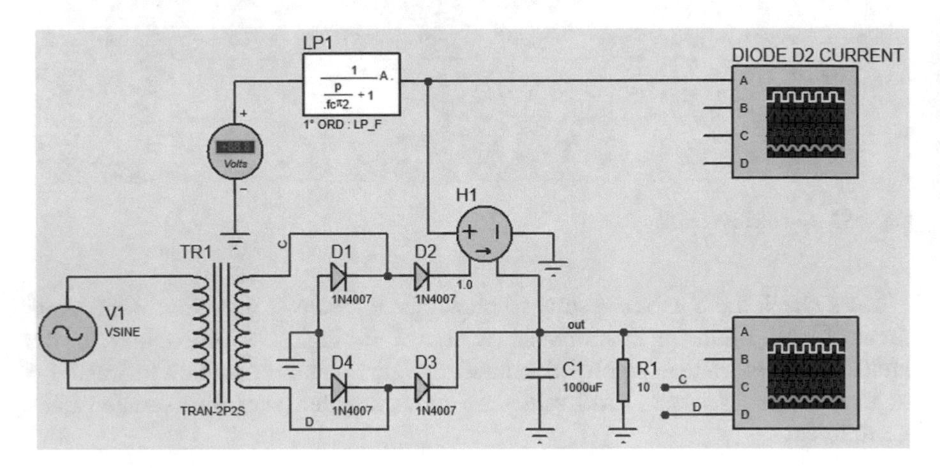

Fig. 2.94 Schematic to measure the average value of the current passed from diode D_2

Edit Component ? X

Part Reference: LP1 Hidden: ☐ OK

Part Value: 1° ORD : LP_F Hidden: ☐ Help

Element: New Cancel

Static gain: 1.0 Hide All

Cut-off frequency [in Hz] 1 Hide All

Other Properties:

☐ Exclude from Simulation ☐ Attach hierarchy module

☐ Exclude from PCB Layout ☐ Hide common pins

☐ Exclude from Current Variant ☐ Edit all properties as text

Fig. 2.95 Settings of low pass filter block

Run the simulation. Simulation result is shown in Fig. 2.96. According to Fig. 2.96, the average value of current is 0.58 A. Obtained result is close to Multisim result.

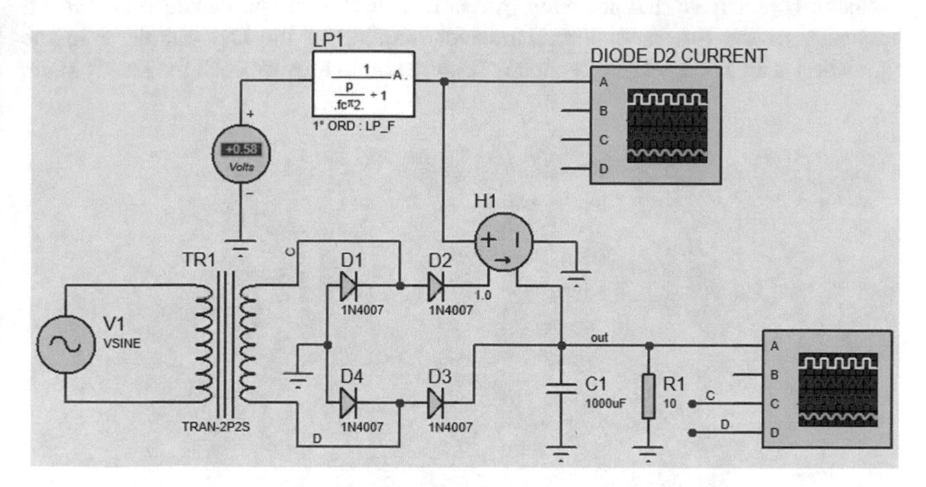

Fig. 2.96 Simulation result

Stop the simulation and double click the low pass filter block. Then decrease the cut-off frequency to 0.01 Hz (Fig. 2.97).

Edit Component

Part Reference: LP1 Hidden:

Part Value: 1° ORD : LP_F Hidden:

Element: New

Static gain: 1.0 Hide All

Cut-off frequency [in Hz] .01 Hide All

Other Properties:

Exclude from Simulation

Exclude from PCB Layout

Exclude from Current Variant

Attach hierarchy module

Hide common pins

Edit all properties as text

OK

Help

Cancel

Fig. 2.97 Settings of low pass filter LP_1

Run the simulation. Simulation result is shown in Fig. 2.98. According to Fig. 2.98, the average value of current is 0.60 A. Obtained result is quite close to Multisim result. Note that accuracy of result increases as you decrease the cut-off frequency of the low pass filter. However, reading of the DC voltmeter in the simulation with lower cut-off frequency requires more time to reach the steady state.

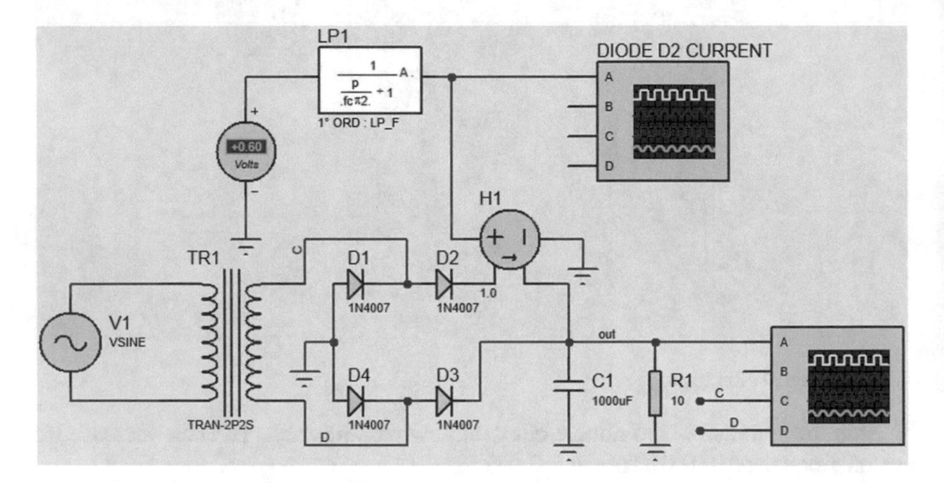

Fig. 2.98 Simulation result

2.16 Example 15: Bridge Block

Full wave rectifier shown in Fig. 2.99 used four discrete diodes to rectify the AC voltage. You can use the schematic shown in Fig. 2.100 as well. In this schematic, instead of using four discrete diodes, one bridge block (Fig. 2.101) is used.

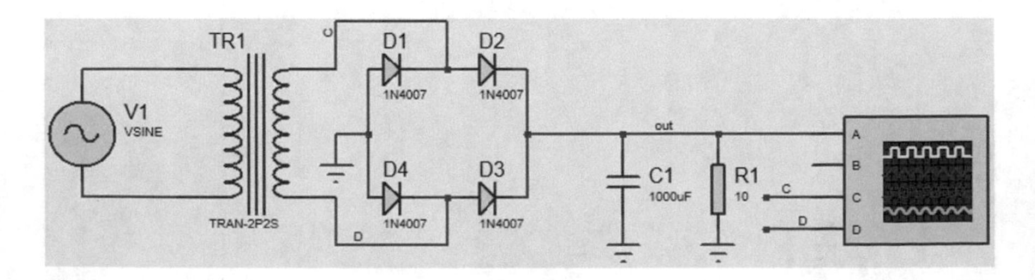

Fig. 2.99 Schematic for Example 15

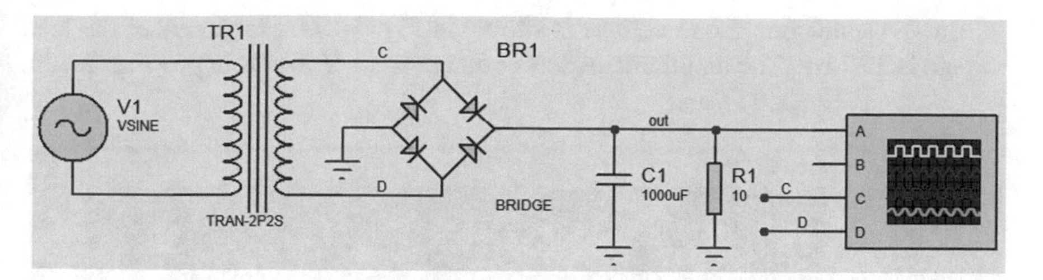

Fig. 2.100 You can use a diode bridge block instead of four discrete diode

Pick Devices

Keywords:

bridge

Match whole words? ☐

Show only parts with models? ☐

Fig. 2.101 Searching for bridge block

2.17 Example 16: Fourier Analysis of Output Voltage of Full Wave Rectifier

In this example, we want to measure the amplitude of harmonics available in the load voltage for the circuit shown in Fig. 2.102.

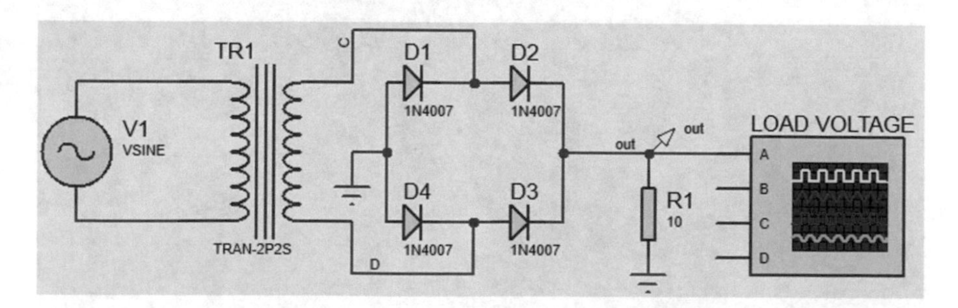

Fig. 2.102 Schematic for Example 16

Run the simulation. Load voltage is shown in Fig. 2.103. Frequency of the load voltage is 120 Hz. The amplitude of load voltage is 15 V according to Fig. 2.104.

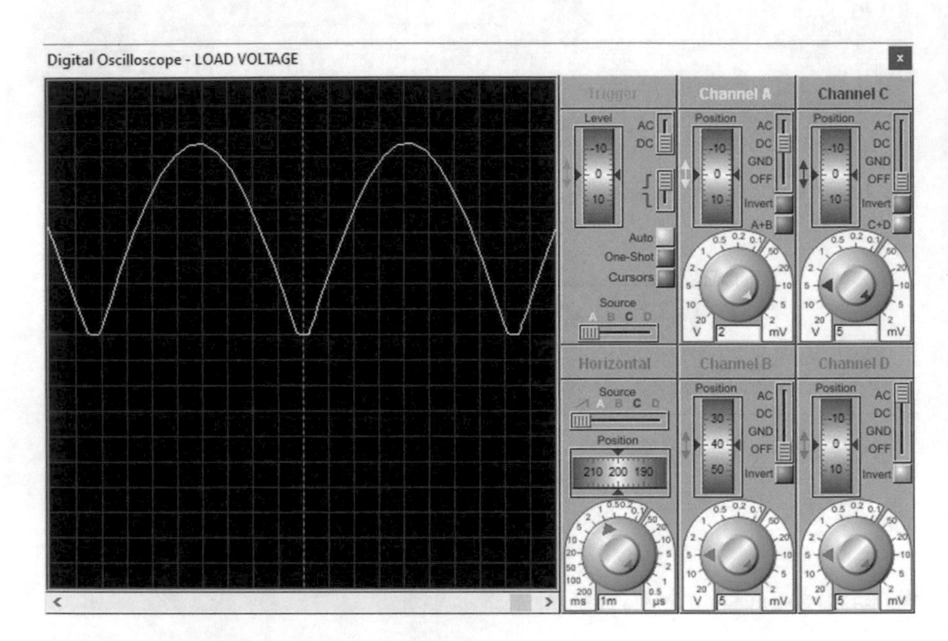

Fig. 2.103 Simulation result

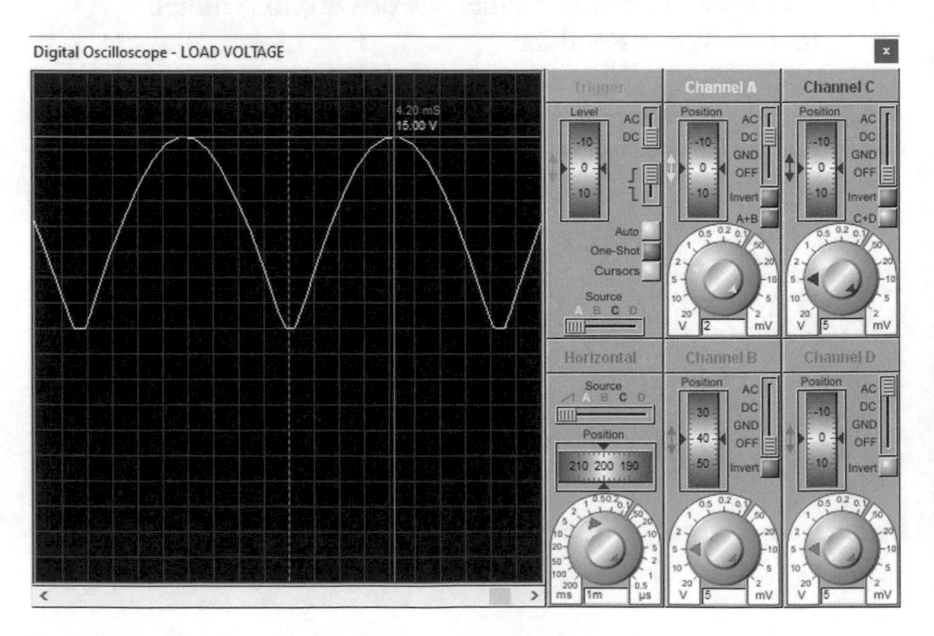

Fig. 2.104 Measurement of output voltage peak

Add a Fourier graph to the schematic (Fig. 2.105).

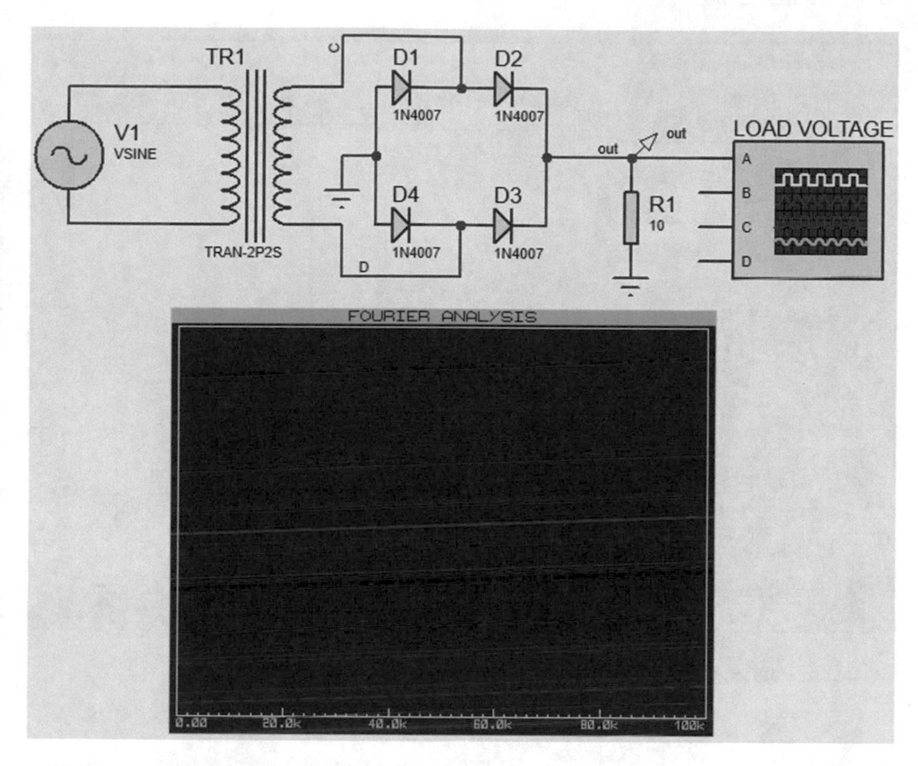

Fig. 2.105 Addition of Fourier graph to the schematic

Double click on the Fourier graph and do the settings similar to Fig. 2.106. Then click the OK button. After clicking the OK button, the schematic changes to what is shown in Fig. 2.107.

Edit Fourier Analysis Graph ? X

Graph title: FOURIER ANALYSIS
Start time: 0
Stop time: 320m
Max Frequency: 1.2k
Resolution: 120 Window None
Left Axis Label: Freq.
Right Axis Label: Output Voltage Harmonics
Options
Y Scale in dBs: ☐
Initial DC solution: ☑
Always simulate? ☑
Log netlist(s)? ☐
SPICE Options
Set Y-Scales
User defined properties:
OK Cancel

Fig. 2.106 Fourier graph settings

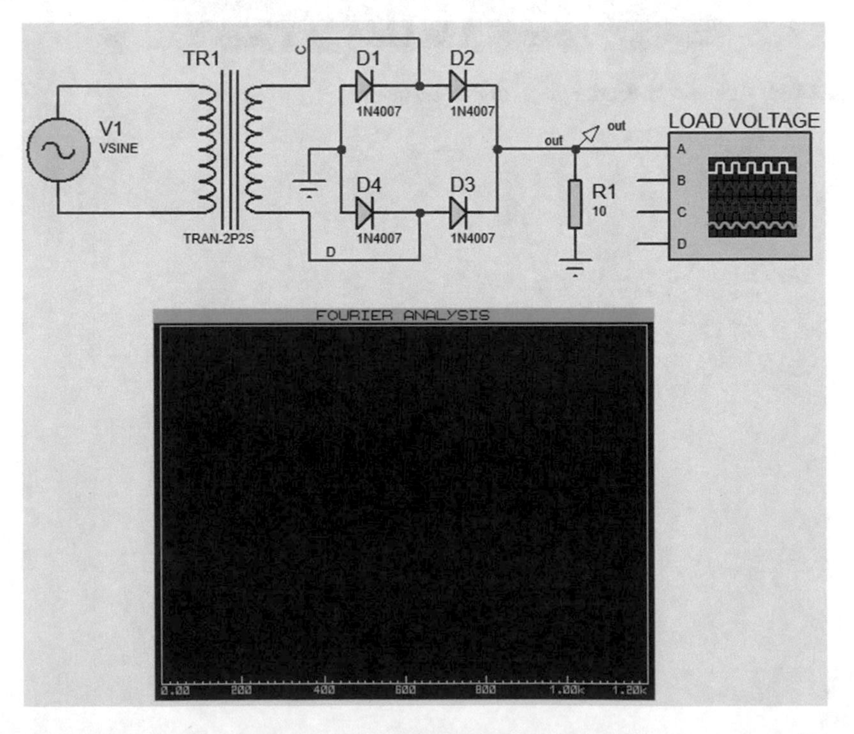

Fig. 2.107 Maximum frequency of Fourier graph is changed to 1.2 kHz

The settings in Fig. 2.106 ask Proteus to take the waveform in the [0, 320 ms] time interval and calculate the Fourier coefficients up to 10th harmonic (1.2 kHz).

Drag and drop the voltage probe "out" onto the top left corner of the Fourier graph (Fig. 2.108).

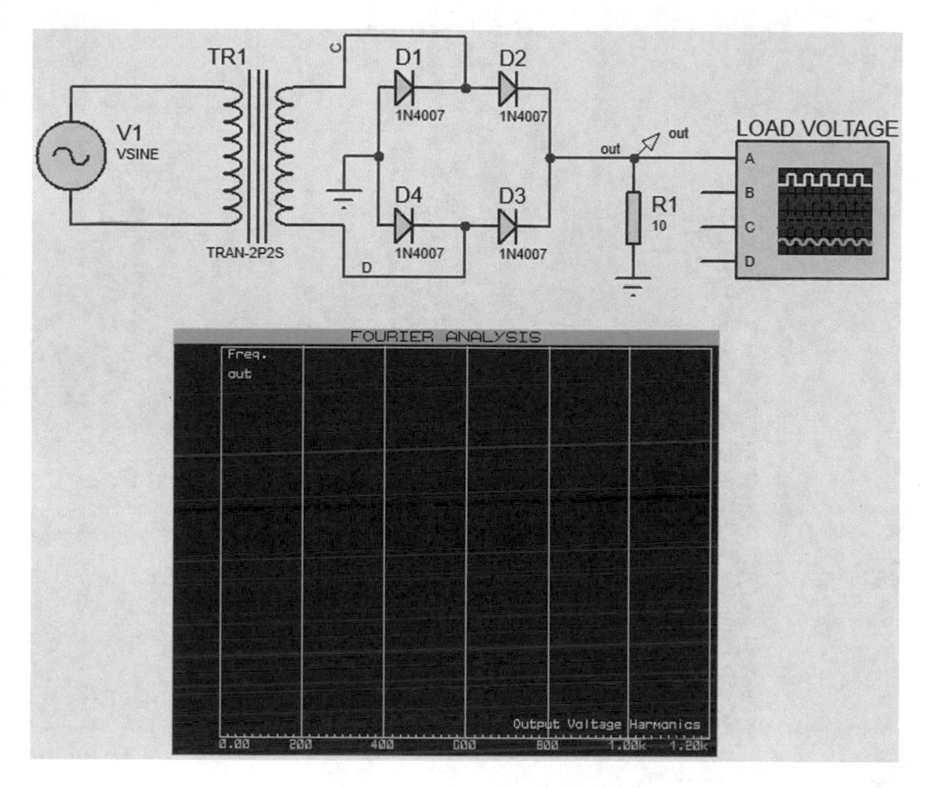

Fig. 2.108 Voltage probe out is dropped onto the Fourier graph

Press the space bar key of your keyboard to run the simulation. Simulation result is shown in Fig. 2.109. Maximize the graph and use the cursor to read the amplitude of harmonics. Amplitudes are shown in Table 2.1.

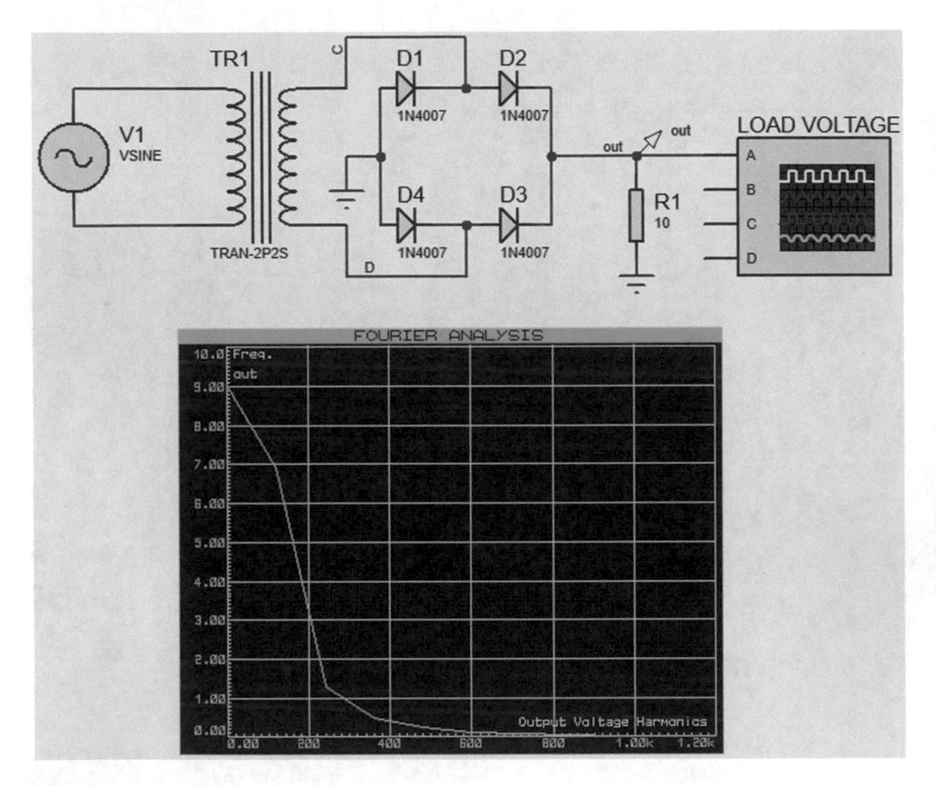

Fig. 2.109 Simulation result

Table 2.1 Amplitudes measured by Proteus

Freq.	Amplitude (V)
DC	9.01
120 Hz	6.93
240 Hz	1.29
360 Hz	0.49
480 Hz	0.233
600 Hz	0.121
720 Hz	0.0648
840 Hz	0.0333
960 Hz	0.0148
1080 Hz	0.00586
1200 Hz	0.00679

Let's check the obtained result. Fourier series of a full wave rectified sine wave is shown in Fig. 2.110. Table 2.2 compares the values suggested by this formula with the values measured by Proteus. Note that corresponding values are quite close together.

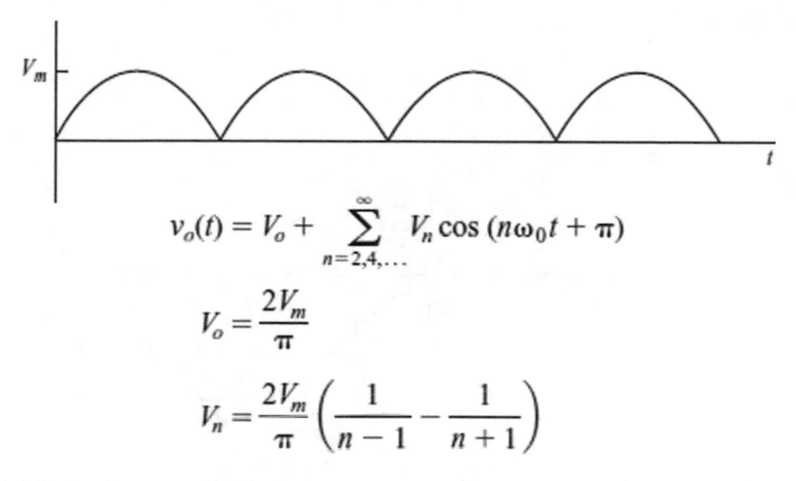

Fig. 2.110 Fourier series of full wave rectified sinusoidal signal

Table 2.2 Comparison between Proteus result and theoretical calculations

Freq.	Proteus amplitude (V)	Theoretical amplitude (V)
DC	9.01	9.54
120 Hz	6.93	6.37
240 Hz	1.29	1.27
360 Hz	0.49	0.55
480 Hz	0.233	0.30
600 Hz	0.121	0.19
720 Hz	0.0648	0.13
840 Hz	0.0333	0.10
960 Hz	0.0148	0.07
1080 Hz	0.00586	0.06
1200 Hz	0.00679	0.05

2.18 Example 17: Harmonic Content of a Triangular Waveform

In this example, we want to see the harmonic content of the triangular waveform shown in Fig. 2.111. $t_1 = 0.25$ ms, $t_2 = 0.5$ ms, $t_3 = 0.75$ ms, $t_4 = 1$ ms and amplitude is 1 V. This waveform can be defined with the aid of File generator block (Fig. 2.112).

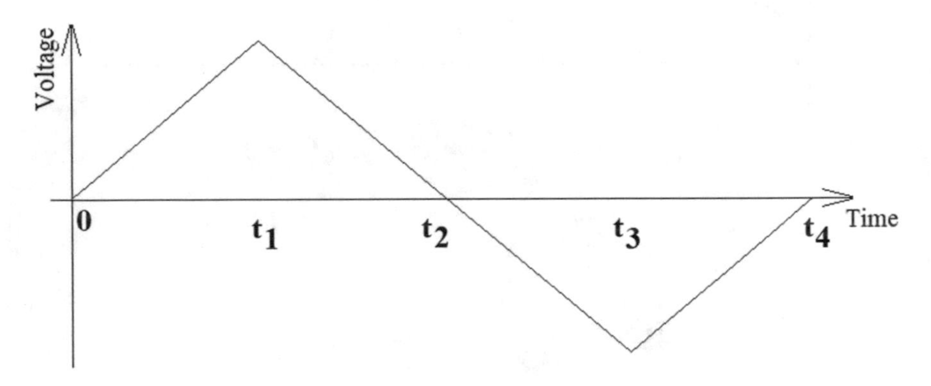

Fig. 2.111 Triangular wave

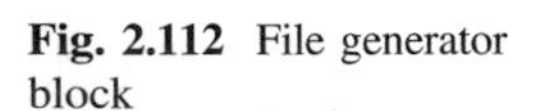

Fig. 2.112 File generator block

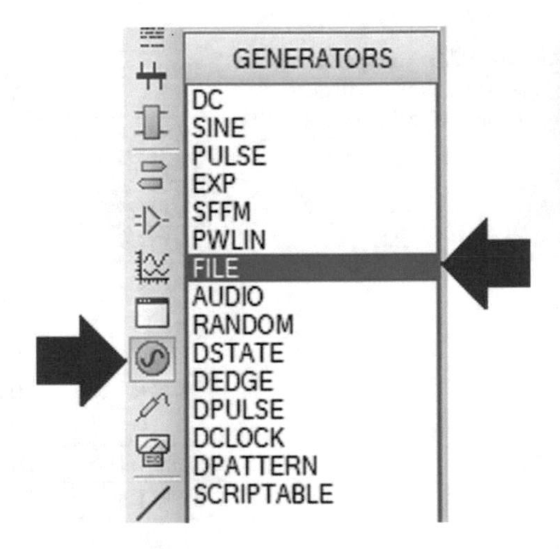

The triangular waveform of this example can be defined with the aid of text file shown in Figs. 2.113 or 2.114. First column shows the time, and second column shows the corresponding voltage.

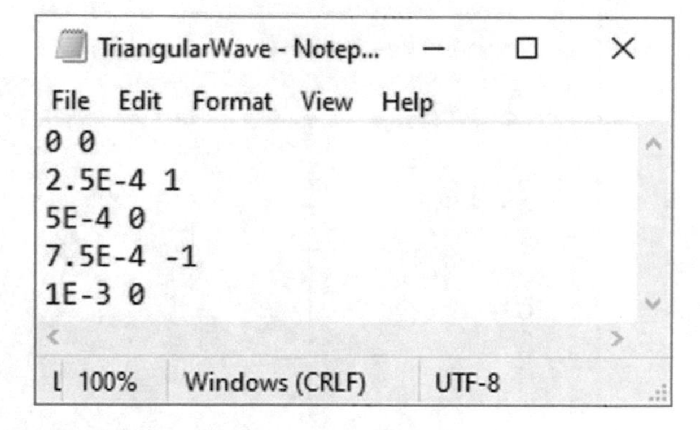

Fig. 2.113 Content of TriangularWave.txt

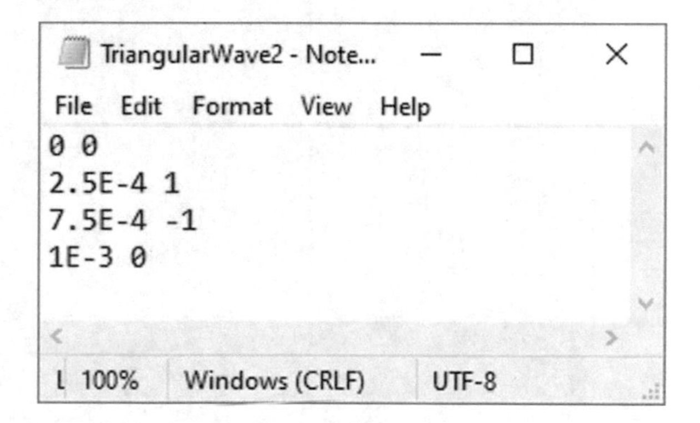

Fig. 2.114 Content of TriangularWave2.txt

Draw the schematic shown in Fig. 2.115.

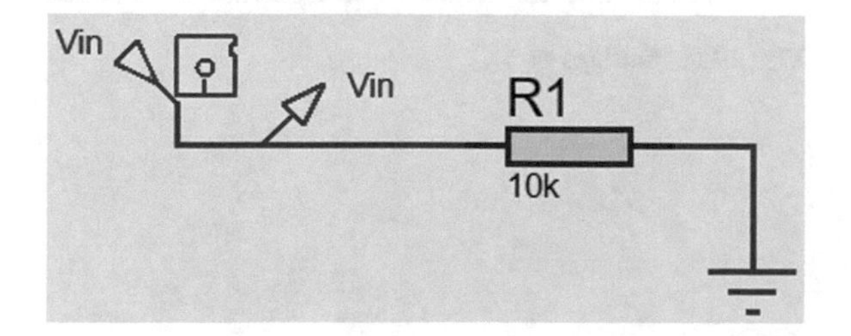

Fig. 2.115 Schematic of Example 17

Double click on the File generator "V_{in}" and use the Browse button to select the text file that defines the triangular waveform (Fig. 2.116). Then click the OK button.

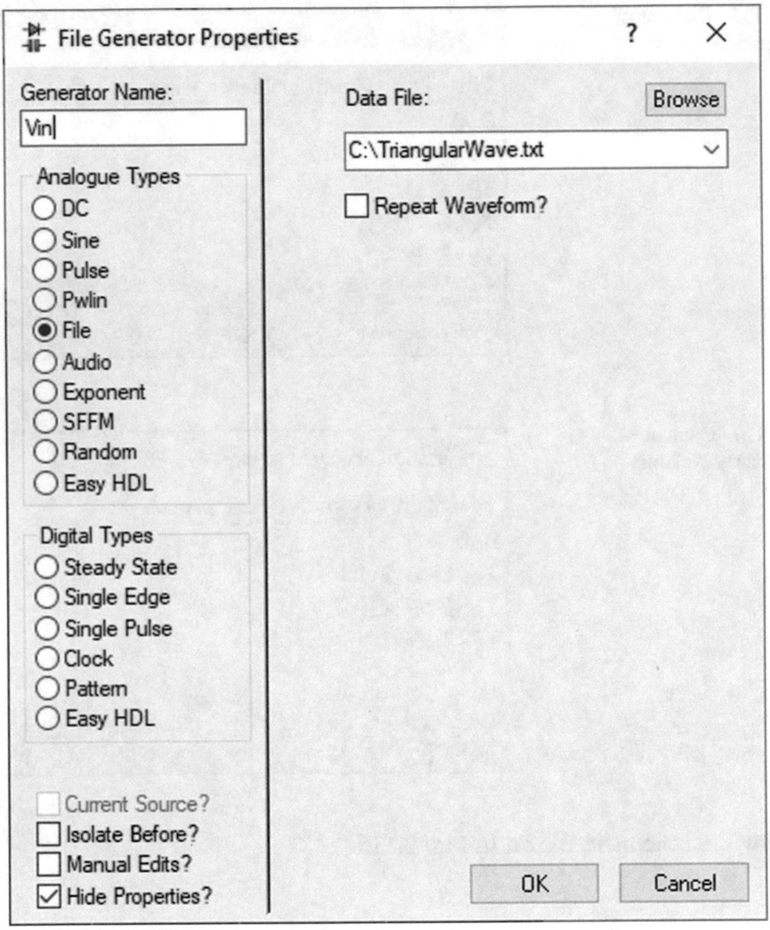

Fig. 2.116 Settings of V_{in}

The waveform generated by the Graph generator is shown in Fig. 2.117.

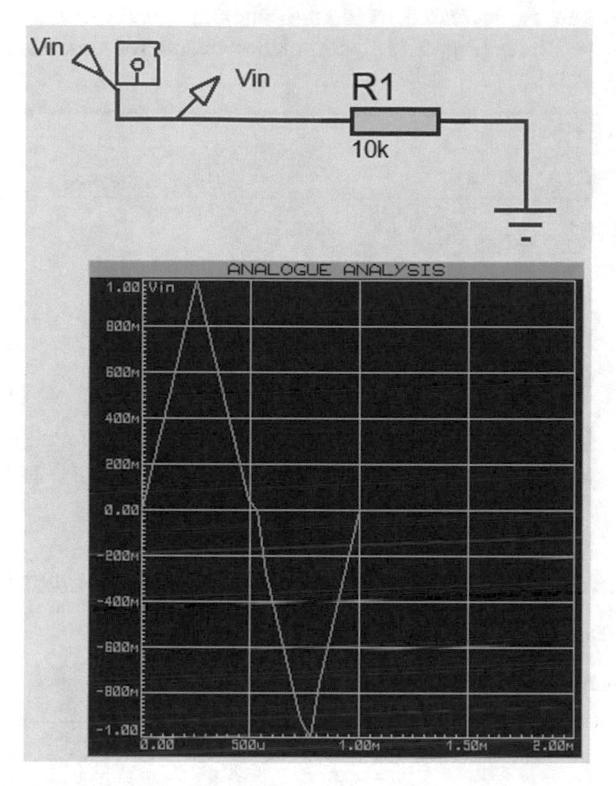

Fig. 2.117 Simulation result

The waveform shown in Fig. 2.117 is a little bit distorted. You can obtain a better triangular waveform by decreasing the time step of simulation. The time step of simulation can be decreased by double clicking on the Analogue graph and entering T_{max} = 0.1 μs (Fig. 2.118). Simulation result for time step of 0.1 μs is shown in Fig. 2.119.

Edit Transient Graph

Graph title: ANALOGUE ANALYSIS
Start time: 0
Stop time: 2m
Left Axis Label:
Right Axis Label:
Options
Initial DC solution: ☑
Always simulate: ☑
Log netlist(s): ☐
SPICE Options
Set Y-Scales
User defined properties:
Tmax=0.1u
OK Cancel

Fig. 2.118 Settings of analogue graph

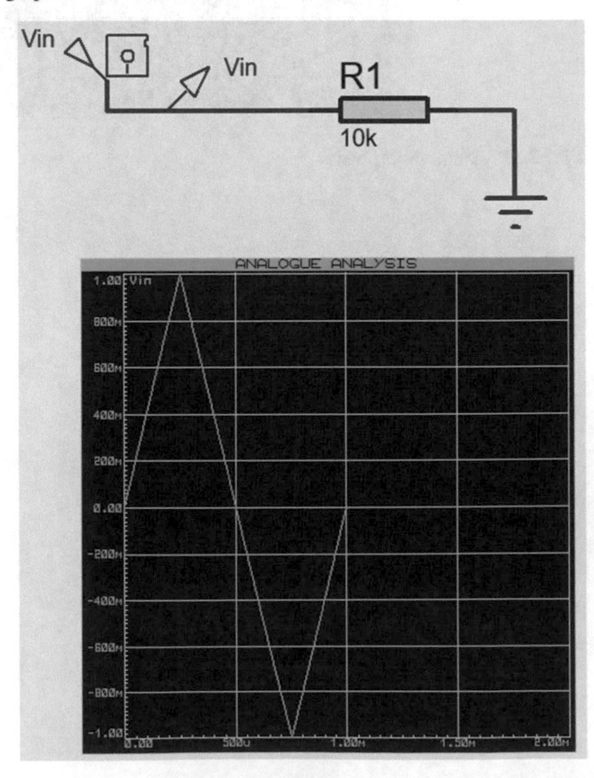

Fig. 2.119 Simulation result

Remove the Analogue graph and add a Fourier graph to the schematic (Fig. 2.120).

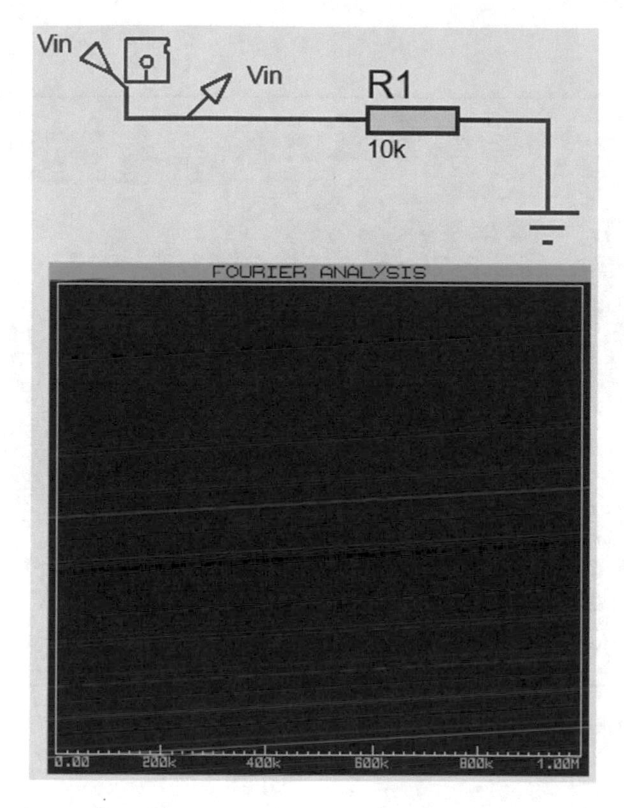

Fig. 2.120 Fourier graph is added to the schematic

Double click on the Fourier graph and do the settings similar to Fig. 2.121. Then click the OK button. The settings in Fig. 2.121 ask Proteus to take the waveform in the [0, 1 ms] time interval and calculate the Fourier coefficients up to 10th harmonic (10 kHz).

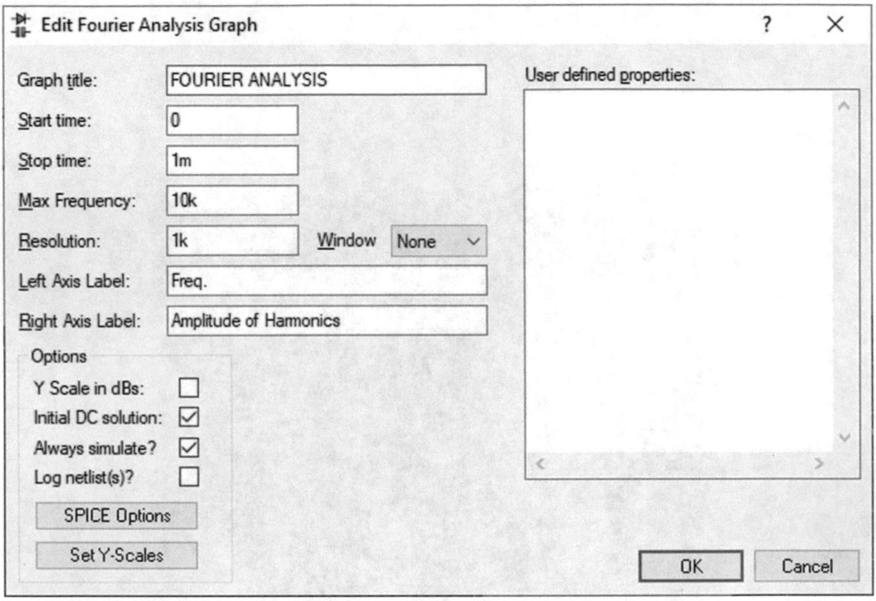

Fig. 2.121 Fourier graph settings

Drag and drop the voltage probe "V_{in}" onto the top left corner of the Fourier graph (Fig. 2.122).

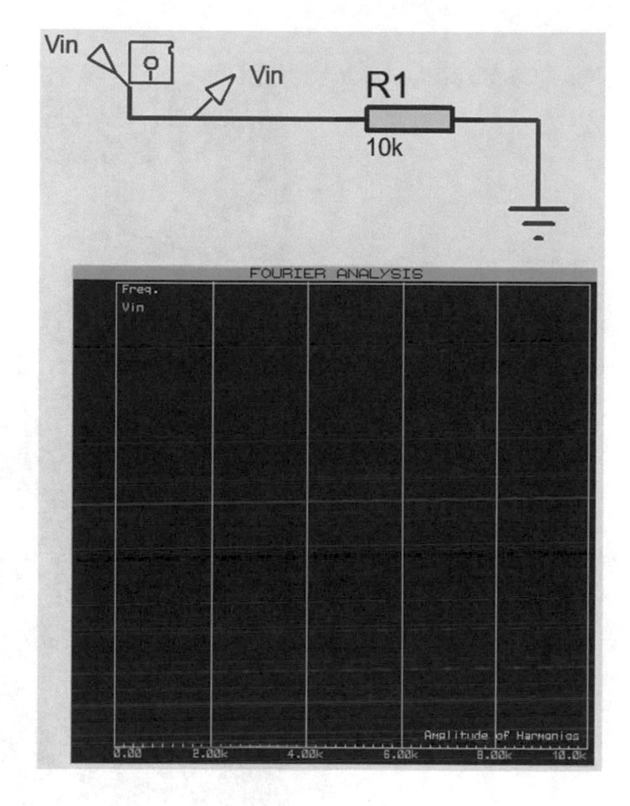

Fig. 2.122 Voltage probe V_{in} is dropped onto the Fourier graph

Run the simulation by pressing the space bar key of your keyboard. Simulation result is shown in Fig. 2.123. Maximize the graph and use the cursor to read the amplitude of harmonics. Amplitudes are shown in Table 2.3.

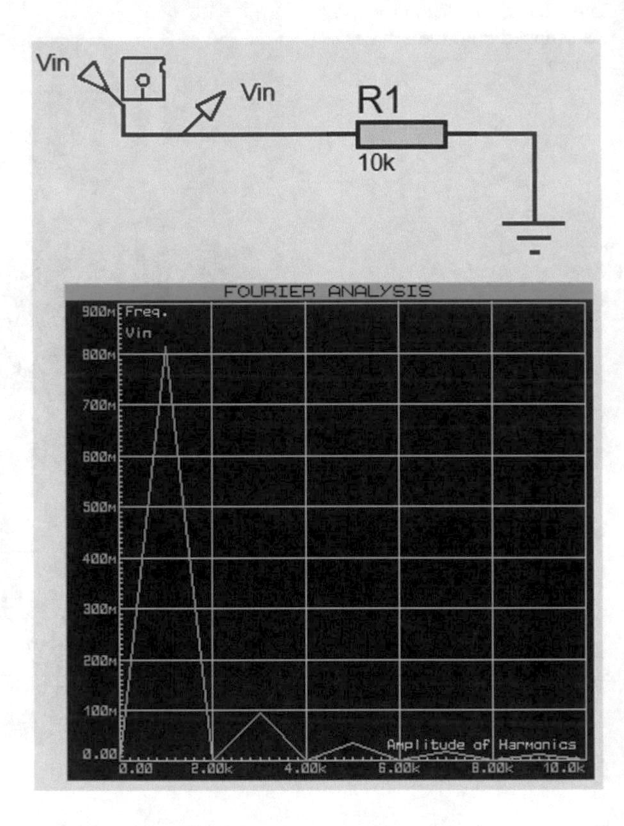

Fig. 2.123 Simulation result

Table 2.3 Amplitudes measured by Proteus

Freq.	Proteus amplitude (mV)
DC	0
1 kHz	811
2 kHz	2.1
3 kHz	93.7
4 kHz	0.94
5 kHz	35.5
6 kHz	1
7 kHz	19.4
8 kHz	0.84
9 kHz	12.3
10 kHz	0.87

Let's check the obtained result. Following MATLAB code calculates the amplitude of first 10 harmonics for a triangular wave with amplitude of 1 V and frequency of 1 kHz. Output of code is shown in Fig. 2.124.

```
clc
A=1;
T=1e-3;

w=2*pi/T;
m=A/(T/4);
syms t
f1=m*t;
f2=-m*(t-(T/2));
f3=m*(t-T);

for n=[1:10]
b=2/T*eval(int(f1*sin(n*w*t),t,0,T/4)+int(f2*sin(n*w*t),t,T/4,3*T/4)+int(f3
*sin(n*w*t),t,3*T/4,T));
S1=strcat('Harmonic number:',string(n));
S2=strcat(S1,', Amplitude in mV:',string(1000*abs(b)));
disp(S2)
end
```

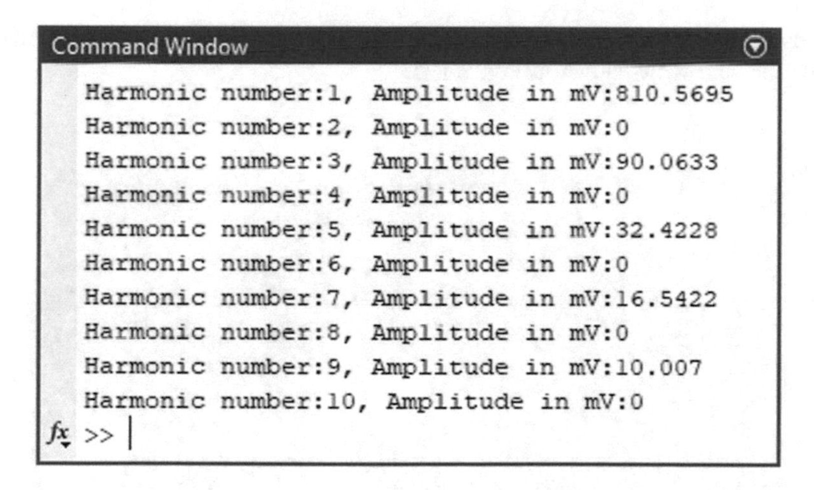

Fig. 2.124 Output of MATLAB code

Table 2.4 compares the values calculated by MATLAB (Fig. 2.124) and the values measured by Proteus. Note that corresponding values are quite close together.

Table 2.4 Comparison between Proteus result and theoretical calculations

Freq.	Proteus amplitude (mV)	MATLAB result (mV)
DC	0	0
1 kHz	811	810.57
2 kHz	2.1	0
3 kHz	93.7	90.06
4 kHz	0.94	0
5 kHz	35.5	32.42
6 kHz	1	0
7 kHz	19.4	16.54
8 kHz	0.84	0
9 kHz	12.3	10.00
10 kHz	0.87	0

2.19 Example 18: Voltage Regulator (I)

In this example, we want to simulate a simple zener diode-based voltage regulator. Draw the schematic shown in Fig. 2.125.

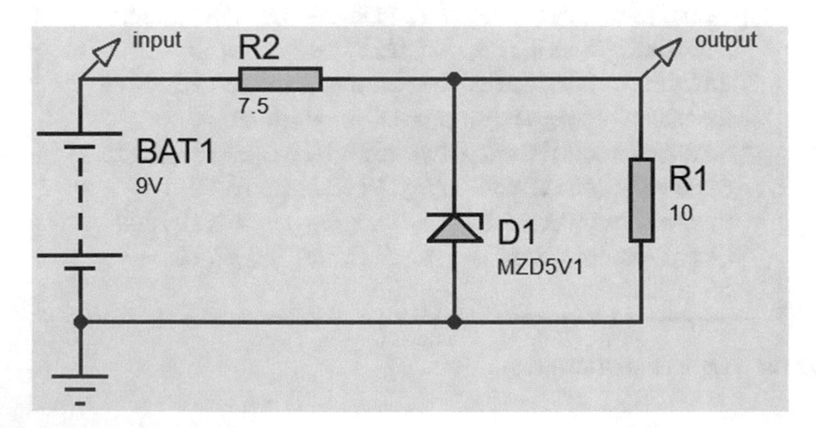

Fig. 2.125 Schematic of Example 18

Run the simulation. Simulation result is shown in Fig. 2.126. Load voltage is 5.1056 V.

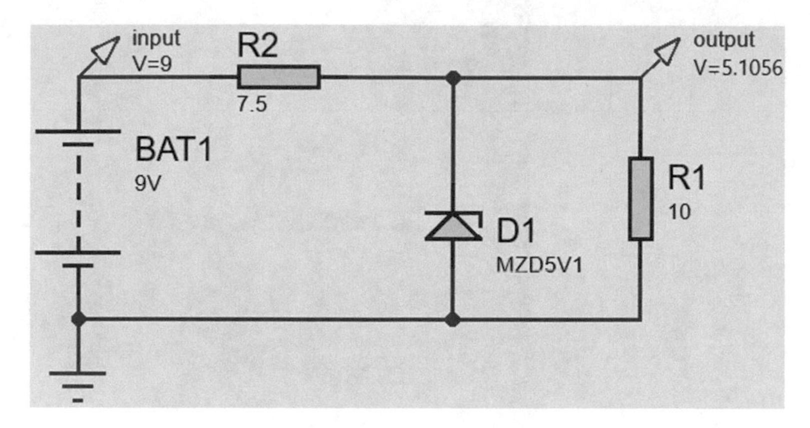

Fig. 2.126 Simulation result

Let's measure the current drawn from input voltage source. According to Fig. 2.127, the current drawn from the input voltage source is 0.519254 A.

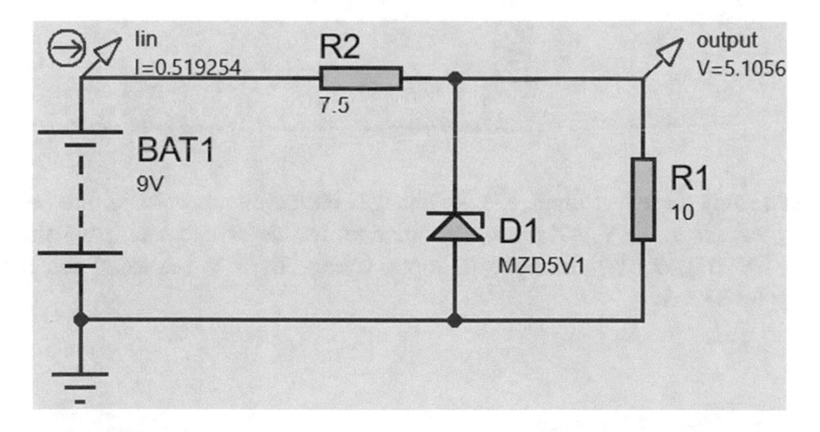

Fig. 2.127 Simulation result

Let's calculate the efficiency of the circuit. According to the calculations shown in Fig. 2.128, the efficiency of the circuit is 55.7791%.

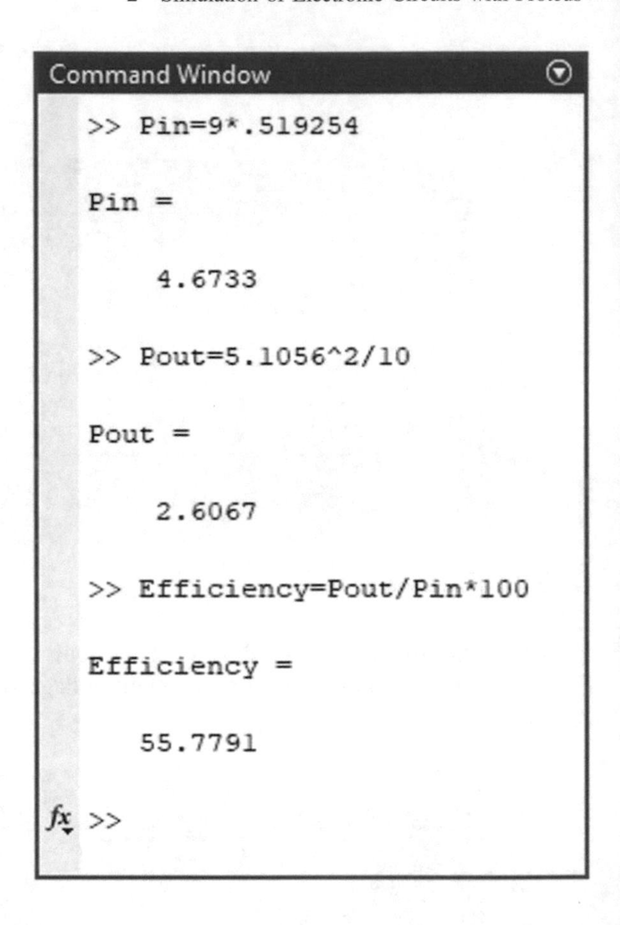

Fig. 2.128 MATLAB calculations

Let's study the effect input voltage changes on the output voltage. Increase the battery voltage to 10 V and run the simulation. In this case, the output voltage is 5.18864 V (Fig. 2.129). Increase in input voltage by 1 V increased the output voltage by 83 mV.

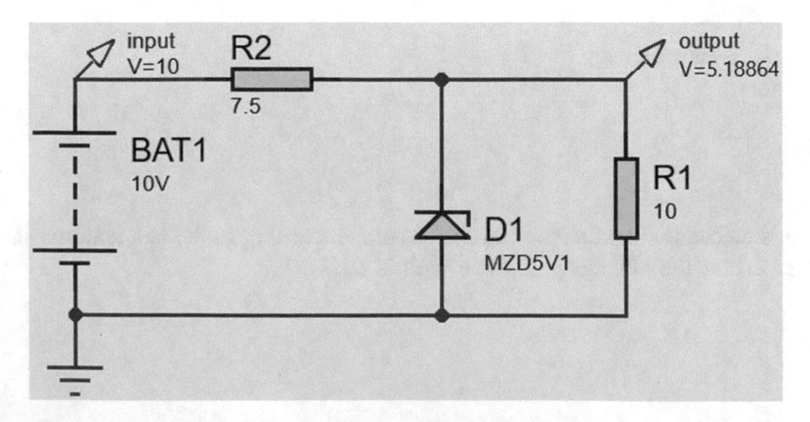

Fig. 2.129 Simulation result

2.20 Example 19: Voltage Regulator (II)

In the previous example, we studied a simple zener diode-based voltage regulator. In this example, we want to use the 7805 linear voltage regulator to generate +5 V in output. Draw the schematic shown in Fig. 2.130.

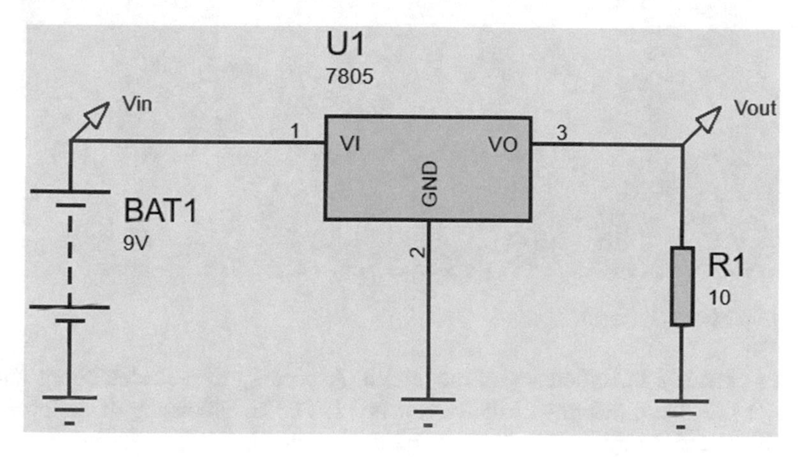

Fig. 2.130 Schematic of Example 19

Run the simulation. Simulation result is shown in Fig. 2.131. Output voltage is 4.97229 V.

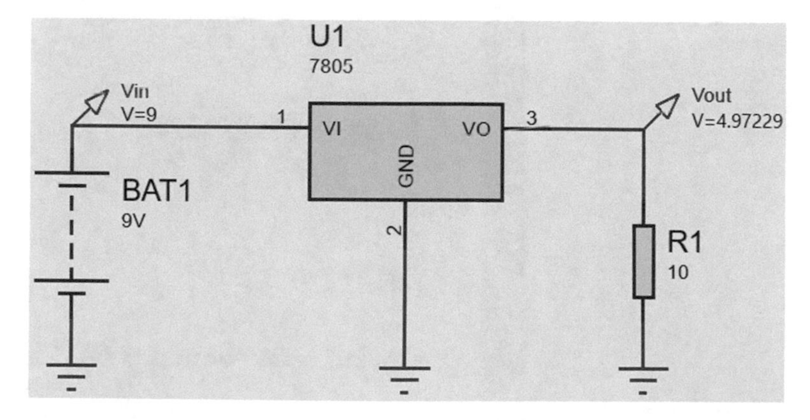

Fig. 2.131 Simulation result

Let's measure the input current. According to Fig. 2.132, the current drawn from the input voltage source is 0.501471 A.

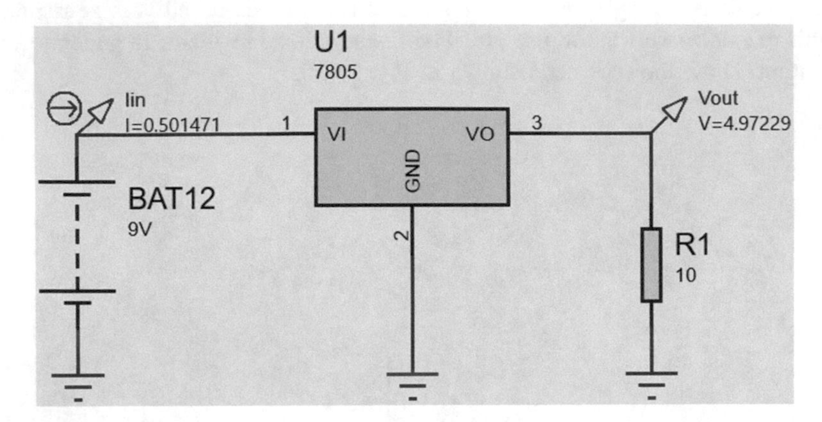

Fig. 2.132 Simulation result

Let's calculate the efficiency of the circuit. According to the calculations shown in Fig. 2.133, the efficiency of the circuit is 54.7803%. Efficiency of the circuit is close to the efficiency of the circuit in Example 18.

Fig. 2.133 MATLAB calculations

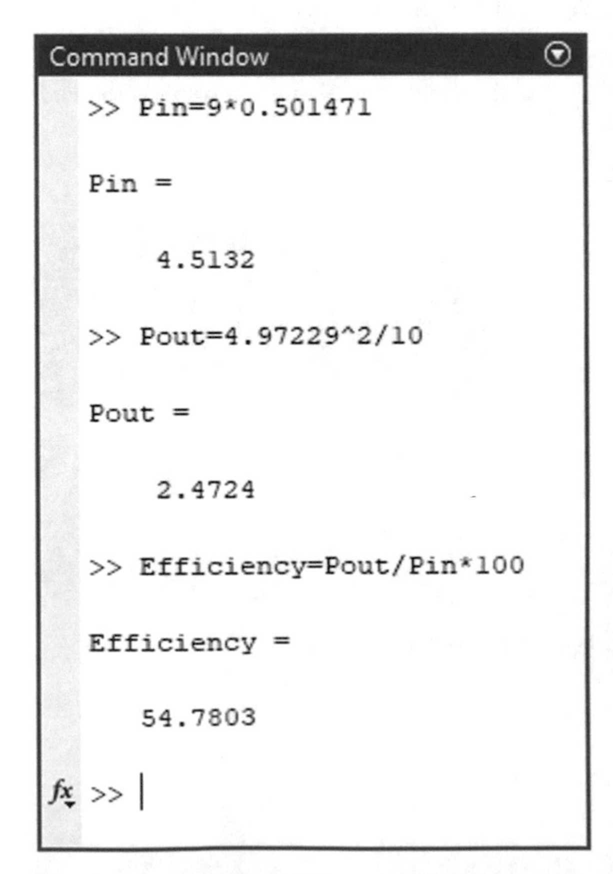

Let's study the effect input voltage changes on the output voltage. Increase the battery voltage to 10 V and run the simulation. In this case, the output voltage is 4.97269 V (Fig. 2.134). Increase of input voltage by 1 V increased the output voltage by 0.4 mV. Increase in the output in current example is quite smaller than increase in the output in previous example. This is expected since 7805 uses the closed loop feedback control to keep the output voltage constant.

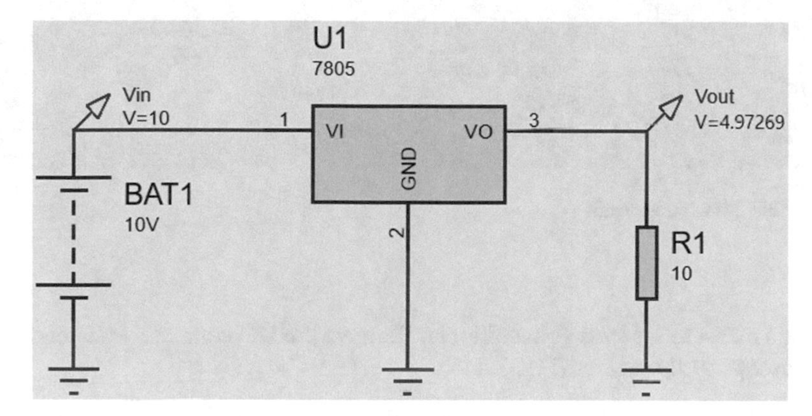

Fig. 2.134 Simulation result

2.21 Example 20: Voltage Regulator (III)

In this example, we study a regulated a power supply. Connect the 7805 to output of Example 12 (Fig. 2.135).

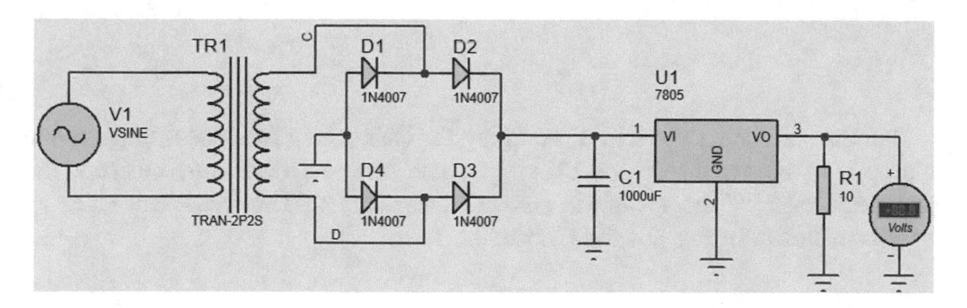

Fig. 2.135 Schematic of Example 20

Run the simulation. The DC voltmeter shows the average value of load voltage (Fig. 2.136).

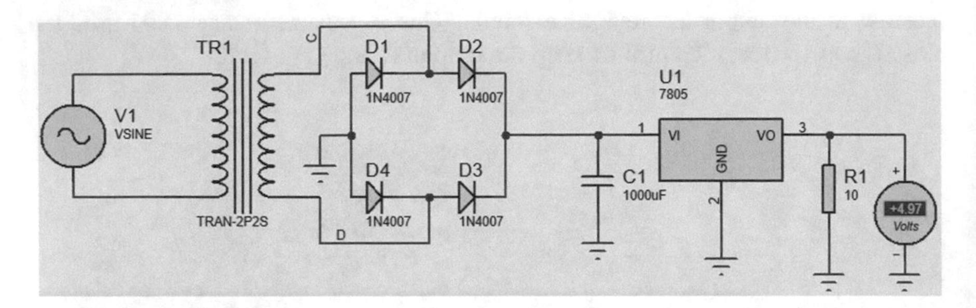

Fig. 2.136 Simulation result

Let's measure the load voltage ripple. Remove the DC voltmeter and connect an oscilloscope to the output (Fig. 2.137).

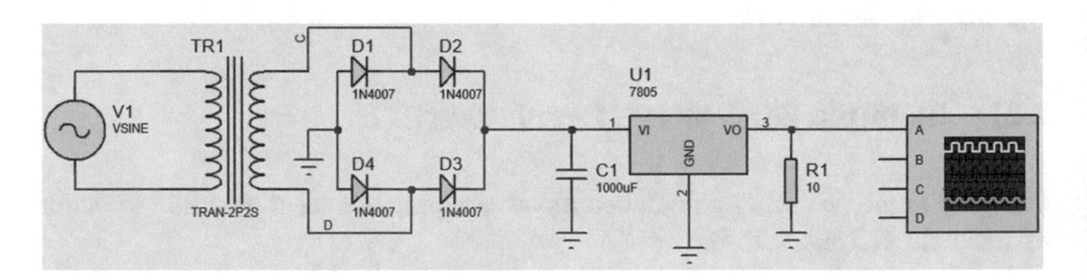

Fig. 2.137 Measurement of load voltage ripple

Run the simulation. Select AC coupling for channel *A*. The load voltage ripple appears on the screen (Fig. 2.138). You can use the cursor to measure the peak-peak value of the waveform shown on the screen. The peak-peak value of waveform shown in Fig. 2.138 is 1.3 mV.

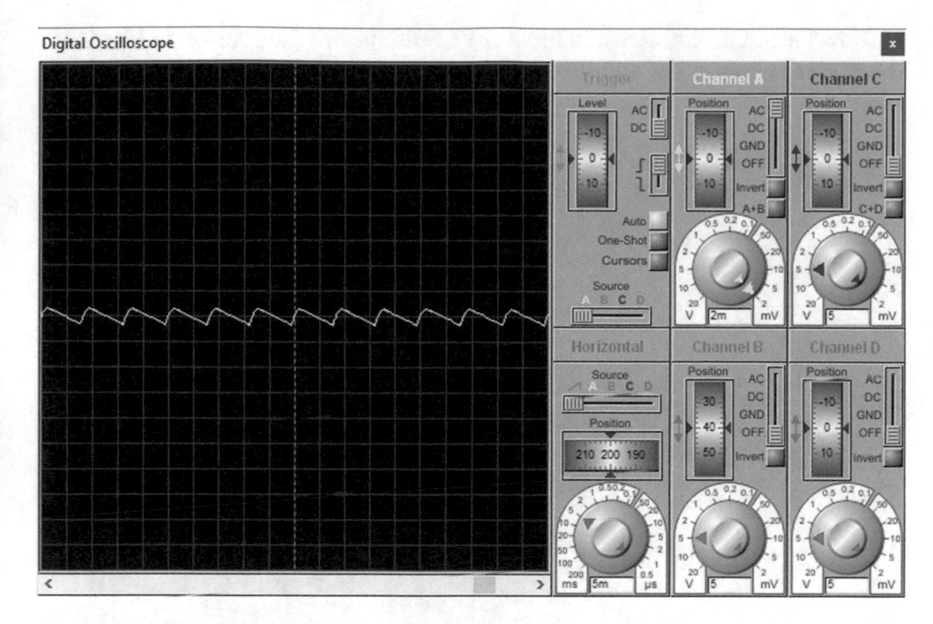

Fig. 2.138 Simulation result

2.22 Example 21: Common Emitter Amplifier

In this example, we want to simulate the circuit shown in Fig. 2.139.

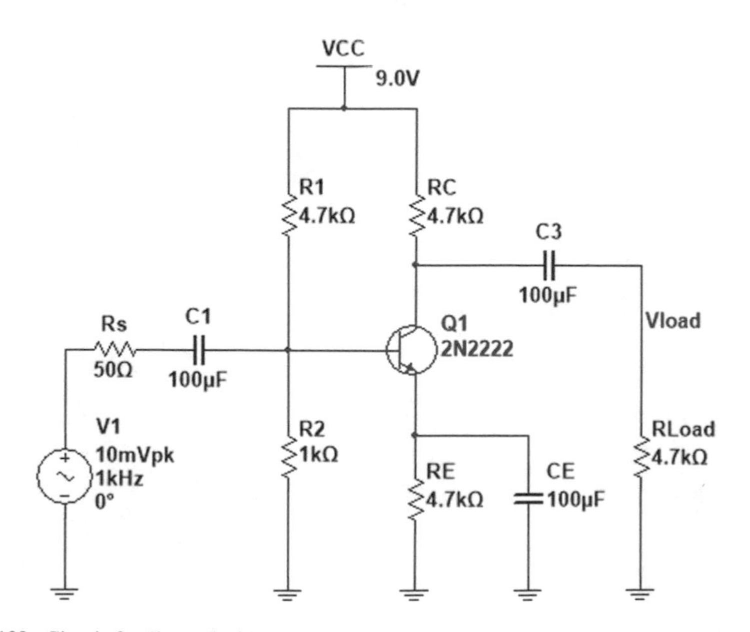

Fig. 2.139 Circuit for Example 21

Draw the schematic shown in Fig. 2.140. Settings of V_1 are shown in Fig. 2.141. The power supply is not connected to the circuit yet.

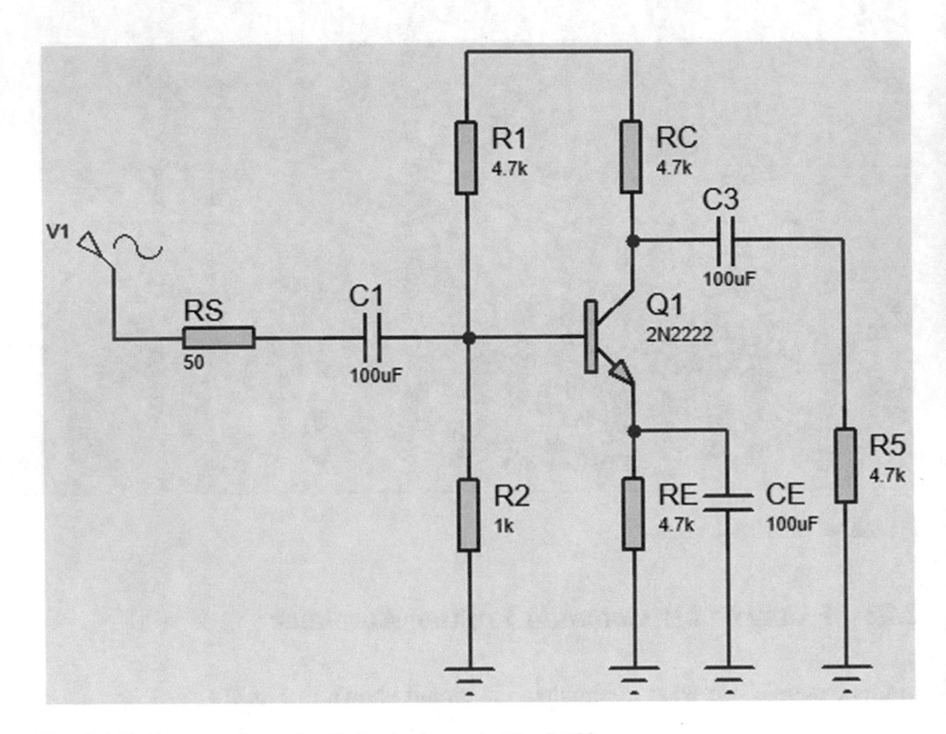

Fig. 2.140 Proteus schematic of circuit shown in Fig. 2.139

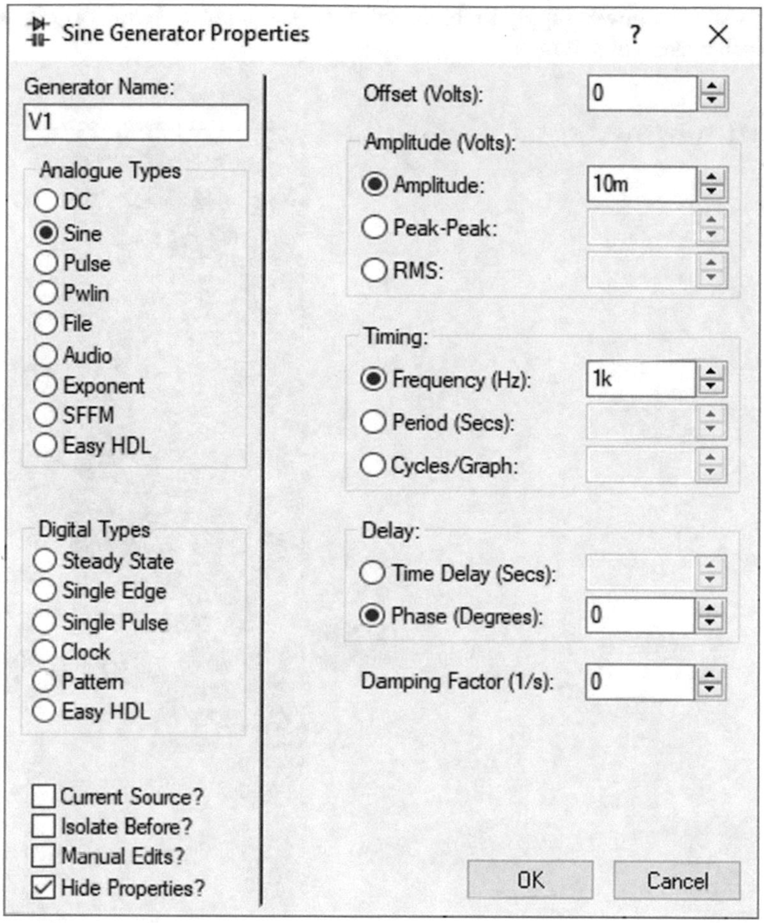

Sine Generator Properties ? X

Generator Name:
V1

Analogue Types
DC
Sine
Pulse
Pwlin
File
Audio
Exponent
SFFM
Easy HDL

Digital Types
Steady State
Single Edge
Single Pulse
Clock
Pattern
Easy HDL

Current Source?
Isolate Before?
Manual Edits?
Hide Properties?

Offset (Volts): 0

Amplitude (Volts):
Amplitude: 10m
Peak-Peak:
RMS:

Timing:
Frequency (Hz): 1k
Period (Secs):
Cycles/Graph:

Delay:
Time Delay (Secs):
Phase (Degrees): 0

Damping Factor (1/s): 0

OK Cancel

Fig. 2.141 Settings of $V1$

Let's add the power supply to the schematic. Add a power terminal (Fig. 2.142) to the schematic (Fig. 2.143).

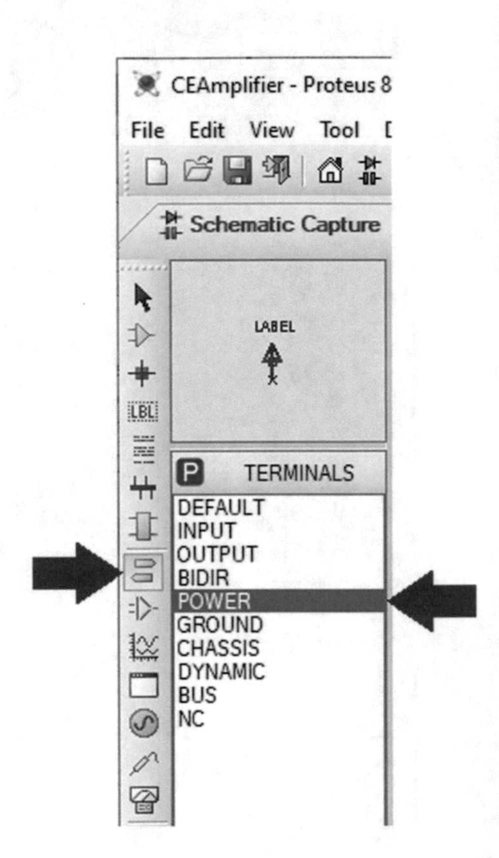

Fig. 2.142 Power block

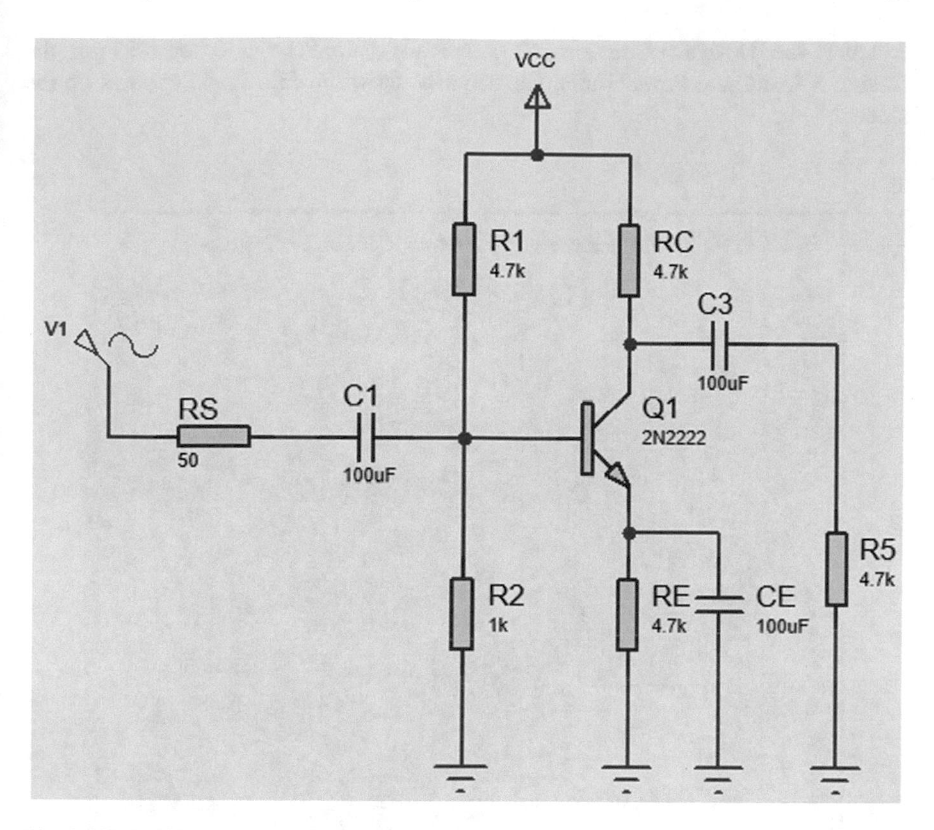

Fig. 2.143 VCC is connected to the drawn schematic

Click the Design > Configure Power Rails (Fig. 2.144). After clicking the Design > Configure Power Rails, the window shown in Fig. 2.145 appears on the screen.

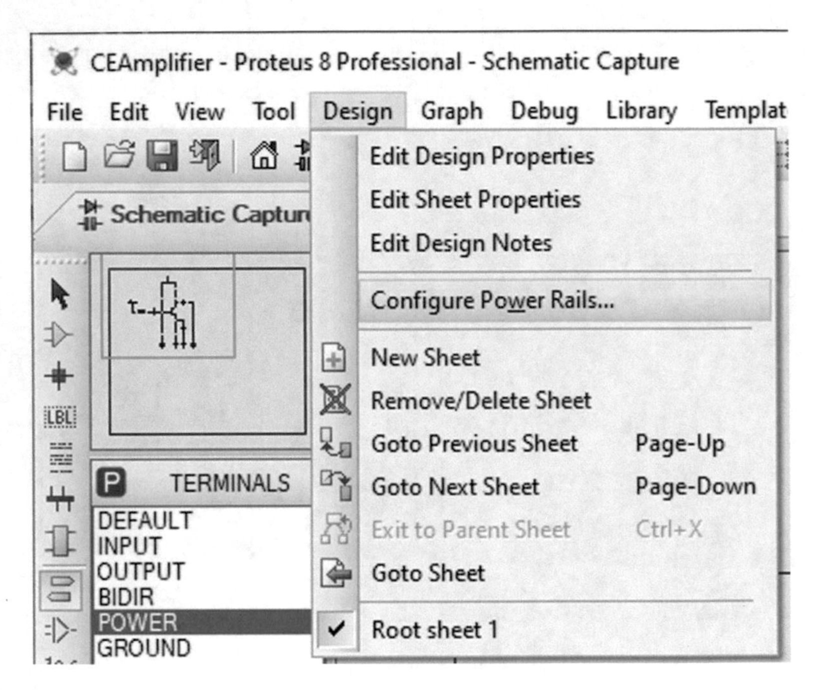

Fig. 2.144 Design > configure power rails

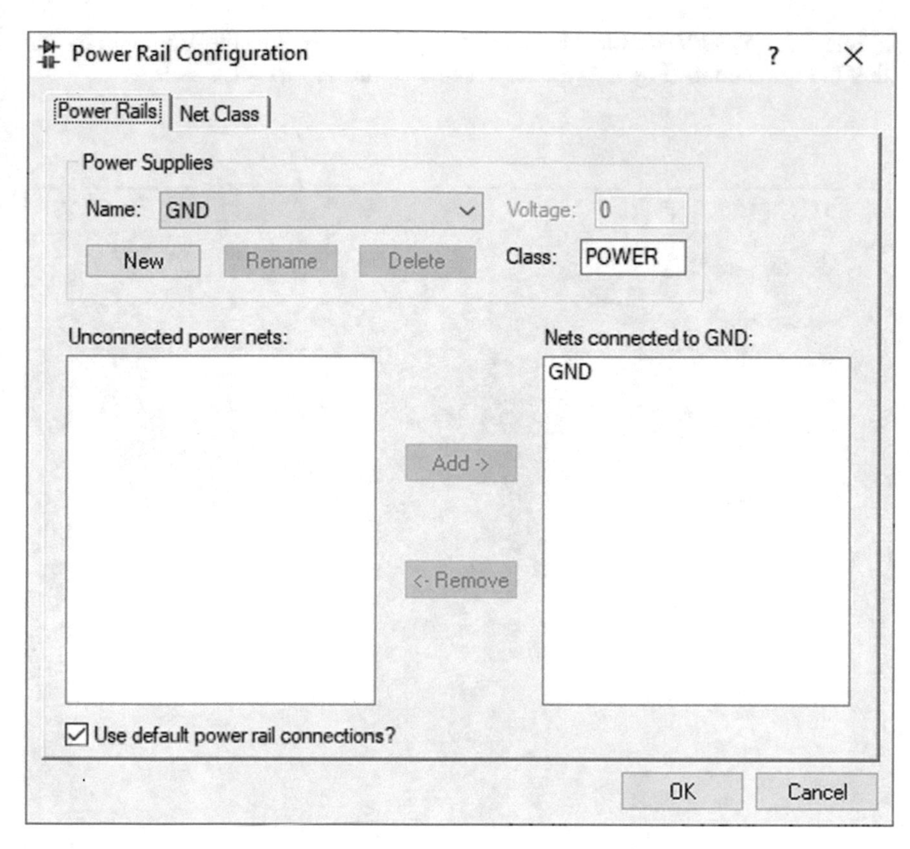

Fig. 2.145 Power rail configuration window

Select the VCC/VDD for the Name box. Then enter 9 to the Voltage box and click the OK button (Fig. 2.146). Now Proteus knows that VCC is 9 V.

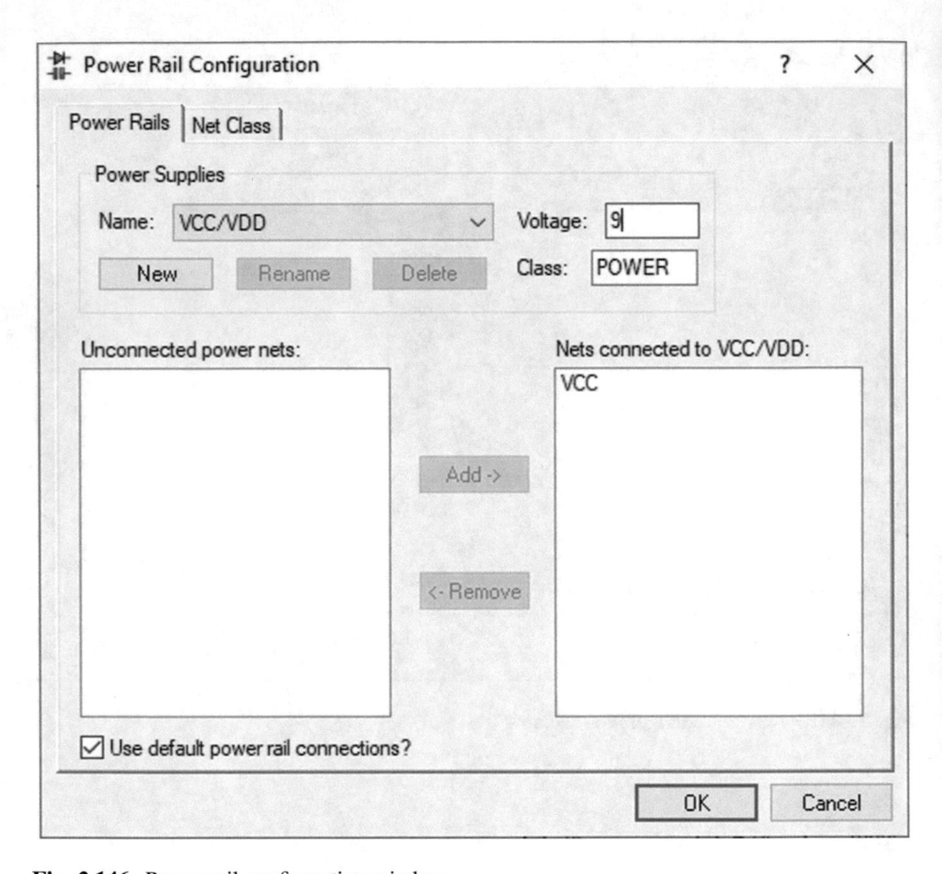

Fig. 2.146 Power rail configuration window

Let's do a DC analysis and measure the operating point of the transistor. Add voltage probes VB, VC, VE and current probe IE to the schematic (Fig. 2.147). Note that there is no need to input signal in measurement of DC operating point. That is why the connection between V_1 and circuit is removed.

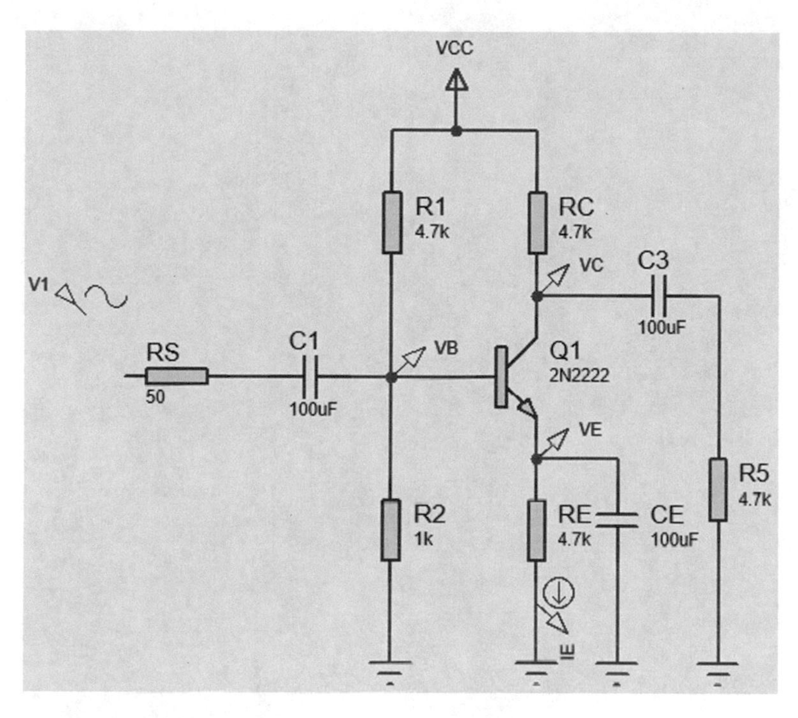

Fig. 2.147 Connection between V_1 and the rest of circuit is removed

Run the simulation. Simulation result is shown in Fig. 2.148. According to Fig. 2.148, $V_{CE} = 8.03815 - 0.966264 = 7.071886$ V and $I_C \approx I_E = 0.205588$ mA.

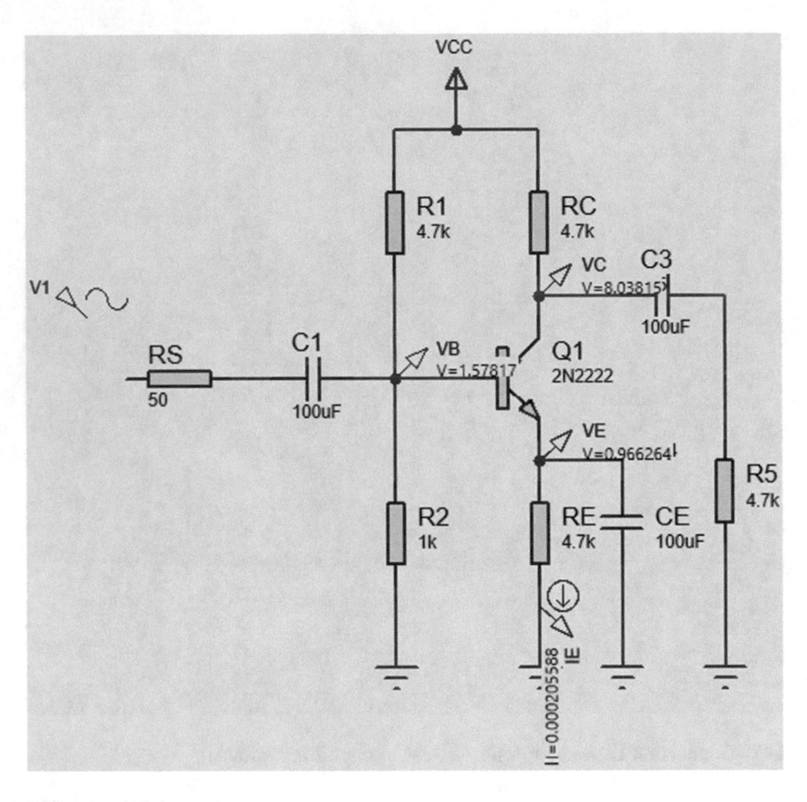

Fig. 2.148 Simulation result

Let's measure the voltage gain of the amplifier. This can be done with the aid of schematic shown in Fig. 2.149. The voltage gain of the amplifier at frequency of 1 kHz is 17.1146 according to the calculation shown in Fig. 2.150.

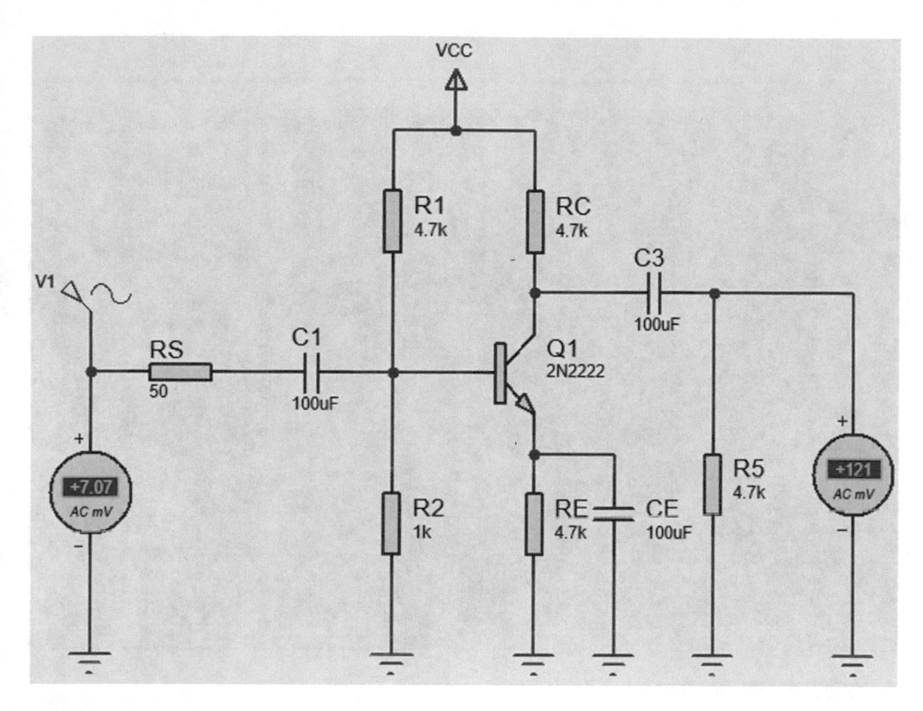

Fig. 2.149 Simulation result

Fig. 2.150 MATLAB calculations

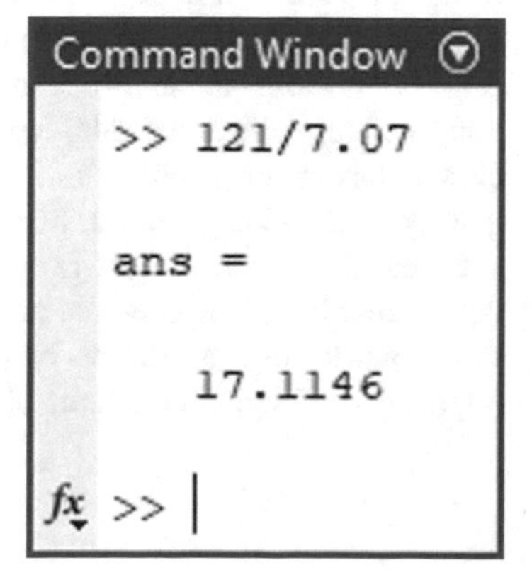

Let's use an oscilloscope to see the input and output signals (Fig. 2.151). Note that output signal is not symmetric and it is distorted. This is expected since this amplifier does not use feedback. If you need an output with very low distortion, then you need to use some form of feedback.

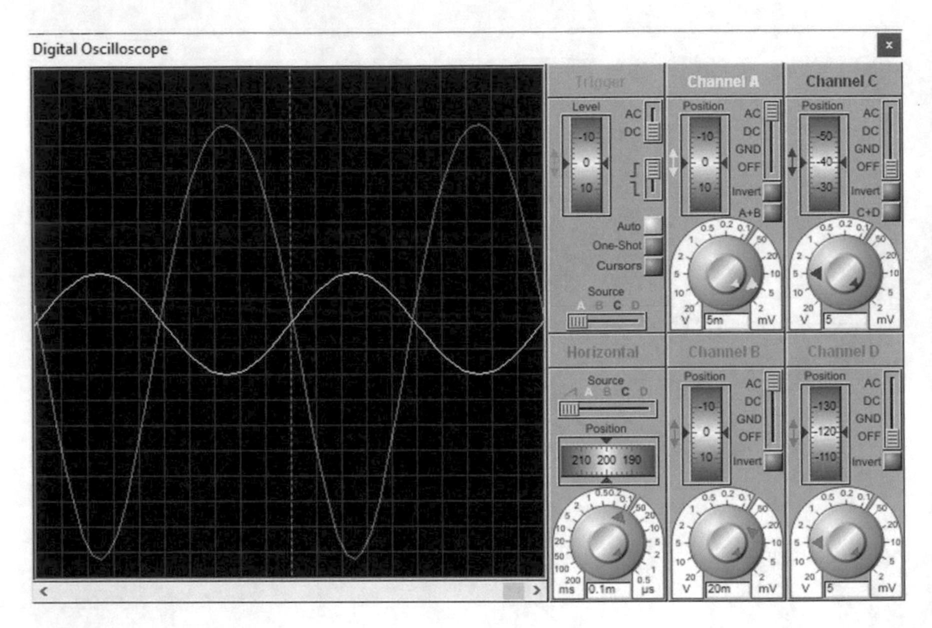

Fig. 2.151 Simulation result

2.23 Example 22: Signal Generator Block

In the previous example, we used a sine generator block (Fig. 2.152) to apply a sinusoidal input to the circuit. In this example, we will use a signal generator block (Fig. 2.153) to generate the input for the circuit. The signal generator block permits you to change the amplitude, frequency and waveform of the signal applied to the circuit during the simulation. Note that you can't change the amplitude and frequency of the sine generator block after running the simulation. Another difference between these two blocks is type of waveforms they can generate. The signal generator block can generate square, saw tooth, triangular and sinusoidal waveforms while sine generator block generates sinusoidal waveform only. Output impedance of signal generator block is 0 Ω.

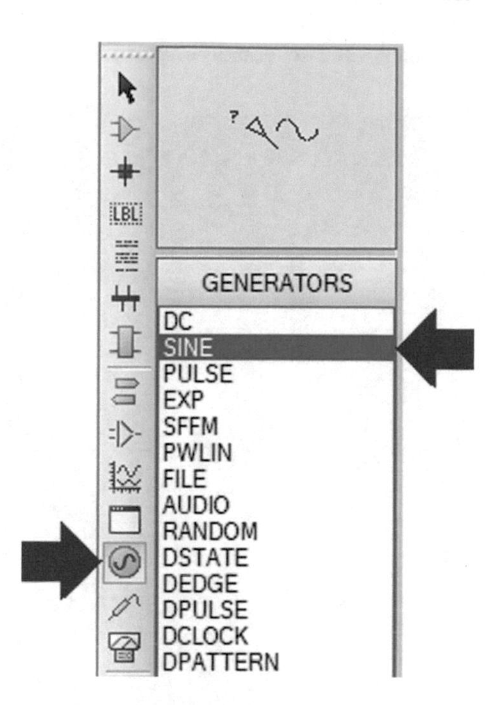

Fig. 2.152 Sine block

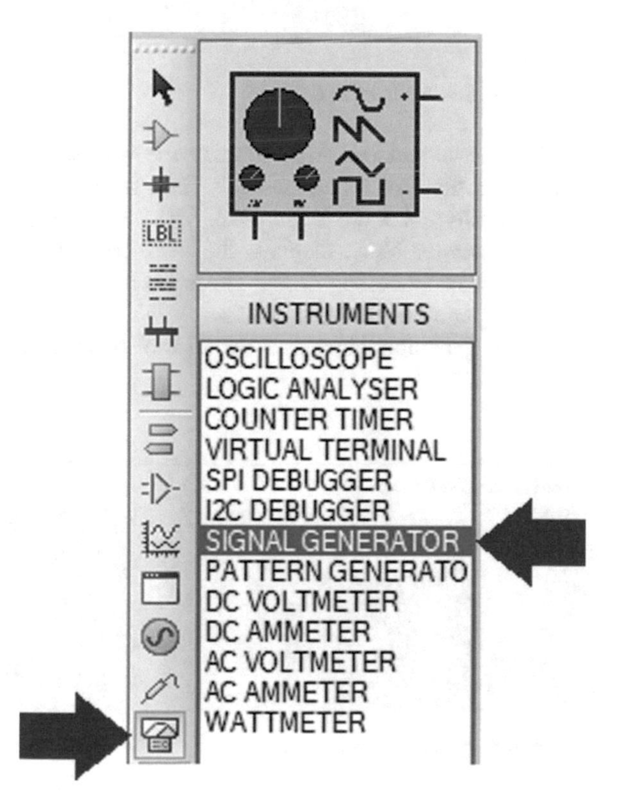

Fig. 2.153 Signal generator block

Change the schematic of previous example to what is shown in Fig. 2.154.

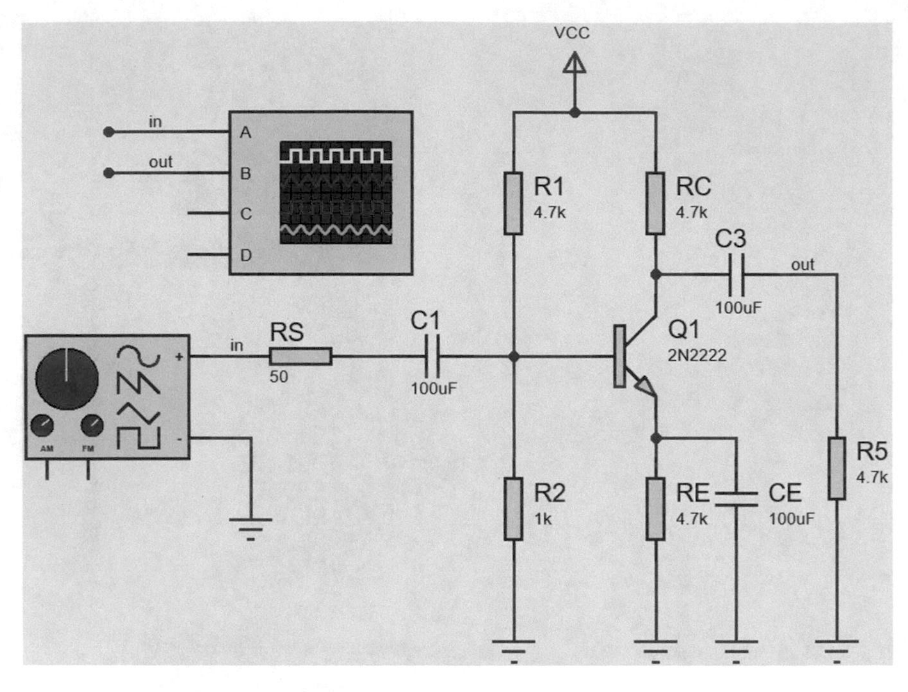

Fig. 2.154 Schematic of Example 22

Run the simulation. After running the simulation, the signal generator window appears on the screen automatically (Fig. 2.155). If it doesn't appear on the screen automatically, click the Debug > 3. VSM Signal Generator (Fig. 2.156). Note that signal generator block displays the peak-peak value and frequency of generated signal. For instance, Fig. 2.155 generates a sinusoidal signal with peak-peak value of 10 mV and frequency of 1 kHz.

Signal generator block is a useful tool to measure the maximum swing of the output.

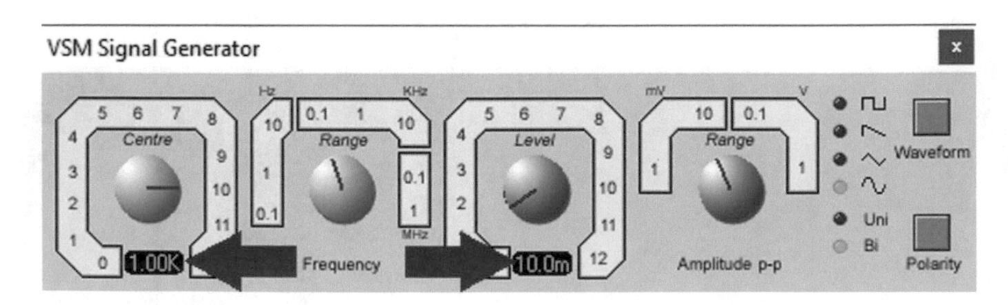

Fig. 2.155 Signal generator settings

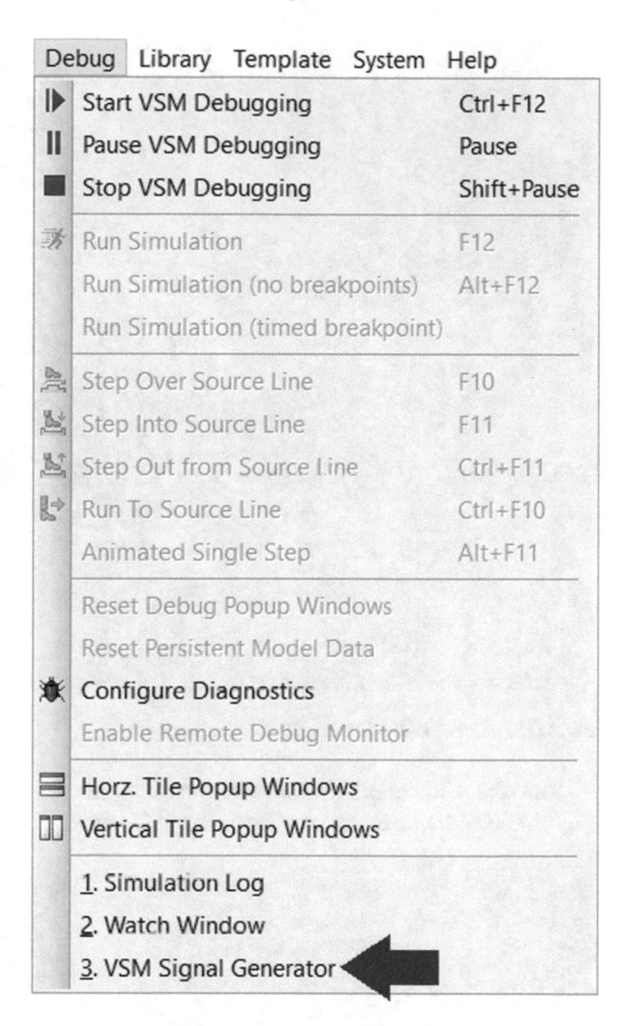

Fig. 2.156 Debug > 3. VSM signal generator

2.24 Example 23: Input Impedance of Common Emitter Amplifier

In this example, we want to measure the input impedance of the amplifier of Example 21 at 1 kHz. Add an AC ammeter to the schematic (Fig. 2.157).

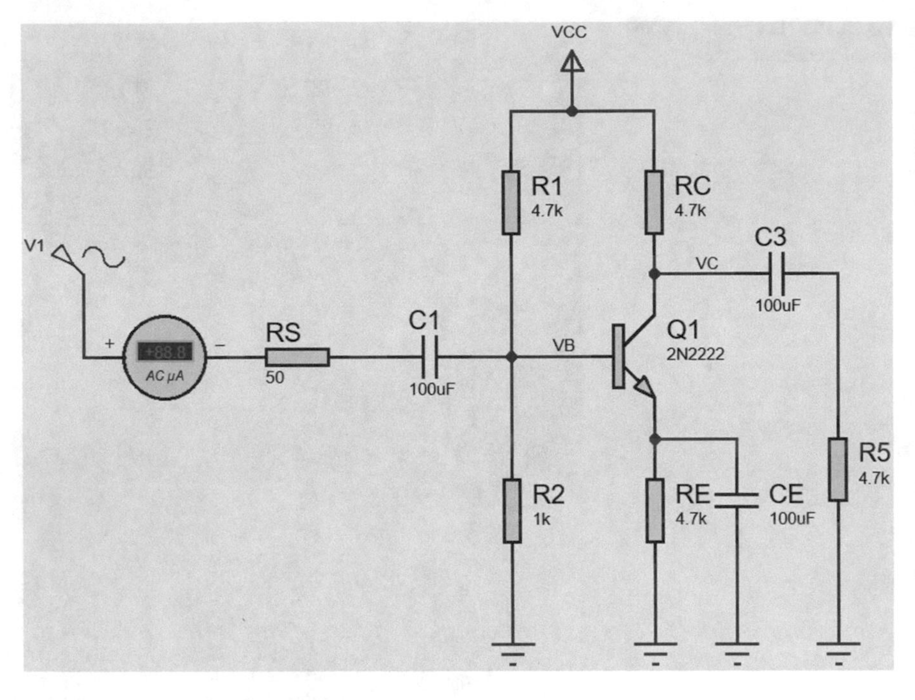

Fig. 2.157 Schematic of Example 23

Run the simulation. Simulation result is shown in Fig. 2.158. According to Fig. 2.158, the current drawn from the AC source V_1 is 8.30 μA.

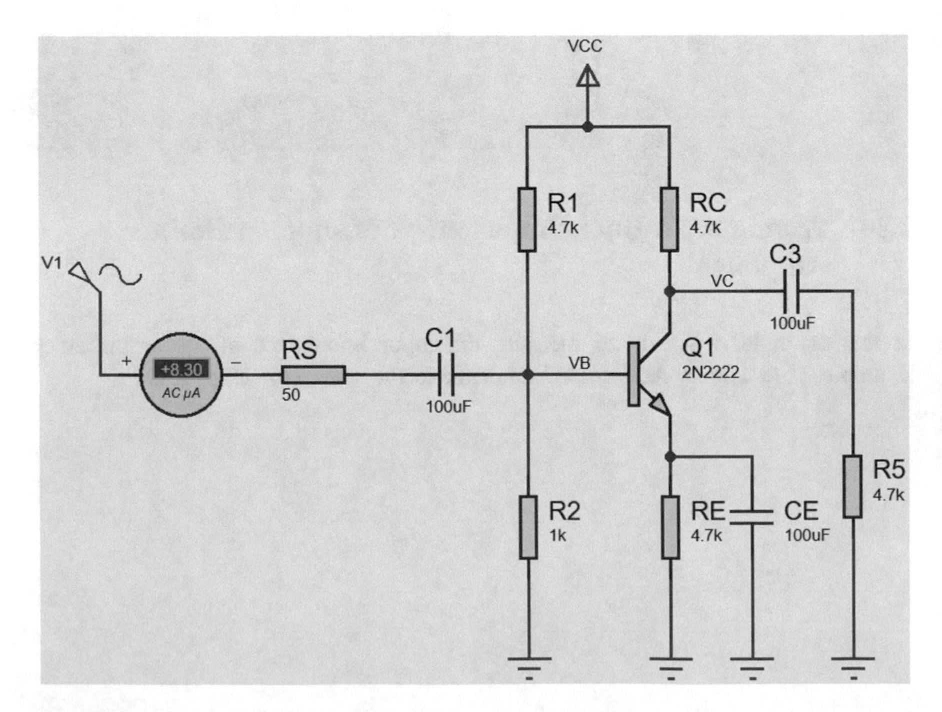

Fig. 2.158 Simulation result

According to the calculations shown in Fig. 2.159, the impedance seen by AC source V_1 at 1 kHz is 851.9359 Ω.

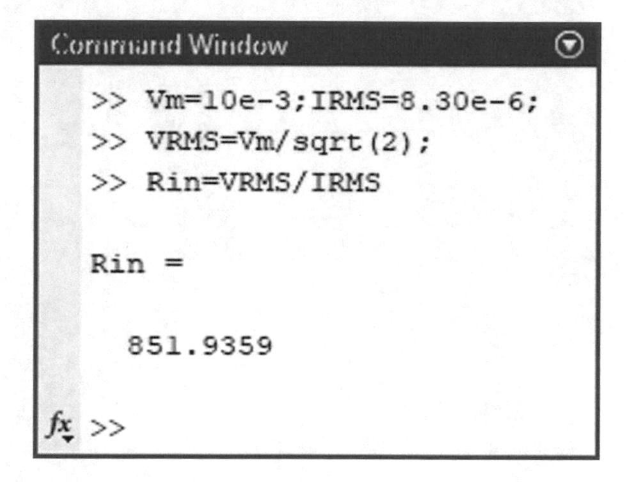

Fig. 2.159 MATLAB commands

2.25 Example 24: Frequency Response of Input Impedance of Common Emitter Amplifier

In the previous example, we measured the input impedance of the common emitter amplifier at 1 kHz. In this example, we want to obtain the frequency response of input impedance of common emitter amplifier of Example 21.

Let's start. Draw the schematic shown in Fig. 2.160.

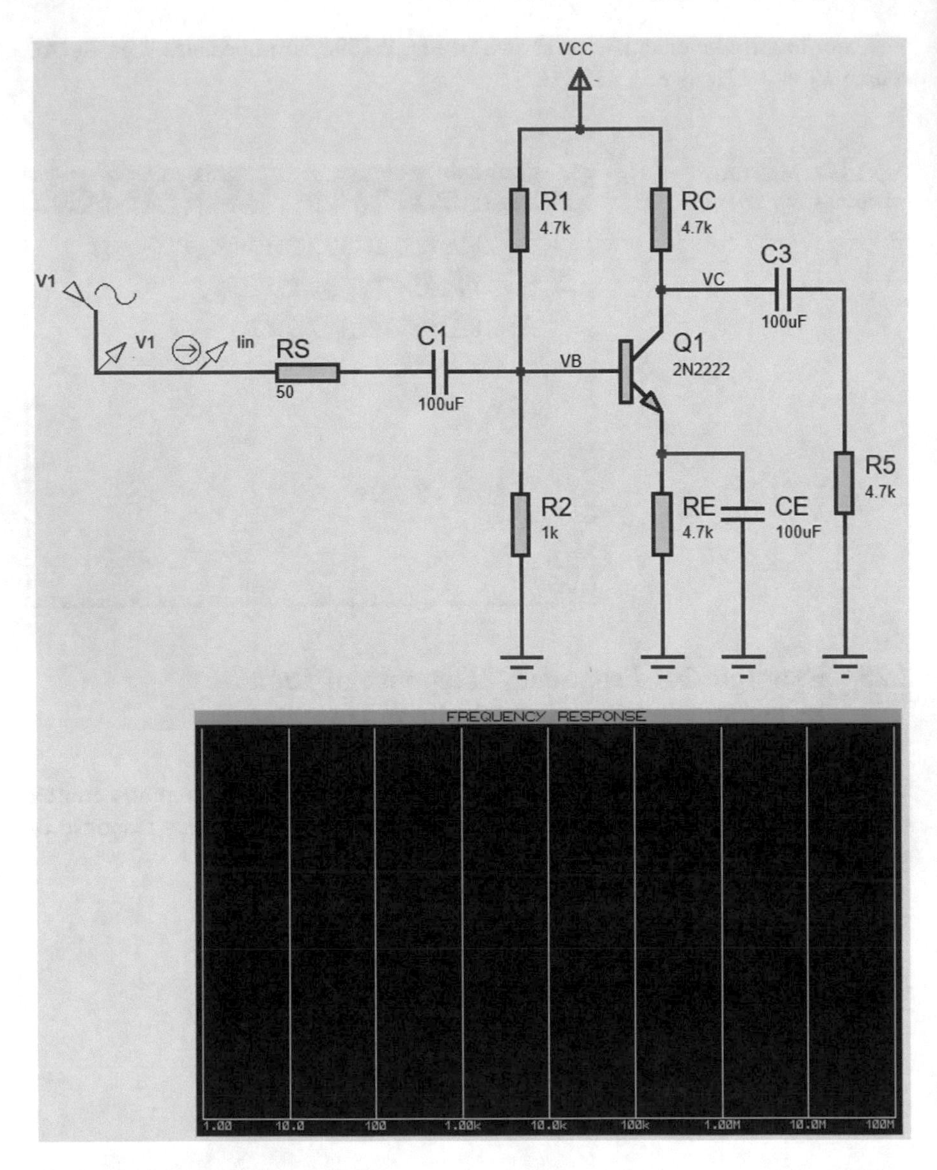

Fig. 2.160 Schematic of Example 24

Right click on the Frequency graph and click the Add Traces (Fig. 2.161). After clicking the Add Traces, Add Phasor Trace window appears on the screen. Do the settings similar to Fig. 2.162 and click the OK button. These settings draw the magnitude graph for us.

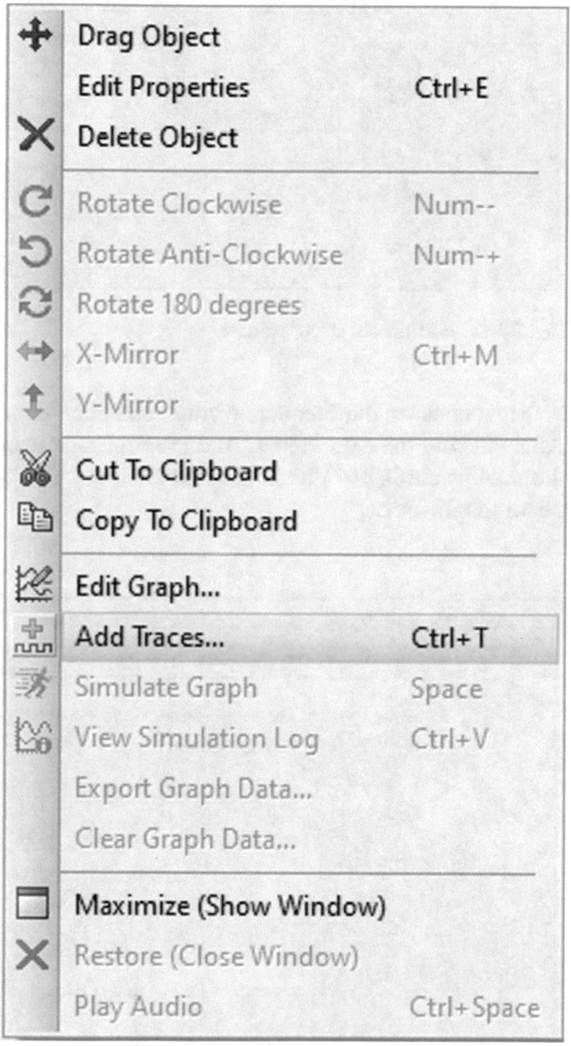

Fig. 2.161 Add traces command

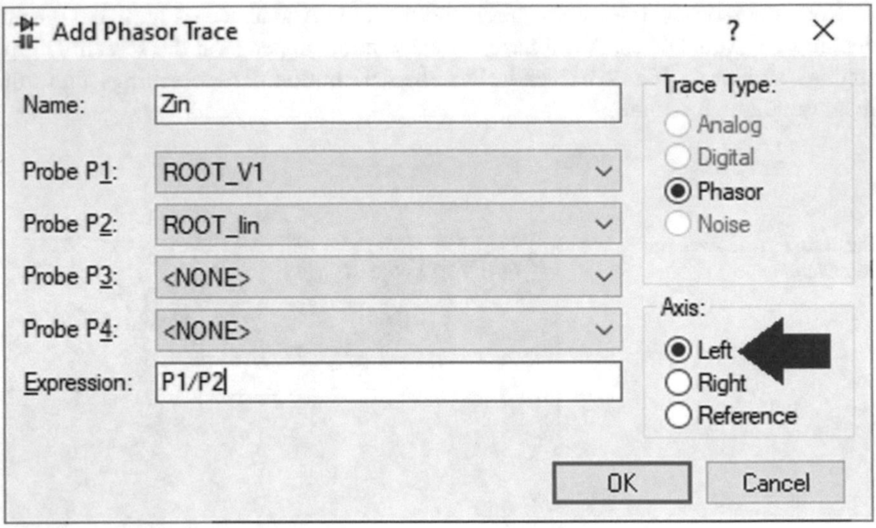

Fig. 2.162 Add phasor trace window

Right click on the Frequency graph and click the Add Traces (Fig. 2.161) again. After clicking the Add Traces, Add Phasor Trace window appears on the screen. Do the settings similar to Fig. 2.163 and click the OK button. These settings draw the phase graph for us.

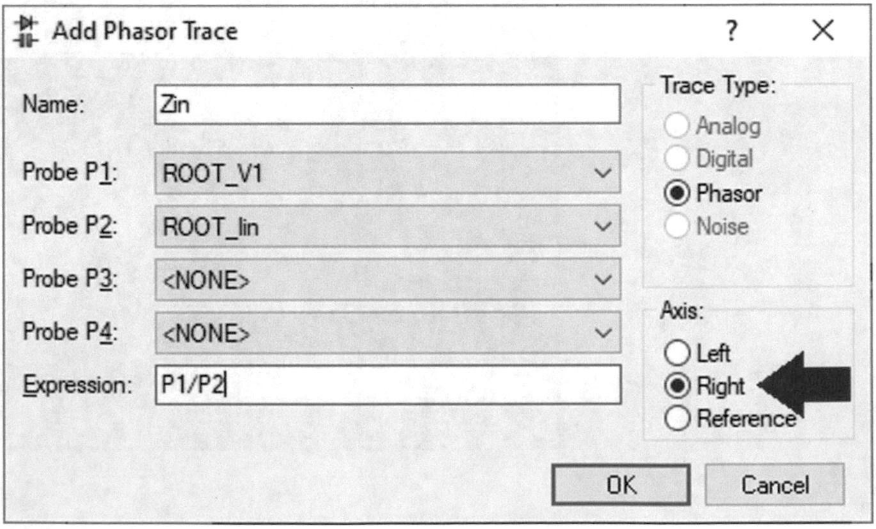

Fig. 2.163 Add phasor trace window

Double click the Frequency graph and do the settings similar to Fig. 2.164.

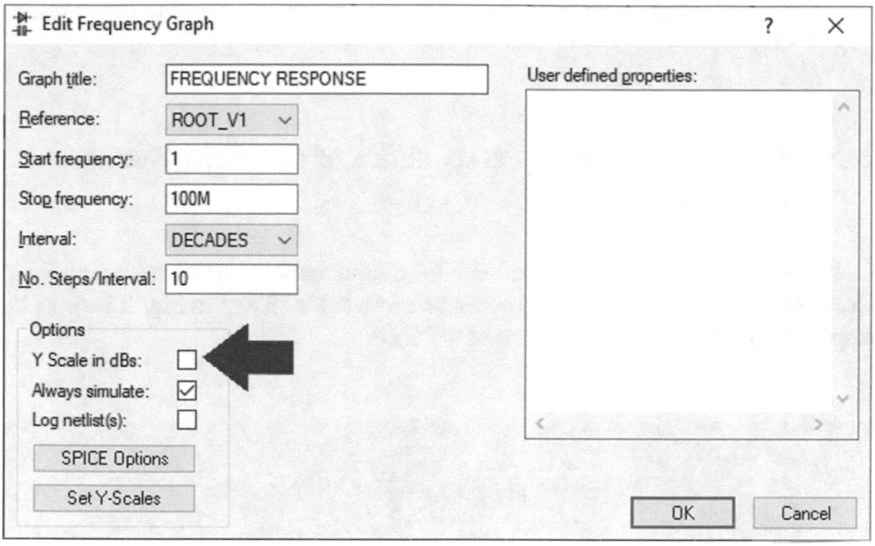

Fig. 2.164 Settings of frequency graph

Press the space bar key of your keyboard to run the simulation. Simulation result is shown in Fig. 2.165.

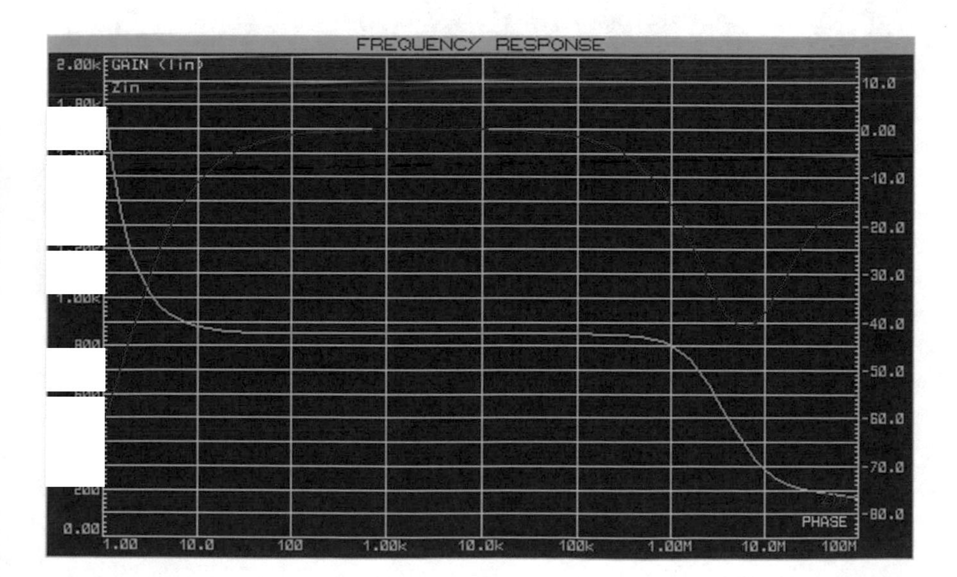

Fig. 2.165 Simulation result

Let's check the obtained result. Values of these two graphs at 1 kHz are 851 Ω and −0.143°. So, the input impedance can be written as $851e^{-0.143j} \approx 851$. Obtained result is quite close to the value we obtained in the previous example (Fig. 2.159).

2.26 Example 25: Output Impedance of Common Emitter Amplifier

In this example, we want to measure the output impedance of the amplifier of Example 21 at 1 kHz. Change the schematic to what is shown in Fig. 2.166. V_1 has amplitude of 10 mV and frequency of 1 kHz.

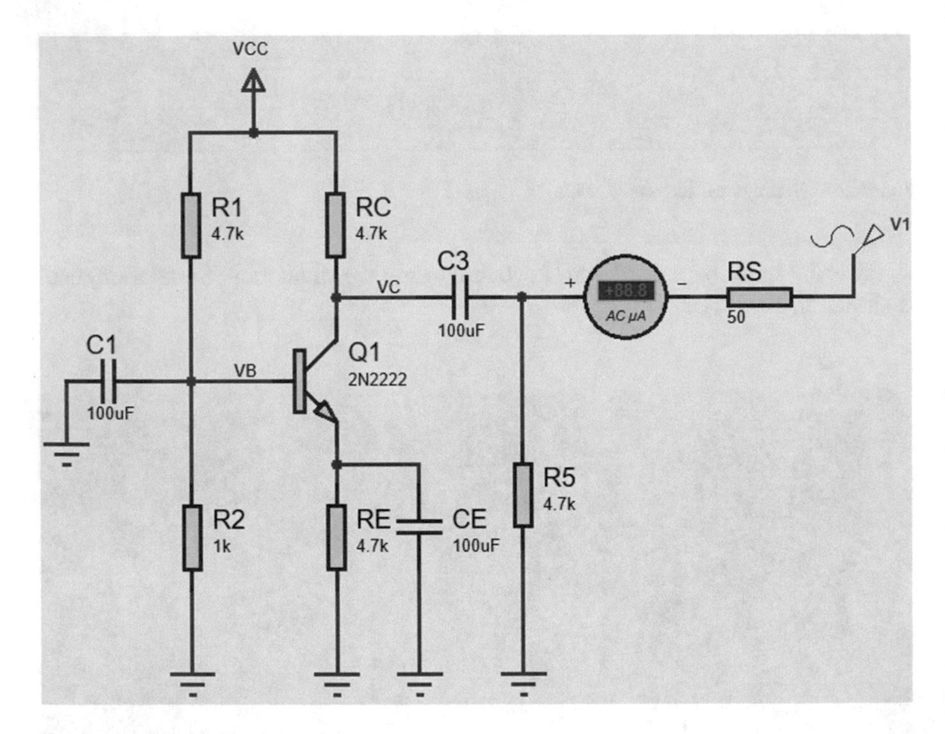

Fig. 2.166 Schematic of Example 25

Run the simulation. Simulation result is shown in Fig. 2.167. The current drawn from V_1 is 2.9 μA.

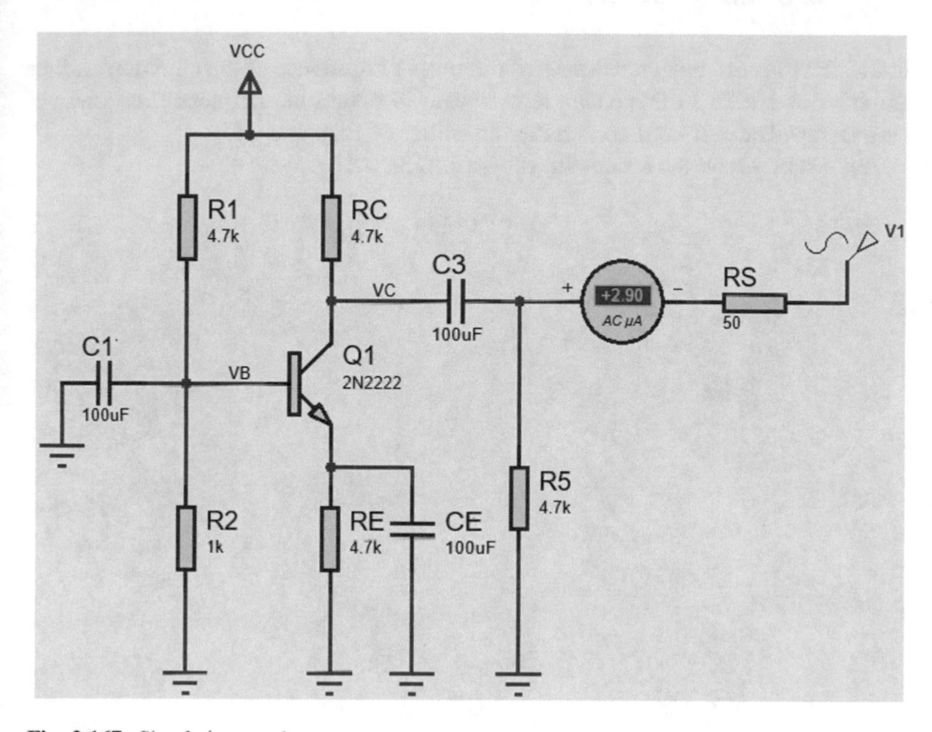

Fig. 2.167 Simulation result

According to the calculations shown in Fig. 2.168, the impedance seen by V_1 is 2.4383 kΩ.

Fig. 2.168 MATLAB calculations

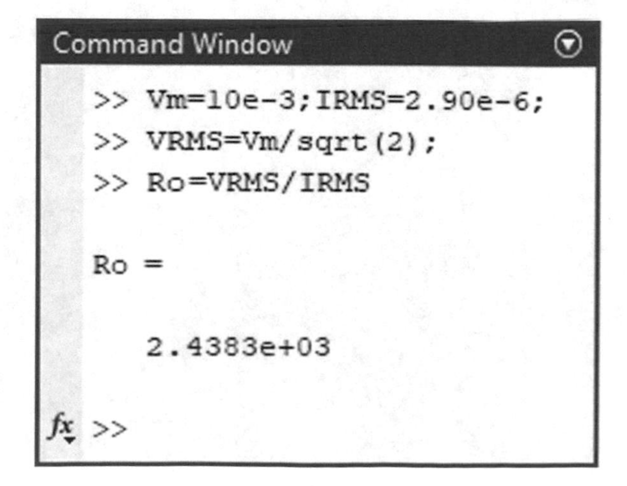

2.27 Example 26: Frequency Response of Output Impedance of Common Emitter Amplifier

In the previous example, we measured the output impedance of the common emitter amplifier at 1 kHz. In this example, we want to obtain the frequency response of output impedance of common emitter amplifier of Example 21.

Let's start. Draw the schematic shown in Fig. 2.169.

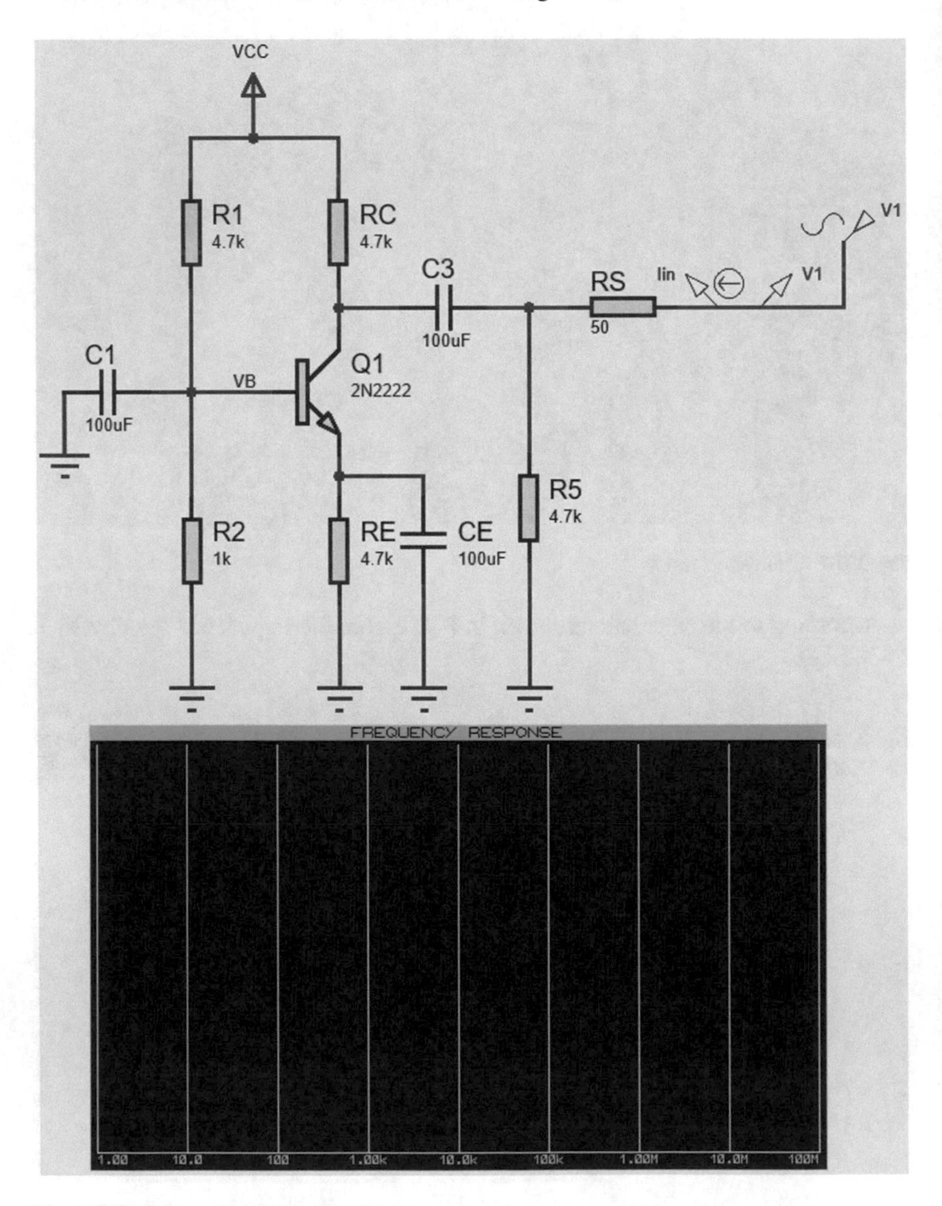

Fig. 2.169 Schematic of Example 26

Right click on the Frequency graph and click the Add Traces (Fig. 2.170). After clicking the Add Traces, Add Phasor Trace window appears on the screen. Do the settings similar to Fig. 2.171 and click the OK button. These settings draw the magnitude graph for us.

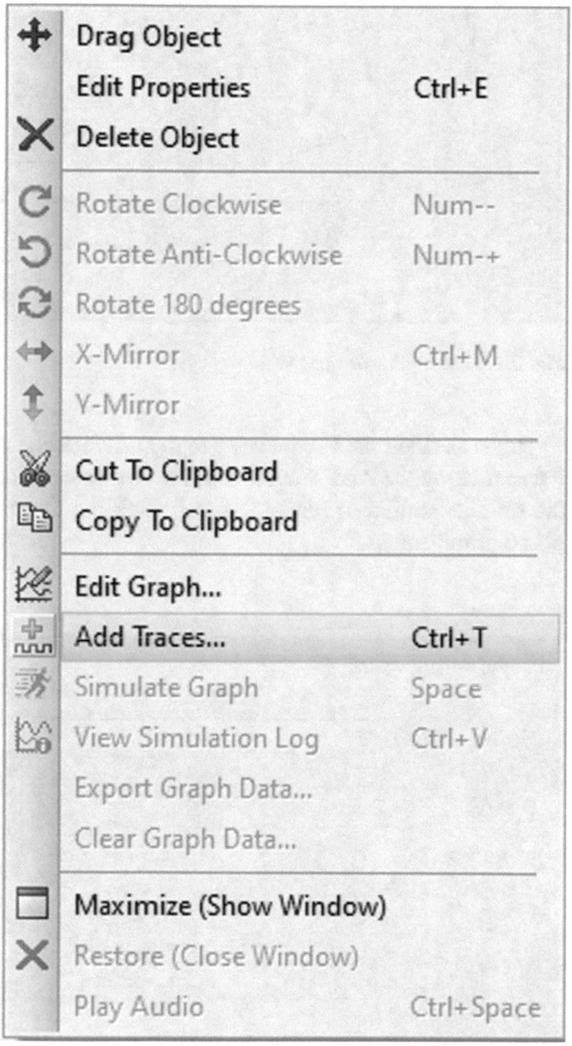

Fig. 2.170 Add traces command

Drag Object
Edit Properties Ctrl+E
Delete Object
Rotate Clockwise Num--
Rotate Anti-Clockwise Num-+
Rotate 180 degrees
X-Mirror Ctrl+M
Y-Mirror
Cut To Clipboard
Copy To Clipboard
Edit Graph...
Add Traces... Ctrl+T
Simulate Graph Space
View Simulation Log Ctrl+V
Export Graph Data...
Clear Graph Data...
Maximize (Show Window)
Restore (Close Window)
Play Audio Ctrl+Space

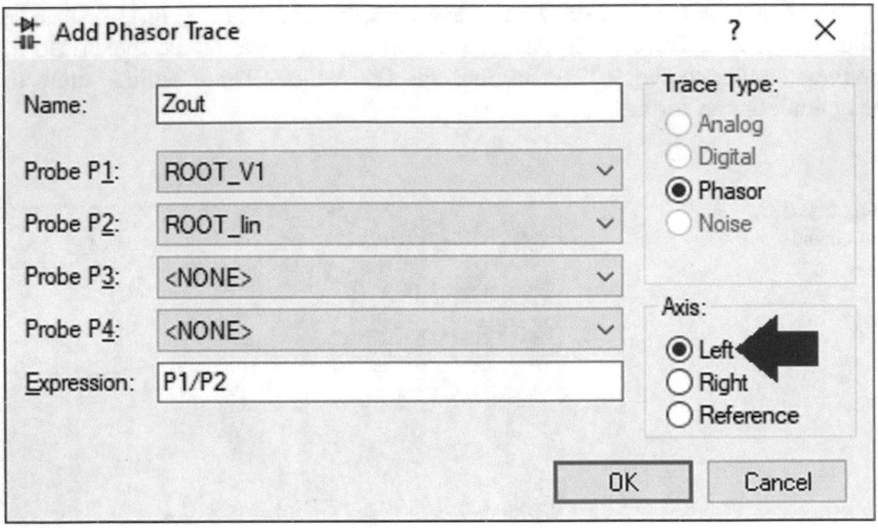

Fig. 2.171 Add phasor trace window

Right click on the Frequency graph and click the Add Traces (Fig. 2.170) again. After clicking the Add Traces, Add Phasor Trace window appears on the screen. Do the settings similar to Fig. 2.172 and click the OK button. These settings draw the phase graph for us.

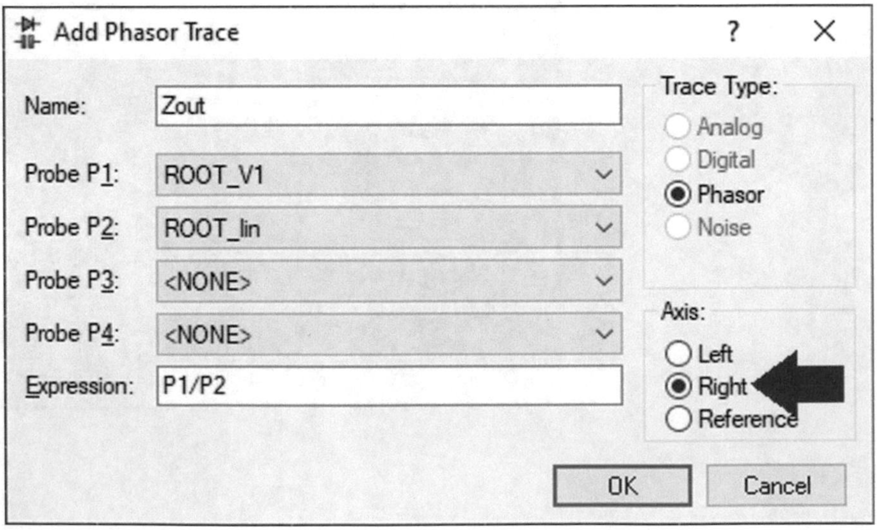

Fig. 2.172 Add phasor trace window

Double click the Frequency graph and do the settings similar to Fig. 2.173.

Edit Frequency Graph ? X
Graph title: FREQUENCY RESPONSE
Reference: ROOT_V1
Start frequency: 1
Stop frequency: 100M
Interval: DECADES
No. Steps/Interval: 10
Options
Y Scale in dBs:
Always simulate:
Log netlist(s):
SPICE Options
Set Y-Scales
User defined properties:
OK Cancel

Fig. 2.173 Settings of frequency graph

Press the space bar key of your keyboard to run the simulation. Simulation result is shown in Fig. 2.174.

Let's check the obtained result. Values of these two graphs at 1 kHz are 2.39 kΩ and −0.0172°. So, the input impedance can be written as $2390e^{-0.0172j} \approx 2.39$ kΩ. Obtained result is quite close to the value we obtained in the previous example (Fig. 2.168).

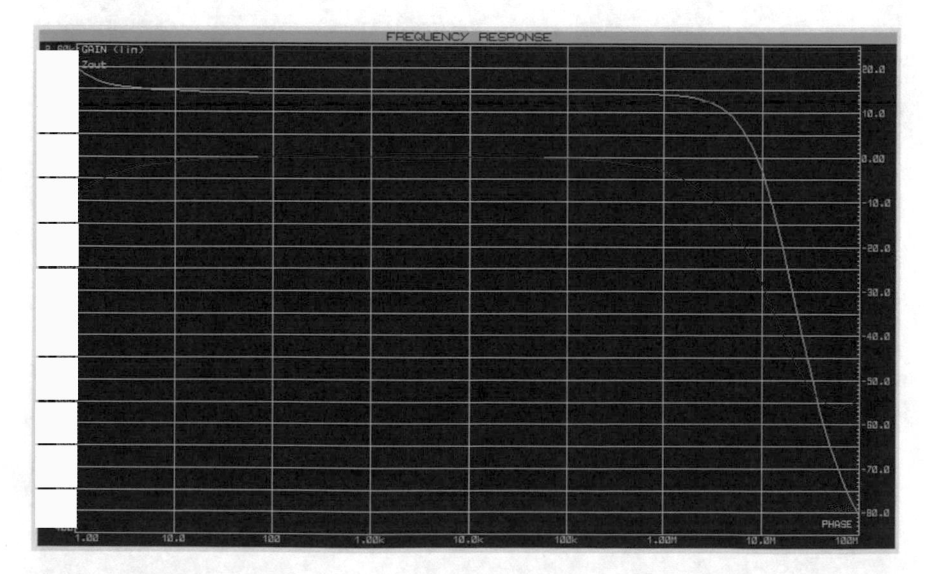

Fig. 2.174 Simulation result

2.28 Example 27: Frequency Response of Amplifier

In this example, we want to obtain the frequency response of the common emitter amplifier of Example 21. Draw the schematic shown in Fig. 2.175. Settings of the Frequency graph are shown in Fig. 2.176.

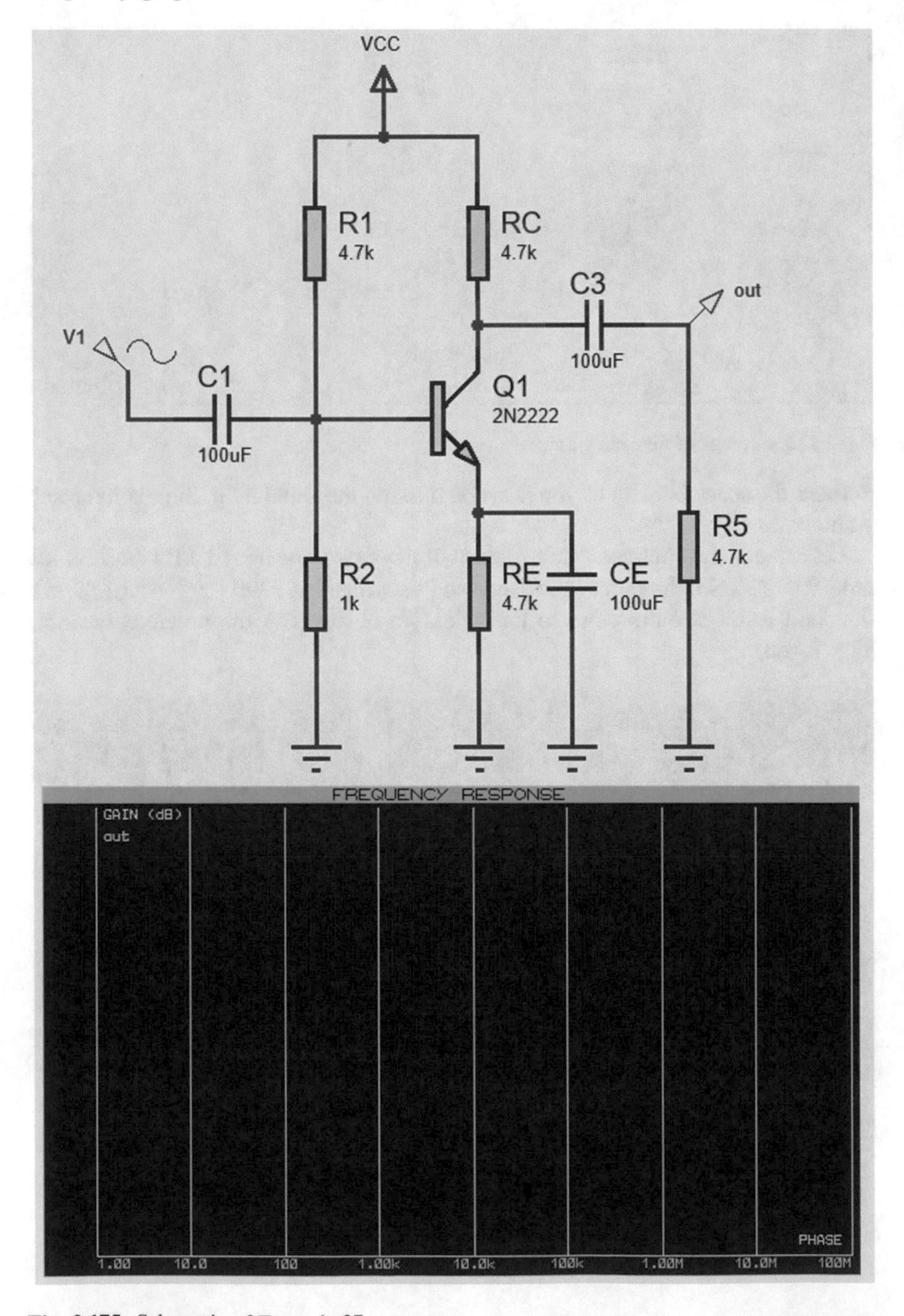

Fig. 2.175 Schematic of Example 27

Edit Frequency Graph

Graph title: FREQUENCY RESPONSE

Reference: ROOT_V1

Start frequency: 1

Stop frequency: 100M

Interval: DECADES

No. Steps/Interval: 10

Options

Y Scale in dBs: ☑

Always simulate: ☑

Log netlist(s): ☐

SPICE Options

Set Y-Scales

User defined properties:

OK Cancel

Fig. 2.176 Settings of frequency response graph

Run the simulation by pressing the space bar key of your keyboard. Simulation result is shown in Fig. 2.177. What you see on the screen is the graph of $\left|\frac{V_{out}(j\omega)}{V_{V_1}(j\omega)}\right|$. V_{out} and V_{V_1} show the voltage of node "out" and "V_1" (Fig. 2.175), respectively.

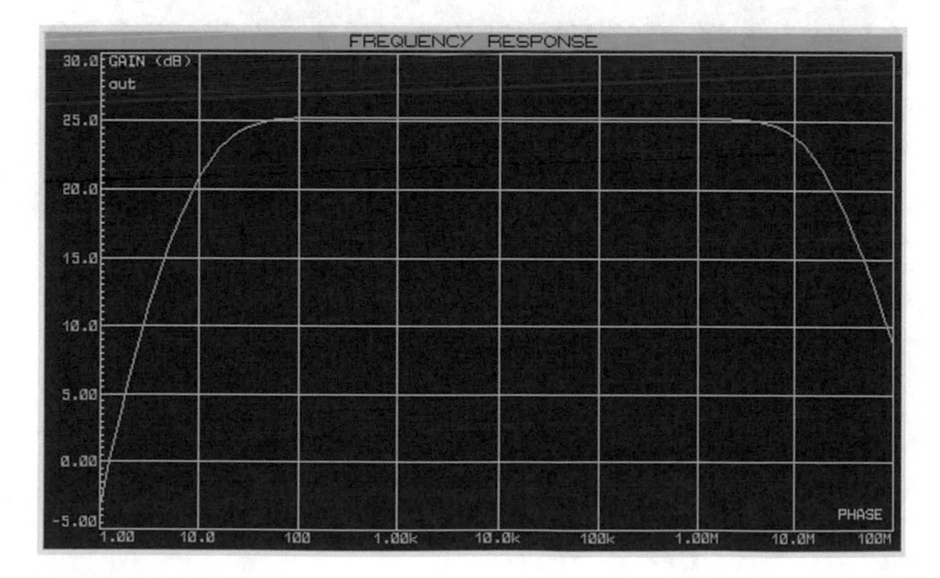

Fig. 2.177 Simulation result

Let's measure the mid-band gain. According to Fig. 2.178, the mid-band gain is 25.3 dB. 25.3 dB is associated with gain of 18.4077 (Fig. 2.179).

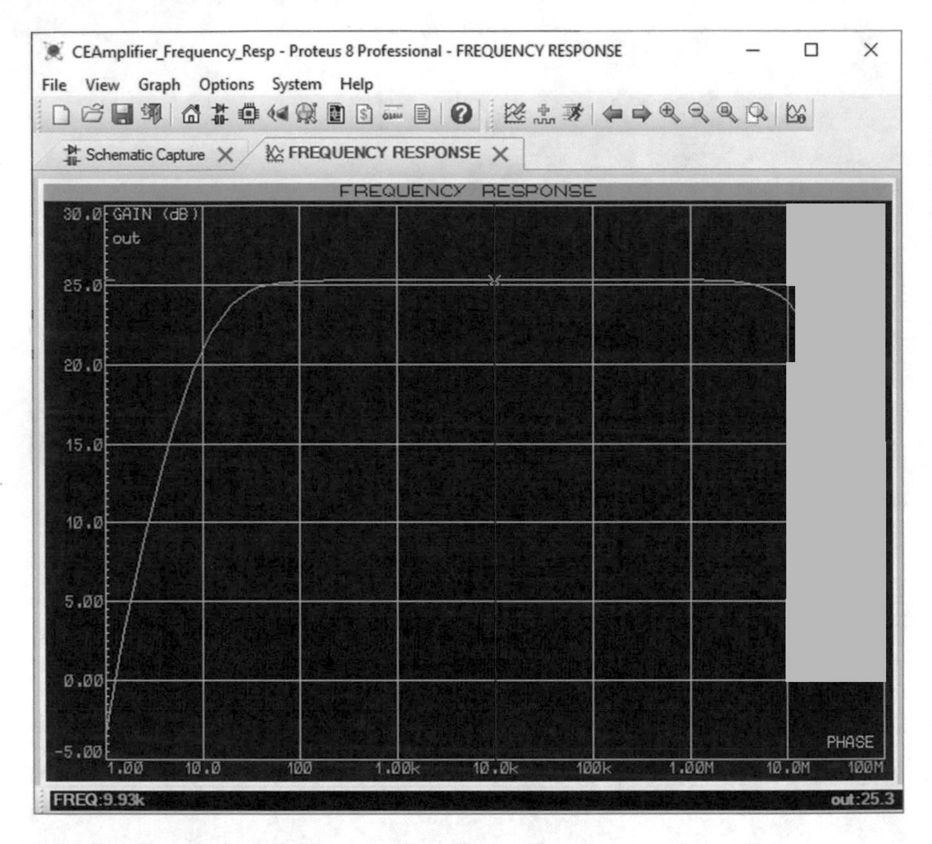

Fig. 2.178 Measurement of midband gain

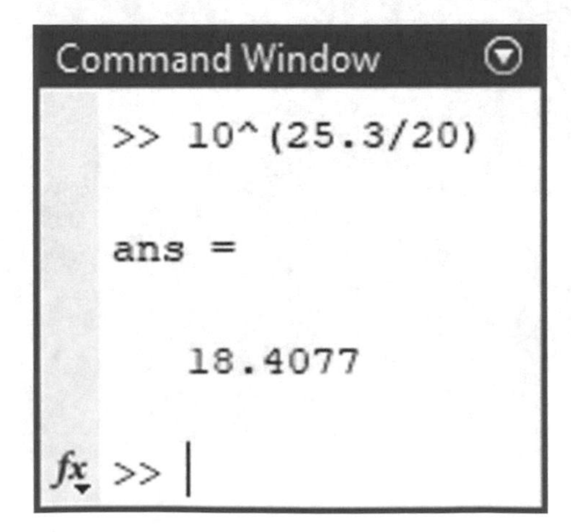

Fig. 2.179 MATLAB calculation

Let's measure the −3 dB cut-off frequencies. We need to search for frequencies which gain decreases to 25.3–3 = 22.3 dB. According to Fig. 2.180, lower cut-off frequency is 13.4 Hz. According to Fig. 2.181, upper cut-off frequency is 16.1 MHz. So, band width (BW) equals to $f_{-3\ \text{dB},H} - f_{-3\ \text{dB},L} = 16.1\ \text{MHz} - 13.1\ \text{Hz} \approx 16.1\ \text{MHz}$.

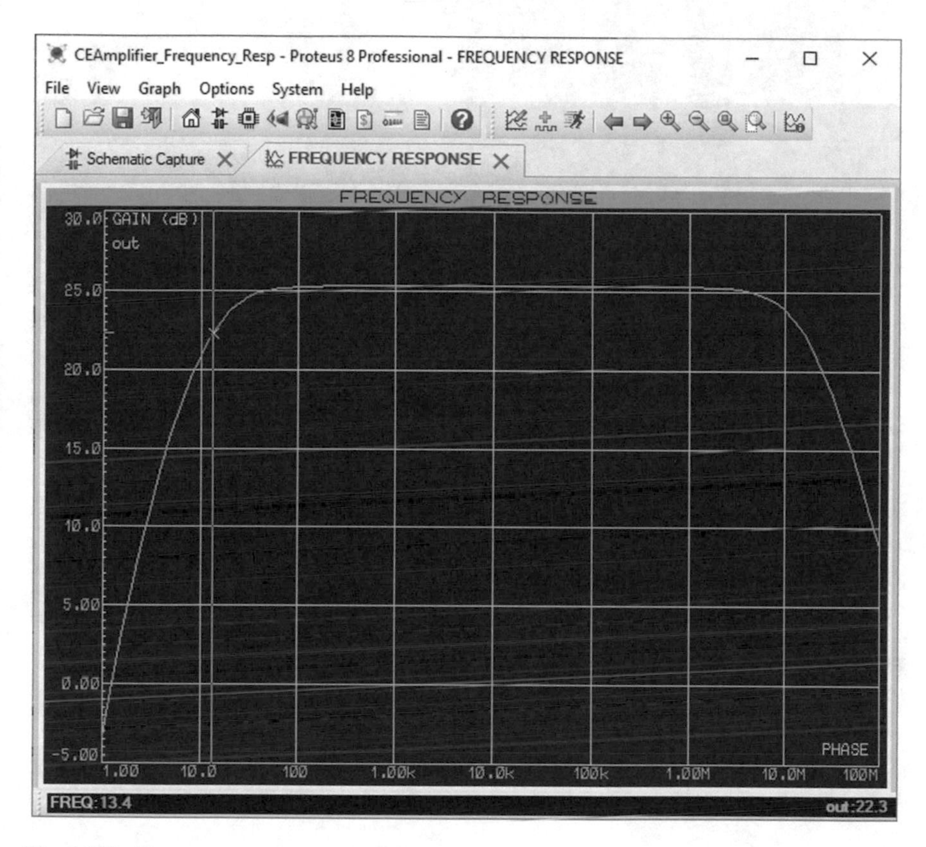

Fig. 2.180 Measurement of low −3 dB frequency

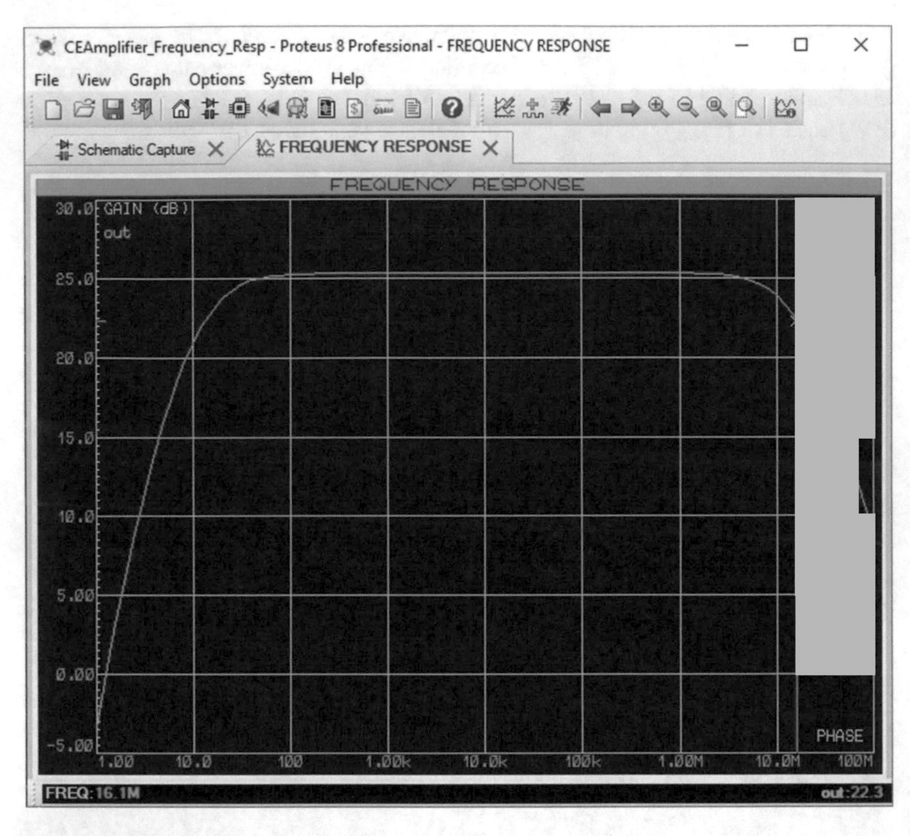

Fig. 2.181 Measurement of high −3 dB frequency

It is very difficult to obtain the bandwidth of 16.1 MHz for a common emitter amplifier in real world. Assume that we have 100 pF of stray capacitance between the base and collector (Fig. 2.182). Frequency response of this amplifier is shown in Fig. 2.183. Note that mid-band gain didn't change and it is equal to 25.3 dB.

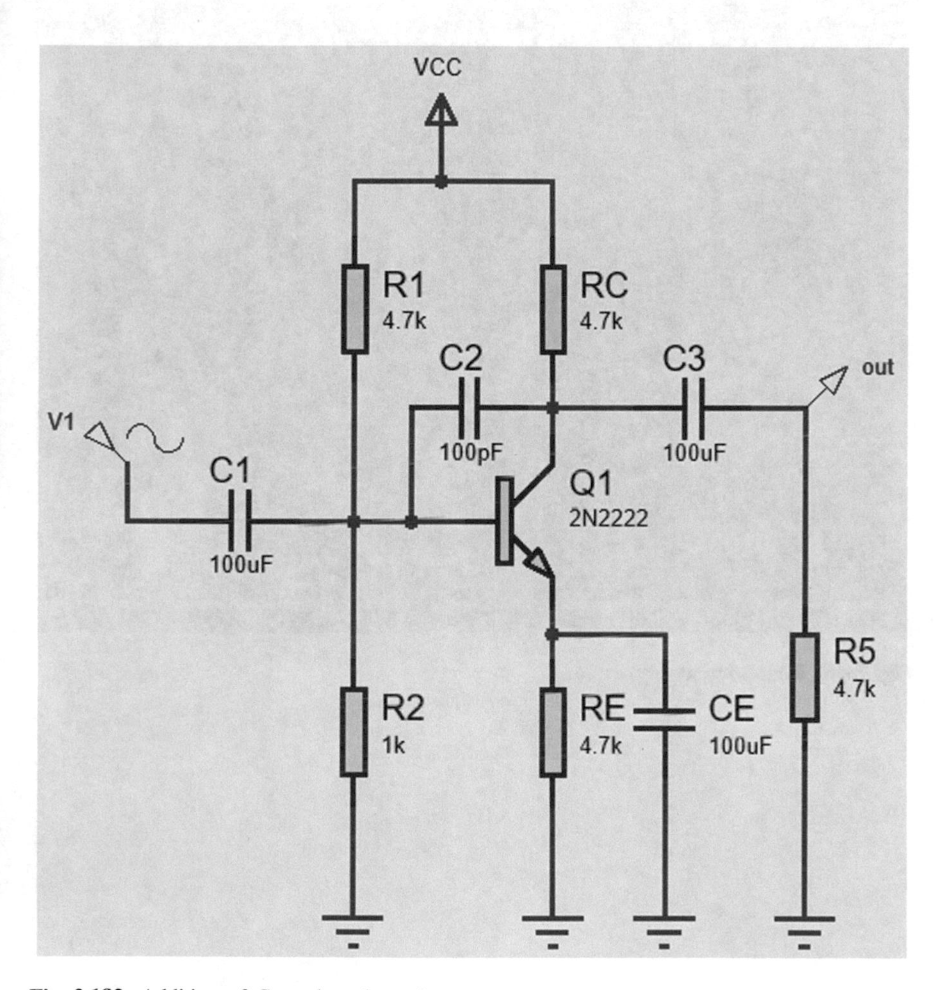

Fig. 2.182 Addition of C_2 to the schematic

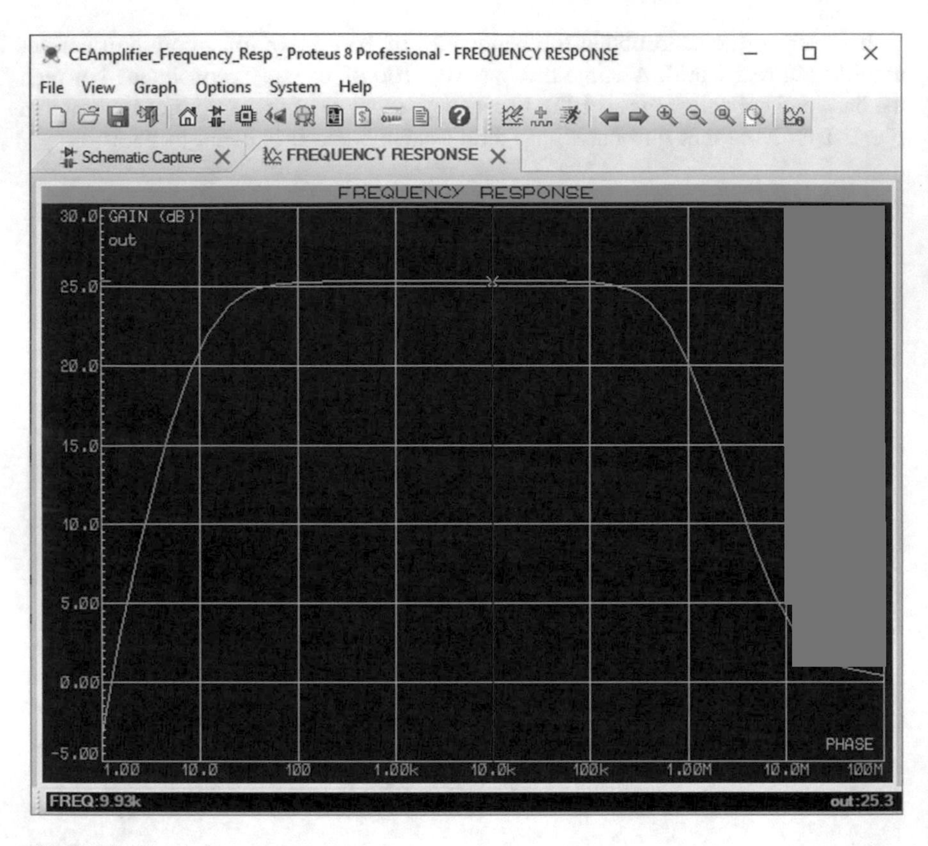

Fig. 2.183 Simulation result

Let's measure the −3 dB cut-off frequencies. According to Figs. 2.184 and 2.185, these frequencies are 13.4 Hz and 659 kHz. Note that upper cut-off frequency decreased considerably. The bandwidth in this case equals to $f_{-3\text{ dB},H} - f_{-3\text{ dB},L} = 659\text{ kHz} - 13.4\text{ Hz} \approx 659\text{ kHz}$.

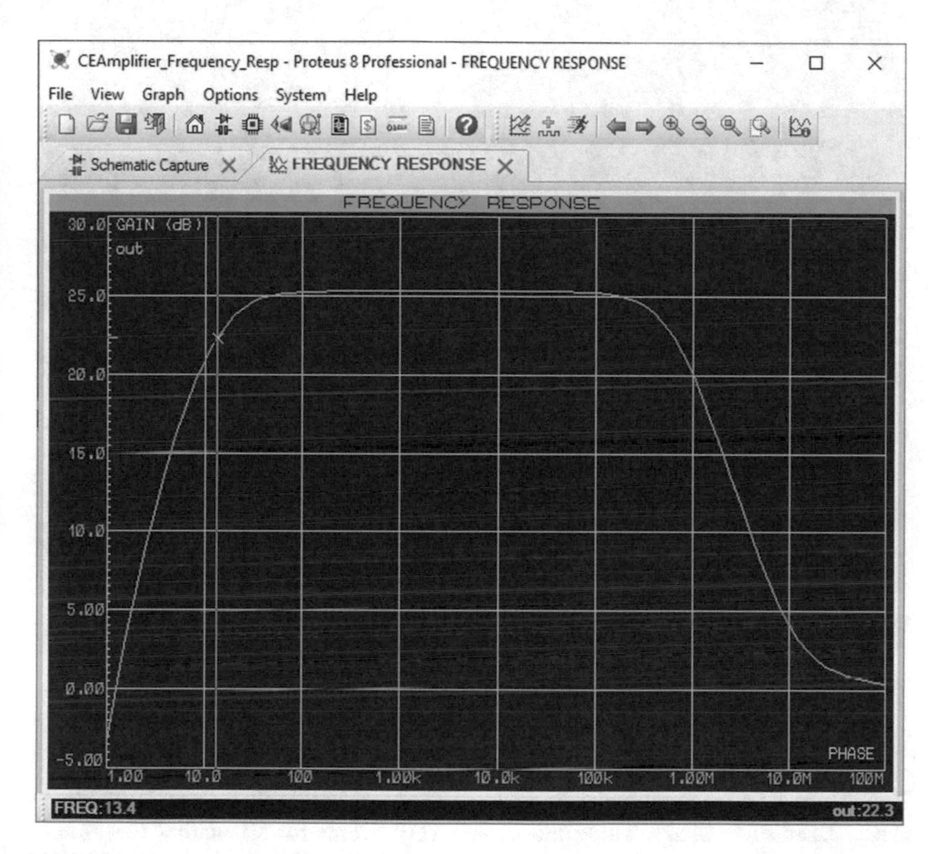

Fig. 2.184 Measurement of low −3 dB frequency

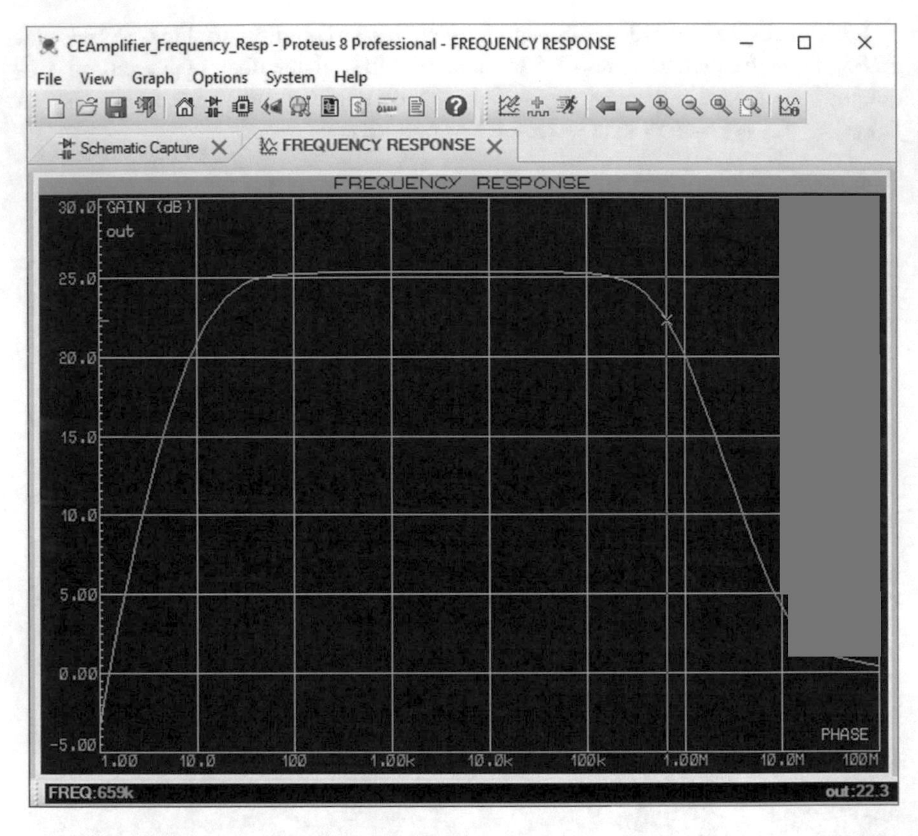

Fig. 2.185 Measurement of high −3 dB frequency

2.29 Example 28: Modeling Custom Semiconductor Devices

You can make a custom bipolar transistor with the aid of "NPN" and "PNP" blocks (Figs. 2.186 and 2.187). These blocks are very useful for simulating the circuits shown in textbooks. Because the schematics in textbooks doesn't contain any specific part number, and they usually use custom hypothetical semiconductor components.

Fig. 2.186 Search for NPN transistor

Pick Devices
Keywords:
NPN
Match whole words?
Show only parts with models?

Pick Devices

Keywords:

PNP

Match whole words?

Show only parts with models?

Fig. 2.187 Search for PNP transistor

For instance, assume that we want to model a NPN transistor with current gain (β or h_{FE}) of 100 and Early voltage of 50 V. Add a NPN block to the schematic (Fig. 2.188).

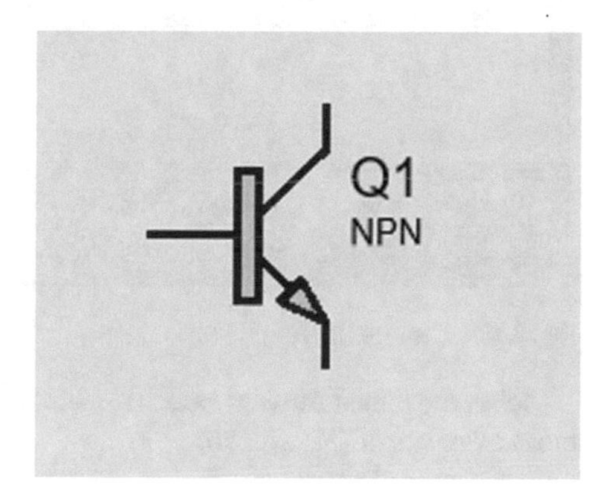

Fig. 2.188 NPN transistor symbol

Double click the transistor. This opens the Edit Component window (Fig. 2.189).

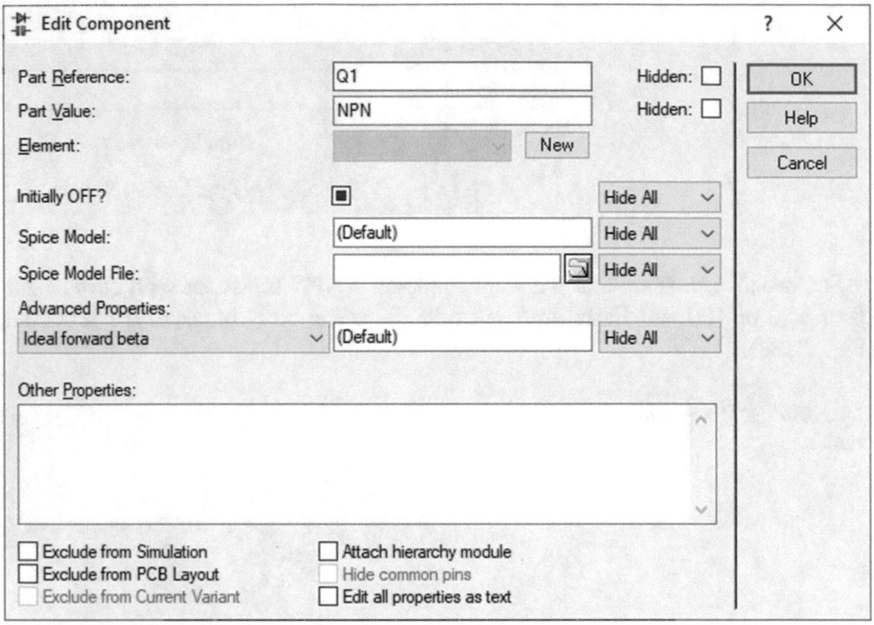

Fig. 2.189 Settings of Q_1

Select the "Ideal forward beta" for Advanced Properties box and enter 100 to the box in front of it (Fig. 2.190).

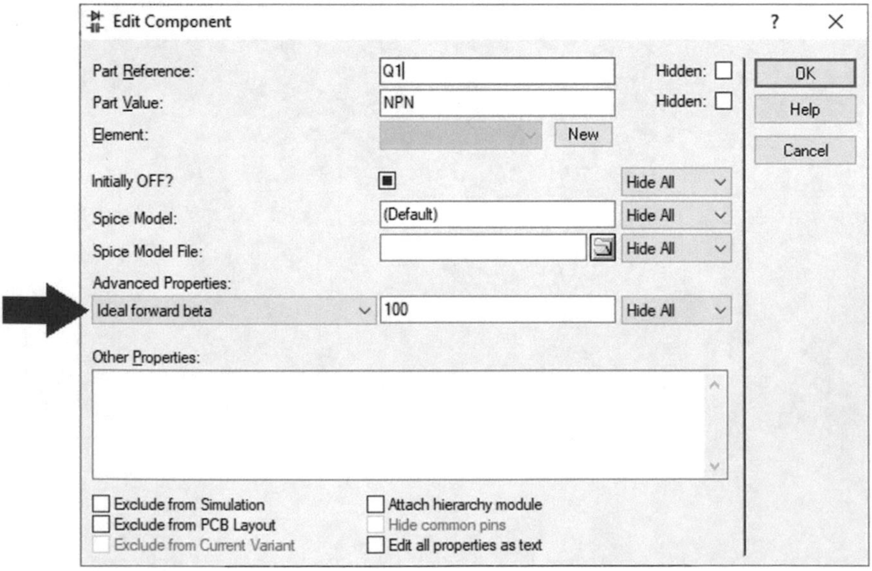

Fig. 2.190 Ideal forward beta determines the current gain of the transistor

Select the "Forward Early voltage" for Advanced Properties box and enter 50 to the box in front of it (Fig. 2.191). Then click the OK button to close the Edit Component window. Now the transistor Q_1 has current gain of 100 and Early voltage of 50 V.

Edit Component ? ×
Part Reference: Q1 Hidden:
Part Value: NPN Hidden:
Element: New
OK
Help
Cancel
Initially OFF? Hide All
Spice Model: (Default) Hide All
Spice Model File: Hide All
Advanced Properties:
Forward Early voltage 50 Hide All
Other Properties:
Exclude from Simulation
Exclude from PCB Layout
Exclude from Current Variant
Attach hierarchy module
Hide common pins
Edit all properties as text

Fig. 2.191 Forward Early voltage determines the early voltage of the transistor

You can make custom MOSFET and JFET transistors as well. These types of transistors can be modeled with the aid of NMOSFET, NMOSFET3, PMOSFET, PMOSFET3, NJFET and PJFET blocks.

Difference between NMOSFET and NMOSFET3 is shown in Fig. 2.192. Substrate of NMOSFET is free while substrate of NMOSFET3 is connected to the source terminal.

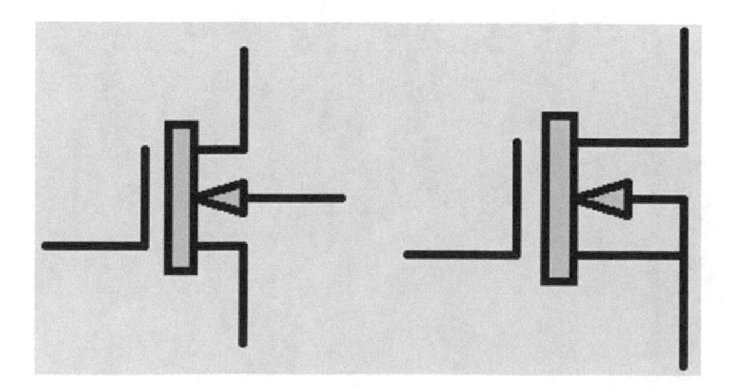

Fig. 2.192 Symbols of NMOSFET (left) and NMOSFET3 (right)

Difference between PMOSFET and PMOSFET3 is shown in Fig. 2.193. Substrate of PMOSFET is free while substrate of PMOSFET3 is connected to the source terminal.

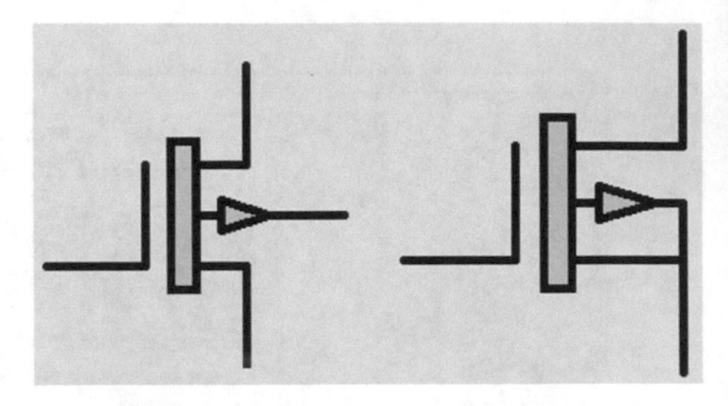

Fig. 2.193 Symbols of PMOSFET (left) and PMOSFET3 (right)

Custom diodes can be modeled with the aid of DIODE block (Fig. 2.194). All of the aforementioned blocks can be found in the "Modeling Primitive" section (Fig. 2.195).

Pick Devices

Keywords:

DIODE

Match whole words?

Show only parts with models?

Fig. 2.194 Search for diode

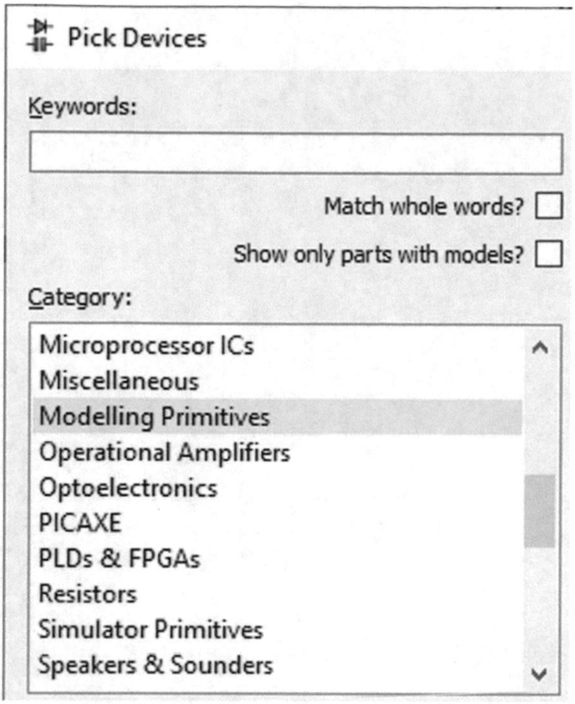

Fig. 2.195 Modeling Primitives

2.30 Example 29: Bill of Material

Proteus can generate the list of components that are used in the schematic. This is very useful when you want to make the circuit. Because the generated list shows all of the components that you need to buy. In this example, we will learn how to generate the list of components for a given schematic.

Consider the schematic shown in Fig. 2.196.

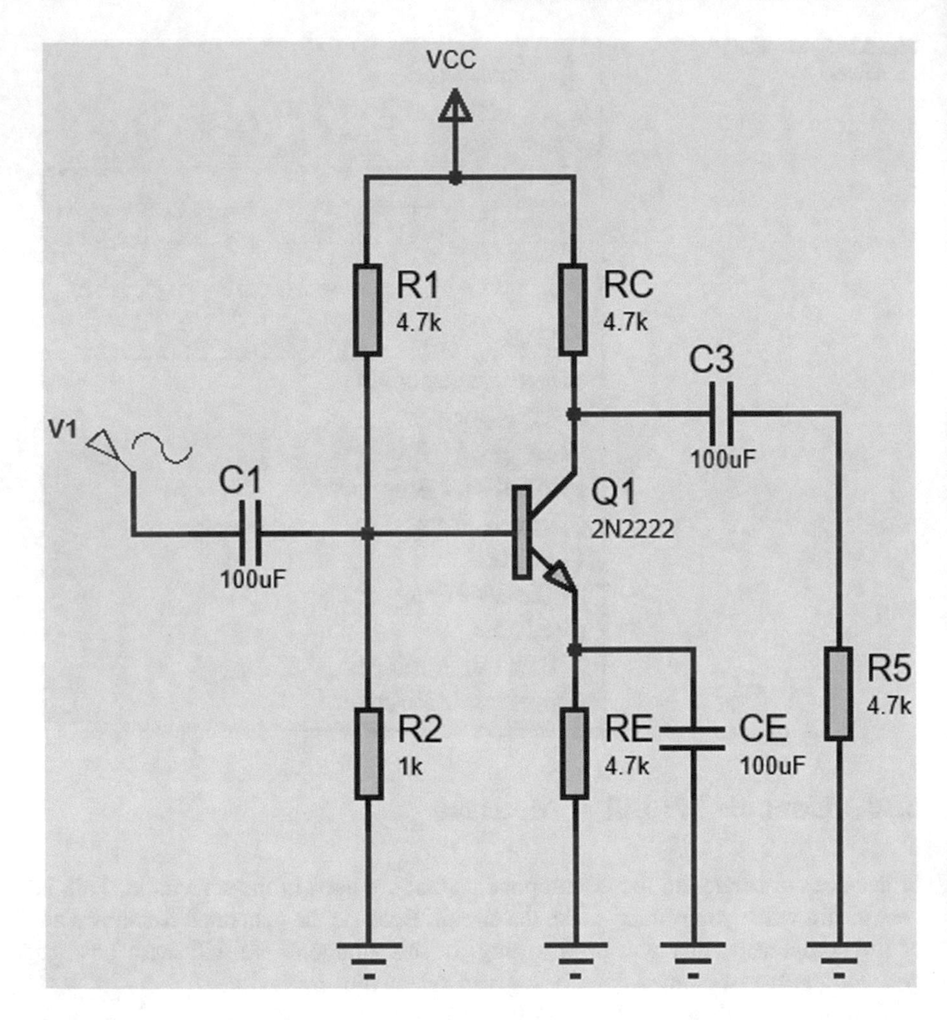

Fig. 2.196 Schematic for Example 29

Click the Bill of Materials icon (Fig. 2.197). After clicking the Bill of Material icon, the report shown in Fig. 2.198 appears on the screen.

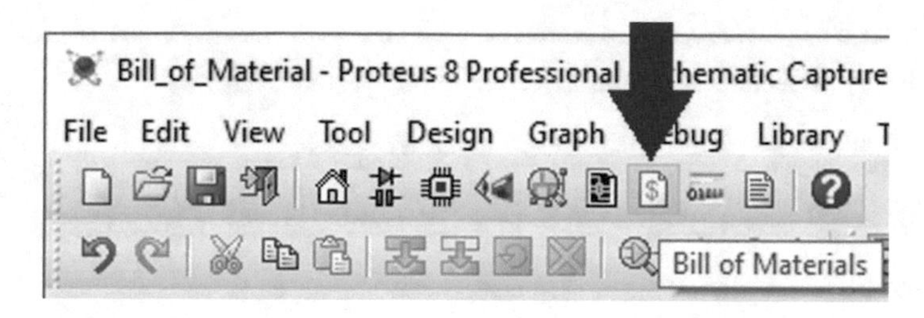

Fig. 2.197 Bill of materials icon

Bill Of Materials for Bill_of_Material

Design Title Bill_of_Material
Author
Document Number
Revision
Design Created
Design Last Modified
Total Parts In Design 9

Quantity	References	Value	Stock Code	Unit Cost
0 Modules				
Quantity	References	Value	Stock Code	Unit Cost
Sub-totals:				£0.00
2 Capacitors				
Quantity	References	Value	Stock Code	Unit Cost
2	C1,C3	100uF		
Sub-totals:				£0.00
3 Resistors				
Quantity	References	Value	Stock Code	Unit Cost
2	R1,R5	4.7k		
1	R2	1k		
Sub-totals:				£0.00
0 Integrated Circuits				
Quantity	References	Value	Stock Code	Unit Cost
Sub-totals:				£0.00
1 Transistors				
Quantity	References	Value	Stock Code	Unit Cost
1	Q1	2N2222		
Sub-totals:				£0.00
0 Diodes				
Quantity	References	Value	Stock Code	Unit Cost
Sub-totals:				£0.00
3 Miscellaneous				
Quantity	References	Value	Stock Code	Unit Cost
1	CE	100uF		
2	RC,RE	4.7k		
Sub-totals:				£0.00

Fig. 2.198 List of used materials in Fig. 2.196

You can print or save the report as a Pdf file by clicking the icons shown in Fig. 2.199.

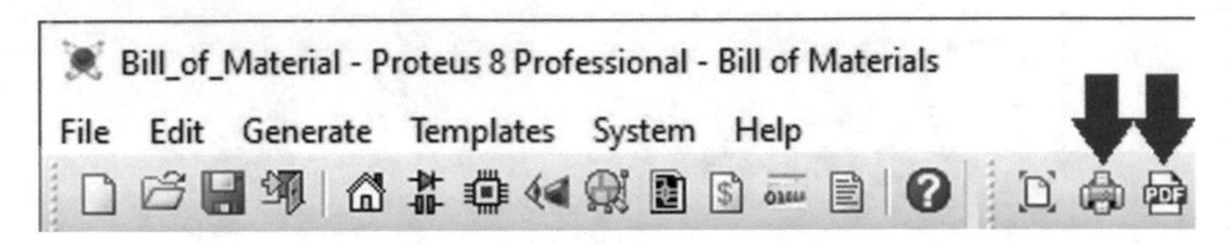

Fig. 2.199 Icons to make pdf file of used components

2.31 Example 30: Common Mode Rejection Ratio (CMRR) of Difference Amplifier

In this example, we want to measure the CMRR of a difference amplifier. Consider the difference amplifier shown in Fig. 2.200. The output voltage of this amplifier is $V_{\text{out}} = \left(\frac{R_1 + R_2}{R_3 + R_4}\right)\frac{R_4}{R_1}V_2 - \frac{R_2}{R_1}V_1$. When $R_1 = R_3$ and $R_2 = R_4$, $V_{\text{out}} = \frac{R_2}{R_1}(V_2 - V_1)$.

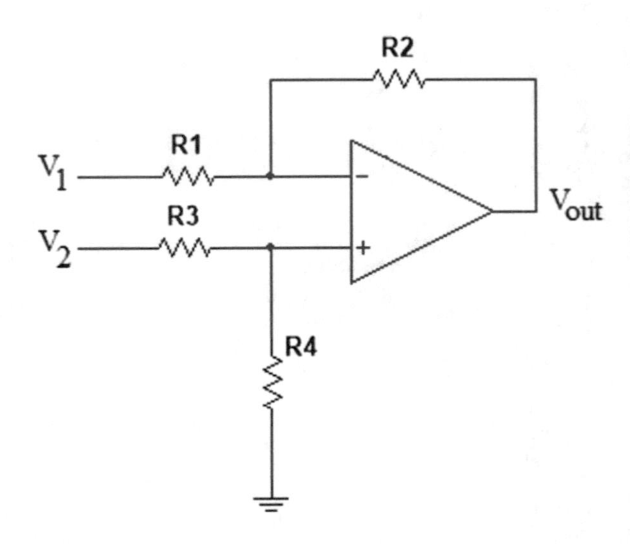

Fig. 2.200 Difference amplifier

Assume that we want to measure the CMRR of the circuit shown in Fig. 2.201.

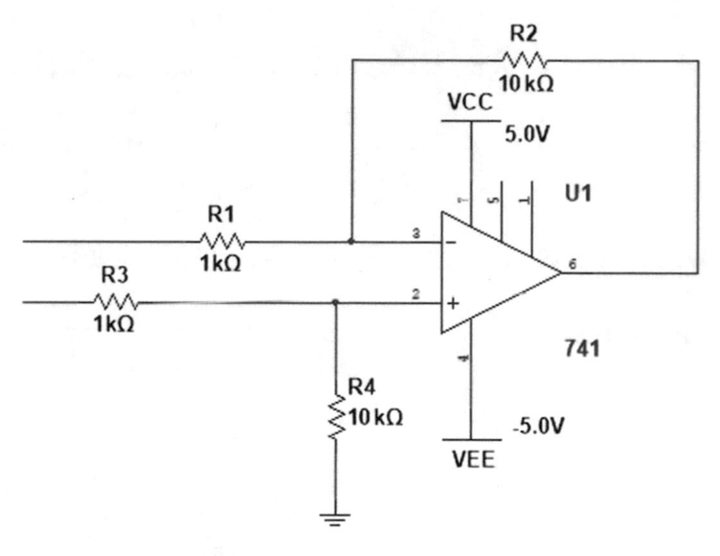

Fig. 2.201 Circuit for Example 30

Draw the circuit in Proteus (Fig. 2.202). You can add the 741 op-amp to the schematic by searching for 741 in the Pick Devices window (Fig. 2.203).

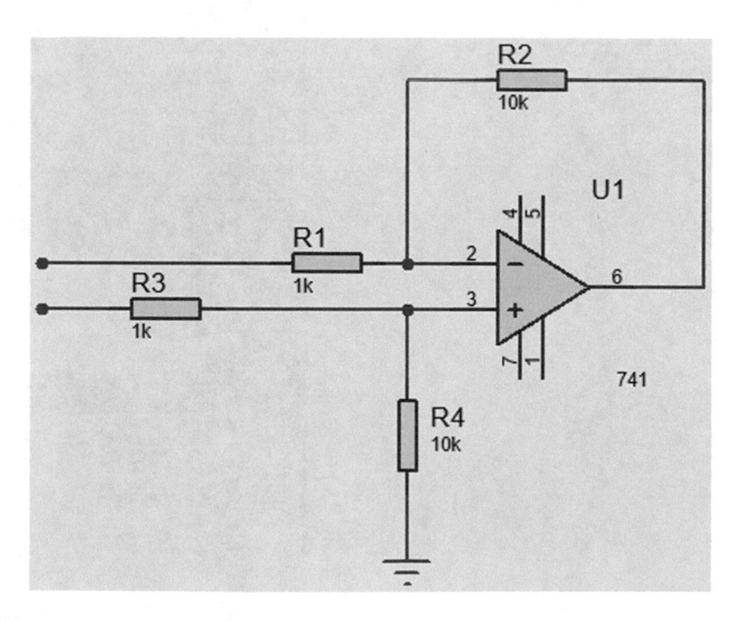

Fig. 2.202 Proteus equivalent of Fig. 2.201

Fig. 2.203 Search for 741

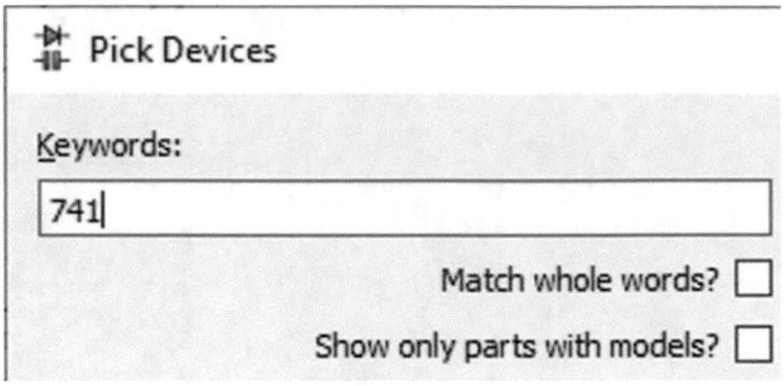

Pick Devices

Keywords:

741

Match whole words?

Show only parts with models?

Add two Power blocks (Fig. 2.204) to the schematic (Fig. 2.205).

Fig. 2.204 Power block

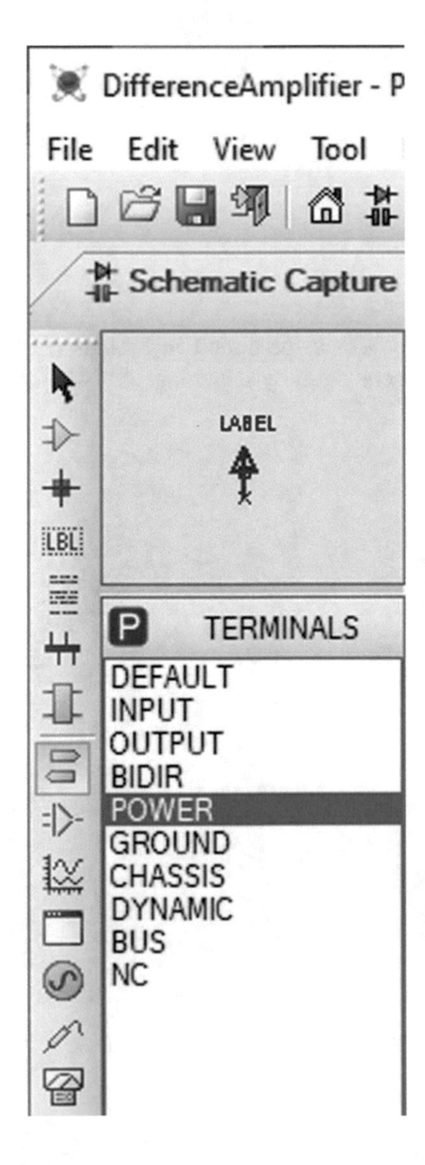

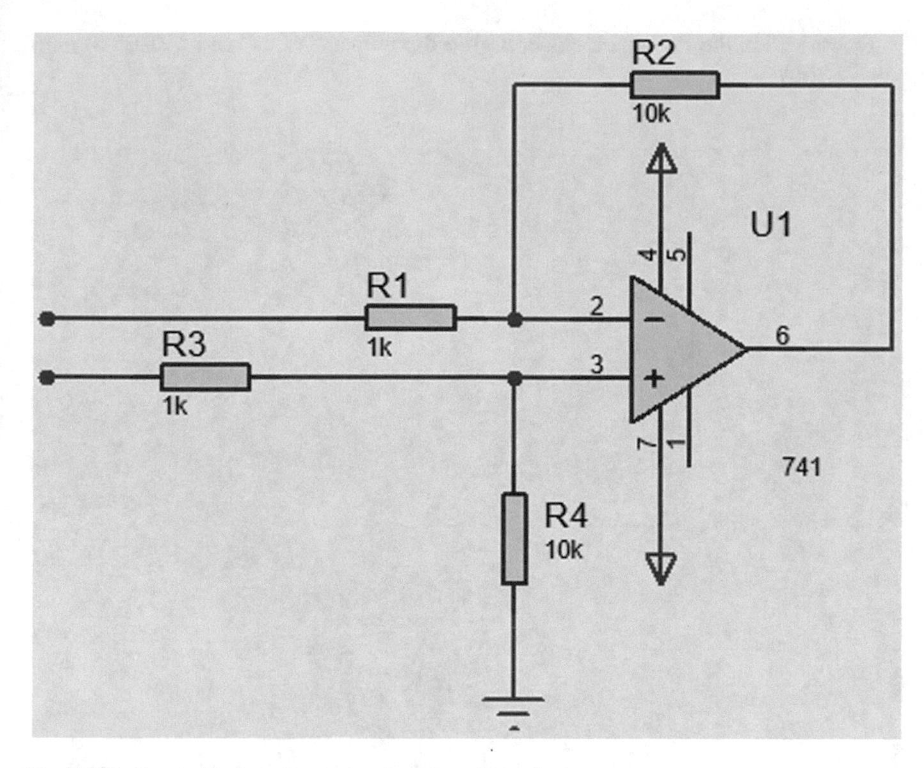

Fig. 2.205 Power blocks are connected to the schematic

Double click the Power blocks and give the names "VCC" and "VEE" to them (Fig. 2.206).

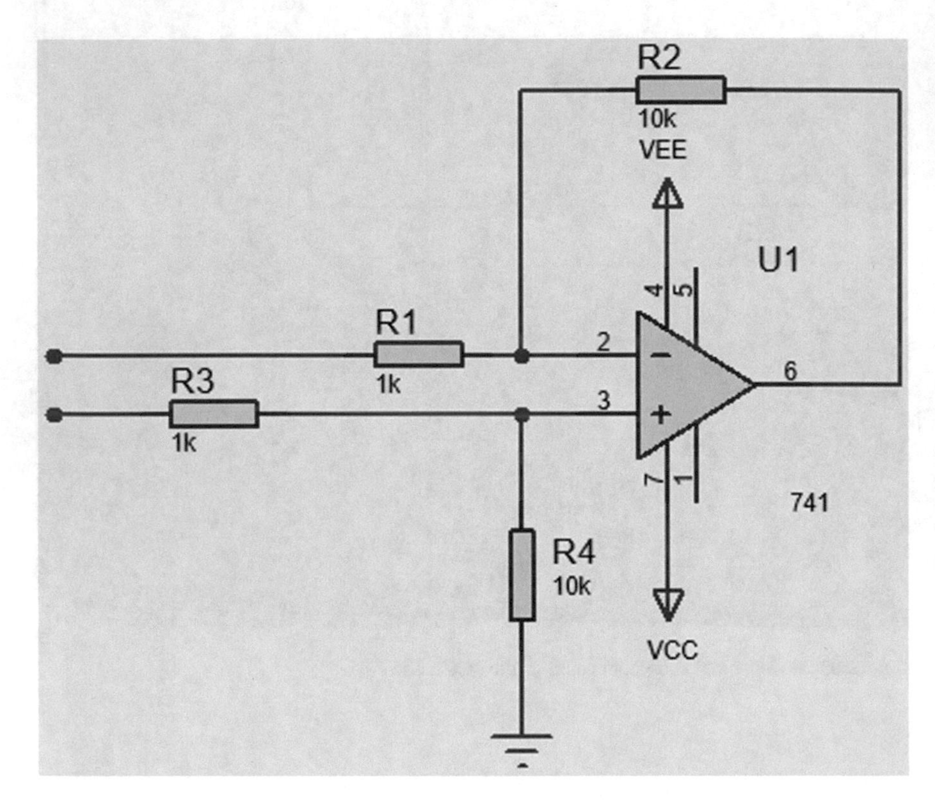

Fig. 2.206 Named VCC and VEE are given to used power blocks

Click the Design > Configure Power Rails (Fig. 2.207). This opens the Power Rail Configuration window for you.

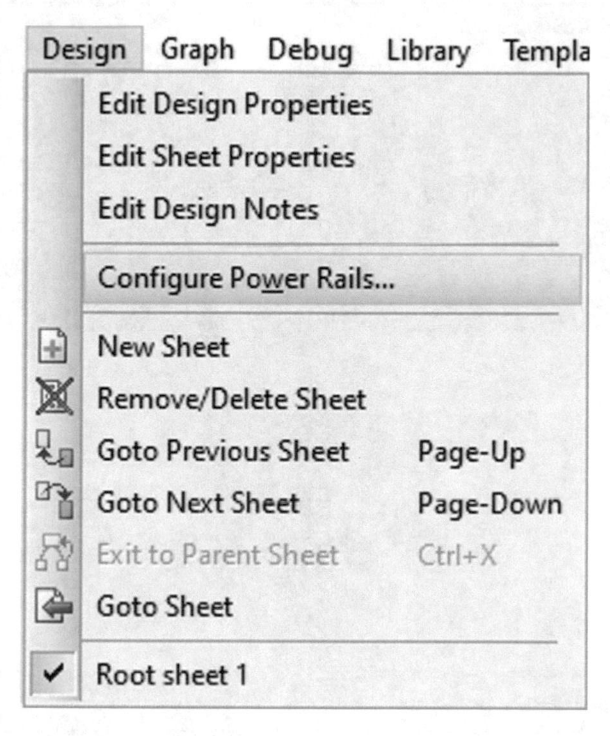

Fig. 2.207 Design > configure power rails

Select the VCC/VDD and enter 5 into the Voltage box (Fig. 2.208).

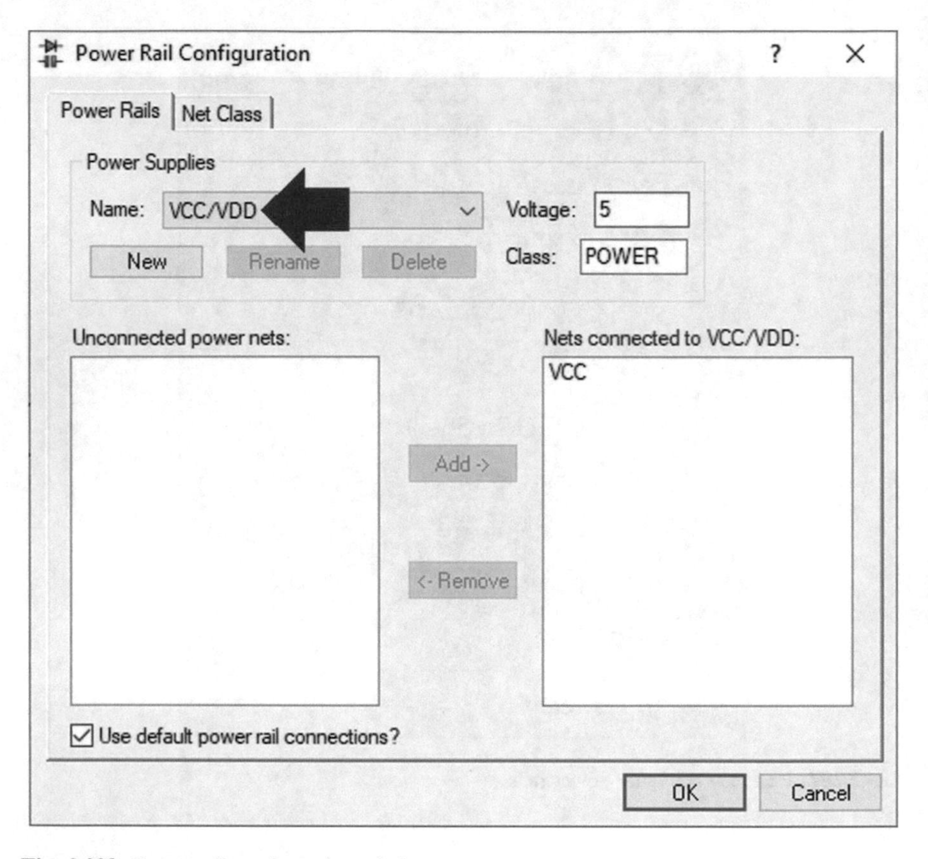

Fig. 2.208 Power rail configuration window

Select the VEE and enter -5 into the Voltage box (Fig. 2.209). Then click the OK button to close the Power Rail Configuration window.

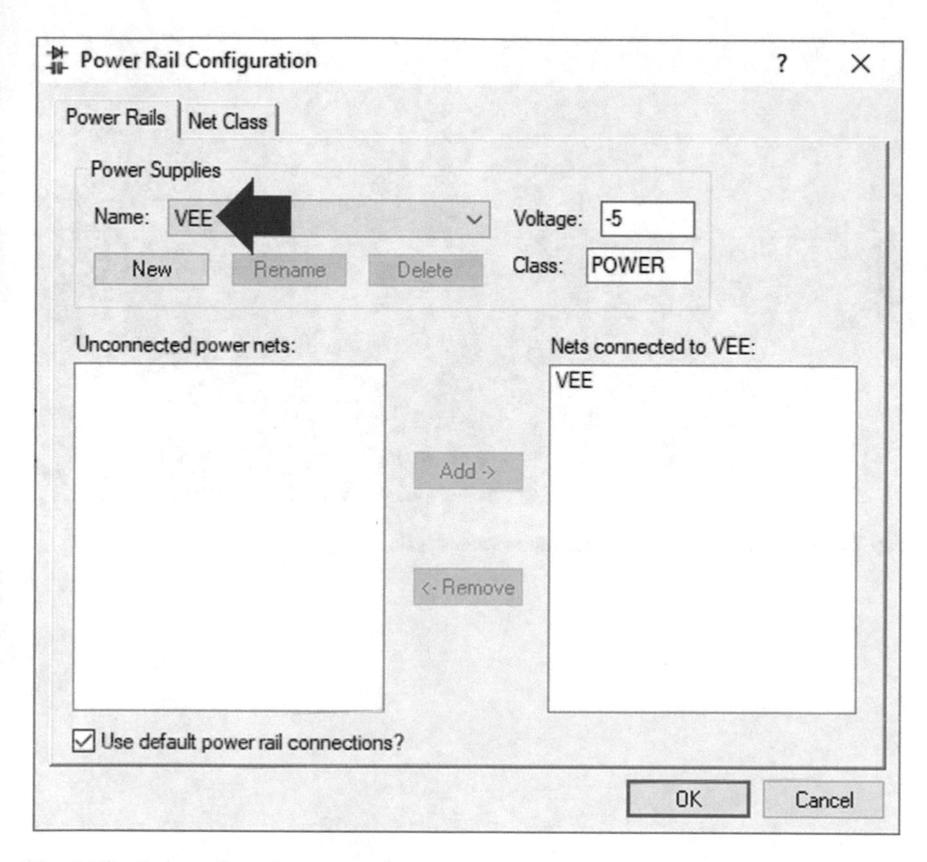

Fig. 2.209 Power rail configuration window

Let's measure the common mode voltage gain. Change the schematic to what is shown in Fig. 2.210. Settings of V_{in} are shown in Fig. 2.211.

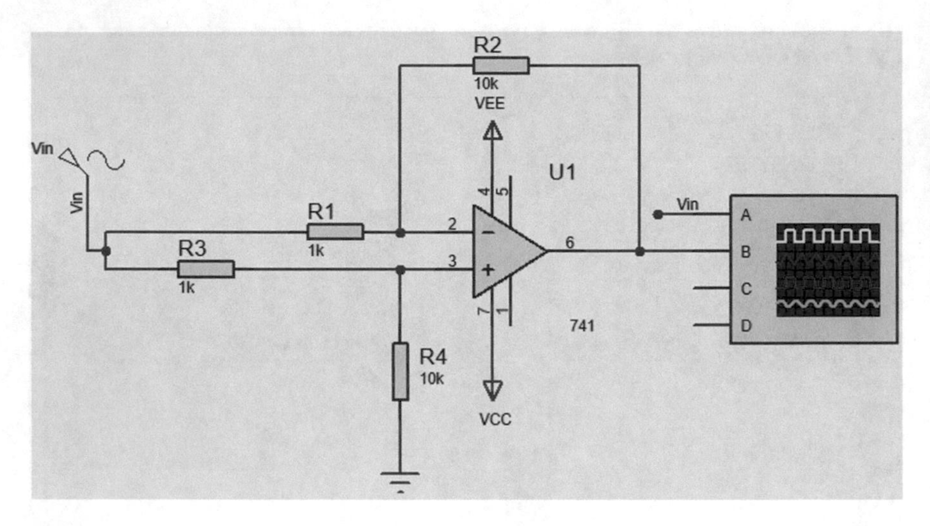

Fig. 2.210 Measurement of common mode voltage gain

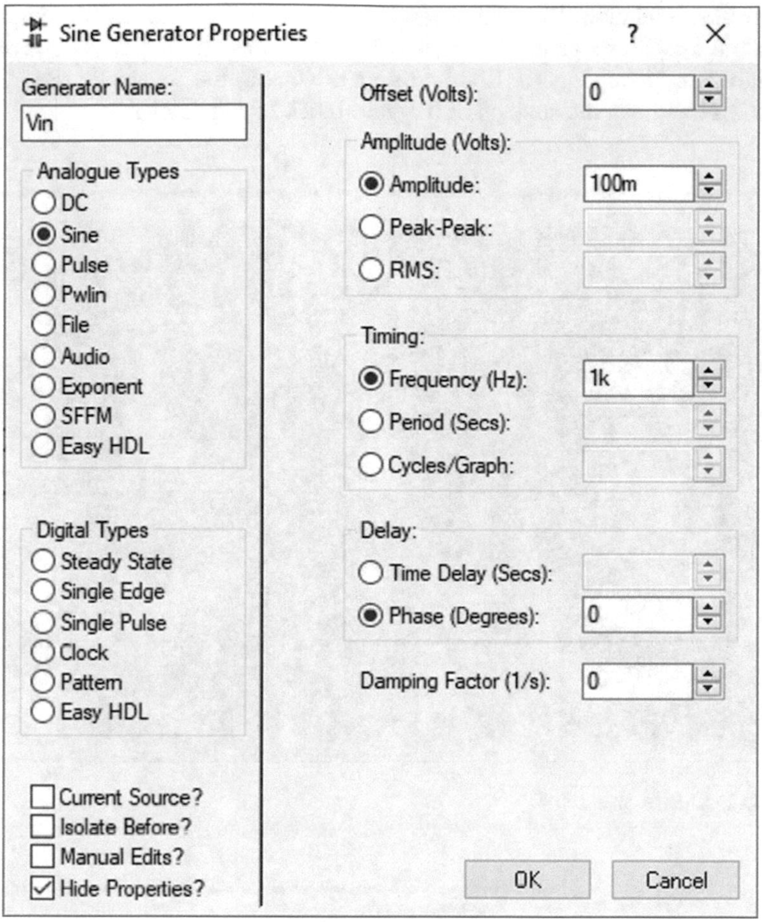

Fig. 2.211 Settings of V_{in}

Run the simulation. Simulation result is shown in Fig. 2.212. The output that enters to the oscilloscope is very small and we can't measure it easily. Let's amplify it with a gain block (Fig. 2.213). Add a gain block to the schematic (Fig. 2.214). Figure 2.215 shows the settings of the gain block in Fig. 2.214.

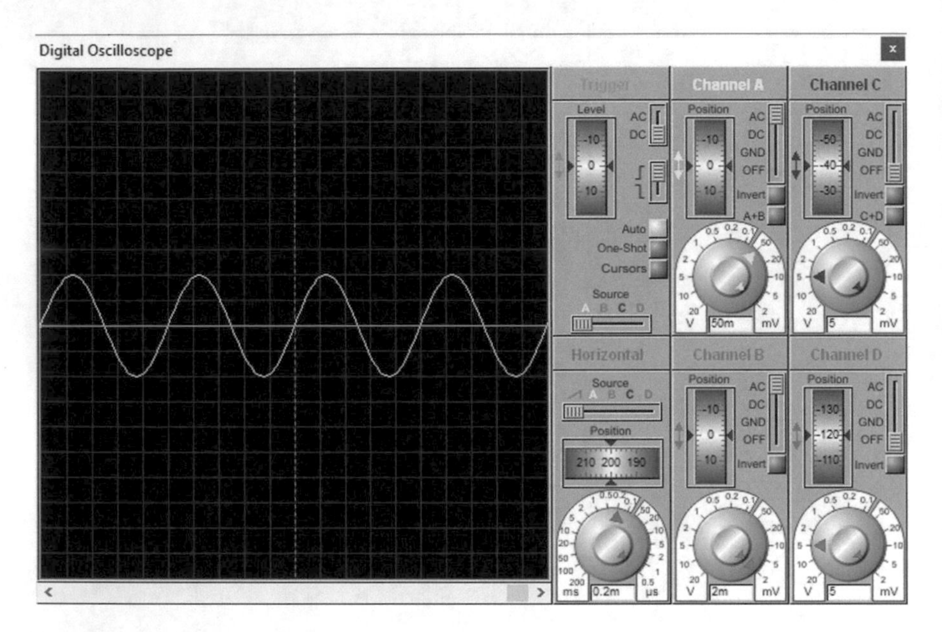

Fig. 2.212 Simulation result

Fig. 2.213 Search for OP: GAIN

Pick Devices

Keywords:

op : gain

Match whole words?

Show only parts with models?

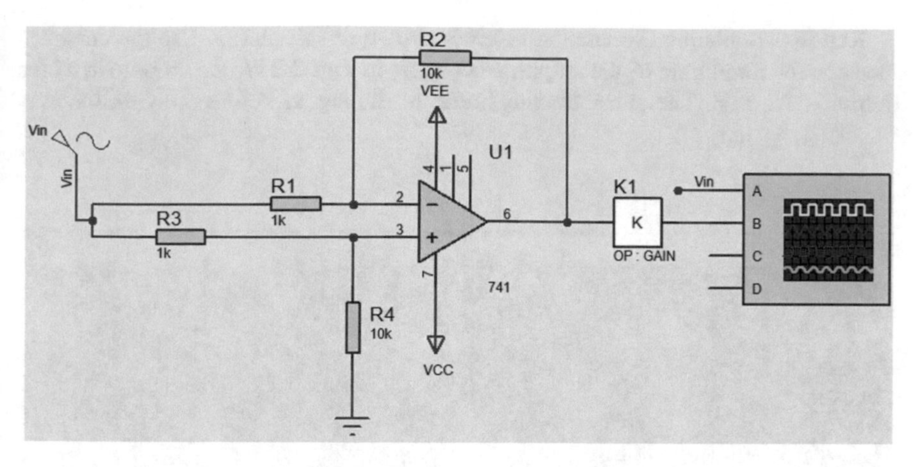

Fig. 2.214 Addition of gain block to the schematic

Edit Component ? ×

Part Reference: K1 Hidden: ☐ OK

Part Value: OP : GAIN Hidden: ☐ Help

Element: New Cancel

Amplification coefficient: 100 Hide All

Other Properties:

☐ Exclude from Simulation ☐ Attach hierarchy module

☐ Exclude from PCB Layout Hide common pins

Exclude from Current Variant ☐ Edit all properties as text

Fig. 2.215 Settings of gain block K_1

Run the simulation. Simulation result is shown in Fig. 2.216. Use the cursor to measure the amplitude of the output. According to Fig. 2.217, the amplitude of the output is 3.1 mV. Therefore, the amplitude of the signal before the gain block is $\frac{3.1\ \text{mV}}{100} = 31\ \mu\text{V}$.

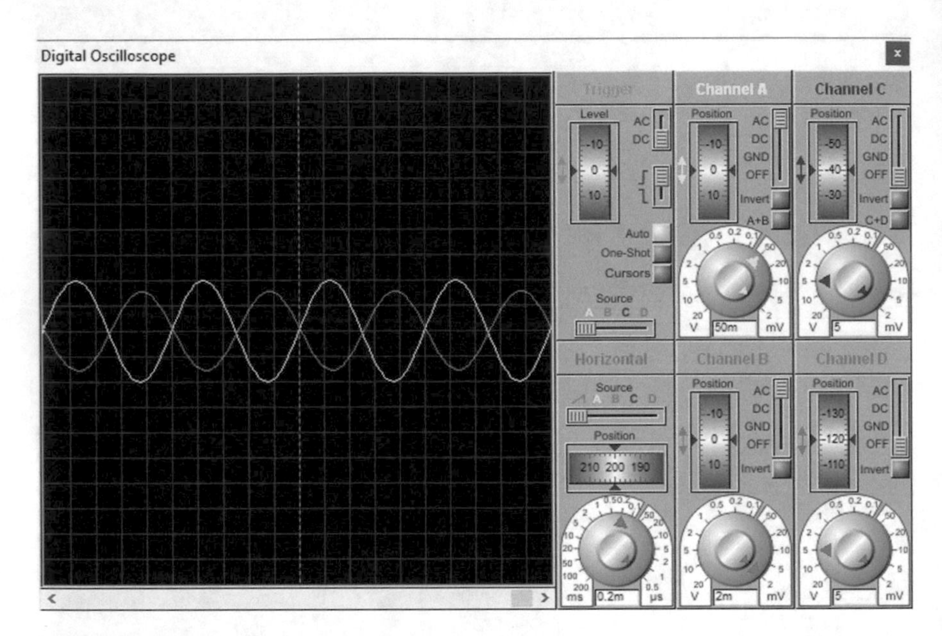

Fig. 2.216 Simulation result

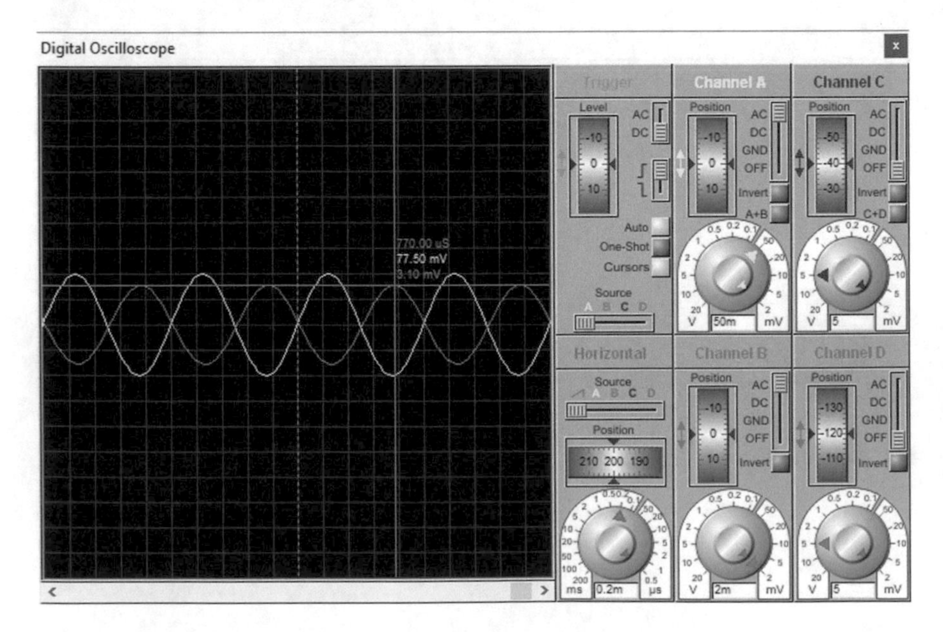

Fig. 2.217 Simulation result

The common mode gain is 3.1×10^{-4} according to the calculation in Fig. 2.218.

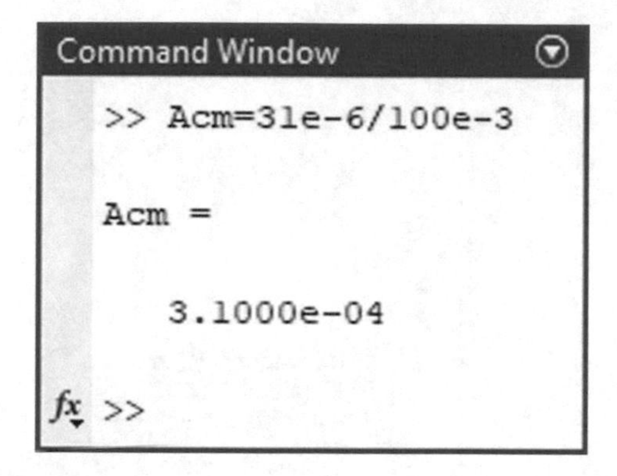

Fig. 2.218 MATLAB commands

Let's measure the differential mode gain of the circuit. Change the schematic to what is shown in Fig. 2.219. Settings of V_{in1} and V_{in2} are shown in Figs. 2.220 and 2.221, respectively. Note that V_{in1} and V_{in2} are out of phase.

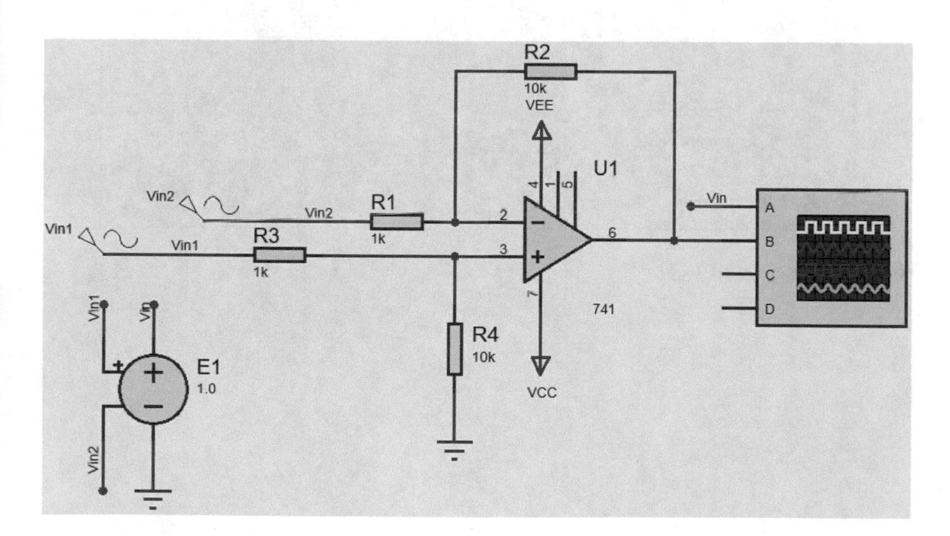

Fig. 2.219 Measurement of differential mode gain

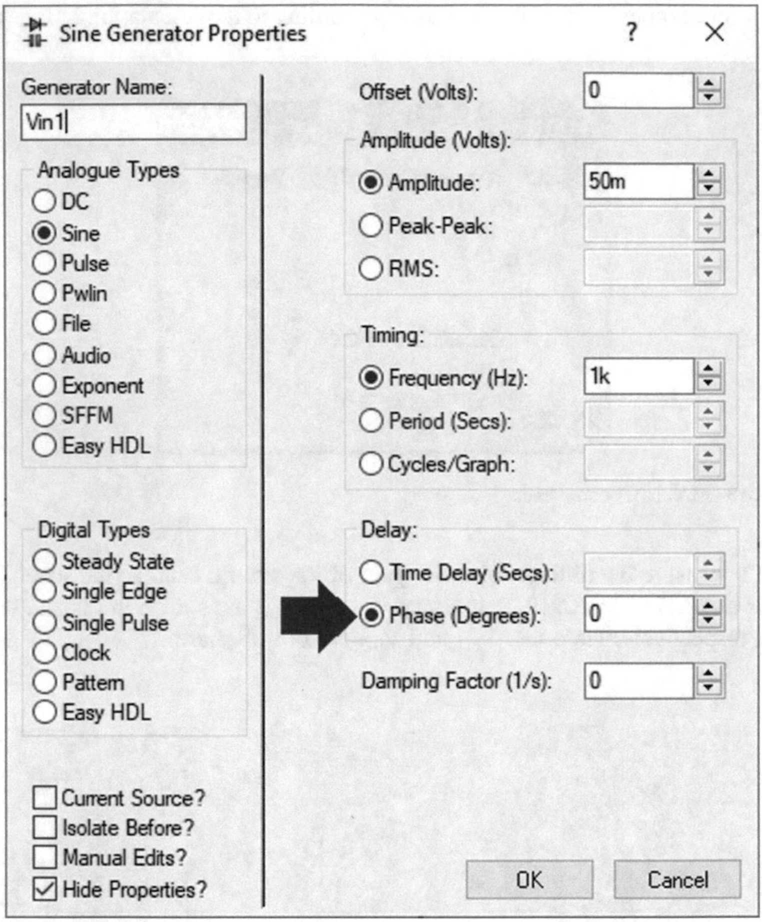

Fig. 2.220 Settings of V_{in1}

Sine Generator Properties
Generator Name: Vin2
Analogue Types: DC, Sine (selected), Pulse, Pwlin, File, Audio, Exponent, SFFM, Easy HDL
Digital Types: Steady State, Single Edge, Single Pulse, Clock, Pattern, Easy HDL
Current Source?
Isolate Before?
Manual Edits?
Hide Properties? (checked)
Offset (Volts): 0
Amplitude (Volts):
Amplitude: 50m (selected)
Peak-Peak:
RMS:
Timing:
Frequency (Hz): 1k (selected)
Period (Secs):
Cycles/Graph:
Delay:
Time Delay (Secs):
Phase (Degrees): 180 (selected)
Damping Factor (1/s): 0
OK
Cancel

Fig. 2.221 Settings of V_{in2}

Run the simulation. Simulation result is shown in Fig. 2.222. Use the cursor to measure the amplitude of the output. According to Fig. 2.223, the amplitude of the output is 990 mV.

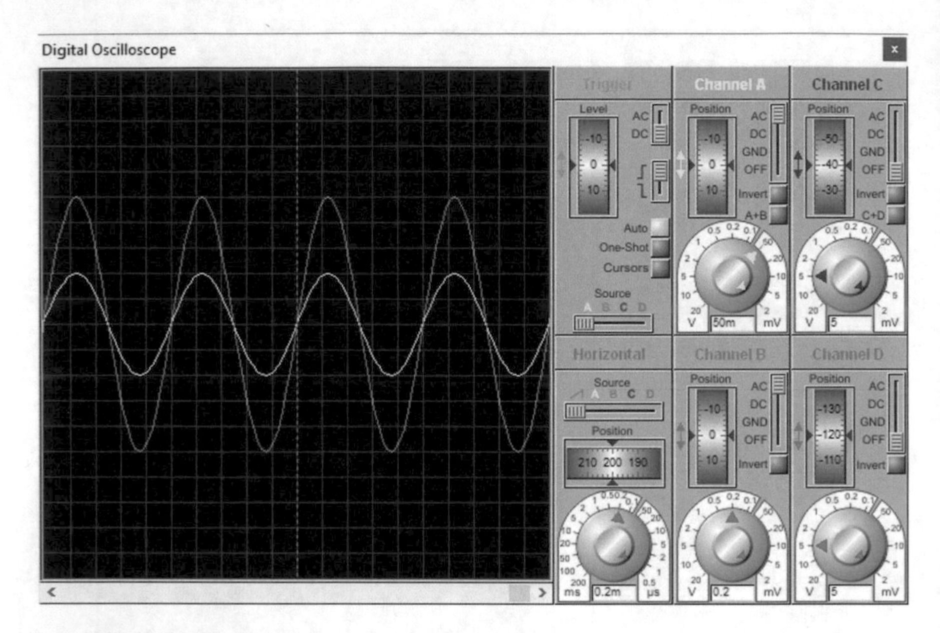

Fig. 2.222 Simulation result

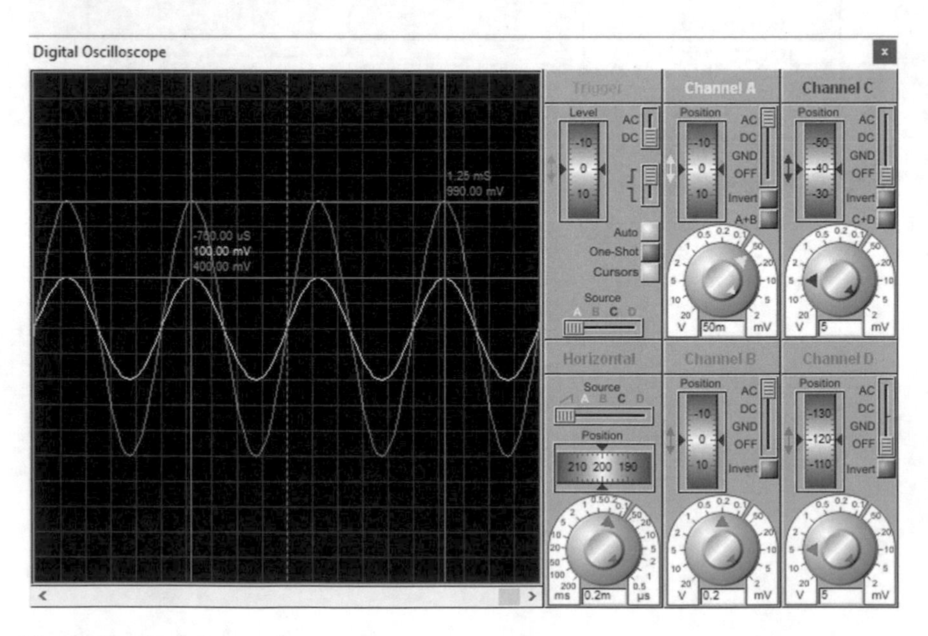

Fig. 2.223 Simulation result

The differential mode gain is 9.9 according to the calculation in Fig. 2.224.

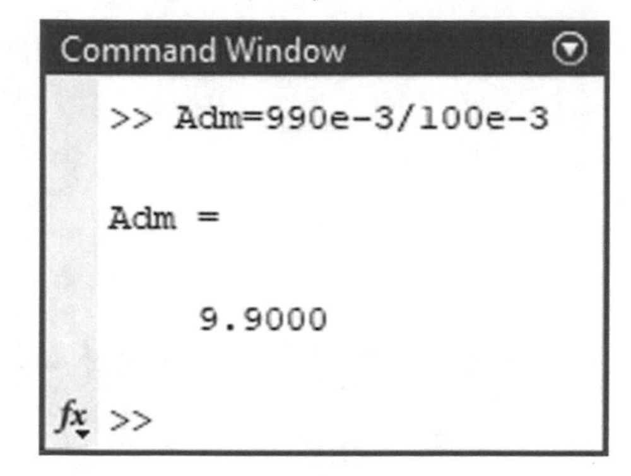

Fig. 2.224 MATLAB calculation

Let's calculate the CMRR. CMRR is 90.0855 dB according to the calculation in Fig. 2.225.

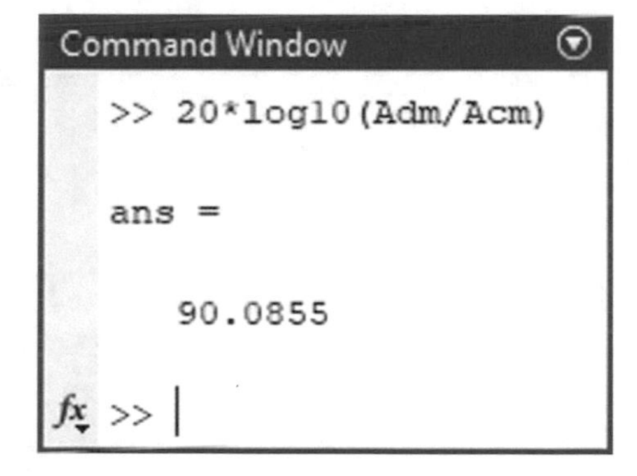

Fig. 2.225 MATLAB calculation

2.32 Example 31: CMRR of Differential Pair

We want to measure the CMRR of differential pair shown in Fig. 2.226. The two collector resistances are accurate to within ± 1%. So, 9.9 kΩ ≤ RC1 ≤ 10.1 kΩ and 9.9 kΩ ≤ RC2 ≤ 10.1 kΩ. Note that output is taken from the collector terminals of the transistors.

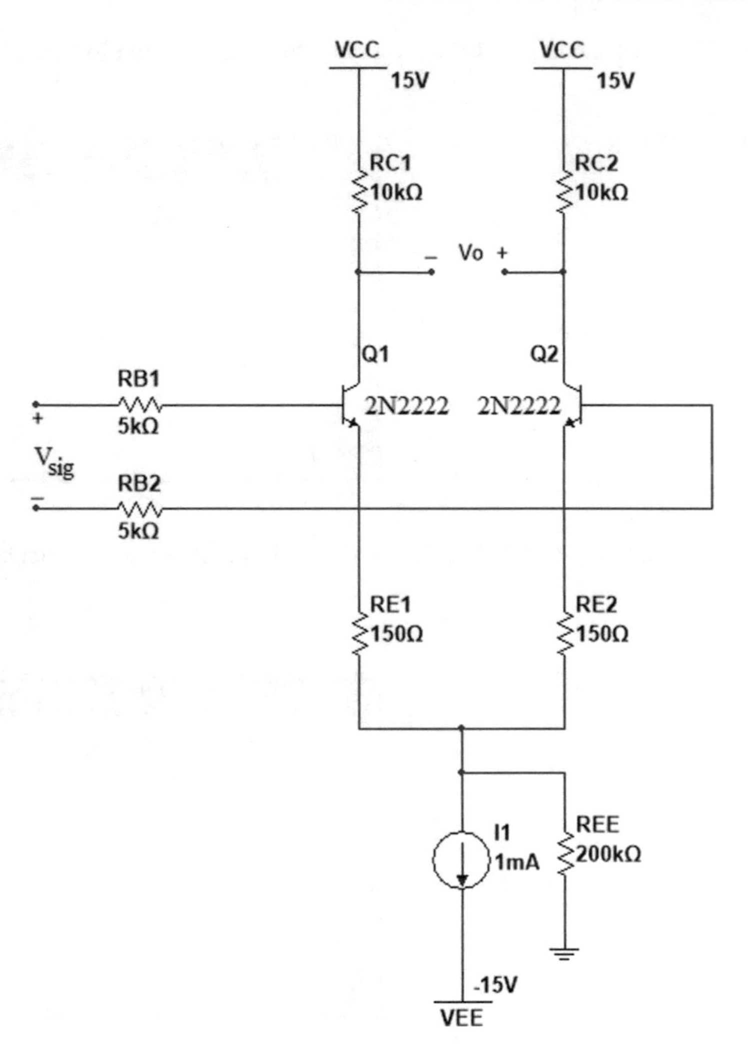

Fig. 2.226 Circuit for Example 31

Draw the circuit in Proteus (Fig. 2.227). The schematic shown in Fig. 2.227 uses a “csource” block (Fig. 2.228) to simulate the DC current source I_1 in Fig. 2.226.

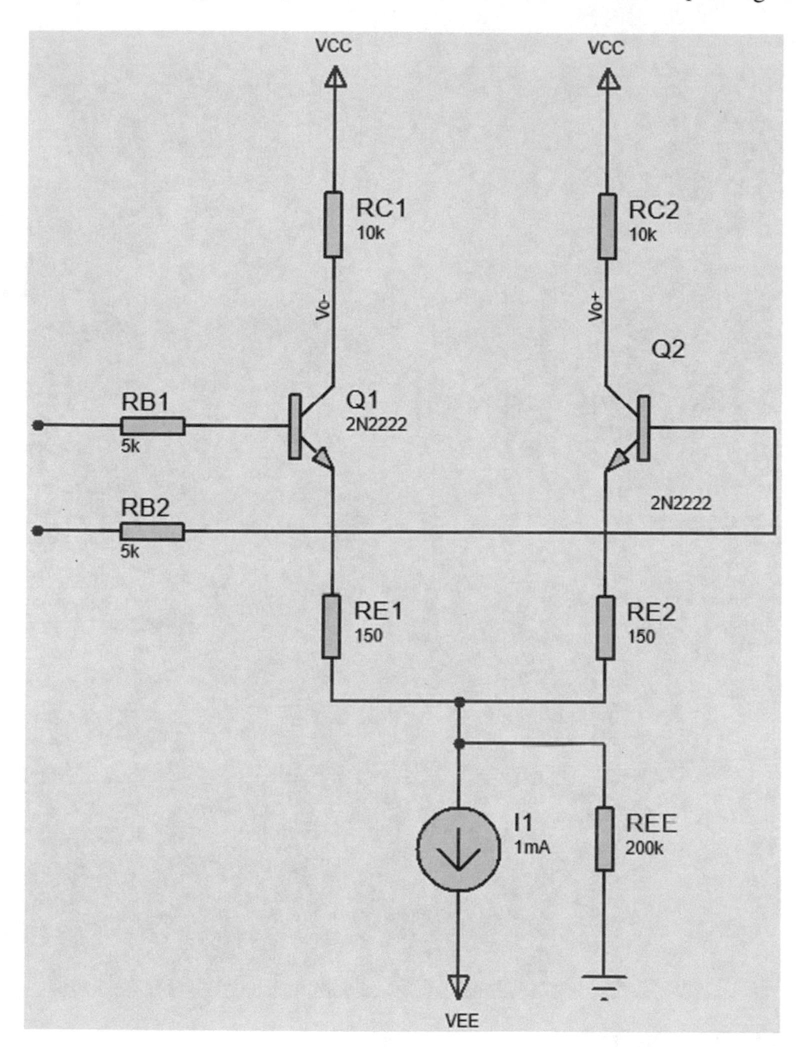

Fig. 2.227 Proteus equivalent of Fig. 2.226

Fig. 2.228 Search for current source block

Pick Devices

Keywords:

csource

Match whole words?

Show only parts with models?

Let's do a DC operating point analysis. According to Fig. 2.229, the transistors are biased at $V_{CE} = 10.0398 - (-0.644679) = 10.6845$ V and $I_C \approx I_E = 0.4982$ mA.

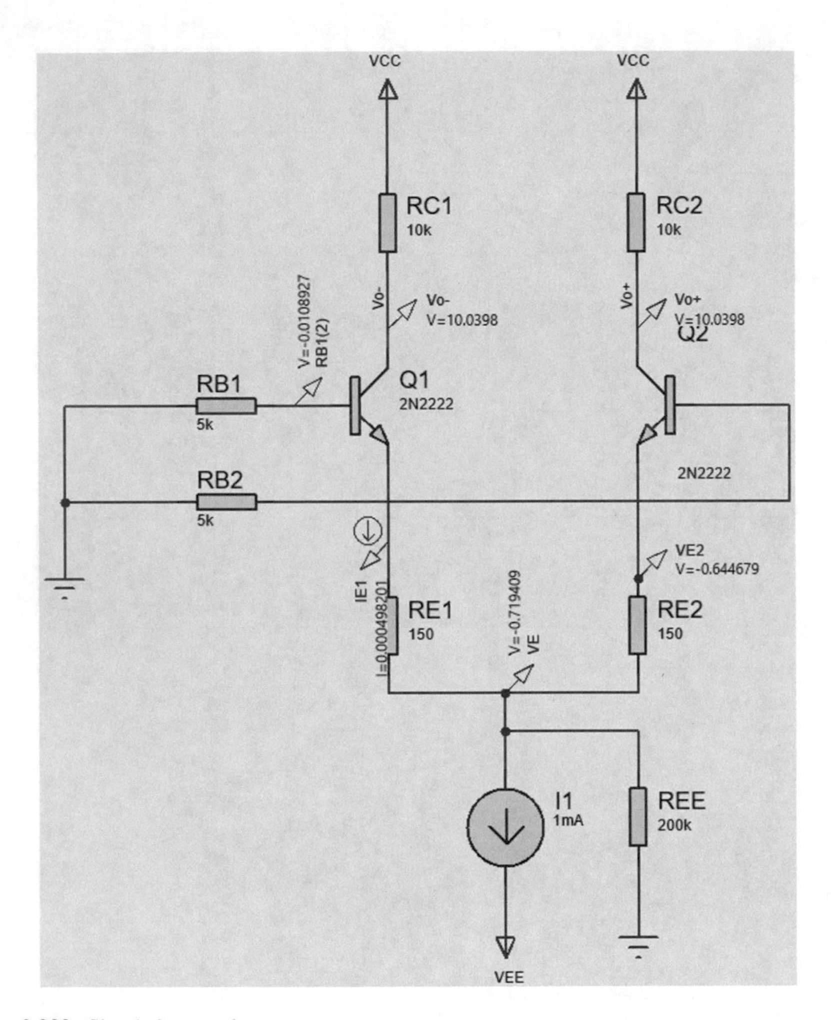

Fig. 2.229 Simulation result

Let's measure the differential mode gain of the circuit. Change the schematic to what is shown in Fig. 2.230. Settings of $V_{\text{in}1}$ and $V_{\text{in}2}$ are shown in Figs. 2.231 and 2.232, respectively. Note that $V_{\text{in}1}$ and $V_{\text{in}2}$ are out of phase. Voltage controlled voltage sources E_1 and E_2 act as voltage sensor and measure input $(V_{\text{in}1} - V_{\text{in}2})$ and output $(V_{o+} - V_{o-})$, respectively.

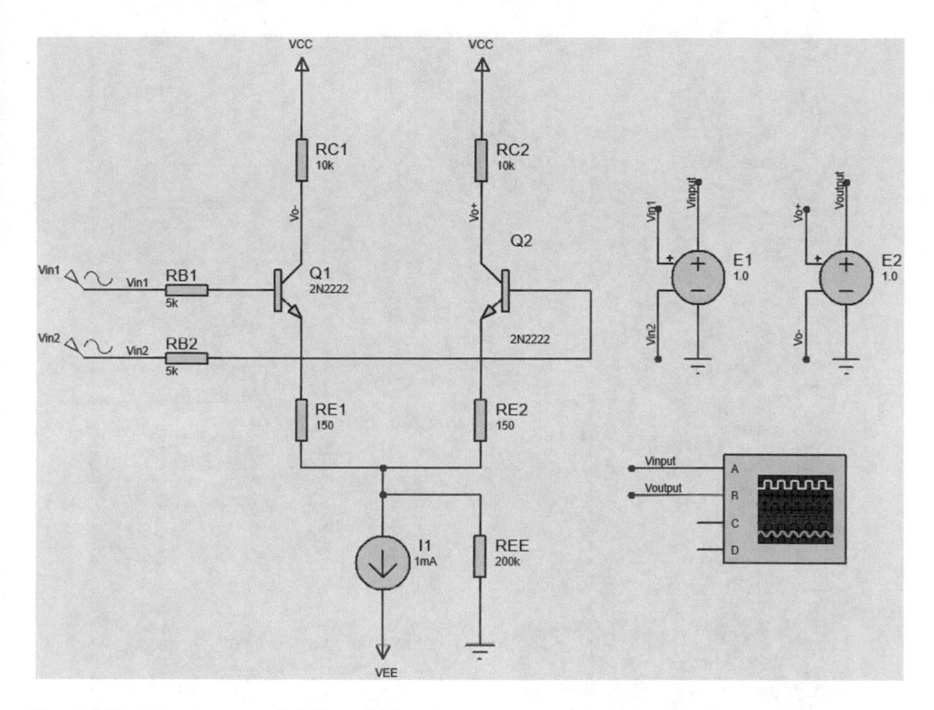

Fig. 2.230 Measurement of differential mode gain

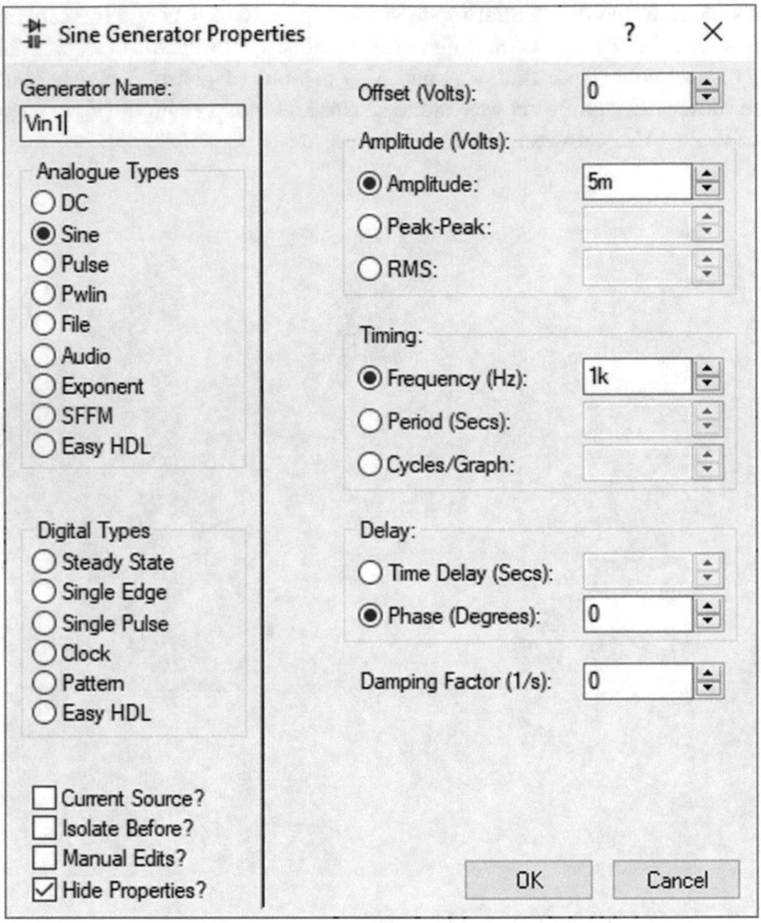

Fig. 2.231 Settings of V_{in1}

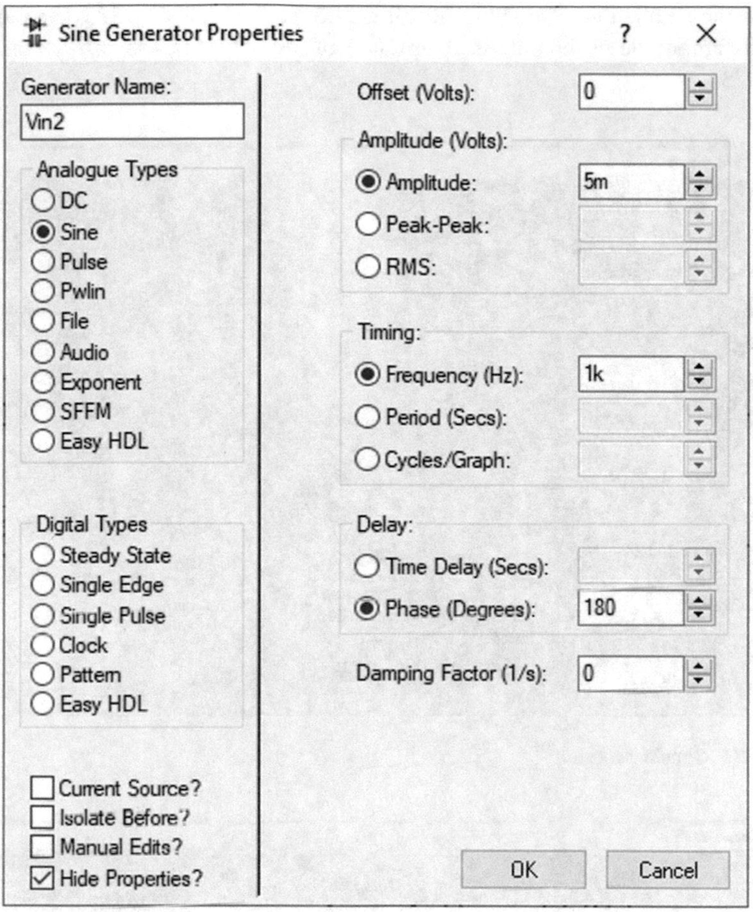

Fig. 2.232 Settings of V_{in2}

Run the simulation. Simulation result is shown in Fig. 2.233. Use the cursor to read the amplitude of the output. Amplitude of the output is 435 mV according to Fig. 2.234.

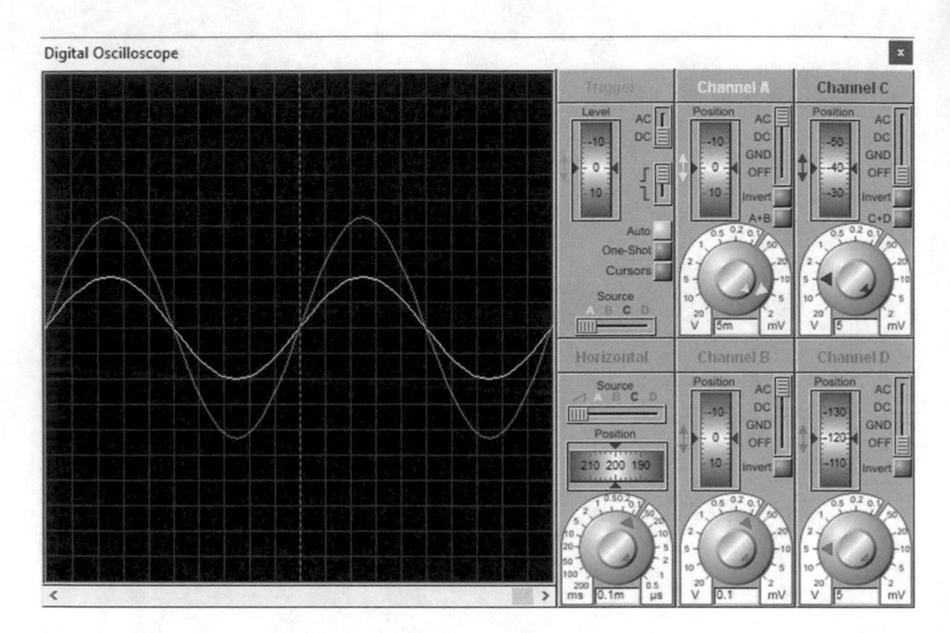

Fig. 2.233 Simulation result

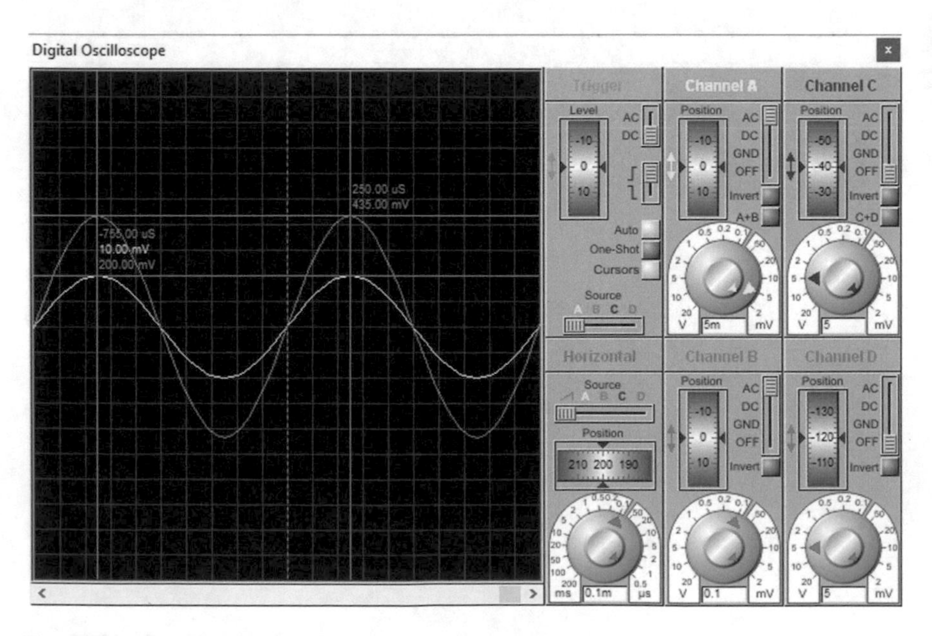

Fig. 2.234 Measurement of peak values

The differential mode gain is 43.5 according to the calculation in Fig. 2.235.

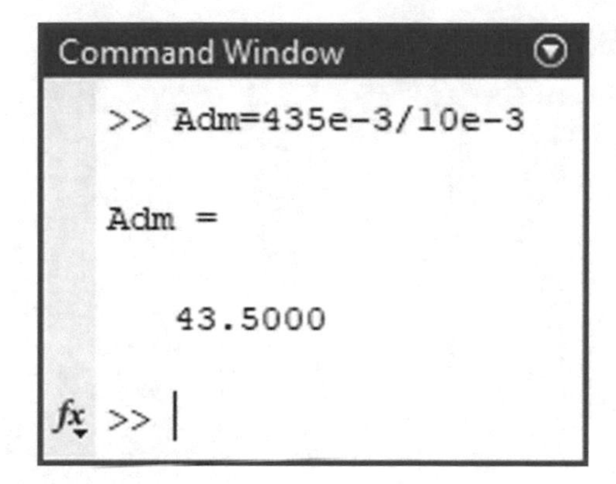

Fig. 2.235 MATLAB calculation

Let's measure the worst-case common mode gain with the assumption that the two collector resistances are accurate to within ±1%. Double click the resistor RC_1 and change it to 9.9 kΩ. Double click the resistor RC_2 and change it to 10.1 kΩ (Fig. 2.236). Settings of V_{in} is shown in Fig. 2.237.

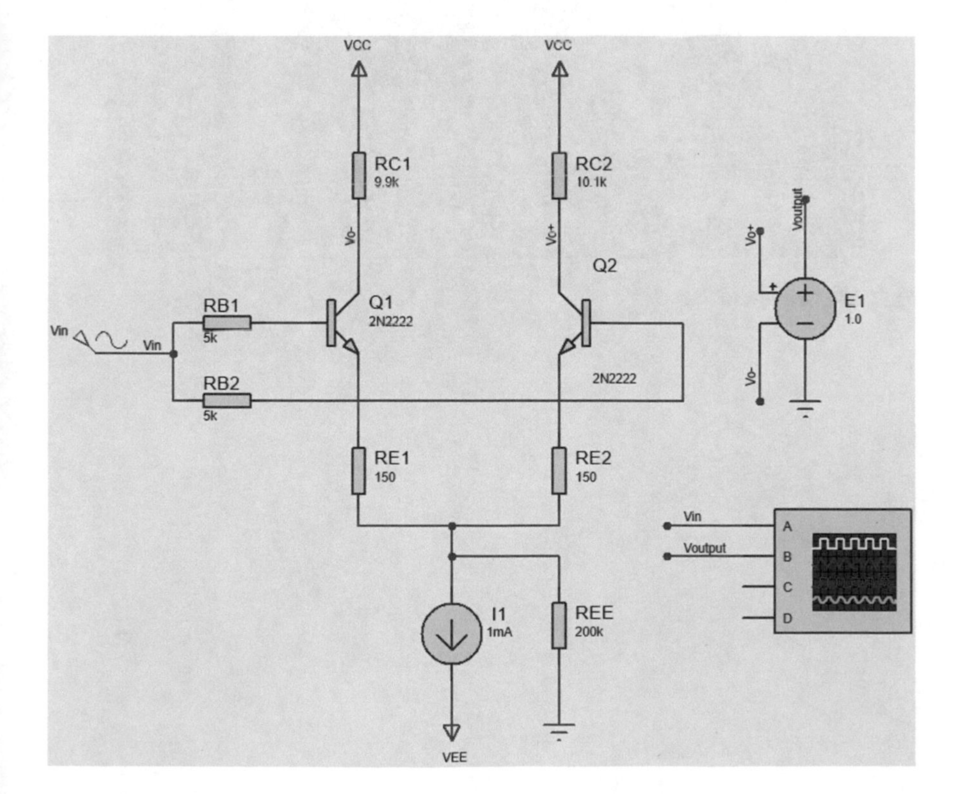

Fig. 2.236 Measurement of worst-case common mode gain

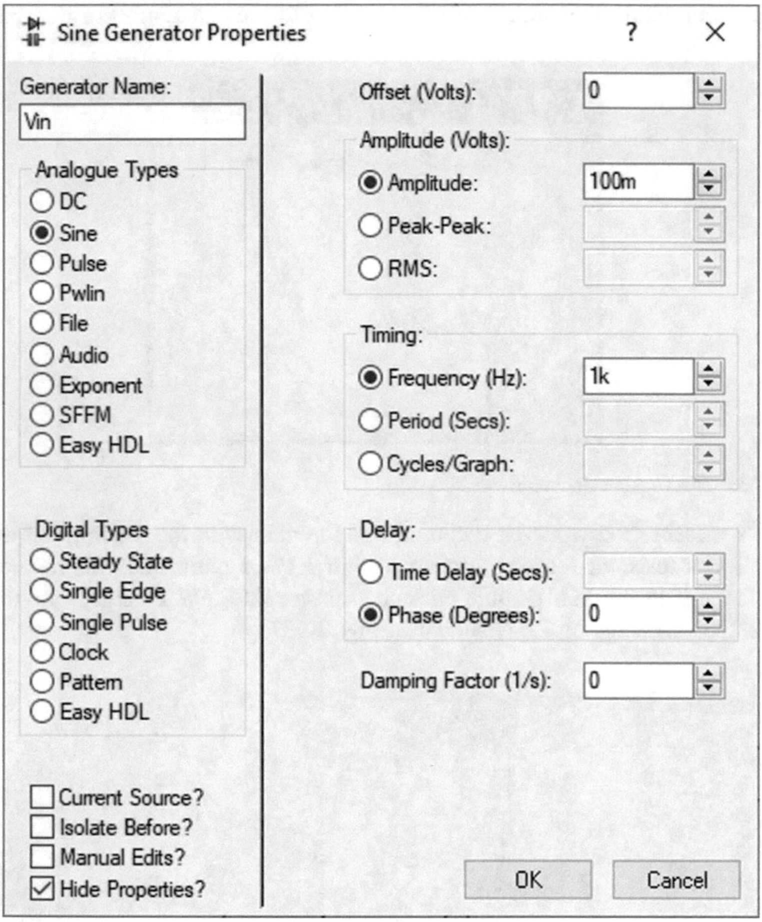

Fig. 2.237 Settings of V_{in}

Run the simulation. Simulation result is shown in Fig. 2.238. Note that output is very small and we can't measure it. Let's amplify the output signal before entering the oscilloscope. This can be done easily by double clicking the voltage controlled voltage source E_1 and increasing the value of Voltage Gain to 100 (Fig. 2.239).

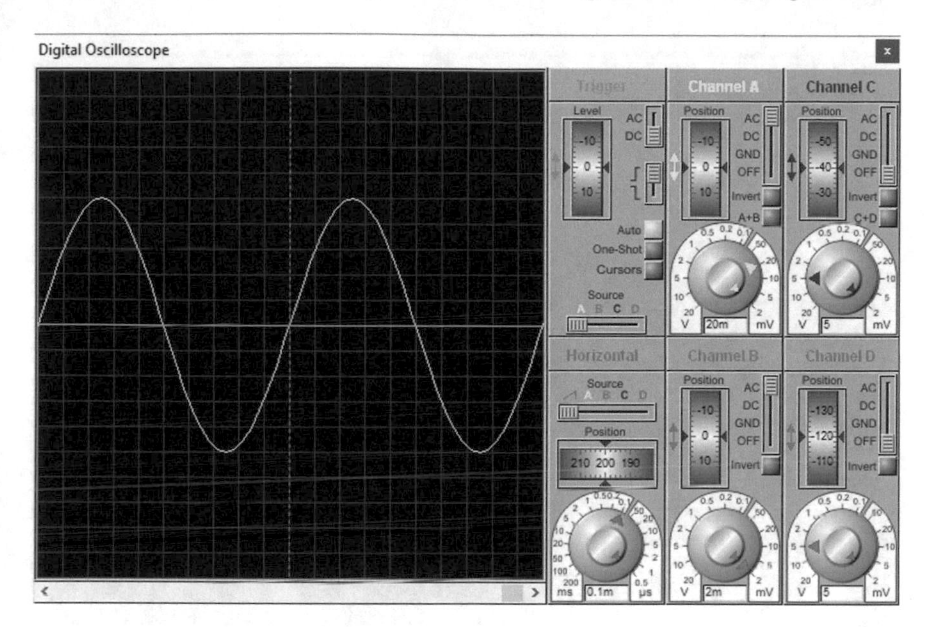

Fig. 2.238 Simulation result

Edit Component

Part Reference: E1 Hidden: ☐

Voltage Gain: 100 Hidden: ☐

Element: New

Other Properties:

OK

Help

Cancel

☐ Exclude from Simulation

☐ Exclude from PCB Layout

☐ Exclude from Current Variant

☐ Attach hierarchy module

☐ Hide common pins

☐ Edit all properties as text

Fig. 2.239 Settings of E_1

Run the simulation. Simulation result is shown in Fig. 2.240. Now you can do the measurement. Amplitude of the output is 4.5 mV according to Fig. 2.241. So, the amplitude of the signal that entered the voltage controlled voltage source E_1 is $\frac{4.5\,\text{mV}}{100} = 45\ \mu\text{V}$.

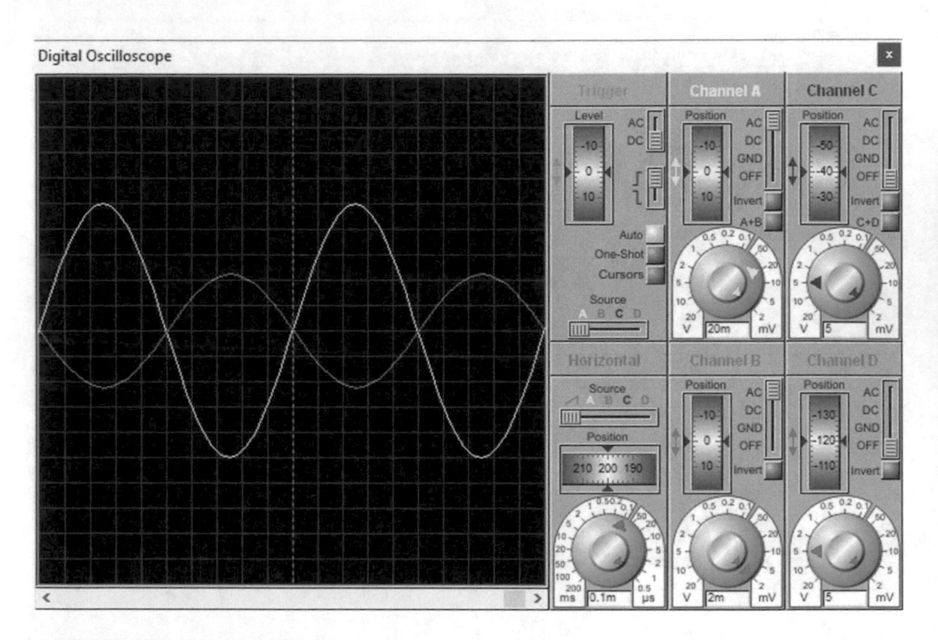

Fig. 2.240 Simulation result

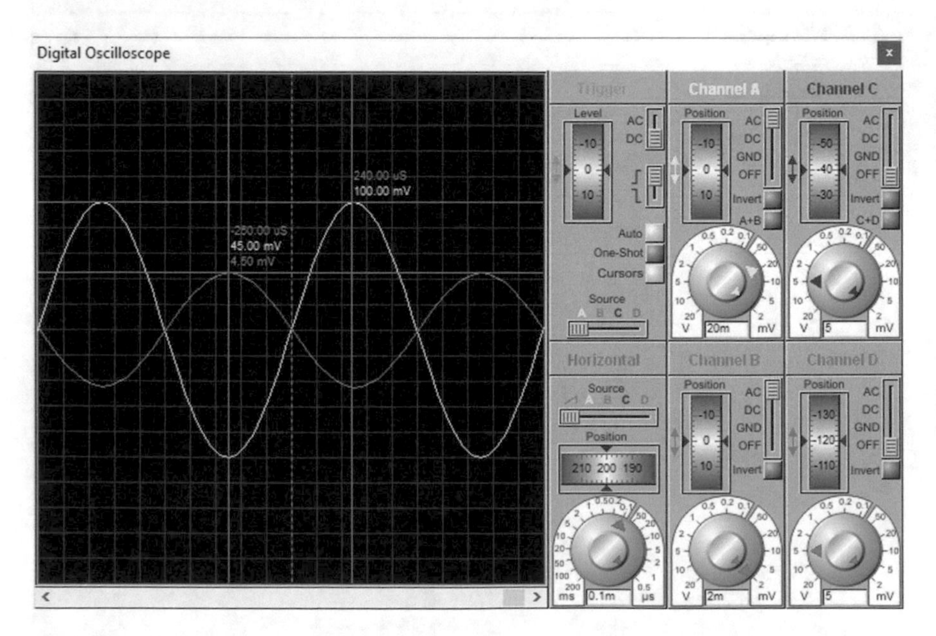

Fig. 2.241 Measurement of peak values

The worst-case common mode gain is 4.5×10^{-4} according to Fig. 2.242. Note that this value shows the maximum value of common mode gain for 1% resistors. If more accurate resistors are used, then the worst common mode gain decreases.

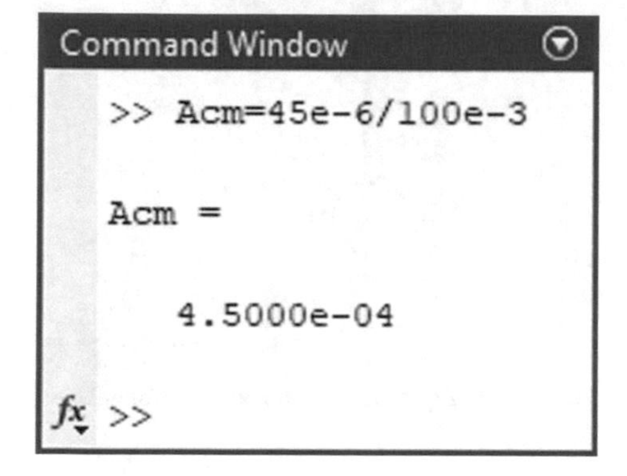

Fig. 2.242 MATLAB calculation

Let's calculate the CMRR. CMRR is 99.7055 dB according to the calculation in Fig. 2.243.

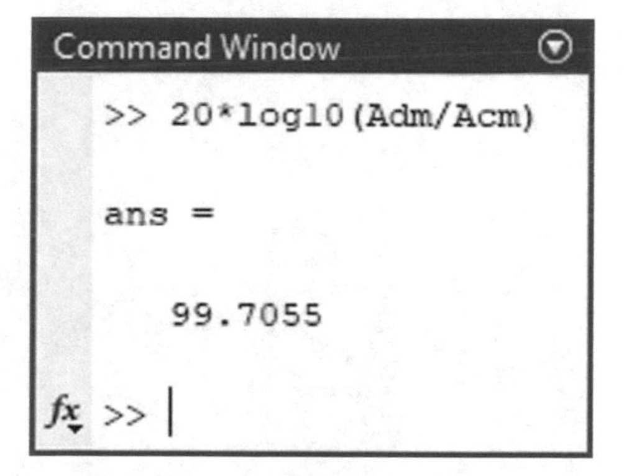

Fig. 2.243 MATLAB calculation

2.33 Example 32: Measurement of Differential Mode Input Impedance of Differential Pair

In this example, we want to measure the differential mode input impedance of previous example. The equivalent circuit for differential inputs is shown in Fig. 2.244.

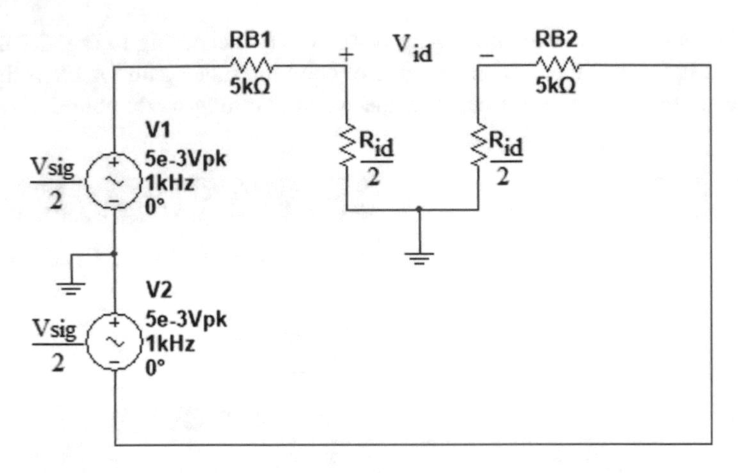

Fig. 2.244 Equivalent circuit for differential mode input impedance of Example 31

Draw the schematic shown in Fig. 2.245. Settings of V_{in1} and V_{in2} are shown in Figs. 2.246 and 2.247, respectively.

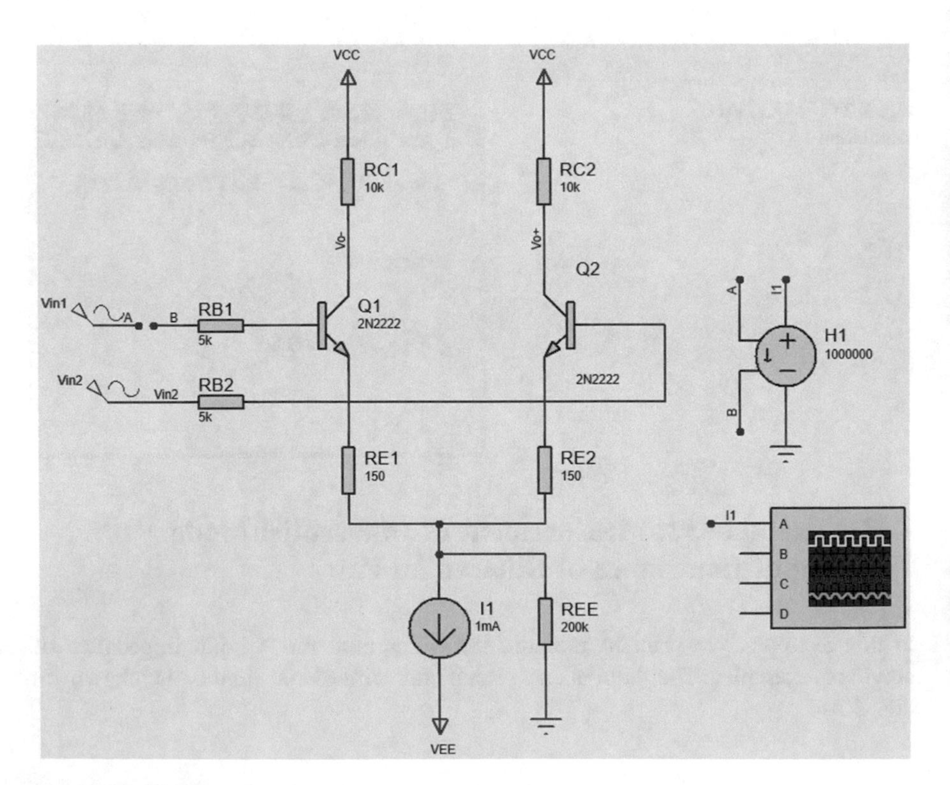

Fig. 2.245 Schematic for Example 32

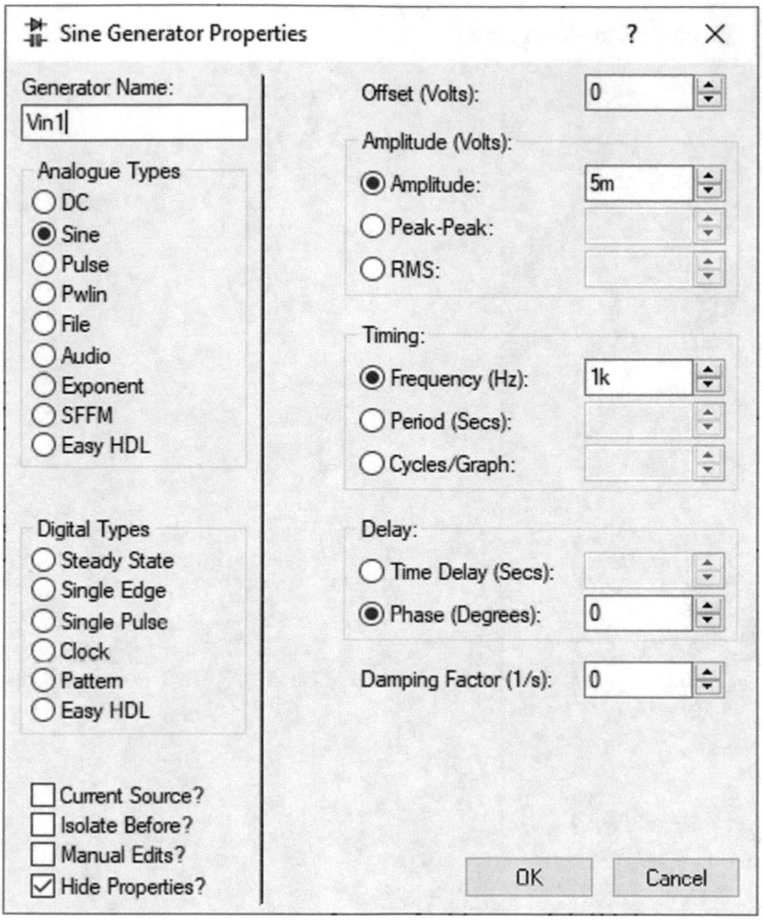

Sine Generator Properties

Generator Name:	Vin1
Analogue Types	DC, (•) Sine, Pulse, Pwlin, File, Audio, Exponent, SFFM, Easy HDL
Digital Types	Steady State, Single Edge, Single Pulse, Clock, Pattern, Easy HDL
Current Source?	☐
Isolate Before?	☐
Manual Edits?	☐
Hide Properties?	☑
Offset (Volts):	0
Amplitude (Volts): (•) Amplitude:	5m
Peak-Peak:	
RMS:	
Timing: (•) Frequency (Hz):	1k
Period (Secs):	
Cycles/Graph:	
Delay: Time Delay (Secs):	
(•) Phase (Degrees):	0
Damping Factor (1/s):	0

OK Cancel

Fig. 2.246 Settings of V_{in1}

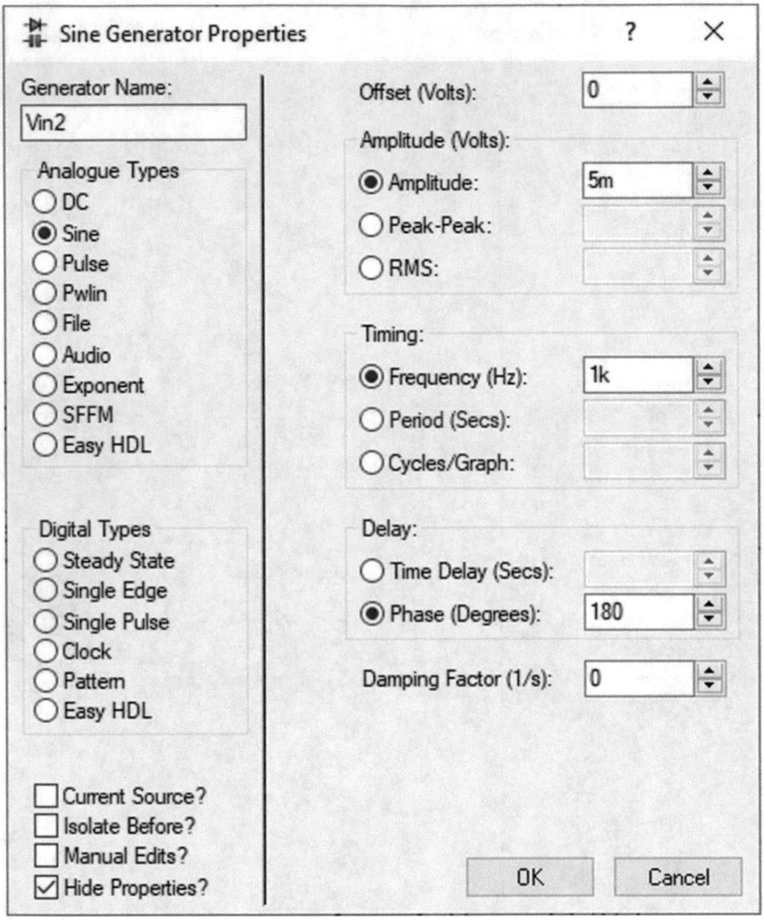

Sine Generator Properties ? X

Generator Name:
Vin2

Analogue Types
DC
Sine
Pulse
Pwlin
File
Audio
Exponent
SFFM
Easy HDL

Digital Types
Steady State
Single Edge
Single Pulse
Clock
Pattern
Easy HDL

Current Source?
Isolate Before?
Manual Edits?
Hide Properties?

Offset (Volts): 0

Amplitude (Volts):
Amplitude: 5m
Peak-Peak:
RMS:

Timing:
Frequency (Hz): 1k
Period (Secs):
Cycles/Graph:

Delay:
Time Delay (Secs):
Phase (Degrees): 180

Damping Factor (1/s): 0

OK Cancel

Fig. 2.247 Settings of V_{in2}

Run the simulation. Simulation result is shown in Fig. 2.248. Select the AC coupling for channel *A* and decrease the Volt/Div of channel *A* to see the AC component better (Fig. 2.249).

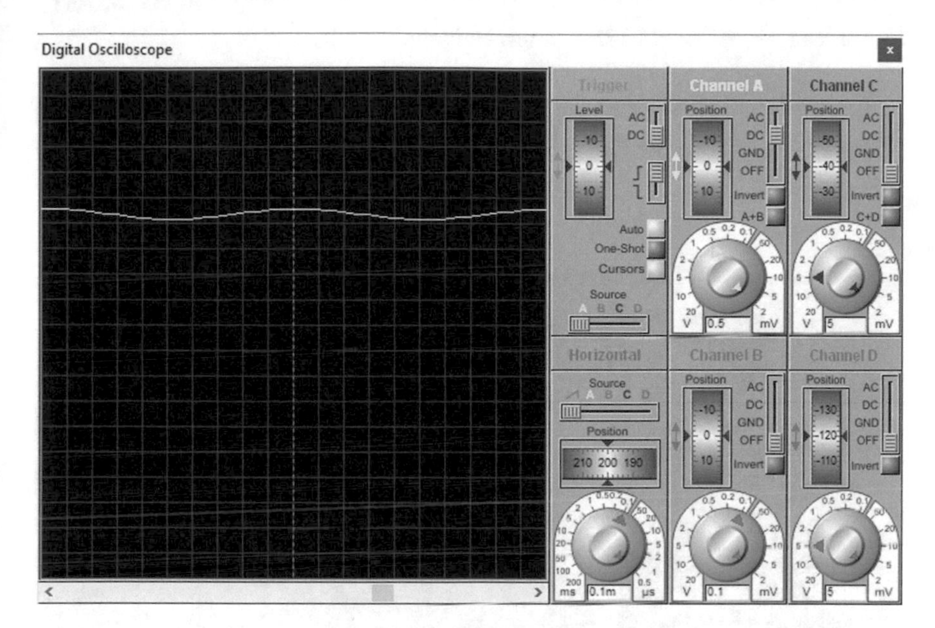

Fig. 2.248 Simulation result

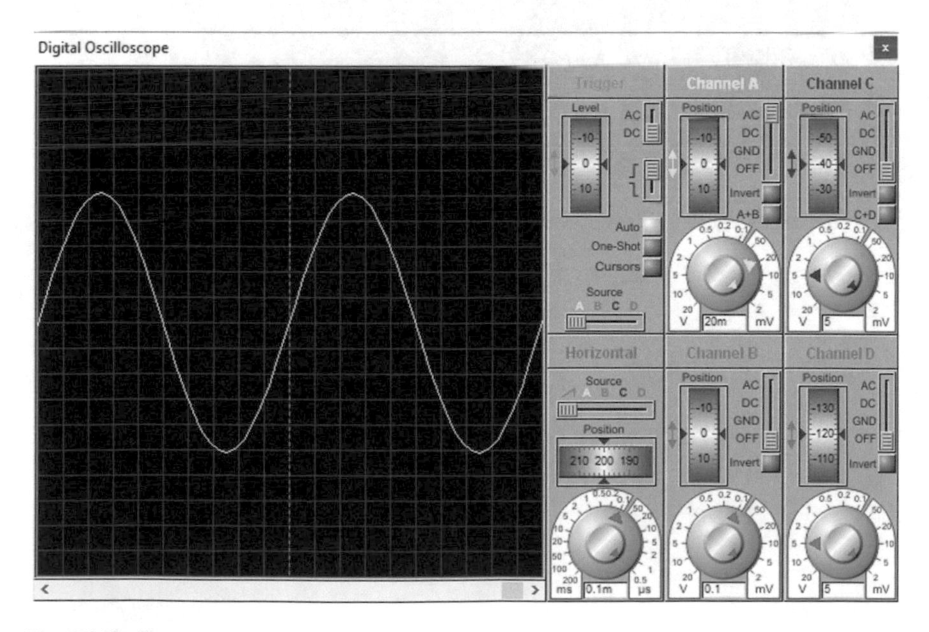

Fig. 2.249 Simulation result

Use the cursor to measure to measure the amplitude of the signal. Amplitude of the signal is 102 mV according to Fig. 2.250. Transimpedance of H_1 is 1000000 Ω. Therefore, amplitude and RMS values of the current drawn from V_{in1} is $\frac{102\ \text{mV}}{1{,}000{,}000\ \Omega} = 102\ \ \text{nA}$ and $\frac{102\ \text{nA}}{\sqrt{2}} = 72.125\ \text{nA}$, respectively.

The resistance seen by V_{in1} is $\frac{\frac{5\ \text{mV}}{\sqrt{2}}}{72.125\ \text{nA}} = 49.02\ k\Omega$. According to Fig. 2.244, the resistance seen by V_{in1} is $R_{B1} + \frac{R_{\text{id}}}{2} = 5\ \text{k}\Omega + \frac{R_{\text{id}}}{2}$. Therefore, $R_{\text{id}} = 88.039\ \text{k}\Omega$.

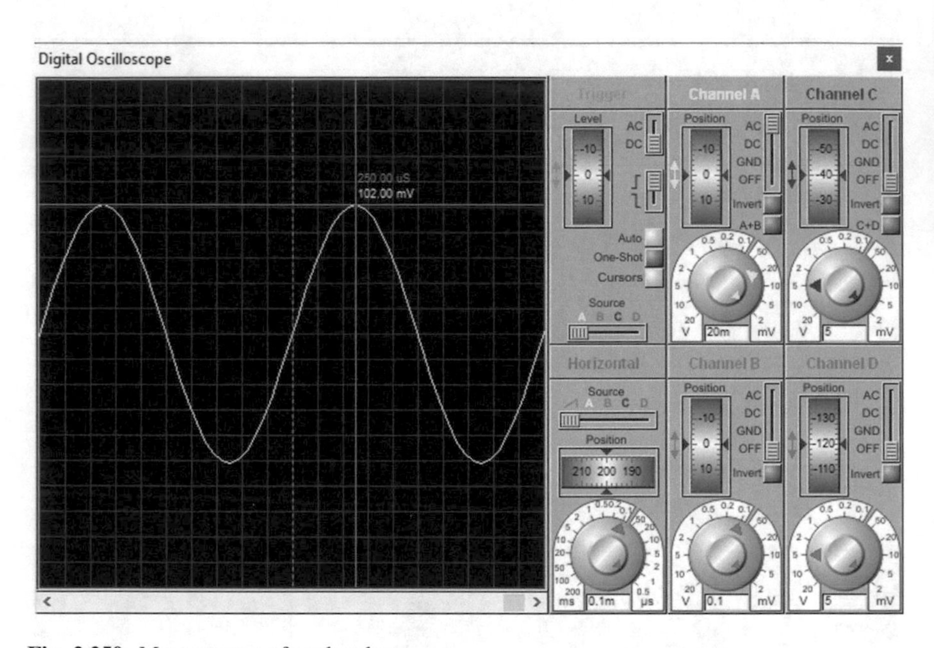

Fig. 2.250 Measurement of peak value

2.34 Example 33: Astable Oscillator with 555

In this example, we want to simulate an astable oscillator with 555 IC. Draw the schematic shown in Fig. 2.251. 555 IC can be added to the schematic by searching for "555" in the Pick Devices window (Fig. 2.252).

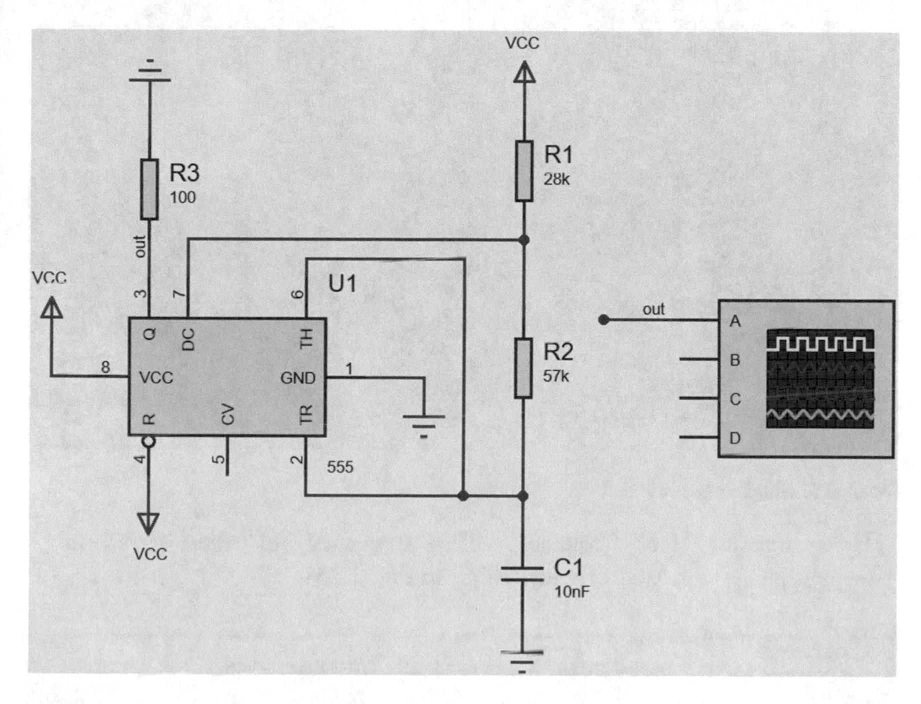

Fig. 2.251 Schematic of Example 33

Fig. 2.252 Search for 555 IC

Pick Devices

Keywords:

555

Match whole words?

Show only parts with models?

Run the simulation. Simulation result is shown in Fig. 2.253.

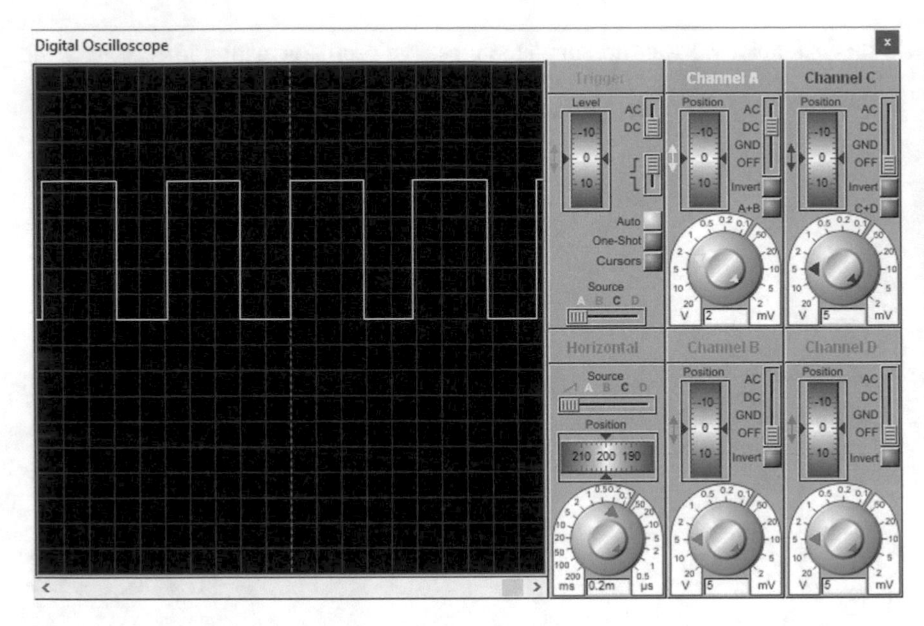

Fig. 2.253 Simulation result

Let's measure the frequency. The frequency of the waveform is $\frac{1}{-950\ \mu\text{S}-(-1.93\ \text{ms})}=1.0204$ kHz according to Fig. 2.254.

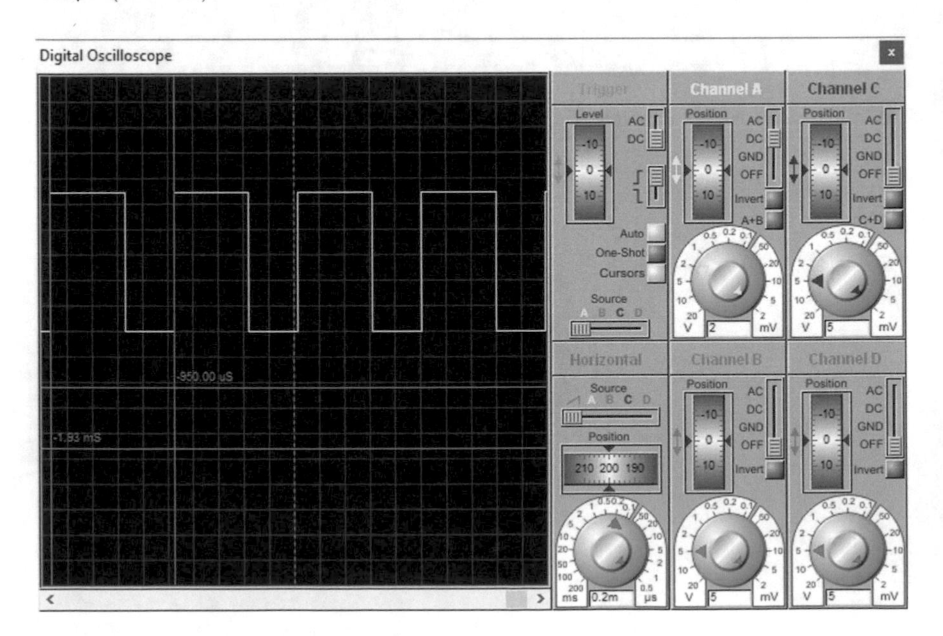

Fig. 2.254 Measurement of period

Width of high portion of waveform is 620 μs–30 μs = 590 μs according to Fig. 2.255.

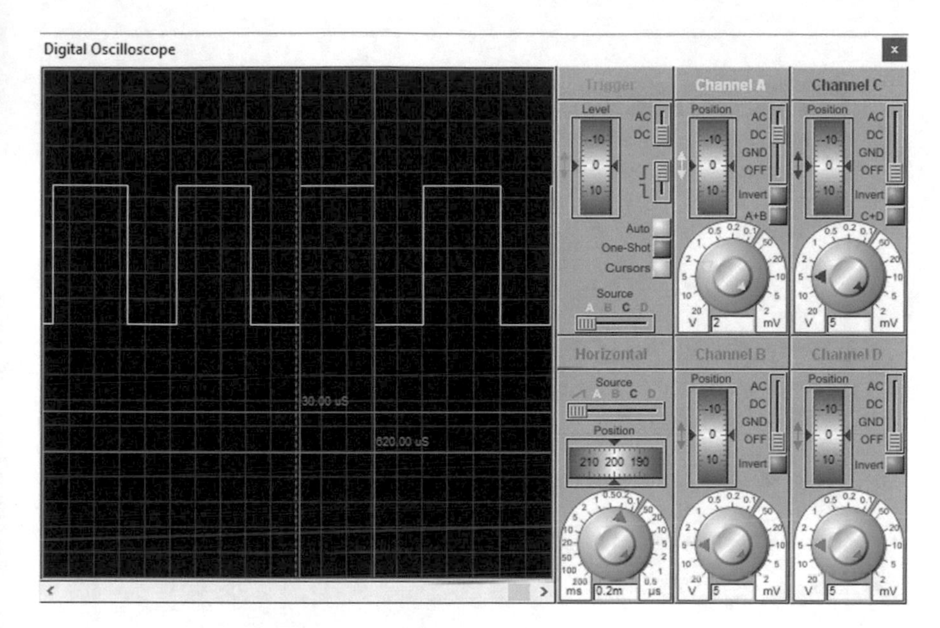

Fig. 2.255 Measurement of duration of high portion of signal

Width of high portion of waveform is 1.01 ms–620 μs = 390 μs according to Fig. 2.256.

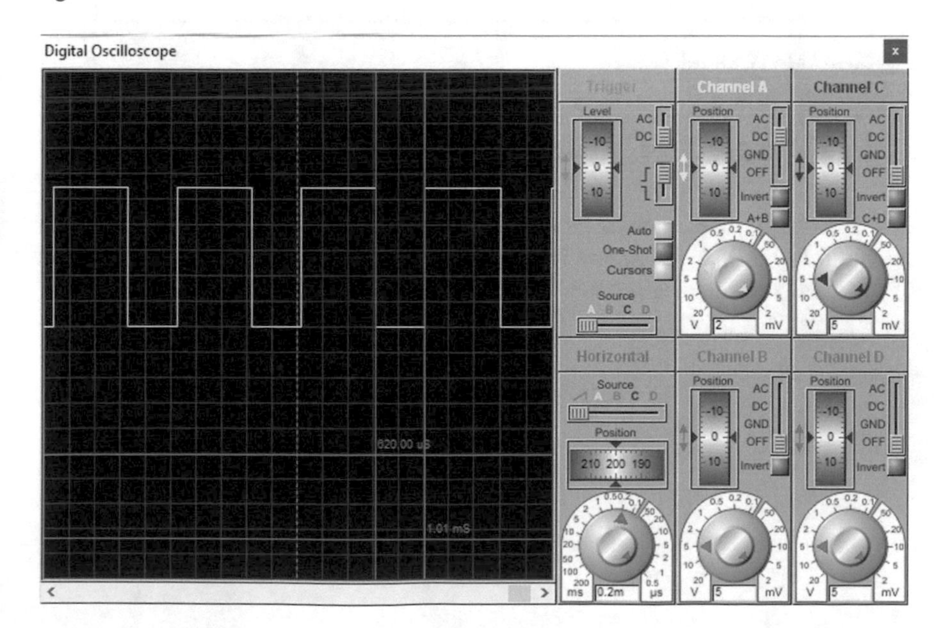

Fig. 2.256 Measurement of duration of low portion of signal

Let's check the obtained results. The calculation shown in Fig. 2.257 shows that Proteus results are correct.

```
Command Window
>> R1=28e3;R2=57e3;C=10e-9;
>> HighDuration_us=0.693*(R1+R2)*C/1e-6

HighDuration_us =

  589.0500

>> LowDuration_us=0.693*R2*C/1e-6

LowDuration_us =

  395.0100

>> frequency=1/((HighDuration_us+LowDuration_us)*1e-6)

frequency =

   1.0162e+03

fx >> |
```

Fig. 2.257 MATLAB calculations

2.35 Example 34: Colpitts Oscillator

In this example, we want to simulate a Colpitts oscillator. Draw the schematic shown in Fig. 2.258. According to the calculation shown in Fig. 2.259, frequency of this oscillator must be around 71.18 kHz.

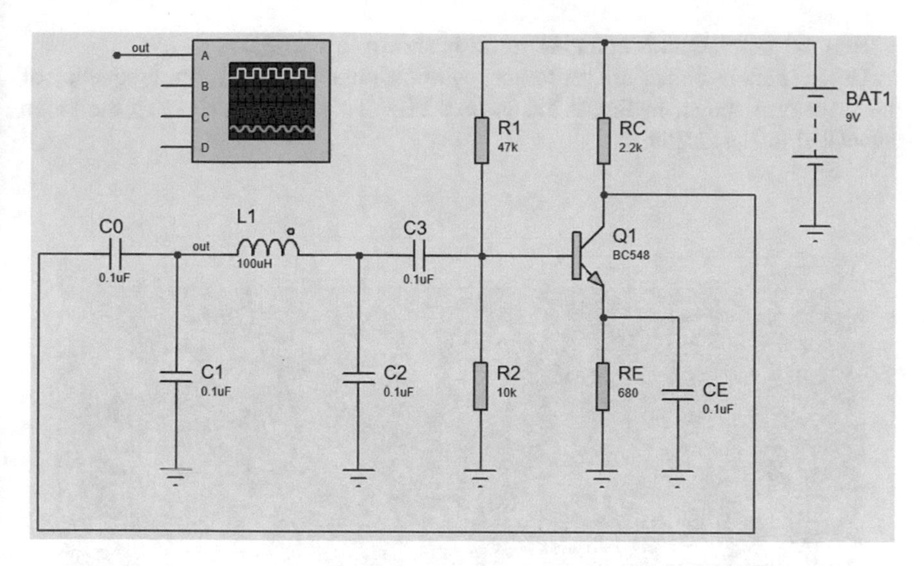

Fig. 2.258 Schematic for Example 34

Fig. 2.259 MATLAB calculations

```
Command Window
>> L=100e-6;C1=0.1e-6;C2=0.1e-6;
>> Ceq=C1*C2/(C1+C2);
>> f=1/(2*pi*sqrt(L*Ceq))

f =

   7.1176e+04

fx >>
```

Run the simulation. Simulation result is shown in Fig. 2.260.

Use the cursor to measure the frequency of the obtained waveform. Frequency of the waveform shown in Fig. 2.260 is 68.5 kHz. This value is close to the value calculated in Fig. 2.259.

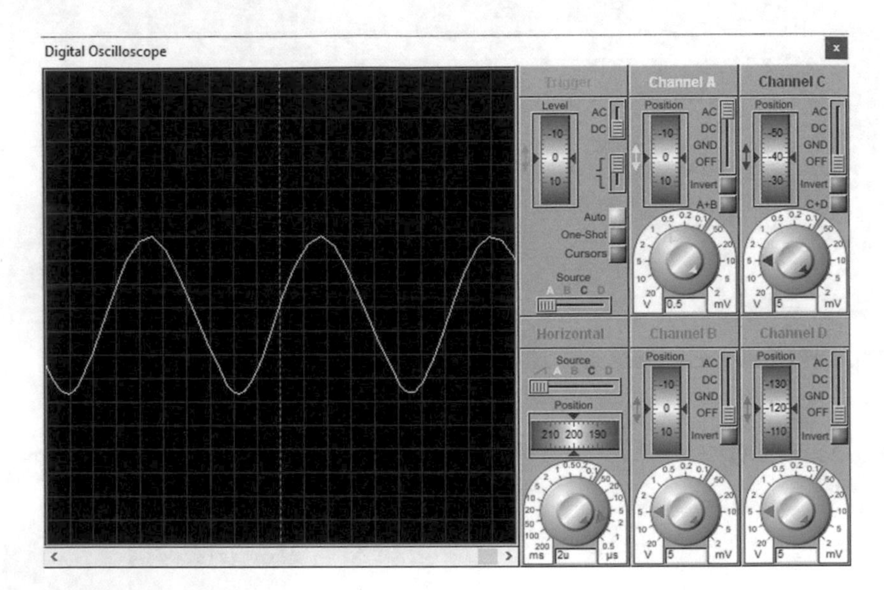

Fig. 2.260 Simulation result

2.36 Example 35: Total Harmonic Distortion (THD) of Colpitts Oscillator

In this example, we want to measure the THD of Colpitts oscillator of previous example. Add a Fourier graph to the schematic of previous example (Fig. 2.261). Settings of Fourier graph are shown in Fig. 2.262.

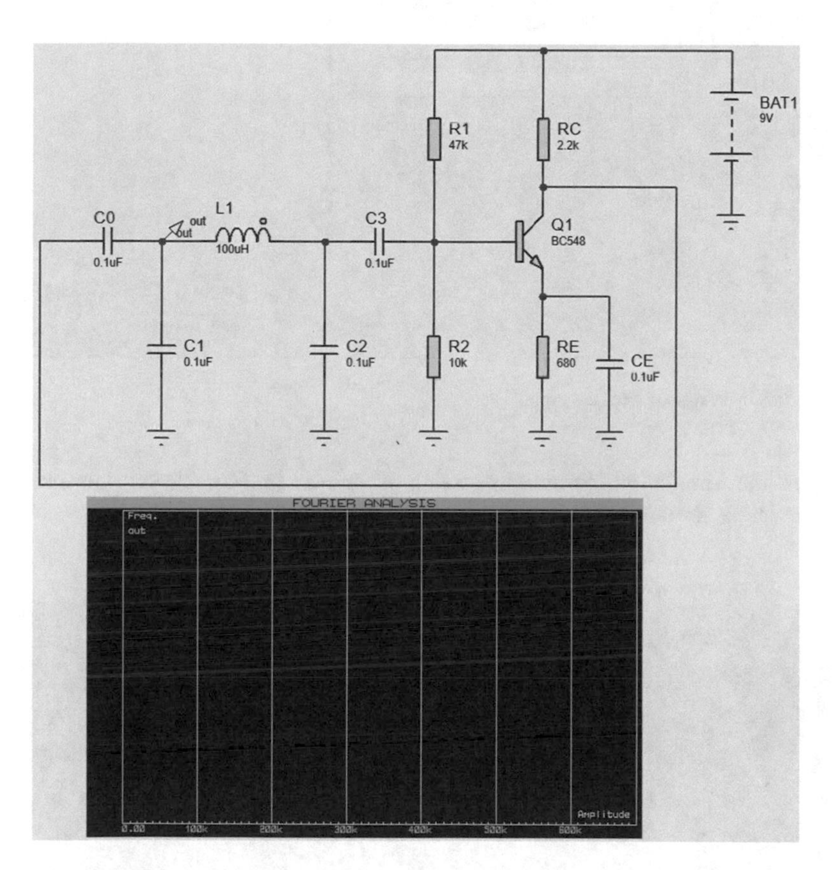

Fig. 2.261 Schematic for Example 35

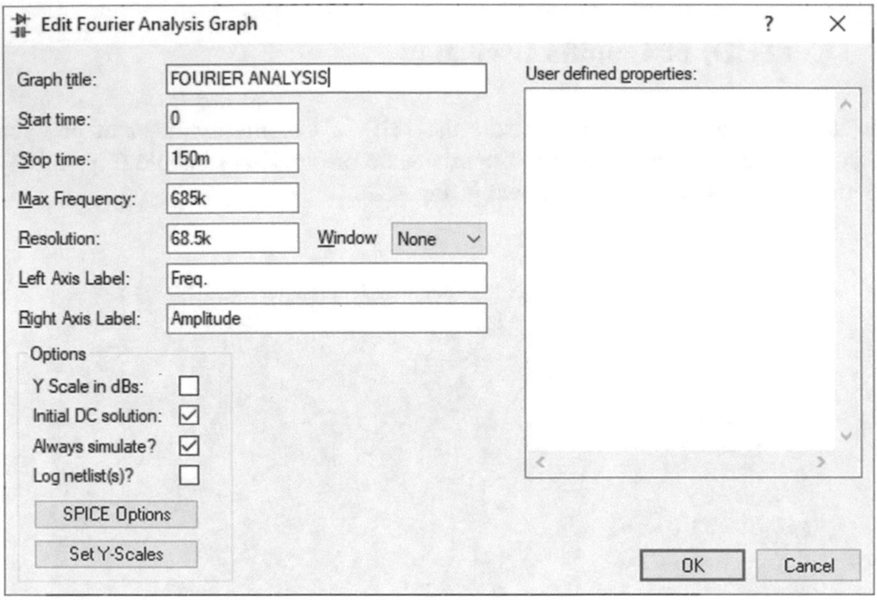

Fig. 2.262 Settings of Fourier graph

Run the simulation. Simulation result is shown in Fig. 2.263. Amplitude of harmonics is shown in Table 2.5.

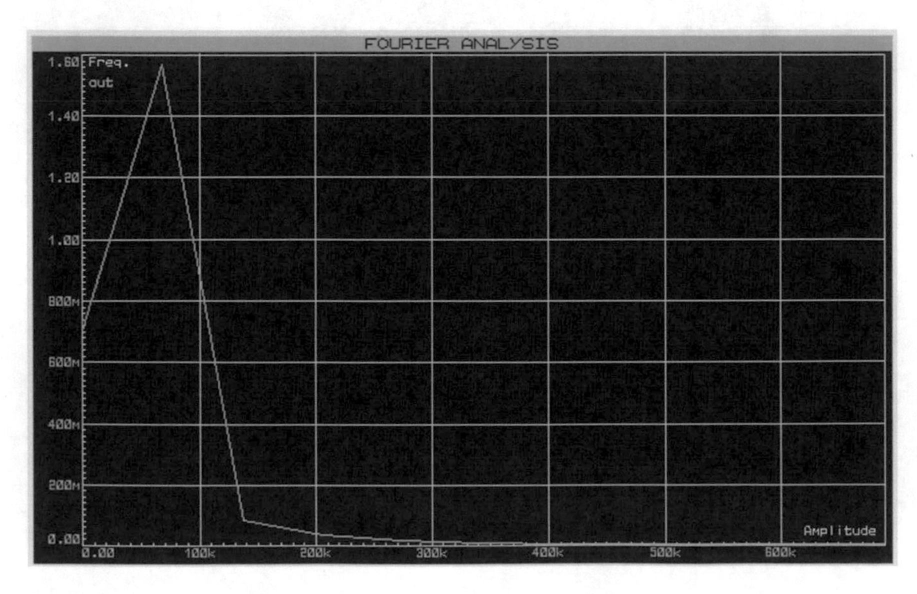

Fig. 2.263 Simulation result

Table 2.5 Amplitude of harmonics in Fig. 2.263

Harmonic number	Amplitude (mV)
DC	700
Fundamental	1570
2nd	86.4
3rd	37.1
4th	16.8
5th	7.75
6th	4.51
7th	3.55
8th	3.24
9	2.31
10	2.59

THD can be calculated with the aid of $\frac{\sqrt{\sum_{h=2}^{N} V_{h,\mathrm{RMS}}^2}}{V_{1,\mathrm{RMS}}}$ formula. THD is 0.0612 or 6.12% according to the calculation shown in Fig. 2.264.

```
Command Window
>> V1RMS=1570e-3/sqrt(2);
>> VhRMS=[86.4 37.1 16.8 7.75 4.51 3.55 3.24 2.31 2.59]*1e-3/sqrt(2);
>> THD=sqrt(VhRMS*VhRMS')/V1RMS

THD =

    0.0612

fx >>
```

Fig. 2.264 MATLAB commands

2.37 Example 36: Wien Bridge Oscillator

In this example, we want to simulate a Wien bridge oscillator. Draw the schematic shown in Fig. 2.265 (VDD = 5 V and VEE = −5 V). According to the calculation shown in Fig. 2.266, frequency of this oscillator must be around 1.59 kHz.

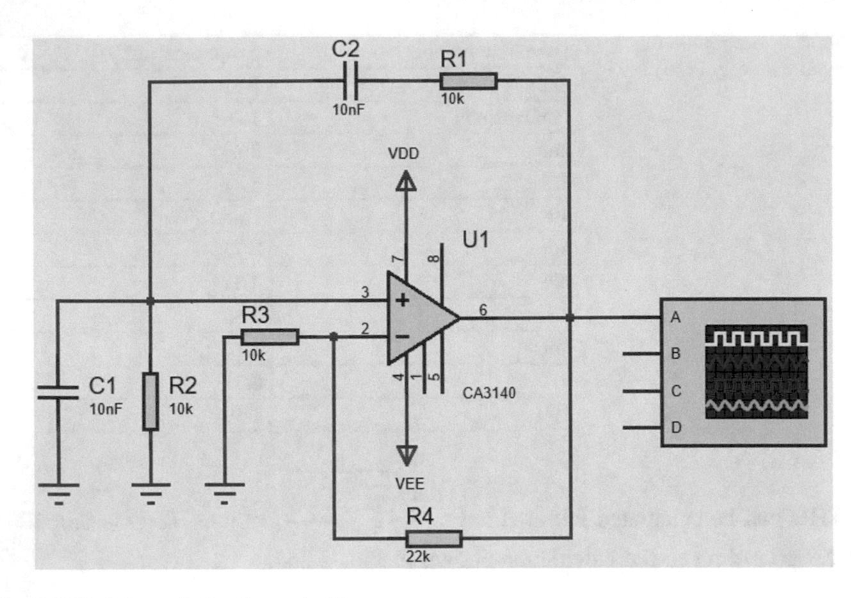

Fig. 2.265 Schematic for Example 36

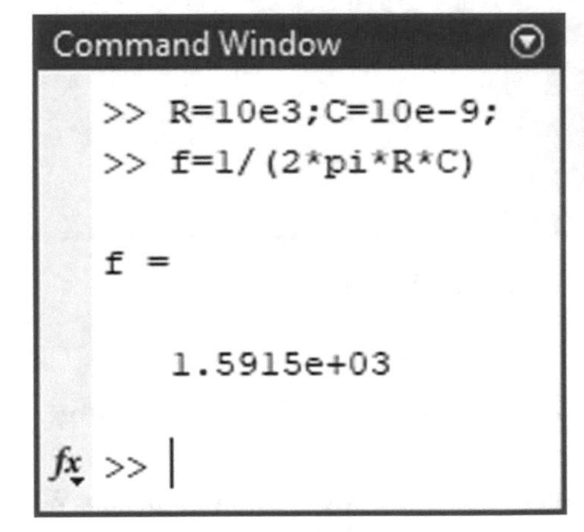

Fig. 2.266 MATLAB calculations

Run the simulation. Simulation result is shown in Fig. 2.267. Note that there is no oscillation in the output. The problem can be solved by addition of initial condition to the capacitor in the feedback path.

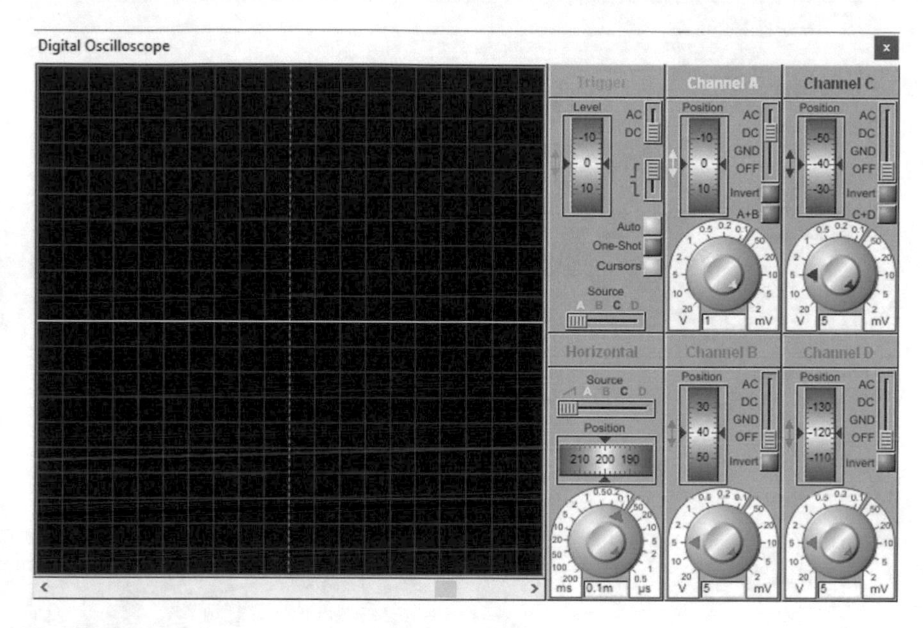

Fig. 2.267 Simulation result

Double click the capacitor C_2 and add the "PRECHARGE = 0" to the Other Properties box (Fig. 2.268) and click the OK button.

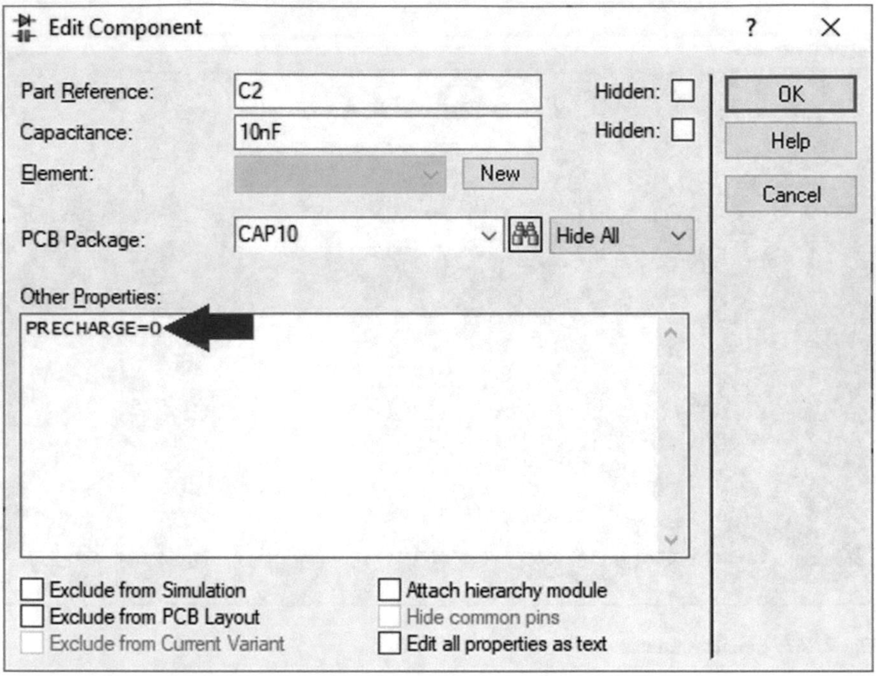

Edit Component ? X

Part Reference: C2 Hidden: ☐ OK

Capacitance: 10nF Hidden: ☐ Help

Element: New Cancel

PCB Package: CAP10 Hide All

Other Properties:

PRECHARGE=0

☐ Exclude from Simulation ☐ Attach hierarchy module

☐ Exclude from PCB Layout ☐ Hide common pins

☐ Exclude from Current Variant ☐ Edit all properties as text

Fig. 2.268 Settings of capacitor C_2

Run the simulation. Simulation result is shown in Fig. 2.269.

Use the cursor to measure the frequency of the obtained waveform. Frequency of the waveform shown in Fig. 2.269 is 1.54 kHz. This value is quite close to the value calculated in Fig. 2.266.

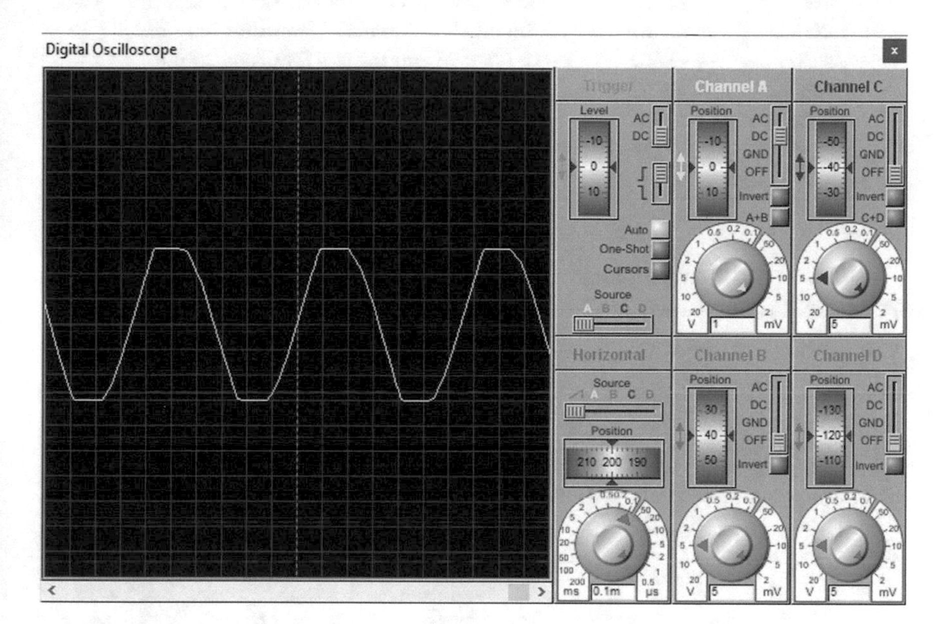

Fig. 2.269 Simulation result

2.38 Example 37: Optocoupler Block

Optocoupler or optoisolator is an electronic component that transfers electrical signals between two isolated circuits by using light. When you search for “optoisolator” in the Pick Devices window (Fig. 2.270), the results shown in Fig. 2.271 appear. These are available optoisolators in Proteus environment.

Fig. 2.270 Search for optocoupler

Pick Devices

Keywords:

optoisolator

Match whole words?

Show only parts with models?

Showing local results: 8

Device	Library	Description
4N25	OPTO	6-Pin DIP Optoisolators Transistor Output.
4N26	OPTO	6-Pin DIP Optoisolators Transistor Output.
4N27	OPTO	6-Pin DIP Optoisolators Transistor Output.
4N28	OPTO	6-Pin DIP Optoisolators Transistor Output.
4N35	OPTO	6-Pin DIP Optoisolators Transistor Output.
4N36	OPTO	6-Pin DIP Optoisolators Transistor Output.
4N37	OPTO	6-Pin DIP Optoisolators Transistor Output.
MOC30XX	OPTO	Zero Cross Optoisolators TRIAC Driver Output

Fig. 2.271 Available optocouplers

Let's study an example. Draw the schematic shown in Fig. 2.272 (VCC = 5 V). Settings of V_{in} are shown in Fig. 2.273.

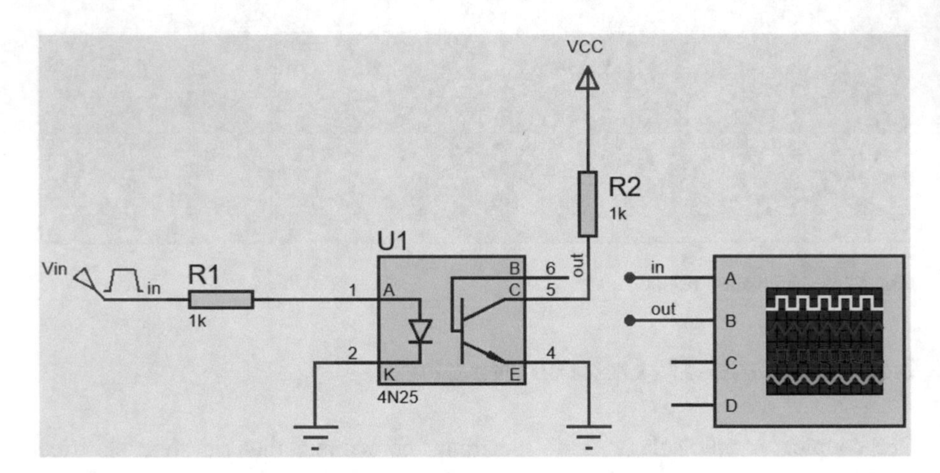

Fig. 2.272 Schematic of Example 37

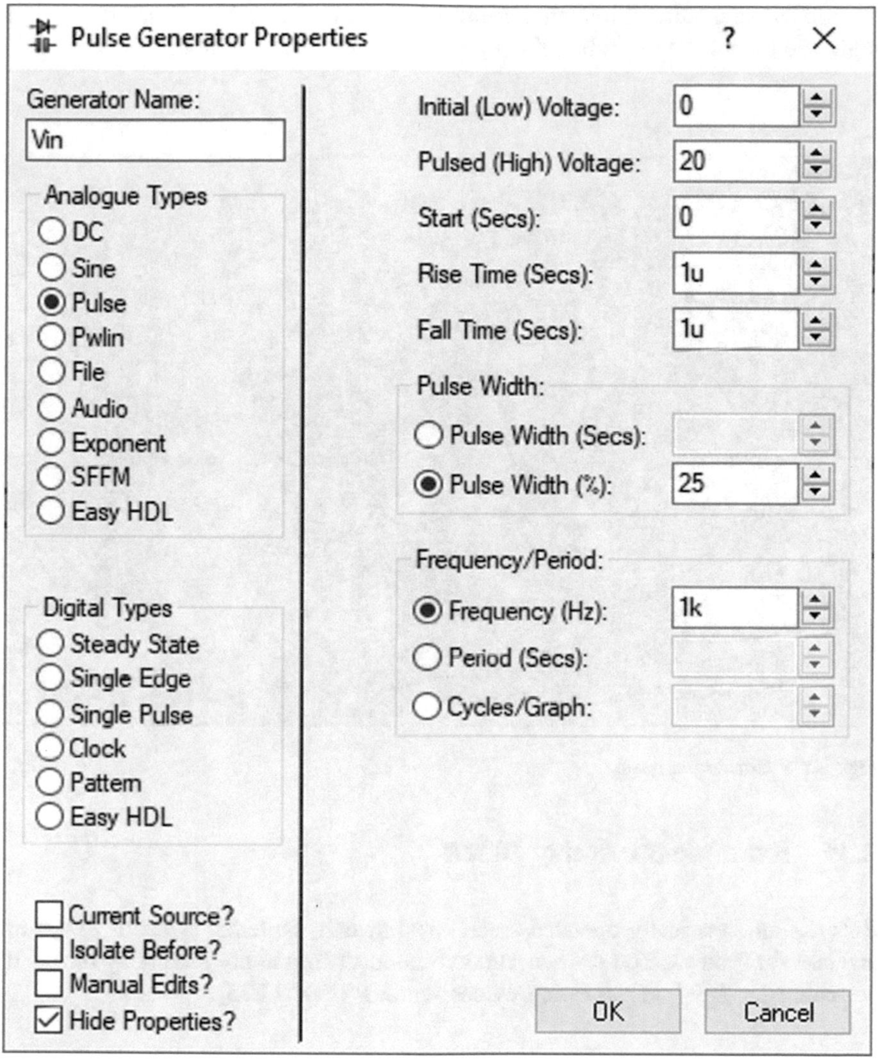

Pulse Generator Properties

Generator Name: Vin

Analogue Types
- DC
- Sine
- Pulse (selected)
- Pwlin
- File
- Audio
- Exponent
- SFFM
- Easy HDL

Digital Types
- Steady State
- Single Edge
- Single Pulse
- Clock
- Pattern
- Easy HDL

- [ ] Current Source?
- [ ] Isolate Before?
- [ ] Manual Edits?
- [x] Hide Properties?

Initial (Low) Voltage:	0
Pulsed (High) Voltage:	20
Start (Secs):	0
Rise Time (Secs):	1u
Fall Time (Secs):	1u

Pulse Width:
- Pulse Width (Secs):
- Pulse Width (%): 25 (selected)

Frequency/Period:
- Frequency (Hz): 1k (selected)
- Period (Secs):
- Cycles/Graph:

OK Cancel

Fig. 2.273 Setting of V_{in}

Run the simulation. Simulation result is shown in Fig. 2.274. When the input is high, the output is low. When the input is low, the output is high.

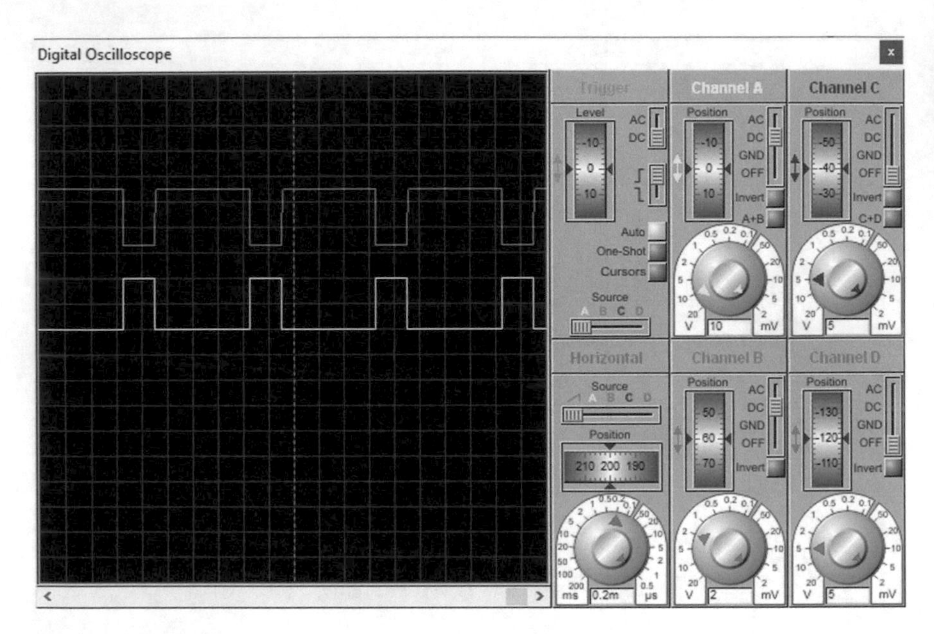

Fig. 2.274 Simulation result

2.39 Example 38: Relay Block

Relay is an electrically operated mechanical switch. Different types of relays are available in Proteus. You can see some of the most commonly used relay blocks by searching for "rly-" in the Pick Devices window (Fig. 2.275).

Fig. 2.275 Search for relay block

Pick Devices

Keywords:

rly-

Match whole words?

Show only parts with models?

Let's study an example. Draw the schematic shown in Fig. 2.276. RL_1 is a "relay" block (Fig. 2.277).

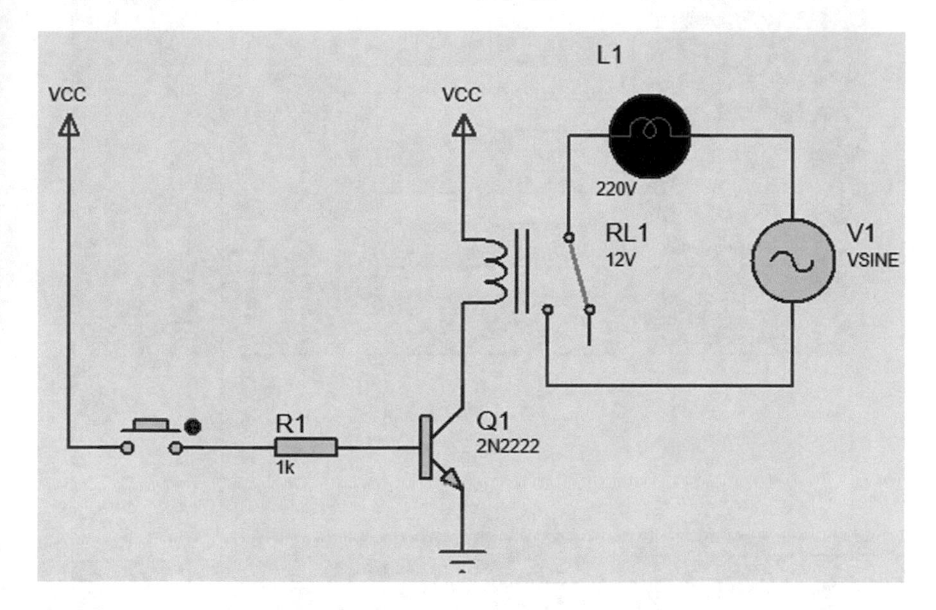

Fig. 2.276 Schematic for Example 38

Fig. 2.277 Search for relay

Pick Devices
Keywords:
relay
Match whole words?
Show only parts with models?

Settings of RL_1, L_1 and V_1 blocks are shown in Fig. 2.278, 2.279 and 2.280, respectively.

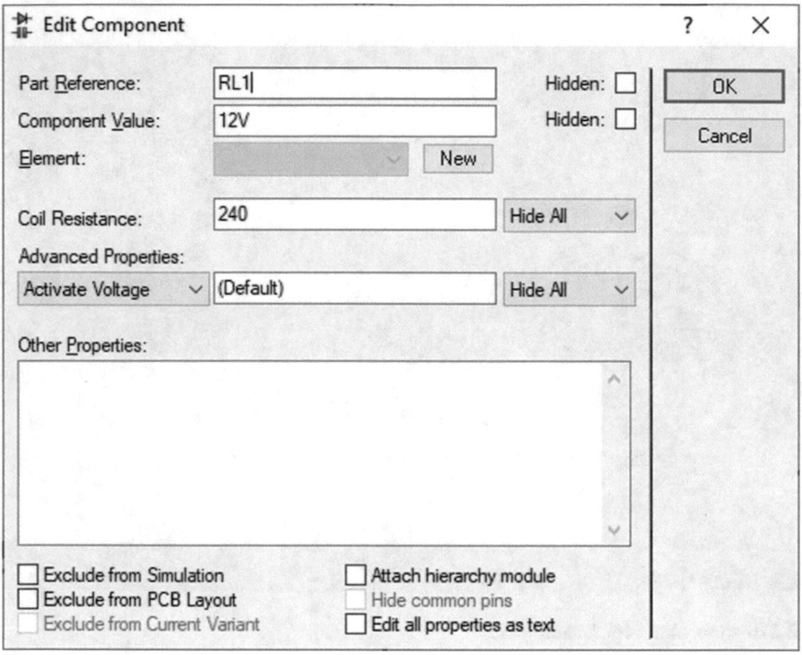

Edit Component ? X

Part Reference: RL1 Hidden: ☐ OK

Component Value: 12V Hidden: ☐ Cancel

Element: New

Coil Resistance: 240 Hide All

Advanced Properties:

Activate Voltage (Default) Hide All

Other Properties:

☐ Exclude from Simulation ☐ Attach hierarchy module

☐ Exclude from PCB Layout ☐ Hide common pins

☐ Exclude from Current Variant ☐ Edit all properties as text

Fig. 2.278 Settings of relay RL_1

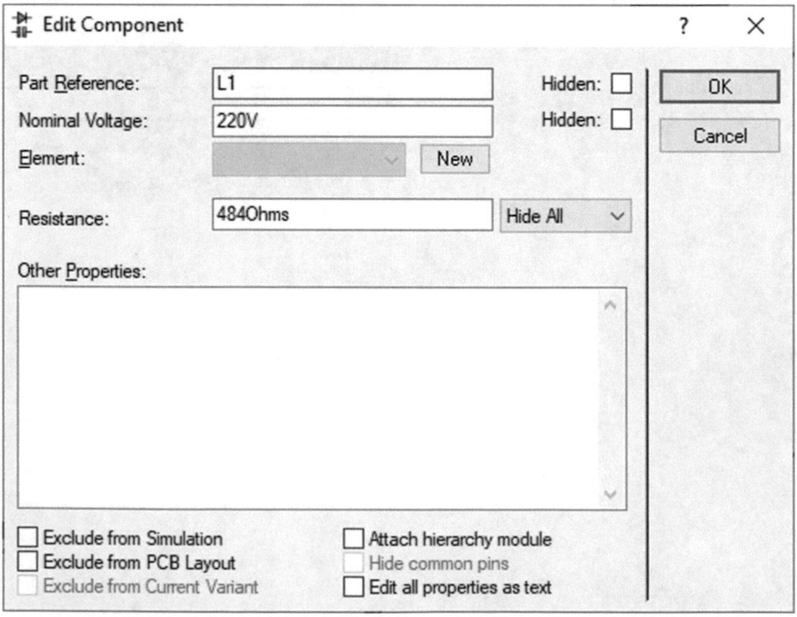

Edit Component ? X

Part Reference: L1 Hidden: ☐ OK

Nominal Voltage: 220V Hidden: ☐ Cancel

Element: New

Resistance: 484Ohms Hide All

Other Properties:

☐ Exclude from Simulation ☐ Attach hierarchy module

☐ Exclude from PCB Layout ☐ Hide common pins

☐ Exclude from Current Variant ☐ Edit all properties as text

Fig. 2.279 Settings of lamp L_1

Edit Component

Part Reference: V1 Hidden: ☐

Part Value: VSINE Hidden: ☐

Element: New

DC Offset: 0 Hide All

Amplitude: 311 Hide All

Frequency: 50 Hide All

Time Delay: 0 Hide All

Damping Factor: 0 Hide All

Other Properties:

☐ Exclude from Simulation ☐ Attach hierarchy module

☐ Exclude from PCB Layout ☐ Hide common pins

☐ Exclude from Current Variant ☐ Edit all properties as text

OK Cancel

Fig. 2.280 Setting of sinusoidal voltage source V_1

Run the simulation. When you press the push button, the lamp turns on (Fig. 2.281). When push button is released, the lamp turns off (Fig. 2.282).

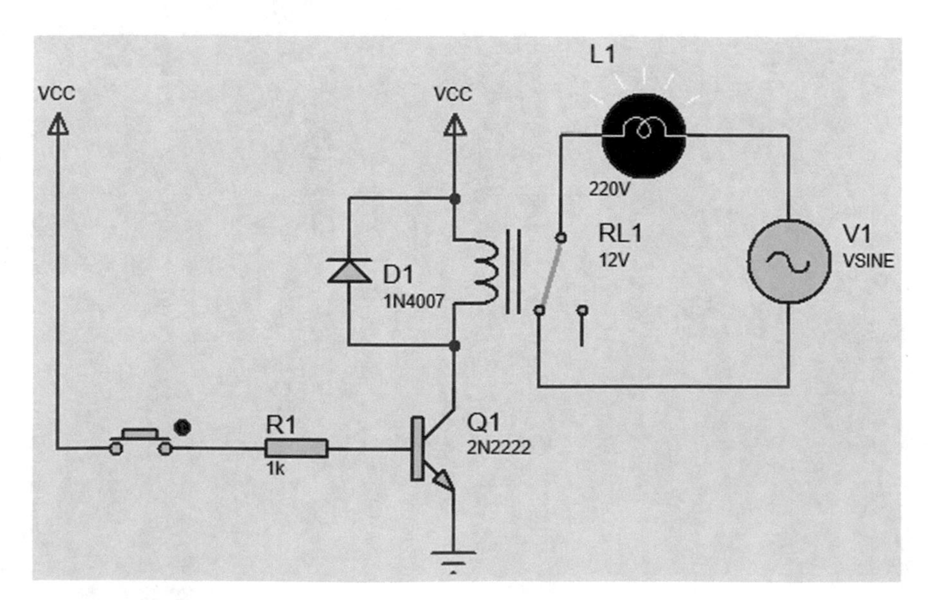

Fig. 2.281 Simulation result (lamp is on)

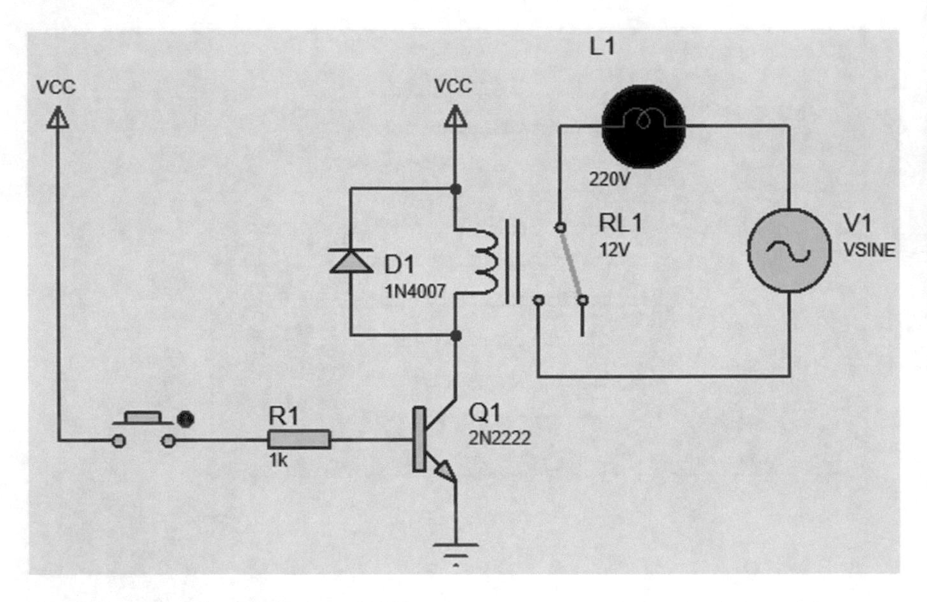

Fig. 2.282 Simulation result (lamp is off)

Note that in real-world circuits you need to add a free-wheeling diode (diode D_1 in Fig. 2.283) to protect the transistor from high voltage that is generated in the coil of relay during the turn off process. In Proteus, the coil of relay is purely resistive and has no inductive nature. Value of relay coil resistance is determined in the "Coil Resistance" box (Fig. 2.278). You can put an inductor in series with the relay coil to simulate the inductive nature of the relay coil.

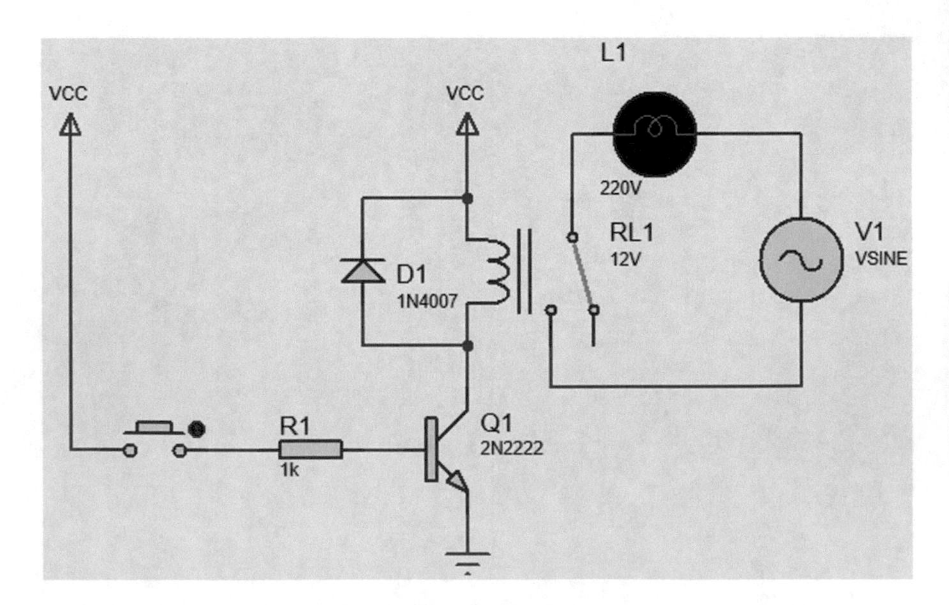

Fig. 2.283 Addition of freewheeling diode to the circuit

2.40 Example 39: Simulation of Control Systems

Closed loop control systems can be simulated in Proteus. Let's study an example. Draw the schematic shown in Fig. 2.284. The blocks that are used in this schematic are shown in Figs. 2.285, 2.286, 2.287 and 2.288. Blocks H_1 and C_1 show the plant and controller, respectively. Note that Proteus use the variable "*p*" to show the Laplace transform complex variable. However, control theory textbooks generally use variable "*s*" to show Laplace transform complex variable.

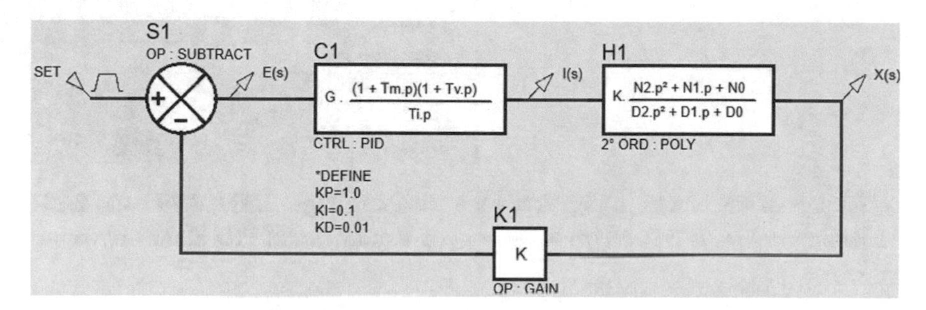

Fig. 2.284 Schematic of Example 39

Fig. 2.285 Search for second-order transfer function

Pick Devices

Keywords:

2 ORD : POLY

Match whole words? ☐

Show only parts with models? ☐

Fig. 2.286 Search for PID controller

Pick Devices

Keywords:

CTRL PID

Match whole words? ☐

Show only parts with models? ☐

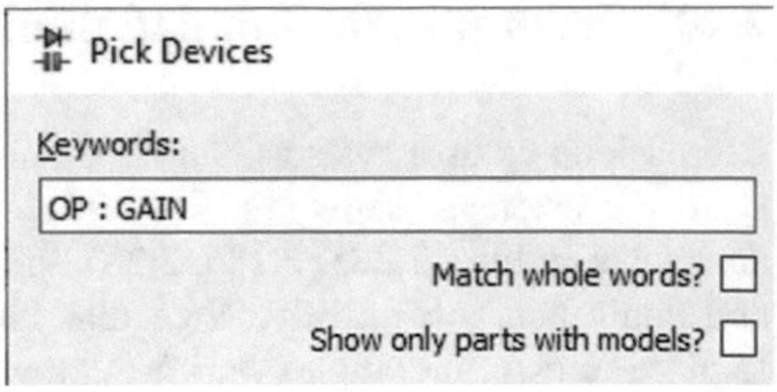

Fig. 2.287 Search for gain block

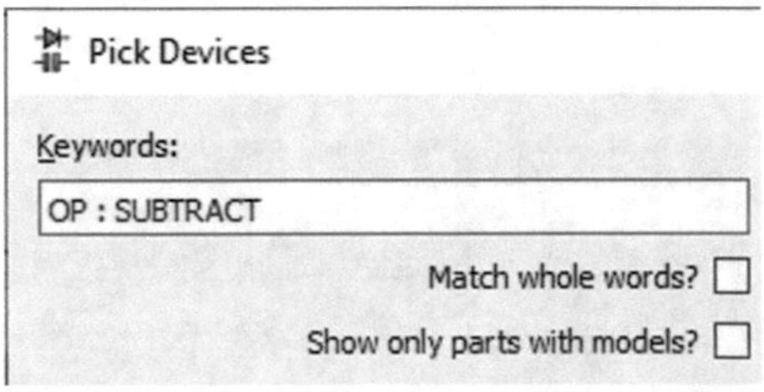

Fig. 2.288 Search for subtract block

Settings of the blocks in Fig. 2.284 are shown in Figs. 2.289, 2.290 and 2.291. According to Fig. 2.289, $H_1(p) = \frac{1}{2p^2 + 8p + 72}$. Parameters of PID block are defined below.

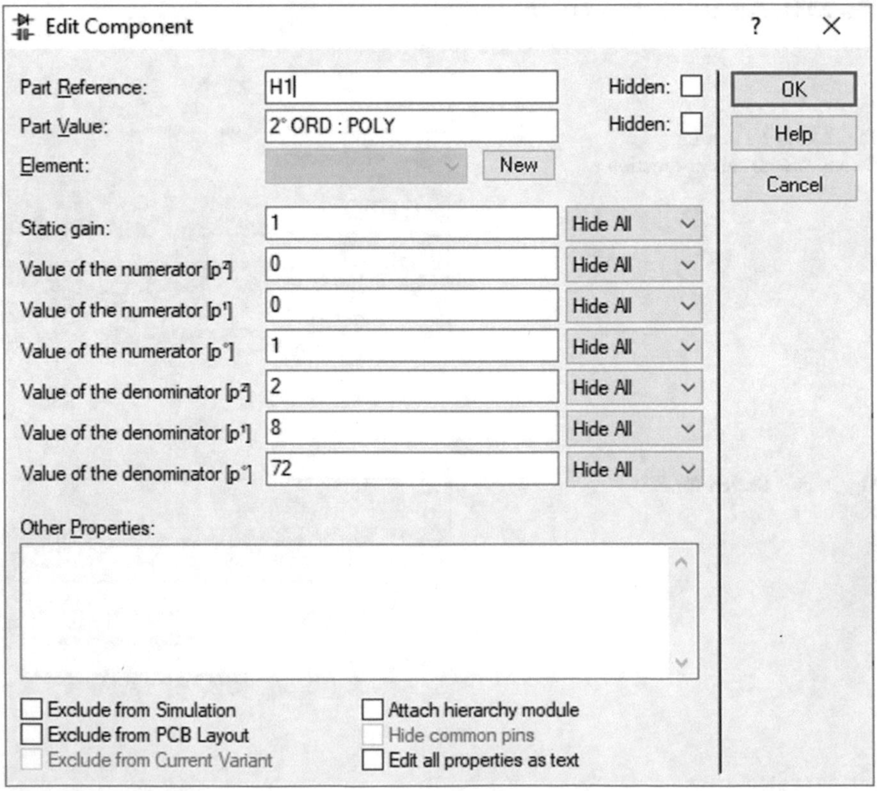

Fig. 2.289 Settings of H_1

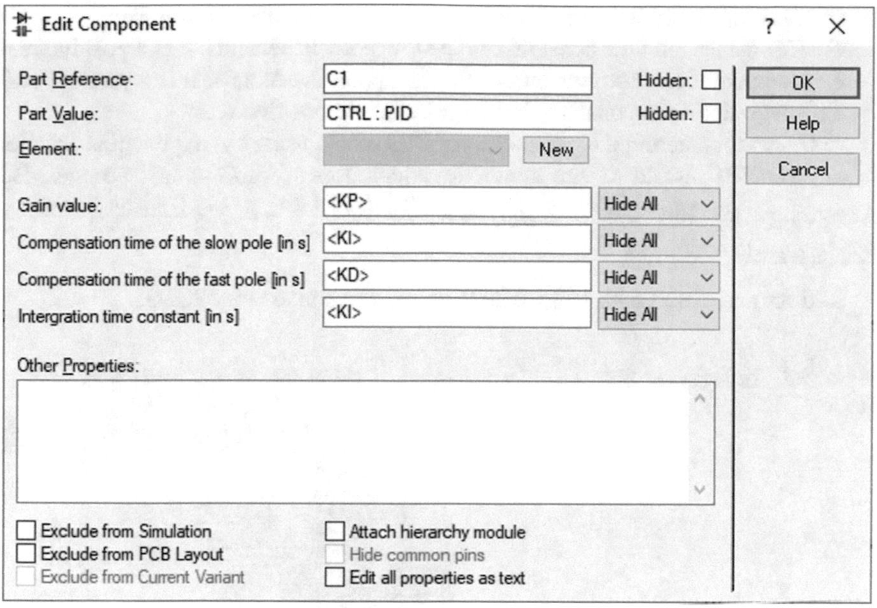

Edit Component

Part Reference: C1 Hidden:

Part Value: CTRL : PID Hidden:

Element: New

Gain value: <KP> Hide All

Compensation time of the slow pole [in s] <KI> Hide All

Compensation time of the fast pole [in s] <KD> Hide All

Intergration time constant [in s] <KI> Hide All

Other Properties:

Exclude from Simulation

Exclude from PCB Layout

Exclude from Current Variant

Attach hierarchy module

Hide common pins

Edit all properties as text

OK

Help

Cancel

Fig. 2.290 Settings of C_1

Edit Component

Part Reference: K1 Hidden:

Part Value: OP : GAIN Hidden:

Element: New

Amplification coefficient: 1.0 Hide All

Other Properties:

Exclude from Simulation

Exclude from PCB Layout

Exclude from Current Variant

Attach hierarchy module

Hide common pins

Edit all properties as text

OK

Help

Cancel

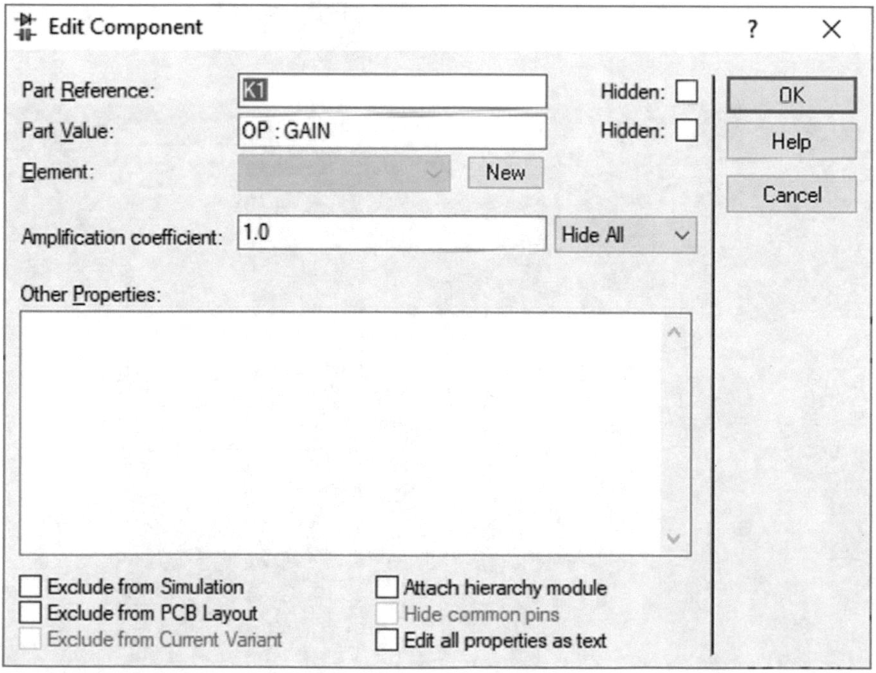

Fig. 2.291 Settings of K_1

G: Gain value, without unit. It is a positive or negative real number.

T_m: Compensation time constant of the slow pole, in seconds. It is a positive real.

T_v: Compensation time constant of the fast pole, in seconds. It is a positive real.

T_i: Integration time constant, in seconds. It is a positive real.

Note that for entering the parameters of PID block, using the exponential notation (e.g., 5.02E+3) instead of the symbolic notation (e.g., 5.02 k) is recommended. According to Fig. 2.290. $C_1(p) = K_P \frac{(1+K_I p)(1+K_D p)}{K_I p} = 1 \times \frac{(1+0.1p)(1+0.01p)}{0.1p} = \frac{0.01p^2+1.1p+10}{p}$.

Add an analogue graph (Fig. 2.292) to the schematic (Fig. 2.293).

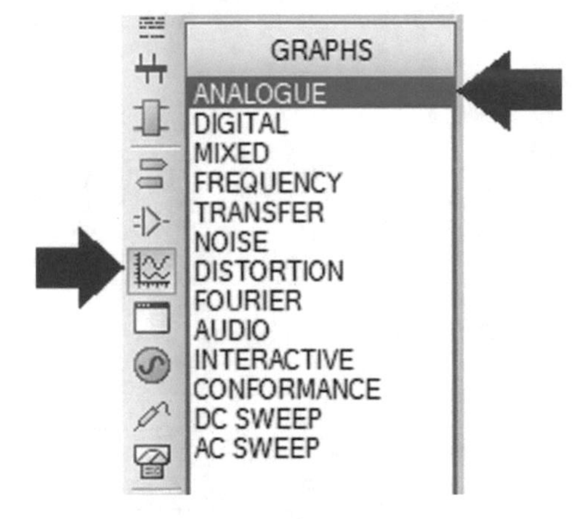

Fig. 2.292 Analogue graph block

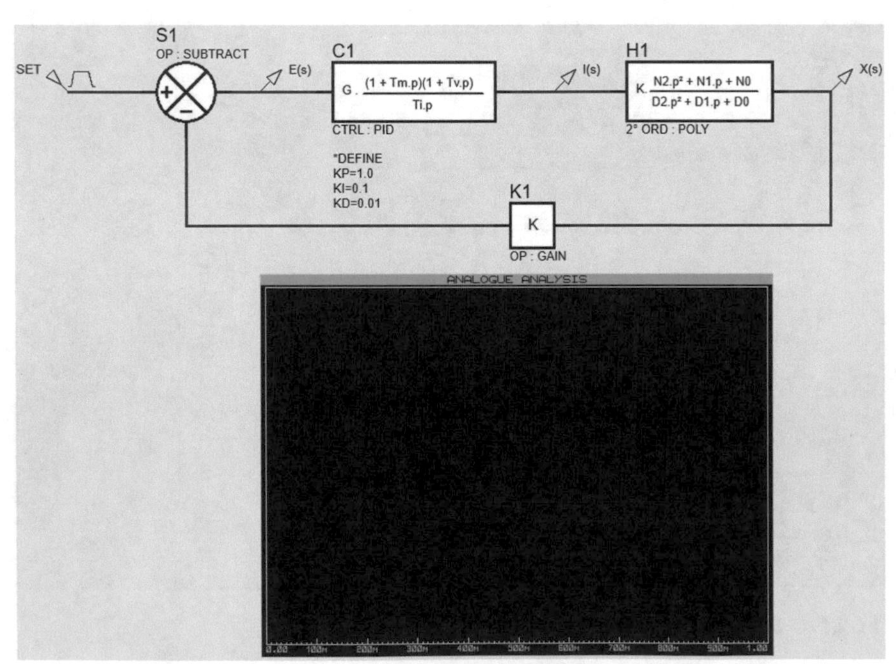

Fig. 2.293 Analogue graph block is added to the schematic

Double click the analogue graph in Fig. 2.293 and change the settings to what is shown in Fig. 2.294.

Edit Transient Graph

Graph title: TEMPERATURE CONTROLLER

Start time: 0

Stop time: 100

Left Axis Label: T(C)

Right Axis Label:

Options

Initial DC solution: ☑

Always simulate: ☑

Log netlist(s): ☐

SPICE Options

Set Y-Scales

User defined properties:

ABSTOL=1e-10

OK Cancel

Fig. 2.294 Edit transient graph window

Drag and drop the pulse generator "SET" and voltage probe "X(s)" on to the analogue graph (Fig. 2.295).

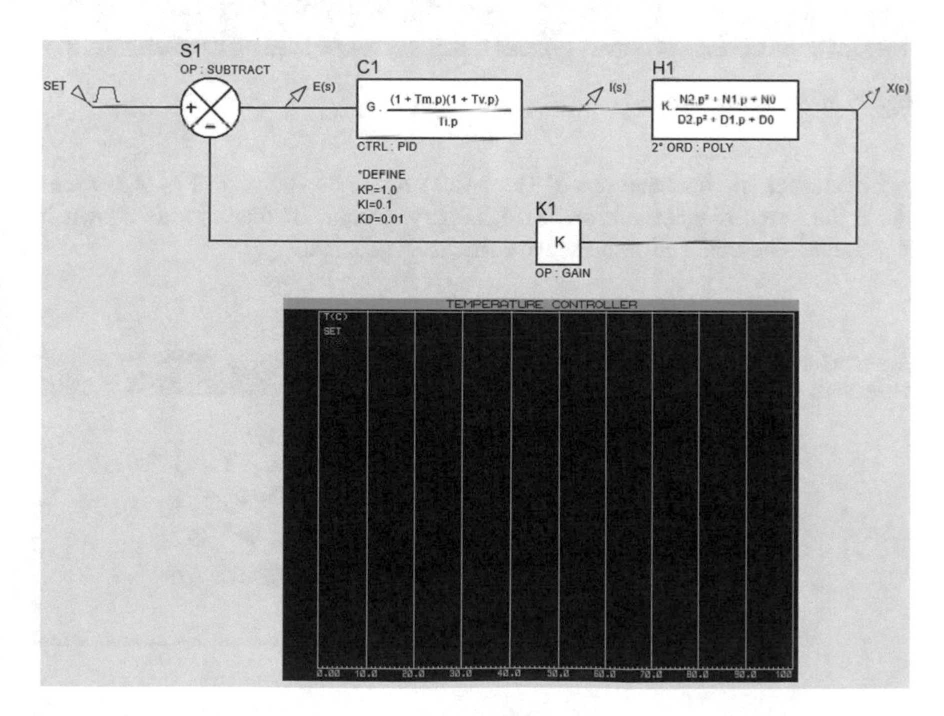

Fig. 2.295 SET and $X(S)$ signals are added to the graph

Run the simulation by pressing the Space bar key of your keyboard. Simulation result is shown in Fig. 2.296.

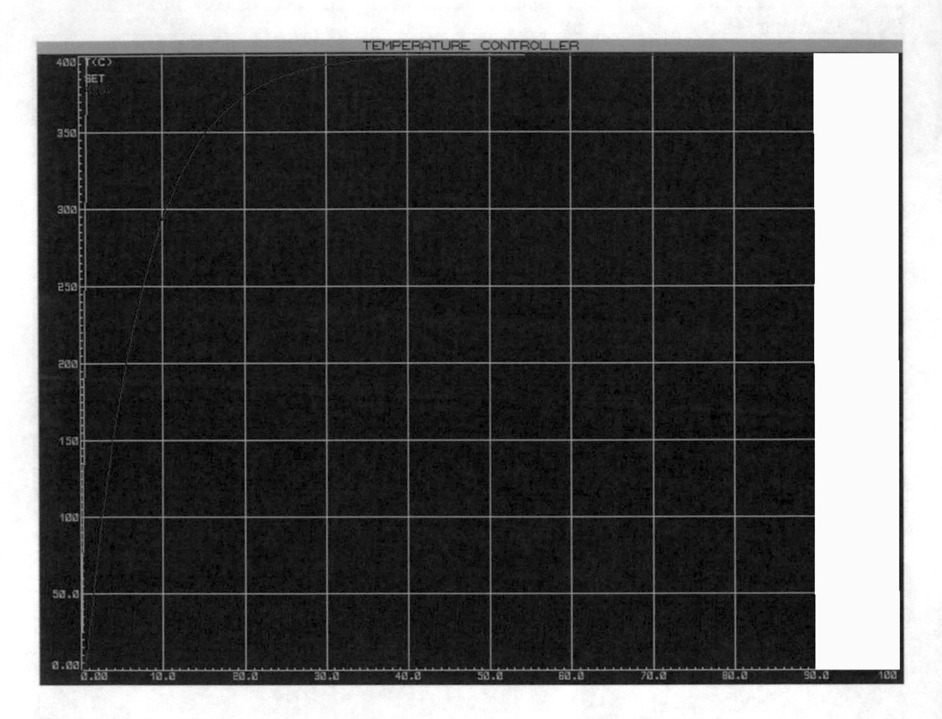

Fig. 2.296 Simulation result

Let's check the obtained result. The MATLAB code shown in Fig. 2.297 simulates the block diagram shown in Fig. 2.284. Output of this code is shown in Fig. 2.298. The obtained result is the same as Fig. 2.296.

Fig. 2.297 MATLAB calculations

Command Window

```
>> s=tf('s');
>> C1=(0.01*s^2+1.1*s+10)/s;
>> H1=1/(2*s^2+8*s+72);
>> CL=feedback(C1*H1,1);
>> step(400*CL),grid on
>>
```

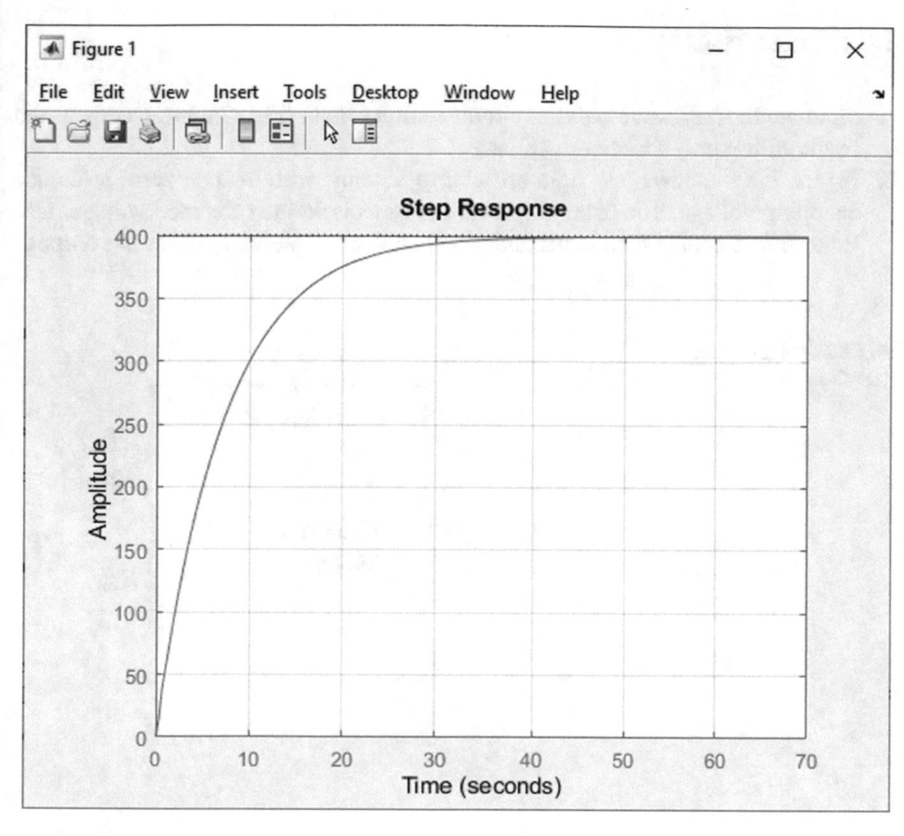

Fig. 2.298 Output of MATLAB code

2.41 Exercises

1. Simulate the half wave rectifier circuit with RL load (Fig. 2.299). Compare the result with purely resistive load case.
2. Figure 2.300 shows an op-amp clamp circuit with a non-zero reference clamping voltage. The clamping level is at precisely the reference voltage. Use Proteus to simulate the circuit and see the effect of Voltage source on output.

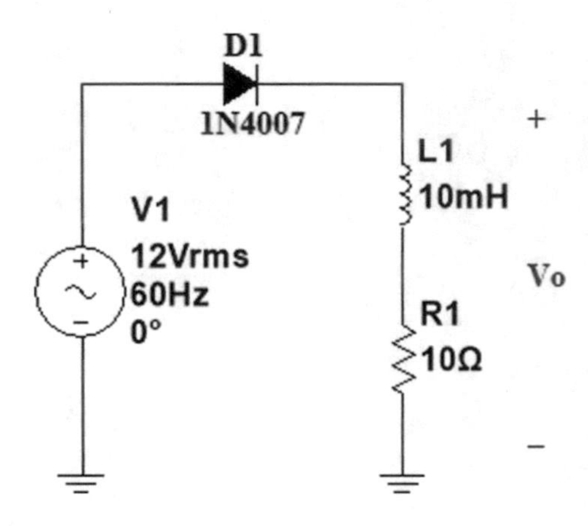

Fig. 2.299 Circuit for Exercise 1

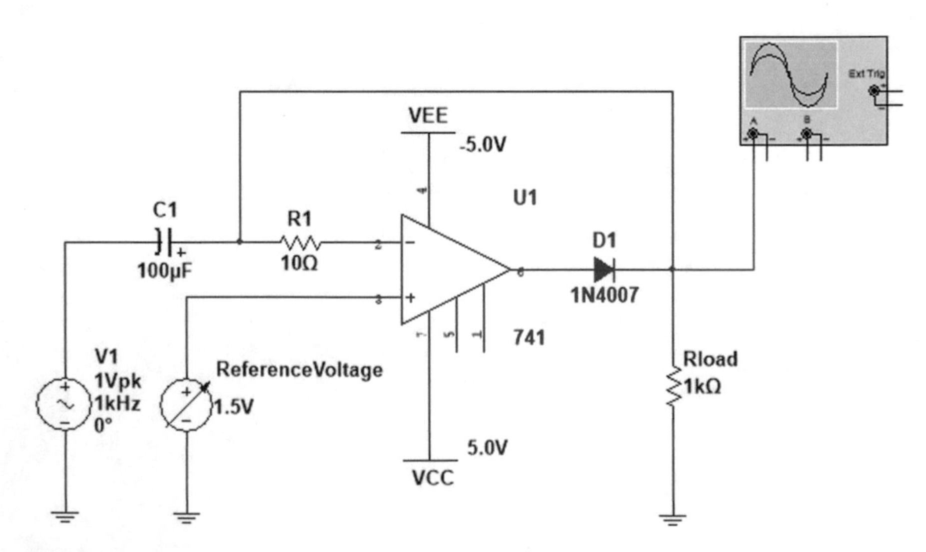

Fig. 2.300 Circuit for Exercise 2

3. Measure the maximum output voltage swing for the circuit shown in Fig. 2.301.

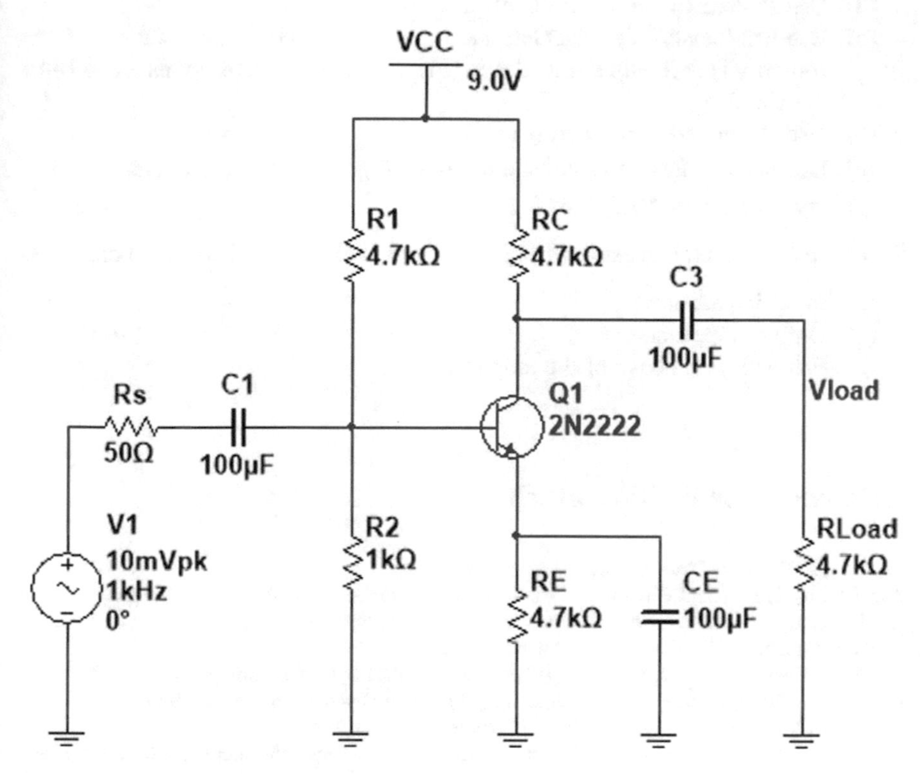

Fig. 2.301 Circuit for Exercise 3

4. Assume the amplifier shown in Fig. 2.302.

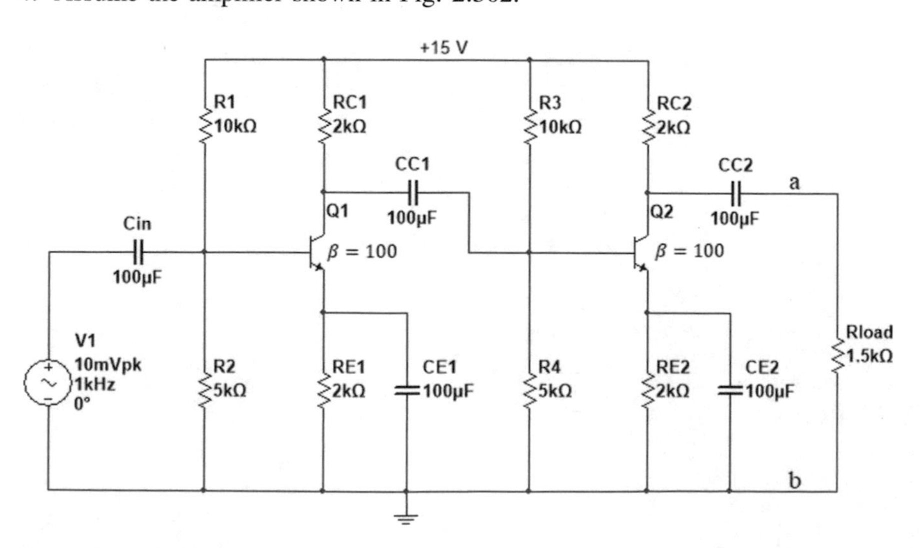

Fig. 2.302 Circuit for Exercise 4

(a) Use hand analysis to calculate the operating point of the circuit.
(b) Use Proteus to verify results of part (a).
(c) Use hand analysis to calculate the input impedance (impedance seen from source V1) and output impedance (impedance seen from points a and b) of the circuit.
(d) Use Proteus to verify part (c).
(e) Use hand analysis to calculate the overall gain ($\frac{V_{ab}}{V_1}$) of the circuit.
(f) Use Proteus to verify part (e).

5. Assume that both transistors in Fig. 2.302 are 2N2222. Use Proteus to draw the:

(a) Input impedance.
(b) Output impedance.
(c) Frequency response of the amplifier.

References for Further Study

Asadi F (2020) State-Space Control Systems, Springer
Asadi F (2022) Essential Circuit Analysis using NI Multisim and MATLAB, Springer
Asadi F (2022) Essential Circuit Analysis using LTspice, Springer
Asadi F, Eguchi K (2021) Electronic Measurements, Springer
Asadi F, Bolanos E, Rodrigues J (2019) Feedback Control Systems, Springer
Rashid MH (2016) Microelectronic circuits: analysis and design.Cengage Learning
Razavi B (2021) Fundamentals of microelectronics, 3rd edn. Wiley
Sedra A, Smith K, Carusone TC, Gaudet V (2019) Microelectronic circuits, 8th edn. Oxford University Press

Chapter 3
Simulation of Digital Circuits with Proteus®

Abstract This chapter shows how a digital circuit can be analyzed in Proteus. In this chapter you will learn how to simulate combinational and sequential digital circuits.

Keywords Full adder · Half adder · Decade counter · Frequency divider circuit · Binary counter · Decade counter · Microcontroller · Boolean logic

3.1 Introduction

In this chapter, you will learn how to analyze digital circuits in Proteus. The theory behind the studied circuits can be found in any standard digital textbook (Mano and Ciletti 2018; Nelson et al. 1995; Floy 2014; Marcovitz 2009). Similar to previous chapters, doing some hand calculations for the given circuits and comparing the hand analysis results with Proteus results are recommended.

3.2 Example 1: Full Adder Circuit

We want to simulate a full adder circuit. The truth table of full adder is shown in Table 3.1.

F. Asadi, *Essential Circuit Analysis Using Proteus®*, Energy Systems in Electrical Engineering, https://doi.org/10.1007/978-981-19-4353-9_3

Table 3.1 Truth table for a full adder

Inputs			Outputs	
A	*B*	*C* (Carry in)	Sum	Carry
0	0	0	0	0
0	0	1	1	0
0	1	0	1	0
0	1	1	0	1
1	0	0	1	0
1	0	1	0	1
1	1	0	0	1
1	1	1	1	1

The equation of Sum and Carry outputs can be written as Sum $= A \oplus B \oplus C$ and Carry $= AB + AC + BC$. Let's simulate these Boolean functions. We need AND, XOR and OR gates to simulate these Boolean functions (Figs. 3.1, 3.2 and 3.3).

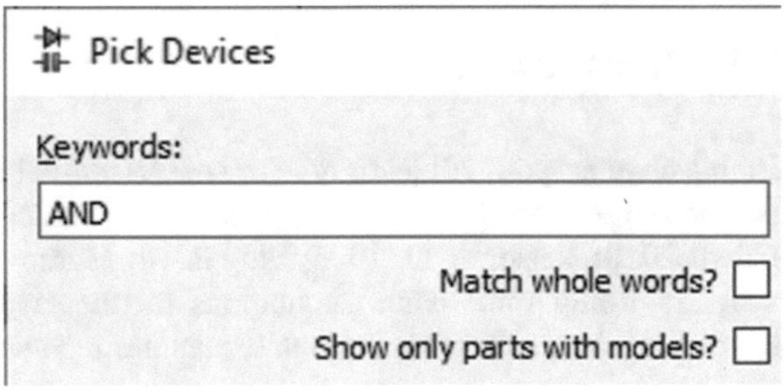

Fig. 3.1 Searching for AND gate

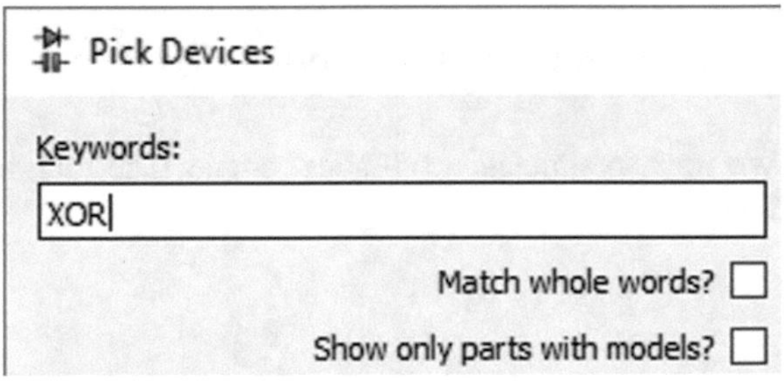

Fig. 3.2 Searching for XOR gate

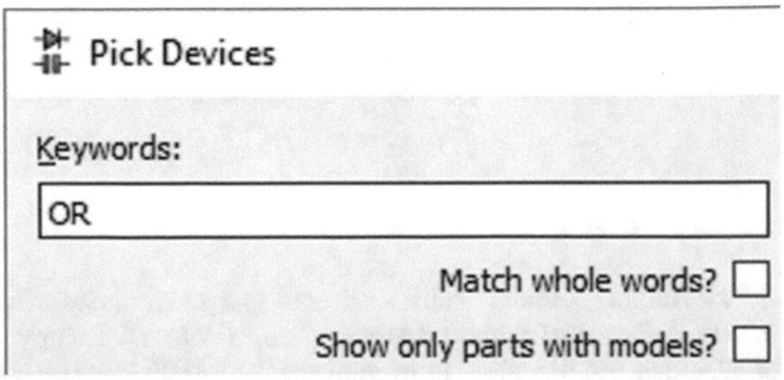

Fig. 3.3 Searching for OR gate

Note that the blocks in Figs. 3.1, 3.2 and 3.3 have two inputs. Proteus has gates with more than two inputs as well. For instance, AND_5 (Fig. 3.4) is an AND gate with five inputs.

Pick Devices

Keywords:

AND_5

Match whole words?

Show only parts with models?

Fig. 3.4 Searching for five-input AND gate

Draw the schematic shown in Fig. 3.5.

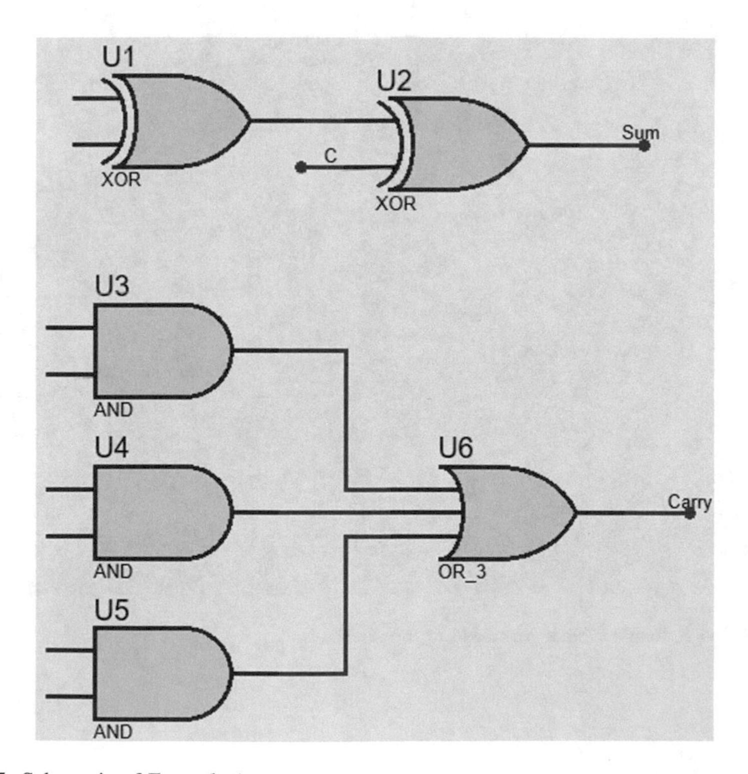

Fig. 3.5 Schematic of Example 1

Now, we need a block to generate the inputs. The logic toggle block (Fig. 3.6) can be used for this purpose.

Pick Devices

Keywords:

LOGICTOGGLE

Match whole words?

Show only parts with models?

Fig. 3.6 Searching for logic toggle block

Add a logic toggle block to the schematic and complete it (Fig. 3.7).

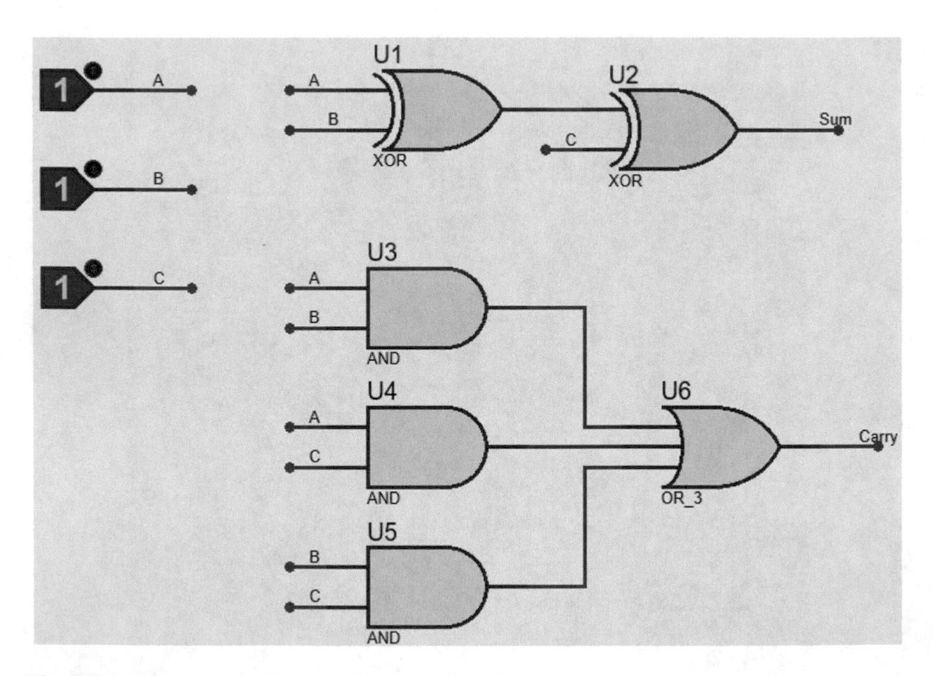

Fig. 3.7 Logic toggle blocks are added to the schematic

A single-pole double-through switch (Fig. 3.8) is another way to apply the input to the circuit (Fig. 3.9). In Fig. 3.9, $V_{CC} = 5$ V.

Pick Devices

Keywords:

SW-SPDT

Match whole words? ☐

Show only parts with models? ☐

Fig. 3.8 Searching for SW-SPDT block

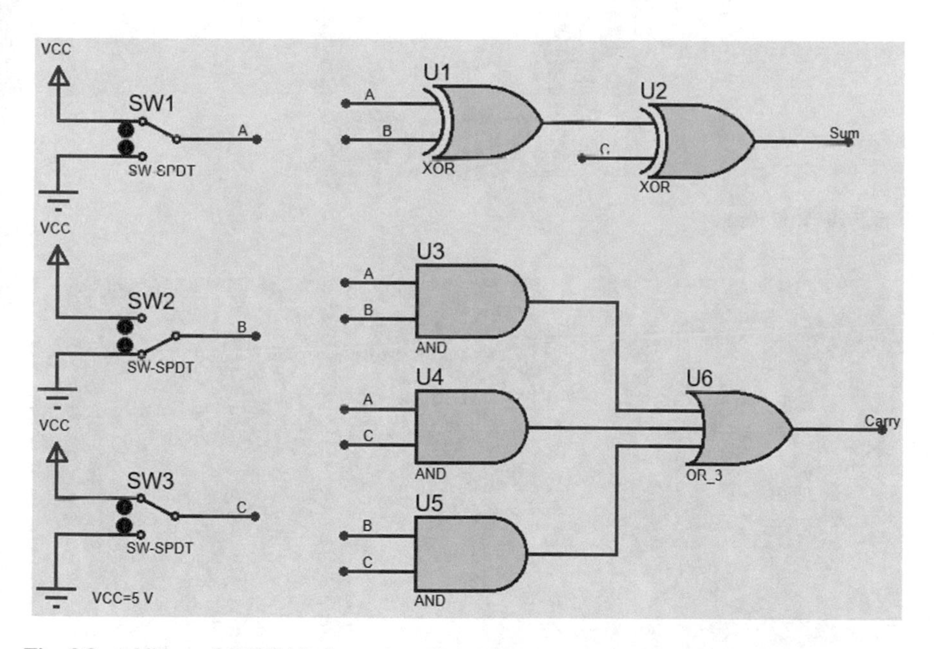

Fig. 3.9 Addition of SPDT blocks to the schematic

Run the simulation (Fig. 3.10). After running the simulation, some squares are added to the pins and show their logical status (Fig. 3.11). The red color shows logical 1 and the blue color shows logical 0. Click the logic toggle block to apply different inputs to the circuit. Compare the Proteus outputs with Table 3.1.

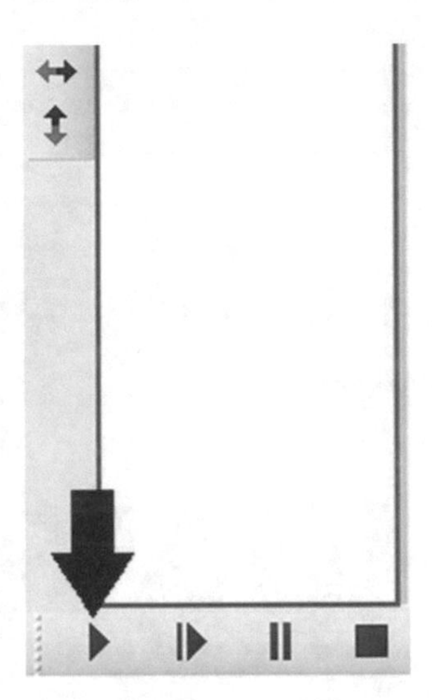

Fig. 3.10 Run icon

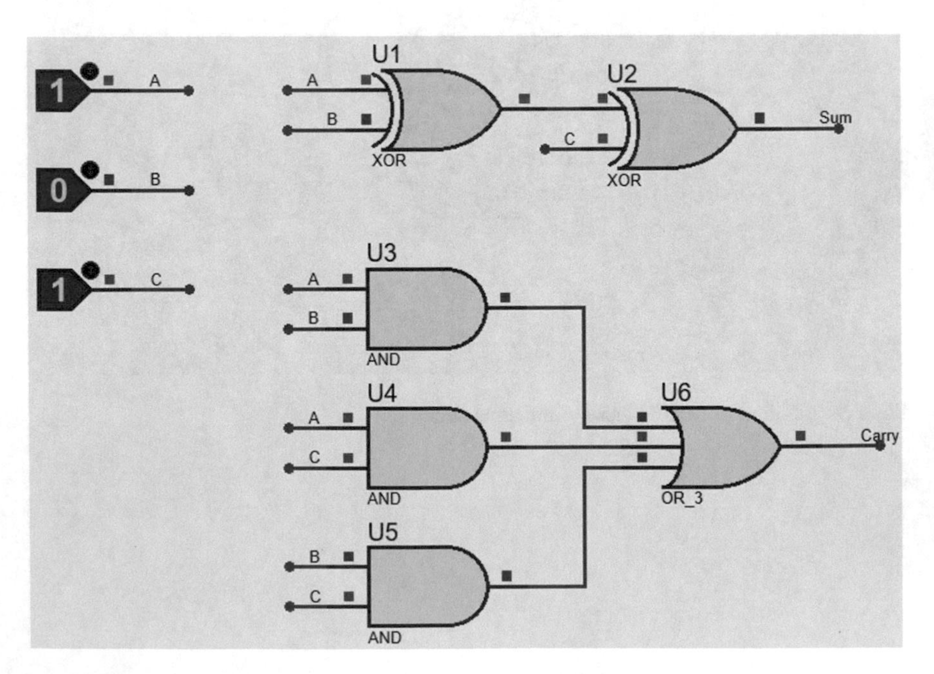

Fig. 3.11 Simulation result

If you can't see the squares after running the simulation, click the System > Set Animation Options (Fig. 3.12). This opens the window shown in Fig. 3.13. Then check the Show Logic State of Pins box.

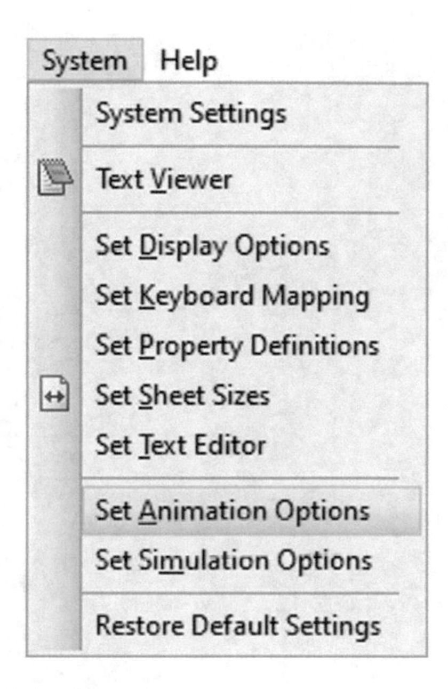

Fig. 3.12 System > Set Animation options

Animated Circuits Configuration
Simulation Speed
Frames per Second: 20
Timestep per Frame: 50m
Single Step Time: 50m
Max. SPICE Timestep: 25m
Step Animation Rate: 4
Voltage/Current Ranges
Maximum Voltage: 6
Current Threshold: 1u
Animation Options
Show Voltage & Current on Probes?
Show Logic State of Pins?
Show Wire Voltage by Colour?
Show Wire Current with Arrows?
SPICE Options
OK
Cancel

Fig. 3.13 Animated circuits configuration window

DC voltmeters can be used to see the outputs as well (Fig. 3.14). You can use LEDs to see the status of outputs as well.

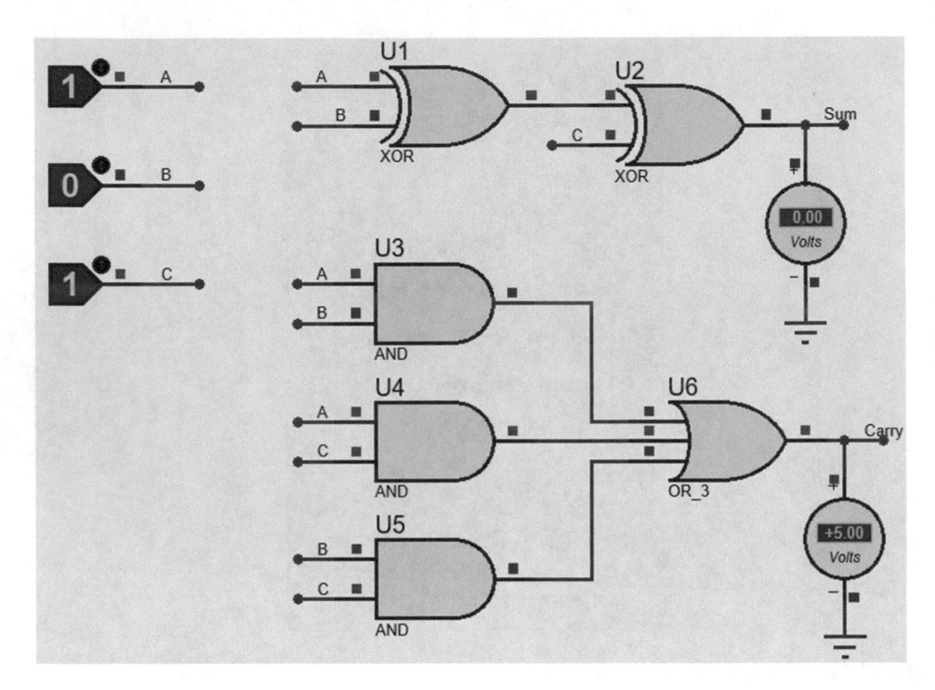

Fig. 3.14 DC voltmeter can be used as an indicator

3.3 Example 2: Logic Probe Block

In the previous example, we learned two ways to see the logical status of a pin. In this example, we will learn another way to see the logical status of a pin.

Proteus has a block called logic probe (Fig. 3.15) which can be used to show the logical status of a pin.

Fig. 3.15 Searching for logic probe

Let's see how this block works. Open the schematic of the previous example and add the logic probe block to the outputs of circuit (Fig. 3.16).

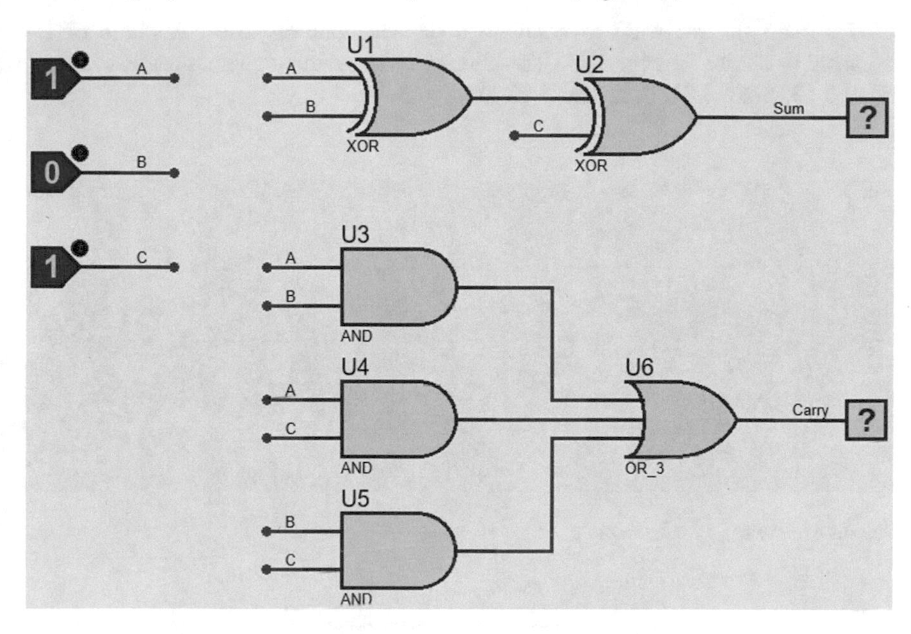

Fig. 3.16 Logic probe block is added to the schematic

Run the simulation. The logic probe block shows the logical status of outputs with 0 and 1 (Fig. 3.17), where 0 means low output and 1 means high output.

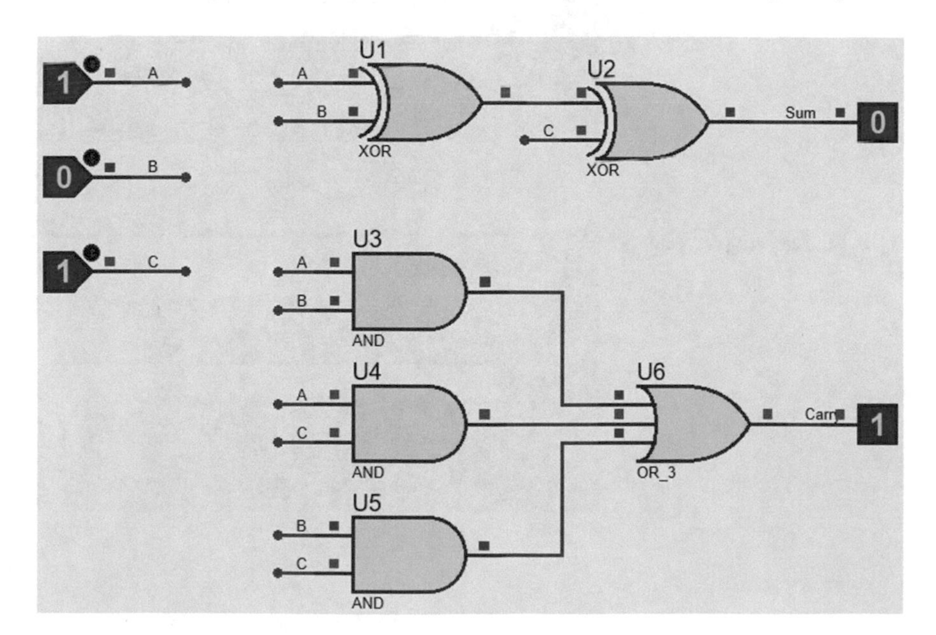

Fig. 3.17 Simulation probe shows the status of digital signals

3.4 Example 3: Decade Counter

In this example, we want to simulate a decade counter. The schematic of this example is shown in Fig. 3.18. The blocks used in this schematic are shown in Figs. 3.19, 3.20, 3.21, 3.22 and 3.23 ($V_{CC} = 5$ V).

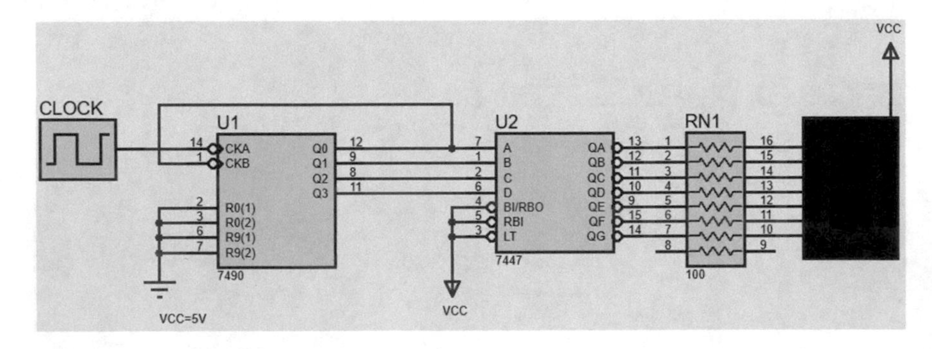

Fig. 3.18 Schematic for Example 3

Fig. 3.19 Searching for clock block

Pick Devices

Keywords:

CLOCK

Match whole words?

Show only parts with models?

Fig. 3.20 Searching for 7490 IC

Pick Devices

Keywords:

7490

Match whole words?

Show only parts with models?

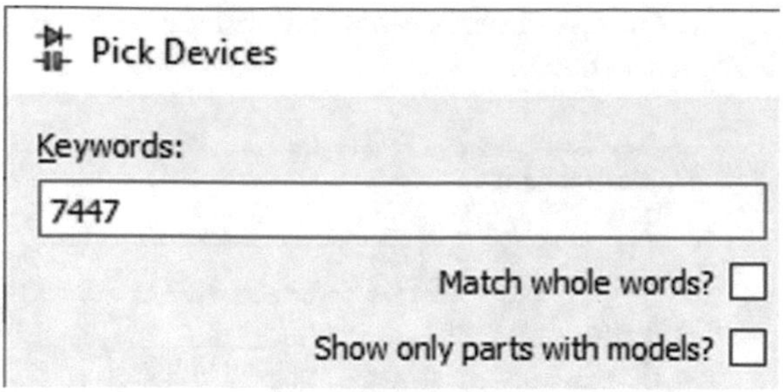

Fig. 3.21 Searching for 7447 IC

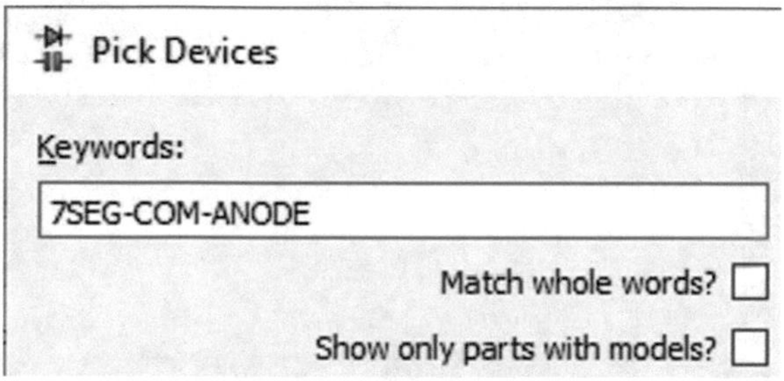

Pick Devices

Keywords:

7447

Match whole words?

Show only parts with models?

Fig. 3.22 Searching for a common anode seven-segment display

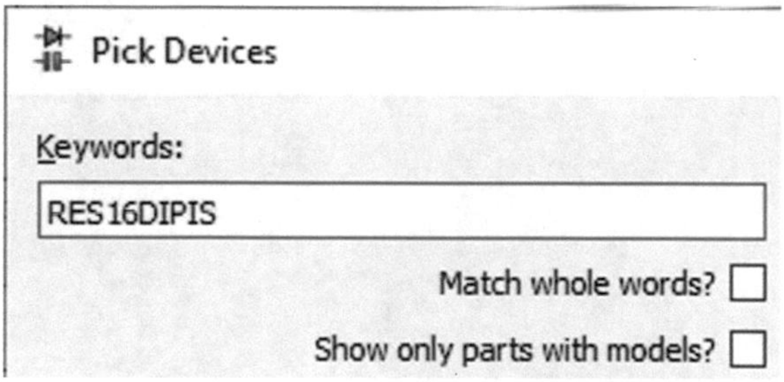

Pick Devices

Keywords:

7SEG-COM-ANODE

Match whole words?

Show only parts with models?

Fig. 3.23 16-pin resistor array

Pick Devices

Keywords:

RES16DIPIS

Match whole words?

Show only parts with models?

The settings of clock block and resistive array RN1 are shown in Figs. 3.24 and 3.25, respectively.

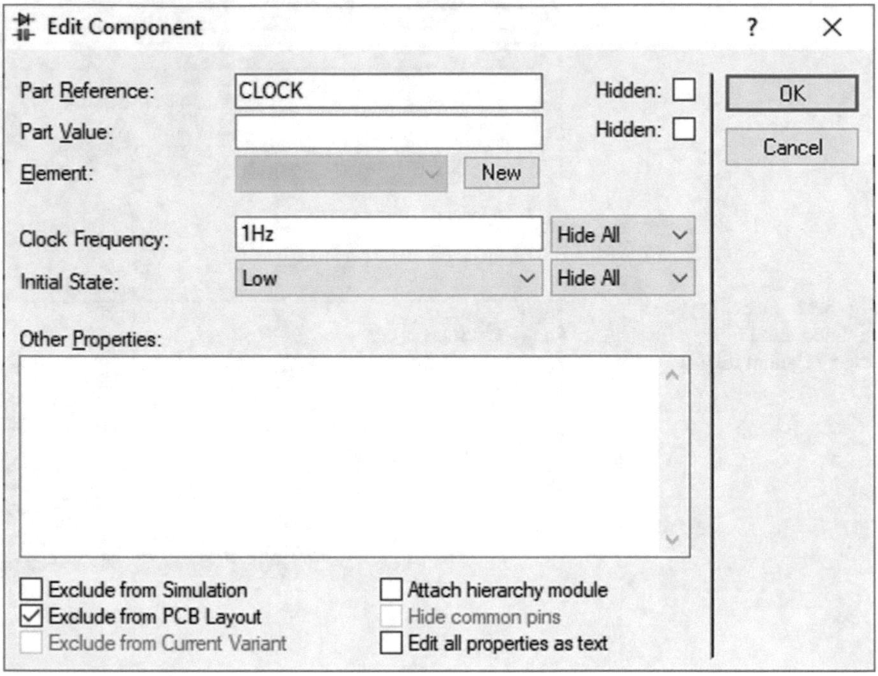

Fig. 3.24 Settings of Clock block

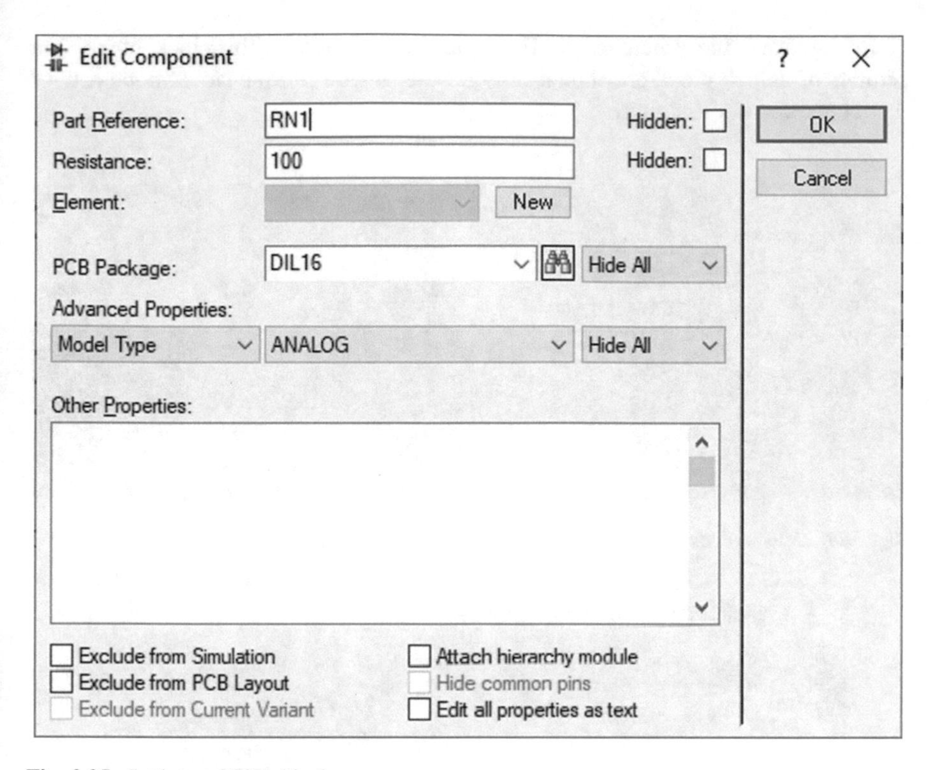

Fig. 3.25 Settings of RN1 block

Run the simulation. After running the simulation, the counter starts to count.

3.5 Example 4: Dclock Block

In the previous example, we used the clock block to generate the required clock signal. The required clock pulse can be generated with the "dclock" generator (Fig. 3.26) as well.

Fig. 3.26 Dclock block

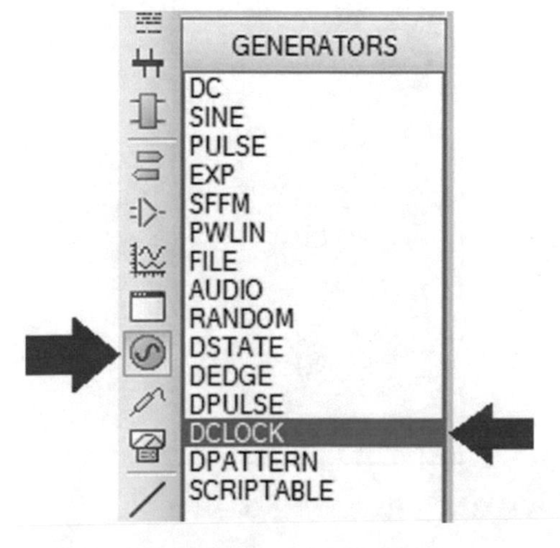

In Fig. 3.27, the schematic of Example 3 is redrawn with dclock block. The settings of dclock block are shown in Fig. 3.28. If you run the simulation, you see that counter starts to count.

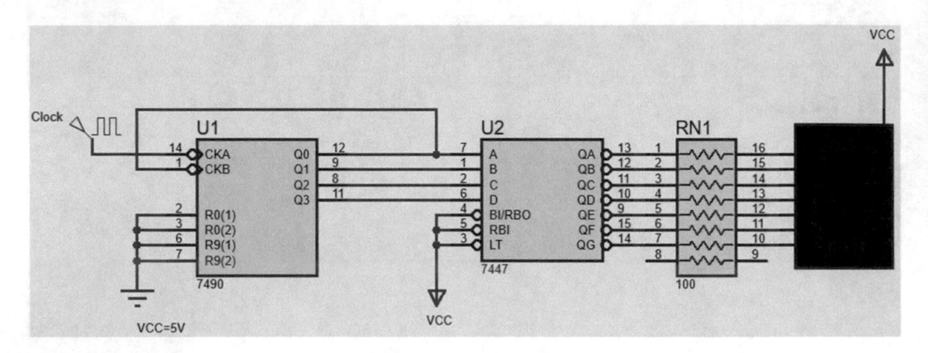

Fig. 3.27 Dclock block is added to the schematic of Example 3

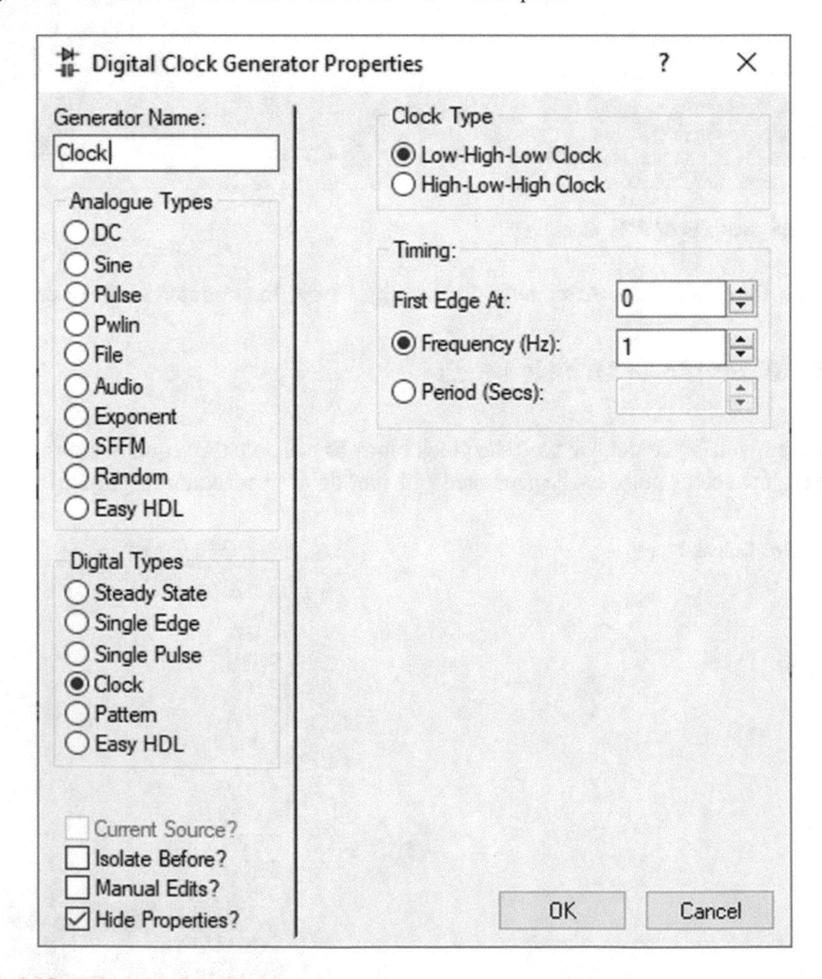

Fig. 3.28 Dclock block settings

3.6 Example 5: Frequency Divider Circuit

In this example, we will simulate a frequency divider circuit. A D flip-flop can be used to divide the frequency of an input square wave by a factor of 2 (Fig. 3.29).

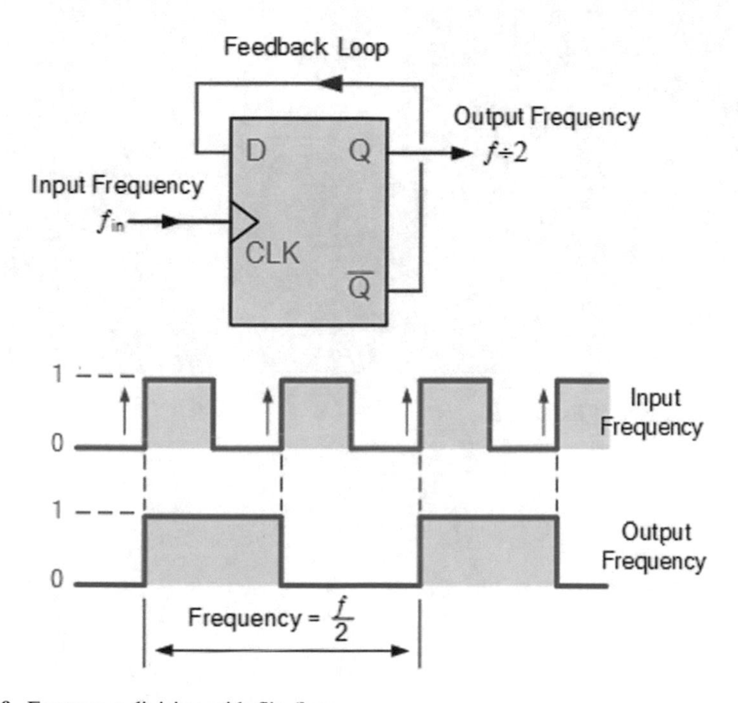

Fig. 3.29 Frequency division with flip-flops

In this example, we will use a 7474 IC. 7474 is composed of two dual-positive edge-triggered D flip-flops (Fig. 3.30).

Fig. 3.30 Inside of 7474 IC

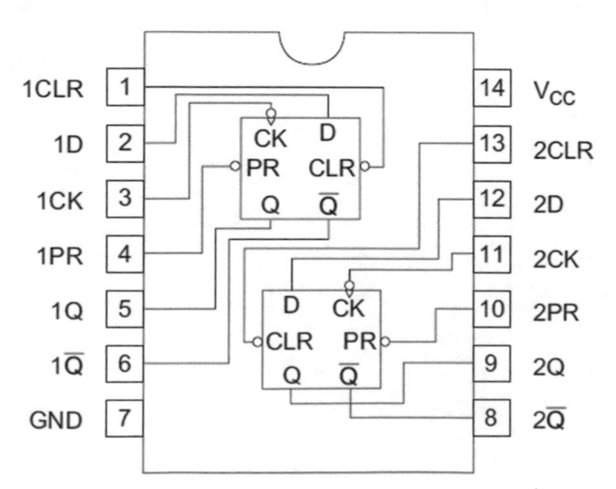

Draw the schematic shown in Fig. 3.31. The settings of V_{in} are shown in Fig. 3.32.

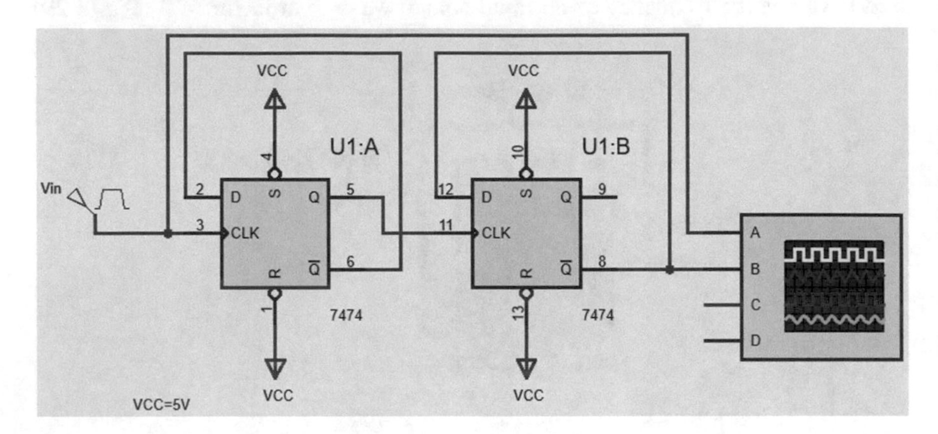

Fig. 3.31 Schematic of Example 5

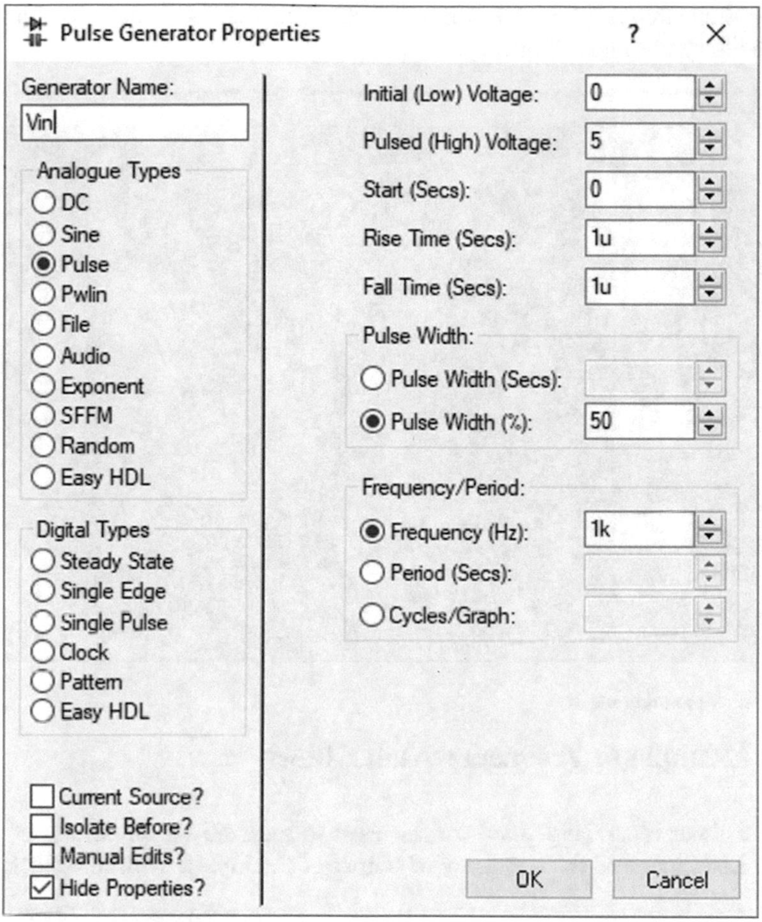

Fig. 3.32 Settings of V_{in}

The simulation result is shown in Fig. 3.33. Note that the frequency of output is one-fourth of the input frequency.

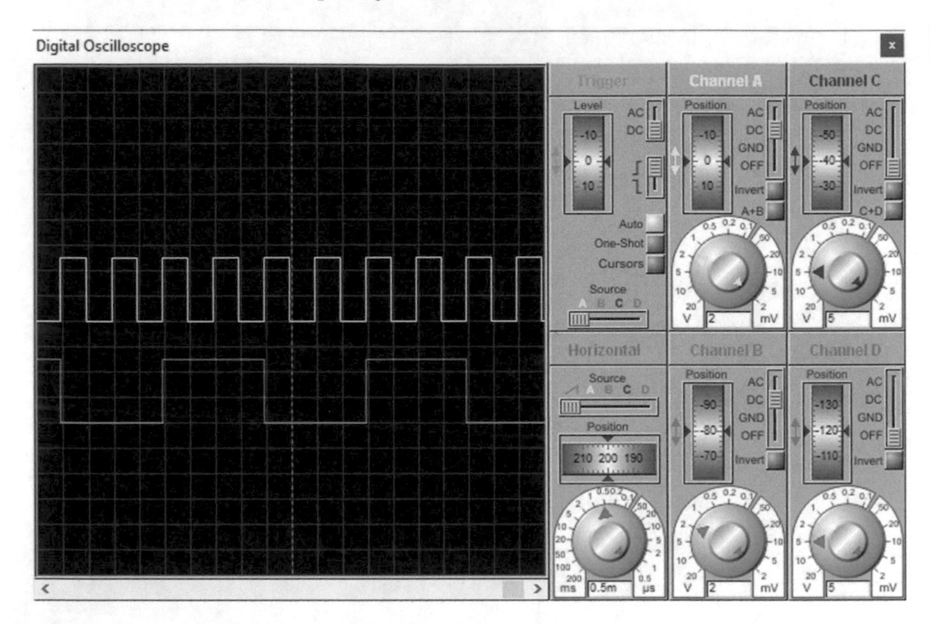

Fig. 3.33 Simulation result

3.7 Example 6: Frequency Meter Block

Counter timer block (Fig. 3.34) can be used to measure the frequency of waveforms. Let's measure the frequency of output of Example 5 with this block.

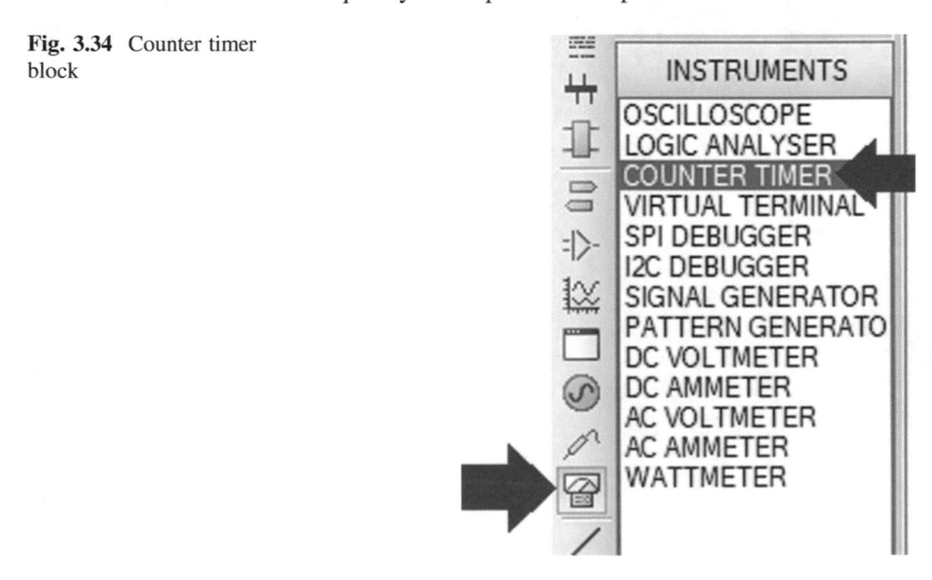

Fig. 3.34 Counter timer block

Connect a counter timer block to the output of Example 5 (Fig. 3.35).

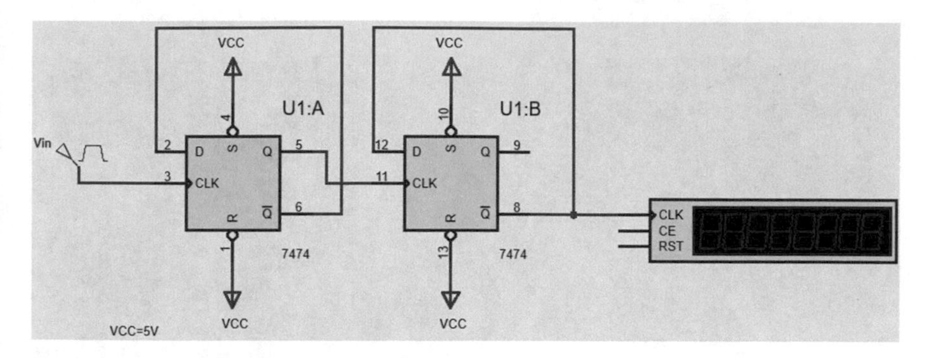

Fig. 3.35 Schematic of Example 6

Double-click the counter timer block and change the operating mode drop-down list to Frequency (Fig. 3.36). Then click the OK button.

Edit Component ? X

Part Reference: Hidden: OK

Part Value: Hidden: Help

Element: New Cancel

Operating Mode: Frequency Hide All

Count Enable Polarity: High Hide All

Reset Edge Polarity: Low-High Hide All

Other Properties:

Exclude from Simulation Attach hierarchy module

Exclude from PCB Layout Hide common pins

Exclude from Current Variant Edit all properties as text

Fig. 3.36 Counter timer settings

Run the simulation. The counter timer block shows the frequency of output (Fig. 3.37).

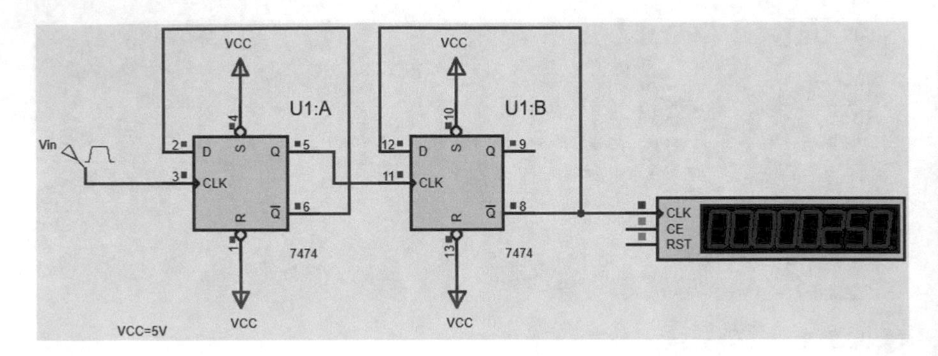

Fig. 3.37 Simulation result

3.8 Example 7: Two-Bit Binary Counter

In this example, we want to simulate a 2-bit binary counter. Draw the schematic shown in Fig. 3.38. The settings of clock are shown in Fig. 3.39.

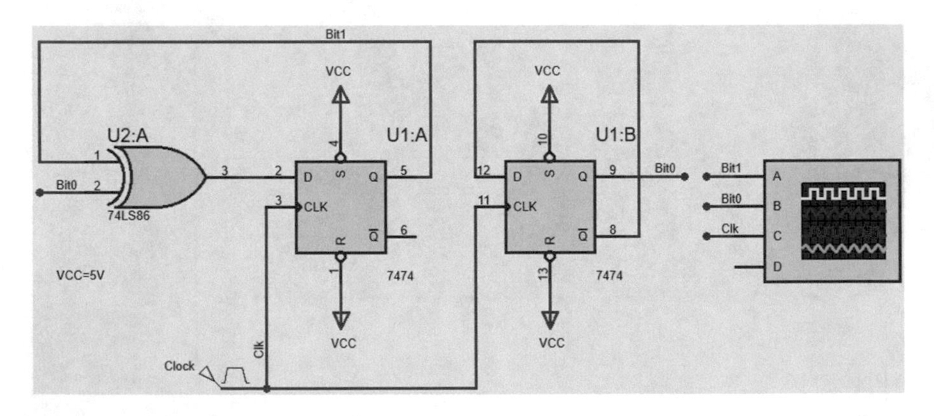

Fig. 3.38 Schematic of Example 7

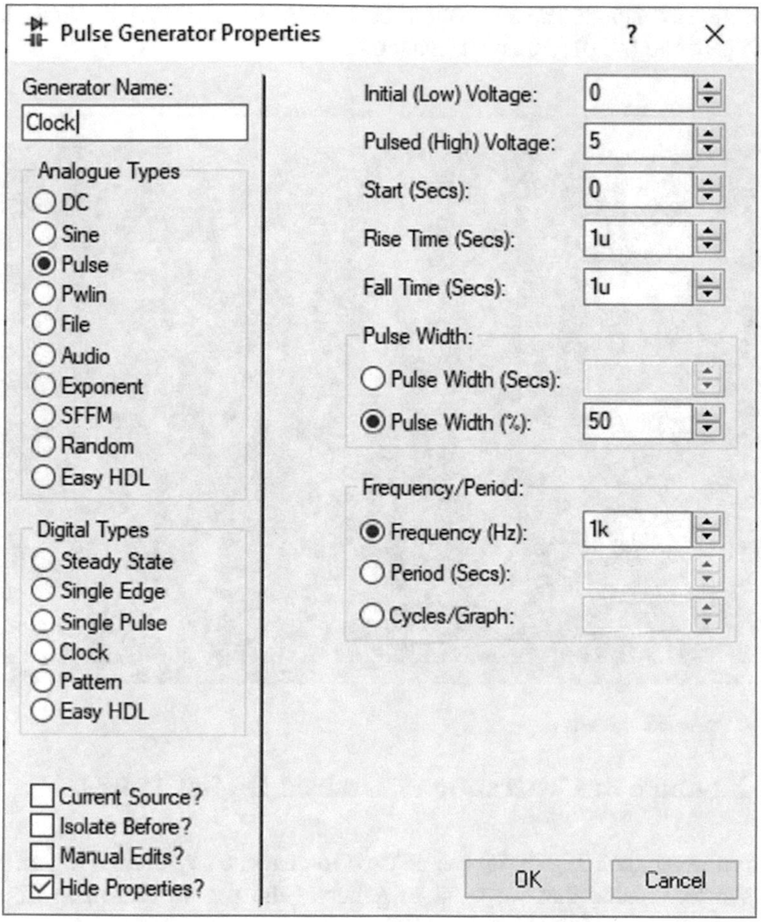

Pulse Generator Properties ? ×

Generator Name:
Clock

Analogue Types
DC
Sine
Pulse
Pwlin
File
Audio
Exponent
SFFM
Random
Easy HDL

Digital Types
Steady State
Single Edge
Single Pulse
Clock
Pattern
Easy HDL

Current Source?
Isolate Before?
Manual Edits?
Hide Properties?

Initial (Low) Voltage: 0
Pulsed (High) Voltage: 5
Start (Secs): 0
Rise Time (Secs): 1u
Fall Time (Secs): 1u

Pulse Width:
Pulse Width (Secs):
Pulse Width (%): 50

Frequency/Period:
Frequency (Hz): 1k
Period (Secs):
Cycles/Graph:

OK Cancel

Fig. 3.39 Settings of clock

Run the simulation. The simulation result is shown in Fig. 3.40. Note that the outputs generate 00, 01, 10 and 11 patterns.

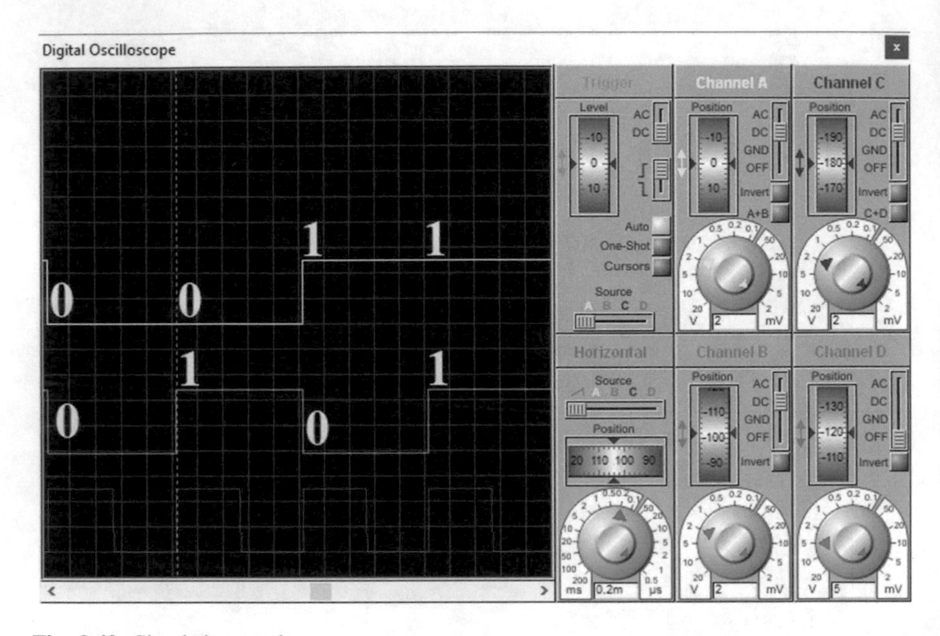

Fig. 3.40 Simulation result

3.9 Example 8: Generating of Desired Digital Pulses

"Dpattern" generator (Fig. 3.41) can be used to generate any desired digital pulses. For instance, assume that we want to generate the pattern shown in Fig. 3.42. Assume that the length of H or L is 100 μs.

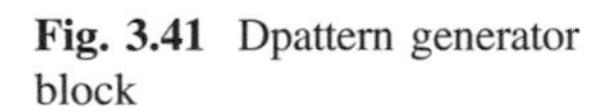

Fig. 3.41 Dpattern generator block

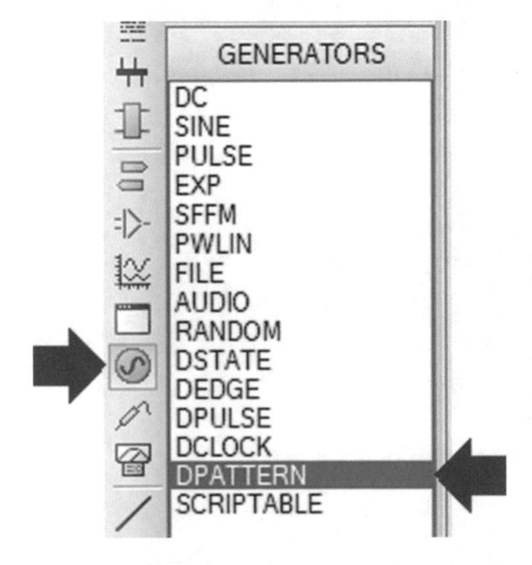

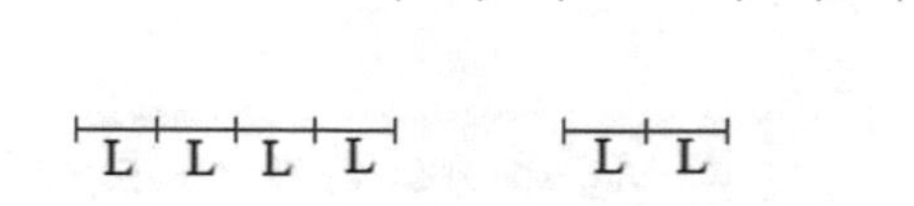

Fig. 3.42 Desired pattern

Draw the schematic shown in Fig. 3.43.

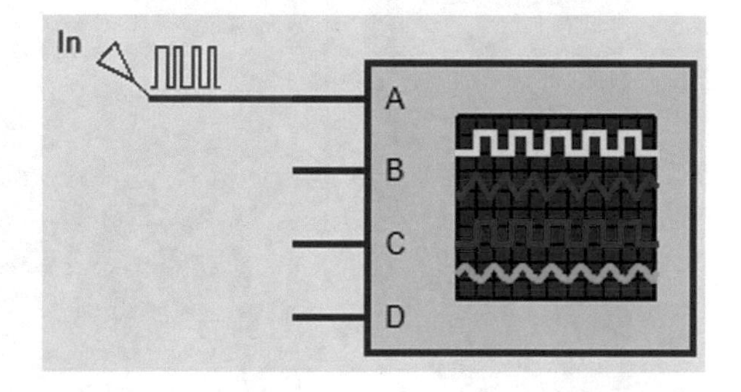

Fig. 3.43 Schematic for Example 8

Double-click the "In" dpattern generator and do the settings similar to Fig. 3.44.

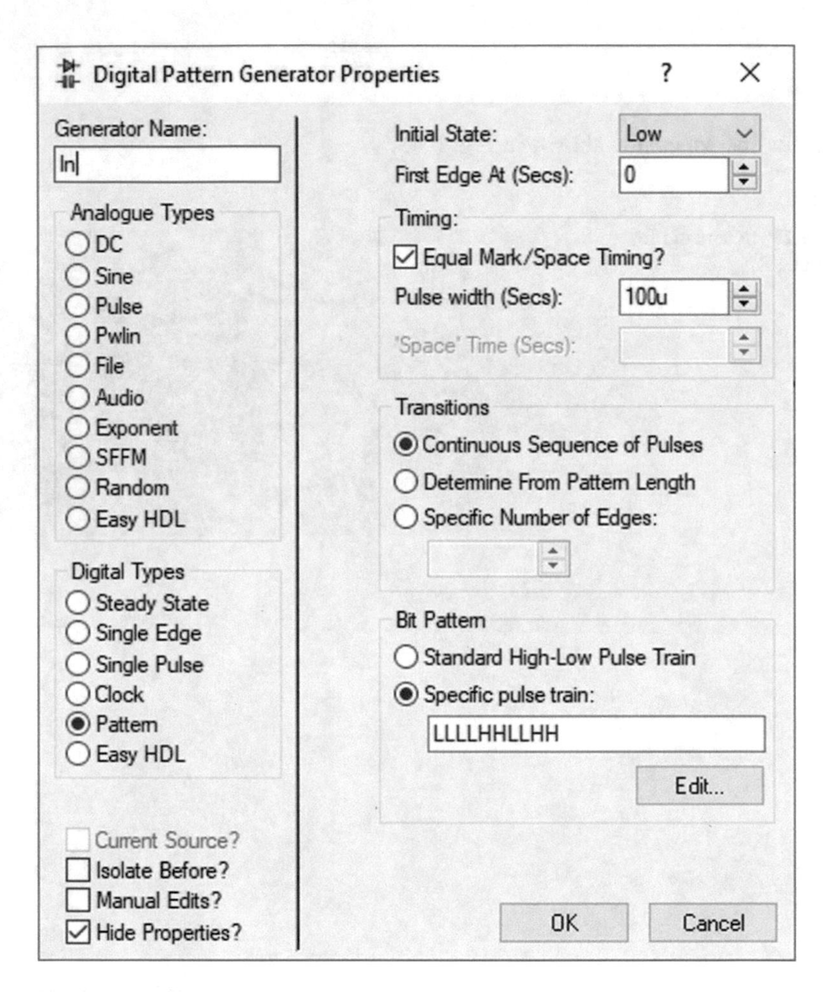

Fig. 3.44 Settings of digital pattern generator

Run the simulation. The simulation result is shown in Fig. 3.45.

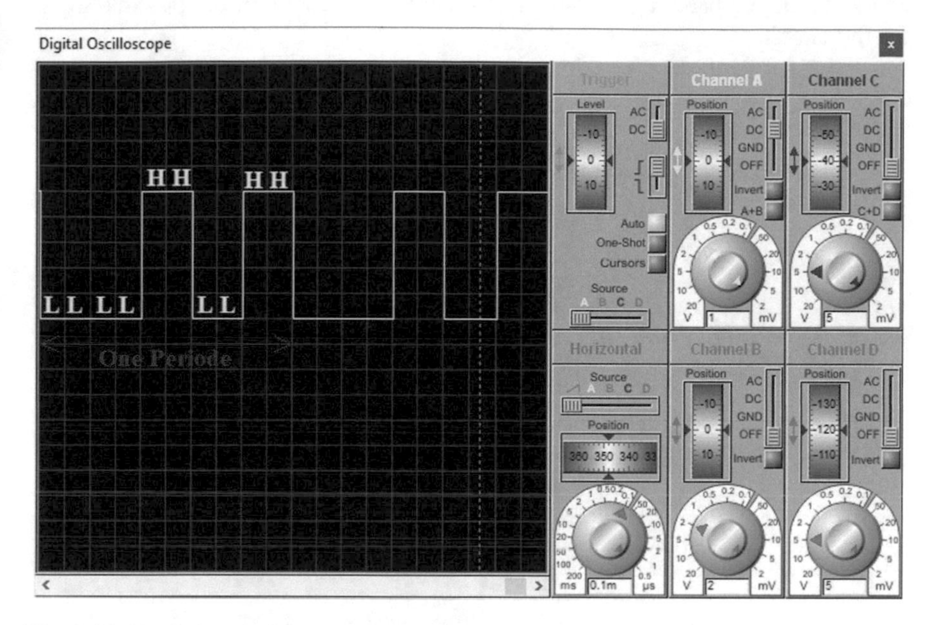

Fig. 3.45 Simulation result

3.10 Example 9: Digital Graph

The digital graph (Fig. 3.46) is a useful way to see the waveforms of a digital circuit. Let's study an example to see how this tool can be used in simulations.

Fig. 3.46 Digital block

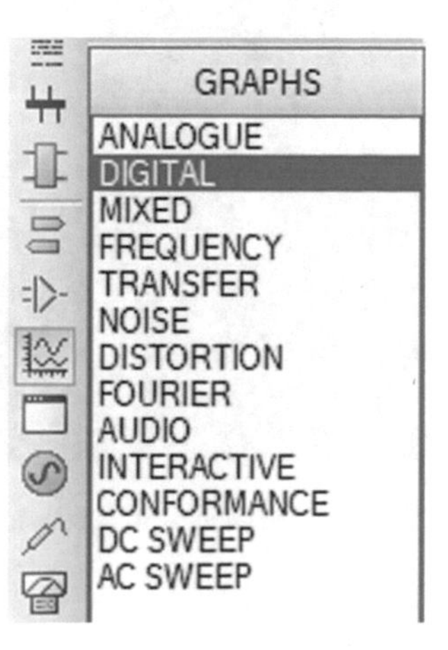

Open the schematic of Example 1 and change it to what is shown in Fig. 3.47. The settings of digital pattern generators A, B and C are shown in Figs. 3.48, 3.49 and 3.50. Digital pattern generators A, B and C generate all of the possible inputs for the circuit.

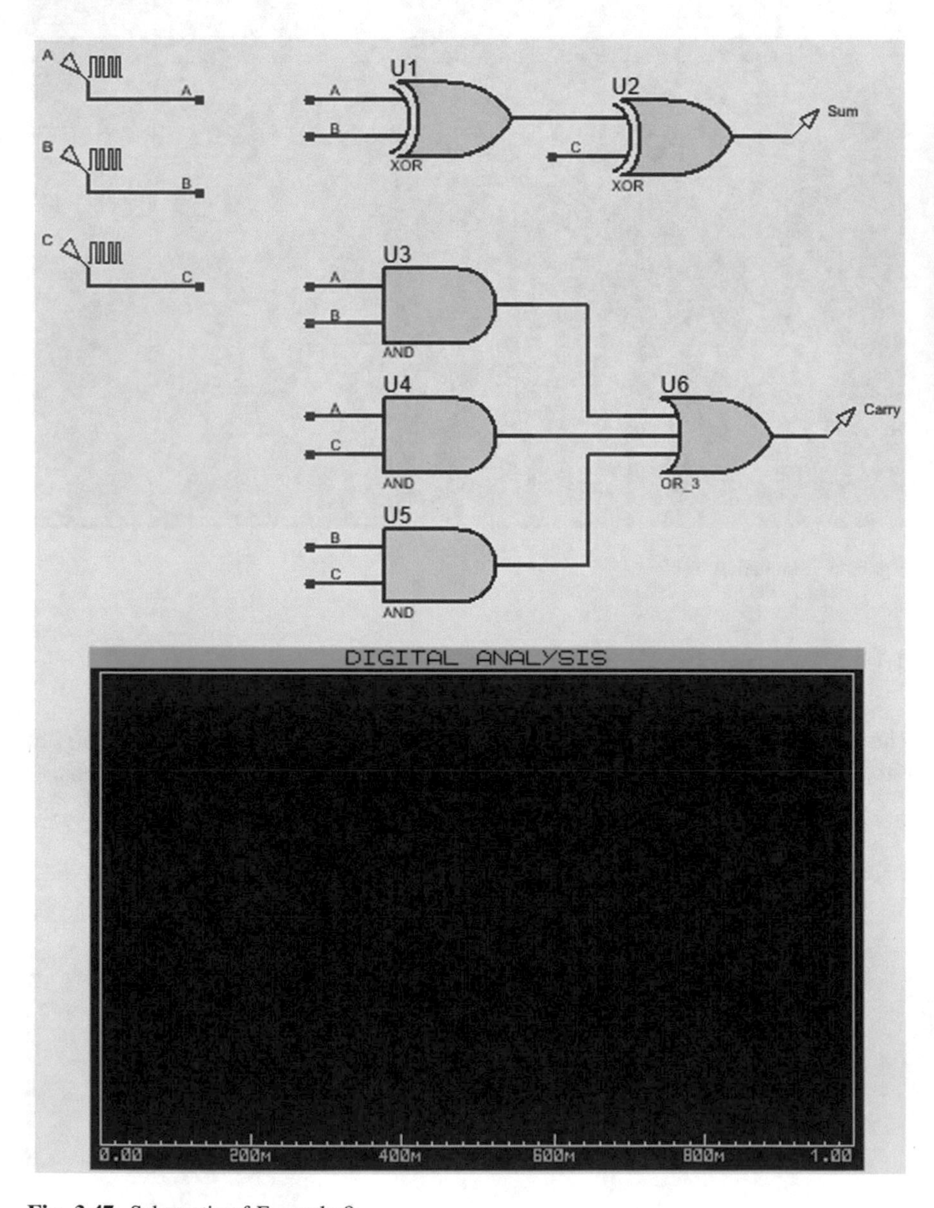

Fig. 3.47 Schematic of Example 9

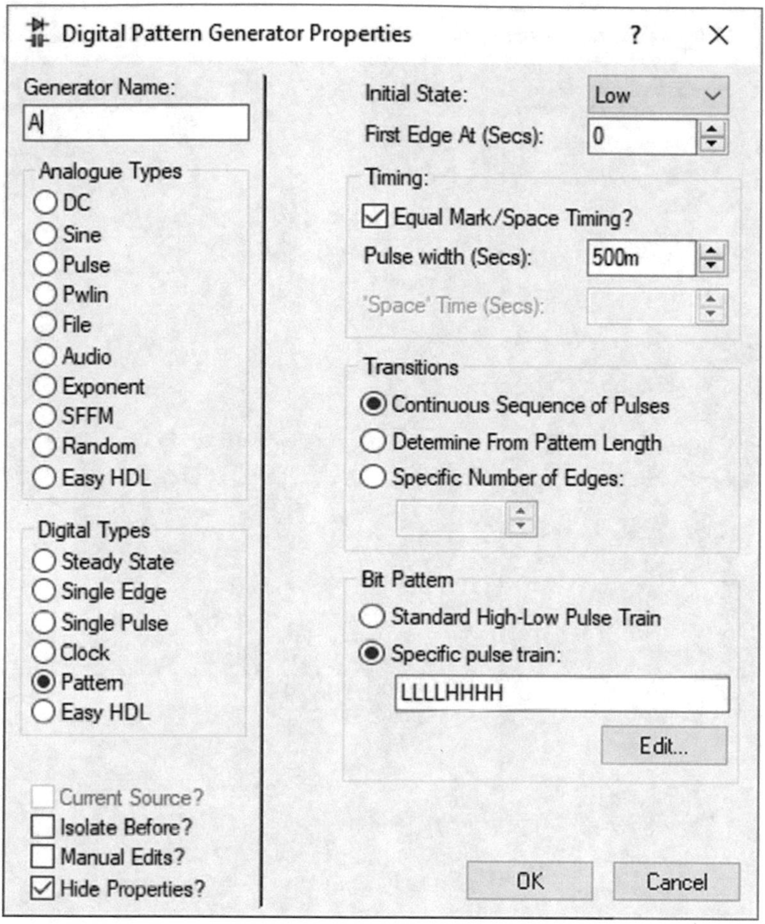

Fig. 3.48 Digital pattern generator A settings

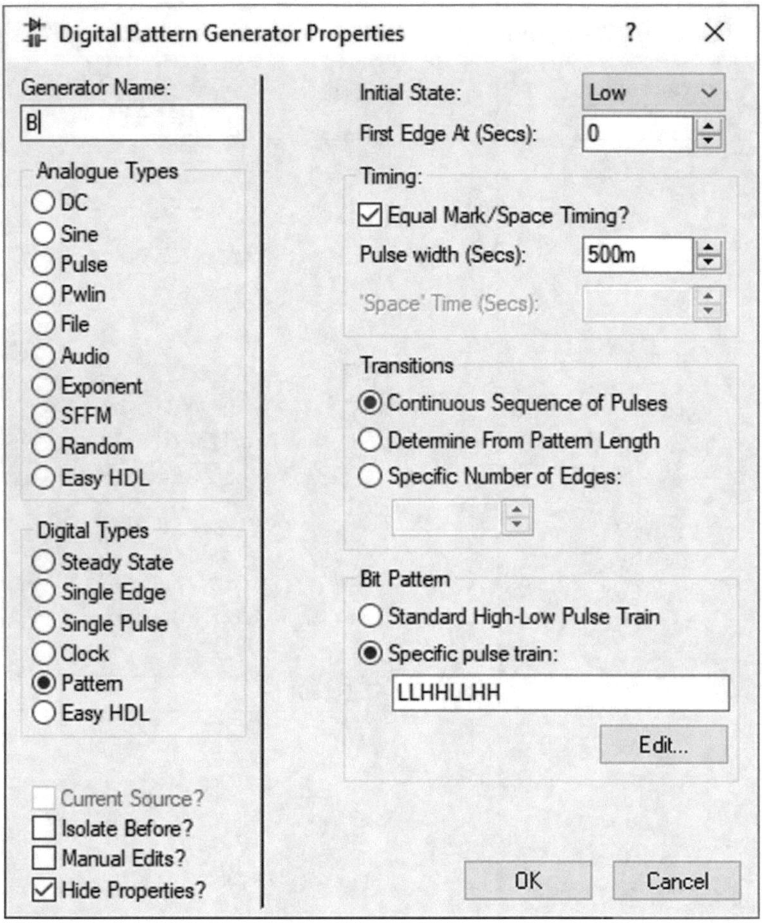

Digital Pattern Generator Properties ? ×

Generator Name:
B

Analogue Types
DC
Sine
Pulse
Pwlin
File
Audio
Exponent
SFFM
Random
Easy HDL

Digital Types
Steady State
Single Edge
Single Pulse
Clock
Pattern
Easy HDL

Current Source?
Isolate Before?
Manual Edits?
Hide Properties?

Initial State: Low
First Edge At (Secs): 0

Timing:
Equal Mark/Space Timing?
Pulse width (Secs): 500m
'Space' Time (Secs):

Transitions
Continuous Sequence of Pulses
Determine From Pattern Length
Specific Number of Edges:

Bit Pattern
Standard High-Low Pulse Train
Specific pulse train:
LLHHLLHH
Edit...

OK Cancel

Fig. 3.49 Digital pattern generator B settings

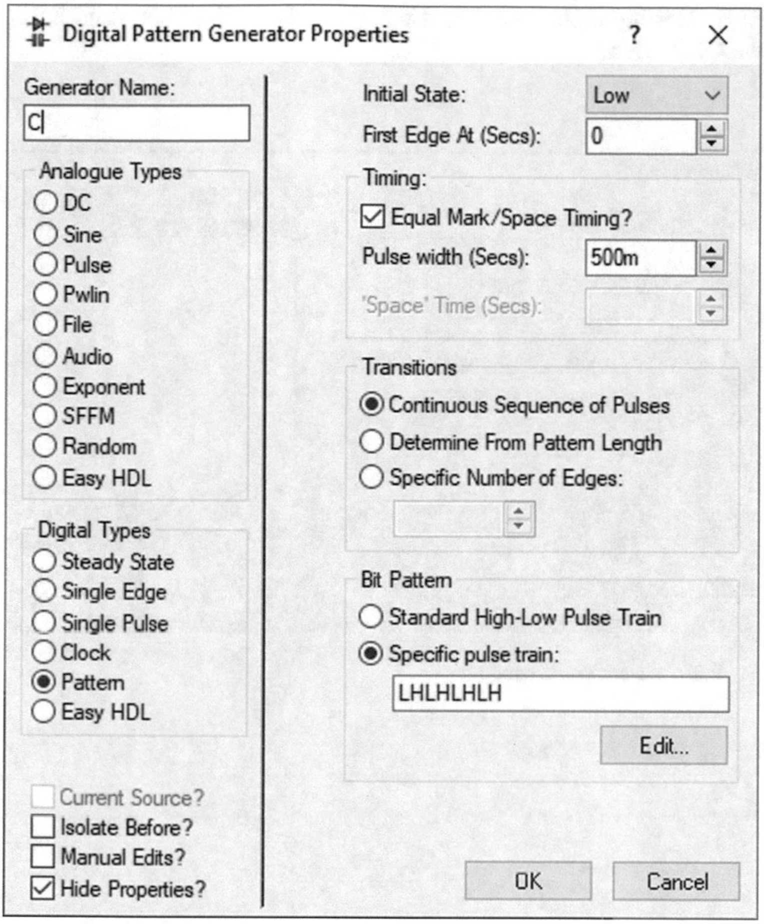

Digital Pattern Generator Properties ? X

Generator Name:
C

Analogue Types
DC
Sine
Pulse
Pwlin
File
Audio
Exponent
SFFM
Random
Easy HDL

Digital Types
Steady State
Single Edge
Single Pulse
Clock
Pattern
Easy HDL

Current Source?
Isolate Before?
Manual Edits?
Hide Properties?

Initial State: Low
First Edge At (Secs): 0

Timing:
Equal Mark/Space Timing?
Pulse width (Secs): 500m
'Space' Time (Secs):

Transitions
Continuous Sequence of Pulses
Determine From Pattern Length
Specific Number of Edges:

Bit Pattern
Standard High-Low Pulse Train
Specific pulse train:
LHLHLHLH
Edit...

OK Cancel

Fig. 3.50 Digital pattern generator C settings

Double-click the digital graph and enter 4 to Stop Time box (Fig. 3.51). Then click the OK button.

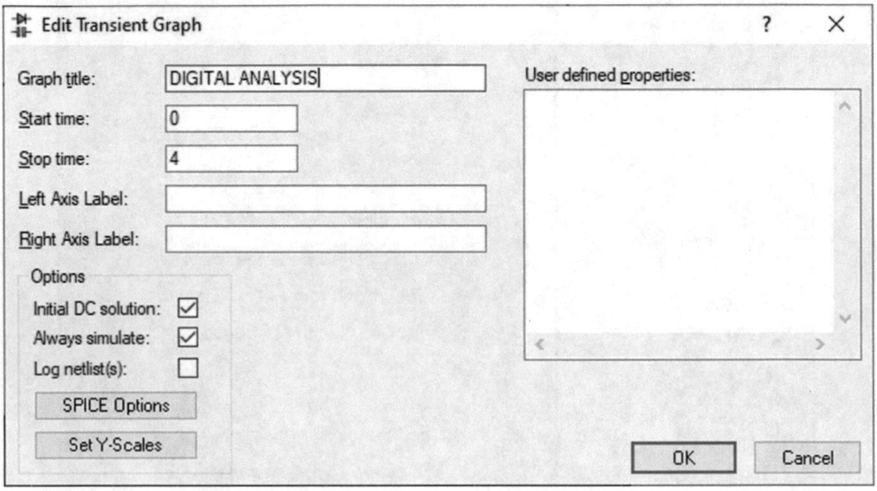

Fig. 3.51 Edit transient graph window

Drag and drop the digital pattern generator *A*, *B*, *C*, and voltage probes Sum and Carry onto the digital graph (Fig. 3.52).

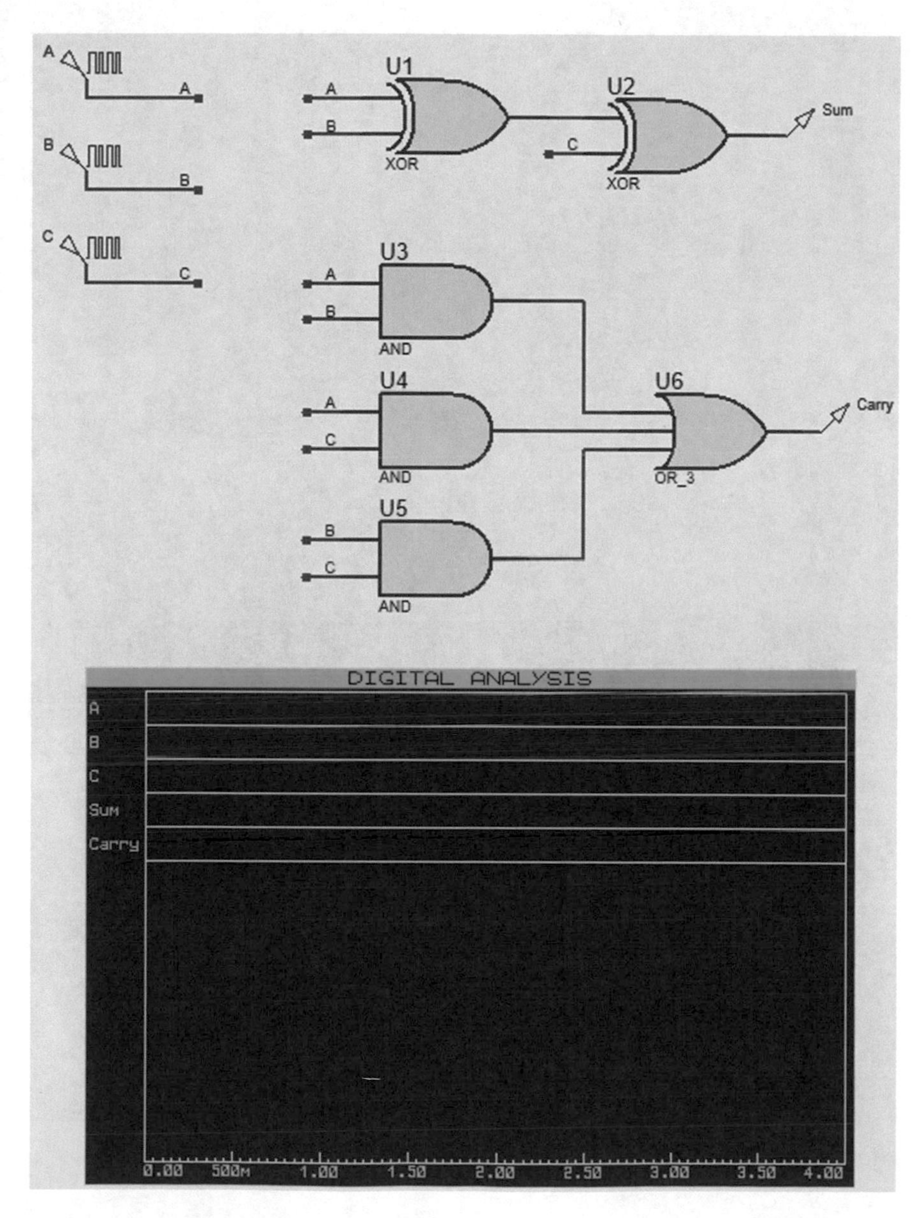

Fig. 3.52 Desired waveforms are added to the digital graph

Run the simulation. The simulation result is shown in Fig. 3.53. Now, you can study the waveforms in order to see whether or not the circuit works correctly.

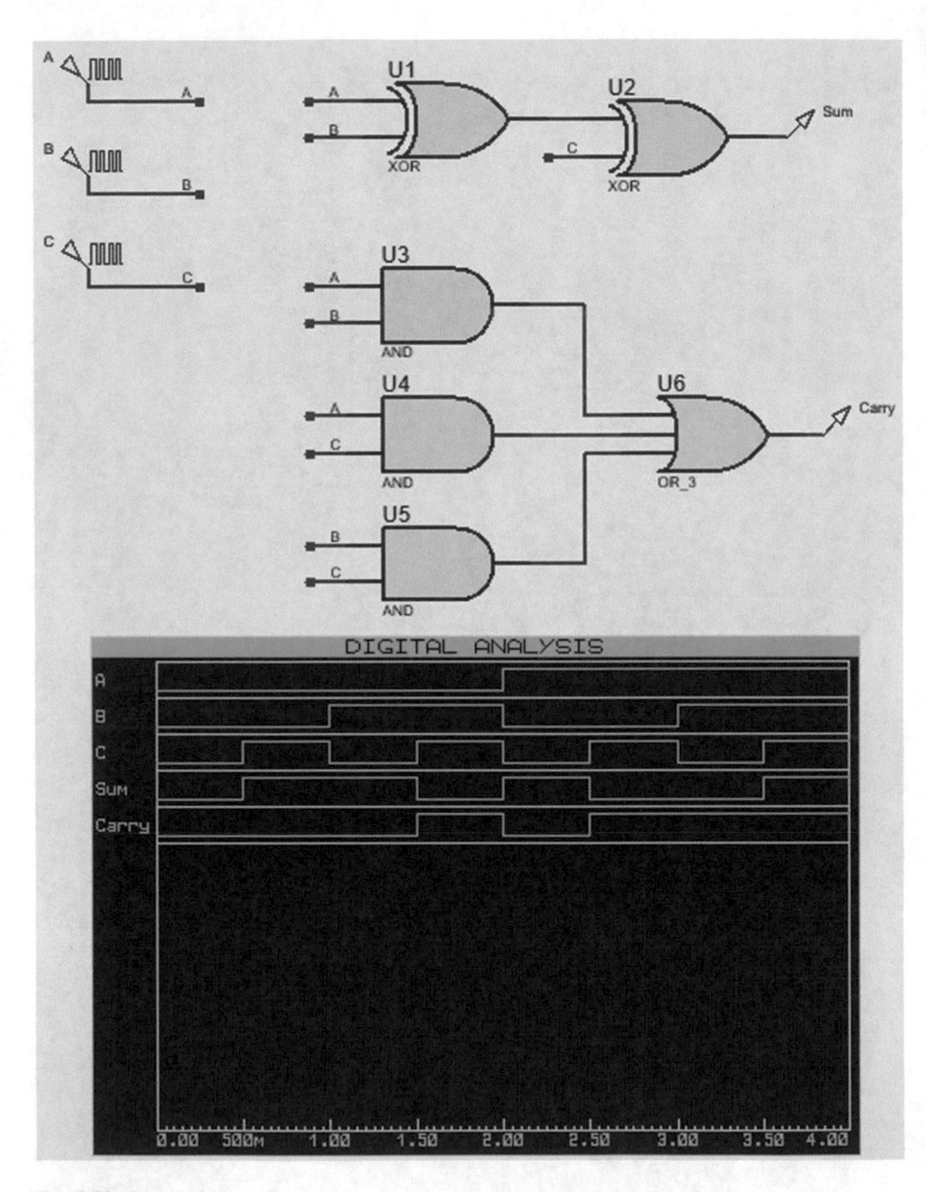

Fig. 3.53 Simulation result

3.11 Example 10: Boolean Block

The Boolean block (Fig. 3.54) permits you to implement a given Boolean function easily. Let's study an example. Consider the truth table given in Table 3.2. *A*, *B* and *C* show the inputs and *O* shows the output.

Using the sum of product $O = A'B'C' + A'B'C + AB'C + ABC'$. Let's use the Bool block to simulate the obtained Boolean function.

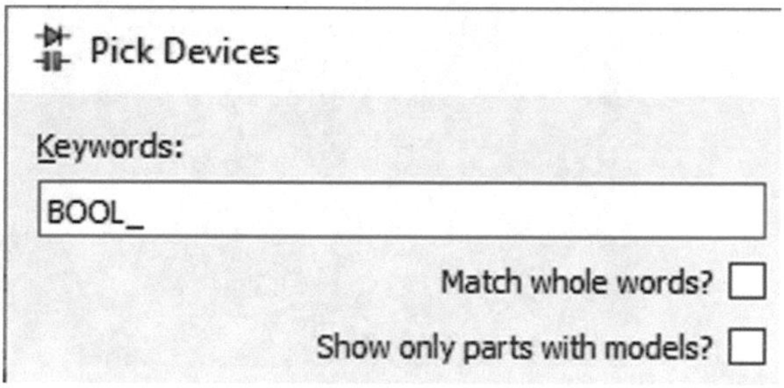

Fig. 3.54 Search for Boolean block

Table 3.2 Truth table of Example 10

A	*B*	*C*	O
0	0	0	1
0	0	1	1
0	1	0	0
0	1	1	0
1	0	0	0
1	0	1	1
1	1	0	1
1	1	1	0

Draw the schematic shown in Fig. 3.55. U1 is a Boolean block with three inputs (Fig. 3.56). Settings of U1 and D1 are shown in Figs. 3.57 and 3.58, respectively. Note that $(!A\&!B\&!C)|(A\&!B\&!C)|(A\&!B\&C)|(A\&B\&!C)$ which is entered into the Boolean Expression box (Fig. 3.57) is the translation of $A'B'C' + A'B'C + AB'C + ABC'$. & shows the logical and operation, | shows the logical or operation and ! shows the logical negation operation.

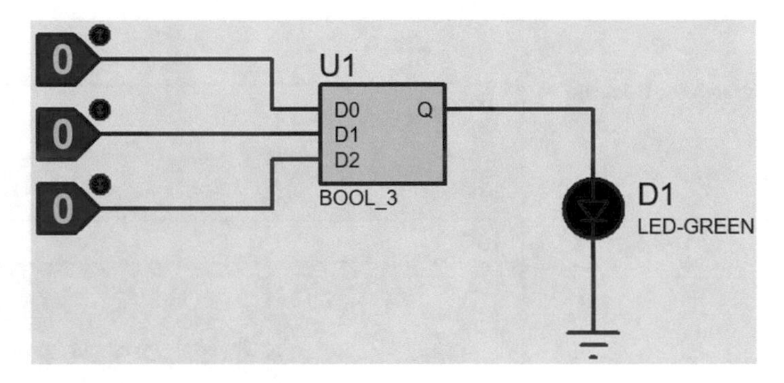

Fig. 3.55 Schematic of Example 10

Fig. 3.56 Three-input Boolean block

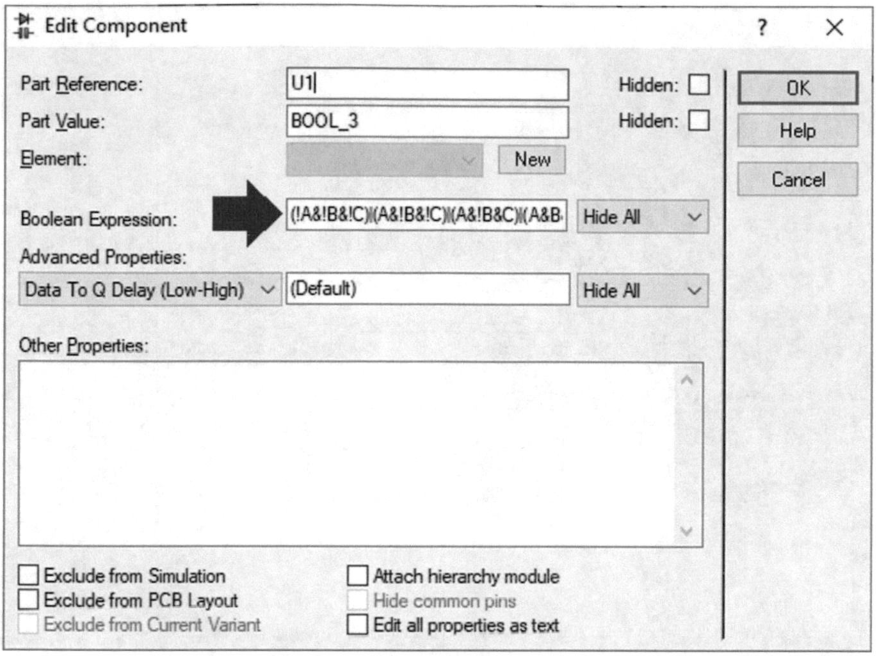

Edit Component ? X
Part Reference: U1 Hidden:
Part Value: BOOL_3 Hidden:
Element: New
Boolean Expression: (!A&!B&!C)|(A&!B&!C)|(A&!B&C)|(A&B Hide All
Advanced Properties:
Data To Q Delay (Low-High) (Default) Hide All
Other Properties:
OK
Help
Cancel
Exclude from Simulation
Exclude from PCB Layout
Exclude from Current Variant
Attach hierarchy module
Hide common pins
Edit all properties as text

Fig. 3.57 Settings of U1 Boolean block

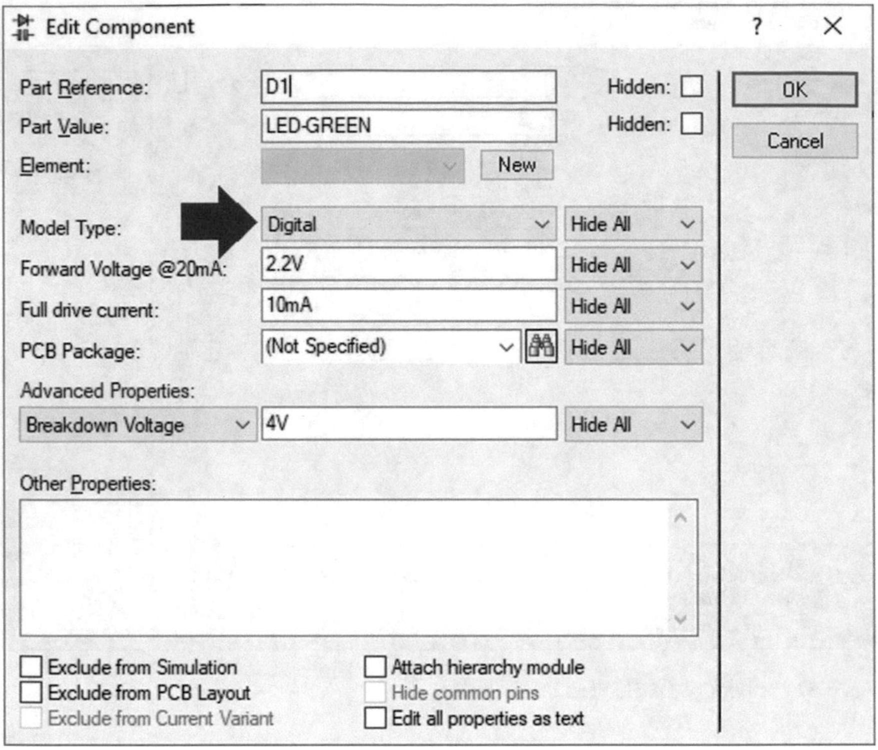

Edit Component ? ×

Part Reference: D1 Hidden: ☐ OK
Part Value: LED-GREEN Hidden: ☐ Cancel
Element: New

Model Type: Digital Hide All
Forward Voltage @20mA: 2.2V Hide All
Full drive current: 10mA Hide All
PCB Package: (Not Specified) Hide All
Advanced Properties:
Breakdown Voltage 4V Hide All

Other Properties:

☐ Exclude from Simulation ☐ Attach hierarchy module
☐ Exclude from PCB Layout ☐ Hide common pins
☐ Exclude from Current Variant ☐ Edit all properties as text

Fig. 3.58 Settings of LED

Run the simulation and apply different inputs in order to ensure that the block generates the expected outputs.

If you click the Help button in Fig. 3.57, a document of the Boolean block appears on the screen (Fig. 3.59). Study this page to obtain more information about the block.

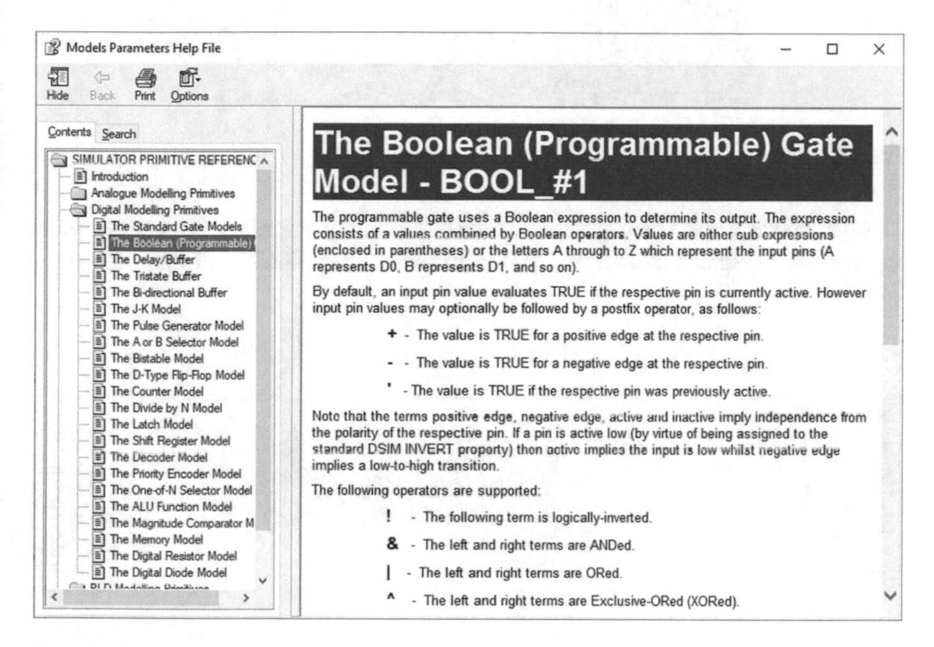

Fig. 3.59 The Boolean section of Proteus help

3.12 Example 11: Bus

Bus Mode icon (Fig. 3.60) permits you to add a bus to your schematic and make your schematic more understandable and less crowded.

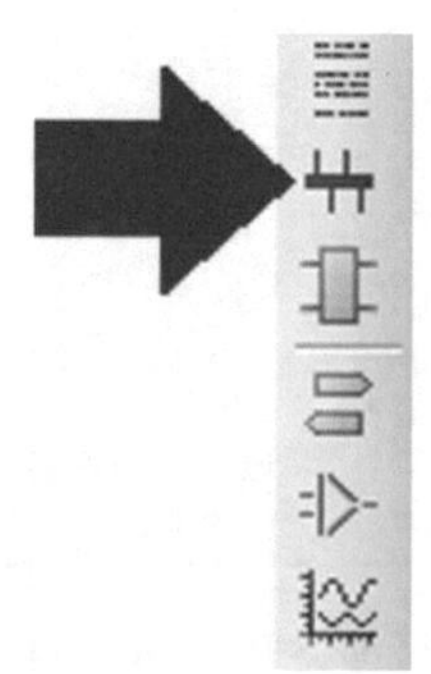

Fig. 3.60 Bus Mode icon

Proteus has a sample simulation which contains the bus block. Let's take a look at that sample simulation. Click the File > Open Sample Project (Fig. 3.61).

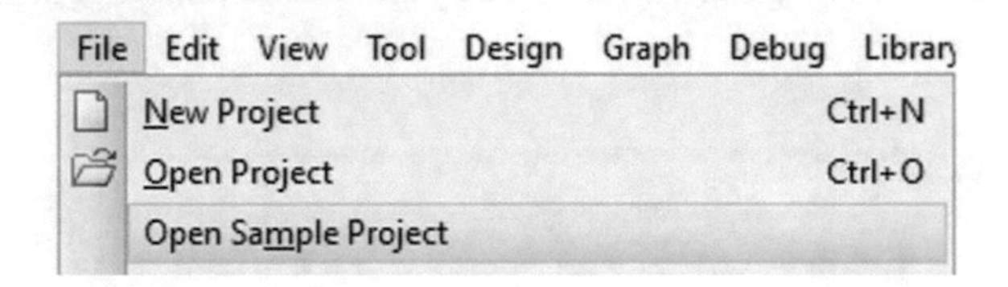

Fig. 3.61 File > Open Sample Project

After clicking the Open Sample Project, the Sample Project Browser window appears on the screen. Search for "8051 external" (Fig. 3.62) and open the result (Fig. 3.63).

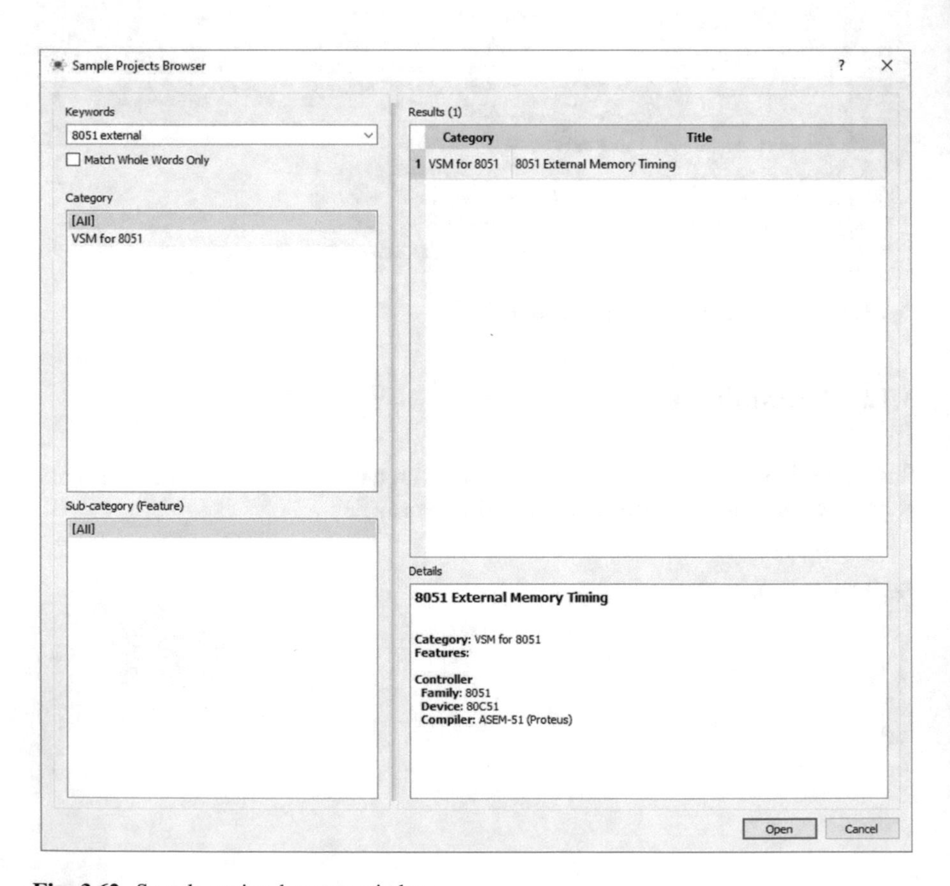

Fig. 3.62 Sample project browser window

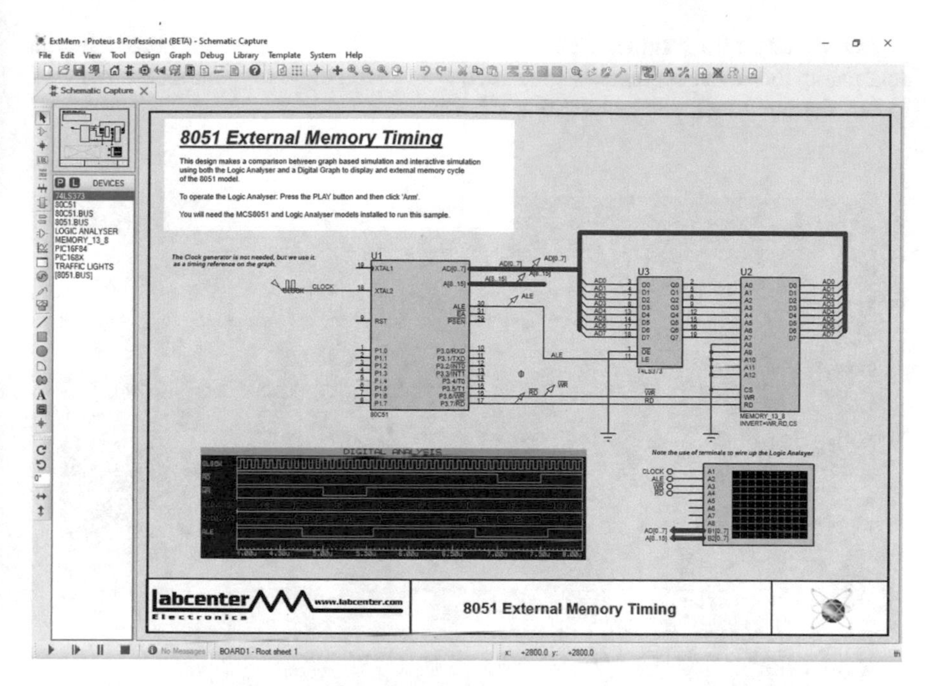

Fig. 3.63 Schematic found for 8051 external

The presence of bus block in Fig. 3.64 makes the schematic easy to understand and tidy. If you try to draw the schematic shown in Fig. 3.64 without the bus block, you will obtain a crowded and difficult-to-understand schematic.

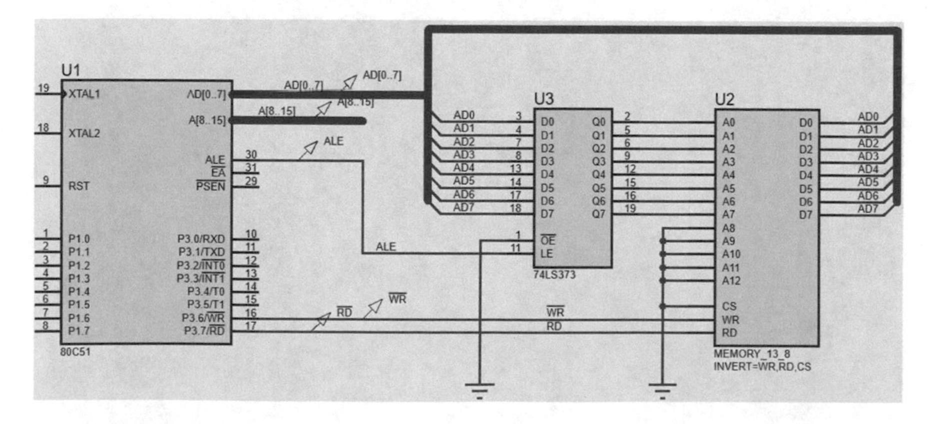

Fig. 3.64 Bus block makes the schematic tidy

Let's study an example. Consider the schematic shown in Fig. 3.65. In this schematic, the seven-segment is controlled with seven logic toggle blocks. We want to use the bus block and redraw this schematic.

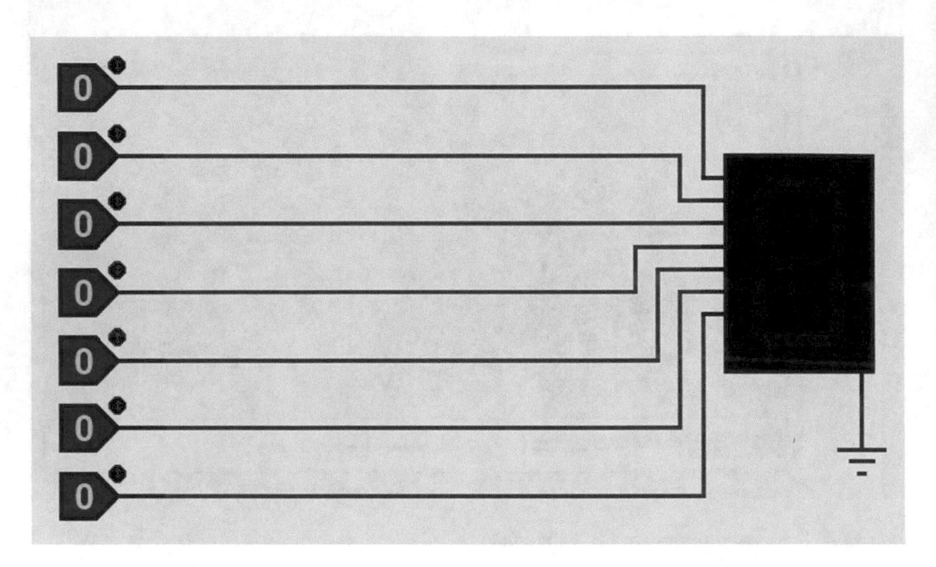

Fig. 3.65 Schematic of Example 11

Remove the wires (Fig. 3.66).

Fig. 3.66 Wires are removed from Fig. 3.65

Click the Bus Mode icon (Fig. 3.60). Then click on the schematic and (without holding down the mouse button) draw the bus. Double-click when you want to finish drawing the bus (Fig. 3.67).

Fig. 3.67 Bus is added to the schematic

Connect the component terminals to the bus (Fig. 3.68).

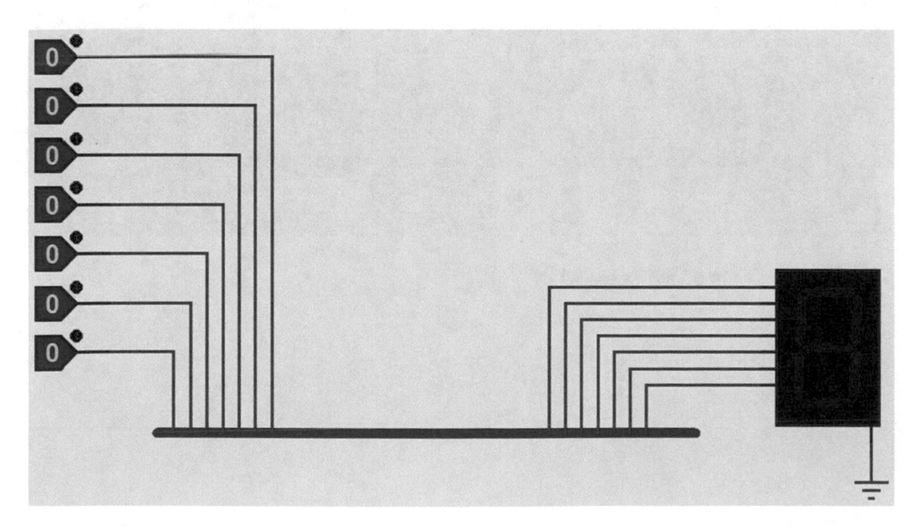

Fig. 3.68 Connecting the components to the bus

Use the Wire Label Mode icon (Fig. 3.69) to add a label to wires (Fig. 3.70). Wires with the same labels are connected together.

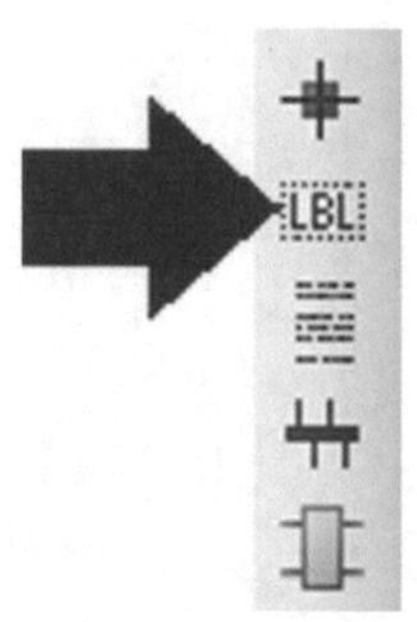

Fig. 3.69 Wire label mode icon

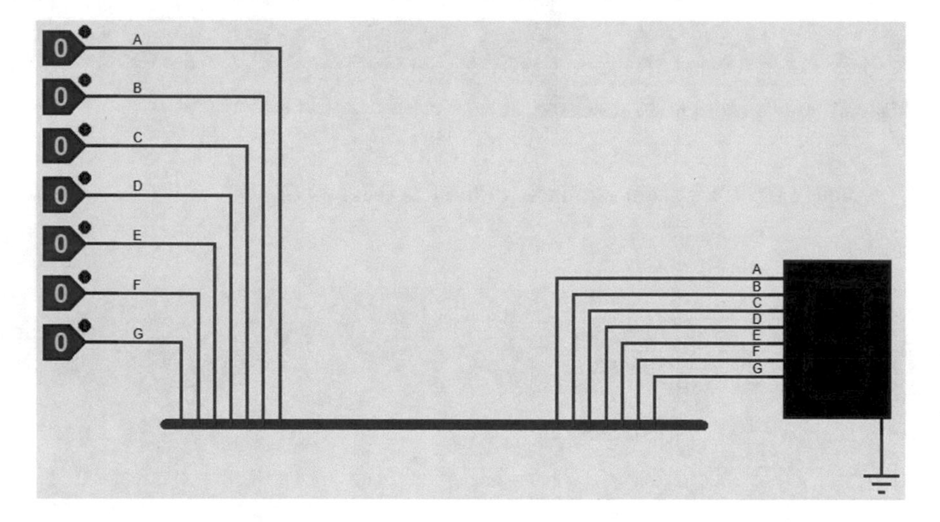

Fig. 3.70 Labels are added to the wires

Run the simulation. Note that you can control the seven segments with the logic toggle blocks. This shows that the transfer of data from logic toggle blocks to the seven segments is done. Both schematics in Figs. 3.65 and 3.71 do the same job. However, the schematic in Fig. 3.71 is more understandable and professional in comparison to the one in Fig. 3.65.

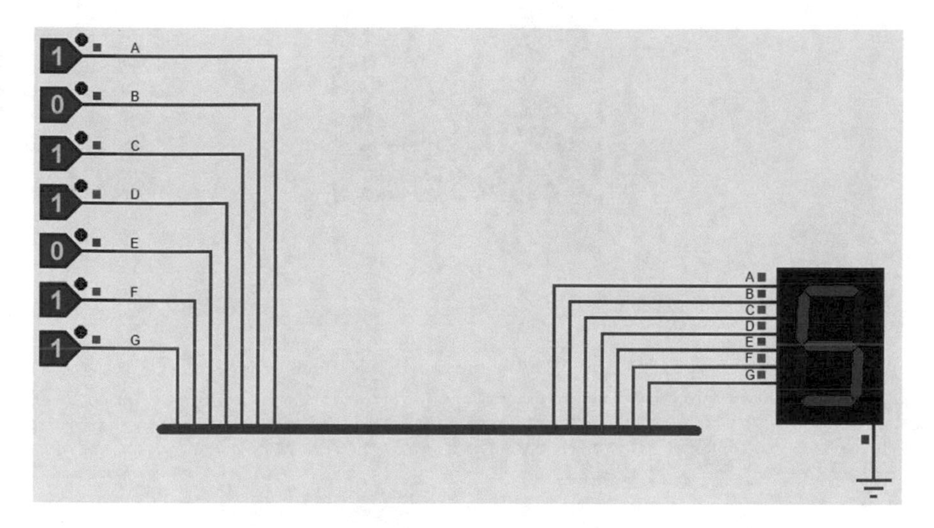

Fig. 3.71 Equivalent of Fig. 3.65 with bus block

3.13 Example 12: Simulation of Circuits Contains a Microcontroller

Proteus can simulate circuits which contain a microcontroller as well. Click the P_{icon} (Fig. 3.72) and then type the microcontroller name in the Keywords box (Fig. 3.73) in order to see whether or not a microcontroller is supported by Proteus.

In this example, we want to simulate a simple code written for the Arduino UNO board. Arduino UNO board uses an ATmega 328P microcontroller. This microcontroller is supported by Proteus.

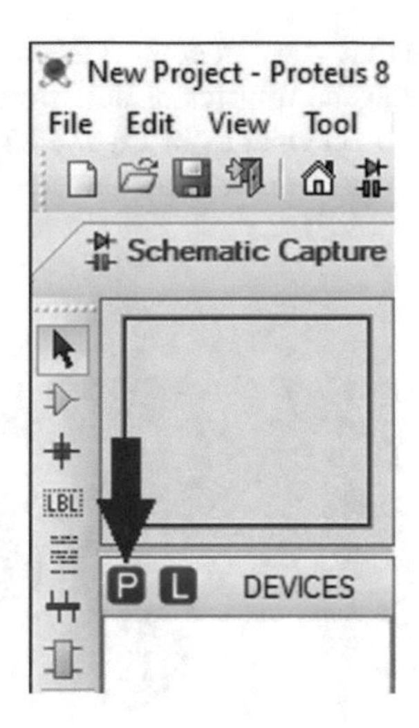

Fig. 3.72 P_{icon}

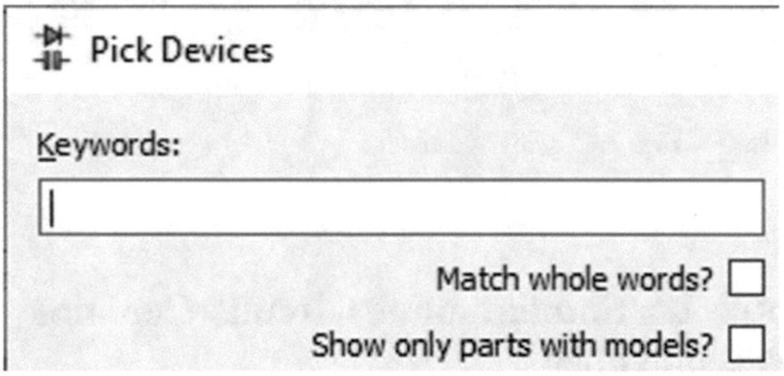

Fig. 3.73 After clicking the P_{icon}, pick devices window appears

Let's start. Open the Arduino IDE and type the program shown in Fig. 3.74. The Arduino IDE can be downloaded from https://www.arduino.cc/en/software. The code in Fig. 3.74 turns on the Arduino UNO board's LED and waits for 1 s. Then it turns off the LED and waits for 1 s. This process is repeated.

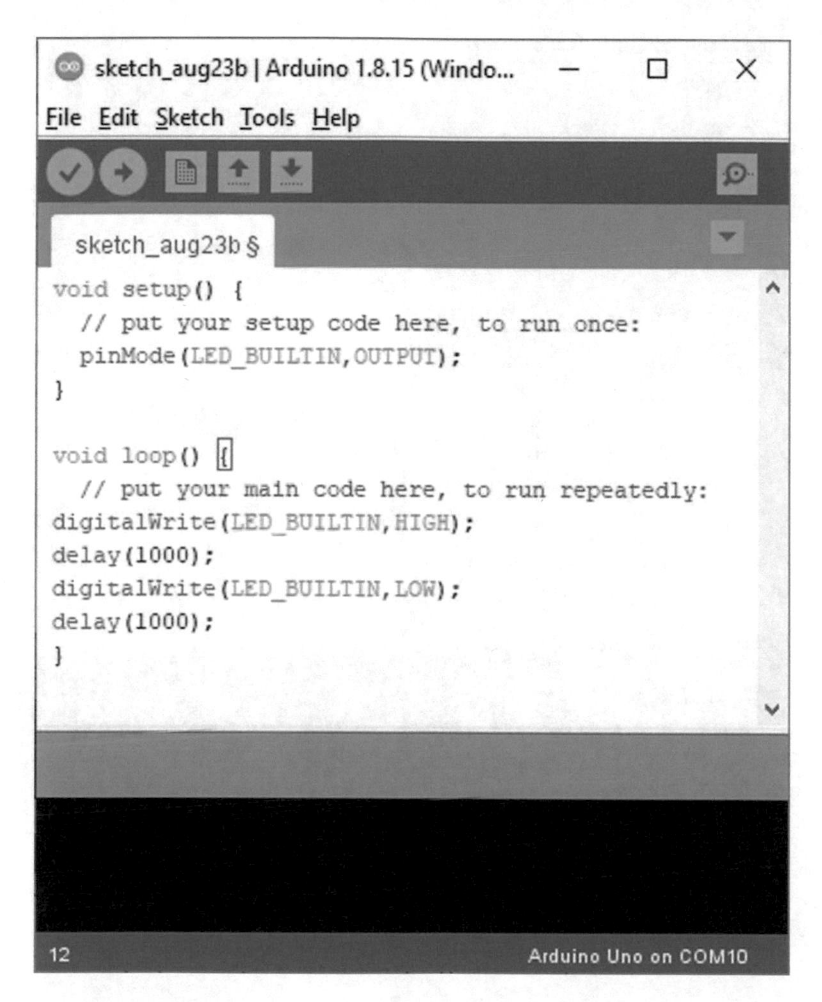

Fig. 3.74 Sample Arduino code

Save the written code in the desired path by clicking the File > Save (Fig. 3.75). Here, the code is saved with the name LED_Blink.

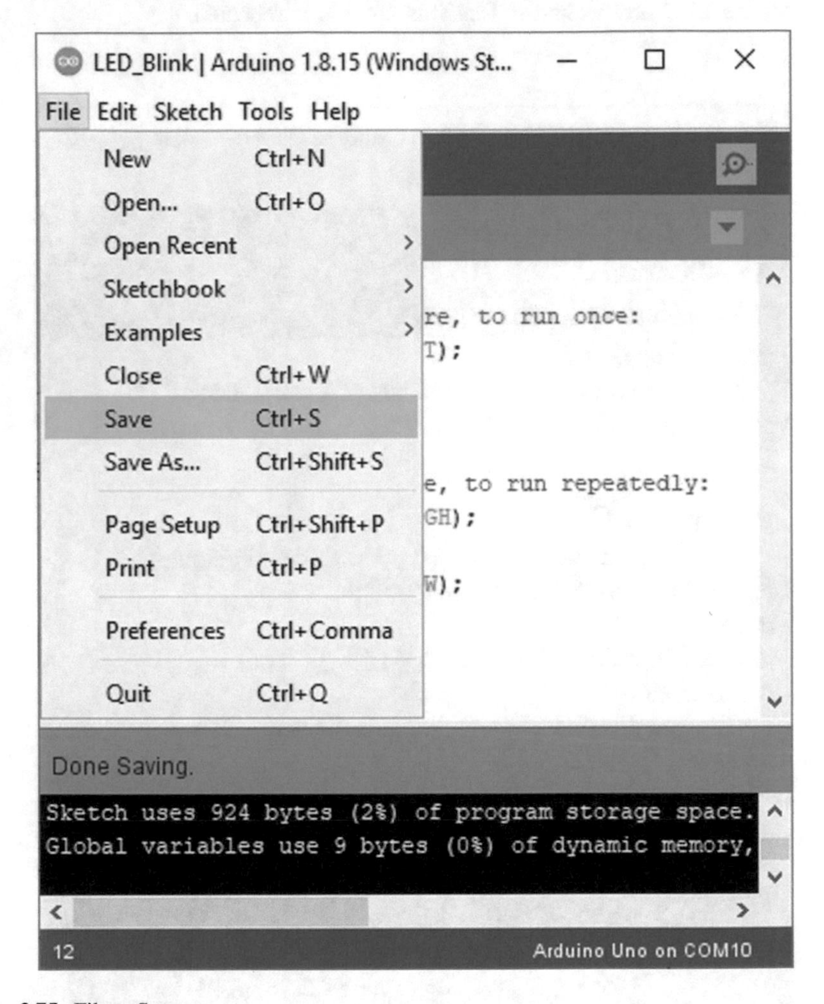

Fig. 3.75 File > Save

Use the Sketch > Export compiled Binary to generate the .hex file (Fig. 3.76). After clicking the Export compiled Binary, two .hex files will be added to the directory in which you saved the code (Fig. 3.77). The LED_Blink.ino.standard.hex will be used by the Proteus schematic.

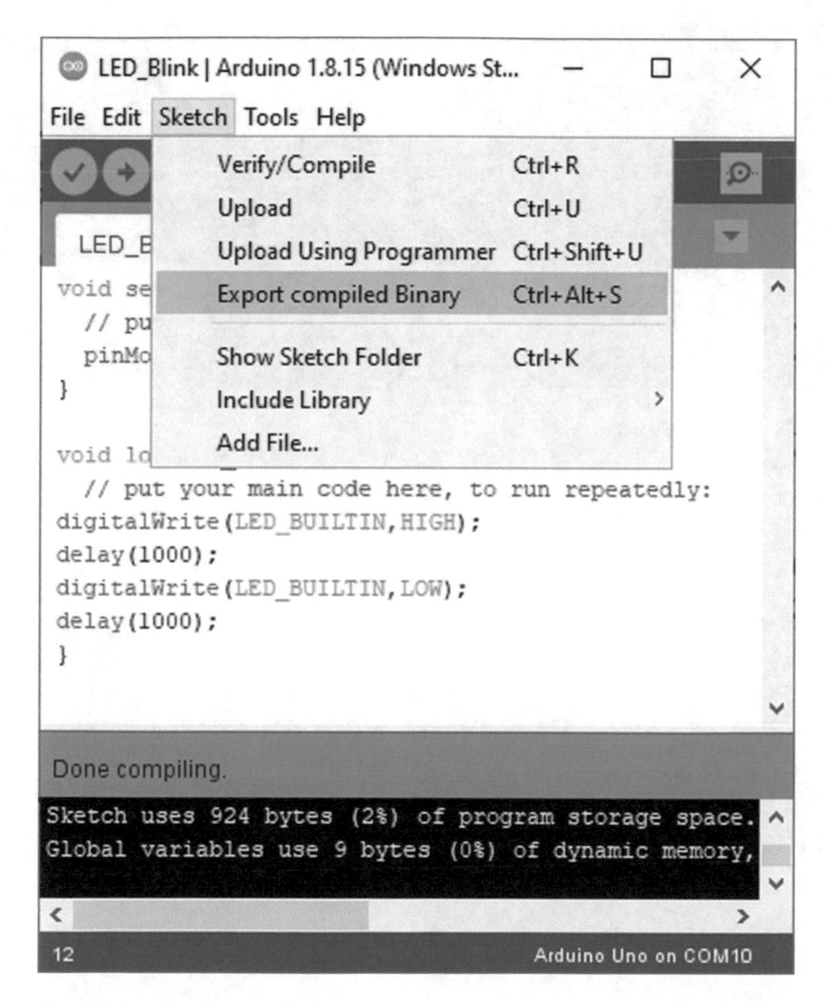

Fig. 3.76 Sketch > Export compiled binary

LED_Blink	23/08/2021 13:19	INO File	1 KB
LED_Blink.ino.standard.hex	23/08/2021 13:20	HEX File	3 KB
LED_Blink.ino.with_bootloader.standard....	23/08/2021 13:20	HEX File	4 KB

Fig. 3.77 Generated hex file

Draw the schematic shown in Fig. 3.78 (V_{CC} = 5 V). Settings of ATMEGA328P and crystal X1 are shown in Figs. 3.79 and 3.80, respectively.

Run the simulation. The LED starts to blink.

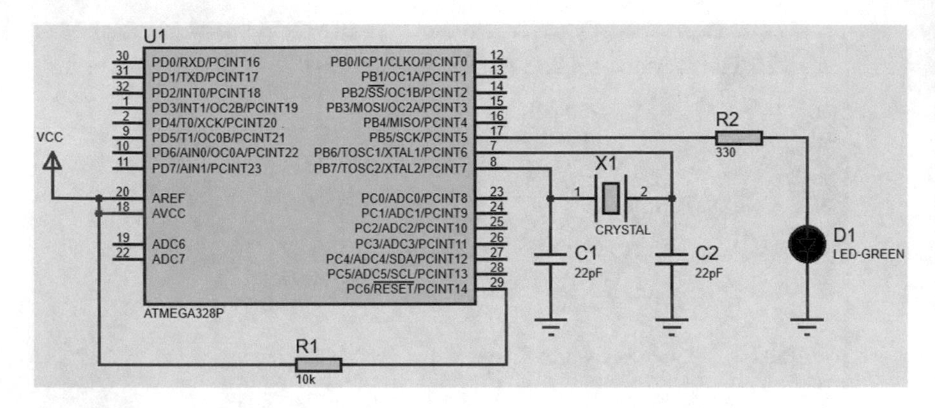

Fig. 3.78 Proteus schematic

Edit Component ? X

Part Reference:	U1	Hidden: ☐	OK
Part Value:	ATMEGA328P	Hidden: ☐	Help
Element:		New	Data
PCB Package:	QFP80P900X900X120-32	Hide All	Hidden Pins
Program File:	LED_Blink\LED_Blink.ino.stanc	Hide All	Edit Firmware
CLKDIV8 (Divide clock by 8)	(1) Unprogrammed	Hide All	Cancel
CKOUT (Clock output)	(1) Unprogrammed	Hide All	
RSTDISBL (External reset disable)	(1) Unprogrammed	Hide All	
WDTON (Watchdog Timer Always On)	(1) Unprogrammed	Hide All	
BOOTRST (Select reset vector)	(1) Unprogrammed	Hide All	
CKSEL Fuses:	(1111) Ext. Crystal 8.0-MHz	Hide All	
Boot Loader Size:	(00) 1024 words. Starts at 0x1CI	Hide All	
SUT Fuses:	(10)	Hide All	

Advanced Properties:

Clock Frequency	16MHz	Hide All

Other Properties:

☐ Exclude from Simulation
☐ Exclude from PCB Layout
☐ Exclude from Current Variant
☐ Attach hierarchy module
☐ Hide common pins
☐ Edit all properties as text

Fig. 3.79 Settings of ATMEGA328P microcontroller

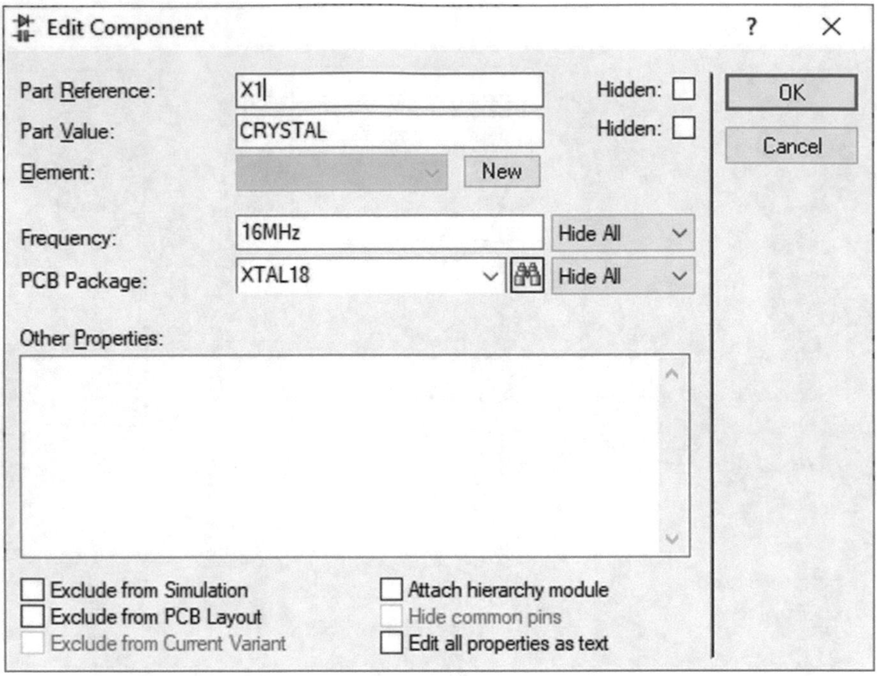

Edit Component ? ×

Part Reference: X1 Hidden: ☐ OK

Part Value: CRYSTAL Hidden: ☐ Cancel

Element: New

Frequency: 16MHz Hide All

PCB Package: XTAL18 Hide All

Other Properties:

☐ Exclude from Simulation ☐ Attach hierarchy module

☐ Exclude from PCB Layout ☐ Hide common pins

☐ Exclude from Current Variant ☐ Edit all properties as text

Fig. 3.80 Settings of crystal X1

3.14 Exercises

1. Simulate the $F = (A + B)(A'B + AC' + B'C)$ Boolean function in Proteus.
2. (a) Simulate an 8×1 multiplexer in Proteus and test it.

   (b) Simulate a 4-bit shift register in Proteus and test it. Use the D flip-flop.
3. Simulate a 4-bit Johnson counter in Proteus and test it.
4. Assume the state diagram shown in Fig. 3.81.

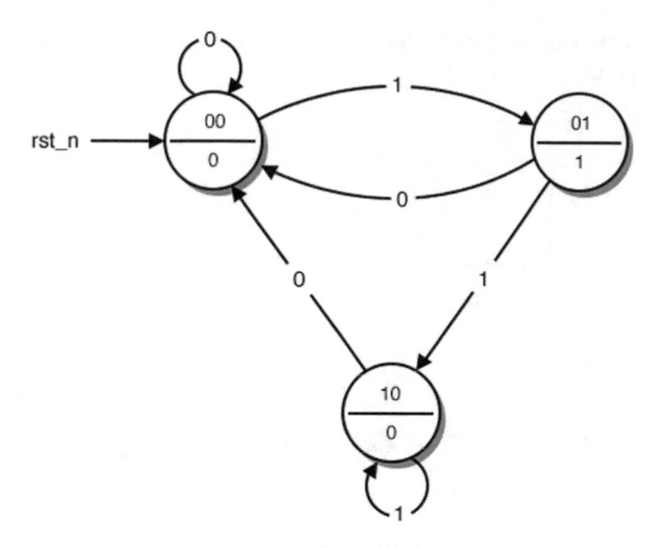

Fig. 3.81 State diagram for Exercise 4

The state table of this state diagram is shown in Table 3.3.

Table 3.3 State table of Fig. 3.81

Current state		Input I	Next state		Outputs Y
A	B		A_{next}	B_{next}	
0	0	0	0	0	0
0	0	1	0	1	0
0	1	0	0	0	1
0	1	1	1	0	1
1	0	0	0	0	0
1	0	1	1	0	0
1	1	0	X	X	X
1	1	1	X	X	X

(a) Use hand analysis to ensure that the circuit in Fig. 3.82 implements the given state diagram.

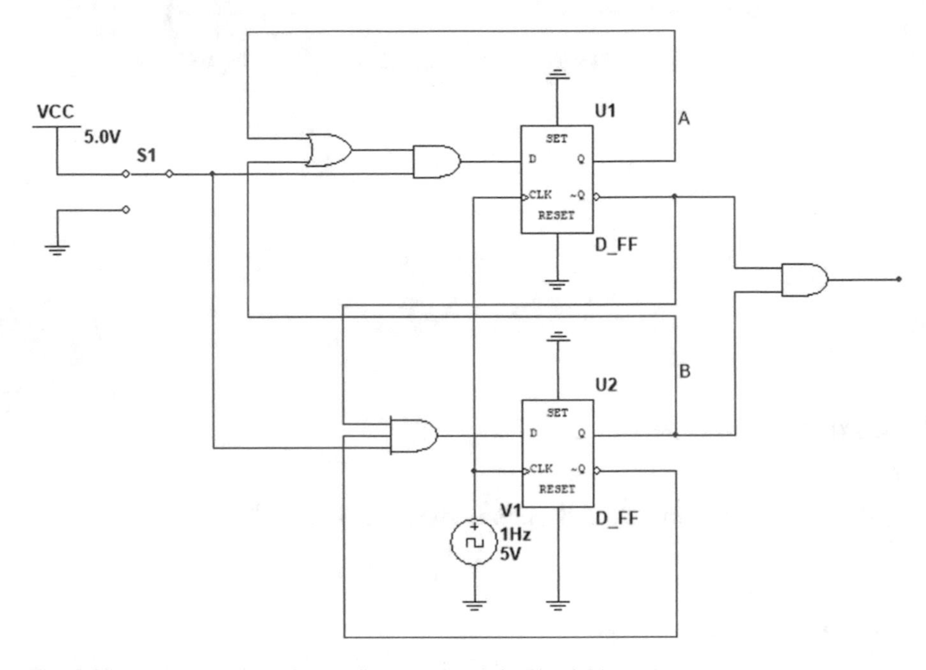

Fig. 3.82 Implementation of state diagram shown in Fig. 3.81

(b) Simulate the circuit shown in Fig. 3.82 with Proteus and ensure that it works similar to Table 3.3.
(c) Draw the JK flip-flop implementation for the given state diagram and test it in Proteus.

References for Further Study

Asadi F, Pongswatd S (2022) Programming the ARM Cortex-M4-based STM32F4 Microcontrollers with Simulink, Springer
Floy T (2014) Digital fundaments, 11th edn. Pearson
Harris D, Harris S (2012) Digital design and computer architecture. Morgan Kaufmann
Mano M, Ciletti MD (2018) Digitla design, 6th edn. Pearson
Marcovitz AB (2009) Introduction to logic design, 3rd edn. Mc-Graw Hill
Nelson V, Nagle HT, Irvin HT, Carrol BD (1995) Digital logic circuit analysis and design. Pearson

Chapter 4
Simulation of Power Electronics Circuits with Proteus®

Abstract This chapter shows how a power electronics circuit can be analyzed in Proteus. In this chapter you will learn how to simulate DC–DC, AC–AC, AC–DC and DC–AC converters.

Keywords Buck · Boost · Buck-Boost · DC–DC converter · DC chopper · Continuous conduction mode · Discontinuous conduction mode · CCM · DCM · Rectifier · Inverter · Single phase inverter · Three phase inverter · Efficiency · Dimmer

4.1 Introduction

In this chapter, you will learn how to analyze power electronics circuits in Proteus. The theory behind the studied circuits can be found in any standard power electronics textbook (Hart 2010; Mohan et al. 2007; Rashid 2013; Erikson and Maksimovic 2020). Similar to previous chapters, doing some hand calculations for the given circuits and comparing the hand analysis results with Proteus results are recommended.

4.2 Example 1: Buck Converter Circuit

In this example, we want to simulate a buck converter circuit. Draw the schematic as shown in Fig. 4.1. Pulse generator VGS and voltage-controlled voltage source E_1 turn on and off the MOSFTE Q_1. Settings of the VGS are shown in Fig. 4.2. According to the settings shown in Fig. 4.2, the switching frequency is 25 kHz and the duty cycle of the pulse applied to the gate source of MOSFET is 60%.

F. Asadi, *Essential Circuit Analysis Using Proteus®*, Energy Systems in Electrical Engineering, https://doi.org/10.1007/978-981-19-4353-9_4

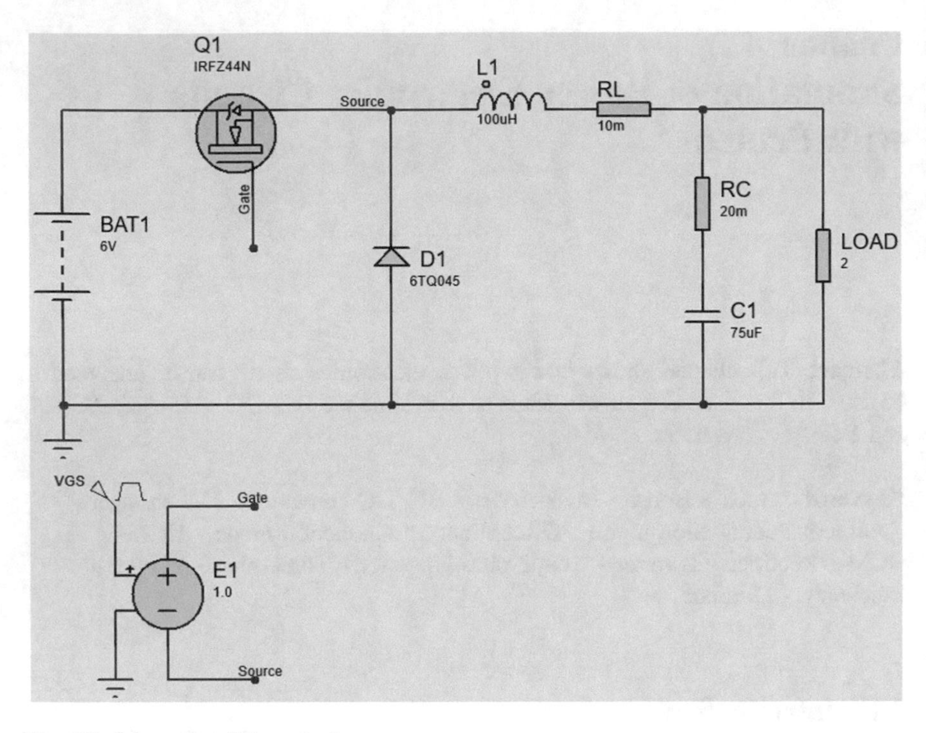

Fig. 4.1 Schematic of Example 1

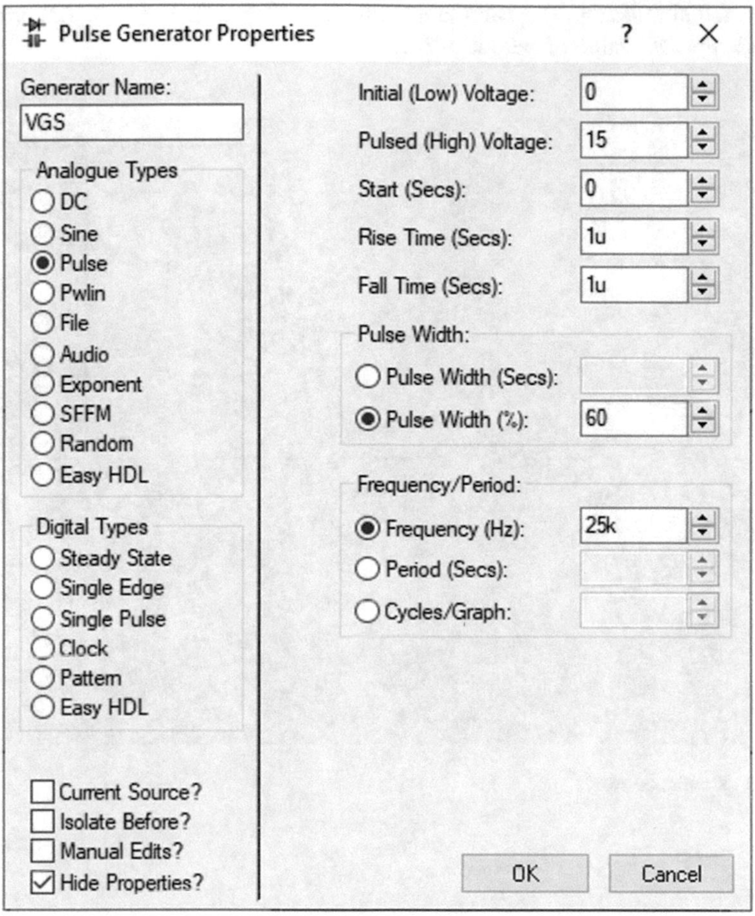

Pulse Generator Properties

Generator Name: VGS

Analogue Types: DC, Sine, Pulse (selected), Pwlin, File, Audio, Exponent, SFFM, Random, Easy HDL

Digital Types: Steady State, Single Edge, Single Pulse, Clock, Pattern, Easy HDL

Current Source? (unchecked)
Isolate Before? (unchecked)
Manual Edits? (unchecked)
Hide Properties? (checked)

Initial (Low) Voltage: 0
Pulsed (High) Voltage: 15
Start (Secs): 0
Rise Time (Secs): 1u
Fall Time (Secs): 1u

Pulse Width:
Pulse Width (Secs):
Pulse Width (%): 60 (selected)

Frequency/Period:
Frequency (Hz): 25k (selected)
Period (Secs):
Cycles/Graph:

OK Cancel

Fig. 4.2 Settings of VGS

Run the simulation. The Simulation result is shown in Fig. 4.3. According to Fig. 4.3, the DC value of output is 3.63 V.

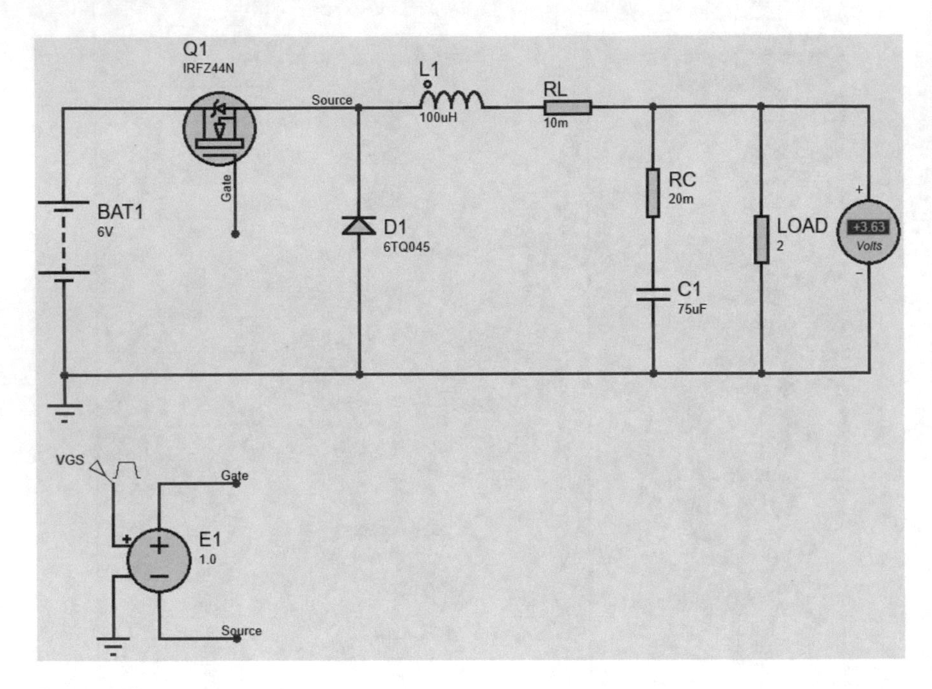

Fig. 4.3 Simulation result

Let's see the output changes in the [0, 5 ms] time interval. Use the schematic shown in Fig. 4.4 for this purpose. The simulation result is shown in Fig. 4.5.

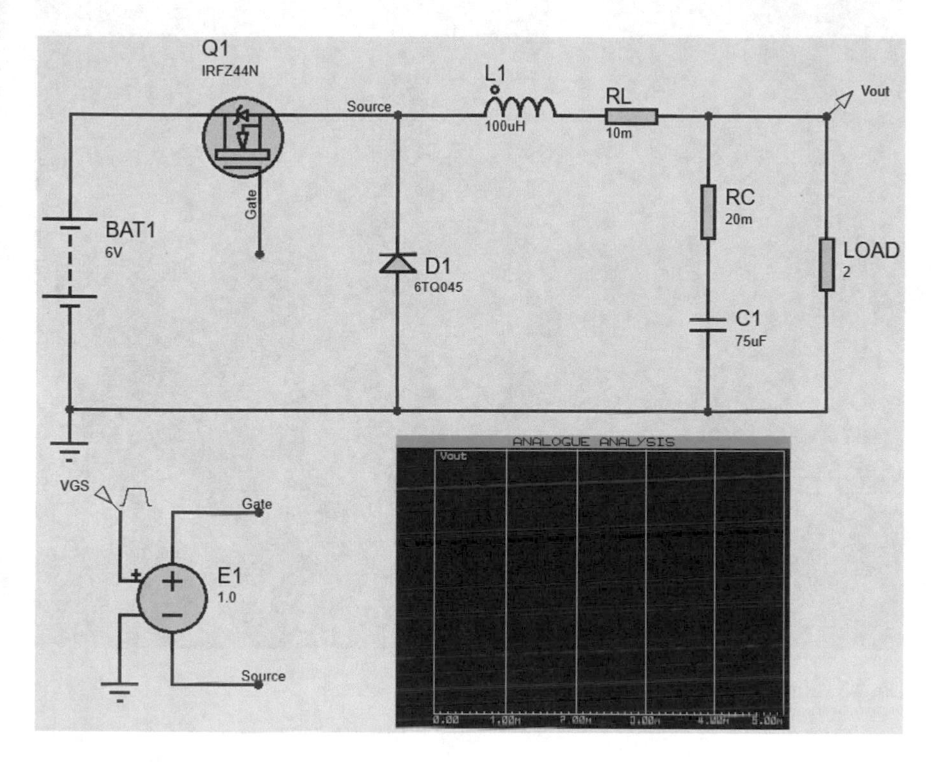

Fig. 4.4 Studying the circuit behavior for [0, 5 ms] interval

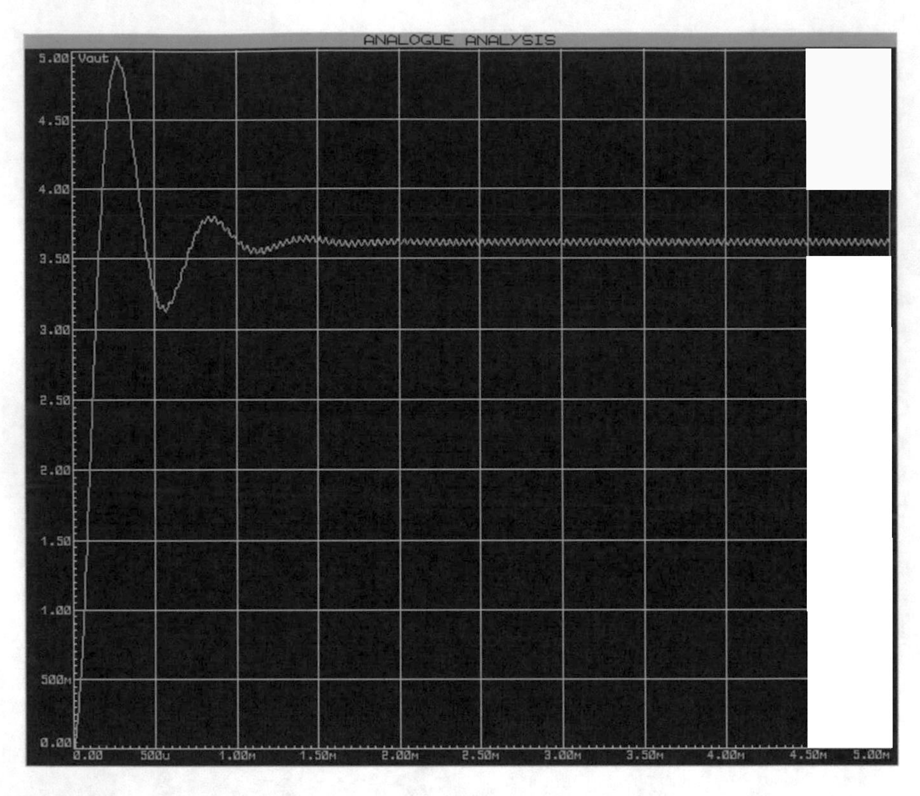

Fig. 4.5 Simulation result

Let's measure the output voltage ripple. In order to do this, zoom into the steady-state portion of the graph (Fig. 4.6). According to Fig. 4.6, the output voltage ripple is $V_{max} - V_{min} = 3.64–3.60 = 0.04$ V = 40 mV.

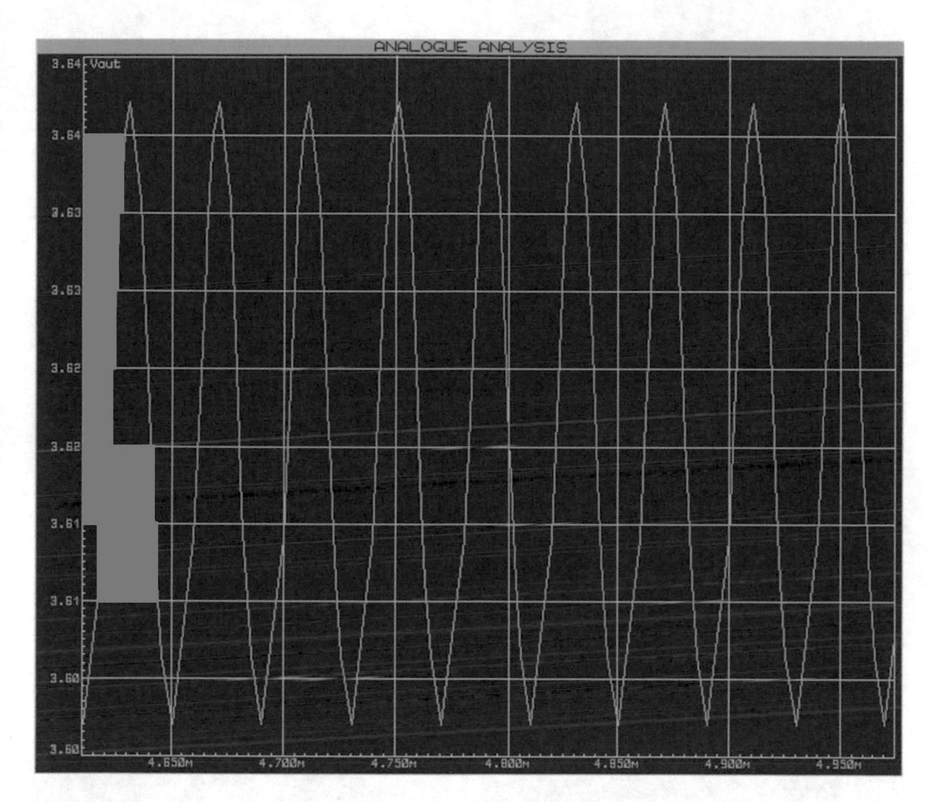

Fig. 4.6 Steady-state waveform for inductor current

4.3 Example 2: Operating Mode of Converter

In this example, we want to determine the operating mode [Continuous Conduction Mode (CCM) or Discontinuous Conduction Mode (DCM)] of the converter of Example 1. We need to observe the inductor current in order to determine the operating mode of converter. The Schematic shown in Figs. 4.7 or 4.8 helps us to observe the inductor current.

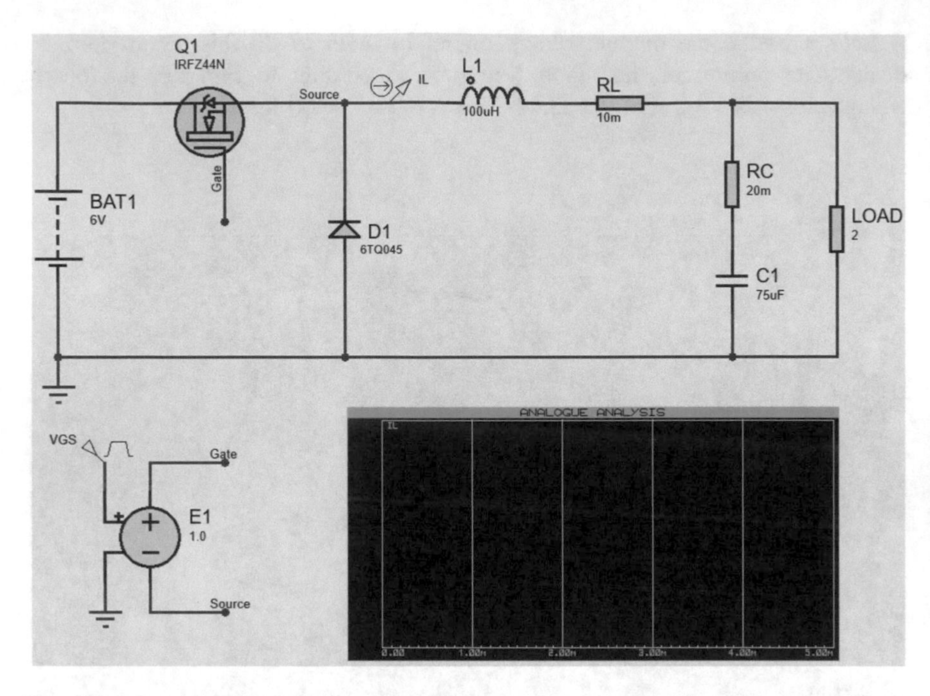

Fig. 4.7 Schematic of Example 2

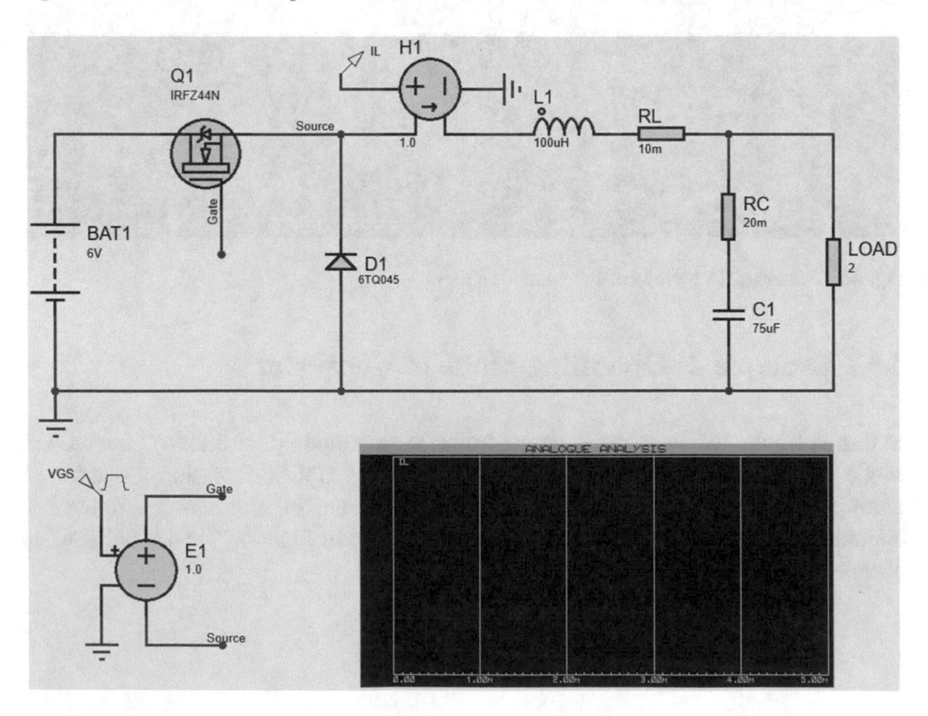

Fig. 4.8 Equivalent schematic for Fig. 4.7

Run the simulation. The simulation result is shown in Fig. 4.9. Steady-state portion of the waveform is shown in Fig. 4.10. According to Fig. 4.10, the inductor current is always positive. So, the converter is operated in Continuous Conduction Mode (CCM).

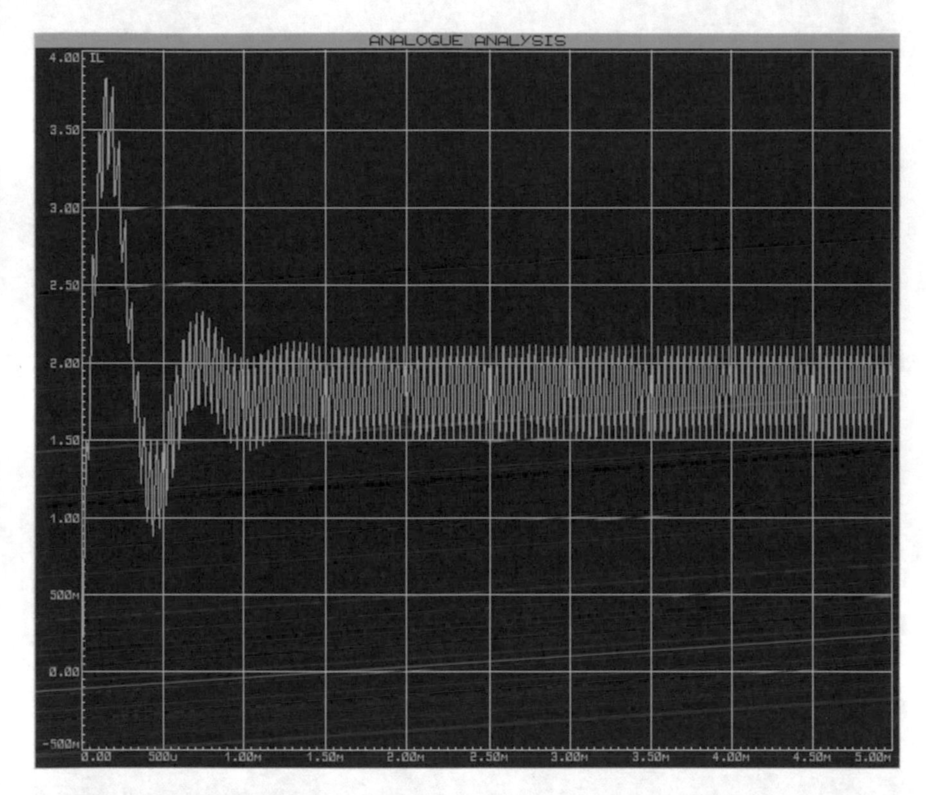

Fig. 4.9 Inductor current waveform

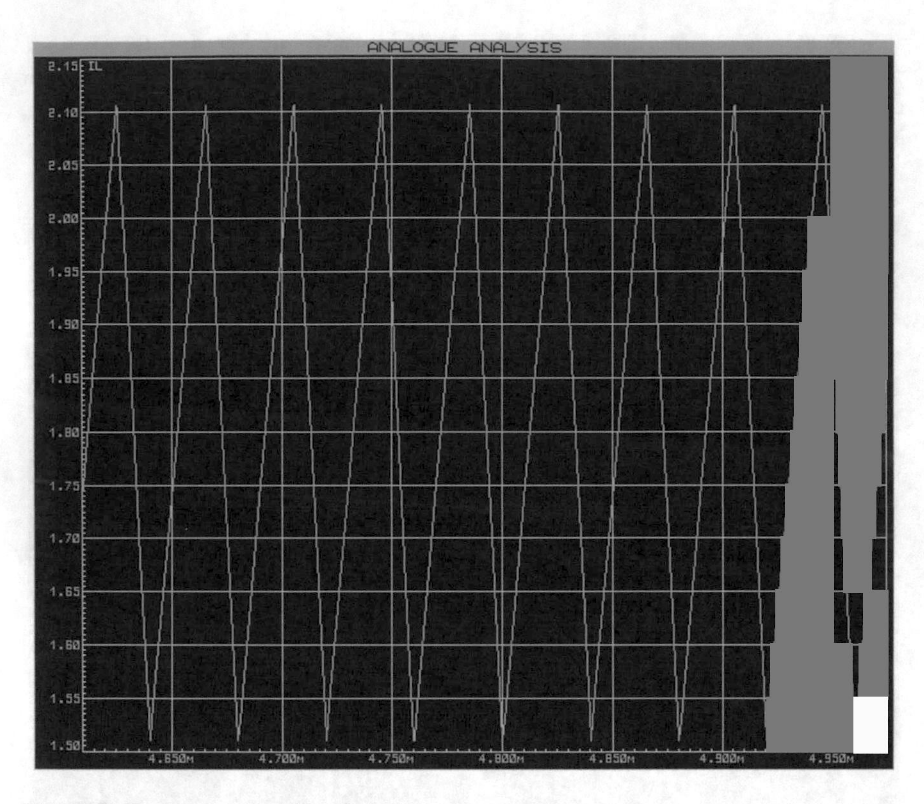

Fig. 4.10 Steady-state waveform for inductor current

Now stop the simulation and increase the output load to 25 Ω. Run the simulation with a 25 Ω load. The simulation result is shown in Fig. 4.11.

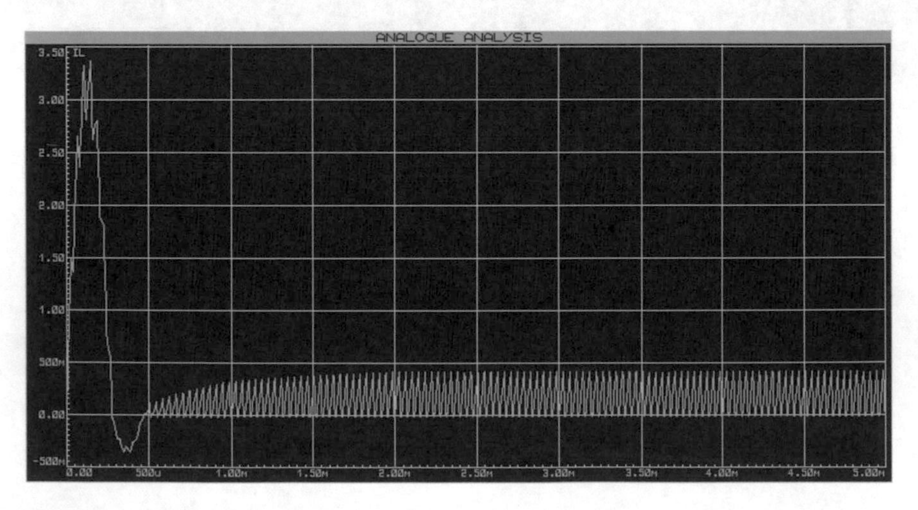

Fig. 4.11 Inductor current waveform

Let's take a look at the steady-state waveform. Steady-state portion of the waveform is shown in Fig. 4.12. Note that the minimum of steady-state current reached zero amps. So, the converter is operated in discontinuous conduction mode for 25 Ω load.

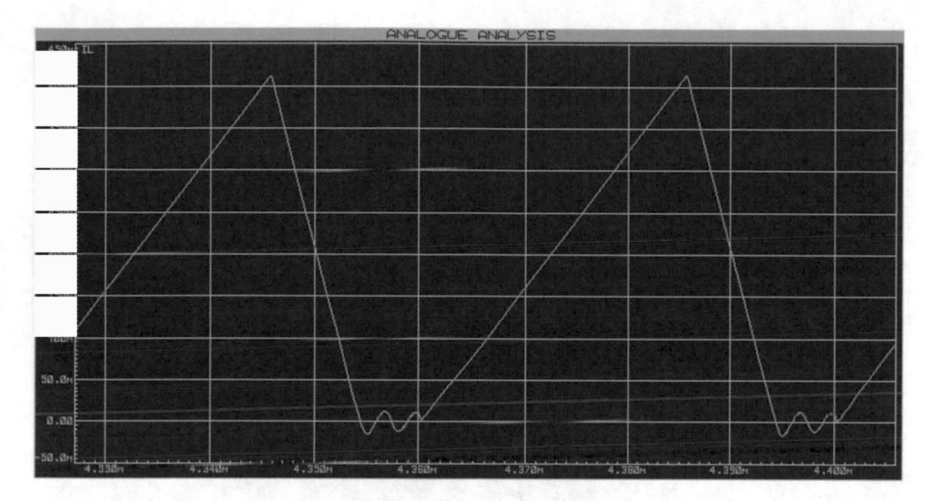

Fig. 4.12 Steady-state waveform for inductor current

4.4 Example 3: Efficiency of the Converter

In this example, we want to calculate the efficiency of the converter. According to Fig. 4.13, the average power consumed by the output load is 6.5522 W.

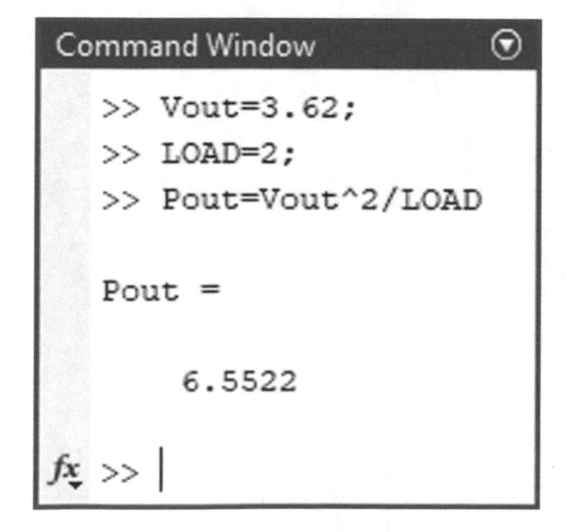

Fig. 4.13 MATLAB calculations

It's time to measure the average input power. The average input power is the product of input voltage into the average value of current drawn from the source. The schematic shown in Fig. 4.14 helps us to see the current drawn from the input voltage source.

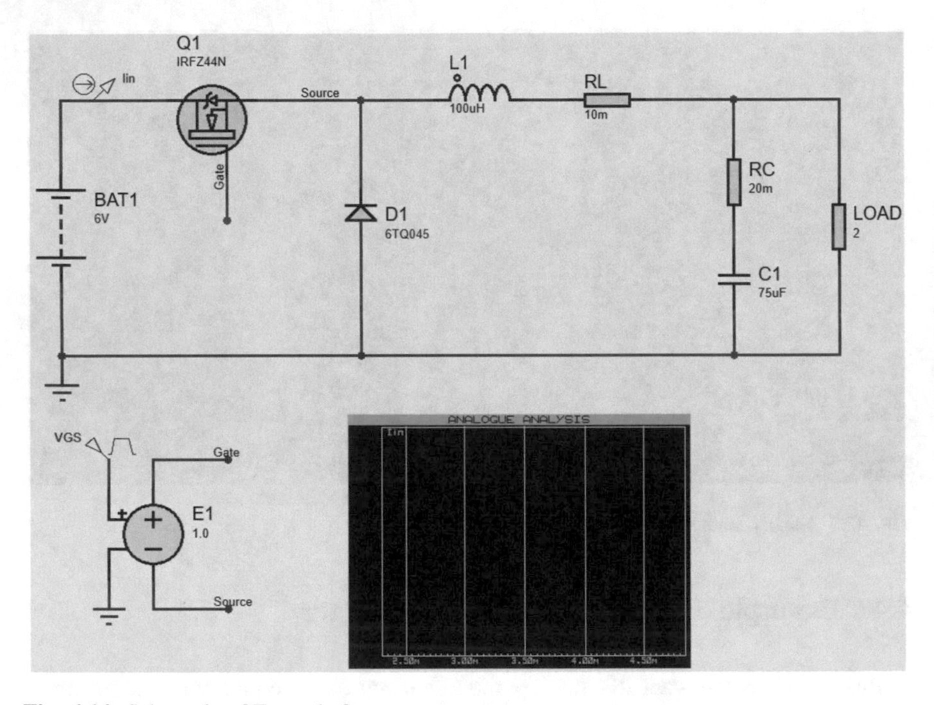

Fig. 4.14 Schematic of Example 3

Run the simulation. The simulation result is shown in Fig. 4.15 and the steady-state waveform is shown in Fig. 4.16.

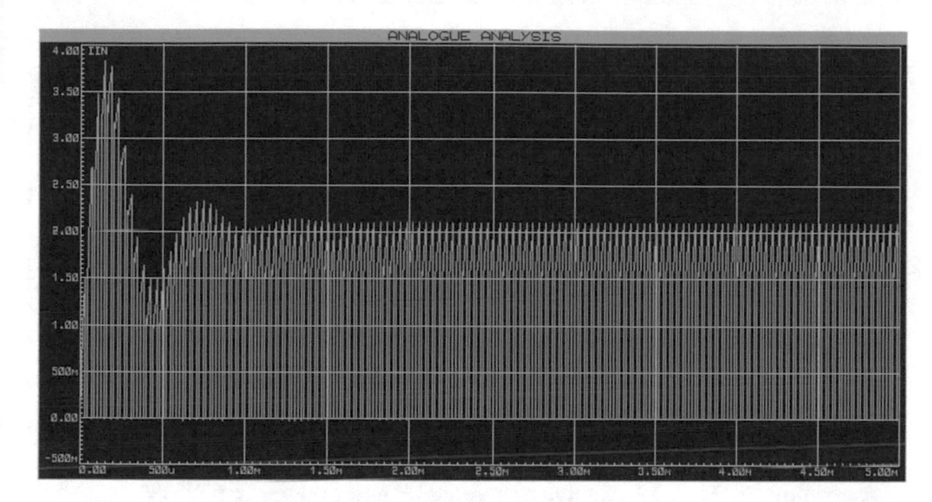

Fig. 4.15 Simulation result

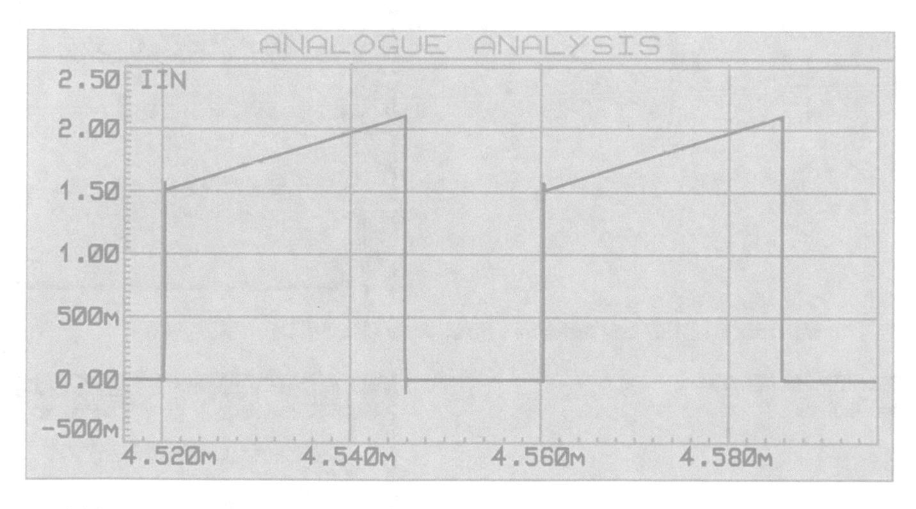

Fig. 4.16 Steady-state waveform for inductor current

According to Fig. 4.16, the waveform of the current drawn from input voltage source is trapezoidal and its frequency is 25 kHz. The value of input current at $t = 4.5604$ ms and $t = 4.5857$ ms are 1.51 A and 2.1 A, respectively. The average value of the input current waveform is calculated in Fig. 4.17.

Fig. 4.17 MATLAB calculations

```
Command Window
>> DT=4.5857e-3-4.5604e-3;
>> T=1/25e3;
>> Iin_avg=0.5*(1.51+2.1)*DT/T

Iin_avg =

    1.1417

fx >> |
```

According to Fig. 4.18, the average value of input power is 6.85 W.

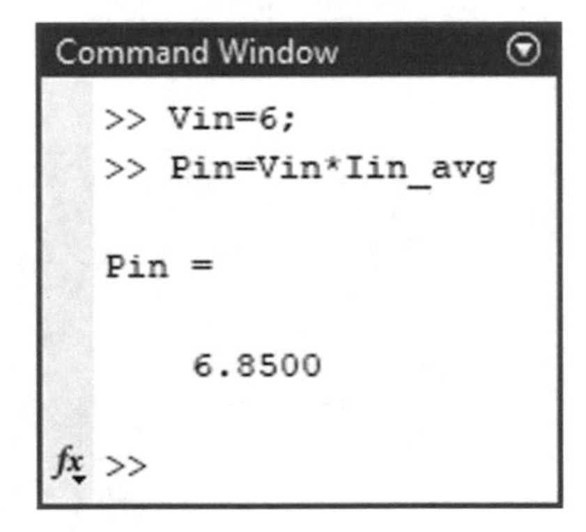

Fig. 4.18 MATLAB calculations

The efficiency of the converter is about 96% (Fig. 4.19).

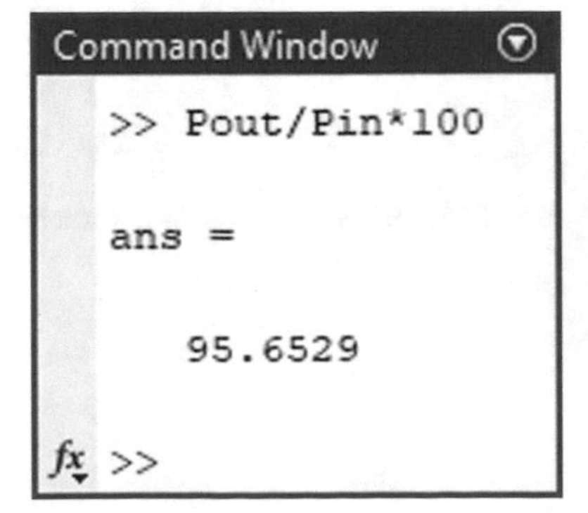

Fig. 4.19 MATLAB calculations

4.5 Example 4: Dimmer Circuit

A dimmer circuit permits you to control the RMS of the voltage which is applied to the load. You can use the dimmer circuit to control the intensity of incandescent light bulbs and small universal motors. In this example, we want to simulate a dimmer circuit. Draw the schematic shown in Fig. 4.20. The settings of V_1 are shown in Fig. 4.21.

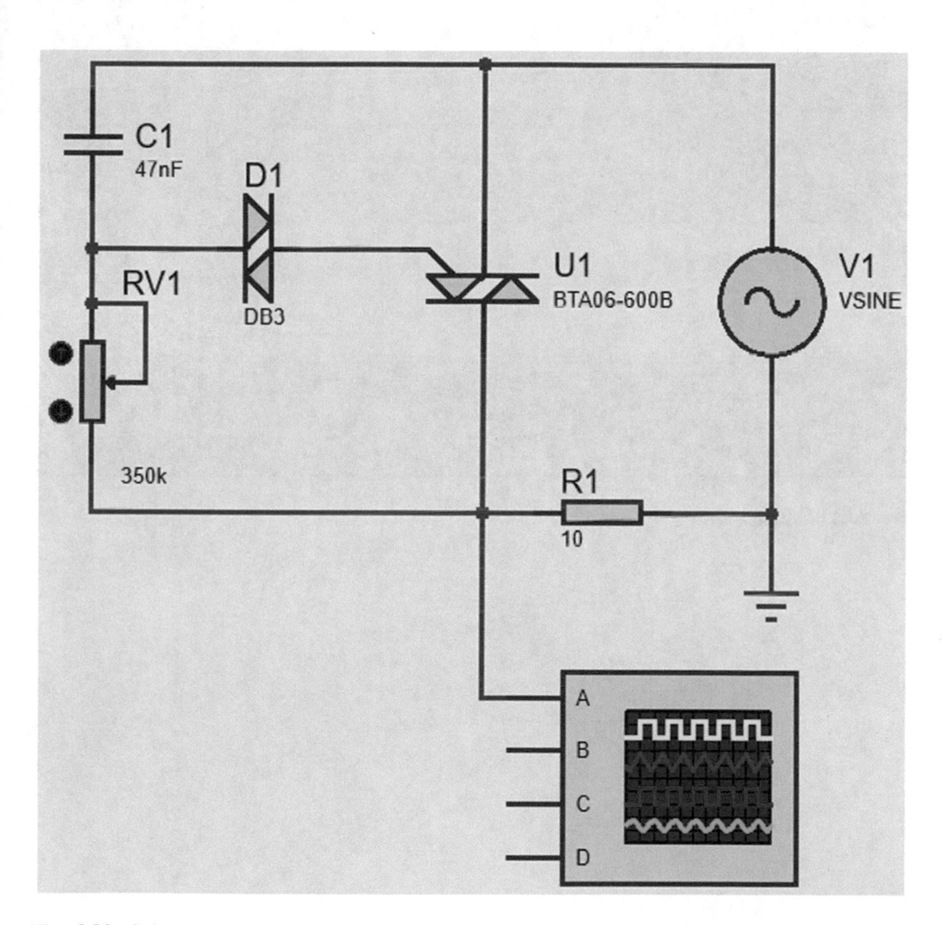

Fig. 4.20 Schematic of Example 4

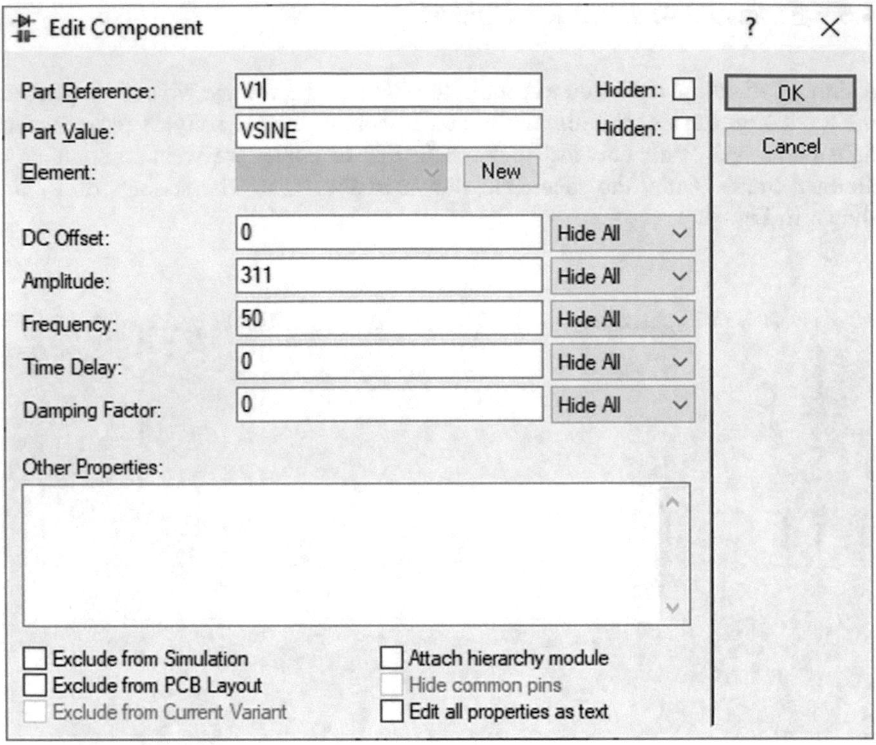

Edit Component

Part Reference: V1 Hidden:
Part Value: VSINE Hidden:
Element: New
DC Offset: 0 Hide All
Amplitude: 311 Hide All
Frequency: 50 Hide All
Time Delay: 0 Hide All
Damping Factor: 0 Hide All
Other Properties:
Exclude from Simulation
Exclude from PCB Layout
Exclude from Current Variant
Attach hierarchy module
Hide common pins
Edit all properties as text
OK
Cancel

Fig. 4.21 Settings of V_1

Run the simulation. When the slider is in the bottom position (Fig. 4.22), the load voltage waveform is almost equal to the source waveform (Fig. 4.23).

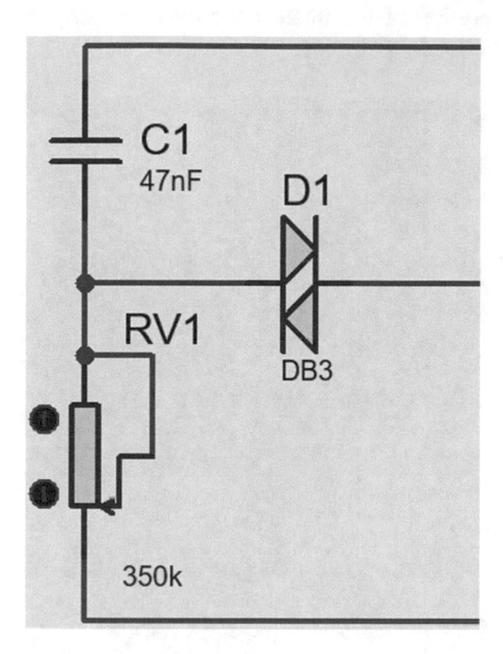

Fig. 4.22 Slider arm is connected to the lower terminal

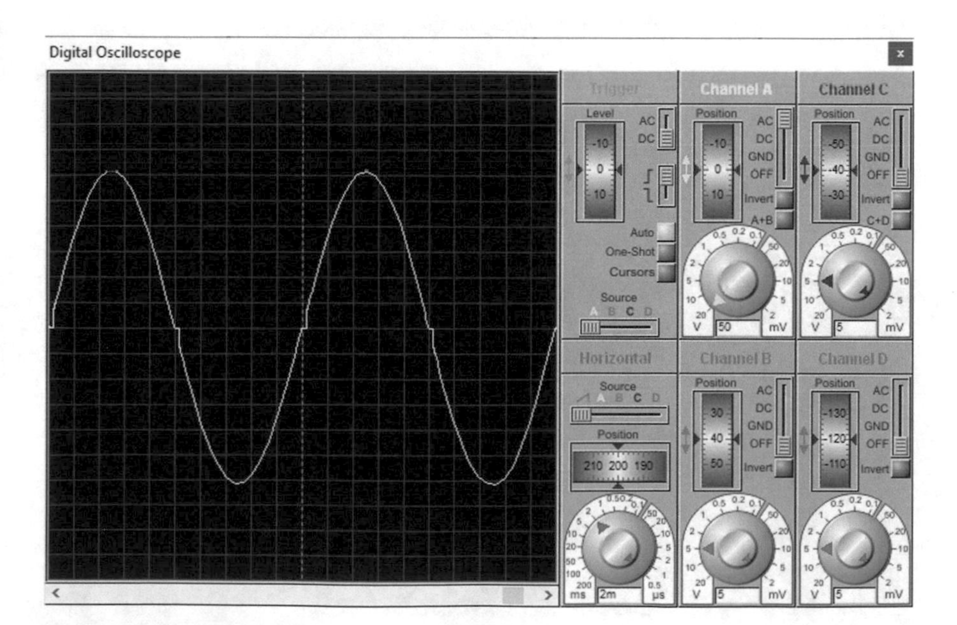

Fig. 4.23 Simulation result

When the slider is in the top position (Fig. 4.24), the load voltage waveform is similar to Fig. 4.25. Note that around half of the input waveform is removed. Removal of some portion of the input waveform is a simple way to control the power applied to the load.

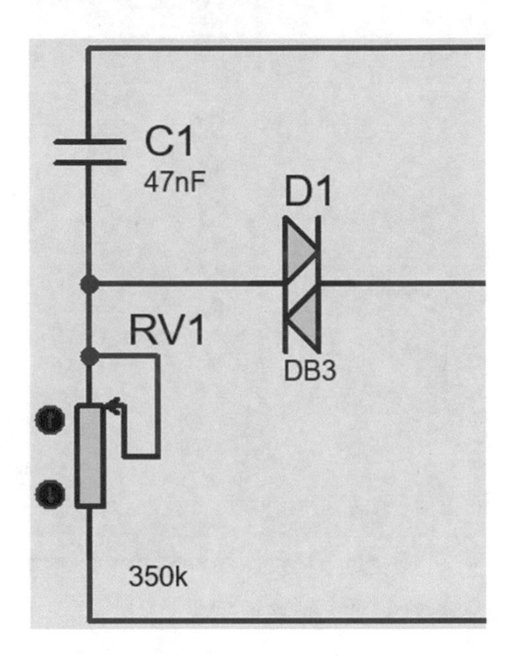

Fig. 4.24 Slider arm is connected to the upper terminal

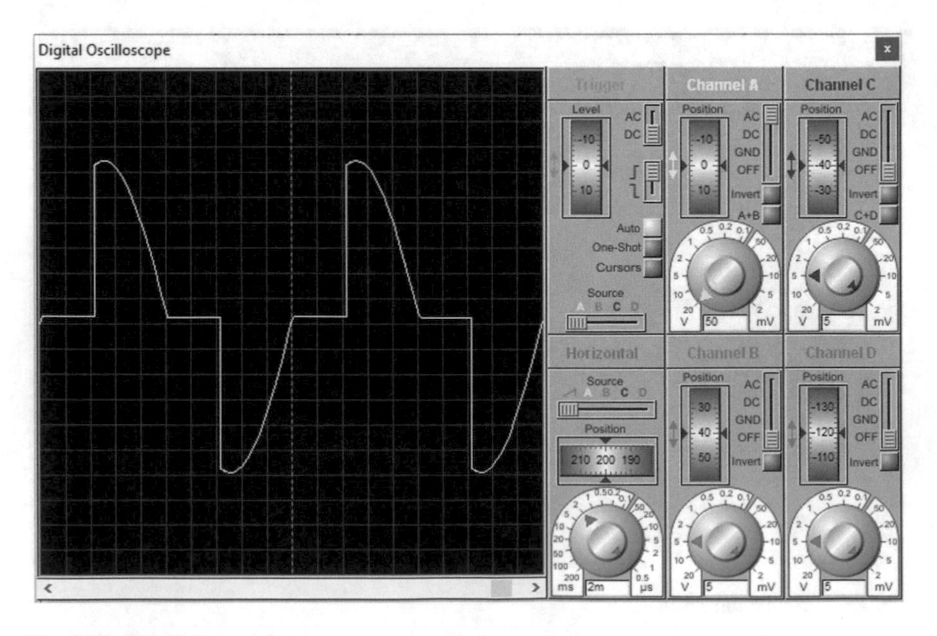

Fig. 4.25 Simulation result

4.6 Example 5: Single-Phase Half Wave Controlled Rectifier

In this example, we want to simulate a single phase half wave controlled (thyristor) rectifier. The schematic of this example is shown in Fig. 4.26. Settings of V_1 and gate signal are shown in Figs. 4.27 and 4.28, respectively. Note that the Start (Secs) box (Fig. 4.28) determines the firing angle of thyristor U_1. According to the settings shown in Fig. 4.28, the firing angle of thyristor U_1 is $\frac{5m}{\frac{1}{50}} = \frac{5m}{20m} \times 360° = 90°$. Note that the frequency of the pulse applied to the gate of the thyristor U_1 is equal to the frequency of the AC source V_1.

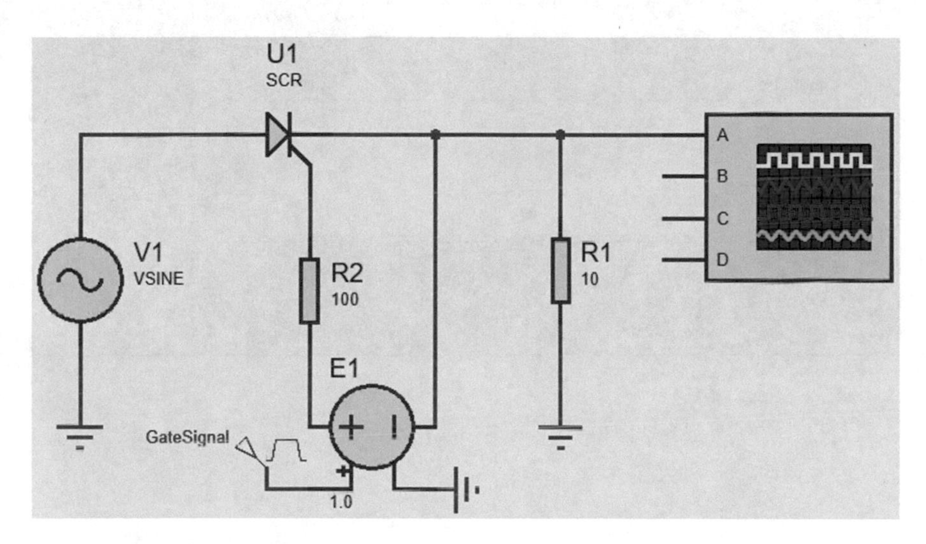

Fig. 4.26 Schematic of Example 5

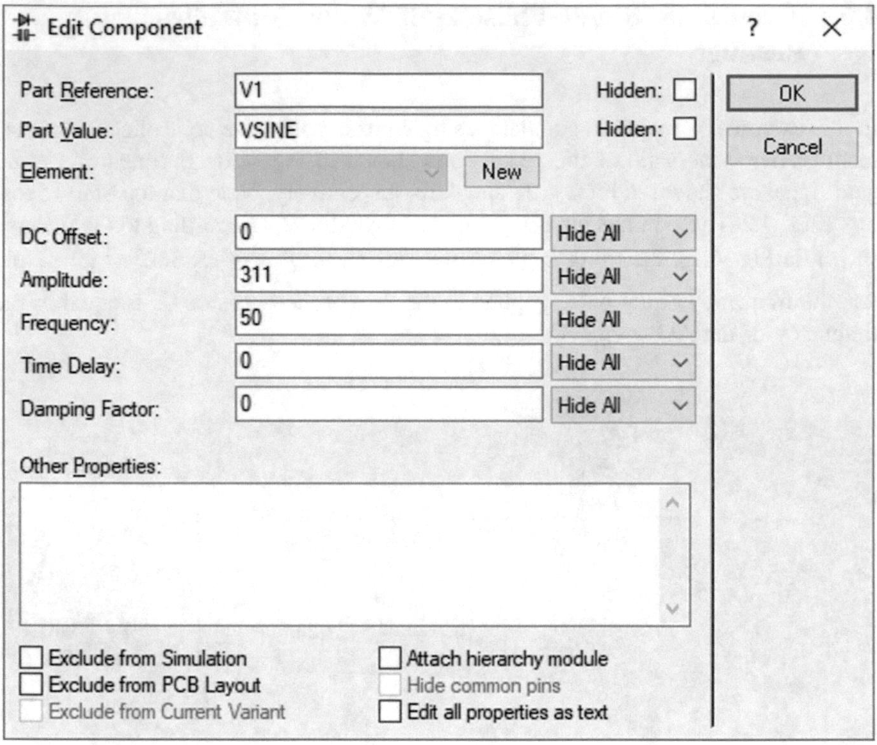

Edit Component

Part Reference:	V1	Hidden: ☐
Part Value:	VSINE	Hidden: ☐
Element:		New
DC Offset:	0	Hide All
Amplitude:	311	Hide All
Frequency:	50	Hide All
Time Delay:	0	Hide All
Damping Factor:	0	Hide All

OK
Cancel

Other Properties:

☐ Exclude from Simulation ☐ Attach hierarchy module
☐ Exclude from PCB Layout ☐ Hide common pins
☐ Exclude from Current Variant ☐ Edit all properties as text

Fig. 4.27 Settings of V_1

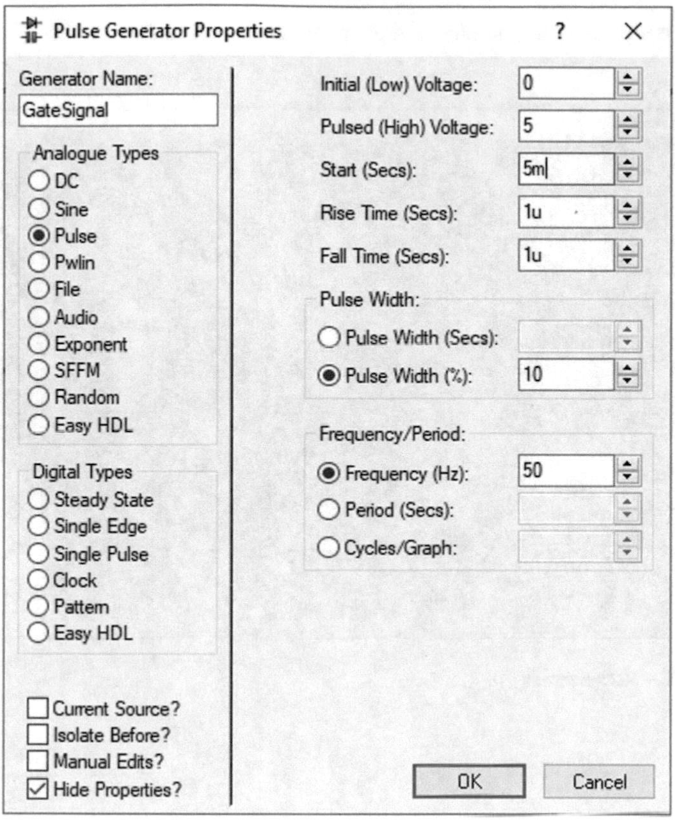

Fig. 4.28 Settings of gate signal

Run the simulation. The simulation result is shown in Fig. 4.29.

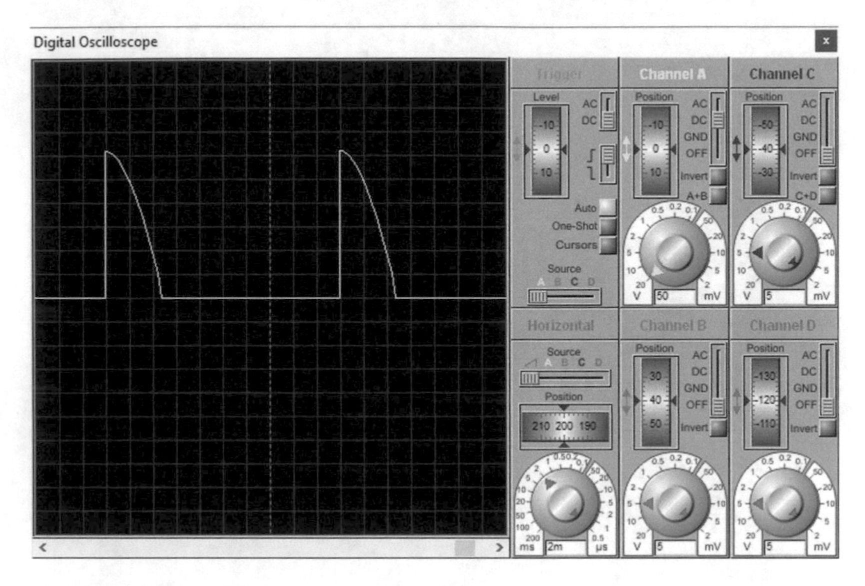

Fig. 4.29 Simulation result

Stop the simulation and double-click the gate signal pulse generator block. Change the start (Secs) to 3 ms (Fig. 4.30); 3 ms produces the firing angle of $\frac{3m}{20m} \times 360^\circ = 54^\circ$. The simulation result for this firing angle is shown in Fig. 4.31.

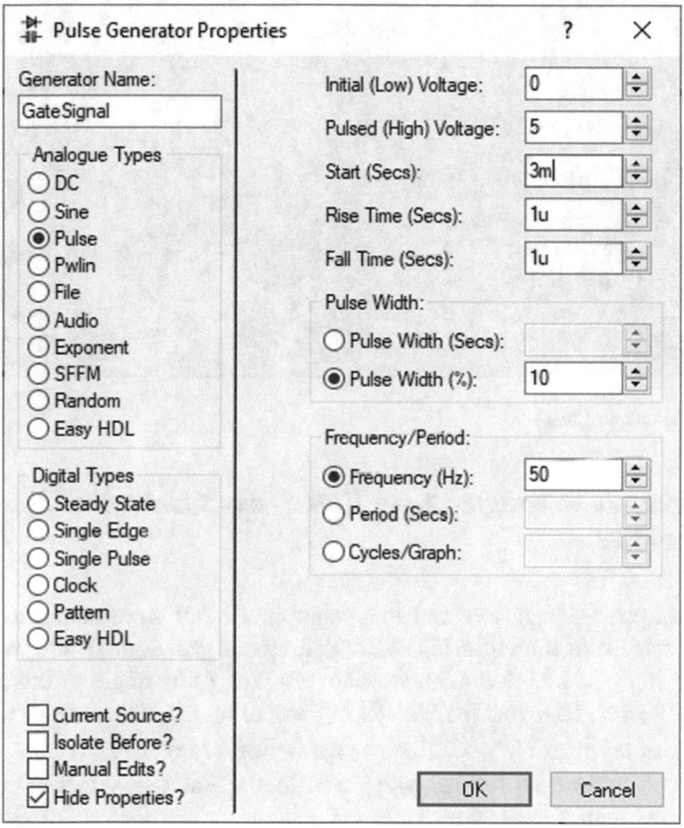

Fig. 4.30 Settings of gate signal

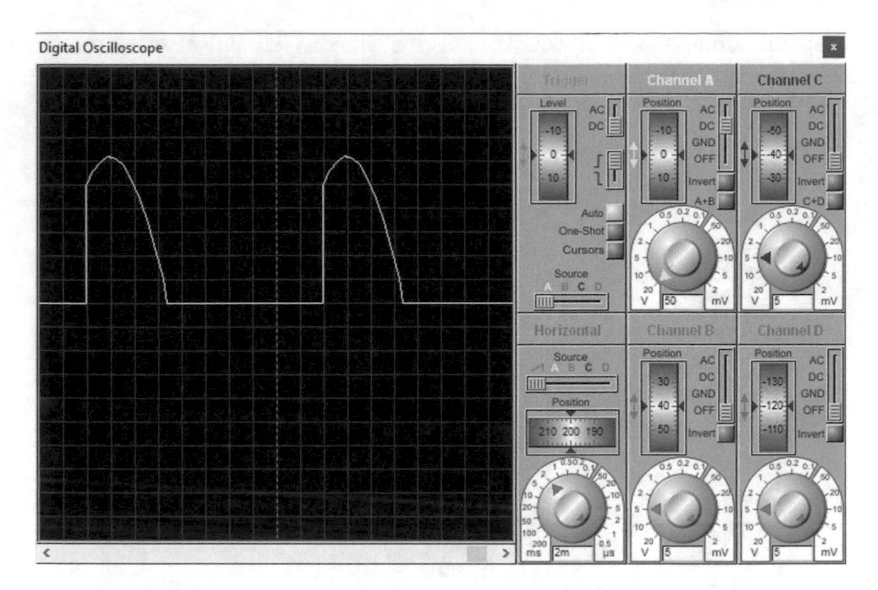

Fig. 4.31 Simulation result

4.7 Example 6: Single-Phase Full Wave Controlled Rectifier

In this example, we want to simulate a single-phase full wave controlled rectifier. Draw the schematic shown in Fig. 4.32. Settings of gate signal1 and gate signal2 are shown in Figs. 4.33 and 4.34, respectively. The firing angle of this rectifier is $\frac{3m}{20m} \times 360^{\circ} = 54^{\circ}$. Note that the Start (Secs) box of thyristor U_2 is equal to the Start (Secs) box of thyristor $U_1 + \frac{T}{2}$. T shows the period of input AC voltage. The Start (Secs) box of thyristor U_1 is filled with 3 ms. So, the Start (Secs) box of thyristor U_2 must be filled with $3\text{ ms} + \frac{20\text{ ms}}{2} = 13\text{ ms}$.

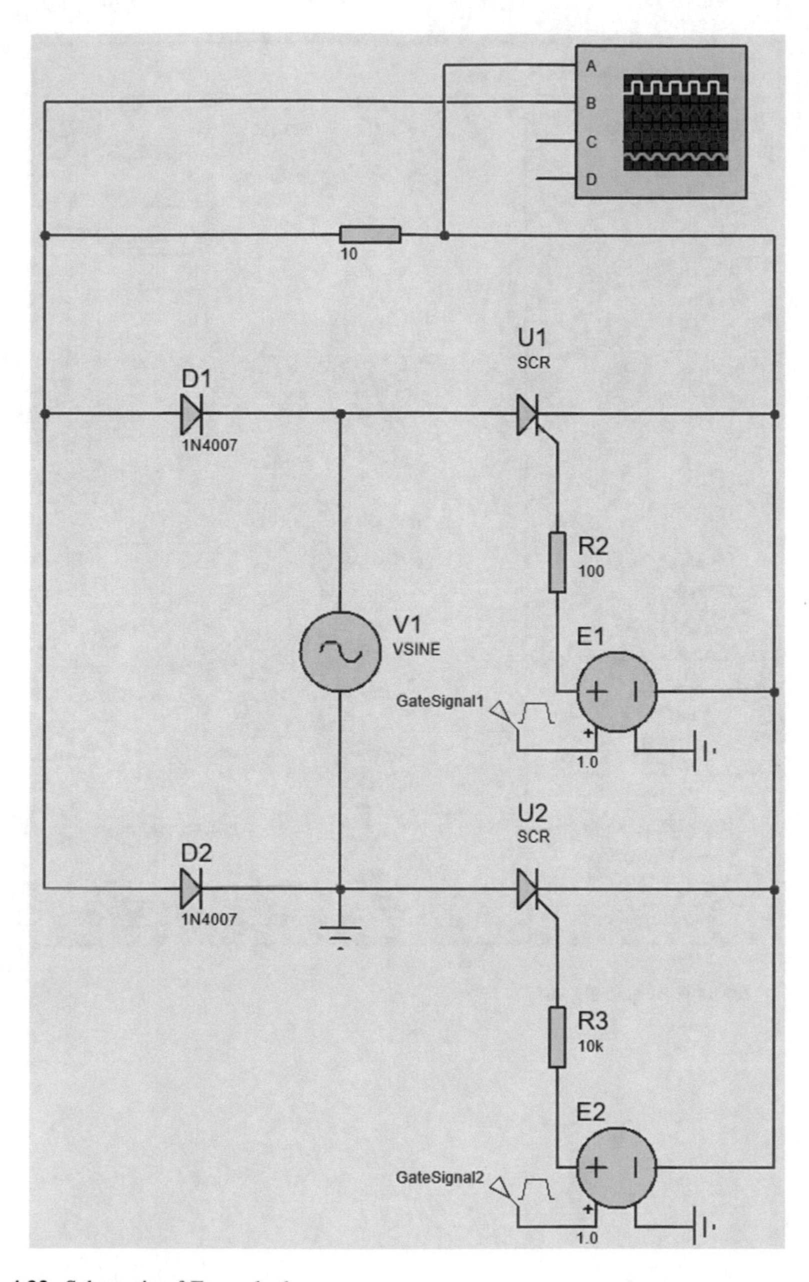

Fig. 4.32 Schematic of Example 6

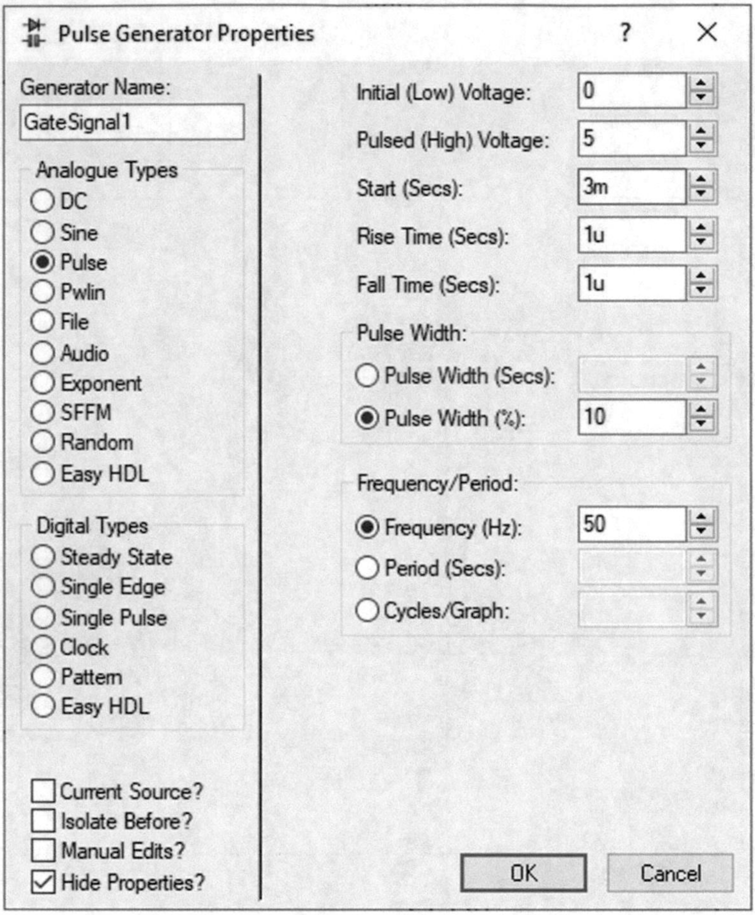

Pulse Generator Properties ? ×

Generator Name:
GateSignal1

Analogue Types
DC
Sine
Pulse
Pwlin
File
Audio
Exponent
SFFM
Random
Easy HDL

Digital Types
Steady State
Single Edge
Single Pulse
Clock
Pattern
Easy HDL

Current Source?
Isolate Before?
Manual Edits?
Hide Properties?

Initial (Low) Voltage: 0
Pulsed (High) Voltage: 5
Start (Secs): 3m
Rise Time (Secs): 1u
Fall Time (Secs): 1u

Pulse Width:
Pulse Width (Secs):
Pulse Width (%): 10

Frequency/Period:
Frequency (Hz): 50
Period (Secs):
Cycles/Graph:

OK Cancel

Fig. 4.33 Settings of gate signal1

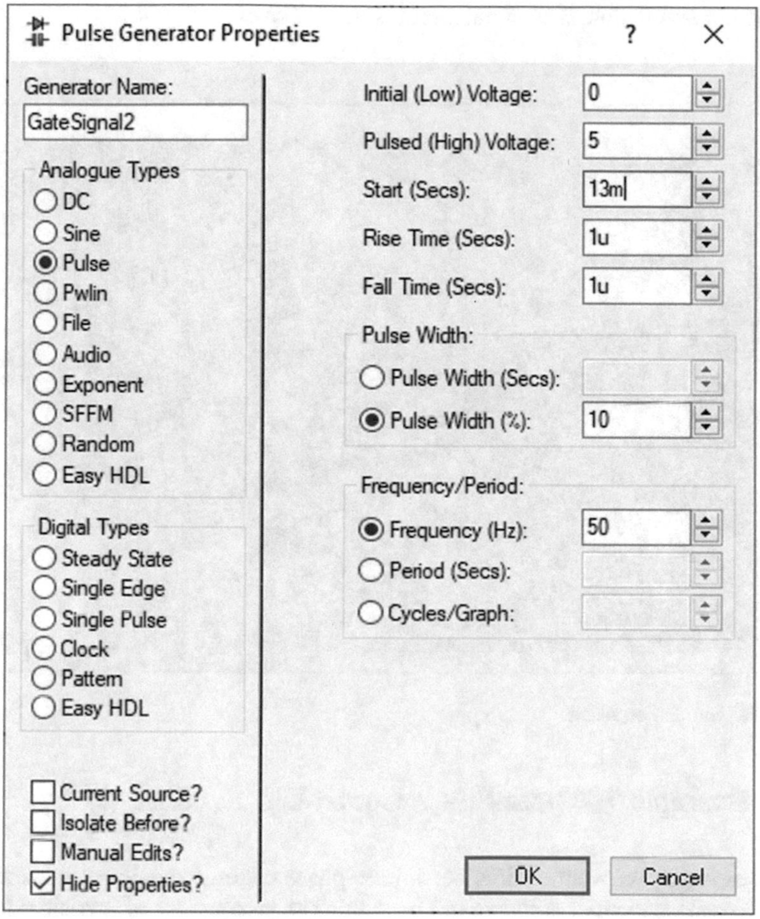

Pulse Generator Properties ? ×

Generator Name:
GateSignal2

Analogue Types
DC
Sine
Pulse
Pwlin
File
Audio
Exponent
SFFM
Random
Easy HDL

Digital Types
Steady State
Single Edge
Single Pulse
Clock
Pattern
Easy HDL

Current Source?
Isolate Before?
Manual Edits?
Hide Properties?

Initial (Low) Voltage: 0
Pulsed (High) Voltage: 5
Start (Secs): 13m
Rise Time (Secs): 1u
Fall Time (Secs): 1u

Pulse Width:
Pulse Width (Secs):
Pulse Width (%): 10

Frequency/Period:
Frequency (Hz): 50
Period (Secs):
Cycles/Graph:

OK Cancel

Fig. 4.34 Settings of gate signal2

Run the simulation. The simulation result is shown in Fig. 4.35.

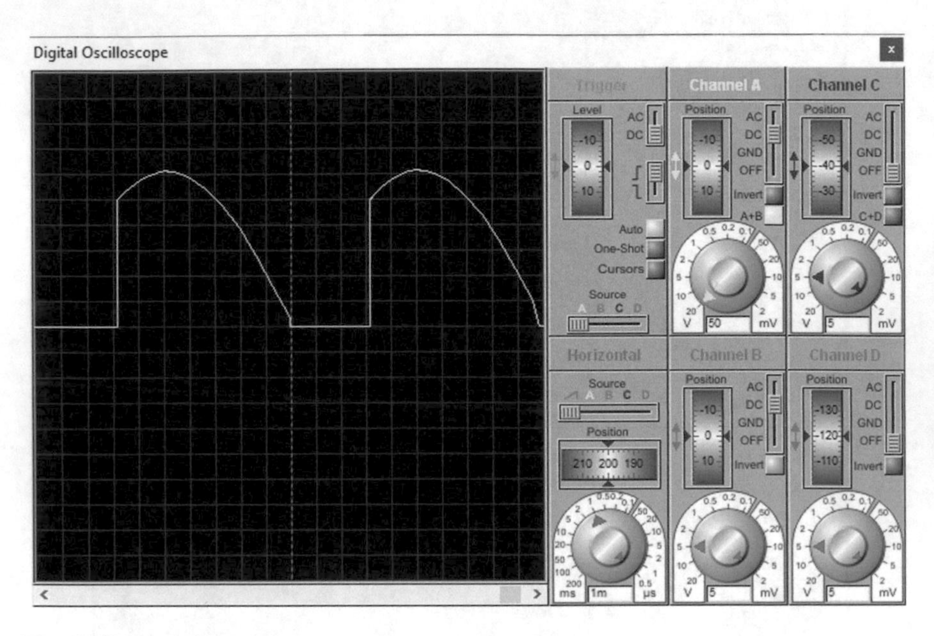

Fig. 4.35 Simulation result

4.8 Example 7: Three-Phase Controlled Rectifier (I)

In this example, we want to simulate a three-phase controlled rectifier. A schematic of a three-phase rectifier is shown in Fig. 4.36. The gate signals of thyristors for two firing angles (0° and 30°) are shown in Fig. 4.37. Note that $T_5 - T_6$ is closed at $t = 0$ and firing angle (α) is measured with respect to $\omega t = \frac{\pi}{6} = 30°$.

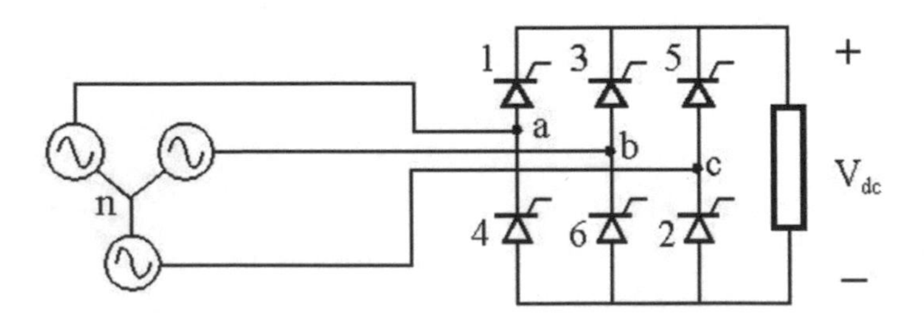

Fig. 4.36 Three-phase controlled rectifier

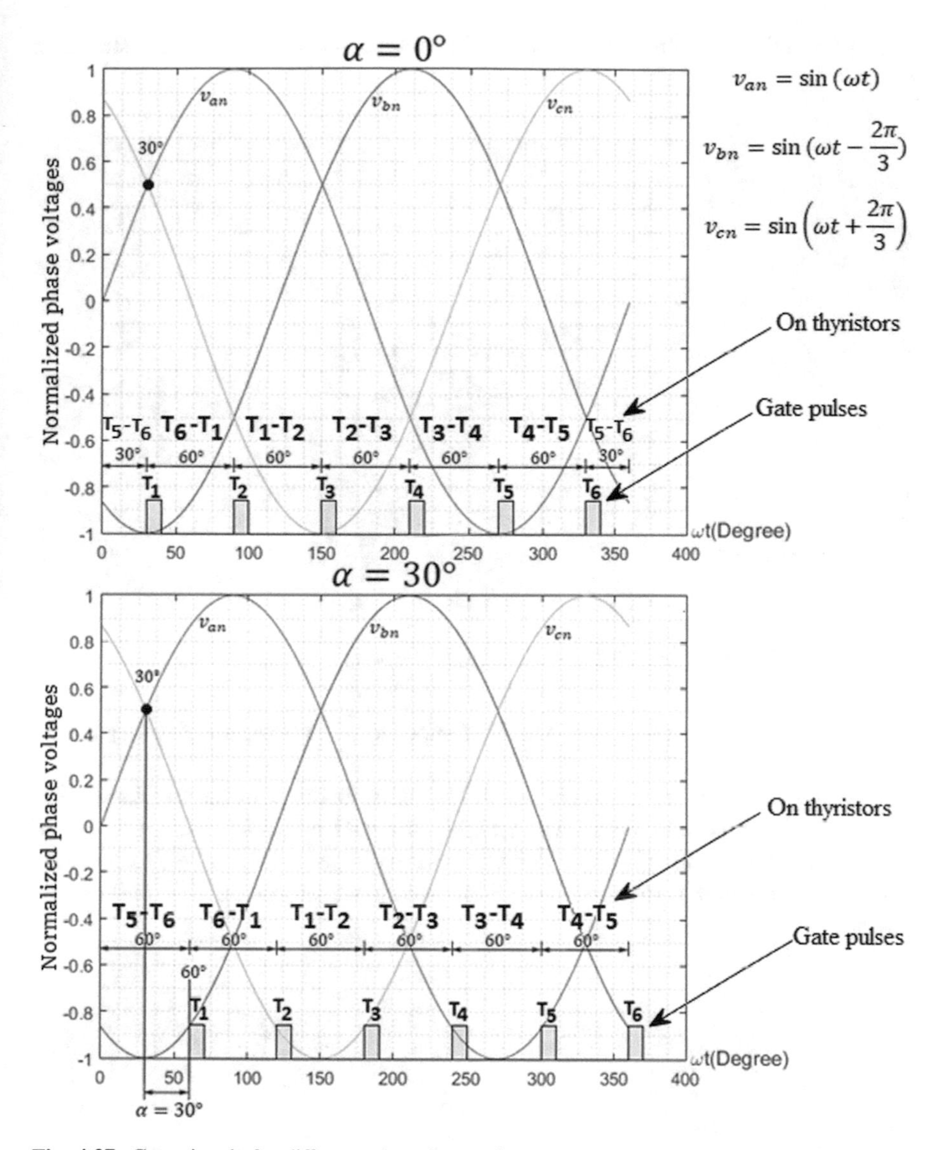

Fig. 4.37 Gate signals for different triggering angles

Let's simulate the three-phase rectifier in Proteus. Draw the schematic shown in Fig. 4.38.

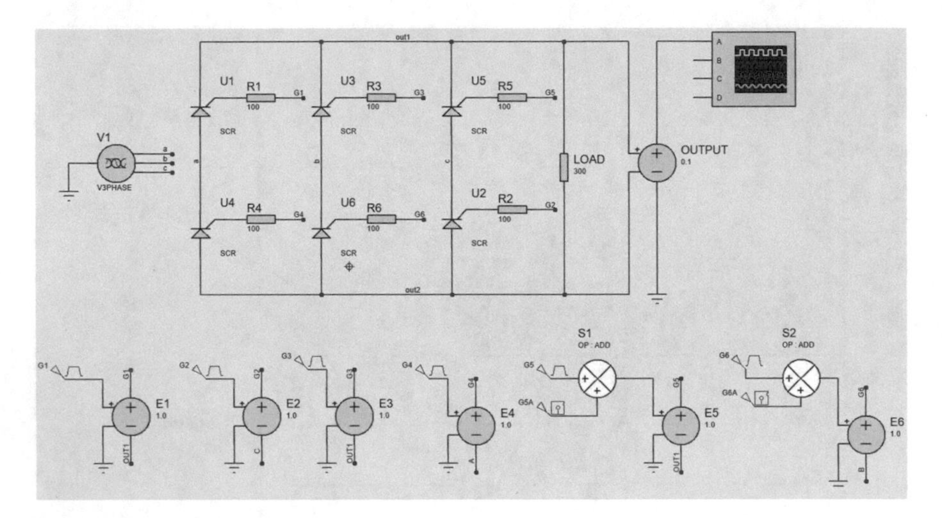

Fig. 4.38 Schematic of Example 7

The schematic shown in Fig. 4.38 is composed of two sections: power circuit and control circuit.

The power circuit section is shown in Fig. 4.39. Settings of the blocks used in the power circuit section are shown in Figs. 4.40, 4.41 and 4.42. The OUTPUT of voltage-controlled voltage source decreases the voltage of load by a factor of 10. Such an attenuation permits us to see the whole waveform on the oscilloscope screen.

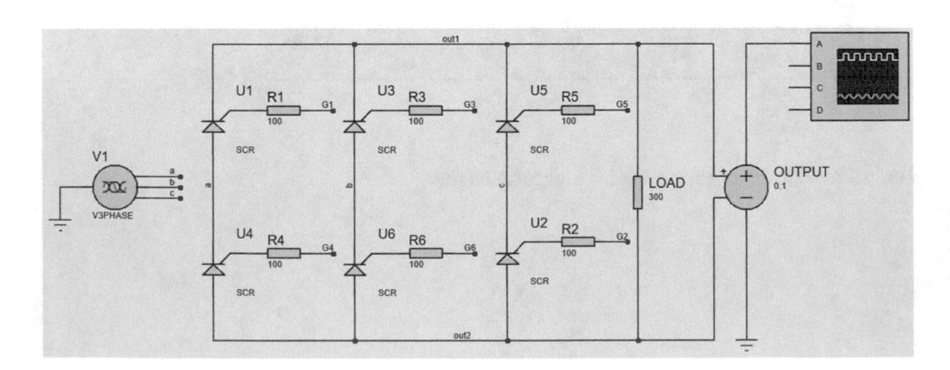

Fig. 4.39 Power circuit

Edit Component ? ×

Part Reference:	V1	Hidden: ☐	OK
Part Value:	V3PHASE	Hidden: ☐	Cancel
Element:		New	
Amplitude (Volts)	220	Hide All	
Amplitude mode:	Rms	Hide All	
Frequency (Hz)	50	Hide All	
Time Delay (sec)	0	Hide All	
Damping Factor:	0	Hide All	
LISA Model File:	V3PHASE	Hide All	

Advanced Properties:

Line resistance (Ohms)	1m	Hide All

Other Properties:

☐ Exclude from Simulation ☐ Attach hierarchy module
☐ Exclude from PCB Layout ☐ Hide common pins
☐ Exclude from Current Variant ☐ Edit all properties as text

Fig. 4.40 Settings of V_1

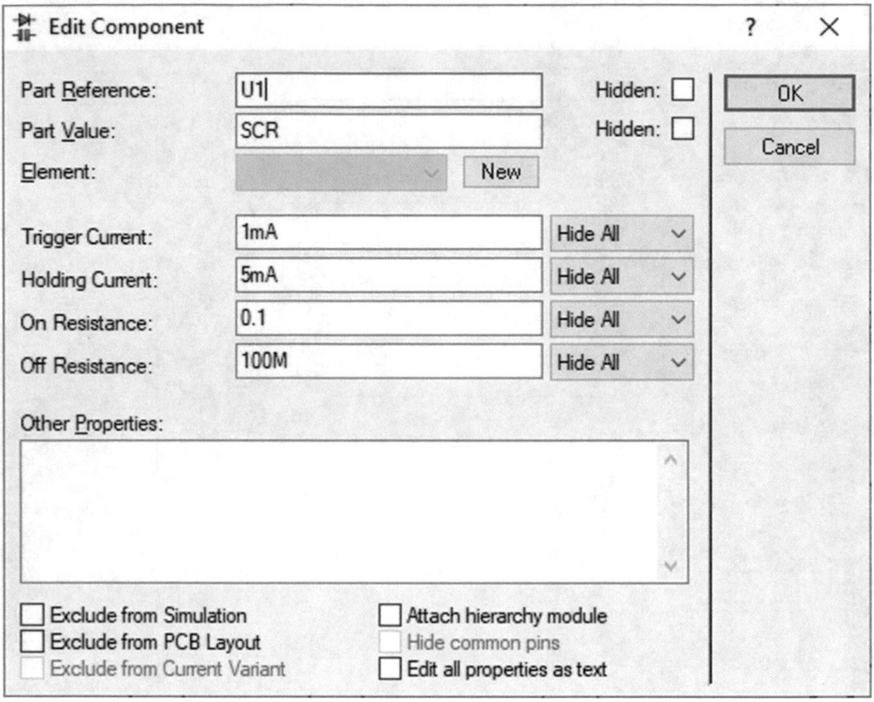

Fig. 4.41 Settings of U_1

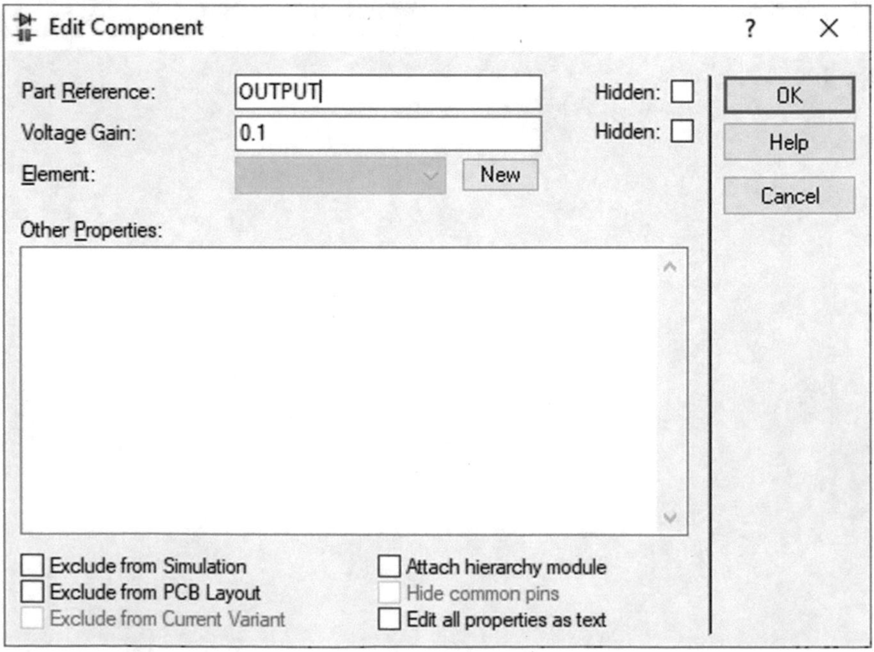

Fig. 4.42 Settings of output

The control circuit section is shown in Fig. 4.43.

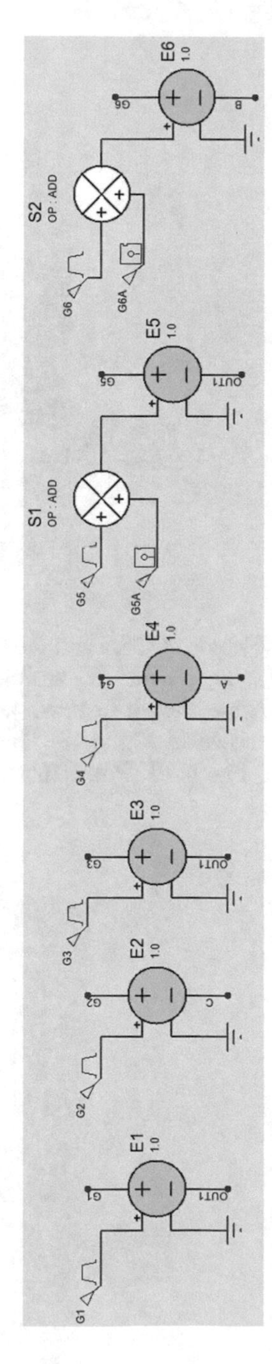

Fig. 4.43 Control circuit section

The required triggering pulse of thyristors is generated with the aid of "pulse" generators (Fig. 4.44) G_1, G_2, G_3, G_4, G_5 and G_6.

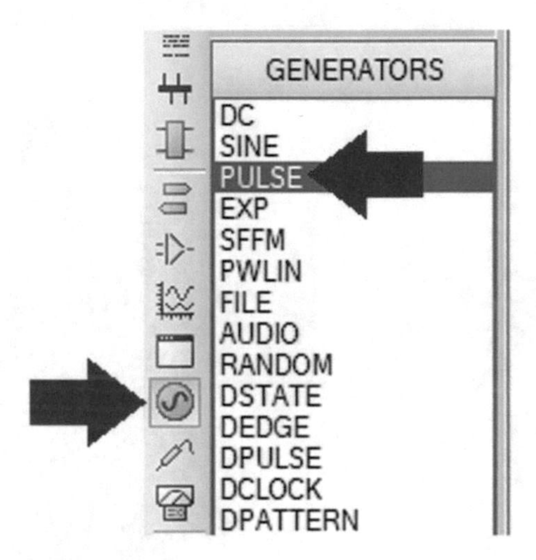

Fig. 4.44 Pulse generator block

The "file" generator (Fig. 4.45) blocks G5A and G6A generate a short pulse to trigger the thyristors U_5 and U_6. So, U_5 and U_6 are closed when the simulation starts. Settings of G5A and G6A are shown in Figs. 4.46 and 4.47, respectively. Contents of InitialPulse.txt are shown in Fig. 4.48. The waveforms generated by G5A and G6A are shown in Fig. 4.49. Note that $t_1 = 1$ μs, $t_2 = 1$ ms and $t_3 = 1.1$ ms.

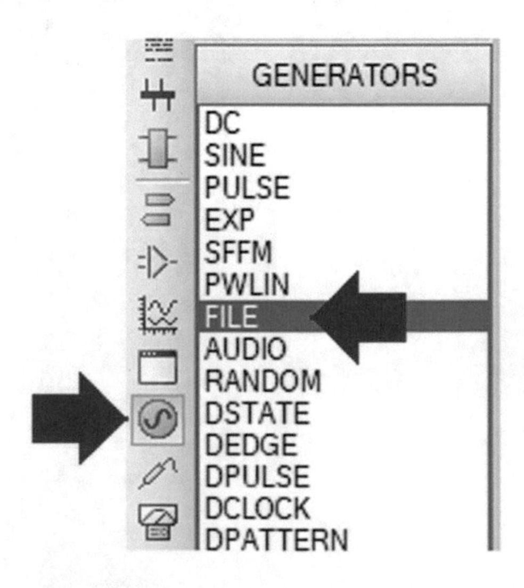

Fig. 4.45 File generator block

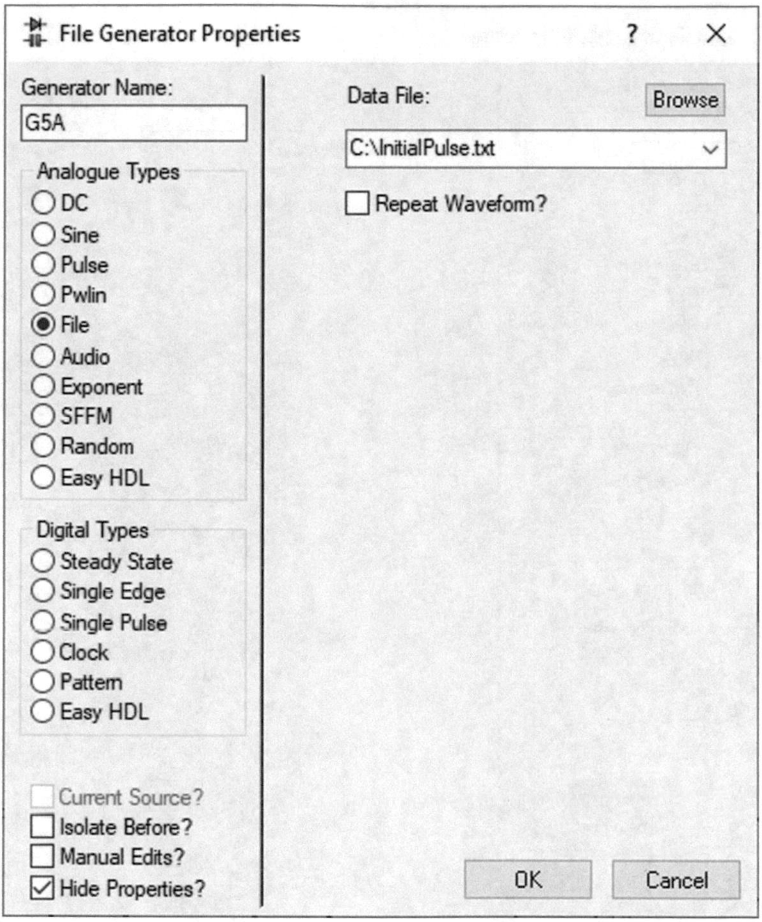

Fig. 4.46 Settings of G5A block

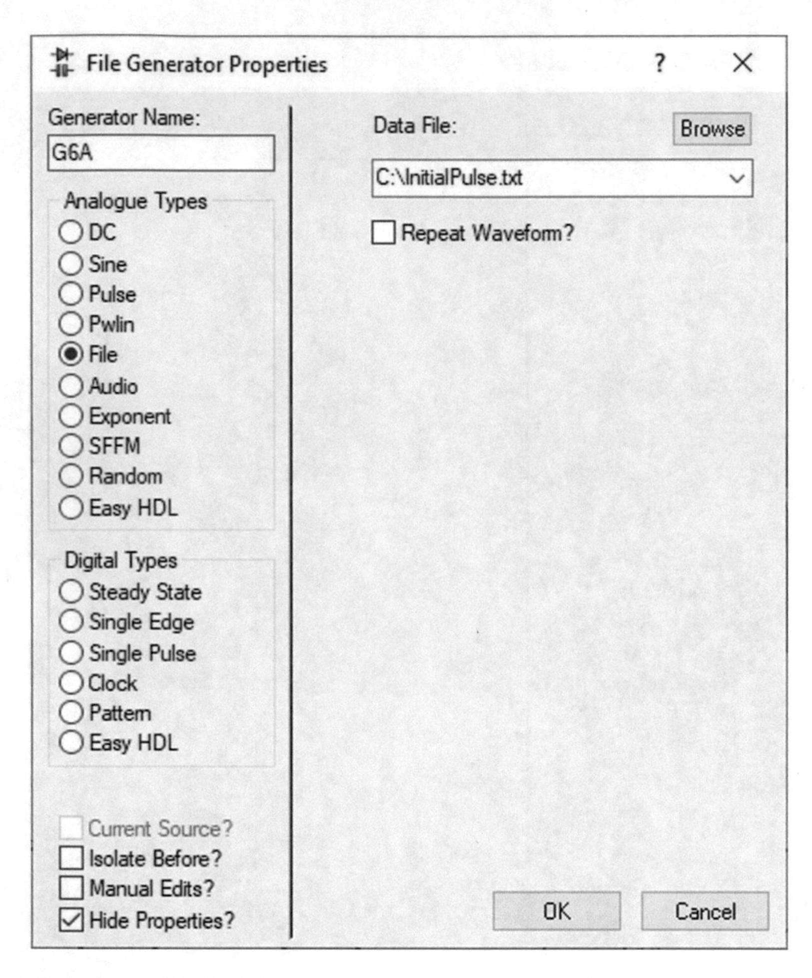

Fig. 4.47 Settings of G6A block

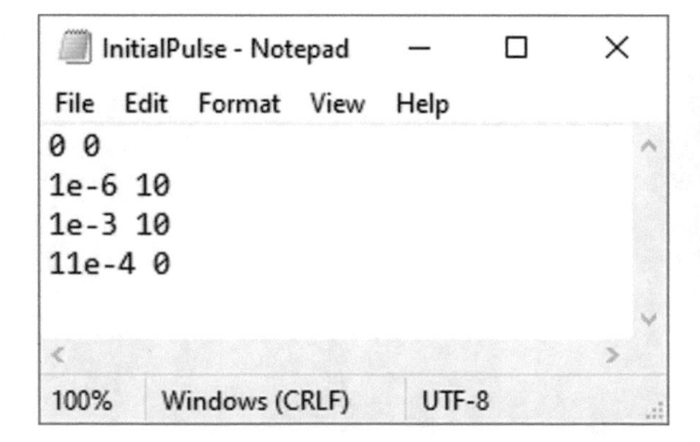

Fig. 4.48 Content of InitialPulse.txt file

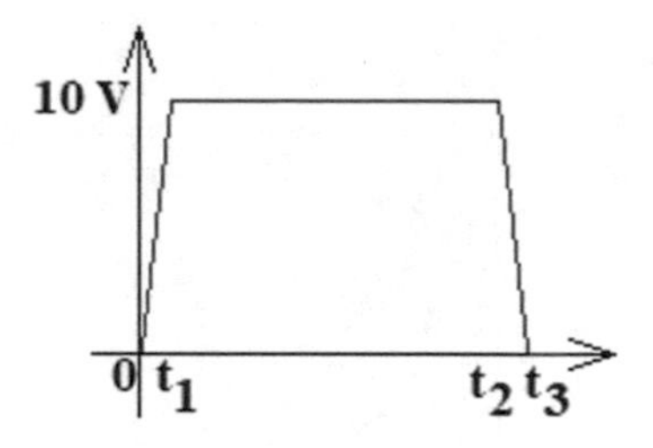

Fig. 4.49 Waveform generated by G5A and G6A

Settings of G_1, G_2, G_3, G_4, G_5 and G_6 are shown in Figs. 4.50, 4.51, 4.52, 4.53, 4.54 and 4.55. These blocks generate short pulses to triggers the thyristor. The width of pulses is controlled by the value entered into the Pulse Width (%) box. The pulse is applied to the gate of the MOSFET at the instant entered into the Start (Secs) box.

Pulse Generator Properties ? X

Generator Name: G1

Analogue Types: DC, Sine, Pulse (selected), Pwlin, File, Audio, Exponent, SFFM, Random, Easy HDL

Digital Types: Steady State, Single Edge, Single Pulse, Clock, Pattern, Easy HDL

Current Source? ☐
Isolate Before? ☐
Manual Edits? ☐
Hide Properties? ☑

Initial (Low) Voltage:	0
Pulsed (High) Voltage:	10
Start (Secs):	2.667m
Rise Time (Secs):	1u
Fall Time (Secs):	1u

Pulse Width:
- Pulse Width (Secs):
- Pulse Width (%) (selected): 5

Frequency/Period:
- Frequency (Hz) (selected): 50
- Period (Secs):
- Cycles/Graph:

OK Cancel

Fig. 4.50 Settings of G_1

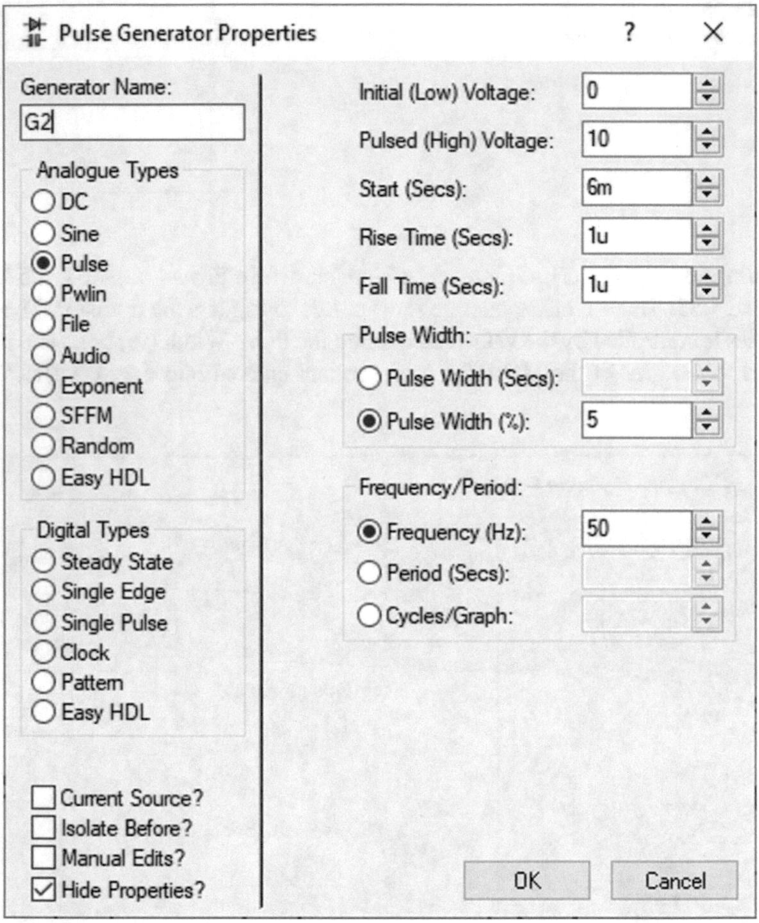

Pulse Generator Properties

Generator Name: G2

Analogue Types: DC, Sine, Pulse (selected), Pwlin, File, Audio, Exponent, SFFM, Random, Easy HDL

Digital Types: Steady State, Single Edge, Single Pulse, Clock, Pattern, Easy HDL

Current Source? Isolate Before? Manual Edits? Hide Properties? (checked)

Initial (Low) Voltage: 0

Pulsed (High) Voltage: 10

Start (Secs): 6m

Rise Time (Secs): 1u

Fall Time (Secs): 1u

Pulse Width: Pulse Width (Secs): / Pulse Width (%): 5 (selected)

Frequency/Period: Frequency (Hz): 50 (selected) / Period (Secs): / Cycles/Graph:

OK Cancel

Fig. 4.51 Settings of G_2

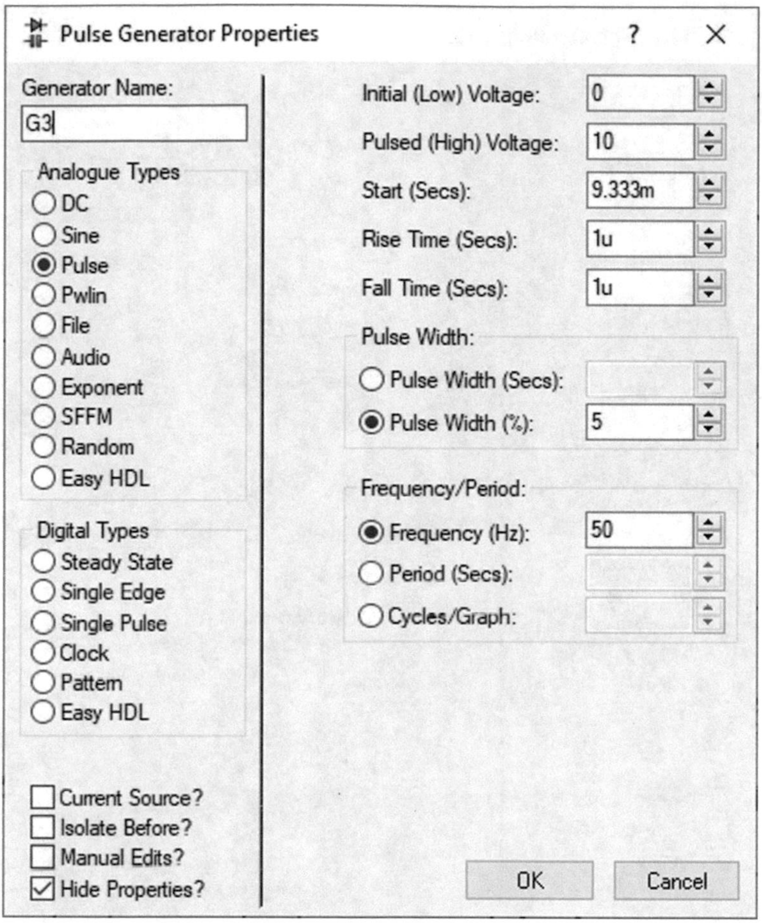

Fig. 4.52 Settings of G_3

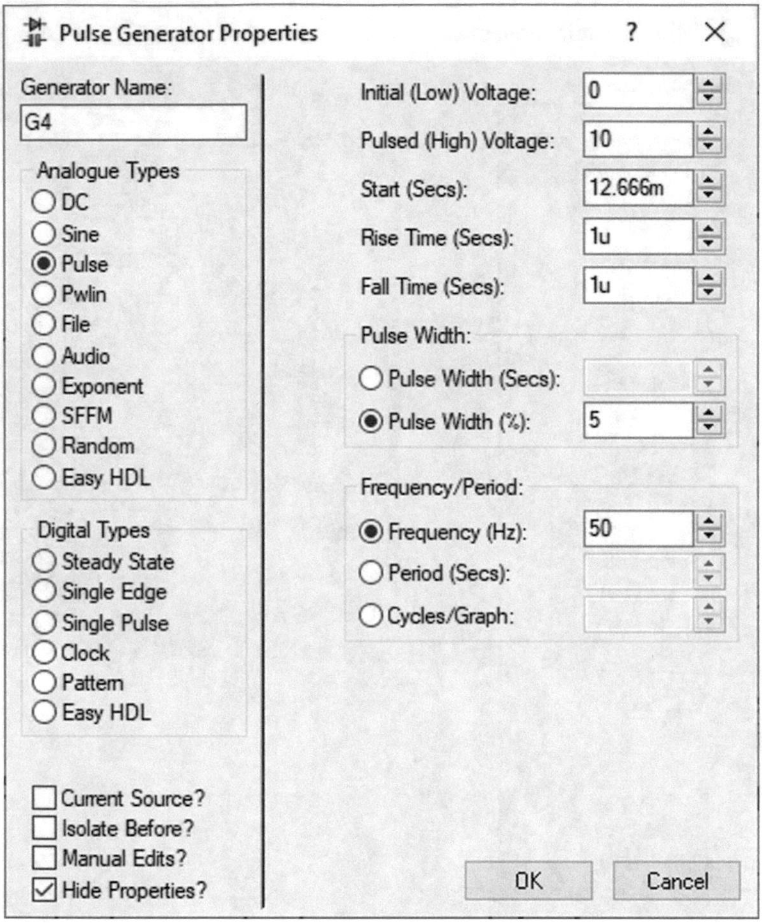

Pulse Generator Properties

Generator Name: G4

Analogue Types: DC, Sine, Pulse (selected), Pwlin, File, Audio, Exponent, SFFM, Random, Easy HDL

Digital Types: Steady State, Single Edge, Single Pulse, Clock, Pattern, Easy HDL

Current Source? / Isolate Before? / Manual Edits? / Hide Properties? (checked)

Initial (Low) Voltage: 0

Pulsed (High) Voltage: 10

Start (Secs): 12.666m

Rise Time (Secs): 1u

Fall Time (Secs): 1u

Pulse Width: Pulse Width (Secs): / Pulse Width (%): 5 (selected)

Frequency/Period: Frequency (Hz): 50 (selected) / Period (Secs): / Cycles/Graph:

OK Cancel

Fig. 4.53 Settings of G_4

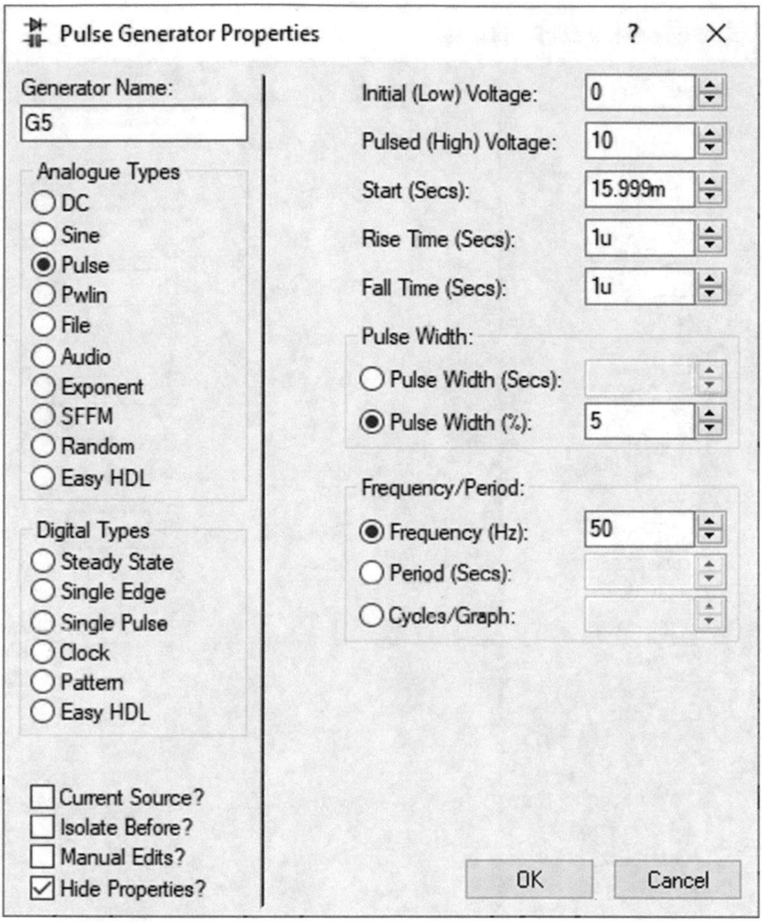

Pulse Generator Properties ? ×

Generator Name: G5

Analogue Types
- DC
- Sine
- (●) Pulse
- Pwlin
- File
- Audio
- Exponent
- SFFM
- Random
- Easy HDL

Digital Types
- Steady State
- Single Edge
- Single Pulse
- Clock
- Pattern
- Easy HDL

- [ ] Current Source?
- [ ] Isolate Before?
- [ ] Manual Edits?
- [x] Hide Properties?

Initial (Low) Voltage: 0
Pulsed (High) Voltage: 10
Start (Secs): 15.999m
Rise Time (Secs): 1u
Fall Time (Secs): 1u

Pulse Width:
- Pulse Width (Secs):
- (●) Pulse Width (%): 5

Frequency/Period:
- (●) Frequency (Hz): 50
- Period (Secs):
- Cycles/Graph:

OK Cancel

Fig. 4.54 Settings of G_5

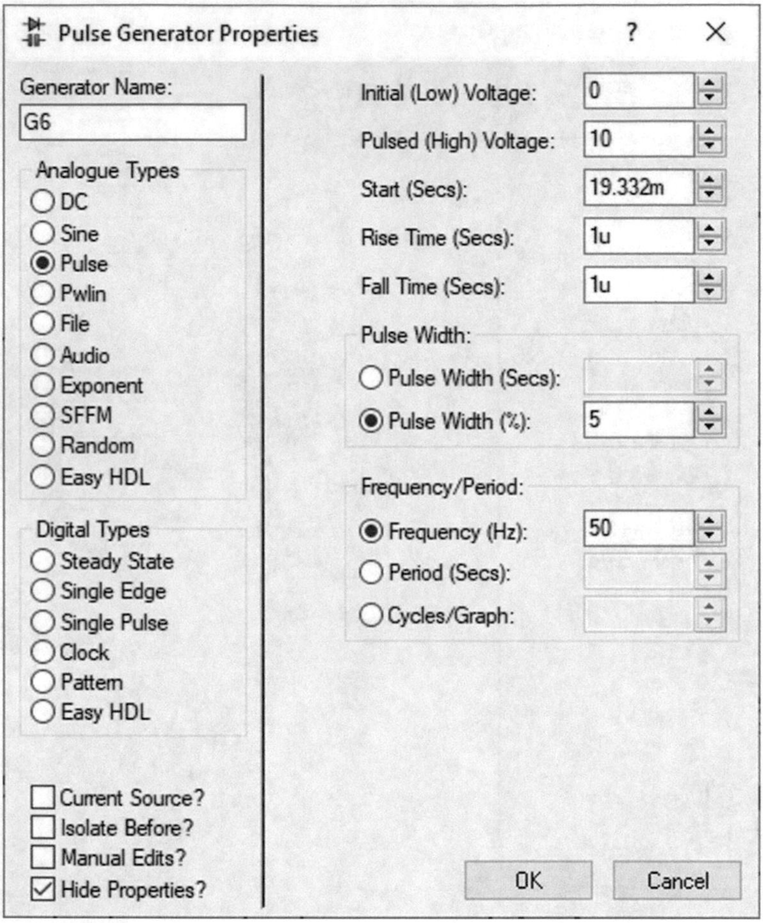

Pulse Generator Properties ? ×

Generator Name: G6

Analogue Types
- DC
- Sine
- Pulse (selected)
- Pwlin
- File
- Audio
- Exponent
- SFFM
- Random
- Easy HDL

Digital Types
- Steady State
- Single Edge
- Single Pulse
- Clock
- Pattern
- Easy HDL

- [ ] Current Source?
- [ ] Isolate Before?
- [ ] Manual Edits?
- [x] Hide Properties?

Initial (Low) Voltage: 0
Pulsed (High) Voltage: 10
Start (Secs): 19.332m
Rise Time (Secs): 1u
Fall Time (Secs): 1u

Pulse Width:
- Pulse Width (Secs):
- Pulse Width (%): 5 (selected)

Frequency/Period:
- Frequency (Hz): 50 (selected)
- Period (Secs):
- Cycles/Graph:

OK Cancel

Fig. 4.55 Settings of G_6

The values entered into Start (Secs) boxes in Figs. 4.50, 4.51, 4.52, 4.53, 4.54 and 4.55 are calculated with the aid of the following MATLAB code. The firing angle for settings shown in Figs. 4.50, 4.51, 4.52, 4.53, 4.54 and 4.55 is 18°.

```
clc
clear all

f=input('Enter the frequency of AC supply (enter 50 or 60): ');
if (f==50)
    TD=1.667;
    DT=3.333;
end
if (f==60)
    TD=1.389;
    DT=2.778;
end
FA=input('Firing angle (in Degrees): ');
T=1000*FA/f/360;

G1=TD+T;
G2=TD+T+DT;
G3=TD+T+2*DT;
G4=TD+T+3*DT;
G5=TD+T+4*DT;
G6=TD+T+5*DT;


str0=['NOTE: Values are in milli seconds'];
str1=['Start (Secs) box of G1: ', num2str(G1)];
str2=['Start (Secs) box of G2: ', num2str(G2)];
str3=['Start (Secs) box of G3: ', num2str(G3)];
str4=['Start (Secs) box of G4: ', num2str(G4)];
str5=['Start (Secs) box of G5: ', num2str(G5)];
str6=['Start (Secs) box of G6: ', num2str(G6)];

disp(str0)
disp(str1)
disp(str2)
disp(str3)
disp(str4)
disp(str5)
disp(str6)
```

Run the simulation. The simulation for firing angle of 18° is shown in Fig. 4.56.

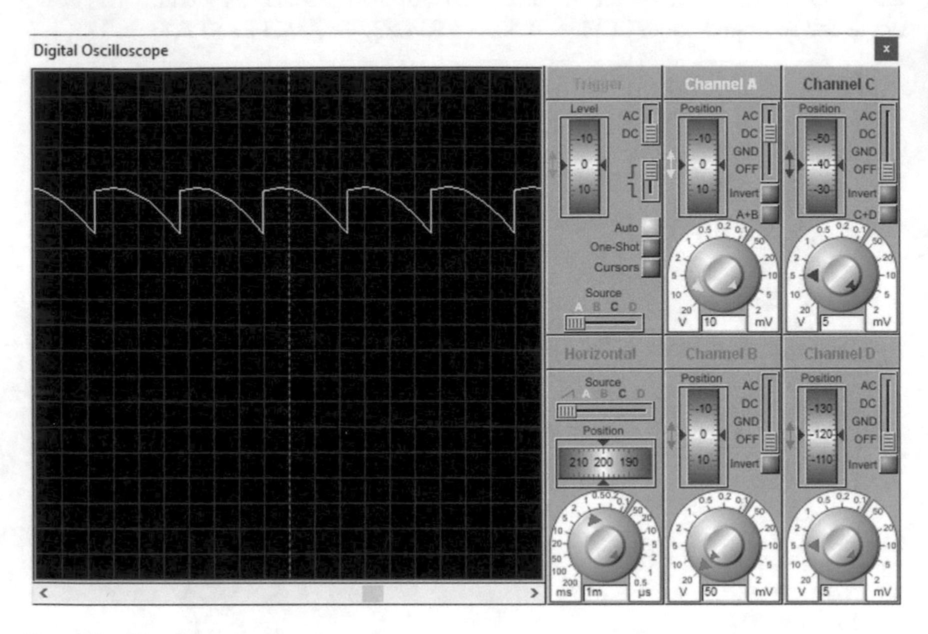

Fig. 4.56 Simulation result

4.9 Example 8: Three-Phase Controlled Rectifier (II)

In the previous example, we simulated a three-phase controlled rectifier. In this example, we want to measure the average value (DC component) of load voltage. Connect channel B to output (Fig. 4.57).

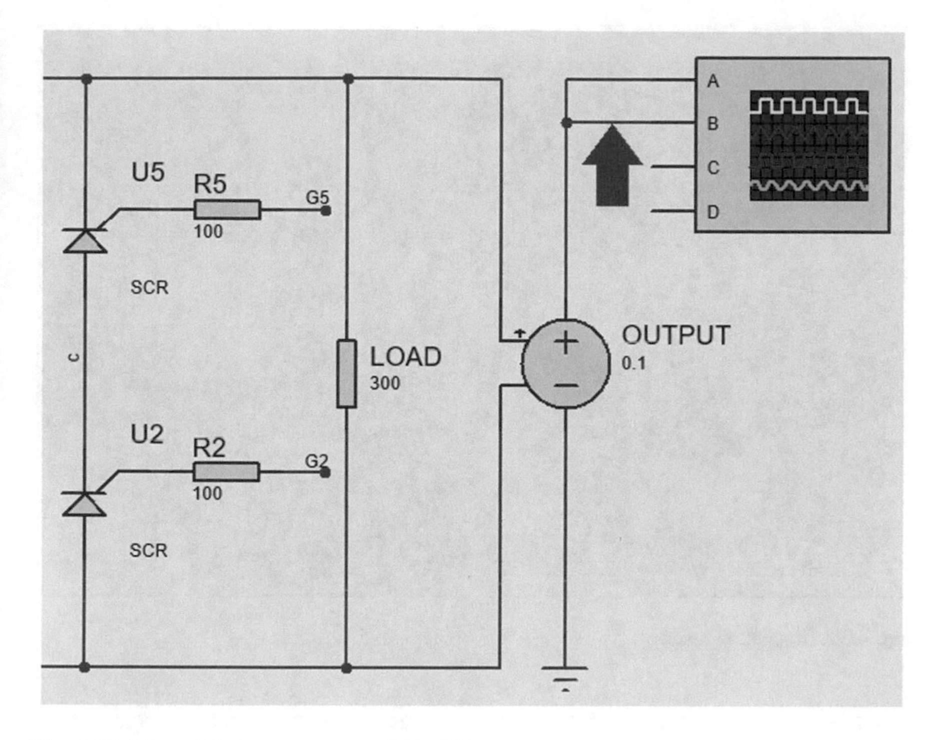

Fig. 4.57 Channel B is connected to OUTPUT

Run the simulation. Select the DC coupling for channel A and AC coupling for channel B. Click the Invert button of channel B and A + B button of channel A (Fig. 4.58). The waveform shown on the screen is the DC component of load voltage. According to Fig. 4.58, the DC component is 485 V. Note that we attenuated the load voltage by a factor of 10 before applying the signal to the oscilloscope. Therefore, the oscilloscope reading must be multiplied by 10 when you want to read the oscilloscope.

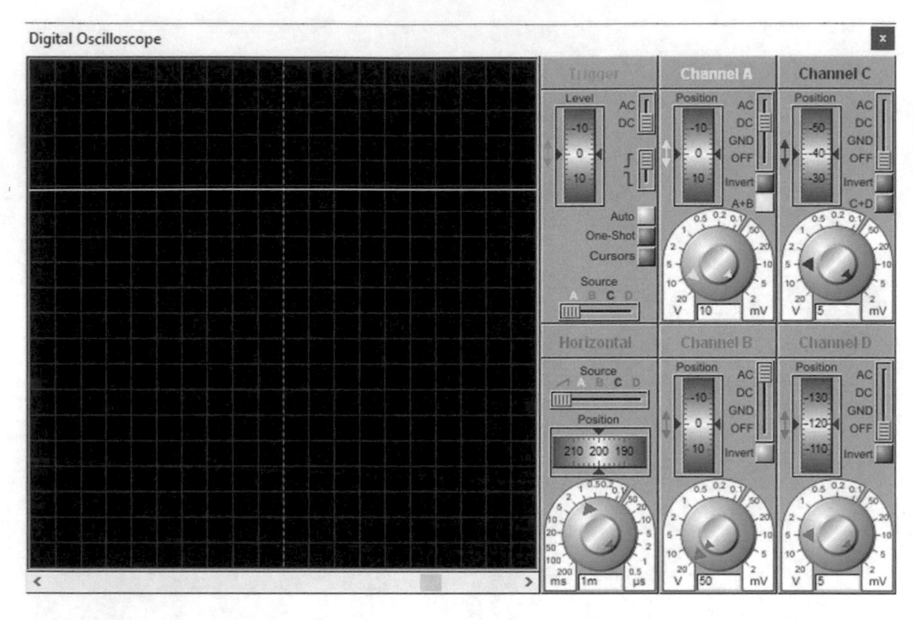

Fig. 4.58 Simulation result

Let's check the obtained result. The average value of output voltage can be calculated with the aid of $V_o = 1.35V_{LL}\cos(\alpha)$ formula. V_o, V_{LL} and α show the average value of output voltage, the RMS of line-line voltage that supplies the rectifier and the firing angle, respectively. According to the calculations shown in Fig. 4.59, the output voltage must be around 489.24 V. The obtained value is quite close to the Proteus value.

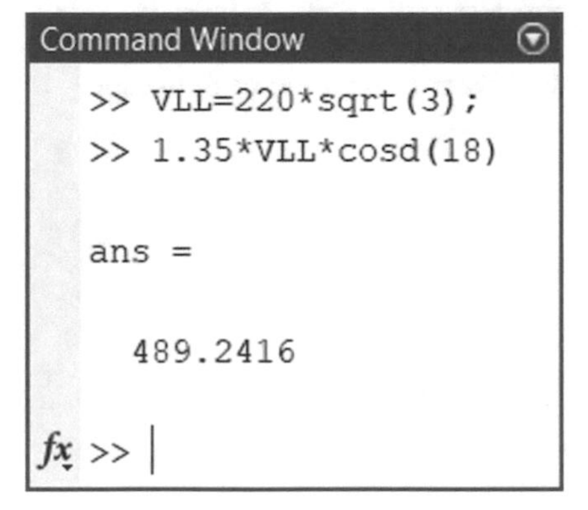

Fig. 4.59 MATLAB calculations

The following MATLAB code calculates the average and RMS of output voltage of the converter. This code ignores the voltage drop of the thyristors. Variable T determines the firing angle of the thyristors: $\alpha = T \times f \times 360°$, where f shows the frequency of input AC source. For instance, for $T = 1$ ms and $f = 50$ Hz, $\alpha = 18°$. Td and DT are two constants: $\text{Td} = \frac{1}{12f}$ and $\text{DT} = \frac{1}{6f}$. For 50 Hz source, $\text{Td} = 1.667 \times 10^{-3}$ and $\text{DT} = 3.333 \times 10^{-3}$ and, for 60 Hz source, $\text{Td} = 1.389 \times 10^{-3}$ and $\text{DT} = 2.778 \times 10^{-3}$.

```
f=50;
Vm=311;
w=2*pi*f;

syms t
Van=Vm*sin(w*t);
Vbn=Vm*sin(w*t-2*pi/3);
Vcn=Vm*sin(w*t+2*pi/3);

T=1e-3;
Td=1.667e-3;
DT=3.333e-3;
V=Van-Vbn;
Average=eval(1/DT*int(V,t,Td+T,Td+T+DT))
RMS=eval(sqrt(1/DT*int(V^2,t,Td+T,Td+T+DT)))
```

After running the above code, the result shown in Fig. 4.60 is obtained.

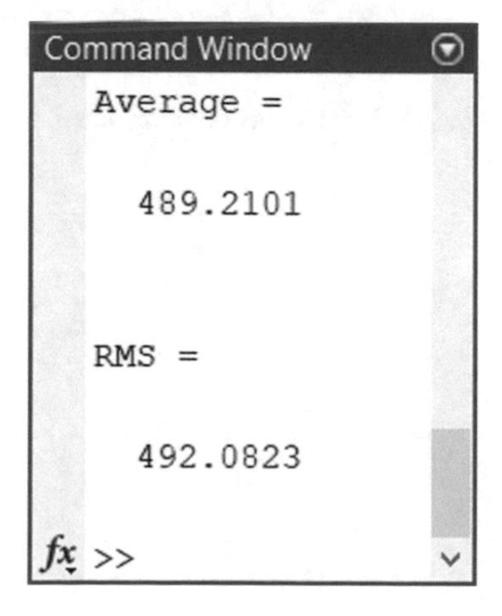

Fig. 4.60 Output of the MATLAB code

Let's connect an AC voltmeter and a DC voltmeter to the output of the circuit. According to Fig. 4.61, the AC voltmeter reading is quite close to the correct value. However, the DC voltmeter reading is not correct.

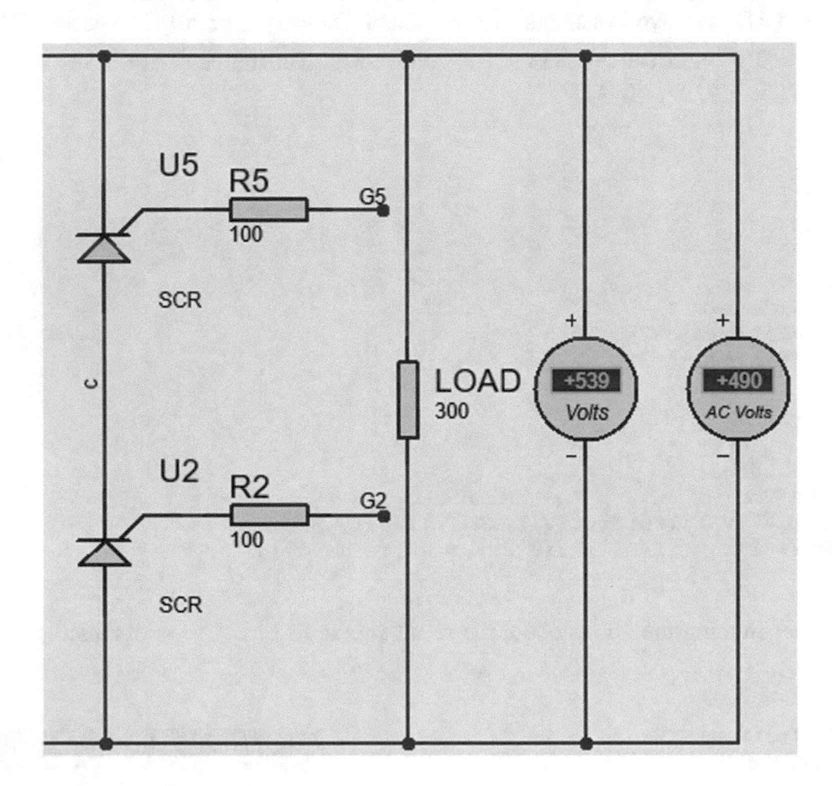

Fig. 4.61 Readings of voltmeters

Let's use a low-pass filter to decrease the amplitude of harmonics reach to the DC voltmeter (Fig. 4.62). This technique is studied in Example 13 in Chap. 2. Settings of low-pass filter are shown in Fig. 4.63. A low-pass filter block can be added to the schematic by searching for "LP_F" in the pick devices window (Fig. 4.64).

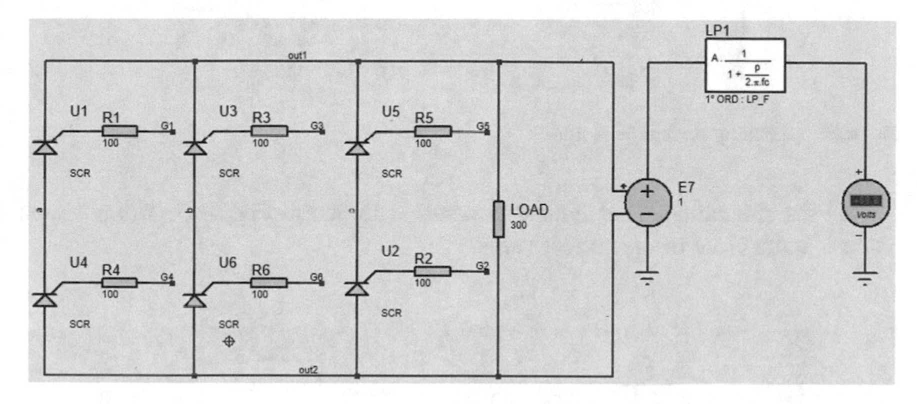

Fig. 4.62 Addition of low-pass filter to the schematic

Edit Component ? ×

Part Reference: LP1 Hidden: ☐ OK

Part Value: 1° ORD : LP_F Hidden: ☐ Help

Element: New Cancel

Static gain: 1.0 Hide All

Cut-off frequency [in Hz] 10 Hide All

Other Properties:

☐ Exclude from Simulation ☐ Attach hierarchy module

☐ Exclude from PCB Layout ☐ Hide common pins

☐ Exclude from Current Variant ☐ Edit all properties as text

Fig. 4.63 Settings of LP1

Pick Devices

Keywords:

LP_F

Match whole words?

Show only parts with models?

Fig. 4.64 Searching for low-pass filter

Run the simulation. The simulation result is shown in Fig. 4.65. The obtained result is quite close to the correct value.

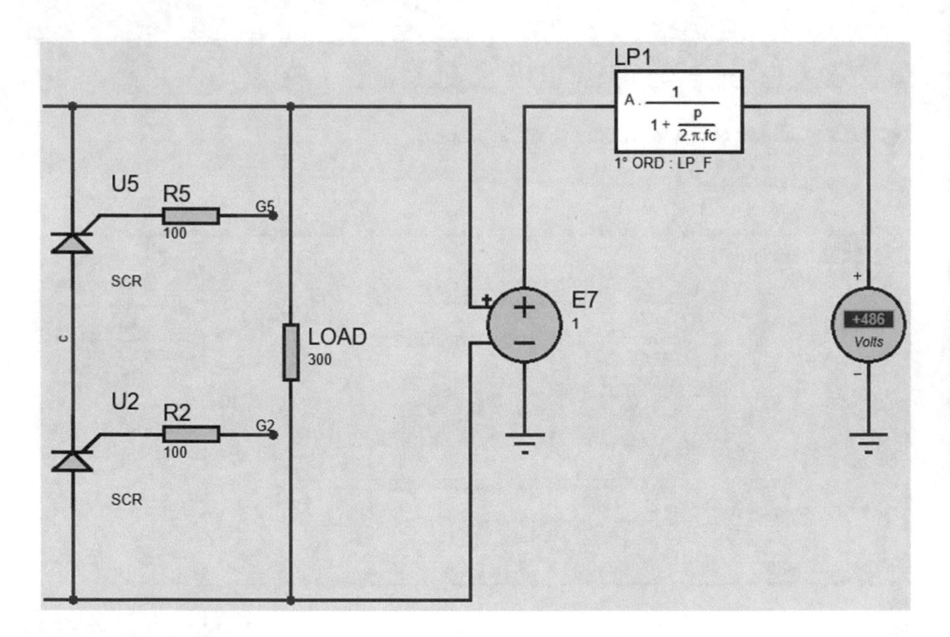

Fig. 4.65 Readings of voltmeter

4.10 Example 9: Three-Phase Controlled Rectifier (III)

We studied two ways to measure the average value of load voltage in the previous example. We will study another method to measure the average value of output voltage in this example. Add two voltage probes to the load (Fig. 4.66).

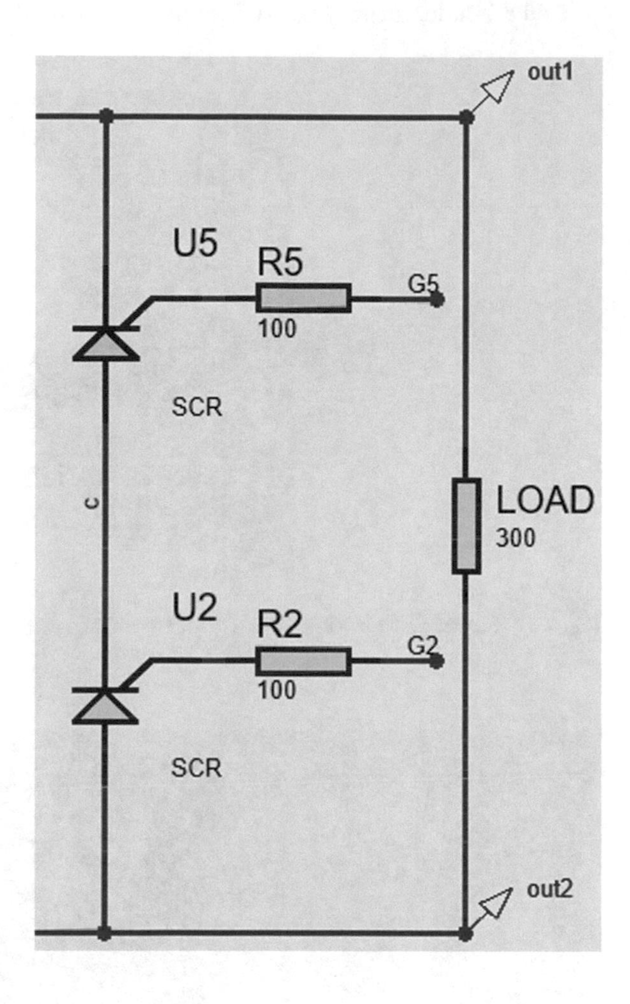

Fig. 4.66 Two voltage probes are added to the load

Add a Fourier graph (Fig. 4.67) to the schematic (Fig. 4.68).

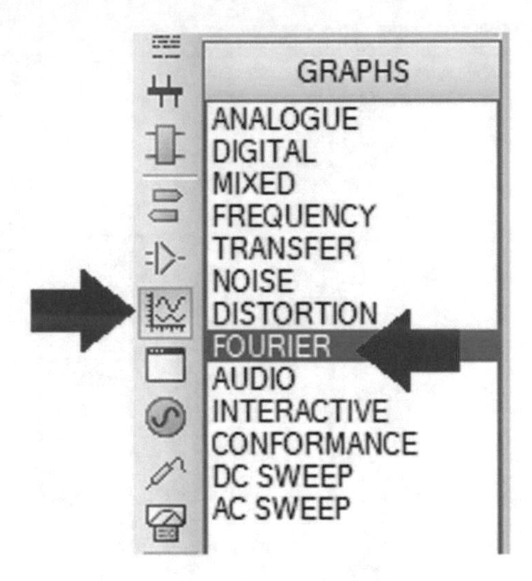

Fig. 4.67 Fourier graph block

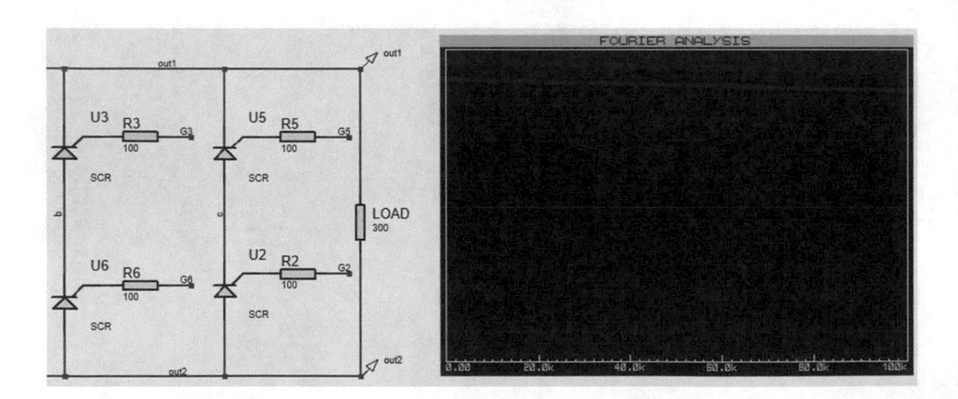

Fig. 4.68 Fourier graph is added to the schematic

Right-click on the Fourier graph and click the Add Traces (Fig. 4.69). After clicking the Add Traces, the window shown in Fig. 4.70 appears. Change the settings to what is shown in Fig. 4.71 and click the OK button.

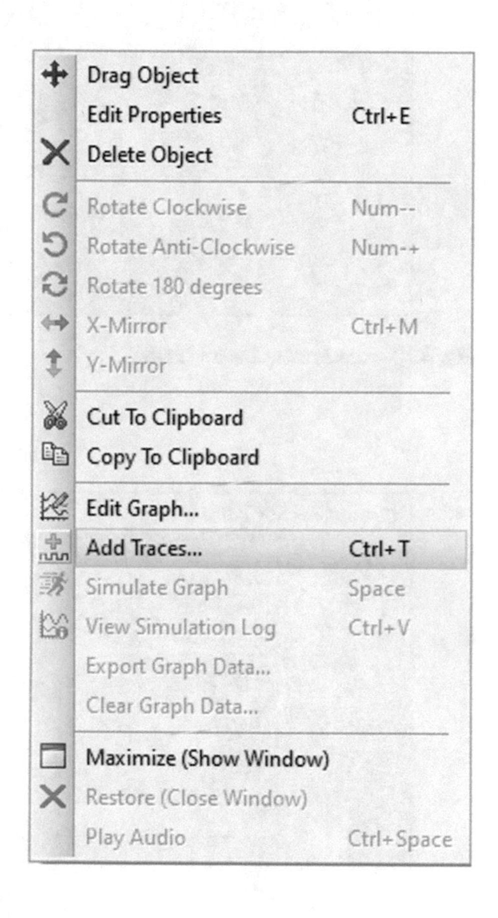

Fig. 4.69 Add traces command

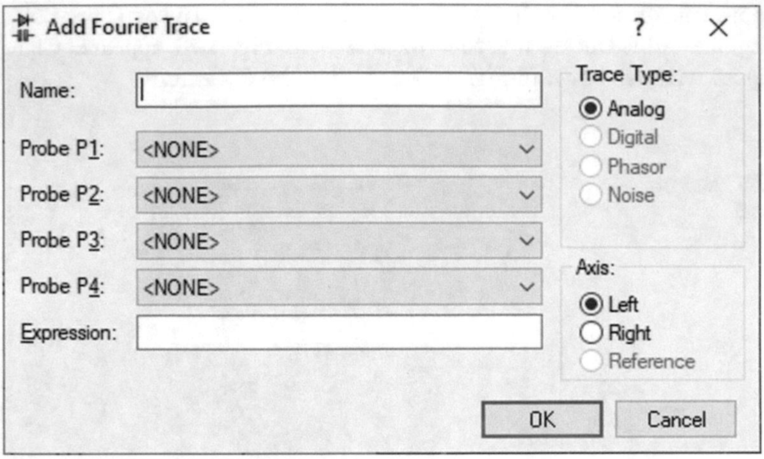

Fig. 4.70 Add Fourier trace window

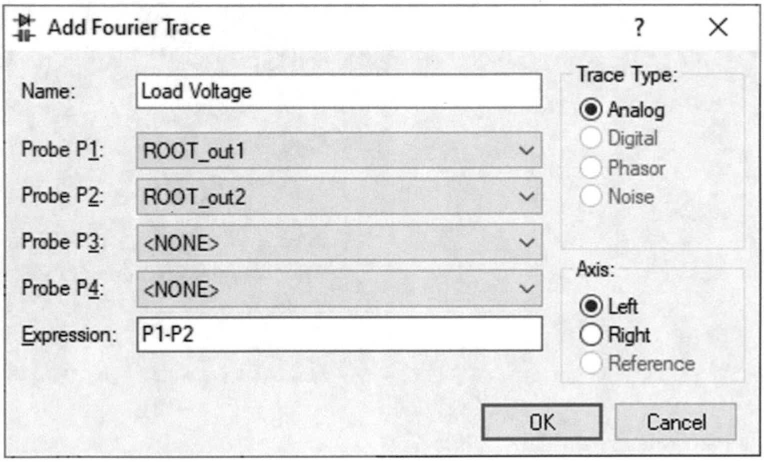

Fig. 4.71 Add Fourier trace window

Double-click the Fourier graph and do the settings similar to Fig. 4.72. Then click the OK button.

Edit Fourier Analysis Graph

Graph title: FOURIER ANALYSIS

Start time: 0

Stop time: 500m

Max Frequency: 500

Resolution: 50 Window None

Left Axis Label:

Right Axis Label:

Options

Y Scale in dBs: ☐

Initial DC solution: ☑

Always simulate? ☑

Log netlist(s)? ☐

SPICE Options

Set Y-Scales

User defined properties:

OK Cancel

Fig. 4.72 Edit Fourier analysis graph window

Run the simulation. The simulation result is shown in Fig. 4.73. Maximize the graph to read it easily. The value of this graph at 0 Hz (DC) is 488 V which is quite close to the correct value shown in Fig. 4.60.

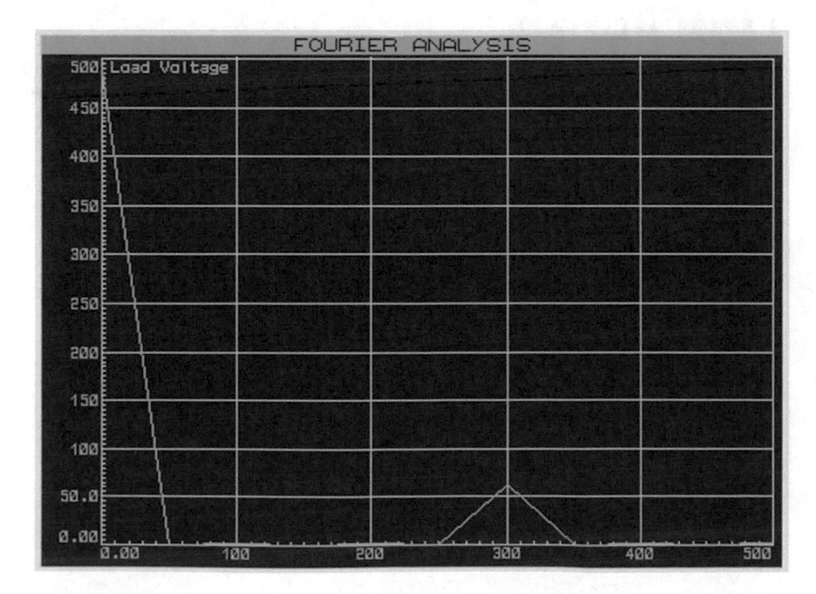

Fig. 4.73 Simulation result

4.11 Example 10: Three-Phase Controlled Rectifier (IV)

Maximum and average values of current passed from the semiconductors are important for the selection of suitable components for the circuit. Let's measure the maximum and average values of current passed from one of the thyristors. Add a current-controlled voltage source to the schematic (Fig. 4.74).

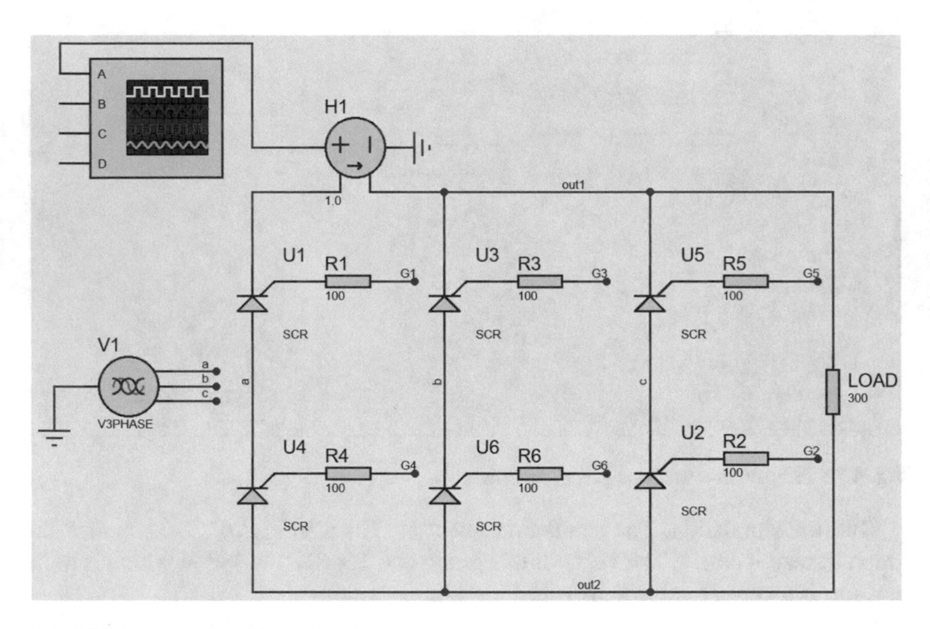

Fig. 4.74 Schematic of Example 10

Run the simulation. The simulation result is shown in Fig. 4.75. Now you can measure the maximum current passed from the device. The maximum is around $9.5 \times 0.2 = 1.9$A.

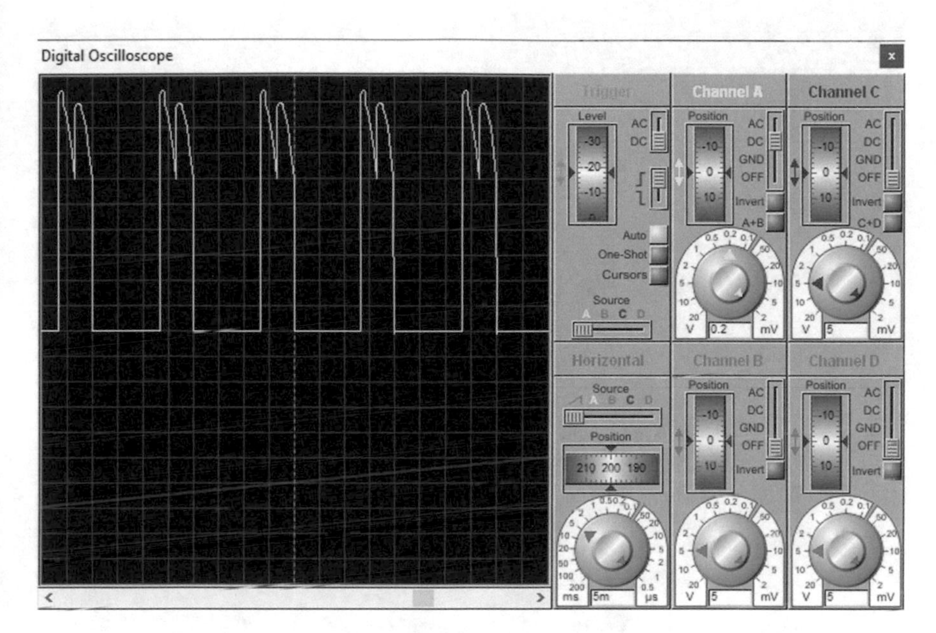

Fig. 4.75 Simulation result

Use the method described in Example 8 to measure the average value of current passed from the device (Fig. 4.76). The result is shown in Fig. 4.77. The average value of current is around 550 mA.

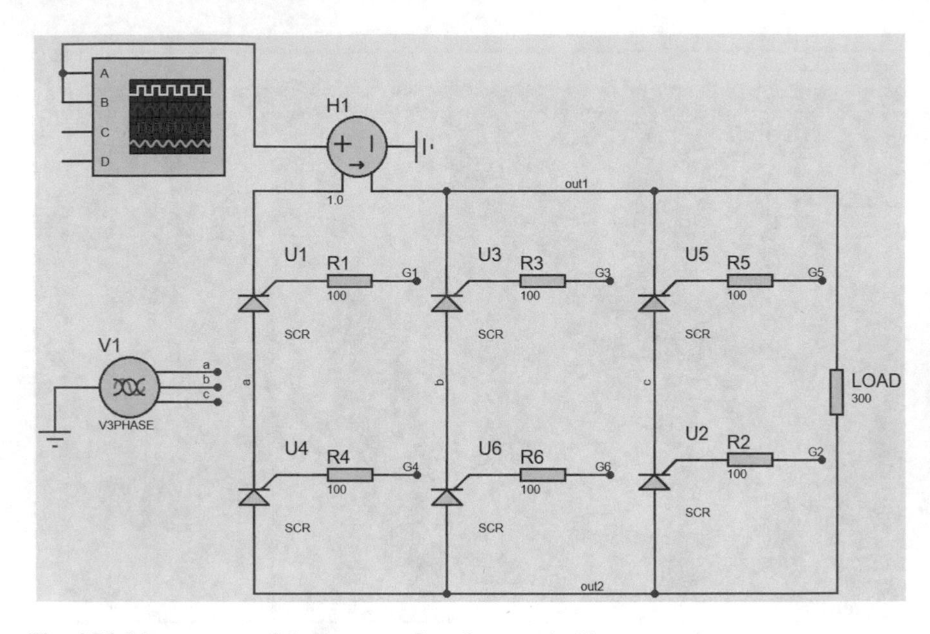

Fig. 4.76 Measurement of the average value of current in U_1

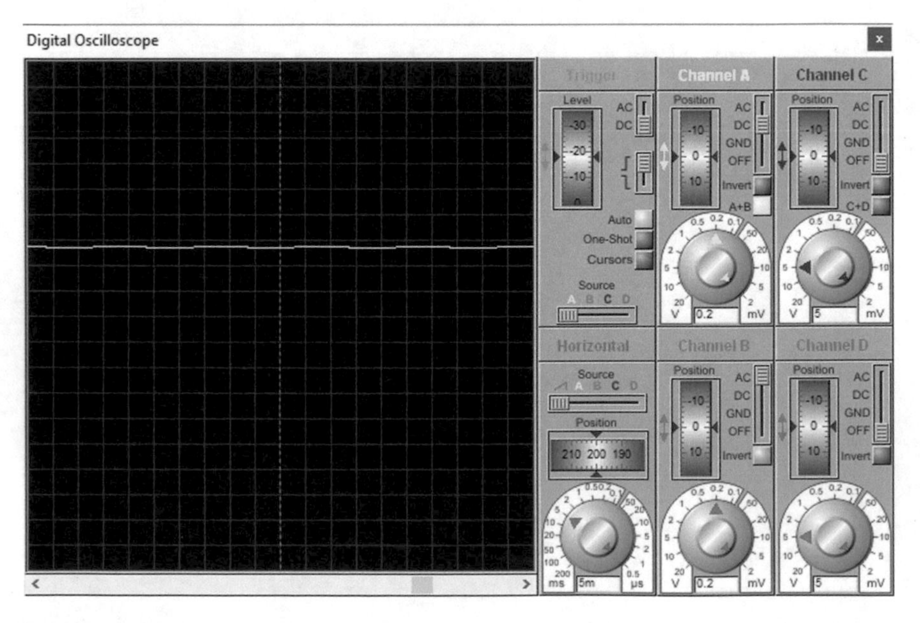

Fig. 4.77 Simulation result

4.12 Example 11: Harmonic Content of Output Voltage of a Rectifier

In this example, we want to see the harmonic content of output voltage of the rectifier shown in Fig. 4.78. The harmonic content of the output voltage can be extracted using the Fourier analysis.

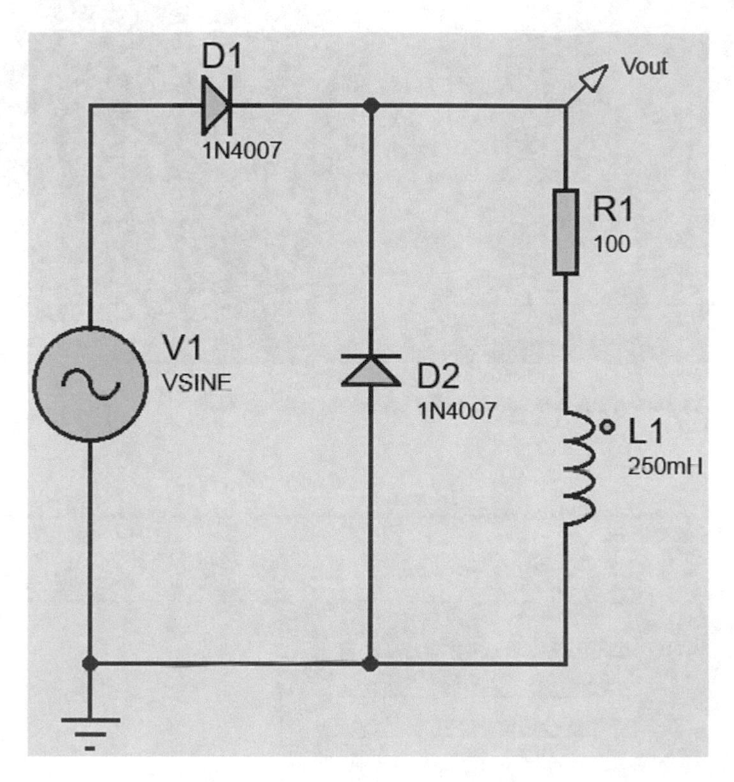

Fig. 4.78 Schematic for Example 11

Let's take a look at the load voltage waveform. Add an analogue graph to the schematic and add voltage probe V_{out} to it (Fig. 4.79). The settings of the added analogue graph are shown in Fig. 4.80.

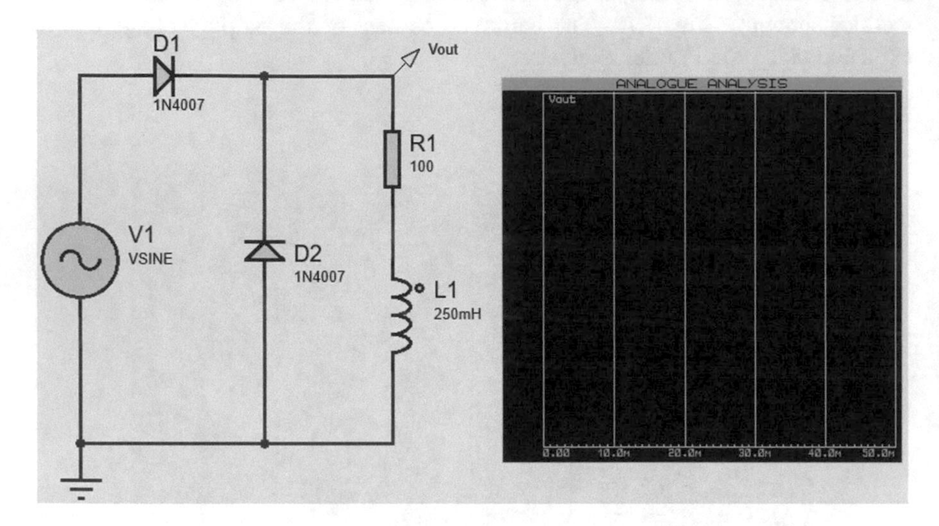

Fig. 4.79 An analog graph is added to the schematic of Fig. 4.78

Edit Transient Graph ? X

Graph title: ANALOGUE ANALYSIS

Start time: 0

Stop time: 50m

Left Axis Label:

Right Axis Label:

Options

Initial DC solution: ☑

Always simulate: ☑

Log netlist(s): ☐

SPICE Options

Set Y-Scales

User defined properties:

OK Cancel

Fig. 4.80 Edit transient graph window

Run the simulation. Figure 4.81 shows the load voltage waveform. The load voltage is a half wave rectified sine wave.

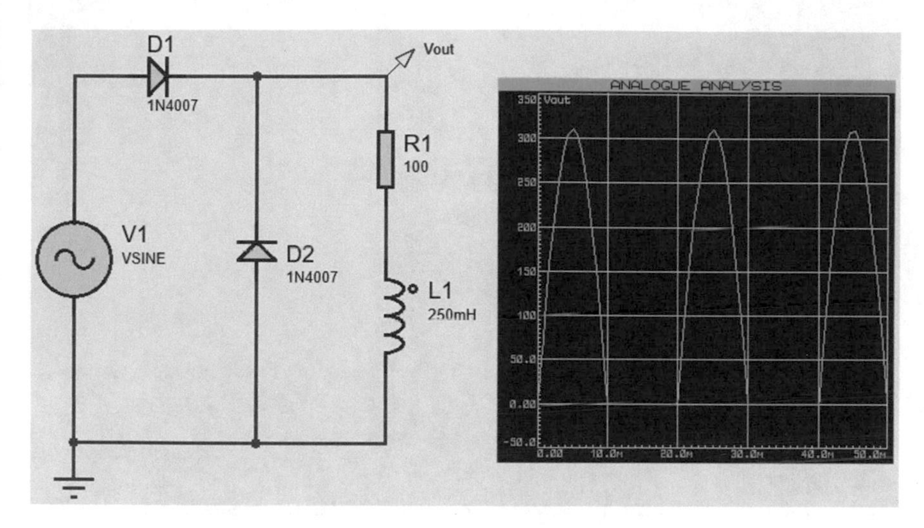

Fig. 4.81 Simulation result

Now remove the analogue analysis graph and add a Fourier graph to the schematic (Fig. 4.82). The settings of the Fourier graph are shown in Fig. 4.83. These settings take the waveform in the [0, 500 ms] interval and use it to calculate the DC, fundamental, 2nd harmonic, 3rd harmonic, 4th harmonic, ..., 10th harmonic amplitudes.

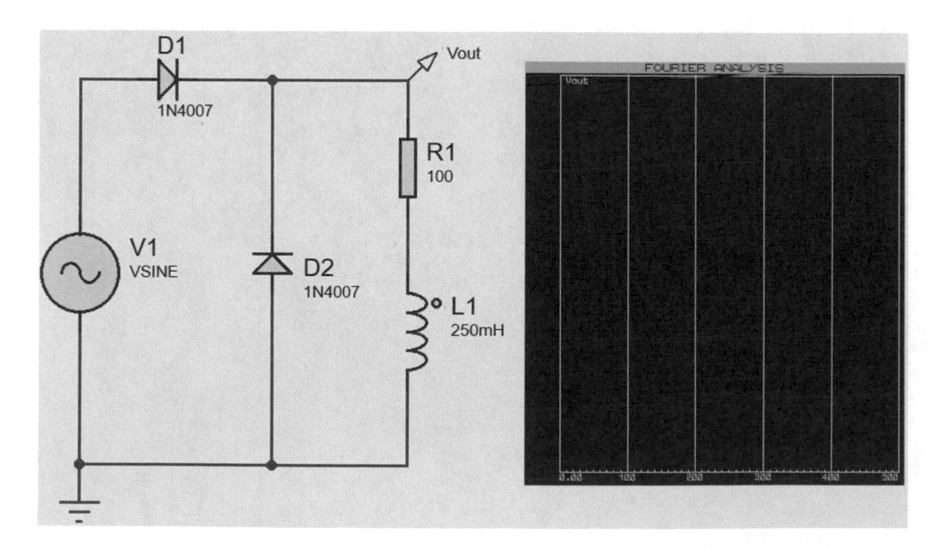

Fig. 4.82 Fourier graph is added to the schematic

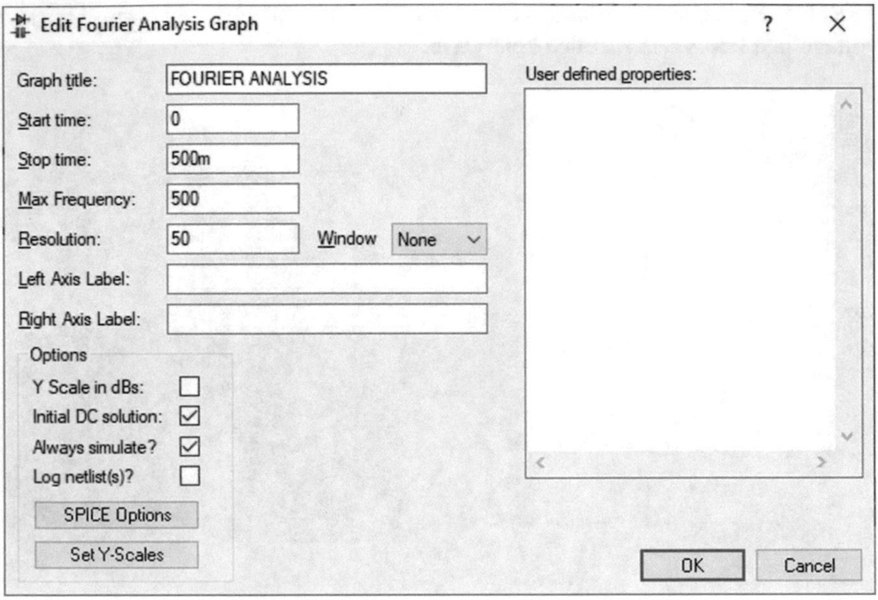

Fig. 4.83 Fourier graph settings

Run the simulation. The simulation result is shown in Fig. 4.84. You can use cursors to read the amplitude of harmonics. The amplitude of harmonics is shown in Table 4.1.

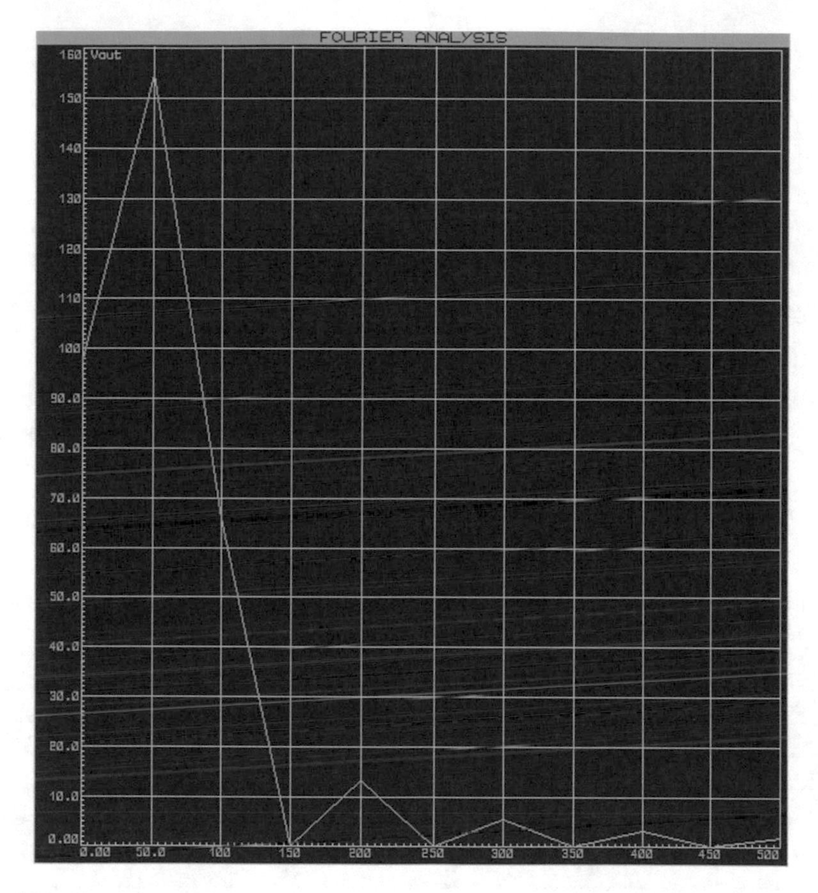

Fig. 4.84 Simulation results

Table 4.1 Amplitude of harmonics in Fig. 4.84

Harmonic	Amplitude (V)
DC component	98
Fundamental	155
2nd	66
3rd	0.07
4th	13.1
5th	0.11
6th	5.6
7th	0.04
8th	3.12
9th	0.04
10th	1.97

Let's check the obtained result. Fourier series of half wave rectified sine wave is shown in Fig. 4.85.

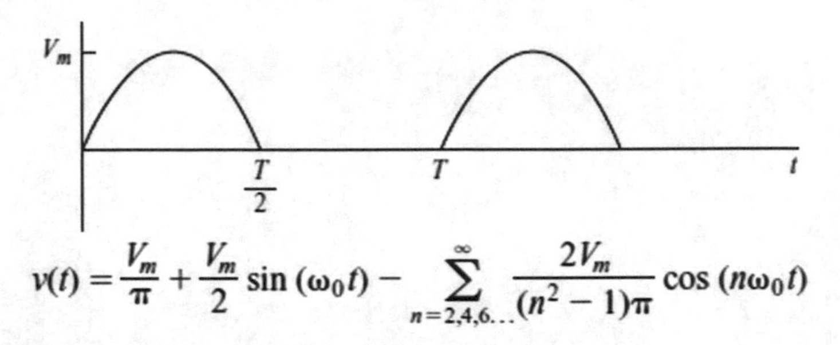

$$v(t) = \frac{V_m}{\pi} + \frac{V_m}{2}\sin(\omega_0 t) - \sum_{n=2,4,6\ldots}^{\infty} \frac{2V_m}{(n^2-1)\pi}\cos(n\omega_0 t)$$

Fig. 4.85 Fourier expansion of half wave rectified sinusoidal signal

Table 4.2 compares the values measured by Proteus with values predicted by the theoretic formula. Note that the values given by Proteus are very close to the values predicted by the theoretic formula in Fig. 4.85. This shows that the Proteus results are correct.

Table 4.2 Comparison of values measured by Proteus with values predicted by the formula in Fig. 4.85

Harmonic	Proteus amplitude (V)	Theoretical amplitude (V)
DC component	98	99.0
Fundamental	155	155.5
2nd	66	66
3rd	0.07	0
4th	13.1	13.2
5th	0.11	0
6th	5.6	5.66
7th	0.04	0
8th	3.12	3.14
9th	0.04	0
10th	1.97	2

4.13 Example 12: Single-Phase Inverter

In this example, we want to simulate a single-phase full bridge inverter (Fig. 4.86) with a unipolar PWM switching technique.

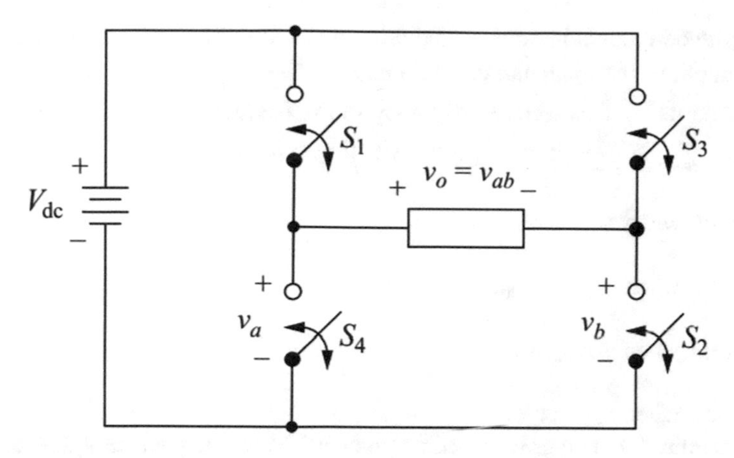

Fig. 4.86 Single-phase full bridge inverter

In a unipolar switching scheme for pulse width modulation, a high-frequency triangular carrier is compared with two low-frequency sinusoidal references (Fig. 4.87). The output $v_o = v_{ab}$ in Fig. 4.86 is switched either from high to zero or from low to zero (Fig. 4.87). That is why it is called unipolar.

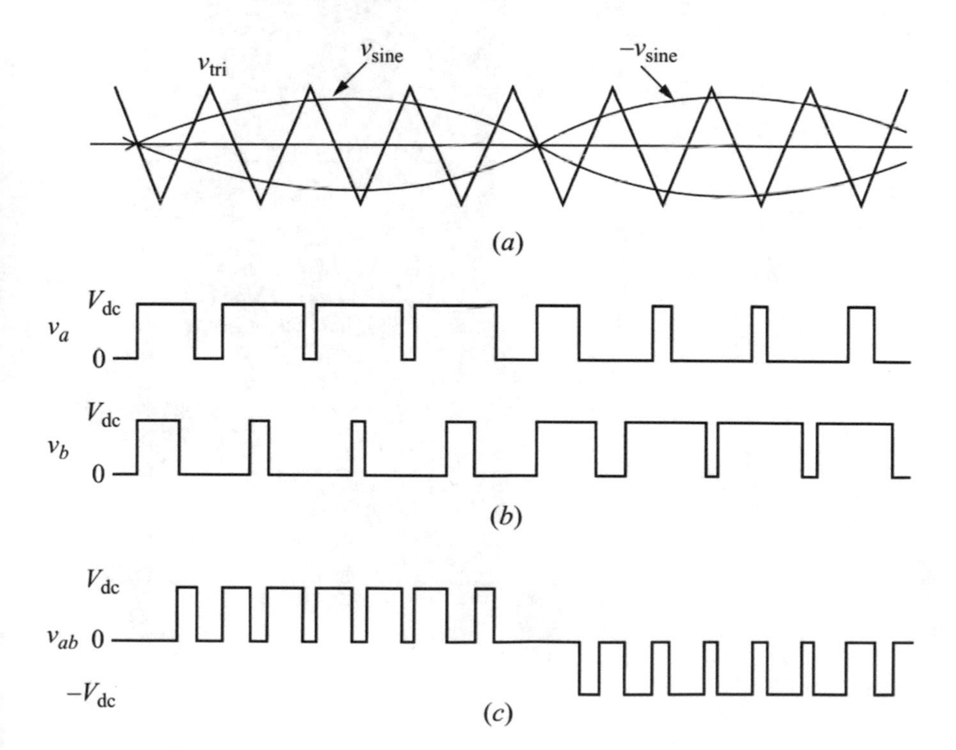

Fig. 4.87 Generation of PWM signals

Amplitude modulation ratio is the ratio of the amplitude of sinusoidal reference to the amplitude of triangular carrier, i.e., $m = \frac{V_{\text{Sinusoidal Reference}}}{V_{\text{Triangular Carrier}}}$. Frequency modulation ratio is the ratio of the carrier frequency to the frequency of sinusoidal reference, i.e., $m_f = \frac{f_{\text{Triangular Carrier}}}{f_{\text{Sinusoidal Reference}}}$. In a unipolar switching scheme,

S_1 is on when $v_{\text{sine}} > v_{\text{tri}}$.

S_2 is on when $-v_{\text{sine}} < v_{\text{tri}}$.

S_3 is on when $-v_{\text{sine}} > v_{\text{tri}}$.

S_4 is on when $v_{\text{sine}} < v_{\text{tri}}$.

The required triangular wave for unipolar PWM can be generated with the aid of a signal generator block (Fig. 4.88).

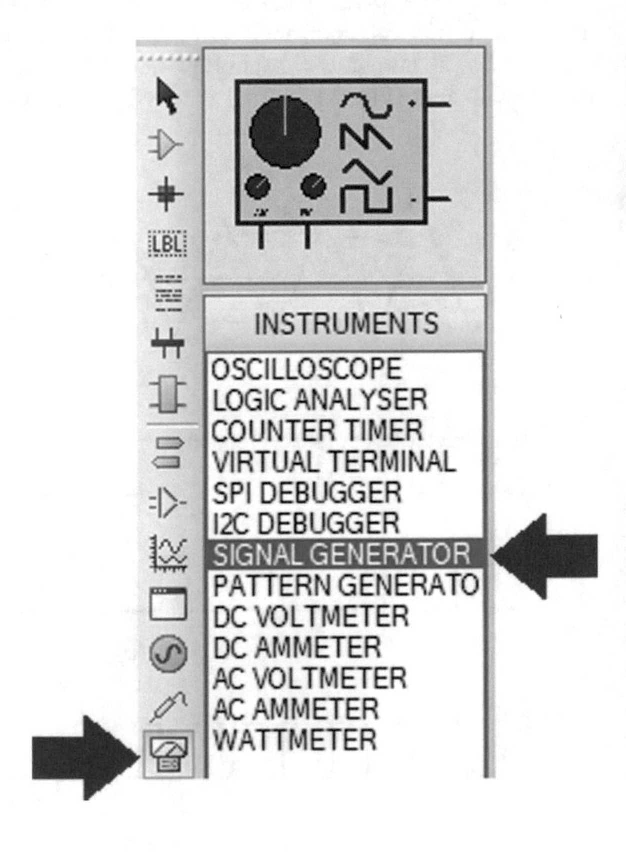

Fig. 4.88 Signal generator block

Draw the schematic shown in Fig. 4.89 and run it. After running the simulation, the signal generator window appears on the screen (Fig. 4.90). Use the controls in this window to generate a triangular wave with a frequency of 1150 Hz and peak-peak amplitude of 2 V (Fig. 4.91).

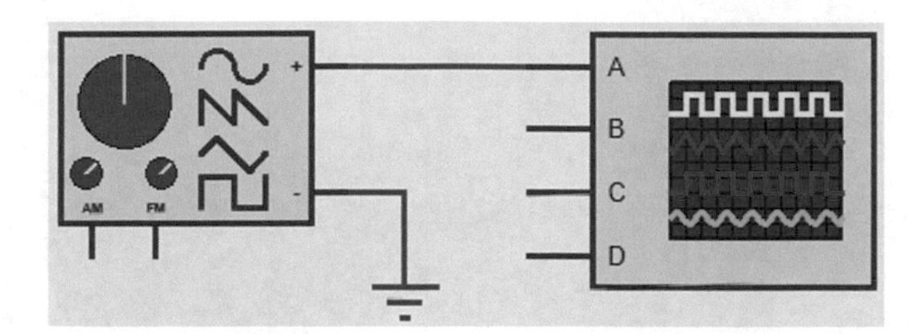

Fig. 4.89 Signal generator is connected to channel A of oscilloscope

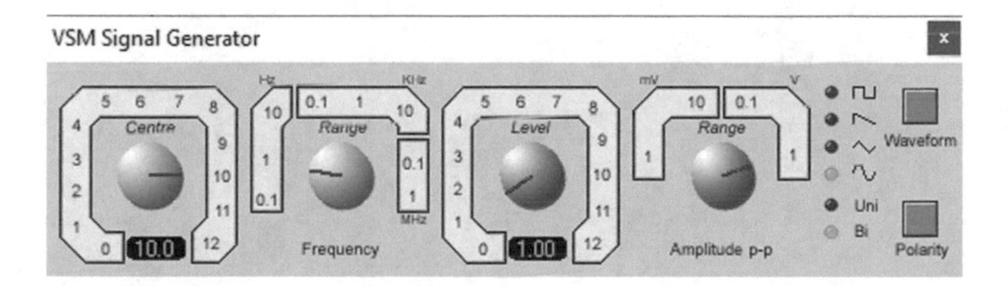

Fig. 4.90 VSM signal generator window

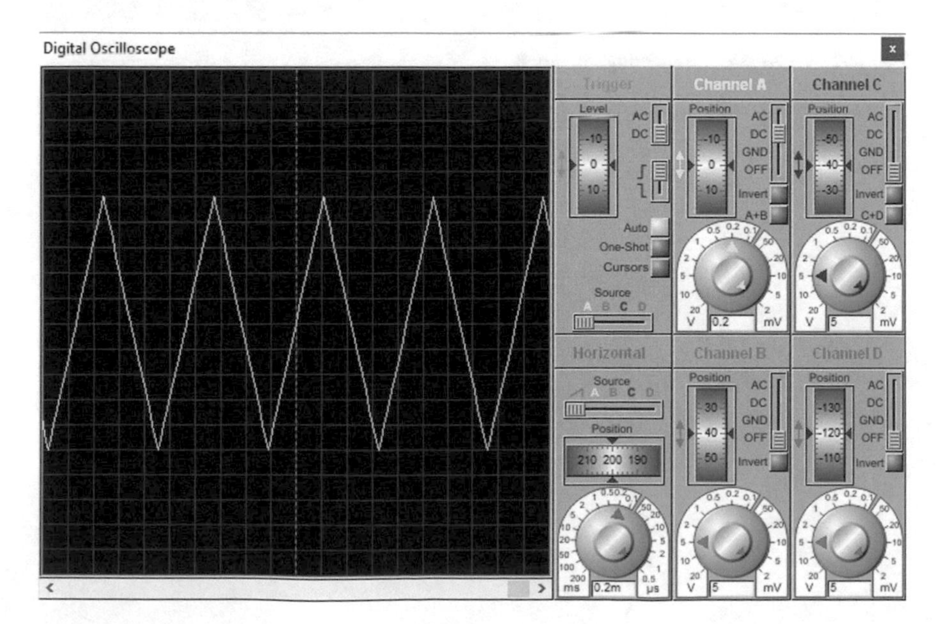

Fig. 4.91 Simulation result

Now, the required carrier is generated. Change the schematic to what is shown in Fig. 4.92.

The schematic shown in Fig. 4.92 is composed of three sections: power circuit, measurement section and control circuit.

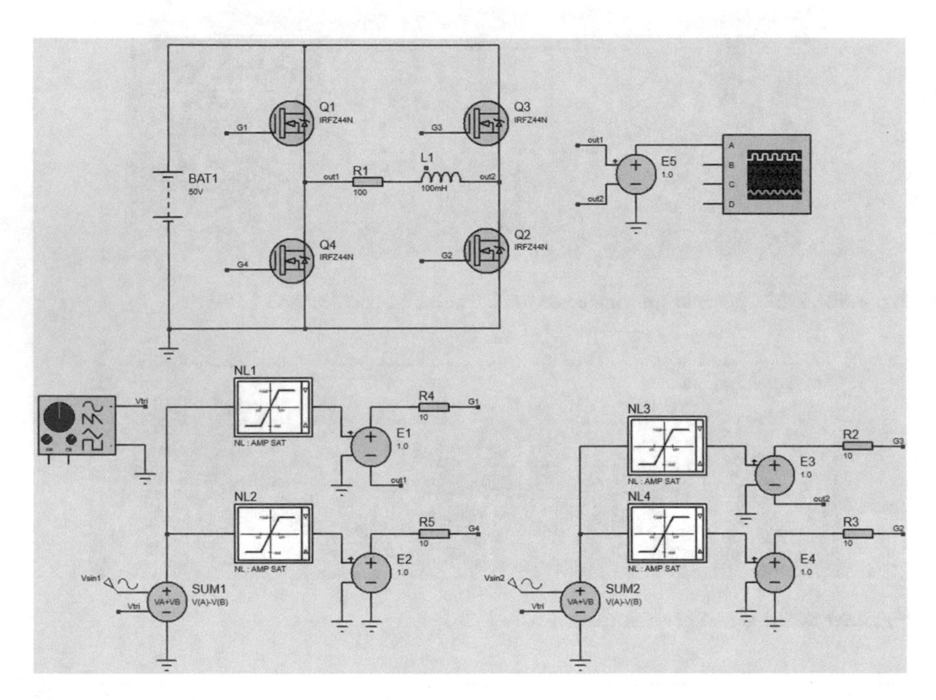

Fig. 4.92 Schematic of full bridge inverter

The power circuit section of the schematic is shown in Fig. 4.93.

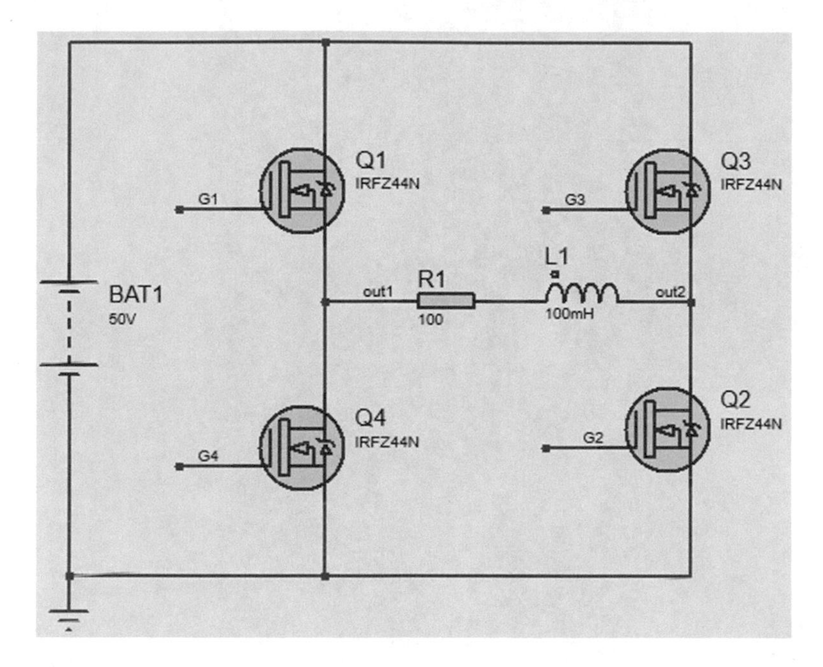

Fig. 4.93 Power circuit

The measurement section of the schematic is shown in Fig. 4.94. This section measures the load voltage.

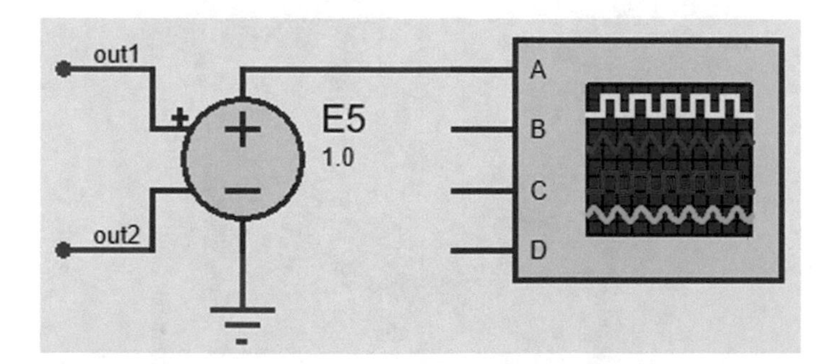

Fig. 4.94 Measurement of load voltage

The Control section is shown in Fig. 4.95. Summer and saturation blocks (Figs. 4.96 and 4.97) are used in this section of schematic.

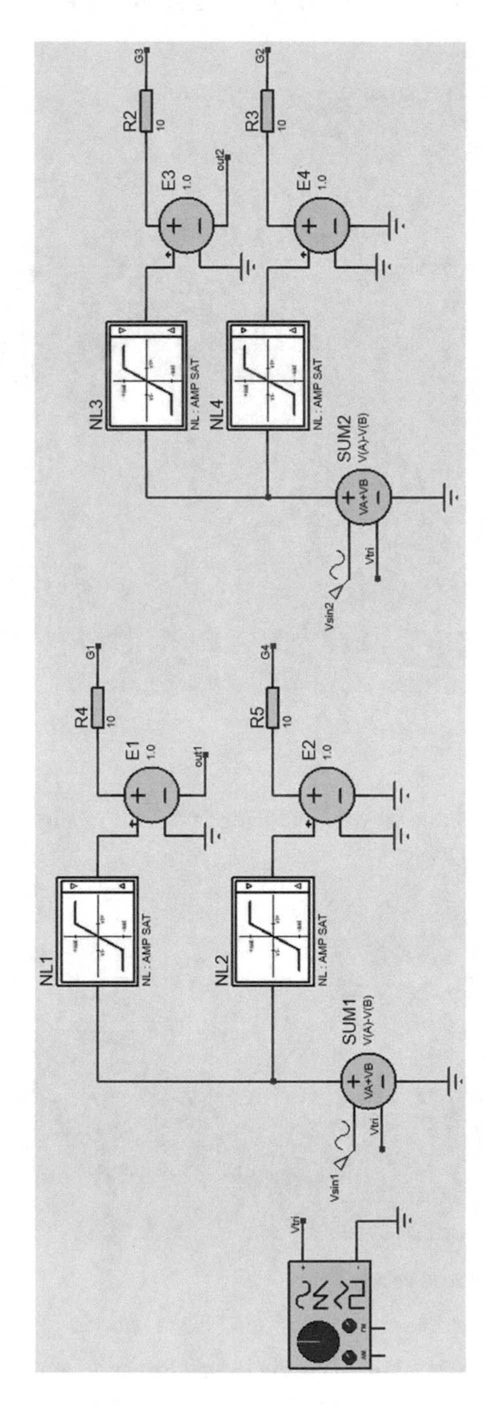

Fig. 4.95 Control section

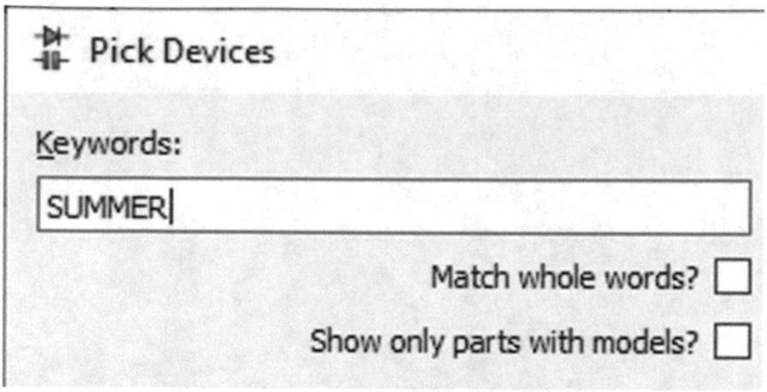

Fig. 4.96 Searching for summer block

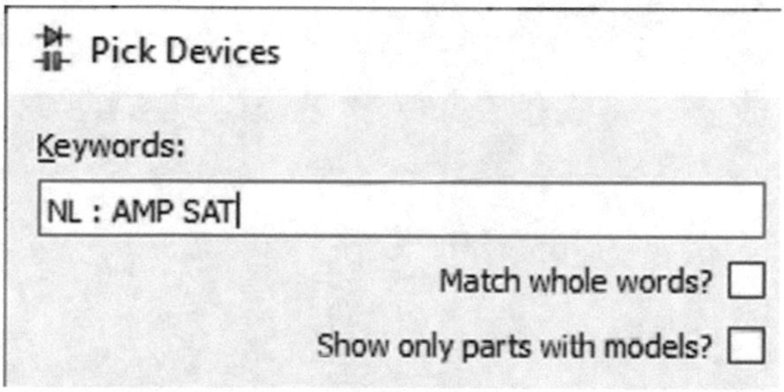

Fig. 4.97 Searching for saturation block

Let's take a closer look at the schematic shown in Fig. 4.95. The section shown in Fig. 4.98 generates a control signal for Q1 and Q4 MOSFETs. The settings of the blocks are shown in Figs. 4.99, 4.100 and 4.101. This section implements the unipolar PWM switching technique. V_{tri} is the triangular waveform shown in Fig. 4.91. V_{sine1} is a sinusoidal reference signal with a frequency of 50 Hz. SUM1 compares V_{sine1} with V_{tri}. When $V_{sin1} > V_{tri}$, the output of SUM1 is positive. When $V_{sin1} < V_{tri}$, the output of SUM1 is negative. When the output of SUM1 is positive, NL_1 and E_1 generate + 12 V between gate and source of Q_1. This forces Q_1 to be on. Q_4 is off in this case. When the output of SUM1 is negative, NL_2 and E_2 generate + 12 V between gate and source of Q_4. This forces Q_4 to be on. Q_1 is off in this case.

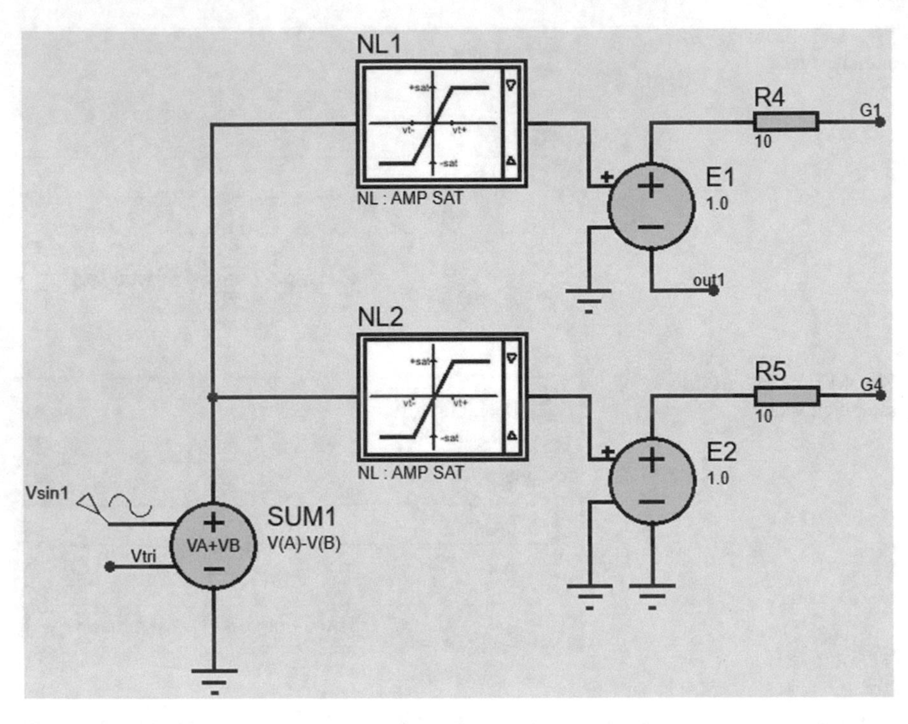

Fig. 4.98 Generation of control signals for Q_1 and Q_4 MOSFETs

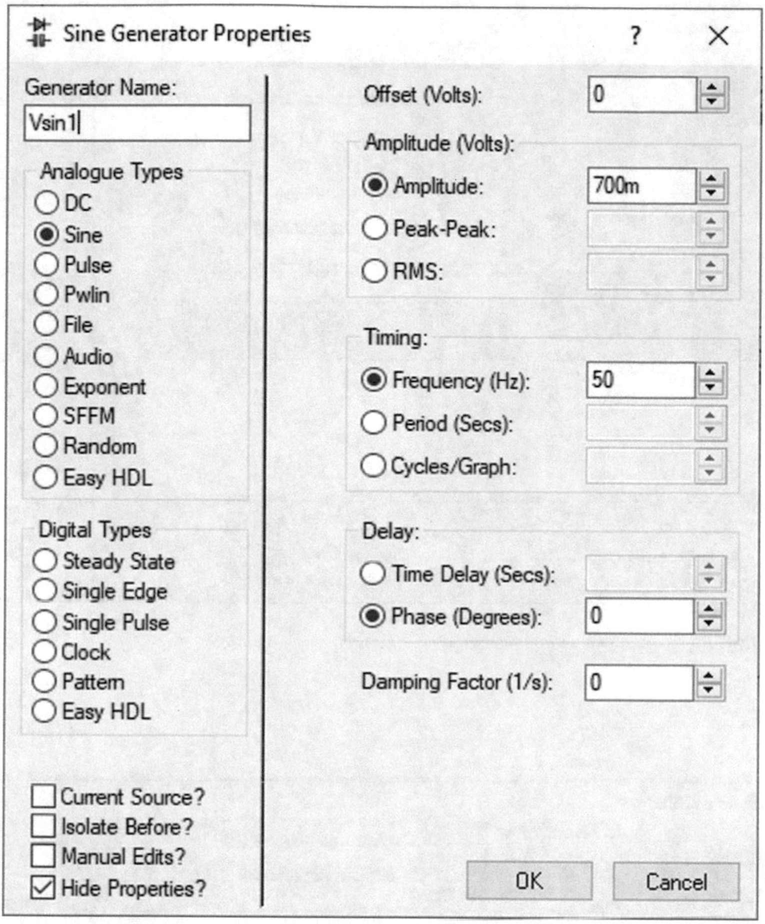

Sine Generator Properties ? X

Generator Name:
Vsin1

Analogue Types
DC
Sine
Pulse
Pwlin
File
Audio
Exponent
SFFM
Random
Easy HDL

Digital Types
Steady State
Single Edge
Single Pulse
Clock
Pattern
Easy HDL

Current Source?
Isolate Before?
Manual Edits?
Hide Properties?

Offset (Volts): 0

Amplitude (Volts):
Amplitude: 700m
Peak-Peak:
RMS:

Timing:
Frequency (Hz): 50
Period (Secs):
Cycles/Graph:

Delay:
Time Delay (Secs):
Phase (Degrees): 0

Damping Factor (1/s): 0

OK Cancel

Fig. 4.99 Settings of $V_{\sin 1}$

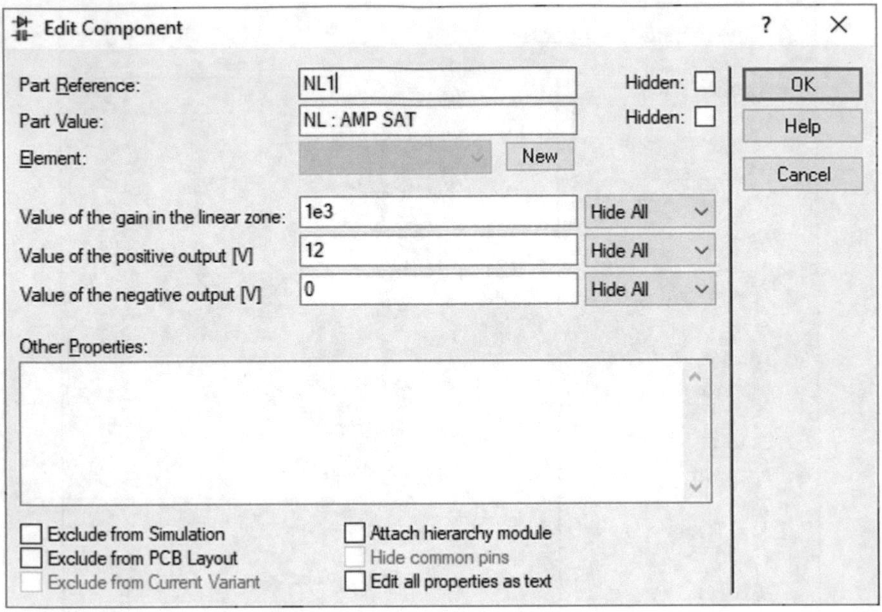

Edit Component

Part Reference: NL1
Part Value: NL : AMP SAT
Element: New
Value of the gain in the linear zone: 1e3 Hide All
Value of the positive output [V] 12 Hide All
Value of the negative output [V] 0 Hide All
Other Properties:
Exclude from Simulation
Exclude from PCB Layout
Exclude from Current Variant
Attach hierarchy module
Hide common pins
Edit all properties as text
Hidden: Hidden: OK Help Cancel

Fig. 4.100 Settings of NL_1

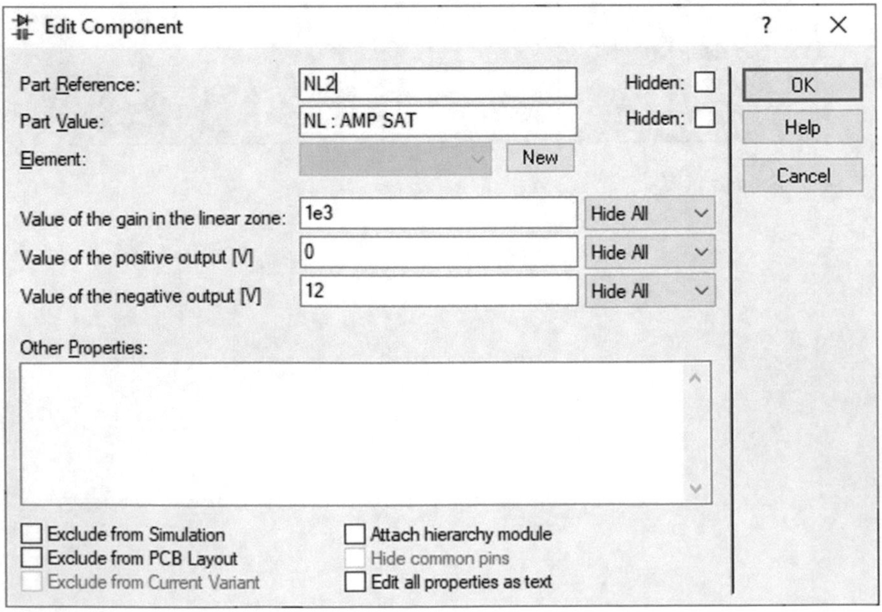

Edit Component

Part Reference: NL2
Part Value: NL : AMP SAT
Element: New
Value of the gain in the linear zone: 1e3 Hide All
Value of the positive output [V] 0 Hide All
Value of the negative output [V] 12 Hide All
Other Properties:
Exclude from Simulation
Exclude from PCB Layout
Exclude from Current Variant
Attach hierarchy module
Hide common pins
Edit all properties as text
Hidden: Hidden: OK Help Cancel

Fig. 4.101 Settings of NL_2

The section shown in Fig. 4.102 generates a control signal for Q_2 and Q_3 MOSFETs. Settings of the blocks are shown in Figs. 4.103, 4.104 and 4.105. This section works similar to Fig. 4.98. V_{sine2} is a sinusoidal reference signal with a frequency of 50 Hz. It has a 180° phase difference with V_{sine1}. SUM2 compares V_{sine2} with V_{tri}. When $V_{sin2} > V_{tri}$, the output of SUM2 is positive. When $V_{sin2} < V_{tri}$, the output of SUM2 is negative. When the output of SUM2 is positive, NL_3 and E_3 generate + 12 V between gate and source of Q_3. This forces Q_3 to be on. Q_2 is off in this case. When the output of SUM2 is negative, NL_4 and E_4 generate + 12 V between gate and source of Q_2. This forces Q_2 to be on. Q_3 is off in this case.

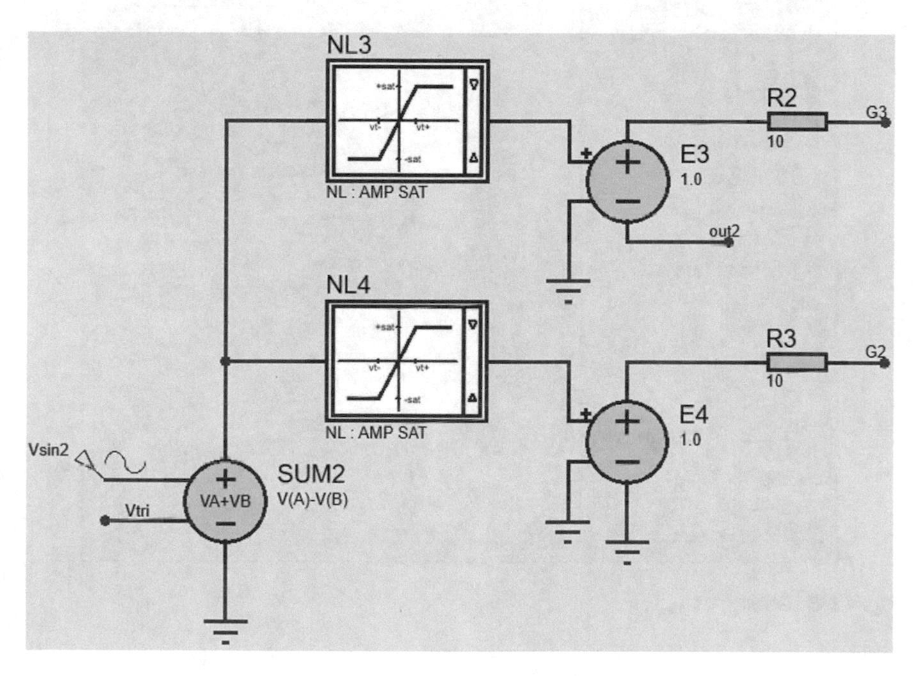

Fig. 4.102 Generation of control signals for Q_3 and Q_2 MOSFETs

Sine Generator Properties
Generator Name: Vsin2
Analogue Types: DC, Sine (selected), Pulse, Pwlin, File, Audio, Exponent, SFFM, Random, Easy HDL
Digital Types: Steady State, Single Edge, Single Pulse, Clock, Pattern, Easy HDL
Current Source? Isolate Before? Manual Edits? Hide Properties? (checked)
Offset (Volts): 0
Amplitude (Volts): Amplitude (selected): 700m; Peak-Peak; RMS
Timing: Frequency (Hz) (selected): 50; Period (Secs); Cycles/Graph
Delay: Time Delay (Secs); Phase (Degrees) (selected): 180
Damping Factor (1/s): 0
OK Cancel

Fig. 4.103 Settings of $V_{\sin 2}$

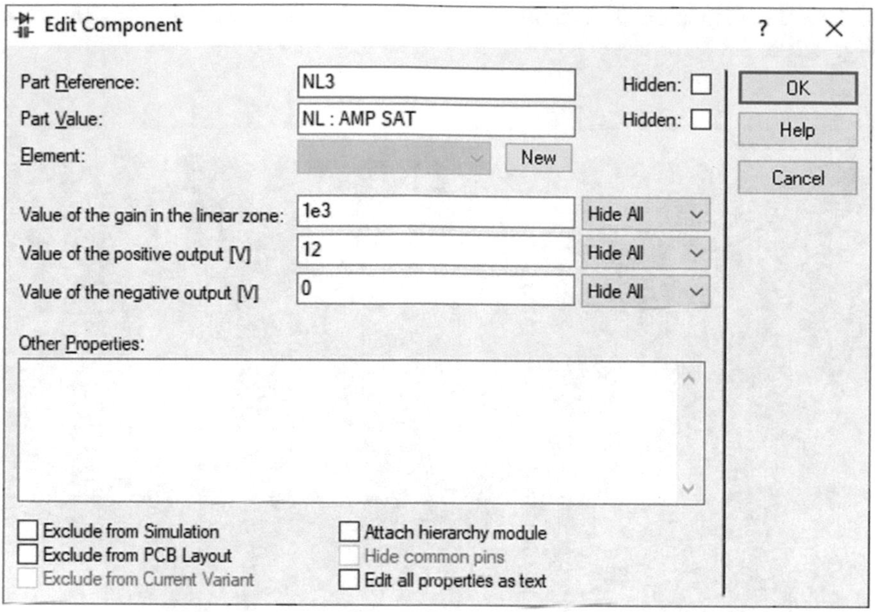

Edit Component ? ×

Part Reference: NL3 Hidden: ☐ OK
Part Value: NL : AMP SAT Hidden: ☐ Help
Element: New Cancel
Value of the gain in the linear zone: 1e3 Hide All
Value of the positive output [V] 12 Hide All
Value of the negative output [V] 0 Hide All
Other Properties:
☐ Exclude from Simulation ☐ Attach hierarchy module
☐ Exclude from PCB Layout Hide common pins
Exclude from Current Variant ☐ Edit all properties as text

Fig. 4.104 Settings of NL_3

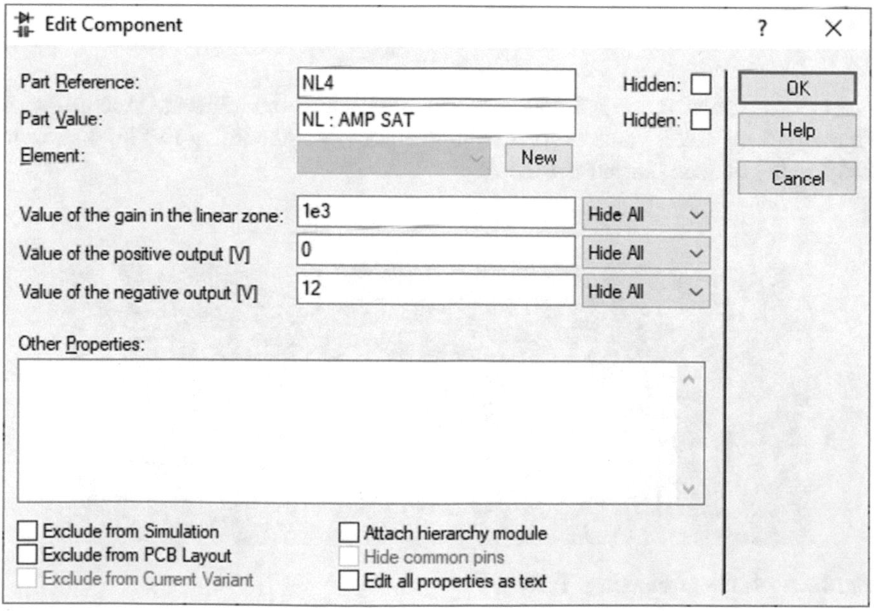

Edit Component ? ×

Part Reference: NL4 Hidden: ☐ OK
Part Value: NL : AMP SAT Hidden: ☐ Help
Element: New Cancel
Value of the gain in the linear zone: 1e3 Hide All
Value of the positive output [V] 0 Hide All
Value of the negative output [V] 12 Hide All
Other Properties:
☐ Exclude from Simulation ☐ Attach hierarchy module
☐ Exclude from PCB Layout Hide common pins
Exclude from Current Variant ☐ Edit all properties as text

Fig. 4.105 Settings of NL_4

Run the simulation. The simulation result (load voltage waveform) is shown in Fig. 4.106.

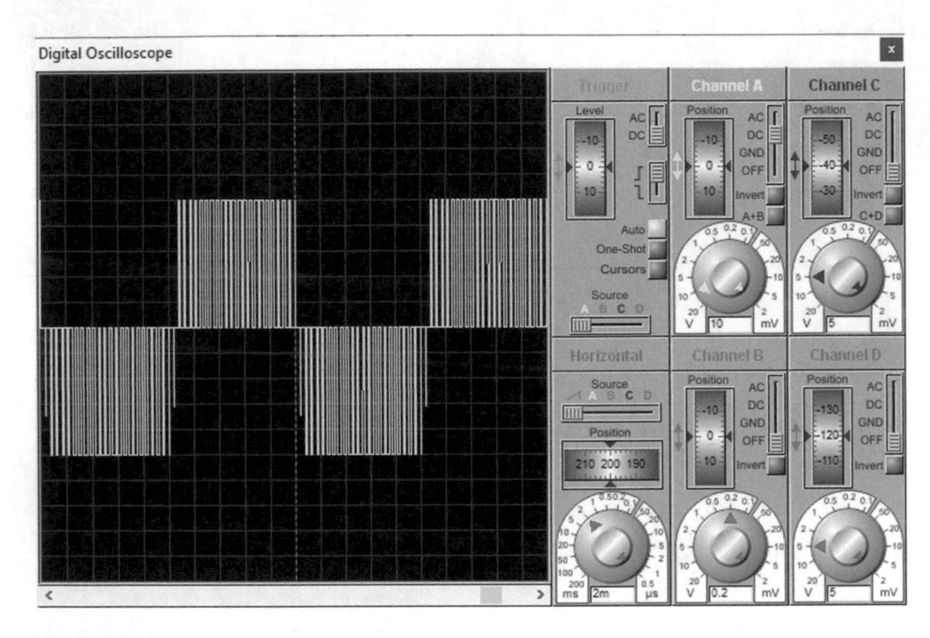

Fig. 4.106 Simulation result

Let's measure the RMS value of load voltage and current. According to Fig. 4.107, the RMS value of load voltage is 33.1 V. According to Fig. 4.108, the RMS value of load current is 0.23 A.

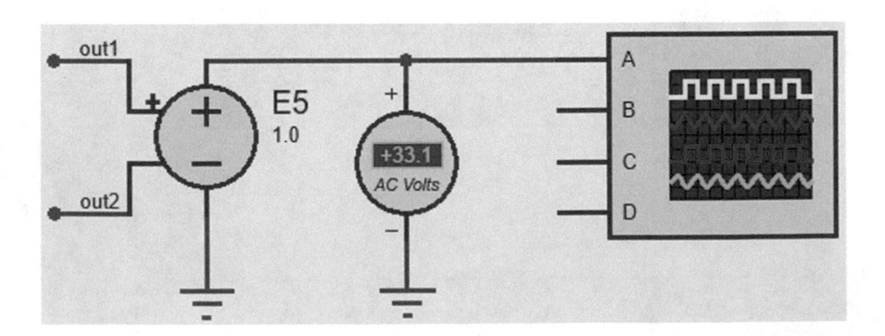

Fig. 4.107 RMS of load voltage is 33.1 V

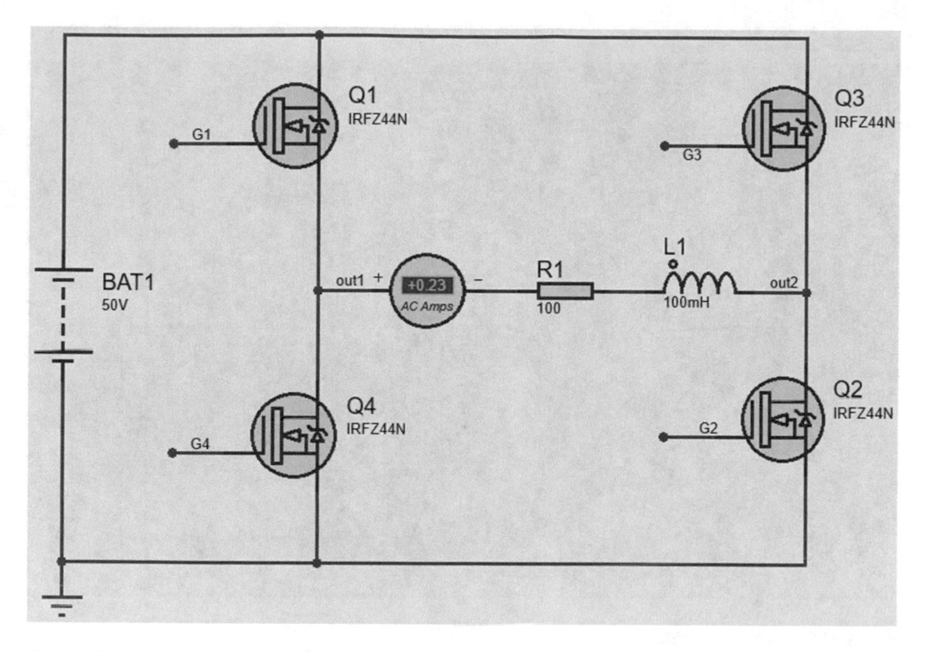

Fig. 4.108 RMS value of load current is 0.23 A

You can use a current-controlled voltage source to see the load current (Fig. 4.109). The load current is shown in Fig. 4.110. Inductor L_1 decreases the amplitude of harmonics in the load current (it acts as a low-pass filter), and this causes the load current to resemble a sinusoidal waveform.

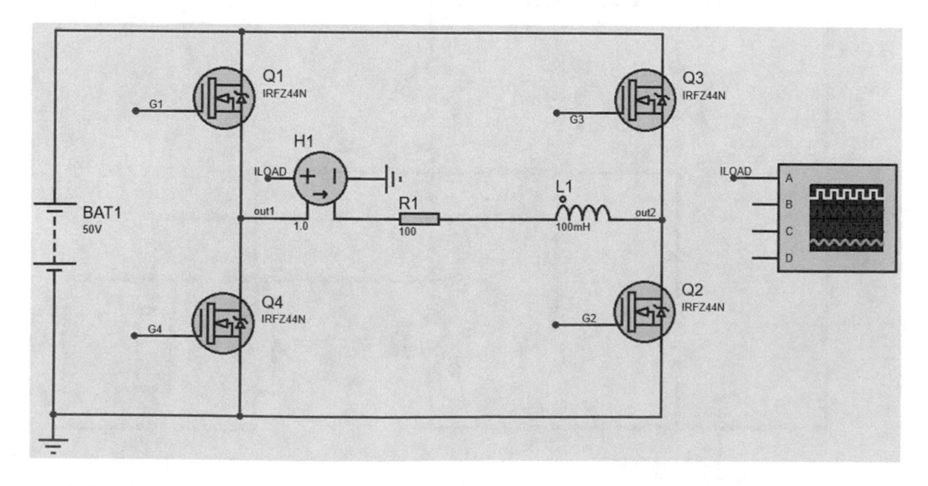

Fig. 4.109 H_1 senses the load current and displays it on the oscilloscope

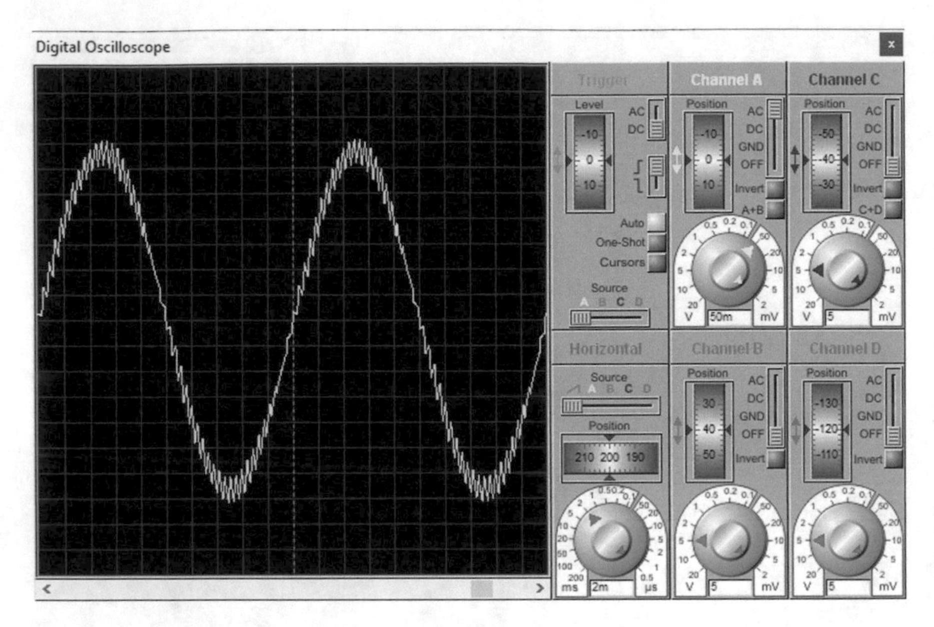

Fig. 4.110 Simulation result

4.14 Example 13: Three-Phase Inverter

In this example we want to simulate a three phase inverter (Fig. 4.111) which uses the Sine-PWM (SPWM) modulation technique.

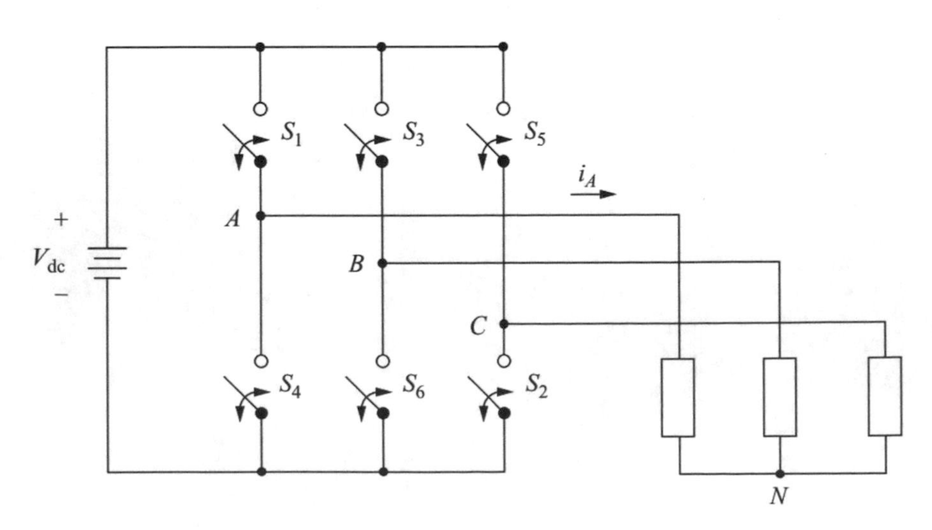

Fig. 4.111 Three-phase inverter circuit

In SPWM, we need three reference signals. The three reference sinusoids are 120° apart (Fig. 4.112). Harmonics will be minimized if the carrier frequency is chosen to be an odd triple multiple of the reference frequency, that is, 3, 9, 15, 21, 27, 33, 39, 45, … times the reference.

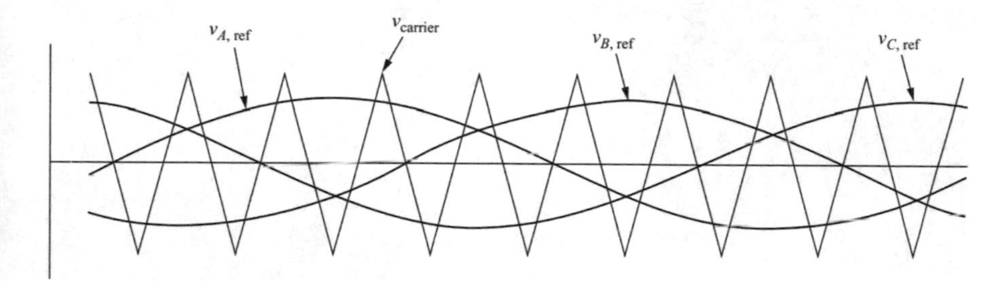

Fig. 4.112 Generation of PWM signals

In the SPWM scheme,

S_1 is on when $v_a > v_{\text{tri}}$.

S_2 is on when $v_c > v_{\text{tri}}$.

S_3 is on when $v_b > v_{\text{tri}}$.

S_4 is on when $v_a < v_{\text{tri}}$.

S_5 is on when $v_c < v_{\text{tri}}$.

S_6 is on when $v_b < v_{\text{tri}}$.

The power circuit of three-phase inverter is shown in Fig. 4.113.

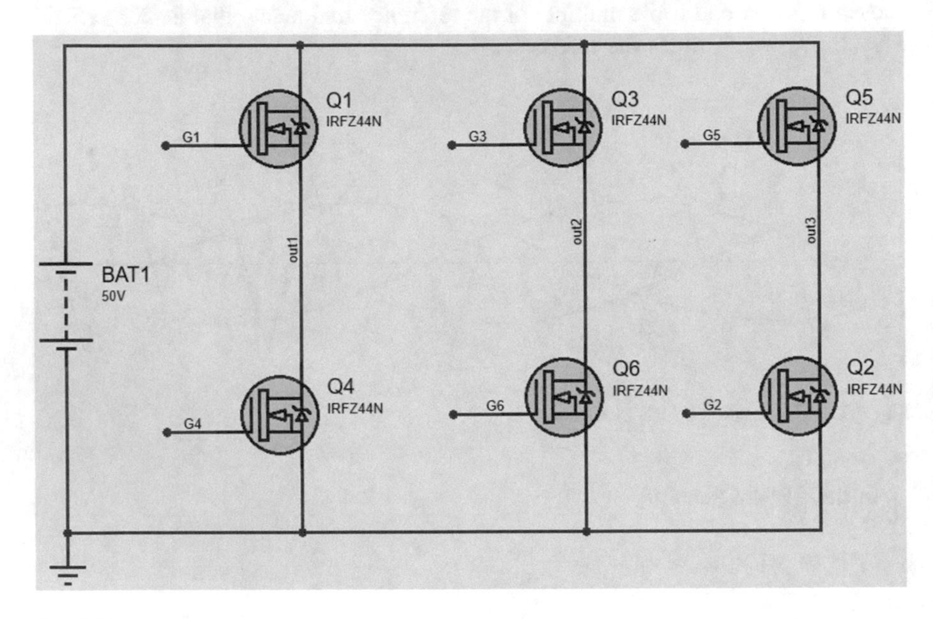

Fig. 4.113 Power circuit

The load of three-phase inverter is shown in Fig. 4.114.

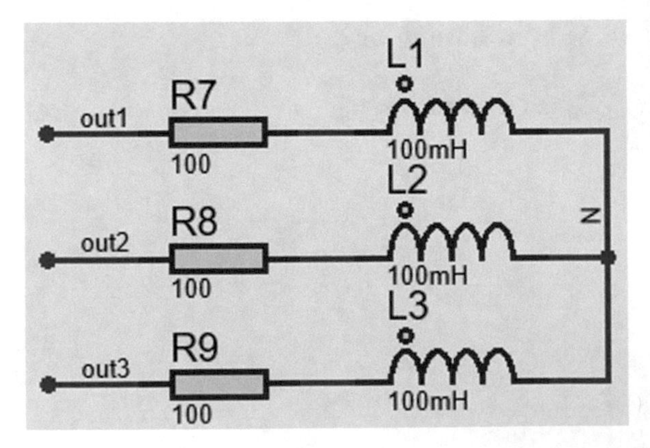

Fig. 4.114 Load of three-phase inverter

The measurement section of three-phase inverter is shown in Fig. 4.115.

The required pulses for MOSFETs are generated with the aid of a schematic shown in Fig. 4.116.

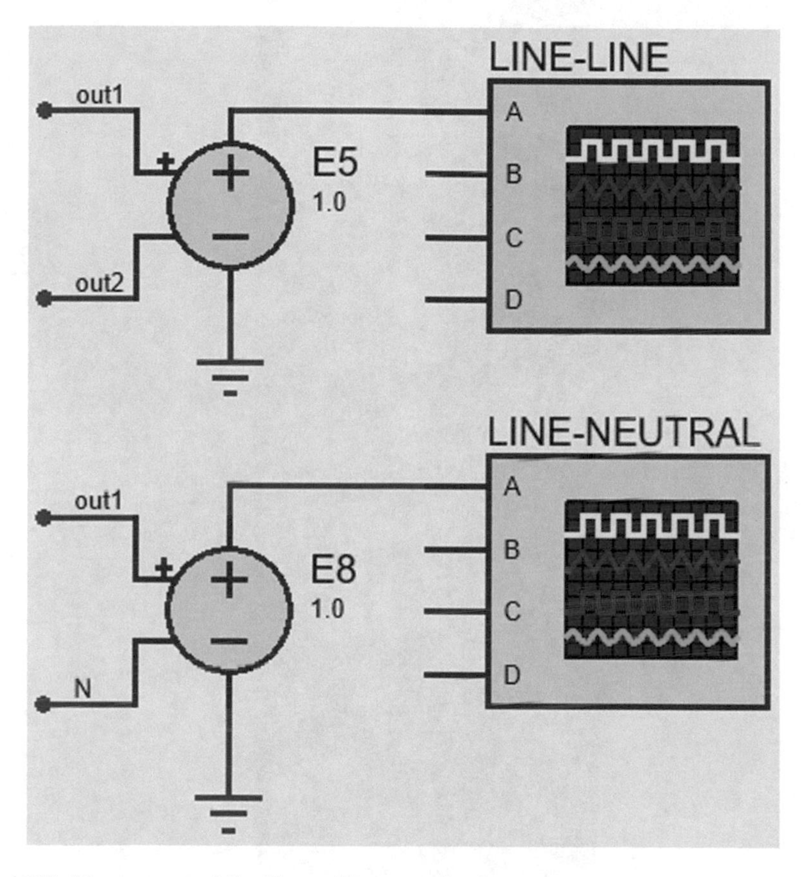

Fig. 4.115 Measurement of line-line and line-neutral voltages

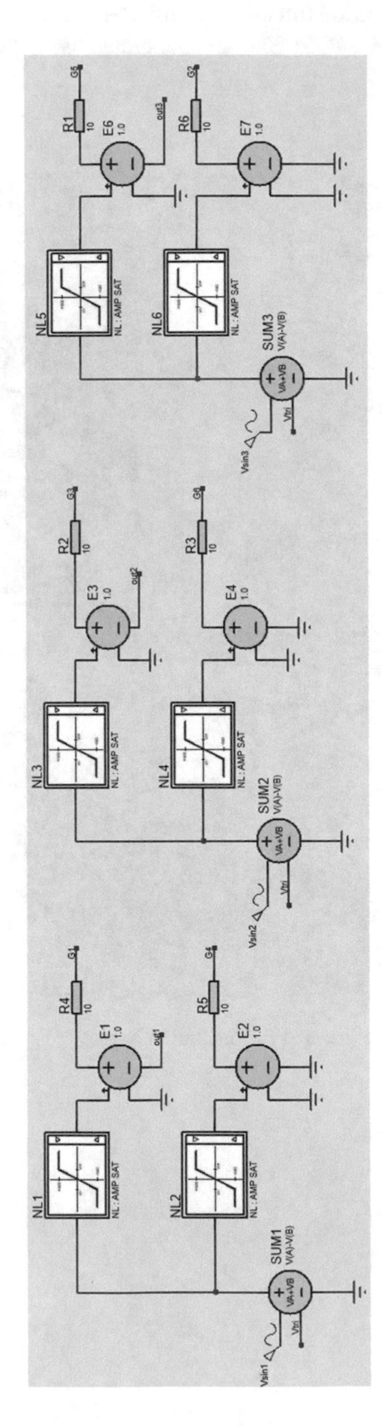

Fig. 4.116 PWM generation section

Let's take a closer look at the schematic shown in Fig. 4.116. The section shown in Fig. 4.117 generates the control signal for MOSFET's Q_1 and Q_4. Settings of NL_1 and NL_2 blocks are shown in Figs. 4.118 and 4.119, respectively. When $V_{sine1} > V_{tri}$, NL_1 and E_1 force Q_1 to be closed. When $V_{sine1} < V_{tri}$, NL_2 and E_2 force Q_4 to be closed. When Q_1 is closed, Q_4 is open. When Q_4 is closed, Q_1 is open.

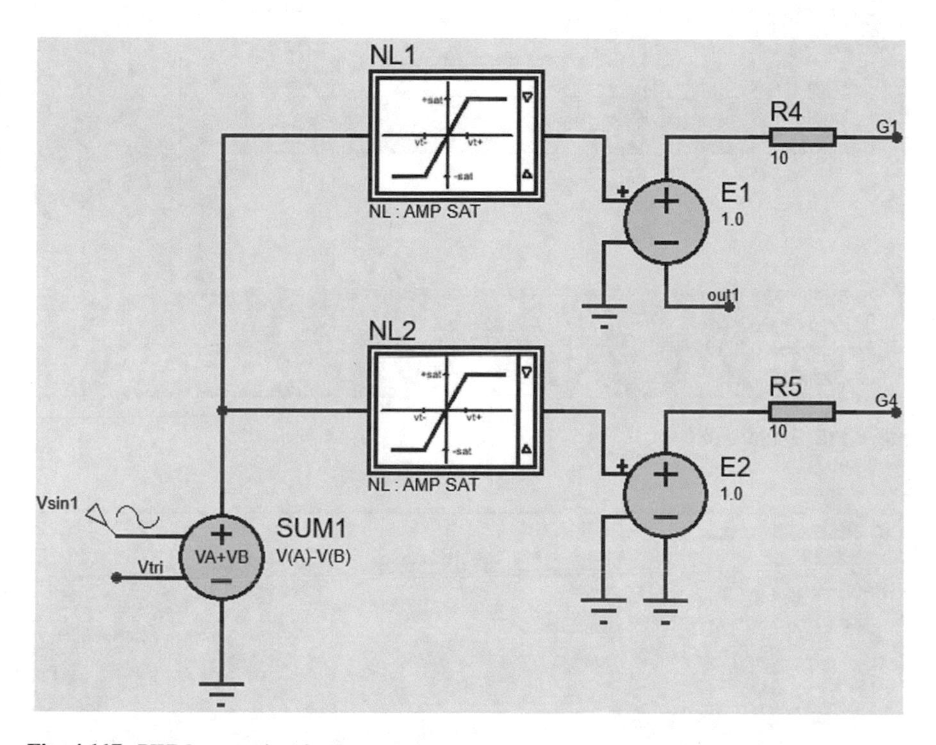

Fig. 4.117 PWM generation for Q_1 and Q_4

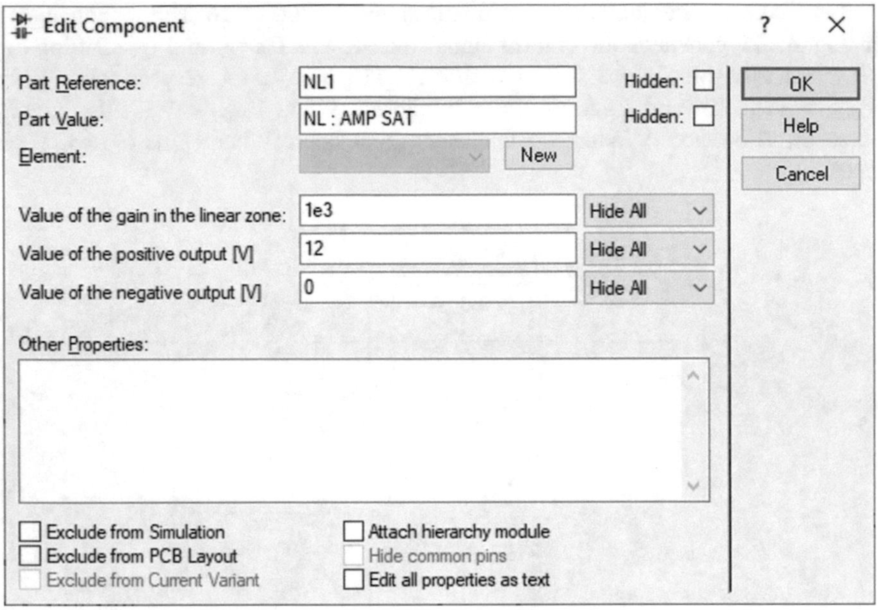

Edit Component ? ×

Part Reference: NL1 Hidden: ☐
Part Value: NL : AMP SAT Hidden: ☐
Element: New

Value of the gain in the linear zone: 1e3 Hide All
Value of the positive output [V] 12 Hide All
Value of the negative output [V] 0 Hide All

Other Properties:

☐ Exclude from Simulation ☐ Attach hierarchy module
☐ Exclude from PCB Layout ☐ Hide common pins
☐ Exclude from Current Variant ☐ Edit all properties as text

OK
Help
Cancel

Fig. 4.118 Settings of NL_1

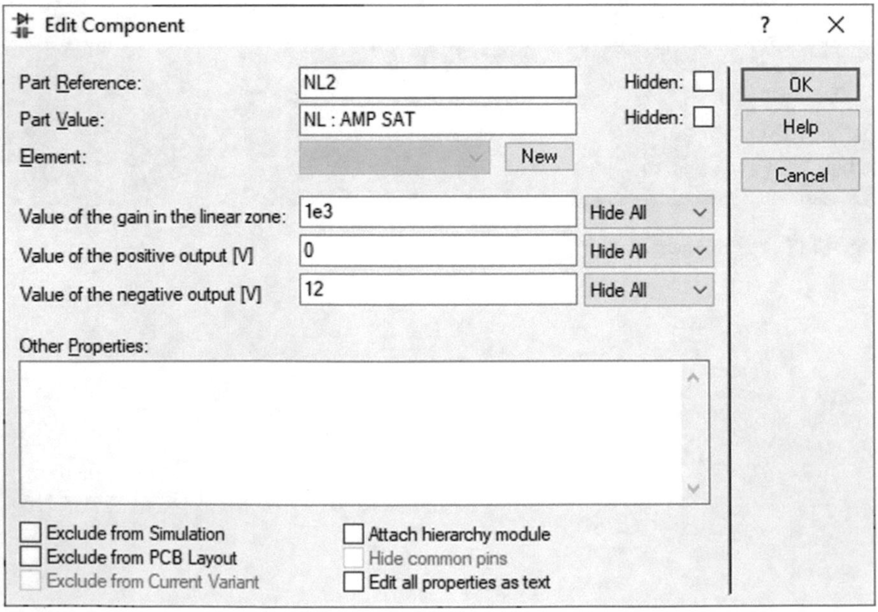

Edit Component ? ×

Part Reference: NL2 Hidden: ☐
Part Value: NL : AMP SAT Hidden: ☐
Element: New

Value of the gain in the linear zone: 1e3 Hide All
Value of the positive output [V] 0 Hide All
Value of the negative output [V] 12 Hide All

Other Properties:

☐ Exclude from Simulation ☐ Attach hierarchy module
☐ Exclude from PCB Layout ☐ Hide common pins
☐ Exclude from Current Variant ☐ Edit all properties as text

OK
Help
Cancel

Fig. 4.119 Settings of NL_2

The section shown in Fig. 4.120 generates the control signal for MOSFETs Q_3 and Q_6. Settings of NL_3 and NL_4 blocks are shown in Figs. 4.121 and 4.122, respectively. When $V_{sine2} > V_{tri}$, NL_3 and E_3 force Q_3 to be closed. When $V_{sine2} < V_{tri}$, NL_4 and E_4 force Q_6 to be closed. When Q_3 is closed, Q_6 is open. When Q_6 is closed, Q_3 is open.

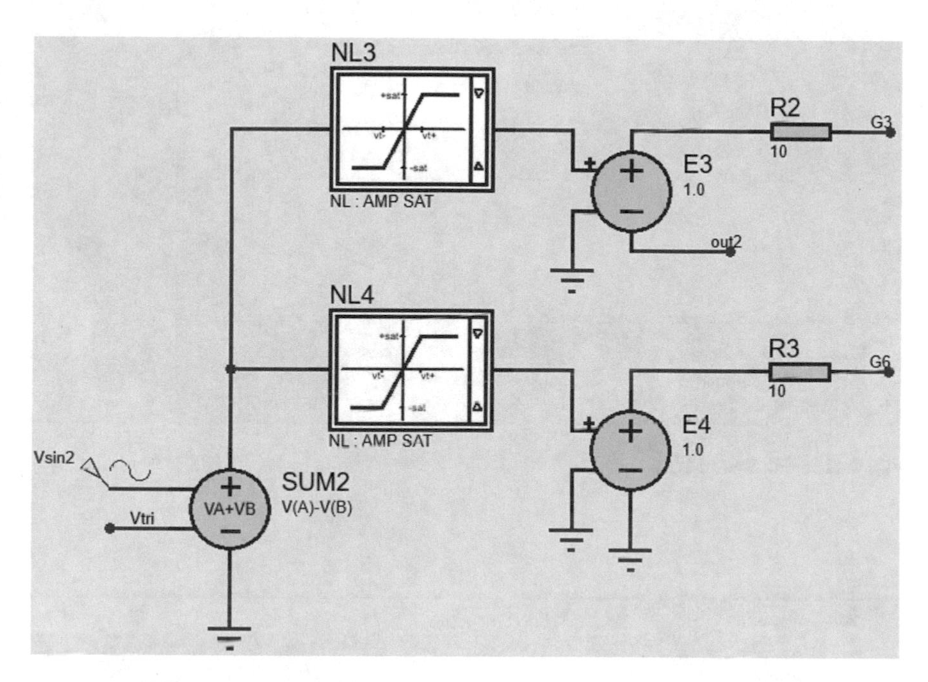

Fig. 4.120 PWM generation for Q_3 and Q_6

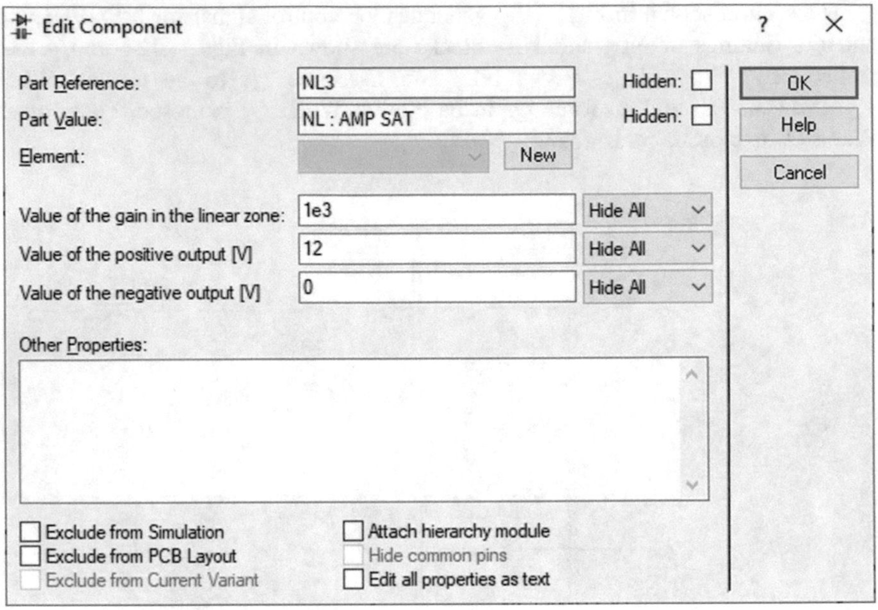

Edit Component

Part Reference: NL3 Hidden: ☐
Part Value: NL : AMP SAT Hidden: ☐
Element: New

Value of the gain in the linear zone: 1e3 Hide All
Value of the positive output [V] 12 Hide All
Value of the negative output [V] 0 Hide All

Other Properties:

☐ Exclude from Simulation ☐ Attach hierarchy module
☐ Exclude from PCB Layout Hide common pins
Exclude from Current Variant ☐ Edit all properties as text

OK Help Cancel

Fig. 4.121 Settings of NL_3

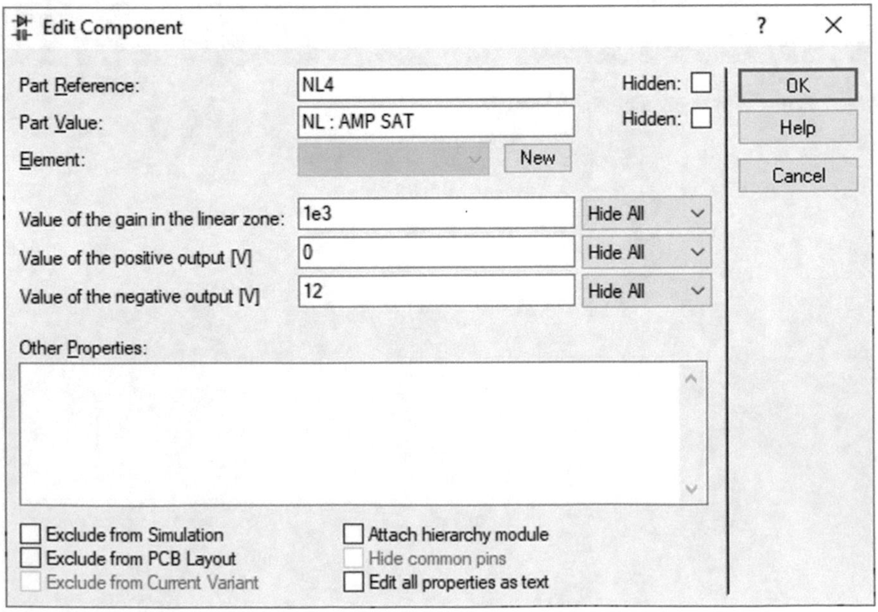

Edit Component

Part Reference: NL4 Hidden: ☐
Part Value: NL : AMP SAT Hidden: ☐
Element: New

Value of the gain in the linear zone: 1e3 Hide All
Value of the positive output [V] 0 Hide All
Value of the negative output [V] 12 Hide All

Other Properties:

☐ Exclude from Simulation ☐ Attach hierarchy module
☐ Exclude from PCB Layout Hide common pins
Exclude from Current Variant ☐ Edit all properties as text

OK Help Cancel

Fig. 4.122 Settings of NL_4

The section shown in Fig. 4.123 generates the control signal for MOSFETs Q_5 and Q_2. Settings of NL_5 and NL_6 blocks are shown in Figs. 4.124 and 4.125, respectively. When $V_{sine3} > V_{tri}$, NL_5 and E_6 force Q_5 to be closed. When $V_{sine3} < V_{tri}$, NL_6 and E_7 force Q_2 to be closed. When Q_5 is closed, Q_2 is open. When Q_2 is closed, Q_5 is open.

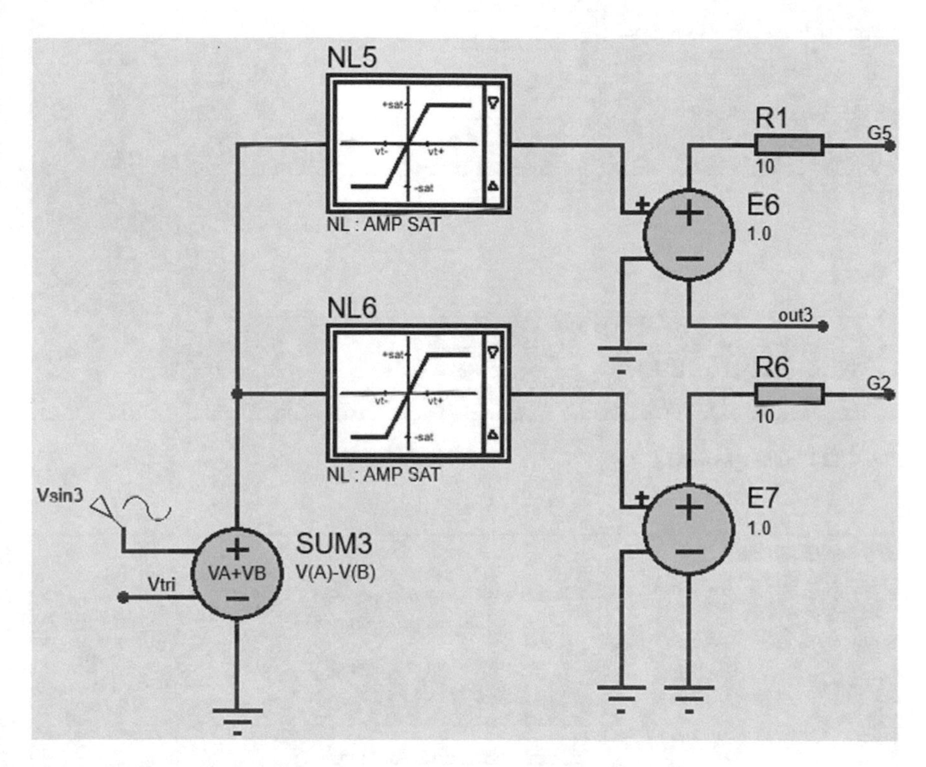

Fig. 4.123 PWM generation for Q_5 and Q_2

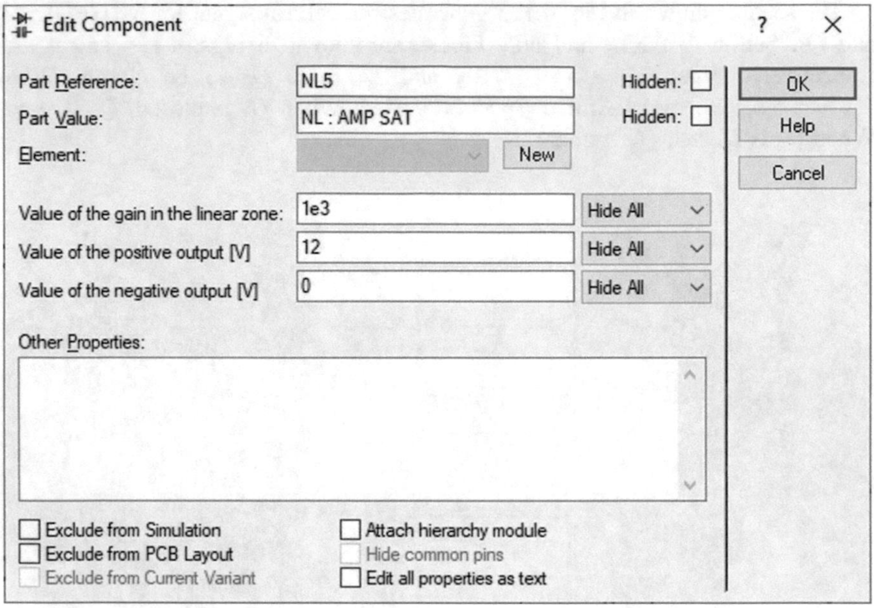

Edit Component

Part Reference: NL5 Hidden:
Part Value: NL : AMP SAT Hidden:
Element: New
OK
Help
Cancel

Value of the gain in the linear zone: 1e3 Hide All
Value of the positive output [V] 12 Hide All
Value of the negative output [V] 0 Hide All

Other Properties:

Exclude from Simulation
Exclude from PCB Layout
Exclude from Current Variant
Attach hierarchy module
Hide common pins
Edit all properties as text

Fig. 4.124 Settings of NL_5

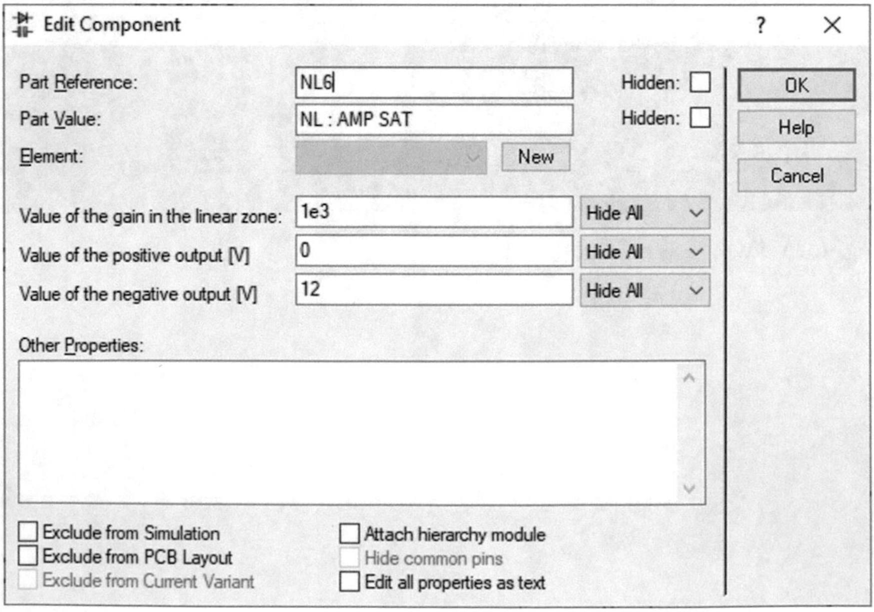

Edit Component

Part Reference: NL6 Hidden:
Part Value: NL : AMP SAT Hidden:
Element: New
OK
Help
Cancel

Value of the gain in the linear zone: 1e3 Hide All
Value of the positive output [V] 0 Hide All
Value of the negative output [V] 12 Hide All

Other Properties:

Exclude from Simulation
Exclude from PCB Layout
Exclude from Current Variant
Attach hierarchy module
Hide common pins
Edit all properties as text

Fig. 4.125 Settings of NL_6

Run the simulation. The simulation results are shown in Figs. 4.126 and 4.127.

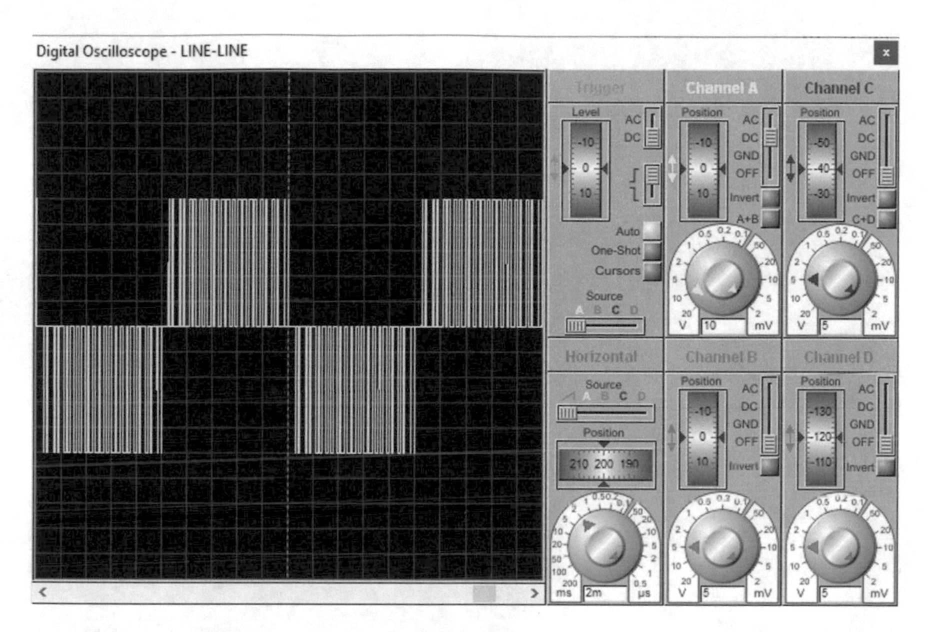

Fig. 4.126 Simulation result (line-line voltage)

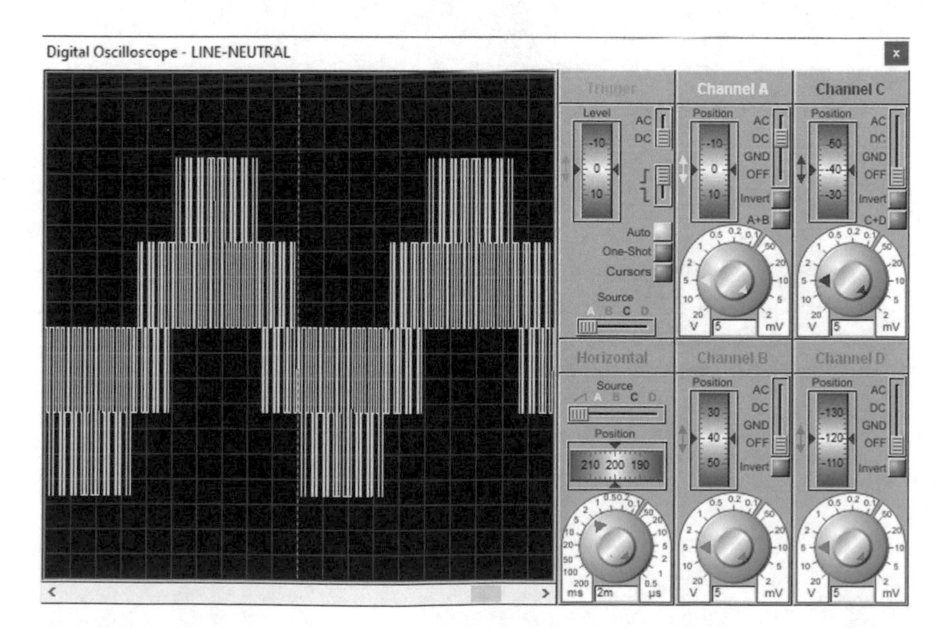

Fig. 4.127 Simulation result (line-neutral voltage)

Let's measure the RMS of output voltage. According to Fig. 4.128, RMS values of line-line voltage and line-neutral voltage are 31.4 V and 17.9 V, respectively.

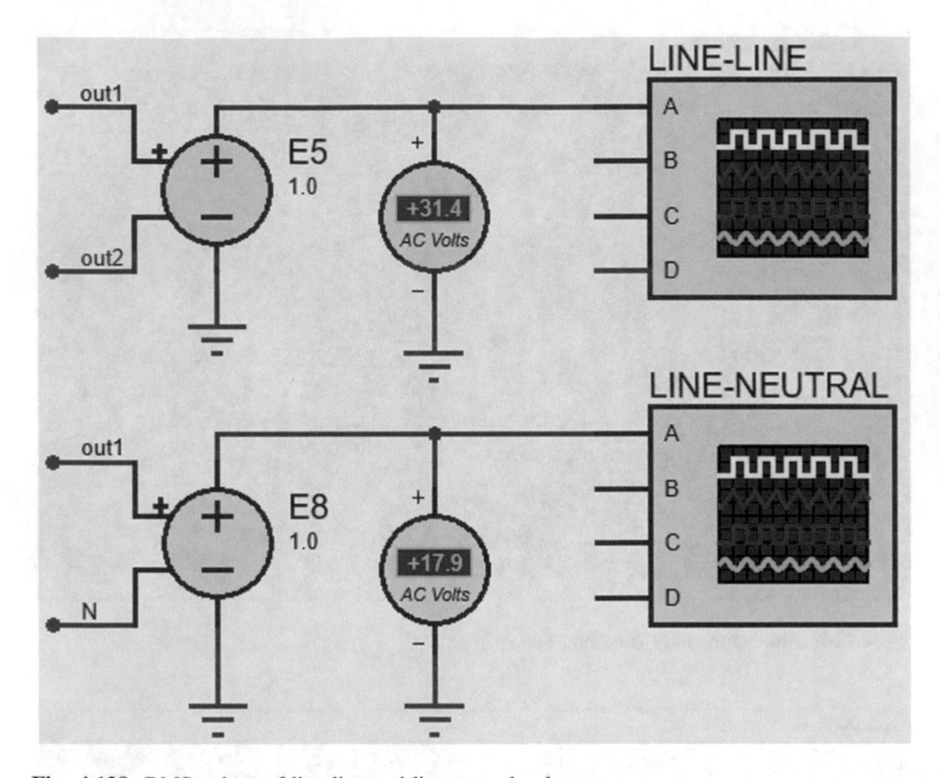

Fig. 4.128 RMS values of line-line and line-neutral voltages

4.15 Exercises

1. A single-phase voltage controller is shown in Fig. 4.129. SCR S1 is triggered at $2k\pi + \alpha$ and SCR S2 is triggered at $(2k + 1)\pi + \alpha$ angles (k = 0, 1, 2, …). The load voltage/current waveforms are shown in Fig. 4.130.
   Use Proteus to simulate the circuit for $\alpha = 30°, 60°$ and 90°.

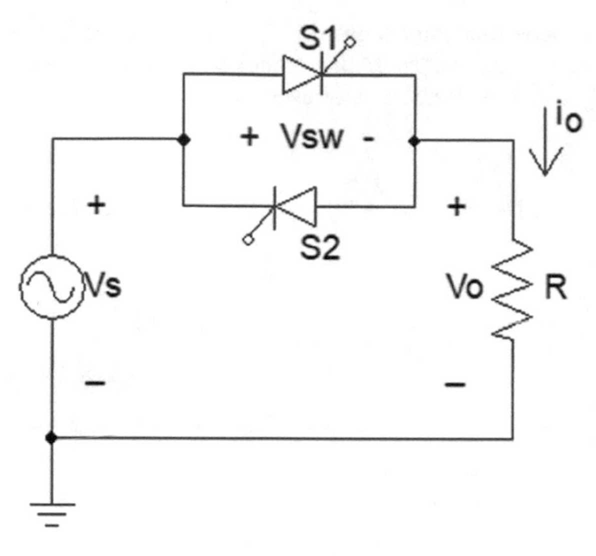

Fig. 4.129 Circuit for Exercise 1

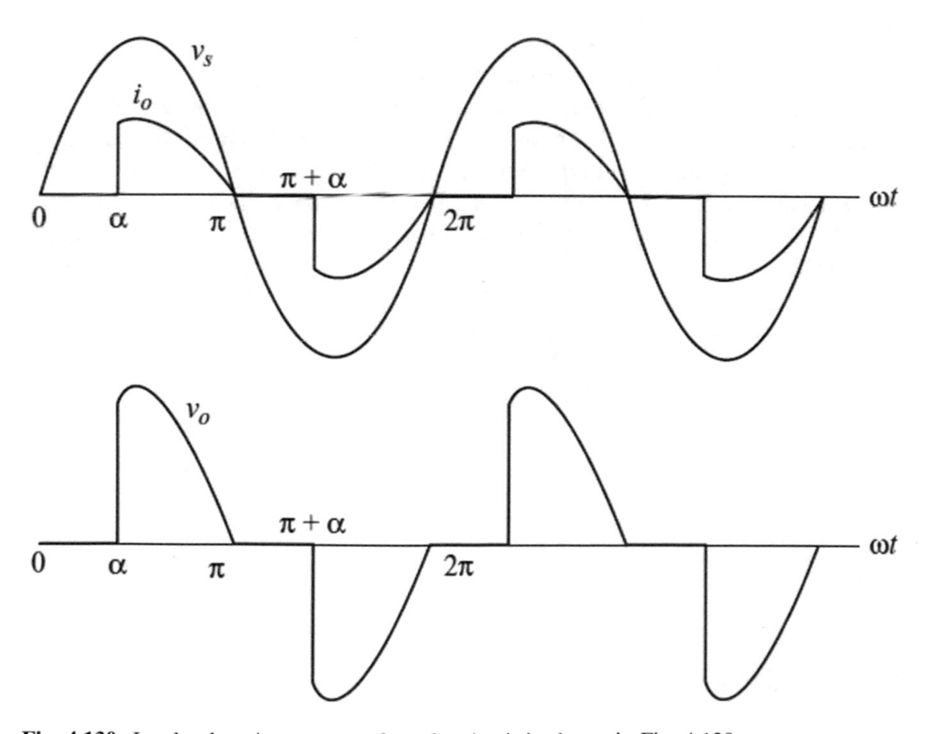

Fig. 4.130 Load voltage/current waveform for circuit is shown in Fig. 4.129

2. Simulate the boost converter shown in Fig. 4.131 with Proteus and measure the output voltage, output voltage ripple, output power and efficiency. Signal S1 has a frequency of 200 kHz and a duty cycle of 0.5.

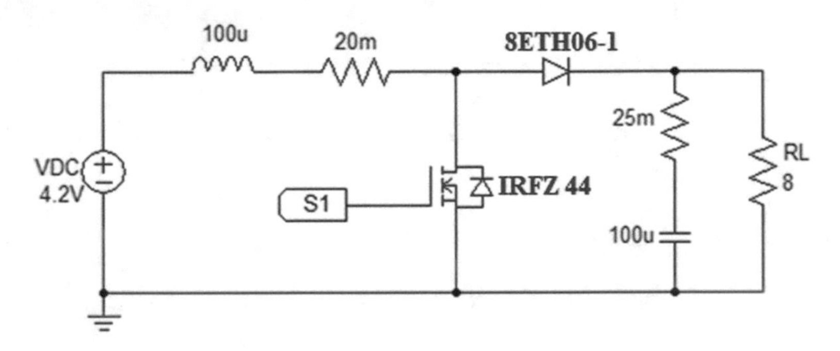

Fig. 4.131 Circuit for Exercise 2

3. Simulate a flyback converter with Proteus.
4. A cascaded multilevel inverter is shown in Fig. 4.132. It is composed of a series connection of two single-phase H bridge inverters. Each of the H bridge inverters can produce an output voltage of $-V_{dc}, 0$ and $+V_{dc}$. So, with a series connection of two H bridge inverters, we can produce $-2V_{dc}, -V_{dc}, 0, +V_{dc}$ and $+2V_{dc}$. Increasing the number of series stages helps to decrease the output of THD.

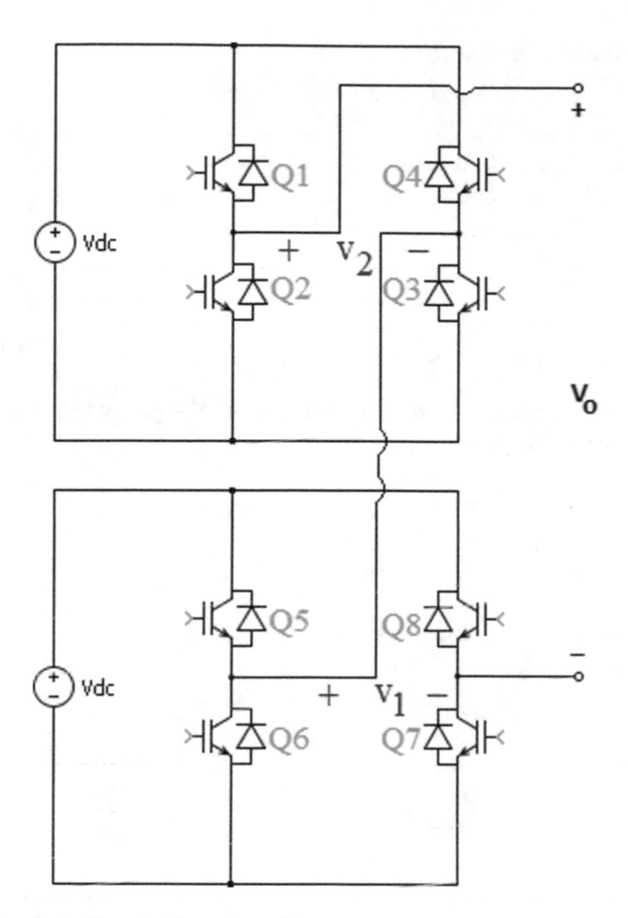

Fig. 4.132 Circuit for Exercise 4

Assume the switching pattern shown in Tables 4.3 and 4.4.

Table 4.3 Switching pattern for Q_1, Q_2, Q_3 and Q_4

Interval	Control signal for Q_1	Control signal for Q_2	Control signal for Q_3	Control signal for Q_4
A: $0 < \omega t < \alpha_2$	1	0	0	1
B: $\alpha_2 < \omega t < \pi - \alpha_2$	1	0	1	0
C: $\pi - \alpha_2 < \omega t < \pi + \alpha_2$	1	0	0	1
D: $\pi + \alpha_2 < \omega t < 2\pi - \alpha_2$	0	1	0	1
E: $2\pi - \alpha_2 < \omega t < 2\pi$	1	0	0	1

Table 4.4 Switching pattern for Q_5, Q_6, Q_7 and Q_8

Interval	Control signal for Q_5	Control signal for Q_6	Control signal for Q_7	Control signal for Q_8
a: $0 < \omega t < \alpha_1$	1	0	0	1
b: $\alpha_1 < \omega t < \pi - \alpha_1$	1	0	1	0
c: $\pi - \alpha_1 < \omega t < \pi + \alpha_1$	1	0	0	1
d: $\pi + \alpha_1 < \omega t < 2\pi - \alpha_1$	0	1	0	1
e: $2\pi - \alpha_1 < \omega t < 2\pi$	1	0	0	1

The switching patterns given in Tables 4.3 and 4.4 produce the voltage waveforms as shown in Fig. 4.133.

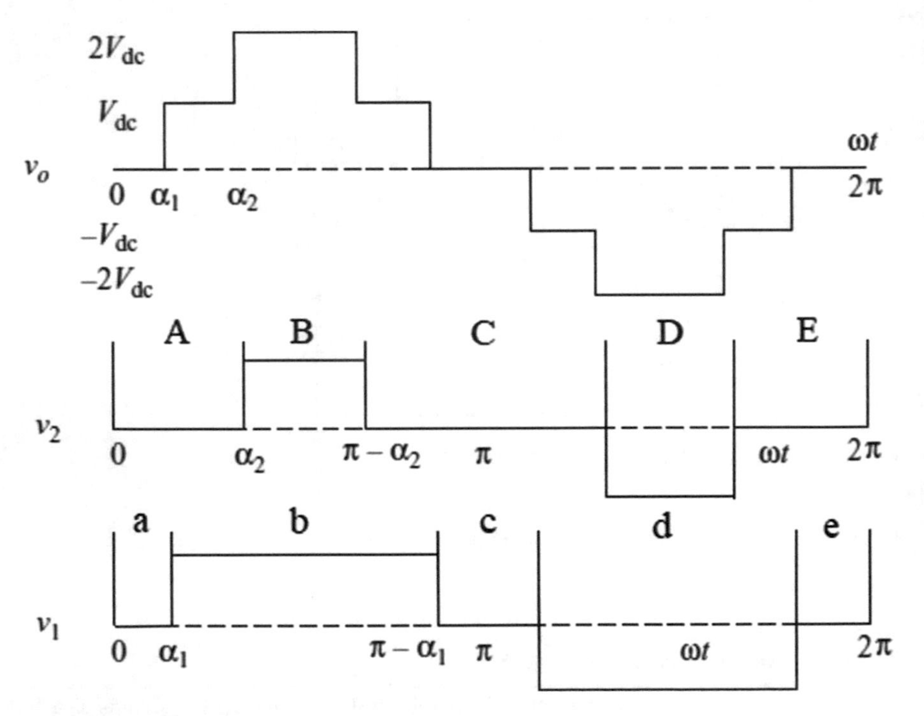

Fig. 4.133 Voltage waveforms for switching patterns are shown in Tables 4.3 and 4.4, respectively

Use Proteus to simulate the circuit for $\alpha_1 = 20°$ and $\alpha_1 = 40°$.

References for Further Study

Asadi F (2018) Robust Control of DC–DC Converters, Springer

Asadi F (2018) Computer Techniques for Dynamic Modeling of DC–DC Power Converters, Springer

Asadi F (2022) Simulation of Power Electronics Circuits with MATLAB/Simulink, Springer

Asadi F, Eguchi K (2018) Dynamics and Control of DC–DC Converters, Springer

Asadi F, Pongswatd S (2019) Modeling Uncertainties in DC–DC Converters, Springer

Erikson RW, Maksimovic D(2020) Fundamentals of power electronics, 3rd edn. Springer

Hart DW (2010) Power electronics. McGraw-Hill

Mohan N, Undeland TM, Robbins WP (2007) Power electronics: converters, applications, and design, 3rd edn. Wiley and Sons

Rashid MH (2013) Power electronics: devices, circuits and applications. Pearson

Index

A
Amplitude modulation ratio, 628

B
Boost converter, 656
Buck converter, 563

C
CMRR of differential pair, 463
Common Mode Rejection Ratio (CMRR), 446
Coupled inductor, 216
Coupling coefficient, 216
Curve Fitting Toolbox, 251
Custom transistor, 438

D
Difference amplifier, 446

F
Fourier series of half wave rectified waveform, 626
Frequency division, 525
Frequency modulation ratio, 628
Full adder circuit, 511

I
Impulse response, 242
Input impedance of differential pair, 475

O
Op-amp clamp circuit, 508

P
Phase difference, 119

R
Resistive circuit, 1

S
Single phase half wave thyristor rectifier, 581
Single phase inverter, 626
Single phase voltage controller, 654
State diagram, 560

T
Thevenin equivalent circuit, 137
Three phase controlled rectifier, 590
Three phase inverter, 642
Truth table, 511

U
Unipolar PWM, 627

W
Worst common mode gain, 475

F. Asadi, *Essential Circuit Analysis Using Proteus®*, Energy Systems in Electrical Engineering, https://doi.org/10.1007/978-981-19-4353-9

Printed in the United States
by Baker & Taylor Publisher Services